The Endoscopic Classification of Representations

Orthogonal and Symplectic Groups

American Mathematical Society

Colloquium Publications
Volume 61

The Endoscopic Classification of Representations

Orthogonal and Symplectic Groups

James Arthur

American Mathematical Society
Providence, Rhode Island

2010 *Mathematics Subject Classification.* Primary 22E55, 22E50, 11R37, 11F66, 58C40.

Supported in part by NSERC Discovery Grant A3483.

For additional information and updates on this book, visit
www.ams.org/bookpages/coll-61

Library of Congress Cataloging-in-Publication Data

Arthur, James, 1944–
The endoscopic classification of representatons : orthogonal and symplectic groups / James Arthur.
pages cm – (American Mathematical Society Colloquium Publications ; volume 61)
Includes bibliographical references and index.
ISBN 978-0-8218-4990-3 (alk. paper)
1. Linear algebraic groups. 2. Class field theory. 3. Algebraic number theory. I. Title.

QA179.A78 2013
512′.4–dc23 2013004694

∞ The paper used in this book is acid-free and falls within the guidelines
established to ensure permanence and durability.
Visit the AMS home page at `http://www.ams.org/`

10 9 8 7 6 5 4 3 2 1 18 17 16 15 14 13

Contents

Foreword: A selective overview

This preface contains a summary of the contents of the volume. We start with a rough description of the main theorems. We then give short descriptions of the contents of the various chapters. At the end, we will add a couple of remarks on the overall structure of the proof, notably our use of induction. The preface can serve as an introduction. The beginning of the actual text, in the form of the first two or three sections of Chapter 1, represents a different sort of introduction. It will be our attempt to motivate what follows from a few basic principles. A reader might consider going directly to these sections after reading the first half of the preface here. One could then return to the more technical second half (on the organization of the volume) only as needed later.

Automorphic representations for $GL(N)$ have been important objects of study for many years. We recall that $GL(N)$, the general linear group of invertible $(N \times N)$-matrices, assigns a group $GL(N, R)$ to any commutative ring R with identity. For example, R could be a fixed number field F, or the ring $\mathbb{A} = \mathbb{A}_F$ of adèles over F. *Automorphic representations* of $GL(N)$ are the irreducible representations of $GL(N, \mathbb{A})$ that occur in the decomposition of its regular representation on $L^2\big(GL(N, F)\backslash GL(N, \mathbb{A})\big)$. This informal definition is made precise in [**L6**], and carries over to any connected reductive group G over F.

The primary aim of the volume is to classify the automorphic representations of special orthogonal and symplectic groups G in terms of those of $GL(N)$. Our main tool will be the stable trace formula for G, which until recently was conditional on the fundamental lemma. The fundamental lemma has now been established in complete generality, and in all of its various forms. In particular, the stabilization of the trace formula is now known for any connected group. However, we will also require the stabilization of twisted trace formulas for $GL(N)$ and $SO(2n)$. Since these have yet to be established, our results will still be conditional.

A secondary purpose will be to lay foundations for the endoscopic study of more general groups G. It is reasonable to believe that the methods we introduce here extend to groups that Ramakrishnan has called *quasiclassical*. These would comprise the largest class of groups whose representations could ultimately be tied to those of general linear groups. Our third goal is expository. In adopting a style that is sometimes more discursive than strictly necessary, we have tried to place at least some of the techniques into perspective. We hope that there will be parts of the volume that are accessible to readers who are not experts in the subject.

Automorphic representations are interesting for many reasons, but among the most fundamental is the arithmetic data they carry. Recall that

$$GL(N,\mathbb{A}) = \widetilde{\prod_v} GL(N,F_v)$$

is a restricted direct product, taken over (equivalence classes of) valuations v of F. An automorphic representation of $GL(N)$ is a restricted direct product

$$\pi = \widetilde{\bigotimes_v} \pi_v,$$

where π_v is an irreducible representation of $GL(N,F_v)$ that is *unramified* for almost all v. We recall that π_v is unramified if v is nonarchimedean, and π_v contains the trivial representation of the hyperspecial maximal compact subgroup $GL(N,\mathfrak{o}_v)$ of integral points in $GL(N,F_v)$. The representation is then parametrized by a semisimple conjugacy class

$$c_v(\pi) = c(\pi_v)$$

in the complex dual group

$$GL(N)^\wedge = GL(N,\mathbb{C})$$

of $GL(N)$. (See [**Bo**, (6.4), (6.5)] for the precise assertion, as it applies to a general connected reductive group G.) It is the relations among the semisimple conjugacy classes $c_v(\pi)$ that will contain the fundamental arithmetic information.

There are three basic theorems for the group $GL(N)$ that together give us a pretty clear understanding of its representations. The first is local, while others, which actually predate the first, are global.

The first theorem is the local Langlands *correspondence* for $GL(N)$. It was established for archimedean fields by Langlands, and more recently for p-adic (which is to say nonarchimedean) fields by Harris, Taylor and Henniart. It classifies the irreducible representations of $GL(N,F_v)$ at all places v by (equivalence classes of) semisimple, N-dimensional representations of the local Langlands group

$$L_{F_v} = \begin{cases} W_{F_v}, & v \text{ archimedean,} \\ W_{F_v} \times SU(2), & \text{otherwise.} \end{cases}$$

In particular, an unramified representation of $GL(N,F_v)$ corresponds to an N-dimensional representation of L_{F_v} that is trivial on the product of $SU(2)$ with the inertia subgroup I_{F_v} of the local Weil group W_{F_v}. It therefore corresponds to a semisimple representation of the cyclic quotient

$$L_{F_v}/I_{F_v} \times SU(2) \cong W_{F_v}/I_{F_v} \cong \mathbb{Z},$$

and hence a semisimple conjugacy class in $GL(N,\mathbb{C})$, as above.

The first of the global theorems is due to Jacquet and Shalika. If π is an irreducible (admissible) representation of $GL(N, \mathbb{A})$, one can form the family of semisimple conjugacy classes

$$c(\pi) = \varinjlim_{S} \{c_v(\pi) = c(\pi_v) : \ v \notin S\}$$

in $GL(N, \mathbb{C})$, defined up to a finite set of valuations S. In other words, $c(\pi)$ is an equivalence class of families, two such families being equivalent if they are equal for almost all v. The theorem of Jacquet and Shalika asserts that if an automorphic representation π of $GL(N)$ is restricted slightly to be *isobaric* [**L7**, §2], it is uniquely determined by $c(\pi)$. This theorem can be regarded as a generalization of the theorem of strong multiplicity one for cuspidal automorphic representations of $GL(N)$.

The other global theorem for $GL(N)$ is due to Moeglin and Waldspurger. It characterizes the automorphic (relatively) discrete spectrum of $GL(N)$ in terms of the set of cuspidal automorphic representations. Since Langlands' general theory of Eisenstein series characterizes the full automorphic spectrum of any group G in terms of discrete spectra, this theorem characterizes the automorphic spectrum for $GL(N)$ in terms of cuspidal automorphic representations. Combined with the first global theorem, it classifies the full automorphic spectrum of $GL(N)$ explicitly in terms of families $c(\pi)$ attached to cuspidal automorphic representations of general linear groups.

Our goal is to generalize these three theorems. As we shall see, there is very little that comes easily. It has been known for many years that the representations of groups other than $GL(N)$ have more structure. In particular, they should separate naturally into L-packets, composed of representations with the same L-functions and ε-factors. This was demonstrated for the group

$$G = SL(2) = Sp(2)$$

by Labesse and Langlands, in a paper [**LL**] that became a model for Langlands's conjectural theory of endoscopy [**L8**], [**L10**].

The simplest and most elegant way to formulate the theory of endoscopy is in terms of the global Langlands group L_F. This is a hypothetical global analogue of the explicit local Langlands groups L_{F_v} defined above. It is thought to be a locally compact extension

$$1 \longrightarrow K_F \longrightarrow L_F \longrightarrow W_F \longrightarrow 1$$

of W_F by a compact connected group K_F. (See [**L7**, §2], [**K3**, §9].) However, its existence is very deep, and could well turn out to be the final theorem in the subject to be proved! One of our first tasks, which we address in §1.4, will be to introduce makeshift objects to be used in place of L_F. For simplicity, however, let us describe our results here in terms of L_F.

Our main results apply to the case that G is a *quasisplit* special orthogonal or symplectic group.* They are stated as three theorems in §1.5. The proof of these theorems will then take up much of the rest of the volume.

Theorem 1.5.1 is the main local result. It contains a local Langlands parametrization of the irreducible representations of $G(F_v)$, for any p-adic valuation v of F, as a disjoint union of finite L-packets Π_{ϕ_v}. These are indexed by local Langlands parameters, namely L-homomorphisms

$$\phi_v : L_{F_v} \longrightarrow {}^L G_v$$

from L_{F_v} to the local L-group ${}^L G_v = \widehat{G} \rtimes \mathrm{Gal}(\overline{F}_v/F_v)$ of G. The theorem includes a way to index the representations in an L-packet Π_{ϕ_v} by linear characters on a finite abelian group $\mathcal{S}_{\phi_v}$ attached to ϕ_v. Since similar results for archimedean valuations v are already known from the work of Shelstad, we obtain a classification of the representations of each local group $G(F_v)$.

Theorem 1.5.1 also contains a somewhat less precise description of the representations of $G(F_v)$ that are local components of automorphic representations. These fall naturally into rather different packets Π_{ψ_v}, indexed according to the conjectures of [**A8**] by L-homomorphisms

$$\psi_v : L_{F_v} \times SU(2) \longrightarrow {}^L G_v, \tag{1}$$

with bounded image. The theorem includes the assertion, also conjectured in [**A8**], that the representations in these packets are all unitary.

Theorem 1.5.2 is the main global result. As a first approximation, it gives a rough decomposition

$$R_{\mathrm{disc}} = \bigoplus_{\psi} R_{\mathrm{disc},\psi} \tag{2}$$

of the representation R_{disc} of $G(\mathbb{A})$ on the automorphic discrete spectrum $L^2_{\mathrm{disc}}\big(G(F)\backslash G(\mathbb{A})\big)$. The indices can be thought of as L-homomorphisms

$$\psi : L_F \times SU(2) \longrightarrow {}^L G \tag{3}$$

of bounded image that do not factor through any proper parabolic subgroup of the global L-group ${}^L G = \widehat{G} \rtimes \mathrm{Gal}(\overline{F}/F)$. They have localizations (1) defined by conjugacy classes of embeddings $L_{F_v} \subset L_F$, or rather the makeshift analogues of such embeddings that we formulate in §1.4. The localizations ψ_v of ψ are unramified at almost all v, and consequently lead to a family

$$c(\psi) = \varinjlim_{S} \{c_v(\psi) = c(\psi_v) : v \notin S\}$$

of equivalence classes of semisimple elements in ${}^L G$. The rough decomposition (2) is of interest as it stands. It implies that G has no embedded eigenvalues, in the sense of unramified Hecke operators. In other words, the family $c(\psi)$ attached to any global parameter ψ in the decomposition of the

*It is understood here, and in everything that follows, that G is "classical", in the sense that it is not an outer twist of the split group $SO(8)$ by a triality automorphism (of order 3).

automorphic discrete spectrum is distinct from any family obtained from the continuous spectrum. This follows from the nature of the parameters ψ in (3), and the application of the theorem of Jacquet-Shalika to the natural image of $c(\psi)$ in the appropriate complex general linear group.

Theorem 1.5.2 also contains a finer decomposition

$$R_{\mathrm{disc},\psi} = \bigoplus_{\pi} m_\psi(\pi)\,\pi, \tag{4}$$

for any global parameter ψ. The indices π range over representations in the global packet of ψ, defined as a restricted direct product of local packets provided by Theorem 1.5.1. The multiplicities $m_\psi(\pi)$ are given by an explicit reciprocity formula in terms of the finite abelian groups $\mathcal{S}_{\psi_v}$, and their global analogue $\mathcal{S}_\psi$. We thus obtain a decomposition of the automorphic discrete spectrum of G into irreducible representations of $G(\mathbb{A})$. We shall say that a parameter $\phi = \psi$ is *generic* if it is trivial on the factor $SU(2)$. The representations $\pi \in \Pi_\phi$, with ϕ generic and $m_\phi(\pi) \neq 0$, are the constituents of the automorphic discrete spectrum that are expected to satisfy the analogue for G of Ramanujan's conjecture. It ψ is not generic, the formula for $m_\psi(\pi)$ has an extra ingredient. It is a sign character ε_ψ on $\mathcal{S}_\psi$, defined (1.5.6) in terms of symplectic root numbers. That the discrete spectrum should be governed by objects of such immediate arithmetic significance seems quite striking.

Theorem 1.5.2 has application to the question of multiplicity one. Suppose that π is an irreducible constituent of the automorphic discrete spectrum of G that also lies in some generic global packet Π_ϕ. We shall then show that the multiplicity of π in the discrete spectrum equals 1 unless $\widehat{G} = SO(2n,\mathbb{C})$, in which case the multiplicity is either 1 or 2, according to an explicit condition we shall give. In particular, if G equals either $SO(2n+1)$ or $Sp(2n)$, the automorphic representations in the discrete spectrum that are expected to satisfy Ramanujan's conjecture all have multiplicity 1. Local results of Moeglin [**M4**] on nontempered p-adic packets suggest that similar results hold for all automorphic representations in the discrete spectrum.

Theorems 1.5.1 and 1.5.2 are founded on the proof of several cases of Langlands' principle of functoriality. In fact, our basic definitions will be derived from the functorial correspondence from G to $GL(N)$, relative to the standard representation of ${}^LG$ into $GL(N,\mathbb{C})$. Otherwise said, our construction of representations of G will be formulated in terms of representations of $GL(N)$. The integer N of course equals $2n$, $2n+1$ and $2n$ as G ranges over the groups $SO(2n+1)$, $Sp(2n)$ and $SO(2n)$ in the three infinite families B_n, C_n and D_n, with dual groups $\widehat{G}$ being equal to $Sp(2n,\mathbb{C})$, $SO(2n+1,\mathbb{C})$ and $SO(2n,\mathbb{C})$, respectively. The third case $SO(2n)$, which includes quasisplit outer twists, is complicated by the fact that it is really the nonconnected group $O(2n)$ that is directly tied to $GL(N)$. This is what is responsible for the failure of multiplicity one described above. It is also the reason we have not yet specified the implicit equivalence relation for the local and global parameters (1) and (3). Let us now agree that they are to be taken up to

$\widehat{G}$-conjugacy if G equals $SO(2n+1)$ or $Sp(2n)$, and up to conjugacy by $O(2n,\mathbb{C})$, a group whose quotient

$$\widetilde{\mathrm{Out}}_N(G) = O(2n,\mathbb{C})/SO(2n,\mathbb{C})$$

acts by outer automorphism on $\widehat{G} = SO(2n,\mathbb{C})$, if G equals $SO(2n)$. This is the understanding on which the decomposition (2) holds.

In the text, we shall write $\widetilde{\Psi}(G_v)$ for the set of equivalence classes of local parameters (1). The packet $\widetilde{\Pi}_{\psi_v}$ attached to any $\psi_v \in \widetilde{\Psi}(G_v)$ will then be composed of $\widetilde{\mathrm{Out}}_N(G_v)$-orbits of equivalence classes of irreducible representations (with $\widetilde{\mathrm{Out}}_N(G)$ being trivial in case G equals $SO(2n+1)$ or $Sp(2n)$). We will write $\widetilde{\Psi}(G)$ for the set of equivalence classes of general global parameters ψ, and $\widetilde{\Psi}_2(G)$ for the subset of classes with the supplementary condition of (3). The packet of any $\psi \in \widetilde{\Psi}(G)$ will then be a restricted direct product

$$\widetilde{\Pi}_\psi = \widetilde{\bigotimes_v} \widetilde{\Pi}_{\psi_v}$$

of local packets. It is this set that corresponds to an (isobaric) automorphic representation of $GL(N)$. In particular, the global packets $\widetilde{\Pi}_\psi$, rather than the individual (orbits of) representations π in $\widetilde{\Pi}_\psi$, are the objects that retain the property of strong multiplicity one from $GL(N)$. Similarly, the global packets $\widetilde{\Pi}_\psi$ attached to parameters $\psi \in \widetilde{\Psi}_2(G)$ retain the qualitative properties of automorphic discrete spectrum of $GL(N)$. They come with a sort of Jordan decomposition, in which the semisimple packets correspond to the generic global parameters ψ, and contain the automorphic representations that are expected to satisfy the G-analogue of Ramanujan's conjecture. In view of these comments, we see that Theorem 1.5.2 can be regarded as a simultaneous analogue for G of both of the global theorems for $GL(N)$.

Theorem 1.5.3 is a global supplement to Theorem 1.5.2. Its first assertion applies to global parameters $\phi \in \widetilde{\Psi}(G)$ that are both generic and simple, in the sense that they correspond to cuspidal automorphic representations π_ϕ of $GL(N)$. Theorem 1.5.3(a) asserts that the dual group $\widehat{G}$ is orthogonal (resp. symplectic) if and only if the symmetric square L-function $L(s,\pi_\phi,S^2)$ (resp. the skew-symmetric square L-function $L(s,\pi_\phi,\Lambda^2)$) has a pole at $s=1$. Theorem 1.5.3(b) asserts that the Rankin-Selberg ε-factor $\varepsilon\big(\frac{1}{2},\pi_{\phi_1}\times\pi_{\phi_2}\big)$ equals 1 for any pair of generic simple parameters $\phi_i \in \widetilde{\Phi}(G_i)$ such that $\widehat{G}_1$ and $\widehat{G}_2$ are either both orthogonal or both symplectic. These two assertions are automorphic analogues of well known properties of Artin L-functions and ε-factors. They are interesting in their own right. But they are also an essential part of our induction argument. We will need them in Chapter 4 to interpret the terms in the trace formula attached to compound parameters $\psi \in \widetilde{\Psi}(G)$.

This completes our summary of the main theorems. The first two sections of Chapter 1 contain further motivation, for the global Langlands group

L_F in §1.1, and the relations between representations of G and $GL(N)$ in §1.2. In §1.3, we will recall the three basic theorems for $GL(N)$. Section 1.4 is given over to our makeshift substitutes for global Langlands parameters, while §1.5 contains the formal statements of the theorems.

As might be expected, the three theorems will have to be established together. The unified proof will take us down a long road, which starts in Chapter 2, and crosses much diverse territory before coming to an end finally in §8.2. The argument is ultimately founded on harmonic analysis, represented locally by orbital integrals and characters, and globally by the trace formula. This of course is at the heart of the theory of endoscopy. We refer the reader to the introductory remarks of individual sections, where we have tried to offer guidance and motivation. We shall be content here with a minimal outline of the main stages.

Chapter 2 is devoted to local endoscopy. It contains a more precise formulation (Theorem 2.2.1) of the local Theorem 1.5.1. This provides for a canonical construction of the local packets $\widetilde{\Pi}_{\psi_v}$ in terms of twisted characters on $GL(N)$. Chapter 2 also includes the statement of Theorem 2.4.1, which we call the local intertwining relation. This is closely related to Theorem 1.5.1 and its refinement Theorem 2.2.1, and from a technical standpoint, can be regarded as our primary local result. It includes a delicate construction of signs, which will be critical for the interpretation of terms in the trace formula.

Chapter 3 is devoted to global endoscopy. We will recall the discrete part of the trace formula in §3.1, and its stabilization in §3.2. We are speaking here of those spectral terms that are linear combinations of automorphic characters, and to which all of the other terms are ultimately dedicated. They are the only terms from the trace formula that will appear explicitly in this volume. In §3.5, we will establish criteria for the vanishing of coefficients in certain identities (Proposition 3.5.1, Corollary 3.5.3). We shall use these criteria many times throughout the volume in drawing conclusions from the comparison of discrete spectral terms.

In general, we will have to treat three separate cases of endoscopy. They are represented respectively by pairs (G, G'), where G is one of the groups to which Theorems 1.5.1, 1.5.2 and 1.5.3 apply and G' is a corresponding endoscopic datum, pairs $\big(\widetilde{G}(N), G\big)$ in which $\widetilde{G}(N)$ is the twisted general linear group $\widetilde{G}^0(N) = GL(N)$ and G is a corresponding twisted endoscopic datum, and pairs $(\widetilde{G}, \widetilde{G}')$ in which $\widetilde{G}$ is a twisted even orthogonal group $\widetilde{G}^0 = SO(2n)$ and $\widetilde{G}'$ is again a corresponding twisted endoscopic datum. The first two cases will be our main concern. However, the third case $(\widetilde{G}, \widetilde{G}')$ is also a necessary part of the story. Among other things, it is forced on us by the need to specify the signs in the local intertwining relation. For the most part, we will not try to treat the three cases uniformly as cases of the general theory of endoscopy. This might have been difficult, given that we have to deduce many local and global results along the way. At any rate, the separate treatment of the three cases gives our exposition a more concrete flavour, if at the expense of some possible sacrifice of efficiency.

In Chapter 4, we will study the comparison of trace formulas. Specifically, we will compare the contribution (4.1.1) of a parameter ψ to the discrete part of the trace formula with the contribution (4.1.2) of ψ to the corresponding endoscopic decomposition. We begin with the statement of Theorem 4.1.2, which we call the stable multiplicity formula. This is closely related to Theorem 1.5.2, and from a technical standpoint again, is our primary global result. Together with the global intertwining relation (Corollary 4.2.1), which we state as a global corollary of Theorem 2.4.1, it governs how individual terms in trace formulas are related. Chapter 4 represents a standard model, in the sense that if we grant the analogues of the two primary theorems for general groups, it explains how the terms on the right hand sides of (4.1.1) and (4.1.2) match. This is discussed heuristically in Sections 4.7 and 4.8. However, our real aim is in the opposite direction. It is to derive Theorems 2.4.1 and 4.1.2 for our groups G *from* the standard model, and whatever else we can bring to bear on the problem. This is the perspective of Sections 4.3 and 4.4. In §4.5, we combine the analysis of these sections with a general induction hypothesis to deduce the stable multiplicity formula and the global intertwining relation for many ψ. Section 4.6 contains the proof of two critical sign lemmas that are essential ingredients of the parallel Sections 4.3 and 4.4.

Chapter 5 is the center of the volume. It is a bridge between the global discussion of Chapters 3 and 4 and the local discussion of Chapters 6 and 7. It also represents a transition from the general comparisons of Chapter 4 to the study of the remaining parameters needed to complete the induction hypotheses. These exceptional cases are the crux of the matter. In §5.2 and §5.3, we shall extract several identities from the standard model, in which we display the possible failure of Theorems 2.4.1 and 4.1.2 as correction terms. Section 5.3 applies to the critical case of a parameter $\psi \in \widetilde{\Psi}_2(G)$, and calls for the introduction of a supplementary parameter ψ_+. In §5.4, we shall resolve the global problems for families of parameters ψ that are assumed to have certain rather technical local properties.

Chapter 6 applies to generic local parameters. It contains a proof of the local Langlands classification for our groups G (modified by the outer automorphism in the case $G = SO(2n)$). We will first have to embed a given local parameter into a family of global parameters with the local constraints of §5.4. This will be the object of Sections 6.2 and 6.3, which rest ultimately on the simple form of the invariant trace formula for G. We will then have to extract the required local information from the global results obtained in §5.4. In §6.4, we will deduce the generic local intertwining relation from its global counterpart in §5.4. Then in §6.5, we will stabilize the orthogonality relations that are known to hold for elliptic tempered characters. This will allow us to quantify the contributions from the remaining elliptic tempered characters, the ones attached to square integrable representations. We will use the information so obtained in §6.6 and §6.7. In these sections, we shall establish Theorems 2.2.1 and 1.5.1 for the remaining "square integrable" Langlands parameters $\phi \in \widetilde{\Phi}_2(G)$. Finally, in §6.8, we shall resolve the various hypotheses taken on at the beginning of Chapter 6.

Chapter 7 applies to nongeneric local parameters. It contains the proof of the local theorems in general. In §7.2, we shall use the construction of §6.2 to embed a given nongeneric local parameter into a family of global parameters, but with local constraints that differ slightly from those of §5.4. We will then deduce special cases of the local theorems that apply to the places v with local constraints. These will follow from the local theorems for generic parameters, established in Chapter 6, and the duality operator of Aubert and Schneider-Stuhler, which we review in §7.1. We will then exploit our control over the places v to derive the local theorems at the localization $\psi = \dot{\psi}_u$ that represents the original given parameter.

We will finish the proof of the global theorems in the first two sections of Chapter 8. Armed with the local theorems, and the resulting refinements of the lemmas from Chapter 5, we will be able to establish almost all of the global results in §8.1. However, there will still to be one final obstacle. It is the case of a simple parameter $\psi \in \widetilde{\Psi}_{\mathrm{sim}}(G)$, which among other things will be essential for a resolution of our induction hypotheses. An examination of this case leads us to the initial impression that it will be resistant to all of our earlier techniques. Fortunately, with further reflection, we will find that there is a way to treat it after all. We will introduce a second supplementary parameter ψ_{++}, which appears ungainly at first, but which, with the support of two rather intricate lemmas, takes us to a successful conclusion.

Section 8.2 is the climax of our long running induction argument, as well as its most difficult point of application. Its final resolution is what brings us to the end of the proof. We will then be free in §8.3 for some general reflections that will give us some perspective on what has been established. In §8.4, we will sharpen our results for the groups $SO(2n)$ in which the outer automorphism creates some ambiguity. We will use the stabilized trace formula to construct the local and global L-packets for generic parameters predicted for these groups by the conjectural theory of endoscopy. In §8.5, we will describe an approximation of a part of the global Langlands group L_F, which is tailored to the classical groups of this volume. It could potentially be used in place of some of the ad hoc global parameters of §1.4 to streamline the statements of the global theorems.

We shall discuss inner forms of orthogonal and symplectic groups in Chapter 9. The automorphic representation theory for inner twists is in some ways easier for knowing what happens in the case of quasisplit groups. In particular, the stable multiplicity formula is already in place, since it applies only to the quasisplit case. However, there are also new difficulties for inner twists, particularly in the local case. We shall describe some of these in Sections 9.1–9.3. We shall then state analogues for inner twists of the main theorems, with the understanding that their proofs will appear elsewhere.

Having briefly summarized the various chapters, we should add some comment on our use of induction. As we have noted, induction is a central part of the unified argument that will carry us from Chapter 2 to Section 8.2. We will have two kinds of hypotheses, both based on the positive integer N that indexes the underlying general linear group $GL(N)$. The first kind includes various ad hoc assumptions, such as those implicit in some of our definitions. For example, the global parameter sets $\widetilde{\Psi}(G)$ defined in §1.4 are based on the inductive application of two "seed" Theorems 1.4.1 and 1.4.2. The second will be the formal induction hypotheses introduced explicitly at the beginning of §4.3, and in more refined form at the beginning of §5.1. They assert essentially that the stated theorems are all valid for parameters of rank less than N. In particular, they include the informal hypotheses implicit in the definitions.

We do not actually have to regard the earlier, informal assumptions as inductive. They really represent implicit appeals to stated theorems, in support of proofs of what amount to corollaries. In fact, from a logical standpoint, it is simpler to treat them as inductive assumptions *only* after we introduce the formal induction hypotheses in §4.3 and §5.1. For a little more discussion of this point, the reader can consult the two parallel *Remarks* following Corollaries 4.1.3 and 4.2.4.

The induction hypotheses of §5.1 are formulated for an abstract family $\widetilde{\mathcal{F}}$ of global parameters. They pertain to the parameters $\psi \in \widetilde{\mathcal{F}}$ of rank less than N, and are supplemented also by a hypothesis (Assumption 5.1.1) for certain parameters in $\widetilde{\mathcal{F}}$ of rank equal to N. The results of Chapter 5 will be applied three times, to three separate families $\widetilde{\mathcal{F}}$. These are the family of generic parameters with local constraints used to establish the local classification of Chapter 6, the family of nongeneric parameters with local constraints used to deduce the local results for nontempered representations in Chapter 7, and the family of all global parameters used to establish the global theorems in the first two sections of Chapter 8. In each of these cases, the assumptions have to be resolved for the given family $\widetilde{\mathcal{F}}$. In the case of Chapter 6, the induction hypotheses are actually imposed in two stages. The local hypothesis at the beginning of §6.3 is needed to construct the family $\widetilde{\mathcal{F}}$, on which we then impose the global part of the general hypothesis of Chapter 5 at the beginning of §6.4. The earlier induction hypotheses of §4.3 apply to general global parameters of rank less than N. They are used in §4.3–§4.6 to deduce the global theorems for parameters that are highly reducible. Their general resolution comes only after the proof of the global theorems in §8.2.

Our induction assumptions have of course to be distinguished from the general condition (Hypothesis 3.2.1) on which our results rely. This is the stabilization of the twisted trace formula for the two groups $GL(N)$ and $SO(2n)$. As a part of the condition, we implicitly include twisted analogues of the two local results that have a role in the stabilization of the ordinary trace formula. These are the orthogonality relations for elliptic tempered characters of [**A10**, Theorem 6.1], and the weak spectral transfer of tempered p-adic characters given by [**A11**, Theorems 6.1 and 6.2]. The stabilization of orthogonality relations in §6.5 requires twisted orthogonality relations for $GL(N)$, as well as those of [**A10**]. It will be an essential part of the local classification in Chapter 6. The two theorems from [**A11**] can be regarded as a partial generalization of the fundamental lemma for the full spherical Hecke algebra [**Hal**]. (Their global proof of course depends on the basic fundamental lemma for the unit, established by Ngo.) We will use them in combination with their twisted analogues in the proofs of Proposition 2.1.1 and Corollary 6.7.4. The first of these gives the image of the twisted transfer of functions from $GL(N)$, which is needed in the proof of Proposition 3.5.1. The second gives a relation among tempered characters that completes the local classification.

There is one other local theorem whose twisted analogues for $GL(N)$ and $SO(2n)$ we shall also have to take for granted. It is Shelstad's strong spectral transfer of tempered archimedean characters, which is to say, her endoscopic classification of representations of real groups. This of course is a major result. Together with its two twisted analogues, it gives the archimedean cases of the local classification in Theorems 2.2.1 and 2.2.4. We shall combine it with a global argument in Chapter 6 to establish the p-adic form of these theorems. The general twisted form of Shelstad's endoscopic classification appears to be within reach. It is likely to be established soon by some extension of recent work by Mezo [**Me**] and Shelstad [**S8**].

Finally, let me include a comment on notation. Because our main theorems require interlocking proofs, which consume a good part of the volume, there is always the risk of losing one's way. Until the end of §8.2, assertions as *Theorems* are generally stated with the understanding that their proofs will usually be taken up much later (unless of course they are simply quoted from some other source). On the other hand, assertions denoted *Propositions*, *Lemmas* or *Corollaries* represent results along the route, for which the reader can expect a timely proof. The theorems stated in §8.4 are not part of the central induction argument. Their proofs, which are formally labeled as such, follow relatively soon after their statements. The theorems stated in §9.4 and §9.5 apply to inner twists. They will be proved elsewhere.

The actual mathematical notation of the volume might appear unconventional at times. I have tried to structure it so as to reflect implicit symmetries in the various objects it represents. With luck, it might help a reader navigate the arguments without necessarily being aware of such symmetries.

The three main theorems of the volume were described in [**A18**, §30]. I gave lecture courses on them in 1994–1995 at the Institute for Advanced

Study and the University of Paris VII, and later in 2000, again at the Institute for Advanced Study. Parts of Chapter 4 were also treated heuristically in the earlier article [**A9**]. In writing this volume, I have added some topics to my original notes. These include the local Langlands classification for $GL(N)$, the treatment of inner twists in Chapter 9 and the remarks on Whittaker models in §8.3. I have also had to fill unforeseen gaps in the notes. For example, I did not realize that twisted endoscopy for $SO(2n)$ was needed to formulate the local intertwining relation. In retrospect, it is probably for the best that this second case of twisted endoscopy does have a role here, since it forces us to confront a general phenomenon in a concrete situation. I have tried to make this point explicit in §2.4 with the discussion surrounding the short exact sequence (2.4.10). In any case, I hope that I have accounted for most of the recent work on the subject, in the references and the text. There will no doubt be omissions. I most regret not being able to describe the results of Moeglin on the structure of p-adic packets $\widetilde{\Pi}_{\psi_v}$ ([**M1**]–[**M4**]). It is clearly an important problem to establish analogues of her results for archimedean packets.

Acknowledgments: I am greatly indebted to the referee, for a close and careful reading of an earlier draft of the manuscript (which went to the end of §8.2), and for many very helpful comments. I also thank Chung Pang Mok, who read a later draft of the manuscript, and offered further helpful suggestions.

CHAPTER 1

Parameters

1.1. The automorphic Langlands group

We begin with some general motivation. We shall review some of the fundamental ideas that underlie the theoretical foundations laid by Langlands. This will help us put our theorems into perspective. It will also lead naturally to a formulation of some of the essential objects with which we need to work.

We take F to be a local or global field of characteristic 0. In other words, F is a finite extension of either the real field $\mathbb{R}$ or a p-adic field $\mathbb{Q}_p$, or it is a finite extension of $\mathbb{Q}$ itself. Suppose that G is a connected reductive algebraic group over F, which to be concrete we take to be a classical matrix group. For example, we could let G be the general linear group

$$G(N) = GL(N)$$

of invertible matrices of rank N over F.

In his original paper [**L2**], Langlands introduced what later became known as the L-group of G. This object is a semidirect product

$${}^LG = \widehat{G} \rtimes \Gamma_F$$

of a complex dual group $\widehat{G}$ of G with the Galois group

$$\Gamma_F = \mathrm{Gal}(\overline{F}/F)$$

of an algebraic closure $\overline{F}$ of F. The action of Γ_F on $\widehat{G}$ (called an L-action) is determined by its action on a based root datum for G and a corresponding splitting for $\widehat{G}$, according to the general theory of algebraic groups. (See [**K3**, §1.1–§1.3].) It factors through the quotient $\Gamma_{E/F}$ of Γ_F attached to any finite Galois extension $E \supset F$ over which G splits. We sometimes formulate the L-group by the simpler prescription

$${}^LG = {}^LG_{E/F} = \widehat{G} \rtimes \Gamma_{E/F},$$

since this suffices for many purposes. If $G = G(N)$, for example, the action of Γ_F on $\widehat{G}$ is trivial. Since $\widehat{G}$ is just the complex general linear group $GL(N, \mathbb{C})$ in this case, one can often work with the restricted form

$${}^LG = \widehat{G} = GL(N, \mathbb{C})$$

of the L-group.

Langlands' conjectures [**L2**] predicate a fundamental role for the L-group in the representation theory of G. Among other things, Langlands conjectured the existence of a natural correspondence

$$\phi \longrightarrow \pi$$

between two quite different kinds of objects. The domain consists of (continuous) L-homomorphisms

$$\phi : \Gamma_F \longrightarrow {}^L G,$$

taken up to conjugation by $\widehat{G}$. (An *L-homomorphism* between two groups that fibre over Γ_F is a homomorphism that commutes with the two projections onto Γ_F.) The codomain consists of irreducible representations π of $G(F)$ if F is local, and automorphic representations π of $G(\mathbb{A})$ if F is global, taken in each case up to the usual relation of equivalence of irreducible representations.

Recall that if F is global, the adele ring is defined as a restricted tensor product

$$\mathbb{A} = \mathbb{A}_F = \prod_v F_v$$

of completions F_v of F. In this case, the Langlands correspondence should satisfy the natural local-global compatibility condition. Namely, if ϕ_v denotes the restriction of ϕ to the subgroup Γ_{F_v} of Γ_F (which is defined up to conjugacy), and π is a restricted tensor product

$$\pi = \widetilde{\bigotimes_v} \pi_v, \qquad \phi_v \rightarrow \pi_v,$$

of representations that correspond to these localizations, then π should correspond to ϕ. We refer the reader to the respective articles [**F**] and [**L6**] for a discussion of restricted direct products and automorphic representations.

The correspondence $\phi \rightarrow \pi$, which remains conjectural, is to be understood in the literal sense of the word. For general G, it will not be a mapping. However, in the case $G = GL(N)$, the correspondence should in fact reduce to a well defined, injective mapping. For local F, this is part of what has now been established, as we will recall in §1.3. For global F, the injectivity would be a consequence of the required local-global compatibility condition and the theorem of strong multiplicity one, or rather its generalization in [**JS**] that we will also recall in §1.3. However, the correspondence will very definitely not be surjective.

In our initial attempts at motivation, we should not lose sight of the fact that the conjectural Langlands correspondence is very deep. For example, even though the mapping $\phi \rightarrow \pi$ is known to exist for $G = GL(1)$, it takes the form of the fundamental reciprocity laws of local and global class field theory. Its generalization to $GL(N)$ would amount to a formulation of nonabelian class field theory.

Langlands actually proposed the correspondence $\phi \to \pi$ with the Weil group W_F in place of the Galois group Γ_F. We recall that W_F is a locally compact group, which was defined separately for local and global F by Weil. It is equipped with a continuous homomorphism

$$W_F \longrightarrow \Gamma_F,$$

with dense image. If F is global, there is a commutative diagram

$$\begin{array}{ccc} W_{F_v} & \longrightarrow & \Gamma_{F_v} \\ \downarrow & & \downarrow \\ W_F & \longrightarrow & \Gamma_F \end{array}$$

for any completion F_v of F, with vertical embeddings defined up to conjugation. (See [**T2**].) In the Weil form of the Langlands correspondence, ϕ represents an L-homomorphism from W_F to ${}^LG$. The restriction mapping of continuous functions on Γ_F to continuous functions on W_F is injective. For this reason, the conjectural Langlands correspondence for Weil groups is a generalization of its version for Galois groups.

For $G = GL(N)$, the Weil form of the conjectural correspondence $\phi \to \pi$ again reduces to an injective mapping. (In the global case, one has to take π to be an *isobaric* automorphic representation, a natural restriction introduced in [**L7**] that includes all the representations in the automorphic spectral decomposition of $GL(N)$.) If $G = GL(1)$, it also becomes surjective. However, for nonabelian groups G, and in particular for $GL(N)$, the correspondence will again not be surjective. One of the purposes of Langlands' article [**L7**] was to suggest the possibility of a larger group, which when used in place of the Weil group, would give rise to a bijective mapping for $GL(N)$. Langlands formulated the group as a complex, reductive, proalgebraic group, in the spirit of the complex form of Grothendieck's motivic Galois group.

Kottwitz later pointed out that Langlands' group ought to have an equivalent but simpler formulation as a locally compact group L_F [**K3**]. It would come with a surjective mapping

$$L_F \longrightarrow W_F$$

onto the Weil group, whose kernel K_F should be compact and connected, and (in the optimistic view of some [**A17**]) even simply connected. If F is local, L_F would take the simple form

$$L_F = \begin{cases} W_F, & \text{if } F \text{ is archimedean}, \\ W_F \times SU(2), & \text{if } F \text{ is nonarchimedean}. \end{cases} \tag{1.1.1}$$

In this case, L_F is actually a *split* extension of W_F by a compact, simply connected group (namely, the trivial group $\{1\}$ if F is archimedean and the three-dimensional compact Lie group $SU(2) = SU(2, \mathbb{R})$ if F is p-adic.) If F is global, L_F remains hypothetical. Its existence is in fact one of the deepest problems in the subject. Whatever form it does ultimately take, it ought to fit into a larger commutative diagram

$$\begin{array}{ccccc} L_{F_v} & \longrightarrow & W_{F_v} & \longrightarrow & \Gamma_{F_v} \\ \downarrow & & \downarrow & & \downarrow \\ L_F & \longrightarrow & W_F & \longrightarrow & \Gamma_F \end{array}$$

for any completion F_v, the vertical embedding on the left again being defined up to conjugation.

The hypothetical formal structure of L_F is thus compatible with an extension of the Langlands correspondence from W_F to L_F. This is what Langlands proposed in [**L7**] (for the proalgebraic form of L_F). The extension amounts to a hypothetical correspondence $\phi \to \pi$, in which ϕ now represents an L-homomorphism

$$\phi : \ L_F \longrightarrow {}^LG,$$

taken again up to $\widehat{G}$-conjugacy. Here it is convenient to use the Weil form of the L-group

$${}^LG = \widehat{G} \rtimes W_F,$$

for the action of W_F on $\widehat{G}$ inherited from Γ_F. An L-homomorphism between two groups over W_F is again one that commutes with the two projections. There are some minor conditions on ϕ that are implicit here. For example, since W_F and L_F are no longer compact, one has to require that for any $\lambda \in L_F$, the image of $\phi(\lambda)$ in $\widehat{G}$ be semisimple. If G is not quasisplit, one generally also requires that ϕ be *relevant* to G, in the sense that if its image lies in a parabolic subgroup ${}^LP$ of ${}^LG$, there is a corresponding parabolic subgroup P of G that is defined over F.

Suppose again that $G = GL(N)$. Then the hypothetical extended correspondence $\phi \to \pi$ again reduces to an injective mapping. However, this time it should also be surjective (provided that for global F, we take the image to be the set of isobaric automorphic representations). If F is local archimedean, so that $L_F = W_F$, the correspondence was established (for any G in fact) by Langlands [**L11**]. If F is a local p-adic field, so that $L_F = W_F \times SU(2)$, the correspondence was established for $GL(N)$ by Harris and Taylor [**HT**] and Henniart [**He1**]. For global F, the correspondence for $GL(N)$ is much deeper, and remains highly conjectural. We have introduced it here as a model to motivate the form of the theorems we seek for classical groups.

If G is more general than $GL(N)$, the extended correspondence $\phi \to \pi$ will not reduce to a mapping. It was to account for this circumstance that Langlands introduced what are now called L-packets. We recall that L-packets are supposed to be the equivalence classes for a natural relation that is weaker than the usual notion of equivalence of irreducible representations. (The supplementary relation is called *L-equivalence*, since it is arithmetic in nature, and is designed to preserve the L-functions and ε-factors of representations.) The extended correspondence $\phi \to \pi$ is supposed to project to a well defined mapping $\phi \to \Pi_\phi$ from the set of parameters ϕ to the set of L-packets. For $G = GL(N)$, L-equivalence reduces to ordinary equivalence. The L-packets Π_ϕ then contain one element each, which is the reason that the correspondence $\phi \to \pi$ reduces to a mapping in this case.

The general construction of L-packets is part of Langlands' conjectural theory of endoscopy. It will be a central topic of investigation for this volume. We recall at this stage simply that the L-packet attached to a given ϕ will be intimately related to the centralizer

$$S_\phi = \mathrm{Cent}\big(\mathrm{Im}(\phi), \widehat{G}\big) \tag{1.1.2}$$

in $\widehat{G}$ of the image $\phi(L_F)$ of ϕ, generally through its finite quotient

$$\mathcal{S}_\phi = S_\phi / S_\phi^0 Z(\widehat{G})^\Gamma. \tag{1.1.3}$$

Following standard notation, we have written S_ϕ^0 for the connected component of 1 in the complex reductive group S_ϕ, $Z(\widehat{G})$ for the center of $\widehat{G}$, and $Z(\widehat{G})^\Gamma$ for the subgroup of invariants in $Z(\widehat{G})$ under the natural action of the Galois group $\Gamma = \Gamma_F$. For $G = GL(N)$, the groups S_ϕ will all be connected. Each quotient $\mathcal{S}_\phi$ is therefore trivial. The implication for other groups G is that we will have to find a way to introduce the centralizers S_ϕ, even though we have no hope of constructing the automorphic Langlands group L_F and the general parameters ϕ.

There is a further matter that must also be taken into consideration. Suppose for example that F is global and that $G = GL(N)$. The problem in this case is that the conjectural parametrization of automorphic representations π by N-dimensional representations

$$\phi:\ L_F \longrightarrow \widehat{G} = GL(N, \mathbb{C})$$

is not compatible with the spectral decomposition of $L^2\big(G(F)\backslash G(\mathbb{A})\big)$. If ϕ is irreducible, π is supposed to be a cuspidal automorphic representation. Any such representation is part of the discrete spectrum (taken modulo the center). However, there are also noncuspidal automorphic representations in the discrete spectrum. These come from residues of Eisenstein series, and include for example the trivial one-dimensional representation of $G(\mathbb{A})$. Such automorphic representations will correspond to certain *reducible* N-dimensional representations of L_F. How is one to account for them?

The answer, it turns out, lies in the product of L_F with the supplementary group $SU(2) = SU(2, \mathbb{R})$. The representations in the discrete automorphic spectrum of $GL(N)$ should be attached to *irreducible unitary* N-dimensional representations of this product. The local constituents of these automorphic representations should again be determined by the restriction of parameters, this time from the product $L_F \times SU(2)$ to its subgroups $L_{F_v} \times SU(2)$. Notice that if v is a p-adic valuation, the localization

$$L_{F_v} \times SU(2) = W_{F_v} \times SU(2) \times SU(2)$$

contains two $SU(2)$-factors. Each will have its own distinct role. In §1.3, we shall recall the general construction, and why it is the product $L_F \times SU(2)$ that governs the automorphic spectrum of $GL(N)$.

Similar considerations should apply to a more general connected group G over any F. One would consider L-homomorphisms

$$\psi : \ L_F \times SU(2) \longrightarrow {}^L G,$$

with relatively compact image in $\widehat{G}$ (the analogue of the unitary condition for $GL(N)$). If F is global, the parameters should govern the automorphic spectrum of G. If F is local, they ought to determine corresponding local constituents. In either case, the relevant representations should occur in packets Π_ψ, which are larger and more complicated than L-packets, but which are better adapted to the spectral properties of automorphic representations. These packets should in turn be related to the centralizers

$$S_\psi = \mathrm{Cent}\big(\mathrm{Im}(\psi), \widehat{G}\big)$$

and their quotients

$$\mathcal{S}_\psi = S_\psi / S_\psi^0 Z(\widehat{G})^\Gamma.$$

For $G = GL(N)$, the groups S_ψ remain connected. However, for other classical groups G we might wish to study, we must again be prepared to introduce parameters ψ and centralizers S_ψ without reference to the global Langlands group L_F.

The objects of study in this volume will be orthogonal and symplectic groups G. Our general goal will be to classify the representations of such groups in terms of those of $GL(N)$. In the hypothetical setting of the discussion above, the problem includes being able to relate the parameters ψ for G with those for $GL(N)$. As further motivation for what is to come, we shall consider this question in the next section. We shall analyze the self-dual, finite dimensional representations of a general group Λ_F. Among other things, this exercise will allow us to introduce endoscopic data, the internal objects for G that drive the classification, in concrete terms.

1.2. Self-dual, finite-dimensional representations

We continue to take F to be any local or global field of characteristic 0. For this section, we let Λ_F denote a general, unspecified topological group. The reader can take Λ_F to be one of the groups Γ_F, W_F or L_F discussed in §1.1, or perhaps the product of one of these groups with $SU(2)$. We assume only that Λ_F is equipped with a continuous mapping $\Lambda_F \to \Gamma_F$, with connected kernel and dense image.

We shall be looking at continuous, N-dimensional representations

$$r : \ \Lambda_F \longrightarrow GL(N, \mathbb{C}).$$

Any such r factors through the preimage of a finite quotient of Γ_F. We can therefore replace Λ_F by its preimage. In fact, one could simply take a large finite quotient of Γ_F in place of Λ_F, which for the purposes of the present exercise we could treat as an abstract finite group.

We say that r is *self-dual* if it is equivalent to its contragredient representation

$$r^\vee(\lambda) = {}^t r(\lambda)^{-1}, \qquad \lambda \in \Lambda_F,$$

where $x \to {}^t x$ is the usual transpose mapping. In other words, the equivalence class of r is invariant under the standard outer automorphism

$$\theta(x) = x^\vee = {}^t x^{-1}, \qquad x \in GL(N),$$

of $GL(N)$. This condition depends only on the inner class of θ. It remains the same if θ is replaced by any conjugate

$$\theta_g(x) = g^{-1}\theta(x)g, \qquad g \in GL(N).$$

We shall analyze the self-dual representations r in terms of orthogonal and symplectic subgroups of $GL(N, \mathbb{C})$.

We decompose a given representation r into a direct sum

$$r = \ell_1 r_1 \oplus \cdots \oplus \ell_r r_r,$$

for inequivalent irreducible representations

$$r_k : \Lambda_F \longrightarrow GL(N_k, \mathbb{C}), \qquad 1 \leqslant k \leqslant r,$$

and multiplicities ℓ_k with

$$N = \ell_1 N_1 + \cdots + \ell_r N_r.$$

The representation is self-dual if and only if there is an involution $k \leftrightarrow k^\vee$ on the indices such that for any k, $r_k^\vee$ is equivalent to $r_{k^\vee}$ and $\ell_k = \ell_{k^\vee}$. We shall say that r is *elliptic* if it satisfies the further constraint that for each k, $k^\vee = k$ and $\ell_k = 1$. We shall concentrate on this case.

Assume that r is elliptic. Then

$$r = r_1 \oplus \cdots \oplus r_r,$$

for distinct irreducible, self-dual representations r_i of Λ_F of degree N_i. If i is any index, we can write

$$r_i^\vee(\lambda) = A_i r_i(\lambda) A_i^{-1}, \qquad \lambda \in \Lambda_F,$$

for a fixed element $A_i \in GL(N_i, \mathbb{C})$. Applying the automorphism θ to each side of this equation, we then see that

$$r_i(\lambda) = A_i^\vee r_i^\vee (A_i^\vee)^{-1} = (A_i^\vee A_i) r_i(\lambda)(A_i^\vee A_i)^{-1}.$$

Since r_i is irreducible, the product $A_i^\vee A_i$ is a scalar matrix. We can therefore write

$${}^t A_i = c_i A_i, \qquad c_i \in \mathbb{C}^*.$$

If we take the transpose of each side of this equation, we see further that $c_i^2 = 1$. Thus, c_i equals $+1$ or -1, and the nonsingular matrix A_i is either symmetric or skew-symmetric. The mapping

$$x_i \longrightarrow (A_i^{-1})\, {}^t x_i\, A_i, \qquad x_i \in GL(N_i),$$

represents the adjoint relative to the bilinear form defined by A_i. Therefore $r_i(\lambda)$ belongs to the corresponding orthogonal group $O(A_i, \mathbb{C})$ or symplectic group $Sp(A_i, \mathbb{C})$, according to whether c_i equals $+1$ or -1.

Let us write I_O and I_S for the set of indices i such that c_i equals $+1$ and -1 respectively. We then write

$$r_\varepsilon(\lambda) = \bigoplus_{i \in I_\varepsilon} r_i(\lambda), \qquad \lambda \in \Lambda_F,$$
$$A_\varepsilon = \bigoplus_{i \in I_\varepsilon} A_i,$$

and

$$N_\varepsilon = \sum_{i \in I_\varepsilon} N_i,$$

for ε equal to O or S. Thus A_O is a symmetric matrix in $GL(N_O, \mathbb{C})$, A_S is a skew-symmetric matrix in $GL(N_S, \mathbb{C})$, and r_O and r_S are representations of Λ_F that take values in the respective groups $O(A_O, \mathbb{C})$ and $Sp(A_S, \mathbb{C})$. We have established a canonical decomposition

$$r = r_O \oplus r_S$$

of the self-dual representation r into orthogonal and symplectic components.

It will be only the equivalence class of r that is relevant, so we are free to replace $r(\lambda)$ by its conjugate

$$B^{-1} r(\lambda) B$$

by a matrix $B \in GL(N, \mathbb{C})$. This has the effect of replacing the matrix

$$A = A_O \oplus A_S$$

by ${}^tBAB$. In particular, we could take A_O to be any symmetric matrix in $GL(N_O, \mathbb{C})$, and A_S to be any skew-symmetric matrix in $GL(N_S, \mathbb{C})$. We may therefore put the orthogonal and symplectic groups that contain the images of r_O and r_S into standard form.

It will be convenient to adopt a slightly different convention for these groups. As our standard orthogonal group in $GL(N)$, we take

$$O(N) = O(N, J),$$

where

$$J = J(N) = \begin{pmatrix} 0 & & 1 \\ & \iddots & \\ 1 & & 0 \end{pmatrix}$$

is the "second diagonal" in $GL(N)$. This is a group of two connected components, whose identity component is the special orthogonal group

$$SO(N) = \{x \in O(N) : \det(x) = 1\}.$$

As the standard symplectic group in $GL(N)$, defined for $N = 2N'$ even, we take the connected group

$$Sp(N) = Sp(N, J')$$

for the skew-symmetric matrix

$$J' = J'(N) = \begin{pmatrix} 0 & -J(N') \\ J(N') & 0 \end{pmatrix}.$$

The advantage of this formalism is that the set of diagonal matrices in either $SO(N)$ or $Sp(N)$ forms a maximal torus. Similarly, the set of upper triangular matrices in either group forms a Borel subgroup. The point is that if

$$_t x = J\,{}^t x\, J = J\,{}^t x\, J^{-1}, \qquad x \in GL(N),$$

denotes the transpose of x about the second diagonal, the automorphism

$$\mathrm{Int}(J) \circ \theta : \ x \longrightarrow \ J\theta(x)J^{-1} = {}_t x^{-1}$$

of $GL(N)$ stabilizes the standard Borel subgroup of upper triangular matrices. Notice that there is a related automorphism

$$\widetilde{\theta}(N) = \mathrm{Int}(\widetilde{J}) \circ \theta : \ x \ \longrightarrow \ \widetilde{J}\theta(x)\widetilde{J}^{-1},$$

defined by the matrix

$$\widetilde{J} = \widetilde{J}(N) = \begin{pmatrix} & 0 & & 1 \\ & & -1 & \\ & \cdot^{\cdot^{\cdot}} & & \\ (-1)^{N+1} & & & 0 \end{pmatrix}, \tag{1.2.1}$$

which stabilizes the standard splitting in $GL(N)$ as well. Both of these automorphisms lie in the inner class of θ, and either one could have been used originally in place of θ.

Returning to our discussion, we can arrange that A equals $J(N_O) \oplus J'(N_S)$. It is best to work with the matrix

$$J_{O,S} = J(N_O, N_S) = \begin{pmatrix} 0 & & -J(N_S') \\ & J(N_O) & \\ J(N_S') & & 0 \end{pmatrix}, \qquad N_S = 2N_S',$$

obtained from the obvious embedding of $J(N_O) \oplus J'(N_S)$ into $GL(N, \mathbb{C})$. The associated elliptic representation r from the given equivalence class then maps Λ_F to the corresponding subgroup of $GL(N, \mathbb{C})$, namely the subgroup

$$O(N_O, \mathbb{C}) \times Sp(N_S, \mathbb{C})$$

defined by the embedding

$$(x, y) \ \longrightarrow \ \begin{pmatrix} y_{11} & 0 & y_{12} \\ 0 & x & 0 \\ y_{21} & 0 & y_{22} \end{pmatrix},$$

where y_{ij} are the four $(N_S' \times N_S')$-block components of the matrix $y \in Sp(N_S, \mathbb{C})$.

The symplectic part r_S of r is the simpler of the two. Its image is contained in the *connected* complex group

$$\widehat{G}_S = Sp(N_S, \mathbb{C}).$$

This in turn is the dual group of the split classical group

$$G_S = SO(N_S + 1).$$

The orthogonal part r_O of r is complicated by the fact that its image is contained only in the *disconnected* group $O(N_O, \mathbb{C})$. Its composition with the projection of $O(N_O, \mathbb{C})$ onto the group

$$O(N_O, \mathbb{C})/SO(N_O, \mathbb{C}) \cong \mathbb{Z}/2\mathbb{Z}$$

of components yields a character η on Λ_F of order 1 or 2. Since we are assuming that the kernel of the mapping of Λ_F to Γ_F is connected, η can be identified with a character on the Galois group Γ_F of order 1 or 2. This in turn determines an extension E of F of degree 1 or 2.

Suppose first that N_O is odd. In this case, the matrix $(-I)$ in $O(N_O)$ represents the nonidentity component, and the orthogonal group is a direct product

$$O(N_O, \mathbb{C}) = SO(N_O, \mathbb{C}) \times \mathbb{Z}/2\mathbb{Z}.$$

We write

$$SO(N_O, \mathbb{C}) = \widehat{G}_O,$$

where G_O is the split group $Sp(N_O - 1)$ over F. We then use η to identify the direct product

$${}^L(G_O)_{E/F} = \widehat{G}_O \times \Gamma_{E/F}$$

with a subgroup of $O(N_O, \mathbb{C})$, namely $SO(N_O, \mathbb{C})$ or $O(N_O, \mathbb{C})$, according to whether η has order 1 or 2. We thus obtain an embedding of the (restricted) L-group of G_O into $GL(N_O, \mathbb{C})$.

Assume next that N is even. In this case, the nonidentity component in $O(N)$ acts by an outer automorphism on $SO(N_O)$. We write

$$SO(N_O, \mathbb{C}) = \widehat{G}_O,$$

where G_O is now the corresponding quasisplit orthogonal group $SO(N_O, \eta)$ over F defined by η. In other words, G_O is the split group $SO(N_O)$ if η is trivial, and the nonsplit group obtained by twisting $SO(N_O)$ over E by the given outer automorphism if η is nontrivial. Let $\widetilde{w}(N_O)$ be the permutation matrix in $GL(N_O)$ that interchanges the middle two coordinates, and leaves the other coordinates invariant. We take this element as a representative of the nonidentity component of $O(N_O, \mathbb{C})$. We then use η to identify the semidirect product

$${}^L(G_O)_{E/F} = \widehat{G}_O \rtimes \Gamma_{E/F}$$

with a subgroup of $O(N_O, \mathbb{C})$, namely $SO(N_O, \mathbb{C})$ or $O(N_O, \mathbb{C})$ as before. We again obtain an embedding of the (restricted) L-group of G_O into $GL(N_O, \mathbb{C})$.

We have shown that the elliptic self-dual representation r factors through the embedded subgroup

$$^L G_{E/F} = {}^L(G_O)_{E/F} \times {}^L(G_S)_{E/F}$$

of $GL(N,\mathbb{C})$ attached to a quasisplit group

$$G = G_O \times G_S$$

over F. The group G is called a θ-twisted *endoscopic group* for $GL(N)$. It is determined by r, and in fact by the decomposition $N = N_O + N_S$ and the character $\eta = \eta_G$ (of order 1 or 2) attached to r. The same is true of the L-embedding

$$\xi = \xi_{O,S,\eta} : \ {}^L G = \widehat{G} \rtimes \Gamma_F \hookrightarrow {}^L G(N) = GL(N,\mathbb{C}) \times \Gamma_F,$$

obtained by inflating the embedding above to the full L-groups.

It is convenient to form the semidirect product

$$\widetilde{G}^+(N) = GL(N) \rtimes \langle\theta\rangle = G(N) \rtimes \langle\widetilde{\theta}(N)\rangle,$$

where $\langle\theta\rangle$ and $\langle\widetilde{\theta}(N)\rangle$ are the groups of order 2 generated by the automorphisms θ and $\widetilde{\theta}(N)$. We write

$$\widetilde{G}^0(N) = GL(N) \rtimes 1 = G(N) \rtimes 1$$

for the identity component, which we can of course identify with the general linear group $GL(N)$, and

$$\widetilde{G}(N) = GL(N) \rtimes \theta = G(N) \rtimes \widetilde{\theta}(N) \tag{1.2.2}$$

for the other connected component. Given r, and hence also the decomposition $N = N_O + N_S$, we can form the semisimple element

$$s = s_{O,S} = J_{O,S}^{-1} \rtimes \theta,$$

in the "dual set" $\widehat{\widetilde{G}}(N)$ of complex points $GL(N,\mathbb{C}) \rtimes \theta$. The complex group $\widehat{G} = \widehat{G}_O \times \widehat{G}_S$, attached to r as above, is then the connected centralizer of s in the group

$$\widehat{G}(N) = \widehat{\widetilde{G}}{}^0(N) = GL(N,\mathbb{C}).$$

The triplet (G, s, ξ) is called an *endoscopic datum* for $\widetilde{G}(N)$, since it becomes a special case of the terminology of [**KS**, p. 16] if we replace $\widetilde{G}(N)$ with either of the pairs $\big(\widetilde{G}^0(N), \widetilde{\theta}(N)\big)$ or $\big(G(N), \theta\big) = \big(GL(N), \theta\big)$.

The endoscopic datum (G, s, ξ) we have introduced has the property of being elliptic. This is a consequence of our condition that the original self-dual representation r is elliptic. A general (nonelliptic) endoscopic datum for $\widetilde{G}(N)$ is again a triplet (G, s, ξ), where G is a quasisplit group over F, s is a semisimple element in $\widehat{\widetilde{G}}(N)$ of which $\widehat{G}$ is the connected centralizer in $\widehat{\widetilde{G}}{}^0(N)$, and ξ is an L-embedding of ${}^L G$ into the L-group ${}^L\widetilde{G}^0(N) = {}^L G(N)$ of $GL(N)$. (In the present setting, we are free to take either the Galois or Weil form of the L-groups.) We require that ξ equal the identity on $\widehat{G}$, and that the projection onto $\widehat{\widetilde{G}}^0(N)$ of the image of ξ lie in the full centralizer of s. The endoscopic group G (or datum (G, s, ξ)) for $\widetilde{G}(N)$ then said to be *elliptic* if $Z(\widehat{G})^\Gamma$, the subgroup of elements in the center $Z(\widehat{G})$ of $\widehat{G}$ invariant under the action of the Galois group $\Gamma = \Gamma_F$, is finite.

The notion of *isomorphism* between two general endoscopic data is defined in [**KS**, p. 18]. In the case at hand, it is given by an element g in the dual group $\widehat{G}(N) = GL(N, \mathbb{C})$ whose action by conjugation is compatible in a natural sense with the two endoscopic data. We write

$$\widetilde{\mathrm{Aut}}_N(G) = \mathrm{Aut}_{\widetilde{G}(N)}(G)$$

for the group of isomorphisms of the endoscopic datum G to itself. The main role for this group is in its image

$$\widetilde{\mathrm{Out}}_N(G) = \widetilde{\mathrm{Aut}}_N(G)/\widetilde{\mathrm{Int}}_N(G)$$

in the group of outer automorphisms of the group G over F. (Following standard practice, we often let the endoscopic group G represent a full endoscopic datum (G, s, ξ), or even an isomorphism class of such data.) If G represents one of the elliptic endoscopic data constructed above, $\widetilde{\mathrm{Out}}_N(G)$ is trivial if the integer N_O is odd or zero. In the remaining case that N_O is even and positive, $\widetilde{\mathrm{Out}}_N(G)$ is a group of order 2, the nontrivial element being the outer automorphism induced by the nontrivial connected component of $O(N_O, \mathbb{C})$.

We write

$$\widetilde{\mathcal{E}}(N) = \mathcal{E}\big(\widetilde{G}(N)\big)$$

for the set of isomorphism classes of endoscopic data for $\widetilde{G}(N)$, and

$$\widetilde{\mathcal{E}}_{\mathrm{ell}}(N) = \mathcal{E}_{\mathrm{ell}}\big(\widetilde{G}(N)\big)$$

for the subset of classes in $\widetilde{\mathcal{E}}(N)$ that are elliptic. The data (G, s, ξ), attached to equivalence classes of elliptic, self-dual representations r as above, form a complete set of representatives of $\widetilde{\mathcal{E}}_{\mathrm{ell}}(N)$. The set $\widetilde{\mathcal{E}}_{\mathrm{ell}}(N)$ is thus parametrized by triplets (N_O, N_S, η), where $N_O + N_S = N$ is a decomposition of N into nonnegative integers with N_S even, and $\eta = \eta_G$ is a character of Γ_F of order 1 or 2 with the property that $\eta = 1$ if $N_O = 0$, and $\eta \neq 1$ if $N_O = 2$. (The last constraint is required in order that the datum be elliptic.) The goal of this volume is to describe the representations of the classical groups G in terms of those of $GL(N)$. The general arguments will be inductive. For this reason, the case in which $\widehat{G}$ is either purely orthogonal or purely symplectic will have a special role. Accordingly, we write

$$\widetilde{\mathcal{E}}_{\mathrm{sim}}(N) = \mathcal{E}_{\mathrm{sim}}\big(\widetilde{G}(N)\big)$$

for the set of elements in $\widetilde{\mathcal{E}}_{\mathrm{ell}}(N)$ that are *simple*, in the sense that one of the integers N_O or N_S vanishes. We then have a chain of sets

$$\widetilde{\mathcal{E}}_{\mathrm{sim}}(N) \subset \widetilde{\mathcal{E}}_{\mathrm{ell}}(N) \subset \widetilde{\mathcal{E}}(N), \tag{1.2.3}$$

which are all finite if F is local, and all infinite if F is global.

The elements $G \in \widetilde{\mathcal{E}}(N)$ are usually called twisted endoscopic data, since they are attached to the automorphism θ. We shall have to work also with ordinary (untwisted) endoscopic data, at least for the quasisplit orthogonal and symplectic groups G that represent elements in $\widetilde{\mathcal{E}}_{\mathrm{sim}}(N)$. An endoscopic datum G' for G is similar to what we have described above for $\widetilde{G}(N)$. It amounts to a triplet (G', s', ξ'), where G' is a (connected) quasisplit group over F, s' is a semisimple element in $\widehat{G}$ of which $\widehat{G}'$ is the connected centralizer in $\widehat{G}$, and ξ' is an L-embedding of ${}^LG'$ into ${}^LG$. We again require that ξ' equal the identity on $\widehat{G}'$, and that its image lie in the centralizer of s' in ${}^LG$. (See [**LS1**, (1.2)], a special case of the general definition in [**KS**], which we have specialized further to the case at hand.) There is again the notion of isomorphism of endoscopic data, which allows us to form the associated finite group

$$\mathrm{Out}_G(G') = \mathrm{Aut}_G(G')/\mathrm{Int}_G(G')$$

of outer automorphisms of any given G'. We write $\mathcal{E}(G)$ for the set of isomorphism classes of endoscopic data G' for G, and $\mathcal{E}_{\mathrm{ell}}(G)$ for the subset of data that are elliptic, in the sense that $Z(\widehat{G}')^\Gamma$ is finite. We then have a second chain of sets

$$\mathcal{E}_{\mathrm{sim}}(G) \subset \mathcal{E}_{\mathrm{ell}}(G) \subset \mathcal{E}(G), \tag{1.2.4}$$

where $\mathcal{E}_{\mathrm{sim}}(G) = \{G\}$ is the subset consisting of G alone. Similar definitions apply to groups G that represent more general data in $\widetilde{\mathcal{E}}(N)$.

It is easy to describe the set $\mathcal{E}_{\mathrm{ell}}(G)$, for any $G \in \widetilde{\mathcal{E}}_{\mathrm{sim}}(N)$. It suffices to consider diagonal matrices $s' \in \widehat{G}$ with eigenvalues ± 1. For example, in the first case that $G = SO(N+1)$ and $\widehat{G} = Sp(N, \mathbb{C})$ (with $N = N_S$ even), it is enough to take diagonal matrices of the form

$$s' = \begin{pmatrix} -I_1'' & & 0 \\ & I_2' & \\ 0 & & -I_1'' \end{pmatrix},$$

where I_1'' is the identity matrix of rank $N_1'' = N_1'/2$, and I_2' is the identity matrix of rank N_2'. The set $\mathcal{E}_{\mathrm{ell}}(G)$ is parametrized by pairs (N_1', N_2') of even integers with $0 \leqslant N_1' \leqslant N_2'$ and $N = N_1' + N_2'$. The corresponding endoscopic groups are the split groups

$$G' = SO(N_1'+1) \times SO(N_2'+1),$$

with dual groups

$$\widehat{G}' = Sp(N_1', \mathbb{C}) \times Sp(N_2', \mathbb{C}) \subset Sp(N, \mathbb{C}) = \widehat{G}.$$

The group $\mathrm{Out}_G(G')$ is trivial in this case unless $N_1' = N_2'$, in which case it has order 2.

The other cases are similar. In the second case that $G = Sp(N-1)$ and $\widehat{G} = SO(N, \mathbb{C})$ (with $N = N_O$ odd), $\mathcal{E}_{\mathrm{ell}}(G)$ is parametrized by pairs (N_1', N_2') of nonnegative even integers with $N = N_1' + (N_2'+1)$, and characters

η' on Γ_F with $(\eta')^2 = 1$. The corresponding endoscopic groups are the quasisplit groups

$$G' = SO(N_1', \eta') \times Sp(N_2'),$$

with dual groups

$$\widehat{G}' = SO(N_1', \mathbb{C}) \times SO(N_2' + 1, \mathbb{C}) \subset SO(N, \mathbb{C}) = \widehat{G}.$$

In the third case that $G = SO(N, \eta)$ and $\widehat{G} = SO(N, \mathbb{C})$ (with $N = N_O$ even), $\mathcal{E}_{\mathrm{ell}}(G)$ is parametrized by pairs of even integers (N_1', N_2') with $0 \leqslant N_1' \leqslant N_2'$ and $N = N_1' + N_2'$, and pairs (η_1', η_2') of characters on Γ_F with $(\eta_1')^2 = (\eta_2')^2 = 1$ and $\eta = \eta_1'\eta_2'$. The corresponding endoscopic groups are the quasisplit groups

$$G' = SO(N_1', \eta_1') \times SO(N_2', \eta_2'),$$

with dual groups

$$\widehat{G}' = SO(N_1', \mathbb{C}) \times SO(N_2', \mathbb{C}) \subset SO(N, \mathbb{C}) = \widehat{G}.$$

In the second and third cases, each character η' has to be nontrivial if the corresponding integer N' equals 2, while if $N' = 0$, η' must of course be trivial. In these cases, the group $\mathrm{Out}_G(G')$ has order 2 unless N is even, and $N_1' = N_2'$ and $\eta_1' = \eta_2'$, in which case it is a product of two groups of order 2, or $N_1' = 0$, in which case the group is trivial.

Observe that in the even orthogonal case, where $\widehat{G} = SO(N, \mathbb{C})$ with N even, the endoscopic data $G' \in \mathcal{E}_{\mathrm{ell}}(G)$ will not be able to isolate constituents r_i of r of *odd* dimension. The discrepancy is made up by a third kind of endoscopic datum. These are the twisted endoscopic data for the even orthogonal groups $G = SO(N, \eta)$ in $\widetilde{\mathcal{E}}_{\mathrm{sim}}(N)$. For any such G, let

$$\widetilde{G} = G \rtimes \widetilde{\theta}, \qquad \widetilde{\theta} = \mathrm{Int}\big(\widetilde{w}(N)\big), \tag{1.2.5}$$

be the nonidentity component in the semi-direct product of G with the group of order two generated by the outer automorphism $\widetilde{\theta}$ of $SO(N)$. In this setting a (twisted) endoscopic datum is a triplet $(\widetilde{G}', \widetilde{s}', \widetilde{\xi}')$, where $\widetilde{G}'$ is a quasisplit group over F, $\widetilde{s}'$ is a semisimple element in the "dual set" $\widehat{\widetilde{G}} = \widehat{G} \rtimes \widetilde{\theta}$ of which $\widehat{\widetilde{G}}'$ is the connected centralizer in $\widehat{G}$, and $\widetilde{\xi}'$ is an L-embedding of ${}^L\widetilde{G}'$ into ${}^LG$, all being subject also to further conditions and definitions as above. The subset $\mathcal{E}_{\mathrm{ell}}(\widetilde{G}) \subset \mathcal{E}(\widetilde{G})$ of isomorphism classes of elliptic endoscopic data for $\widetilde{G}$ is parametrized by pairs of *odd* integers $(\widetilde{N}_1', \widetilde{N}_2')$, with $0 \leqslant \widetilde{N}_1' \leqslant \widetilde{N}_2'$ and $N = \widetilde{N}_1' + \widetilde{N}_2'$, and pairs of characters $(\widetilde{\eta}_1', \widetilde{\eta}_2')$ on Γ_F, with $(\widetilde{\eta}_1')^2 = (\widetilde{\eta}_2')^2 = 1$ and $\eta = \widetilde{\eta}_1'\widetilde{\eta}_2'$. The corresponding endoscopic groups are the quasisplit groups

$$\widetilde{G}' = Sp(\widetilde{N}_1' - 1) \times Sp(\widetilde{N}_2' - 1),$$

with dual groups

$$\widehat{\widetilde{G}}' = SO(\widetilde{N}_1', \mathbb{C}) \times SO(\widetilde{N}_2', \mathbb{C}) \subset SO(N, \mathbb{C}) = \widehat{G}.$$

The family $\mathcal{E}_{\mathrm{ell}}(\widetilde{G})$ will have to be part of our analysis. However, its role will be subsidiary to that of the two primary families $\widetilde{\mathcal{E}}_{\mathrm{ell}}(N)$ and $\mathcal{E}_{\mathrm{ell}}(G)$.

We have completed our brief study of elliptic self-dual representations r. Remember that we are regarding these objects as parameters, in the spirit of §1.1. We have seen that a parameter for $GL(N)$ factors into a product of two parameters for quasisplit classical groups. The products are governed by twisted endoscopic data $G \in \widetilde{\mathcal{E}}_{\mathrm{ell}}(N)$. They can be refined further according to ordinary endoscopic data $G' \in \mathcal{E}_{\mathrm{ell}}(G)$. Thus, while the parameters will not be available (for lack of a global Langlands group L_F), the endoscopic data that control many of their properties will be. Before we can study the ramifications of this, we must first formulate a makeshift substitute for the parameters attached to our classical groups. We shall do so in §1.4, after a review in §1.3 of the representations of $GL(N)$ that will serve as parameters for this group.

We have considered only the self-dual representation r that are elliptic, since it is these objects that pertain directly to our theorems. Before going on, we might ask what happens if r is not elliptic. A moment's reflection reveals that any such r factors through subgroups of $GL(N,\mathbb{C})$ attached to *several* data $G \in \widetilde{\mathcal{E}}_{\mathrm{ell}}(N)$, in contrast to what we have seen in the elliptic case. This is because r also factors through a subgroup attached to a datum M in the complement of $\widetilde{\mathcal{E}}_{\mathrm{ell}}(N)$ in $\mathcal{E}(N)$, and because any such M can be identified with a proper Levi subgroup of several G. The analysis of general self-dual representations r is therefore more complicated, though still not very difficult. It is best formulated in terms of the centralizers

$$\widetilde{S}_r(N) = S_r\big(\widetilde{G}(N)\big) = \mathrm{Cent}\big(\mathrm{Im}(r), \widehat{\widetilde{G}}(N)\big)$$

and

$$S_r = S_r(G) = \mathrm{Cent}\big(\mathrm{Im}(r), \widehat{G}\big)$$

of the images of r. We shall return to this matter briefly in §1.4, and then more systematically as part of the general theory of Chapter 4.

1.3. Representations of $GL(N)$

A general goal, for the present volume and beyond, is to classify representations of a broad class of groups in terms of those of general linear groups. What makes this useful is the fact that much of the representation theory of $GL(N)$ is both well understood and relatively simple. We shall review what we need of the theory.

Suppose first that F is local. In this case, we can replace the abstract group Λ_F of the last section by the local Langlands group L_F defined by (1.1.1). The local Langlands classification parametrizes irreducible representations of $GL(N,F)$ in terms of N-dimensional representations

$$\phi:\ L_F \longrightarrow GL(N,\mathbb{C}).$$

Before we state it formally, we should recall a few basic notions.

Given a finite dimensional (semisimple, continuous) representation ϕ of L_F, we can form the local L-function $L(s, \phi)$, a meromorphic function of $s \in \mathbb{C}$. We can also form the local ε-factor $\varepsilon(s, \phi, \psi_F)$, a monomial of the form ab^{-s} which also depends on a nontrivial additive character ψ_F of F. If F is archimedean, we refer the reader to the definition in [**T2**, §3]. If F is p-adic, we extend ϕ analytically to a representation of the product of W_F with the complexification $SL(2, \mathbb{C})$ of the subgroup $SU(2)$ of L_F. We can then form the representation

$$\mu_\phi(w) = \phi\left(w, \begin{pmatrix} |w|^{\frac{1}{2}} & 0 \\ 0 & |w|^{-\frac{1}{2}} \end{pmatrix}\right), \qquad w \in W_F,$$

of W_F, where $|w|$ is the absolute value on W_F, and the nilpotent matrix

$$N_\phi = \log \phi\left(1, \begin{pmatrix} 1 & 1 \\ 0 & 1 \end{pmatrix}\right).$$

The pair $V_\phi = (\mu_\phi, N_\phi)$ gives a representation of the Weil-Deligne group [**T2**, (4.1.3)], for which we define an L-function

$$L(s, \phi) = Z(V_\phi, q_F^{-s})$$

and ε-factor

$$\varepsilon(s, \phi, \psi_F) = \varepsilon(V_\phi, q_F^{-s}),$$

following notation in [**T2**, §4]. (We have written q_F here for the order of the residue field of F.) Of particular interest are the tensor product L-function

$$L(s, \phi_1 \times \phi_2) = L(s, \phi_1 \otimes \phi_2)$$

and ε-factor

$$\varepsilon(s, \phi_1 \times \phi_2, \psi_F) = \varepsilon(s, \phi_1 \otimes \phi_2, \psi_F),$$

attached to any pair of representations ϕ_1 and ϕ_2 of L_F.

We expect also to be able to attach local L-functions $L(s, \pi, r)$ and ε-factors $\varepsilon(s, \pi, r, \psi_F)$ to any connected reductive group G over F, where π ranges over irreducible representations of $G(F)$, and r is a finite dimensional representation of ${}^LG$. For general G, this has been done in only the simplest of cases. However, if G is a product $G(N_1) \times G(N_2)$ of general linear groups, there is a broader theory [**JPS**]. (See also [**MW2**, Appendice].) It applies to any representation $\pi = \pi_1 \times \pi_2$, in the case that r is the standard representation

$$r(g_1, g_2):\ X \longrightarrow g_1 \cdot X \cdot {}^tg_2, \qquad g_i \in G(N_i), \tag{1.3.1}$$

of

$$\widehat{G} = GL(N_1, \mathbb{C}) \times GL(N_2, \mathbb{C})$$

on the space of complex $(N_1 \times N_2)$-matrices X. The theory yields functions

$$L(s, \pi_1 \times \pi_2) = L(s, \pi, r)$$

and

$$\varepsilon(s, \pi_1 \times \pi_2, \psi_F) = \varepsilon(s, \pi, r, \psi_F),$$

known as local Rankin-Selberg convolutions.

The local classification for $GL(N)$ is essentially characterized by being compatible with these local Rankin-Selberg convolutions. It has other important properties as well. Some of these relate to supplementary conditions we can impose on the parameters ϕ as follows.

Suppose for a moment that G is any connected group over F. We write $\Phi(G)$ for the set of $\widehat{G}$-orbits of (semisimple, continuous, G-relevant) L-homomorphisms

$$\phi : \, L_F \longrightarrow {}^LG,$$

and $\Pi(G)$ for the set of equivalence classes of irreducible (admissible) representations of $G(F)$. (See [**Bo**].) These sets come with parallel chains of subsets

$$\Phi_{2,\mathrm{bdd}}(G) \subset \Phi_{\mathrm{bdd}}(G) \subset \Phi(G)$$

and

$$\Pi_{2,\mathrm{temp}}(G) \subset \Pi_{\mathrm{temp}}(G) \subset \Pi(G).$$

In the second chain, $\Pi_{\mathrm{temp}}(G)$ denotes the set of tempered representations in $\Pi(G)$, and

$$\Pi_{2,\mathrm{temp}}(G) = \Pi_2(G) \cap \Pi_{\mathrm{temp}}(G),$$

where $\Pi_2(G)$ is the set of representations in $\Pi(G)$ that are essentially square integrable, in the sense that after tensoring with the appropriate positive character on $G(F)$, they are square integrable modulo the centre of $G(F)$. Recall that $\Pi_{\mathrm{temp}}(G)$ can be described informally as the set of representations $\pi \in \Pi(G)$ that occur in the spectral decomposition of $L^2\big(G(F)\big)$. Similarly, $\Pi_{2,\mathrm{temp}}(G)$ is the set of π that occur in the discrete spectrum (taken modulo the center). In the first chain, $\Phi_{\mathrm{bdd}}(G)$ denotes the set of $\phi \in \Phi(G)$ whose image in ${}^LG$ projects onto a relatively compact subset of $\widehat{G}$, and

$$\Phi_{2,\mathrm{bdd}}(G) = \Phi_2(G) \cap \Phi_{\mathrm{bdd}}(G),$$

where $\Phi_2(G)$ is the set of parameters ϕ in $\Phi(G)$ whose image does not lie in any proper parabolic subgroup ${}^LP$ of ${}^LG$.

In the case $G = G(N) = GL(N)$ of present concern, we write $\Phi(N) = \Phi\big(GL(N)\big)$ and $\Pi(N) = \Pi\big(GL(N)\big)$, and follow similar notation for the corresponding subsets above. Then $\Phi(N)$ can be identified with the set of equivalence classes of (semisimple, continuous) N-dimensional representations of L_F. The subset

$$\Phi_{\mathrm{sim}}(N) = \Phi_2(N)$$

consists of those representations that are irreducible, while $\Phi_{\mathrm{bdd}}(N)$ corresponds to representations that are unitary. On the other hand, the set $\Pi_{\mathrm{unit}}(N)$ of unitary representations in $\Pi(N)$ properly contains $\Pi_{\mathrm{temp}}(N)$, if $N \geqslant 2$. It has an elegant classification [**V2**], [**Tad1**], but our point here is that it is not parallel to the set of N-dimensional representations ϕ that are unitary. We do observe that

$$\Pi_{2,\text{temp}}(N) = \Pi_2(N) \cap \Pi_{\text{unit}}(N) = \Pi_{2,\text{unit}}(N),$$

so the notions of tempered and unitary are the same for square integrable representations.

If F is p-adic, we can write $\Pi_{\text{scusp,temp}}(N)$ (resp. $\Pi_{\text{scusp}}(N)$) for the set of *supercuspidal* representations in $\Pi_{2,\text{temp}}(N)$ (resp. $\Pi_2(N)$). It is also convenient to write $M_{\text{sim,bdd}}(N)$ (resp. $M_{\text{sim}}(N)$) for the set of parameters $\mu = \phi$ in $\Phi_{\text{sim,bdd}}(N)$ (resp. $\Phi_{\text{sim}}(N)$) that are trivial on the second factor $SU(2)$ of L_F. If F is archimedean, it is natural to take $\Pi_{\text{scusp,temp}}(N)$ and $\Pi_{\text{scusp}}(N)$ to be empty unless $GL(N,F)$ is compact modulo the center (which is a silly way of saying that $N = 1$), in which case we can take them to be the corresponding sets $\Pi_{2,\text{temp}}(GL(1)) = \Pi_{\text{temp}}(GL(1))$ and $\Pi_2(GL(1)) = \Pi(GL(1))$. With the parameter sets defined accordingly, we then have two parallel chains

$$M_{\text{sim,bdd}}(N) \subset \Phi_{\text{sim,bdd}}(N) \subset \Phi_{\text{bdd}}(N) \subset \Phi(N), \tag{1.3.2}$$

and

$$\Pi_{\text{scusp,temp}}(N) \subset \Pi_{2,\text{temp}}(N) \subset \Pi_{\text{temp}}(N) \subset \Pi(N), \tag{1.3.3}$$

for our given local field F.

The local classification for $G = GL(N)$ can now be formulated as follows.

THEOREM 1.3.1 (Langlands [**L11**], Harris-Taylor [**HT**], Henniart [**He1**], Scholze [**Sch**]). *There is a unique bijective correspondence $\phi \to \pi$ from $\Phi(N)$ onto $\Pi(N)$ such that*

$$\phi \otimes \chi \longrightarrow \pi \otimes (\chi \circ \det), \tag{i}$$

for any character χ in the group $\Phi(1) = \Pi(1)$,

$$\det \circ \phi \longrightarrow \eta_\pi, \tag{ii}$$

for the central character η_π of π, and

$$\phi^\vee \longrightarrow \pi^\vee, \tag{iii}$$

for the contragredient involutions $\vee$ on $\Phi(N)$ and $\Pi(N)$, and such that if

$$\phi_i \longrightarrow \pi_i, \qquad \phi_i \in \Phi(N_i),\ i = 1, 2,$$

then

$$L(s, \pi_1 \times \pi_2) = L(s, \phi_1 \times \phi_2) \tag{iv}$$

and

$$\varepsilon(s, \pi_1 \times \pi_2, \psi_F) = \varepsilon(s, \phi_1 \times \phi_2, \psi_F). \tag{v}$$

Furthermore, the bijection is compatible with the two chains (1.3.2) and (1.3.3), in the sense that it maps each subset in (1.3.2) onto its counterpart in (1.3.3). □

We now take F to be global. This brings us to the representations that will be the foundation of all that follows. They are the objects in the set

$$\mathcal{A}_{\text{cusp}}(N) = \mathcal{A}_{\text{cusp}}\big(GL(N)\big)$$

of (equivalence classes of) unitary, cuspidal automorphic representations of $GL(N)$.

Assume first that G is an arbitrary connected group over the global field F. To suppress the noncompact part of the center, one often works with the closed subgroup

$$G(\mathbb{A})^1 = \{x \in G(\mathbb{A}) : \ |\chi(x)| = 1, \ \chi \in X^*(G)_F\}$$

of $G(\mathbb{A})$, where $X^*(G)_F$ is the additive group of characters of G defined over F. We recall that $G(F)\backslash G(\mathbb{A})^1$ has finite volume, and that there is a sequence

$$L^2_{\text{cusp}}\big(G(F)\backslash G(\mathbb{A})^1\big) \subset L^2_{\text{disc}}\big(G(F)\backslash G(\mathbb{A})^1\big) \subset L^2\big(G(F)\backslash G(\mathbb{A})^1\big),$$

of embedded, right $G(\mathbb{A})^1$-invariant Hilbert spaces. In particular, the space $L^2_{\text{cusp}}\big(G(F)\backslash G(\mathbb{A})^1\big)$ of cuspidal functions in $L^2\big(G(F)\backslash G(\mathbb{A})^1\big)$ is contained in the subspace $L^2_{\text{disc}}\big(G(F)\backslash G(\mathbb{A})^1\big)$ that decomposes under the action of $G(\mathbb{A})^1$ into a direct sum of irreducible representations. We can then introduce a corresponding chain of subsets of irreducible automorphic representations

$$\mathcal{A}_{\text{cusp}}(G) \subset \mathcal{A}_2(G) \subset \mathcal{A}(G).$$

By definition, $\mathcal{A}_{\text{cusp}}(G)$, $\mathcal{A}_2(G)$ and $\mathcal{A}(G)$ denote the subsets of irreducible *unitary* representations π of $G(\mathbb{A})$ whose restrictions to $G(\mathbb{A})^1$ are irreducible constituents of the respective spaces L^2_{cusp}, L^2_{disc} and L^2. We shall also write $\mathcal{A}^+_{\text{cusp}}(G)$ and $\mathcal{A}^+_2(G)$ for the analogues of $\mathcal{A}_{\text{cusp}}(G)$ and $\mathcal{A}_2(G)$ defined without the condition that π be unitary. (The definition of $\mathcal{A}(G)$ here is somewhat informal, and will be used only for guidance. It can in fact be made precise at the singular points in the continuous spectrum where there might be some ambiguity.)

We specialize again to the case $G = GL(N)$, taken now over the global field F. We write $\mathcal{A}(N) = \mathcal{A}\big(GL(N)\big)$, with similar notation for the corresponding subsets above. Observe that $GL(N,\mathbb{A})^1$ is the group of adelic matrices $x \in GL(N,\mathbb{A})$ whose determinant has absolute value 1. In this case, the set $\mathcal{A}(N)$ is easy to characterize. It is composed of the induced representations

$$\pi = \mathcal{I}_P(\pi_1 \otimes \cdots \otimes \pi_r), \qquad \pi_i \in \mathcal{A}_2(N_i),$$

where P is the standard parabolic subgroup of block upper triangular matrices in $G = GL(N)$ corresponding to a partition $(N_1, \ldots, N_r)$ of N, with the standard Levi subgroup of block diagonal matrices

$$M_P = GL(N_1) \times \cdots \times GL(N_r).$$

This follows from the theory of Eisenstein series [**L1**], [**L5**], [**A1**] (valid for any G), and the fact [**Be**] (special to $G = GL(N)$) that an induced representation $\mathcal{I}_P^G(\sigma)$ is irreducible for any representation σ of $M_P(\mathbb{A})$ that is irreducible and unitary. Let us also write $\mathcal{A}^+(N)$ for the set of induced representations as above, but with the components π_i now taken from the larger sets $\mathcal{A}_2^+(N_i)$. These are global analogues of what are usually called standard representations. They can be reducible at certain points, although they typically remain irreducible. We thus have a chain of sets

$$\mathcal{A}_{\text{cusp}}(N) \subset \mathcal{A}_2(N) \subset \mathcal{A}(N) \subset \mathcal{A}^+(N) \tag{1.3.4}$$

of representations, for our given global field F. This is a rough global analogue of the local sequence (1.3.3). We have dropped the subscript "temp" in the global notation, since the local components of a constituent of $L^2\big(GL(N,F)\backslash GL(N,\mathbb{A})^1\big)$ need not be tempered, and added the superscript $+$ to remind ourselves that $\mathcal{A}^+(N)$ contains full induced representations, rather than irreducible quotients.

There are two fundamental theorems on the automorphic representations of $GL(N)$ that will be essential to us. The first is the classification of automorphic representations by Jacquet and Shalika, while the second is the characterization by Moeglin and Waldspurger of the discrete spectrum in terms of cuspidal spectra. We shall review each in turn.

Suppose again that G is an arbitrary connected group over the global field F. An automorphic representation π of G is among other things, a weakly continuous, irreducible representation of $G(\mathbb{A})$. As such, it can be written as a restricted tensor product

$$\pi = \bigotimes_v \pi_v$$

of irreducible representations of the local groups $G(F_v)$, almost all which are unramified [**F**]. Recall that if π_v is unramified, the group $G_v = G/F_v$ is unramified. This means that F_v is a p-adic field, that G_v is quasisplit, and that the action of W_{F_v} on $\widehat{G}$ factors through the infinite cyclic quotient

$$W_{F_v}/I_{F_v} = \langle \text{Frob}_v \rangle$$

of W_{F_v} by the inertia subgroup I_{F_v} with canonical generator Frob_v. Recall also that the local Langlands correspondence has long existed in this very particular context. The $\widehat{G}$-orbit of homomorphisms

$$\phi_v : \ L_{F_v} \longrightarrow {}^L G_v = \widehat{G} \rtimes \langle \text{Frob}_v \rangle$$

in $\Phi(G_v)$ to which π_v corresponds factors through the quotient

$$L_{F_v}/\big(I_{F_v} \times SU(2)\big) = W_{F_v}/I_{F_v}$$

of L_{F_v}. The resulting mapping

$$\pi_v \longrightarrow c(\pi_v) = \phi_v(\text{Frob}_v)$$

is a bijection from the set of unramified representations of $G(F_v)$ (relative to any given hyperspecial maximal compact subgroup $K_v \subset G(F_v)$) and the set of semisimple $\widehat{G}$-orbits in ${}^L G_v$ that project to the Frobenius generator of W_{F_v}/I_{F_v}. Let $c_v(\pi)$ be the image of $c(\pi_v)$ in ${}^LG$ under the embedding of ${}^LG_v$ into ${}^LG$ that is defined canonically up to conjugation. In this way, the automorphic representation π of G gives rise to a family of semisimple conjugacy classes

$$c^S(\pi) = \{c_v(\pi) : \ v \notin S\}$$

in ${}^LG$, where S is some finite set of valuations of F outside of which G is unramified.

Let $\mathcal{C}^S_{\mathrm{aut}}(G)$ be the set of families

$$c^S = \{c_v : \ v \notin S\}$$

of semisimple conjugacy classes in ${}^LG$ obtained in this way. That is, $c^S = c^S(\pi)$, for some automorphic representation π of G. We define $\mathcal{C}_{\mathrm{aut}}(G)$ to be the set of equivalence classes of such families, c^S and $(c')^{S'}$ being equivalent if c_v equals c'_v for almost all v. We then have a mapping

$$\pi \longrightarrow c(\pi)$$

from the set of automorphic representations of G onto $\mathcal{C}_{\mathrm{aut}}(G)$.

The families $c^S(\pi)$ arise most often in the guise of (partial) global L-functions. Suppose that r is a finite dimensional representation of ${}^LG$ that is unramified outside of S. The corresponding incomplete L-function is given by an infinite product

$$L^S(s,\pi,r) = \prod_{v\notin S} L(s,\pi_v,r_v)$$

of unramified local L-functions

$$L(s,\pi_v,r_v) = \det\big(1 - r(c(\pi_v))q_v^{-s}\big)^{-1}, \qquad v \notin S,$$

which converges for the real part of s large. It is at the ramified places $v \in S$ that the construction of local L-functions and ε-factors poses a challenge. As the reader is no doubt well aware, the completed L-function

$$L(s,\pi,r) = \prod_{v} L(s,\pi_v,r_v)$$

is expected to have analytic continuation as a meromorphic function of $s \in \mathbb{C}$ that satisfies the functional equation

$$L(s,\pi,r) = \varepsilon(s,\pi,r)L(1-s,\pi,r^\vee), \tag{1.3.5}$$

where $r^\vee$ is the contragredient of r, and

$$\varepsilon(s,\pi,r) = \prod_{v\notin S} \varepsilon(s,\pi_v,r_v,\psi_{F_v}).$$

Here, ψ_{F_v} stands for the localization of a nontrivial additive character ψ_F on $\mathbb{A}/F$ that is unramified outside of S.

We return again to the case $G = GL(N)$. The elements in the set

$$\mathcal{C}_{\mathrm{aut}}(N) = \mathcal{C}_{\mathrm{aut}}\big(GL(N)\big)$$

can be identified with families of semisimple conjugacy classes in $GL(N, \mathbb{C})$, defined up to the equivalence relation above. It is convenient to restrict the domain of the corresponding mapping $\pi \to c(\pi)$.

Recall that general automorphic representations for $GL(N)$ can be characterized as the irreducible constituents of induced representations

$$\rho = \mathcal{I}_P(\pi_1 \otimes \cdots \otimes \pi_r), \qquad \pi_i = \mathcal{A}^+_{\mathrm{cusp}}(N_i).$$

(See Proposition 1.2 of [**L6**], which applies to any G.) With this condition, the nonunitary induced representation ρ can have many irreducible constituents. However, it does have a canonical constituent. This is the irreducible representation

$$\pi = \bigotimes_v \pi_v,$$

where $\pi_v \in \Pi(GL(N)_v)$ is obtained from the local Langlands parameter

$$\phi_v = \phi_{1,v} \oplus \cdots \oplus \phi_{r,v}, \qquad \phi_{i,v} \longrightarrow \pi_{i,v},$$

in $\Phi(GL(N)_v)$ by Theorem 1.3.1 (applied to both $GL(N)_v$ and its subgroups $GL(N_i)_v$). *Isobaric* representations are the automorphic representations π of $GL(N)$ obtained in this way. The equivalence class of π does not change if we reorder the cuspidal representations $\{\pi_i\}$. We formalize this property by writing

$$\pi = \pi_1 \boxplus \cdots \boxplus \pi_r, \qquad \pi_i \in \mathcal{A}^+_{\mathrm{cusp}}(N_i), \tag{1.3.6}$$

in the notation [**L7**, §2] of Langlands, where the right hand side is regarded as a formal, unordered direct sum.

THEOREM 1.3.2 (Jacquet-Shalika [**JS**]). *The mapping*

$$\pi = \pi_1 \boxplus \cdots \boxplus \pi_r \longrightarrow c(\pi),$$

from the set of equivalence classes of isobaric automorphic representations π of $GL(N)$ to the set of elements $c = c(\pi)$ in $\mathcal{C}_{\mathrm{aut}}(N)$, is a bijection. □

Global Rankin-Selberg convolutions $L(s, \pi_1 \times \pi_2)$ and ε-factors $\varepsilon(s, \pi_1 \times \pi_2)$ are defined as in the local case. They correspond to automorphic representations $\pi = \pi_1 \times \pi_2$ of a group $G = GL(N_1) \times GL(N_2)$, and the standard representation (1.3.1) of $\widehat{G}$. The analytic behaviour of these functions is now quite well understood [**JPS**], [**MW2**, Appendice]. In particular, $L(s, \pi_1 \times \pi_2)$ has analytic continuation with functional equation (1.3.5). Moreover, if π_1 and π_2 are cuspidal, $L(s, \pi_1 \times \pi_2)$ is an entire function of s unless N_1 equals N_2, and $\pi_2^\vee$ is equivalent to the representation

$$\pi_1(g_1)\,|\det g_1|^{-s_1}, \qquad g_1 \in GL(N_1, \mathbb{A}),$$

for some $s_1 \in \mathbb{C}$, in which case $L(s, \pi_1 \times \pi_2)$ has only a simple pole at $s = s_1 + 1$. It is this property that is used to prove Theorem 1.3.2.

Theorem 1.3.2 can be regarded as a characterization of (isobaric) automorphic representations of $GL(N)$ in terms of simpler objects, families of semisimple conjugacy classes. However, it does not in itself characterize the spectral properties of these representations. For example, it is not a priori clear that the representations in $\mathcal{A}(N)$, or even its subset $\mathcal{A}_2(N)$, are isobaric. The following theorem provides the necessary corroboration.

THEOREM 1.3.3 (Moeglin-Waldspurger [**MW2**]). *The representations $\pi \in \mathcal{A}_2(N)$ are parametrized by the set of pairs (m, μ), where $N = mn$ is divisible by m, and μ is a representation in $\mathcal{A}_{\text{cusp}}(m)$. If P is the standard parabolic subgroup corresponding to the partition $(m, \dots, m)$ of N, and σ_μ is the representation*

$$x \longrightarrow \mu(x_1)|\det x_1|^{\frac{n-1}{2}} \otimes \mu(x_2)|\det x_2|^{\frac{n-3}{2}} \otimes \cdots \otimes \mu(x_n)|\det x_n|^{-\frac{(n-1)}{2}}$$

of the Levi subgroup

$$M_P(\mathbb{A}) \cong \{x = (x_1, \dots, x_n) : \; x_i \in GL(m, \mathbb{A})\},$$

then π is the unique irreducible quotient of the induced representation $\mathcal{I}_P(\sigma_\mu)$. Moreover, the restriction of π to $GL(N, \mathbb{A})^1$ occurs in the discrete spectrum with multiplicity one. □

Consider the representation $\pi \in \mathcal{A}_2(N)$ described in the theorem. For any valuation v, its local component π_v is the Langlands quotient of the local component $\mathcal{I}_P(\sigma_{\mu,v})$ of $\mathcal{I}_P(\sigma_\mu)$. It then follows from the definitions that π_v is the irreducible representation of $GL(N, F_v)$ that corresponds to the Langlands parameter of the induced representation $\mathcal{I}_P(\sigma_{\mu,v})$. In other words, the automorphic representation π is isobaric. We can therefore write

$$\pi = \mu\Big(\tfrac{n-1}{2}\Big) \boxplus \mu\Big(\tfrac{n-3}{2}\Big) \boxplus \cdots \boxplus \mu\Big(-\tfrac{(n-1)}{2}\Big), \tag{1.3.7}$$

in the notation (1.3.6), for cuspidal automorphic representations

$$\mu(i) : \; x_i \longrightarrow \mu(x)|x|^i, \qquad x \in GL(m, \mathbb{A}).$$

Theorem 1.3.3 also provides a description of the automorphic representations in the larger set $\mathcal{A}(N)$. For as we noted above, $\mathcal{A}(N)$ consists of the set of irreducible induced representations

$$\pi = \mathcal{I}_P(\pi_1 \otimes \cdots \otimes \pi_r), \qquad \pi_i \in \mathcal{A}_2(N_i),$$

where $N = N_1 + \cdots + N_r$ is again a partition of N, and P is the corresponding standard parabolic subgroup of $GL(N)$. It is easy to see from its irreducibility that π is also isobaric. We can therefore write

$$\pi = \pi_1 \boxplus \cdots \boxplus \pi_r, \qquad \pi_i \in \mathcal{A}_2(N_i). \tag{1.3.8}$$

Observe that despite the notation, (1.3.8) differs slightly from the general isobaric representation (1.3.6). Its constituents π_i are more complex, since they are not cuspidal, but the associated induced representation is simpler, since it is irreducible.

There is another way to view Theorem 1.3.3. Since F is global, the Langlands group L_F is not available to us. If it were, we would expect its set $\Phi(N)$ of (equivalence classes of) N-dimensional representations

$$\phi : \; L_F \longrightarrow GL(N, \mathbb{C})$$

to parametrize the isobaric automorphic representations of $GL(N)$. A very simple case, for example, is the pullback $|\lambda|$ to L_F of the absolute value on the quotient

$$W_F^{\text{ab}} = GL(1, F)\backslash GL(1, \mathbb{A})$$

of L_F. This of course corresponds to the automorphic character on $GL(1)$ given by the original absolute value. We would expect the subset

$$\Phi_{\text{sim,bdd}}(N) = \Phi_{2,\text{bdd}}(N)$$

of irreducible unitary representations in $\Phi(N)$ to parametrize the unitary cuspidal automorphic representations of $GL(N)$. How then are we to account for the full discrete spectrum $\mathcal{A}_2(N)$? A convenient way to do so is to take the product of L_F with the group $SU(2) = SU(2, \mathbb{R})$.

Let us write $\Psi(N)$ temporarily for the set of (equivalence classes of) *unitary* N-dimension representations

$$\psi : \ L_F \times SU(2) \longrightarrow GL(N, \mathbb{C}).$$

According to Theorem 1.3.3, it would then be the subset

$$\Psi_{\text{sim}}(N) = \Psi_2(N)$$

of irreducible representations in $\Psi(N)$ that parametrizes $\mathcal{A}_2(N)$. Indeed, any $\psi \in \Psi_{\text{sim}}(N)$ decomposes uniquely as a tensor product $\mu \otimes \nu$ of irreducible representations of L_F and $SU(2)$. It therefore gives rise to a pair (m, μ), in which $\mu \in \mathcal{A}_{\text{cusp}}(m)$ represents a unitary cuspidal automorphic representation of $GL(N)$ as well as the corresponding irreducible, unitary, m-dimensional representation of L_F. Conversely, given any such pair, we again identify $\mu \in \mathcal{A}_{\text{cusp}}(m)$ with the corresponding representation of L_F, and we take ν to be the unique irreducible representation of $SU(2)$ of degree $n = Nm^{-1}$.

We are identifying any finite dimensional representation of $SU(2)$ with its analytic extension to $SL(2, \mathbb{C})$. With this convention, we attach an N-dimensional representation

$$\phi_\psi : \ u \longrightarrow \psi\left(u, \begin{pmatrix} |u|^{\frac{1}{2}} & 0 \\ 0 & |u|^{-\frac{1}{2}} \end{pmatrix}\right), \qquad u \in L_F,$$

of L_F to any $\psi \in \Psi(N)$. If $\psi = \mu \otimes \nu$ belongs to the subset $\Psi_{\text{sim}}(N)$, ϕ_ψ decomposes as a direct sum

$$u \longrightarrow \mu(u)|u|^{\frac{n-1}{2}} \oplus \cdots \oplus \mu(u)|u|^{-(\frac{n-1}{2})},$$

to which we associate the induced representation $\mathcal{I}_P(\sigma_\mu)$ of Theoreom 1.3.3. According to the rules of the hypothetical global correspondence from $\Phi(N)$ to isobaric automorphic representations of $GL(N)$, it is the unique irreducible quotient $\pi = \pi_\psi$ of $\mathcal{I}_P(\sigma_\mu)$ that is supposed to correspond to the parameter ϕ_ψ. This representation is unitary, as of course is implicit in Theorem 1.3.3. The mapping $\psi \to \pi_\psi$ is thus an explicit realization of the bijective correspondence from $\Psi_{\mathrm{sim}}(N)$ to $\mathcal{A}_2(N)$ implied by the theorem of Moeglin and Waldspurger (and the existence of the group L_F).

More generally, suppose that ψ belongs to the larger set $\Psi(N)$. Then the isobaric representation π_ψ attached to the parameter $\phi_\psi \in \Phi(N)$ belongs to $\mathcal{A}(N)$. It can be described in the familiar way as the irreducible unitary representation induced from a unitary representation of a Levi subgroup. The mapping $\psi \to \pi_\psi$ becomes a bijection from $\Psi(N)$ to $\mathcal{A}(N)$. In general, the restriction $\psi \to \phi_\psi$ of parameters is an *injection* from $\Psi(N)$ into $\Phi(N)$. The role of the set $\Psi(N)$ we have just defined in terms of the supplementary group $SU(2)$ is thus to single out the subset of $\Phi(N)$ that corresponds to the subset $\mathcal{A}(N)$ of "globally tempered" automorphic representations.

We are also free to form larger sets

$$\Psi^+(N) \supset \Psi(N)$$

and

$$\Psi^+_{\mathrm{sim}}(N) \supset \Psi_{\mathrm{sim}}(N)$$

of representations of $L_F \times SU(2)$ by removing the condition that they be unitary. Any element $\psi \in \Psi^+(N)$ is then a direct sum of irreducible representations $\psi_i \in \Psi^+_{\mathrm{sim}}(N_i)$. The components ψ_i should correspond to automorphic representations $\pi_i = \pi_{\psi_i}$ in $\mathcal{A}^+_2(N_i)$. The corresponding induced representation

$$\rho_\psi = \mathcal{I}_P(\pi_1 \otimes \cdots \otimes \pi_r)$$

then belongs to the set of $\mathcal{A}^+(N)$ of (possibly reducible) representations of $GL(N,\mathbb{A})$ introduced above. Notice that the extended mapping $\psi \to \phi_\psi$ from $\Psi^+(N)$ to $\Phi(N)$ is no longer injective. In particular, ρ_ψ will not in general be equal to the automorphic representation corresponding to ϕ_ψ. However, the mapping $\psi \to \rho_\psi$ will be a bijection from $\Psi^+(N)$ to $\mathcal{A}^+(N)$. In the interests of symmetry, we can also write

$$\Psi_{\mathrm{cusp}}(N) = \Phi_{\mathrm{sim,bdd}}(N)$$

for the set of representations ψ that are trivial on the factor $SU(2)$. This gives us a chain of sets

$$\Psi_{\mathrm{cusp}}(N) \subset \Psi_{\mathrm{sim}}(N) \subset \Psi(N) \subset \Psi^+(N) \tag{1.3.9}$$

that is parallel to (1.3.4). We will then have a bijective correspondence $\psi \to \rho_\psi$ that takes each set in (1.3.9) to its counterpart in (1.3.4).

The global parameter sets in the chain (1.3.9) are hypothetical, depending as they do on the global Langlands group L_F. However, their local analogues are not. They can be defined for the general linear group $G_v(N) = GL(N)_v$ over any completion F_v of F. Replacing L_F by L_{F_v} in the definitions above, we obtain local parameter sets

$$\Psi_{\mathrm{cusp},v}(N) \subset \Psi_{\mathrm{sim},v}(N) \subset \Psi_v(N) \subset \Psi^+_v(N). \tag{1.3.10}$$

We also obtain a restriction mapping $\psi \to \psi_v$ from the hypothetical global set $\Psi^+(N)$ to the local set $\Psi^+_v(N)$. The generalized Ramanujan conjecture for $GL(N)$ asserts that this mapping takes $\Psi(N)$ to the subset $\Psi_v(N)$ of $\Psi_v^+(N)$. However, the conjecture is not known. For this reason, we will be forced to work with the larger local sets $\Psi_v^+(N)$ in our study of global spectra.

Our purpose in introducing the hypothetical families of parameters (1.3.9) has been to persuade ourselves that they correspond to well defined families of automorphic representations (1.3.4). This will inform the discussion of the next section. There we shall revisit the constructions of the last section for orthogonal and symplectic groups, but with the objects (1.3.4) in place of the parameter sets (1.3.9). Notice that the three left hand sets in (1.3.4) contain only isobaric automorphic representations. They therefore correspond bijectively with the three subsets

$$\mathcal{C}_{\text{sim}}(N) \subset \mathcal{C}_2(N) \subset \mathcal{C}(N) \subset \mathcal{C}_{\text{aut}}(N) \tag{1.3.11}$$

of $\mathcal{C}_{\text{aut}}(N)$ under the mapping of Theorem 1.3.2. We have written

$$\mathcal{C}_{\text{sim}}(N) = \mathcal{C}_{\text{cusp}}(N)$$

for the smallest of these sets, in part because it would be bijective with the hypothetical family

$$\Phi_{\text{sim,bdd}}(N) = \Psi_{\text{cusp}}(N)$$

of irreducible unitary, N-dimensional representations of the group L_F. The largest set $\mathcal{C}_{\text{aut}}(N)$ would of course be bijective with the family $\Phi(N)$ of all N-dimensional representations of L_F.

The sets (1.3.11) can all be expressed in terms of the smallest set $\mathcal{C}_{\text{sim}}(N)$ (or rather its analogue $\mathcal{C}_{\text{sim}}(m)$ for $m \leqslant N$). This is a consequence of Theorem 1.3.3, or if one prefers, its embodiment in the left hand three parameter in (1.3.9). The sets $\mathcal{C}_{\text{sim}}(N)$ thus contain all the global information for general linear groups. The notation we have chosen is meant to reflect the role of their elements as the *simple* building blocks. Our ultimate goal is to show that this fundamental role extends to the representations of orthogonal and symplectic groups.

1.4. A substitute for global parameters

We can now resume the discussion from §1.2. We shall use the cuspidal automorphic representations of $GL(N)$ as a substitute for the irreducible N-dimensional representations of the hypothetical global Langlands group L_F. This will allow us to construct objects that ultimately parametrize families of automorphic representations of classical groups.

Assume that F is global. In the last section, we wrote $\Psi_{\text{sim}}(N) = \Psi_2(N)$ temporarily for the set of equivalence classes of irreducible unitary representations of the group $L_F \times SU(2)$. From now on, we let

$$\Psi_{\text{sim}}(N) = \Psi_{\text{sim}}\big(G(N)\big) = \Psi_{\text{sim}}\big(GL(N)\big)$$

stand for the set of formal tensor products

$$\psi = \mu \boxtimes \nu, \qquad \mu \in \mathcal{A}_{\text{cusp}}(m),$$

where $N = mn$ and ν is the unique irreducible representation of $SU(2)$ of degree n. Elements in this set are in bijective correspondence with the set of pairs (m, μ) of Theorem 1.3.3. We have therefore a canonical bijection $\psi \to \pi_\psi$ from $\Psi_{\mathrm{sim}}(N)$ onto the relative discrete, automorphic spectrum $\mathcal{A}_2(N)$ of $GL(N)$. For any such ψ, we set $c(\psi)$ equal to $c(\pi_\psi)$, an equivalence class of families in $\mathcal{C}_2(N)$. Then

$$c(\psi) \sim \{c_v(\psi) : v \notin S\},$$

for some finite set $S \supset S_\infty$ of valuations of F outside of which μ is unramified, and for semisimple conjugacy classes

$$c_v(\psi) = c_v(\mu) \otimes c_v(\nu) = c_v(\mu) q_v^{\frac{n-1}{2}} \oplus \cdots \oplus c_v(\mu) q_v^{-\frac{n-1}{2}}$$

in $GL(N, \mathbb{C})$. Here S_∞ denotes the set of archimedean valuations of F as usual, while q_v is again the order of the residue field of F_v.

In §1.3, we also wrote $\Psi(N)$ temporarily for the set of all unitary representations of $L_F \times SU(2)$. Following the notation (1.3.8), we now let

$$\Psi(N) = \Psi\big(G(N)\big) = \Psi\big(GL(N)\big)$$

denote the set of formal, unordered direct sums

$$\psi = \ell_1\psi_1 \boxplus \cdots \boxplus \ell_r\psi_r, \tag{1.4.1}$$

for positive integers ℓ_k and *distinct* elements $\psi_k = \mu_k \boxtimes \nu_k$ in $\Psi_{\mathrm{sim}}(N_k)$. The ranks here are positive integers $N_k = m_k n_k$ such that

$$N = \ell_1 N_1 + \cdots + \ell_r N_r = \ell_1 m_1 n_1 + \cdots + \ell_r m_r n_r.$$

For any ψ, we take π_ψ to be the irreducible induced representation

$$\mathcal{I}_P\big(\underbrace{\pi_{\psi_1} \otimes \cdots \otimes \pi_{\psi_1}}_{\ell_1} \otimes \cdots \otimes \underbrace{\pi_{\psi_r} \otimes \cdots \otimes \pi_{\psi_r}}_{\ell_r}\big)$$

of $GL(N, \mathbb{A})$, where P is the parabolic subgroup corresponding to the partition

$$\big(\underbrace{N_1, \ldots, N_1}_{\ell_1}, \ldots, \underbrace{N_r, \ldots, N_r}_{\ell_r}\big)$$

of N. We then have a bijection $\psi \to \pi_\psi$ from $\Psi(N)$ onto the full automorphic spectrum $\mathcal{A}(N)$ of $GL(N)$. Again we set $c(\psi)$ equal to the associated class $c(\pi_\psi)$ in $\mathcal{C}(N)$. Then

$$c(\psi) \sim \{c_v(\psi) : v \notin S\},$$

for a finite set $S \supset S_\infty$ outside of which each μ_i is unramified, and semisimple conjugacy classes

$$c_v(\psi) = \big(\underbrace{c_v(\psi_1) \oplus \cdots \oplus c_v(\psi_1)}_{\ell_1}\big) \oplus \cdots \oplus \big(\underbrace{c_v(\psi_r) \oplus \cdots \oplus c_v(\psi_r)}_{\ell_r}\big)$$

in $GL(N, \mathbb{C})$.

Suppose that v is *any* valuation in F. If $\psi = \mu \boxtimes \nu$ belongs to $\Psi_{\mathrm{sim}}(N)$, the local component μ_v of μ is an irreducible representation of $GL(m, F_v)$.

It corresponds to an m-dimensional representation of the local Langlands group L_{F_v}, whose tensor product ψ_v with ν belongs to the set $\Psi_v^+(N)$ of N-dimensional representations of $L_{F_v} \otimes SU(2)$. More generally, if ψ is any element (1.4.1) in the larger set $\Psi(N)$, the direct sum

$$\psi_v = \big(\underbrace{\psi_{1,v} \oplus \cdots \oplus \psi_{1,v}}_{\ell_1}\big) \oplus \cdots \oplus \big(\underbrace{\psi_{r,v} \oplus \cdots \oplus \psi_{r,v}}_{\ell_r}\big)$$

also belongs to the set $\Psi_v^+(N)$. There is thus a mapping $\psi \to \psi_v$ from $\Psi(N)$ to $\Psi_v^+(N)$. We emphasize again that we cannot say that ψ_v lies in the smaller local set $\Psi_v(N)$, since the generalized Ramanujan conjecture for $GL(N)$ has not been established.

We now have a set $\Psi(N)$ of formal parameters to represent the automorphic spectrum $\mathcal{A}(N)$ of $GL(N)$. By Theorem 1.3.2, the elements in $\mathcal{A}(N)$ are faithfully represented by the (equivalence classes of) families $c(\psi)$ in $\mathcal{C}(N)$. As we noted at the end of §1.3, and have seen explicitly here, these families are easily constructed in terms of cuspidal families c. Reiterating the conclusion from §1.3, the sets $\mathcal{C}_{\mathrm{sim}}(N) = \mathcal{C}_{\mathrm{cusp}}(N)$ of cuspidal families are to be regarded as our fundamental data, in terms of which everything we hope to establish for classical groups can be formulated.

We are now ready to formalize the constructions of §1.2. The outer automorphism θ: $x \to x^\vee$ of $G(N) = GL(N)$ acts on the set $\Psi(N)$. It transforms a general parameter (1.4.1) in this set to its contragredient

$$\begin{aligned}\psi^\vee &= \ell_1\psi_1^\vee \boxplus \cdots \boxplus \ell_r\psi_r^\vee \\ &= \ell_1(\mu_1^\vee \boxtimes \nu_1^\vee) \boxplus \cdots \boxplus \ell_r(\mu_r^\vee \boxtimes \nu_r^\vee) \\ &= \ell_1(\mu_1^\vee \boxtimes \nu_1) \boxplus \cdots \boxplus \ell_r(\mu_r^\vee \boxtimes \nu_r),\end{aligned}$$

where $\mu_k^\vee$ is the contragredient of the cuspidal automorphic representation μ_k of $GL(m_k)$. We have written $\nu_k^\vee = \nu_k$, since any irreducible representation of $SU(2)$ is self-dual. The contragredient $\pi_\psi^\vee$ of the associated automorphic representation $\pi_\psi \in \mathcal{A}(N)$ is then equal to $\pi_{\psi^\vee}$. We introduce the subset

$$\widetilde{\Psi}(N) = \Psi\big(\widetilde{G}(N)\big) = \{\psi \in \Psi(N) :\ \psi^\vee = \psi\} \tag{1.4.2}$$

of self-dual parameters in $\Psi(N)$, a family that is naturally associated to the connected component

$$\widetilde{G}(N) = G(N) \rtimes \widetilde{\theta}(N)$$

in $\widetilde{G}(N)^+ = GL(N) \rtimes \langle\theta\rangle$. Elements in this subset correspond to the representations $\pi_\psi \in \mathcal{A}(N)$ that extend to the group $\widetilde{G}(N,\mathbb{A})^+$ generated by $\widetilde{G}(N,\mathbb{A})$. There is also an action

$$c = \{c_v\} \longrightarrow c^\vee = \{c_v^\vee\}$$

of θ on the set of automorphic families $\mathcal{C}(N)$. The subset $\widetilde{\Psi}(N)$ of $\Psi(N)$ corresponds to the subset $\widetilde{\mathcal{C}}(N)$ of self-dual families in $\mathcal{C}(N)$.

It is sometimes convenient to write K_ψ for the set that parametrizes the simple constituents ψ_k in (1.4.1) of a general element $\psi \in \Psi(N)$. We shall usually think of K_ψ as an abstract indexing set, of which (1.4.1) represents a (noncanonical) enumeration. If ψ belongs to the subset $\widetilde{\Psi}(N)$ of $\Psi(N)$, there is an involution $k \to k^\vee$ on K_ψ such that $\psi_{k^\vee} = \psi_k^\vee$ and $\ell_{k^\vee} = \ell_k$. We can then write the indexing set as a disjoint union

$$K_\psi = I_\psi \coprod J_\psi \coprod J_\psi^\vee, \qquad J_\psi^\vee = \{j^\vee : j \in J_\psi\},$$

where I_ψ is the set of fixed points of the involution, and J_ψ is some complementary subset that represents the orbits of order 2. The union of I_ψ and J_ψ of course represents the set $\{K_\psi\}$ of all orbits of the involution on K_ψ.

Having defined the set $\widetilde{\Psi}(N)$, we can form the subset

$$\widetilde{\Psi}_{\mathrm{sim}}(N) = \widetilde{\Psi}(N) \cap \Psi_{\mathrm{sim}}(N),$$

of parameters (1.4.1) in $\widetilde{\Psi}(N)$ that are *simple*, in the sense that $r = 1$ and $\ell_1 = 1$. An element ψ in $\widetilde{\Psi}_{\mathrm{sim}}(N)$ is thus a self-dual parameter $\mu \boxtimes \nu$ in the set $\Psi_{\mathrm{sim}}(N)$. We shall also write

$$\widetilde{\Phi}_{\mathrm{sim}}(N) = \widetilde{\Psi}_{\mathrm{cusp}}(N) = \widetilde{\Psi}(N) \cap \Psi_{\mathrm{cusp}}(N)$$

for the subset of parameters in $\widetilde{\Psi}_{\mathrm{sim}}(N)$ that are generic, in the sense that ν is trivial. A simple generic parameter in $\widetilde{\Psi}(N)$ is thus a self-dual element in the basic set $\mathcal{A}_{\mathrm{cusp}}(N)$ of unitary cuspidal automorphic representations of $GL(N)$. These will be our fundamental building blocks for orthogonal and symplectic groups.

Recall that the term *simple* was also applied in §1.2 to endoscopic data. It was used to designate the subset $\widetilde{\mathcal{E}}_{\mathrm{sim}}(N)$ of elliptic endoscopic data $G \in \widetilde{\mathcal{E}}_{\mathrm{ell}}(N)$ that are not composite. We note here that any G in $\widetilde{\mathcal{E}}_{\mathrm{sim}}(N)$, or indeed in the larger set $\widetilde{\mathcal{E}}_{\mathrm{ell}}(N)$, has the property that the split component of its center is trivial. This means that the group $G(\mathbb{A})^1$ equals the full adelic group $G(\mathbb{A})$.

THEOREM 1.4.1. *Suppose that $\phi \in \widetilde{\Phi}_{\mathrm{sim}}(N)$ is a simple generic (formal) global parameter. Then there is a unique $G_\phi \in \widetilde{\mathcal{E}}_{\mathrm{ell}}(N)$ (taken as an isomorphism class of endoscopic data $(G_\phi, s_\phi, \xi_\phi)$) such that*

$$c(\phi) = \xi_\phi\big(c(\pi)\big),$$

for some representation π in $\mathcal{A}_2(G)$. Moreover, G_ϕ is simple.

The assertion here is quite transparent. From among all the twisted, elliptic endoscopic data $G \in \widetilde{\mathcal{E}}_{\mathrm{ell}}(N)$ for $\widetilde{G}(N)$, there is exactly one that is the source of the conjugacy class data of ϕ. This is of course what we would expect from the discussion of §1.2.

Theorem 1.4.1 is the first of a collection of theorems for orthogonal and symplectic groups, the main body of which will be formulated in the next section. We have stated it here as a "seed theorem", which will be needed to define the objects on which the others depend. All of the theorems will be established together. The proof will be based on a long induction argument, which is to be carried throughout the course of the volume. The implication for Theorem 1.4.1 is that the definitions it yields will ultimately have to be inductive. We remind the reader that our results will also be conditional on the stabilization of the discrete part of the twisted trace formula, a hypothesis we will state formally in §3.2.

Suppose that N is fixed, and that Theorem 1.4.1 holds with N replaced by any integer $m \leqslant N$. What can we say about an arbitrary parameter ψ in the set $\widetilde{\Psi}(N) = \Psi\big(\widetilde{G}(N)\big)$?

We can certainly write

$$\psi = \Big(\bigboxplus_{i \in I_\psi} \ell_i \psi_i\Big) \boxplus \Big(\bigboxplus_{j \in J_\psi} \ell_j (\psi_j \boxplus \psi_{j^\vee})\Big). \tag{1.4.3}$$

If i belongs to I_ψ, we apply Theorem 1.4.1 to the simple generic factor $\mu_i \in \widetilde{\Phi}_{\mathrm{sim}}(m_i)$ of $\psi_i = \mu_i \otimes \nu_i$. This gives a canonical datum $(G_{\mu_i}, s_{\mu_i}, \xi_{\mu_i})$ in $\widetilde{\mathcal{E}}_{\mathrm{sim}}(m_i)$, which it will be convenient to denote by H_i. If j belongs to J_ψ, we simply set $H_j = GL(m_j)$. We thus obtain a connected reductive group H_k over F for any index k in either I_ψ or J_ψ, or equivalently, in the set $\{K_\psi\}$ of orbits of the involution on K_ψ. Let ${}^L H_k$ be the Galois form of its L-group. We can then form the fibre product

$$\mathcal{L}_\psi = \prod_{k \in \{K_\psi\}} ({}^L H_k \longrightarrow \Gamma_F) \tag{1.4.4}$$

of these groups over Γ_F. This is to be our substitute for the global Langlands group in our study of automorphic representations attached to ψ. To make matters slightly more transparent, we have formulated it in algebraic form, as an extension of the profinite group Γ_F by a complex reductive group rather than an extension of W_F by a compact topological group. For this reason, we shall work with the Galois forms of L-groups throughout much of the volume, rather than their Weil form. This leads to no difficulty in the framework of orthogonal and symplectic groups we will study.

If an index $k = i$ in (1.4.1) belongs to I_ψ, we have the standard embedding

$$\widetilde{\mu}_i = \xi_{\mu_i} : \ {}^L H_i \longrightarrow {}^L\big(GL(m_i)\big) = GL(m_i, \mathbb{C}) \times \Gamma_F$$

that is part of the endoscopic datum G_i. If $k = j$ belongs to J_ψ, we define a standard embedding

$$\widetilde{\mu}_j : \ {}^L H_j \longrightarrow {}^L\big(GL(2m_j)\big) = GL(2m_j, \mathbb{C}) \times \Gamma_F$$

by setting

$$\widetilde{\mu}_j(h_j \times \sigma) = (h_j \oplus {}_t h_j^{-1}) \times \sigma, \qquad h_j \in \widehat{H}_j = GL(m_j, \mathbb{C}), \ \sigma \in \Gamma_F.$$

We then define an L-embedding

$$\widetilde{\psi} : \ \mathcal{L}_\psi \times SL(2, \mathbb{C}) \longrightarrow {}^L G(N) = GL(N, \mathbb{C}) \times \Gamma_F$$

as the direct sum

$$(1.4.5) \qquad \widetilde{\psi} = \Big(\bigoplus_{i \in I_\psi} \ell_i(\widetilde{\mu}_i \otimes \nu_i) \Big) \oplus \Big(\bigoplus_{j \in J_\psi} \ell_j(\widetilde{\mu}_j \otimes \nu_j) \Big).$$

Our use of $SL(2,\mathbb{C})$ here in place of $SU(2)$ is purely notational, and is in keeping with our construction of $\mathcal{L}_\psi$ as a complex proalgebraic group. We are of course free to interpret the embedding $\widetilde{\psi}$ also as an N-dimensional representation of $\mathcal{L}_\psi \times SL(2,\mathbb{C})$. With either interpretation, we shall be primarily interested in the equivalence class of $\widetilde{\psi}$, as a $GL(N,\mathbb{C})$ conjugacy class of homomorphisms from $\mathcal{L}_\psi \times SL(2,\mathbb{C})$ to either $GL(N,\mathbb{C})$ or ${}^L G(N)$.

Suppose that G belongs to $\widetilde{\mathcal{E}}_{\mathrm{sim}}(N)$. We write $\widetilde{\Psi}(G)$ for the set of elements $\psi \in \widetilde{\Psi}(N)$ such that $\widetilde{\psi}$ factors through ${}^L G$. By this, we mean that there exists an L-homomorphism

$$\widetilde{\psi}_G : \mathcal{L}_\psi \times SL(2,\mathbb{C}) \longrightarrow {}^L G$$

such that

$$(1.4.6) \qquad \xi \circ \widetilde{\psi}_G = \widetilde{\psi},$$

where ξ is the embedding of ${}^L G$ into ${}^L\big(G(N)\big)$ that is part of the twisted endoscopic datum represented by G. Since $\widetilde{\psi}$ and ξ are to be regarded as $GL(N,\mathbb{C})$-conjugacy classes of homomorphisms, $\widetilde{\psi}_G$ is determined up to the stabilizer in $GL(N,\mathbb{C})$ of its image, a group that contains $\widehat{G}$. The quotient of this group by $\widehat{G}$ equals $\widetilde{\mathrm{O}}\mathrm{ut}_N(G)$, the group of outer automorphisms of the endoscopic datum G, described in §1.2. It is a finite group, which is actually trivial unless G is an even orthogonal group, in which case it equals $\mathbb{Z}/2\mathbb{Z}$. In the latter case though, there can be two $\widehat{G}$-orbits of homomorphisms $\widetilde{\psi}_G$ in the class of $\widetilde{\psi}$. It is for this reason that we write $\widetilde{\Psi}(G)$ in place of the more familiar symbol $\Psi(G)$.

More generally, suppose that G belongs to the larger set $\widetilde{\mathcal{E}}_{\mathrm{ell}}(N)$, or even the full set $\widetilde{\mathcal{E}}(N)$ of endoscopic data for $\widetilde{G}(N)$. As a group over F, G equals a product

$$G = \prod_\iota G_\iota$$

of groups G_ι that can range over (quasisplit) connected orthogonal and symplectic groups and (split) general linear groups. We define the set of parameters for G as the product

$$\widetilde{\Psi}(G) = \prod_\iota \widetilde{\Psi}_\iota(G_\iota),$$

where $\Psi_\iota(G_\iota)$ equals $\widetilde{\Psi}(G_\iota)$ if G_ι is orthogonal or symplectic, and equals $\Psi(G_\iota)$ if G_ι is a general linear group. After a little reflection, we see that an element in $\widetilde{\Psi}(G)$ can be identified with a pair $(\psi, \widetilde{\psi}_G)$, for a parameter $\psi \in \widetilde{\Psi}(N)$ and an L-embedding

$$\widetilde{\psi}_G : \mathcal{L}_\psi \times SL(2,\mathbb{C}) \longrightarrow {}^L G$$

that satisfies (1.4.6), and is defined as a $\widehat{G}$-orbit only up to the action of $\widetilde{\mathrm{O}}\mathrm{ut}_N(G)$. The projection

$$(\psi, \widetilde{\psi}_G) \longrightarrow \psi$$

is not generally injective. This is in contrast to the injective embedding of $\widetilde{\Psi}(G)$ into $\widetilde{\Psi}(N)$ for simple G, which is an implicit aspect of our original definition. (It is also in contrast to the special case of elliptic parameters $\psi \in \widetilde{\Psi}_{\mathrm{ell}}(N)$ and elliptic endoscopic data $G \in \widetilde{\mathcal{E}}_{\mathrm{ell}}(N)$, which we will describe in a moment.) However, we shall still sometimes allow ourselves to denote elements in the more general sets $\widetilde{\Psi}(G)$ by ψ when there is no danger of confusion.

Suppose that ψ is any parameter in $\widetilde{\Psi}(N)$. Let

$$\widetilde{S}_\psi(N) = S_\psi\big(\widetilde{G}(N)\big) = \mathrm{Cent}\big(\mathrm{Im}(\widetilde{\psi}), \widehat{\widetilde{G}}(N)\big)$$

and

$$\widetilde{S}^0_\psi(N) = S_\psi(\widetilde{G}^0(N)\big) = \mathrm{Cent}\big(\mathrm{Im}(\widetilde{\psi}), \widehat{\widetilde{G}}^0(N)\big) = \mathrm{Cent}\big(\mathrm{Im}(\widetilde{\psi}), GL(N, \mathbb{C})\big)$$

be the centralizers of the image

$$\mathrm{Im}(\widetilde{\psi}) = \widetilde{\psi}\big(\mathcal{L}_\psi \times SL(2, \mathbb{C})\big)$$

of $\widetilde{\psi}$ in the respective components $\widehat{\widetilde{G}}(N) = \widehat{\widetilde{G}}^0(N) \rtimes \theta$ and $\widehat{\widetilde{G}}^0(N)$. Then $\widetilde{S}^0_\psi(N)$ is a reductive subgroup of $\widehat{\widetilde{G}}^0(N)$, which acts simply transitively by left or right translation on $\widetilde{S}_\psi(N)$. Both $\widetilde{S}_\psi(N)$ and $\widetilde{S}^0_\psi(N)$ are connected. On the other hand, if G is a datum in $\widetilde{\mathcal{E}}(N)$ and ψ belongs to $\widetilde{\Psi}(G)$, the centralizer

$$S_\psi = S_\psi(G) = \mathrm{Cent}\big(\mathrm{Im}(\widetilde{\psi}_G), \widehat{G}\big)$$

of the image of $\widetilde{\psi}_G$ need not be connected. Its quotient

$$\mathcal{S}_\psi = \mathcal{S}_\psi(G) = S_\psi / S^0_\psi Z(\widehat{G})^{\Gamma_F}$$

is a finite abelian group, which plays an essential role in the theory. Notice that there is a canonical element

$$s_\psi = \widetilde{\psi}_G\left(1, \begin{pmatrix} -1 & 0 \\ 0 & -1 \end{pmatrix}\right)$$

in S_ψ. Its image in $\mathcal{S}_\psi$ (which we denote also by s_ψ) will be a part of the description of nontempered automorphic representations of G. The centralizers $\widetilde{S}_\psi(N)$ and S_ψ are in obvious analogy with the centralizers $\widetilde{S}_r(N)$ and S_r described at the end of §1.2. Their definition settles the question posed in §1.1 of how to introduce centralizers of parameters without recourse to the global Langlands group L_F.

We write

$$\widetilde{\Psi}_{\mathrm{ell}}(N) = \Psi_{\mathrm{ell}}\big(\widetilde{G}(N)\big)$$

for the subset of parameters $\psi \in \widetilde{\Psi}(N)$ such that the indexing set J_ψ is empty, and such that $\ell_i = 1$ for each $i \in I_\psi$. These objects are analogous to the self-dual representations r we called elliptic in §1.2. The discussion of §1.2 applies here without change. It tells us that any $\psi \in \widetilde{\Psi}_{\mathrm{ell}}(N)$ has a unique source in one of the sets $\widetilde{\Psi}(G)$. More precisely, let $\widetilde{\Psi}_2(G)$ be the preimage of $\widetilde{\Psi}_{\mathrm{ell}}(N)$ in $\widetilde{\Psi}(G)$, for any $G \in \widetilde{\mathcal{E}}_{\mathrm{ell}}(N)$. The mapping from $\widetilde{\Psi}(G)$ to $\widetilde{\Psi}(N)$ then takes $\widetilde{\Psi}_2(G)$ injectively onto a subset of $\widetilde{\Psi}_{\mathrm{ell}}(N)$, which we identify with $\widetilde{\Psi}_2(G)$, and $\widetilde{\Psi}_{\mathrm{ell}}(N)$ is then the disjoint union

$$\widetilde{\Psi}_{\mathrm{ell}}(N) = \coprod_{G \in \widetilde{\mathcal{E}}_{\mathrm{ell}}(N)} \widetilde{\Psi}_2(G)$$

of these subsets. We thus have parallel chains of parameter sets

$$\widetilde{\Psi}_{\mathrm{sim}}(N) \subset \widetilde{\Psi}_{\mathrm{ell}}(N) \subset \widetilde{\Psi}(N)$$

and

$$\widetilde{\Psi}_{\mathrm{sim}}(G) \subset \widetilde{\Psi}_2(G) \subset \widetilde{\Psi}(G), \qquad G \in \widetilde{\mathcal{E}}_{\mathrm{ell}}(N), \tag{1.4.7}$$

where $\widetilde{\Psi}_{\mathrm{sim}}(G)$ denotes the intersection of $\widetilde{\Psi}_{\mathrm{sim}}(N)$ with $\widetilde{\Psi}_2(G)$. Observe that $\widetilde{\Psi}_2(G)$ is the subset of parameters $\psi \in \widetilde{\Psi}(G)$ such that the centralizer S_ψ is finite, while $\widetilde{\Psi}_{\mathrm{sim}}(G)$ consists of those ψ such that S_ψ equals the minimal group $Z(\widehat{G})^{\Gamma_F}$.

Suppose that

$$\psi = \psi_1 \boxplus \cdots \boxplus \psi_r$$

belongs to $\widetilde{\Psi}_{\mathrm{ell}}(N)$. How do we determine the group $G \in \widetilde{\mathcal{E}}_{\mathrm{ell}}(N)$ such that ψ lies in $\widetilde{\Psi}_2(G)$? To answer the question, we have to be able to write $\psi = \psi_O \boxplus \psi_S$, where

$$\psi_O = \boxplus_{i \in I_{\psi,O}} \psi_i$$

is the sum of those components of orthogonal type,

$$\psi_S = \boxplus_{i \in I_{\psi,S}} \psi_i$$

is the sum of components of symplectic type, and

$$I_\psi = I_{\psi,O} \coprod I_{\psi,S}$$

is the corresponding partition of the set $I_\psi = K_\psi$ of indices.

Consider a general component

$$\psi_i = \mu_i \boxtimes \nu_i$$

of the given ψ. The irreducible cuspidal element $\mu_i \in \widetilde{\Psi}_{\mathrm{sim}}(m_i)$ has a central character $\eta_i = \eta_{\mu_i}$ of order 1 or 2, which we identify with a character on Γ_F. It also gives rise to a datum $H_i \in \widetilde{\mathcal{E}}_{\mathrm{sim}}(m_i)$, according to Theorem 1.4.1. This provides a complex, connected classical dual group $\widehat{H}_i \subset GL(m_i, \mathbb{C})$. The n_i-dimensional representation ν_i of $SL(2, \mathbb{C})$ determines the complex connected classical group $\widehat{K}_i \subset GL(n_i, \mathbb{C})$ that contains its image. By considering principal unipotent elements, for example, the reader can see that $\widehat{K}_i$ is symplectic when n_i is even, and orthogonal when n_i is odd. The tensor product embedding

$$\widehat{H}_i \times \widehat{K}_i \longrightarrow GL(N_i, \mathbb{C}), \qquad N_i = m_i n_i,$$

is then contained in a canonical classical group $\widehat{G}_i \subset GL(N_i, \mathbb{C})$. In concrete terms, $\widehat{G}_i$ is symplectic if one of $\widehat{H}_i$ or $\widehat{K}_i$ is symplectic and the other is orthogonal, and is orthogonal if $\widehat{H}_i$ and $\widehat{K}_i$ are both of the same type. This allows us to designate ψ_i as either orthogonal or symplectic, and thus gives rise to the required decomposition $\psi = \psi_O \boxplus \psi_S$. As in §1.2, we set

$$N_\varepsilon = \sum_{i \in I_{\psi,\varepsilon}} N_i, \qquad \varepsilon = O, S.$$

Then ψ_S lies in the subset $\widetilde{\Psi}_2(G_S)$ of $\widetilde{\Psi}_{\mathrm{ell}}(N_S)$, for the datum $G_S \in \widetilde{\mathcal{E}}_{\mathrm{sim}}(N_S)$ with dual group $\widehat{G}_S = Sp(N_S, \mathbb{C})$, while ψ_O lies in the subset $\widetilde{\Psi}_2(G_O)$ of $\widetilde{\Psi}_{\mathrm{ell}}(N_O)$, for the datum $G_O \in \widetilde{\mathcal{E}}_{\mathrm{sim}}(N_O)$ with dual group $\widehat{G}_O = SO(N_O, \mathbb{C})$ and character

$$\eta_O = \prod_{i \in I_{\psi,O}} (\eta_i)^{n_i} = \prod_{i \in I_{\psi,O}} \eta_i.$$

The original element ψ therefore lies in $\widetilde{\Psi}_2(G)$, where G is the product datum $G_O \times G_S$ in $\widetilde{\mathcal{E}}_{\mathrm{ell}}(N)$.

More generally, suppose that ψ is a parameter (1.4.3) in the larger set $\widetilde{\Psi}(N)$. Then ψ can have several preimages in the various families $\widetilde{\Psi}(G)$, $G \in \widetilde{\mathcal{E}}_{\mathrm{ell}}(N)$. How does one determine them explicitly? For a given datum $G = G_O \times G_S$, there are several ways to divide the components ψ_k among the two factors G_O and G_S. To describe the possibilities, one can see that it suffices to determine whether a given ψ belongs to the subset $\widetilde{\Psi}(G)$ of $\widetilde{\Psi}(N)$ attached to a fixed simple datum $G \in \widetilde{\mathcal{E}}_{\mathrm{sim}}(N)$.

The dual group $\widehat{G}$ of $G \in \widetilde{\mathcal{E}}_{\mathrm{sim}}(N)$ is purely orthogonal or symplectic. The parameter ψ comes with a partition of its index set K_ψ into subsets I_ψ, J_ψ and $J_\psi^\vee$. If j belongs to J_ψ, the corresponding component $\ell_j(\widetilde{\psi}_j \oplus \widetilde{\psi}_{j^\vee})$ of the embedding $\widetilde{\psi}$ takes values in a factor

$${}^L M_j \cong {}^L\big(\underbrace{GL(N_j) \times \cdots \times GL(N_j)}_{\ell_j}\big)$$

of a Levi subgroup M of ${}^L G$. As we have just seen (at least in the special case that $\psi \in \widetilde{\Psi}_2(G)$), the complementary set I_ψ has its own partition into subsets $I_{\psi,O}$ and $I_{\psi,S}$ of orthogonal and symplectic type. It also indexes characters η_i of order 1 or 2. If an index $i \in I_\psi$ has *even* multiplicity $\ell_i = 2\ell_i'$, the component $\ell_i \widetilde{\psi}_i$ of $\widetilde{\psi}$ takes values in another factor

$${}^L M_i \cong {}^L\big(\underbrace{GL(N_i) \times \cdots \times GL(N_i)}_{\ell_i'}\big)$$

of ${}^{L}M$, whether i belongs to $I_{\psi,O}$ or $I_{\psi,S}$. The indices $j \in J_\psi$ and $i \in I_\psi$ with ℓ_i even therefore impose no constraints. However, if $i \in I_\psi$ with ℓ_i odd, there is a supplementary copy of ψ_i to contend with. In this case ψ_i and $\widehat{G}$ must be of the same type, either both symplectic or both orthogonal. Moreover, if $\widehat{G}$ is orthogonal, these factors also impose a constraint on the character $\eta = \eta_G$ that goes with G. It must satisfy the identity

$$\eta = \prod_{i \in I_{\psi,O}} (\eta_i)^{\ell_i n_i} = \prod_{i \in I_{\psi,O}} (\eta_i)^{\ell_i}.$$

The last two conditions are both necessary and sufficient for ψ to belong to the subset $\widetilde{\Psi}(G)$.

It is also easy to describe the centralizers $S_\psi = S_\psi(G)$. Suppose that $G \in \widetilde{\mathcal{E}}_{\mathrm{sim}}(N)$, and that ψ is a parameter (1.4.3) that lies in the subset $\widetilde{\Psi}(G)$ of $\widetilde{\Psi}$. Every index $j \in J_\psi$ contributes a factor $GL(\ell_j, \mathbb{C})$ to S_ψ. The complementary indexing set has a separate partition

$$I_\psi = I_\psi^+(G) \coprod I_\psi^-(G),$$

where $I_\psi^+(G)$ (resp. $I_\psi^-(G)$) is the set of indices $i \in I_\psi$ such that ψ_i and $\widehat{G}$ are of the same (resp. different) type. In other words, $I_\psi^+(G) = I_{\psi,O}$ and $I_\psi^-(G) = I_{\psi,S}$ if $\widehat{G}$ is orthogonal, while $I_\psi^+(G) = I_{\psi,S}$ and $I_\psi^-(G) = I_{\psi,O}$ if $\widehat{G}$ is symplectic. It follows from what we have just observed that an index $i \in I_\psi^-(G)$ has even multiplicity ℓ_i. It contributes a factor $Sp(\ell_i, \mathbb{C})$ to S_ψ. Each index $i \in I_\psi^+(G)$ contributes a factor $O(\ell_i, \mathbb{C})$ to S_ψ, with the stipulation that a certain product (with multiplicities) of the determinants of these factors be 1. Let

$$\Big(\prod_{i \in I_\psi^+(G)} O(\ell_i, \mathbb{C}) \Big)_\psi^+$$

be the kernel of the character

$$\xi_\psi^+ = \xi_\psi^+(G) : \prod_i g_i \longrightarrow \prod_i (\det g_i)^{N_i}, \qquad g_i \in O(\ell_i, \mathbb{C}),\ i \in I_\psi^+(G).$$

The centralizer S_ψ is then given by

$$(1.4.8)\quad S_\psi = \Big(\prod_{i \in I_\psi^+(G)} O(\ell_i, \mathbb{C}) \Big)_\psi^+ \times \Big(\prod_{i \in I_\psi^-(G)} Sp(\ell_i, \mathbb{C}) \Big) \times \Big(\prod_{j \in J_\psi} GL(\ell_j, \mathbb{C}) \Big).$$

Suppose in particular that ψ belongs to $\widetilde{\Psi}_2(G)$. Then

$$I_\psi^+(G) = I_\psi = \{1, \dots, r\},$$

and each ℓ_i equals 1. In this case, the centralizer specializes to

$$(1.4.9)\qquad S_\psi = \Big(\prod_{i=1}^{r} O(1) \Big)_\psi^+ \cong \begin{cases} (\mathbb{Z}/2\mathbb{Z})^r, & \text{if each } N_i \text{ is even,} \\ (\mathbb{Z}/2\mathbb{Z})^{r-1}, & \text{otherwise.} \end{cases}$$

Similar constructions apply to (ordinary) endoscopic data $G' \in \mathcal{E}(G)$, or for that matter, iterated data $G'' \in \mathcal{E}(G')$, $G''' \in \mathcal{E}(G'')$, and so on. As a group over F, G' is again a product of elementary groups G'_ι for which the sets $\Psi_\iota(G'_\iota)$ have been defined. We define $\widetilde{\Psi}(G')$ to be the corresponding product of these sets. A parameter $\psi' = \psi_{G'}$ in $\widetilde{\Psi}(G')$ can be identified with a pair $(\psi, \widetilde{\psi}')$, for a parameter ψ in $\widetilde{\Psi}(G)$, and an L-embedding

$$\widetilde{\psi}' = \widetilde{\psi}_{G'} : \mathcal{L}_\psi \times SL(2,\mathbb{C}) \longrightarrow {}^L G'$$

that satisfies

$$\xi' \circ \widetilde{\psi}' = \widetilde{\psi}_G, \tag{1.4.10}$$

and is defined as a $\widehat{G}'$-orbit only up to the action of the finite group

$$\widetilde{\mathrm{O}}\mathrm{ut}_N(G') = \mathrm{Out}_G(G') \times \widetilde{\mathrm{O}}\mathrm{ut}_N(G).$$

Here, ξ': ${}^L G' \to {}^L G$ is the L-embedding attached to G'. Notions introduced earlier for ψ have obvious meaning for parameters $\psi' \in \widetilde{\Psi}(G')$ here. For example, we have the centralizer

$$S_{\psi'} = S_{\psi'}(G') = \mathrm{Cent}\big(\mathrm{Im}(\psi'), \widehat{G}'\big)$$

and its abelian quotient

$$\mathcal{S}_{\psi'} = \mathcal{S}_{\psi'}(G') = S_{\psi'}/S^0_{\psi'} Z(\widehat{G}')^{\Gamma_F},$$

for any such ψ'.

Any pair

$$(G', \psi'), \qquad G' \in \mathcal{E}(G),\ \psi' \in \widetilde{\Psi}(G'),$$

gives rise to a second pair

$$(\psi, s), \qquad \psi \in \widetilde{\Psi}(G),\ s \in S_\psi,$$

where $\psi = \psi_G$ is the parameter in $\widetilde{\Psi}(G)$ attached to ψ', and s is the image in $\widehat{G}$ of the semisimple element $s' \in \widehat{G}'$ that is part of the endoscopic datum G'. Conversely, suppose that ψ is any parameter in $\widetilde{\Psi}(G)$, and that s is a semisimple element in S_ψ. Let $\widehat{G}'$ be the connected centralizer of s in $\widehat{G}$, and let s' be the preimage of s in $\widehat{G}'$. The product

$$\mathcal{G}' = \widehat{G}' \cdot \widetilde{\psi}_G\big(\mathcal{L}_\psi \times SL(2,\mathbb{C})\big)$$

of $\widehat{G}'$ with the image of $\widetilde{\psi}_G$ is an L-subgroup of ${}^L G$, for which the identity embedding ξ' is an L-homomorphism. We take G' to be a quasisplit group for which $\widehat{G}'$, with the L-action induced by $\mathcal{G}'$, is a dual group. In the present situation, there is a natural way to identify $\mathcal{G}'$ with ${}^L G'$. The triplet (G', s', ξ') represented by G' is then an endoscopic datum for G', in the restricted sense of §1.2. Since s lies in the centralizer of the image of $\widetilde{\psi}_G$, $\widetilde{\psi}_G$ factors through ${}^L G'$. We obtain an L-embedding $\widetilde{\psi}'$ of $\mathcal{L}_\psi \times SL(2,\mathbb{C})$ into ${}^L G'$ that satisfies (1.4.10), and hence an element $\psi' \in \widetilde{\Psi}(G')$. The pair (ψ, s) thus gives rise to a pair (G', ψ') of the original sort.

The correspondence

$$(G', \psi') \longrightarrow (\psi, s) \tag{1.4.11}$$

is a general phenomenon. It applies to arbitrary endoscopic data, at least in the context of Langlands parameters. In particular, it has an obvious variant for our other primary case in which (G, G') is replaced by $(\widetilde{G}(N), G)$. This leads to a more systematic way to view the discussion of §1.2. In general, the correspondence reduces many questions on the transfer of characters to a study of groups S_ψ. It will be part of the foundations we will be calling the standard model in Chapter 4.

The discussion of this section applies also to our supplementary case, that of a bitorsor $\widetilde{G}$ (1.2.5) for an even orthogonal group $G \in \widetilde{\mathcal{E}}_{\mathrm{sim}}(N)$. We write $\Psi(\widetilde{G}) = \widetilde{\Psi}(\widetilde{G})$ for the subset of $\widetilde{\theta}$-fixed elements in $\widetilde{\Psi}(G)$. These are the parameters in $\widetilde{\Psi}(G)$ such that the $\widetilde{\mathrm{Aut}}_N(G)$-orbit of L-embeddings

$$\widetilde{\psi}_G : \mathcal{L}_\psi \times SL(2, \mathbb{C}) \longrightarrow {}^L G$$

contains only one $\widehat{G}$-orbit, or equivalently, such that $\widetilde{\psi}_G$ factors through the L-embedding

$$\widetilde{\xi}' : {}^L\widetilde{G}' \longrightarrow {}^L G$$

of some twisted endoscopic datum $\widetilde{G}' \in \mathcal{E}_{\mathrm{ell}}(\widetilde{G})$. They comprise the minimal set in a chain

$$\Psi(\widetilde{G}) \subset \widetilde{\Psi}(G) \subset \widetilde{\Psi}(N) \subset \Psi(N)$$

of subsets of our original family of parameters $\Psi(N)$ for $GL(N)$. Observe that the subset

$$\Psi_2(\widetilde{G}) = \widetilde{\Psi}_2(G) \cap \Psi(\widetilde{G})$$

of $\Psi(\widetilde{G})$ consists of those parameters (1.4.1) that lie in $\widetilde{\Psi}_2(G)$, and for which one of the irreducible ranks N_i is *odd*. In particular, the subset

$$\Psi_{\mathrm{sim}}(\widetilde{G}) = \widetilde{\Psi}_{\mathrm{sim}}(G) \cap \Psi(\widetilde{G}) = \widetilde{\Psi}_{\mathrm{sim}}(G) \cap \Psi_2(\widetilde{G})$$

of simple elements in $\Psi(\widetilde{G})$ is empty, since N is even. For any $\psi \in \Psi(\widetilde{G})$, the centralizer

$$\widetilde{S}_\psi = S_\psi(\widetilde{G}) = \mathrm{Cent}\big(\mathrm{Im}(\widetilde{\psi}_G), \widehat{\widetilde{G}}\big)$$

is a bitorsor under S_ψ, whose quotient

$$\widetilde{\mathcal{S}}_\psi = \mathcal{S}_\psi(\widetilde{G}) = \widetilde{S}_\psi / \widetilde{S}_\psi^0 Z(\widehat{G})^{\Gamma_F}$$

is a bitorsor under the quotient $\mathcal{S}_\psi$ of S_ψ. It gives rise to an analogue of the correspondence (1.4.11), with $(\widetilde{G}, \widetilde{G}')$ in place of (G, G').

We have so far devoted this section to global parameters. We have shown that their study is parallel to that of the finite dimensional representations in §1.2. We have also tried to demonstrate that objects attached to any ψ can be computed explicitly, even if the details may be somewhat complicated. The details themselves need not be taken too seriously. When we resume this discussion in Chapter 4, we shall treat the relations among our global parameters in a more systematic fashion.

We shall now discuss the local parameters attached to a completion F_v of F. These are simpler, since they really do represent homomorphisms defined on the group $L_{F_v} \times SU(2)$. In particular, the remarks in §1.3 pertaining to the chain (1.3.10) apply here.

Observe that the local analogue of the assertion of Theorem 1.4.1 is elementary. For we have seen in §1.2 that any parameter $\psi_v \in \widetilde{\Psi}_{\mathrm{sim},v}(N)$, and in particular, any $\psi_v = \phi_v$ in the subset $\widetilde{\Phi}_{\mathrm{sim},v}(N)$ of generic parameters in $\widetilde{\Psi}_{\mathrm{sim},v}(N)$, factors through a unique local endoscopic datum G_v in the set

$$\widetilde{\mathcal{E}}_{\mathrm{ell},v}(N) = \mathcal{E}_{\mathrm{ell}}\big(\widetilde{G}_v(N)\big).$$

Moreover, this datum is simple. In these more concrete terms, all of the global discussion above carries over to the completion F_v of F. In fact, it extends without change to the larger parameter set

$$\widetilde{\Psi}_v^+(N) = \{\psi_v \in \Psi_v^+(N) : \psi_v^\vee = \psi_v\}$$

of self-dual, N-dimensional representations of $L_{F_v} \times SU(2)$ that are not necessarily unitary.

We thus attach chains of parameter sets

$$\widetilde{\Psi}_{\mathrm{sim}}(G_v) \subset \widetilde{\Psi}_2(G_v) \subset \widetilde{\Psi}(G_v) \subset \widetilde{\Psi}^+(G_v) \tag{1.4.12}$$

to local endoscopic data $G_v \in \widetilde{\mathcal{E}}_v(N)$, following the local analogues of the global notation (1.4.7). They consist of L-homomorphisms

$$\psi_v : L_{F_v} \times SU(2) \longrightarrow {}^L G_v,$$

which can be composed with the endoscopic embedding

$$\xi_v = \xi_{G_v} : {}^L G_v \longrightarrow {}^L\big(G_v(N)\big).$$

The resulting mapping from $\widetilde{\Psi}^+(G_v)$ to $\widetilde{\Psi}_v^+(N)$ is injective if G_v is simple, or if ψ_v belongs to the subset $\widetilde{\Psi}_2^+(G_v)$ of $\widetilde{\Psi}^+(G_v)$. This allows us in each of the two cases to identify the domain of the mapping with a subset of $\widetilde{\Psi}_v^+(N)$. For example, if G_v is simple, any parameter set in (1.4.12) equals the intersection of $\widetilde{\Psi}^+(G_v)$ with the relevant set of parameters for $\widetilde{G}_v(N)$ in the chain

$$\widetilde{\Psi}_{\mathrm{sim},v}(N) \subset \widetilde{\Psi}_{\mathrm{ell},v}(N) \subset \widetilde{\Psi}_v(N) \subset \widetilde{\Psi}_v^+(N).$$

For any parameter ψ_v in $\widetilde{\Psi}^+(G_v)$, the local centralizer

$$S_{\psi_v} = S_{\psi_v}(G_v) = \mathrm{Cent}\big(\mathrm{Im}(\psi_v), \widehat{G}_v\big),$$

and its abelian quotient

$$\mathcal{S}_{\psi_v} = \mathcal{S}_{\psi_v}(G_v) = S_{\psi_v}/S_{\psi_v}^0 Z(\widehat{G}_v)^{\Gamma_v},$$

are defined in the usual way. We also have the set $\widetilde{\Psi}^+(G_v')$ attached to any $G_v' \in \mathcal{E}(G_v)$, as well as the local analogue of the correspondence (1.4.11). We shall apply all of these local constructions at will, sometimes with the local field being F itself (and the notation modified accordingly) rather than the completion F_v here of the global field F.

Finally, we will need to define a localization mapping $\psi \to \psi_v$ from the global set $\widetilde{\Psi}(G)$ attached to any $G \in \widetilde{\mathcal{E}}_{\mathrm{sim}}(N)$ to the associated local set $\widetilde{\Psi}^+(G_v)$. It would not be hard to formulate the localization as a mapping from $\widetilde{\Psi}(N)$ to $\widetilde{\Psi}_v^+(N)$, or if we wanted to follow the general remarks for the hypothetical group L_F at the end of the last section, as a mapping between the larger sets $\Psi(N)$ and $\Psi_v^+(N)$. To complete the definition, however, we would need to know that it takes the subset $\widetilde{\Psi}(G)$ of $\widetilde{\Psi}(N)$ to the subset $\widetilde{\Psi}^+(G_v)$ of $\widetilde{\Psi}_v^+(N)$. This property is not elementary. It is a consequence of a second "seed theorem", which we state here as a complement to Theorem 1.4.1, but which like Theorem 1.4.1, will have to be proved inductively at the same time as our other theorems.

THEOREM 1.4.2. *Suppose that $\phi \in \widetilde{\Phi}_{\mathrm{sim}}(N)$ is simple generic, as in Theorem 1.4.1. Then the localization ϕ_v of ϕ at any v, a priori an element in the subset $\widetilde{\Phi}_v(N)$ of generic parameters in $\widetilde{\Psi}_v^+(N)$, lies in the subset $\widetilde{\Phi}(G_{\phi,v})$ of $\widetilde{\Phi}_v(N)$ attached to the localization $G_{\phi,v}$ of the global datum $G_\phi \in \widetilde{\mathcal{E}}_{\mathrm{sim}}(N)$ of Theorem 1.4.1.*

The theorem asserts that the N-dimensional representation ϕ_v of L_{F_v}, which is attached by the local Langlands correspondence to the v-component of the cuspidal automorphic representation of $GL(N)$ given by ϕ, factors through the local endoscopic embedding

$$\xi_{\phi,v} : \ {}^LG_{\phi,v} \longrightarrow {}^LG_v(N).$$

It allows us to identify ϕ_v with an L-homomorphism

$$\phi_v : \ L_{F_v} \longrightarrow {}^LG_{\phi,v}.$$

Here G_ϕ is the endoscopic datum attached to ϕ by Theorem 1.4.1, and ϕ_v is determined as a mapping into ${}^LG_{\phi,v}$ up to the action of the group $\widetilde{\mathrm{Aut}}_N(G_\phi)$ on $\widehat{G}_\phi$. Like Theorem 1.4.1, this theorem will be proved by a long induction argument that includes the proof of our other results.

Suppose that Theorems 1.4.1 and 1.4.2 both hold with N replaced by any integer $m \leqslant N$. Let us consider the group $\mathcal{L}_\psi$ attached to a general parameter (1.4.1) in $\widetilde{\Psi}(N)$, and its relation to the local Langlands group L_{F_v} at v. If ${}^LH_k$ is one of the factors (1.4.4) of $\mathcal{L}_\psi$, H_k represents either a simple endoscopic datum in $\widetilde{\mathcal{E}}_{\mathrm{sim}}(m_k)$, or a general linear group $GL(m_k)$. In the first case, Theorem 1.4.2 gives a conjugacy class of L-homomorphisms

$$\begin{array}{ccc} L_{F_v} & \longrightarrow & \Gamma_{F_v} \\ \downarrow & & \downarrow \\ {}^LH_k & \longrightarrow & \Gamma_F \end{array}, \tag{1.4.13}$$

which is determined up to the action of $\widetilde{\mathrm{Aut}}_{m_k}(H_k)$ on $\widehat{H}_k$. In the second case, we obtain a similar assertion from the local Langlands correspondence for $GL(m_k)$, with $\widetilde{\mathrm{Aut}}_{m_k}(H_k)$ equal to $\mathrm{Int}\big(GL(m_k,\mathbb{C})\big)$. The fibre product (1.4.4) then yields a conjugacy class of L-homomorphisms

$$\begin{array}{ccc} L_{F_v} & \longrightarrow & \Gamma_{F_v} \\ \downarrow & & \downarrow \\ \mathcal{L}_\psi & \longrightarrow & \Gamma_F \end{array} \tag{1.4.14}$$

which is determined up to the L-action of the group

$$\widetilde{\mathrm{Aut}}(\mathcal{L}_\psi) = \prod_k \widetilde{\mathrm{Aut}}_{m_k}(H_k) \tag{1.4.15}$$

on $\mathcal{L}_\psi$.

Suppose that ψ belongs to $\widetilde{\Psi}(G)$, for some $G \in \widetilde{\mathcal{E}}_{\mathrm{sim}}(N)$, or indeed, for any G in the general set $\widetilde{\mathcal{E}}(N)$. It then follows from this discussion that we can identify the localization of ψ with an L-homomorphism

$$\psi_v : \ L_{F_v} \times SU(2) \longrightarrow {}^L G_v,$$

which fits into a larger commutative diagram of L-homomorphisms

$$\begin{array}{ccccc} L_{F_v} \times SU(2) & \xrightarrow{\psi_v} & {}^L G_v & \longrightarrow & \Gamma_{F_v} \\ \downarrow & & \downarrow & & \downarrow \\ \mathcal{L}_\psi \times SL(2,\mathbb{C}) & \xrightarrow{\widetilde{\psi}_G} & {}^L G & \longrightarrow & \Gamma_F. \end{array} \tag{1.4.16}$$

The left hand vertical arrow is given by the mapping of L_{F_v} into $\mathcal{L}_\psi$ in (1.4.14) and the usual embedding of $SU(2)$ into $SL(2,\mathbb{C})$. Since ψ_v is essentially the restriction of the global embedding $\widetilde{\psi}_G$ to the image of $L_{F_v} \times SU(2)$, the global centralizer S_ψ is contained in S_{ψ_v}. It follows that there is a canonical mapping

$$x \longrightarrow x_v, \qquad x \in \mathcal{S}_\psi,$$

of $\mathcal{S}_\psi$ into $\mathcal{S}_{\psi_v}$.

1.5. Statement of three main theorems

We shall now state our main theorems. They apply to the quasisplit orthogonal and symplectic groups we have introduced, and provide what amounts to a classification of the representations of these groups in terms of representations of general linear groups. They will be conditional on the stabilization of the twisted trace formula, an assumption we shall state formally when we come to discuss the relevant part of the trace formula in Chapter 3.

Recall that the groups under consideration represent simple endoscopic data $G \in \widetilde{\mathcal{E}}_{\mathrm{sim}}(N)$, for the torsor $\widetilde{G}(N)$ attached to a general linear group

$G(N) = GL(N)$. By presenting them in standard form in §1.2, we have implicitly attached extra structure to these groups. Each $G \in \widetilde{\mathcal{E}}_{\mathrm{sim}}(N)$ comes with a scheme structure over the ring of integers $\mathfrak{o}_F$. It also comes with a standard Borel subgroup B, with maximal torus T, and a standard splitting for (T, B). We have also identified the group $\widetilde{\mathrm{Out}}_N(G)$ of outer automorphisms of G over F with the subgroup of F-automorphisms of G that preserve the given splitting. This group seems innocuous enough, being of order 1 or 2, but it does play an essential role.

The first theorem concerns the case that F is local. It classifies representations of $G(F)$ in terms of a construction defined by the endoscopic transfer of characters. We shall postpone the formal description of this construction until §2.1, as part of our general discussion of local endoscopic transfer. In the meantime, we shall be content to give a somewhat less precise statement of the theorem.

We are assuming for the moment that F is local. We fix a maximal compact subgroup K_F of the group $G(F)$, which we take to be special if F is p-adic, and hyperspecial if G is unramified over F. The $\mathfrak{o}_F$-scheme structure on G actually leads to a natural choice of K_F. For example, if F is p-adic and G is split, we can take $K_F = G(\mathfrak{o}_F)$. On the other hand, if F is p-adic, but $G = SO(N, \eta_G)$ is quasisplit but not split (so that $N = 2n$ is even), K_F is a little more complicated. (See the diagrams in [**Ti**, §1.16].) In any case, given K_F, we write $\mathcal{H}(G)$ for the corresponding Hecke algebra of smooth, left and right K_F-finite functions of compact support on $G(F)$. We also write $\widetilde{\mathcal{H}}(G)$ for the subalgebra of $\widetilde{\mathrm{Out}}_N(G)$-invariant functions in $\mathcal{H}(G)$. Because our results are ultimately tied to $GL(N)$, they will apply to the representation theory of $\widetilde{\mathcal{H}}(G)$, which is slightly weaker than that of $\mathcal{H}(G)$.

We write $\widetilde{\Pi}(G)$ for the set of orbits of $\widetilde{\mathrm{Out}}_N(G)$ in $\Pi(G)$, which we recall is the set of irreducible representations of $G(F)$. Then $\widetilde{\Pi}(G)$ equals $\Pi(G)$ unless G is an even orthogonal group $SO(2n, \eta_G)$, in which case $\widetilde{\Pi}(G)$ contains orbits of order 2 as well as of order 1. Any element $\pi \in \widetilde{\Pi}(G)$ gives rise to a well defined character

$$f_G(\pi) = \mathrm{tr}\big(\pi(f)\big), \qquad f \in \widetilde{\mathcal{H}}(G),$$

on $\widetilde{\mathcal{H}}(G)$. Similar notation applies to any subset of $\Pi(G)$, such as for example the set $\Pi_{\mathrm{unit}}(G)$ of unitary representations in $\Pi(G)$. We shall state the local theorem in terms of packets that are "sets over $\widetilde{\Pi}_{\mathrm{unit}}(G)$".

By a *set over* S, or an S-*set*, or even an S-*packet*, we mean simply a set S_1 with a fibration

$$S_1 \longrightarrow S$$

over S. Equivalently, S_1 is a set that can be represented as a disjoint union of subsets of S. Any function on S, such as the character $f_G(\pi)$ on $\widetilde{\mathcal{H}}(G)$ in case S equals $\widetilde{\Pi}(G)$, will be identified with its pullback to a function on S_1.

The order

$$m_1 : S \longrightarrow \mathbb{N} \cup \{0, \infty\}$$

of the fibres in S_1 represents a multiplicity function, which makes S_1 into a multiset on S in the sense of [**Z**, p. 169]. If S_1 is multiplicity free, in that every element in S has multiplicity at most 1, it is of course just a subset of S.

THEOREM 1.5.1. *Assume that F is local and that $G \in \widetilde{\mathcal{E}}_{\text{sim}}(N)$.*

(a) *For any local parameter $\psi \in \widetilde{\Psi}(G)$, there is a finite set $\widetilde{\Pi}_\psi$ over $\widetilde{\Pi}_{\text{unit}}(G)$, constructed from ψ by endoscopic transfer, and equipped with a canonical mapping*

$$\pi \longrightarrow \langle \cdot, \pi \rangle, \qquad \pi \in \widetilde{\Pi}_\psi,$$

from $\widetilde{\Pi}_\psi$ into the group $\widehat{\mathcal{S}}_\psi$ of characters on $\mathcal{S}_\psi$ such that $\langle \cdot, \pi \rangle = 1$ if G and π are unramified (relative to K_F).

(b) *If $\phi = \psi$ belongs to the subset $\widetilde{\Phi}_{\text{bdd}}(G)$ of parameters in $\widetilde{\Psi}(G)$ that are trivial on the factor $SU(2)$, the elements in $\widetilde{\Pi}_\phi$ are tempered and multiplicity free, and the corresponding mapping from $\widetilde{\Pi}_\phi$ to $\widehat{\mathcal{S}}_\phi$ is injective. Moreover, every element in $\widetilde{\Pi}_{\text{temp}}(G)$ belongs to exactly one packet $\widetilde{\Pi}_\phi$. Finally, if F is nonarchimedean, the mapping from $\widetilde{\Pi}_\phi$ to $\widehat{\mathcal{S}}_\phi$ is bijective.*

The finite subsets $\widetilde{\Pi}_\phi$ of $\widetilde{\Pi}_{\text{temp}}(G)$ in (b) represent the tempered L-packets for $G(F)$. They are composed of ($\widetilde{\text{O}}\text{ut}_N(G)$-orbits of) tempered representations, which are parametrized by characters in $\widehat{\mathcal{S}}_\phi$. Since the theorem asserts that $\widetilde{\Pi}_{\text{temp}}(G)$ is a disjoint union of these packets, it can be regarded as a classification of the irreducible, tempered representations of $G(F)$.

The more general sets $\widetilde{\Pi}_\psi$ of (a) represent local factors of automorphic representations. They can contain ($\widetilde{\text{O}}\text{ut}_N(G)$-orbits of) nontempered representations, and typically only fibre over $\widehat{\mathcal{S}}_\psi$. It seems likely that these packets are also multiplicity free (that is, subsets of $\widetilde{\Pi}_{\text{unit}}(G)$ rather than multi-subsets). For nonarchimedean F, Moeglin has recently established this fact [**M4**], using some of the properties of the tempered L-packets $\widetilde{\Pi}_\phi$ we will establish in the course of proving the theorem. However, the general mapping from $\widetilde{\Pi}_\psi$ to $\widehat{\mathcal{S}}_\psi$ is still not injective, so the elements in $\widetilde{\Pi}_\psi$ are not parametrized by characters on $\mathcal{S}_\psi$. One could rearrange the definition simply by combining all the elements in $\widetilde{\Pi}_\psi$ that map to a given character. With this equivalent formulation, the elements in $\widetilde{\Pi}_\psi$ would no longer be irreducible, but they would map injectively to $\widehat{\mathcal{S}}_\psi$. This was the point of view in the announcement of [**A18**, §30].

Analogues Π_ϕ of the packets $\widetilde{\Pi}_\phi$ in (b) have been defined in some cases for other groups. If F is archimedean, Langlands [**L11**] classified general representations of any group G in terms of packets. If F is p-adic, Lusztig

[**Lu2**] classified the unipotent representations of adjoint split groups G, and their inner forms, in terms of packets. This followed his earlier work with Kazhdan [**KL**], and established a conjecture he had announced in [**Lu1**]. Lusztig later generalized his construction to general (absolutely simple) adjoint groups [**Lu3**]. Unipotent representations correspond to parameters ϕ whose restriction to the inertia subgroup $I_F \subset W_F$ of L_F is trivial. DeBacker and Reeder [**DR**] have recently formulated another class of representations in terms of packets, namely depth-0 supercuspidal representations in "general position", for unramified p-adic groups G (and their "pure inner forms"). Depth-0 representations correspond to parameters ϕ whose restriction to the wild inertia subgroup $P_F \subset I_F$ of L_F is trivial, while the extra conditions imply among other things that ϕ is also trivial on the subgroup $SU(2)$ of L_F.

We should emphasize that Theorem 1.5.1 asserts implicitly that the various objects can all be defined in a canonical and quite specific way. We will make this precise in the statement of Theorem 2.2.1 from the next section. The main point is that the representations of Theorem 1.5.1 will be given by their characters, and that these in turn will be defined by the endoscopic transfer of twisted characters from general linear groups. This is essential for our understanding of the local constituents of automorphic representations. For the global theorems will be based on the global trace formula, which reduces ultimately to a theory of characters. We therefore need an endoscopic local classification in terms of packets and characters in order to establish the global theory of endoscopy for our groups G, which we recall was proposed in general by Langlands in [**L10**].

It remains an open question whether the various packets that have been constructed for more general groups all satisfy some analogue of the strong form of Theorem 1.5.1. For archimedean F, Shelstad [**S2**]–[**S6**] has shown that the general packets Π_ϕ of the Langlands classification are indeed defined by the endoscopic transfer of characters. To see that they satisfy the stronger condition of Theorem 2.2.1 in the special case here of quasisplit orthogonal and symplectic groups, and therefore represent the implicit packets of Theorem 1.5.1, one also needs the twisted supplements of Shelstad [**S8**] and Mezo [**Me**], together with some further work in progress. We will use these results throughout the volume, and in particular in the (global) proofs of (the local) Theorems 1.5.1 and 2.2.1.

More generally, Adams, Barbasch and Vogan [**ABV**] have constructed general packets Π_ψ by endoscopic transfer, for any real group G. They do not include twisted supplements, so it is not known whether their packets include those implicit in Theorem 1.5.1, although one expects this to be so. It is also not known that the representations in the general packets Π_ψ are unitary. However, the purely local methods of [**ABV**] are entirely different from those here, and are likely to have an important role in future study. (See also the earlier paper [**BV**] of Barbasch and Vogan, which treats the case of complex groups by more elementary means.) For p-adic F, the packets in [**Lu2**] and [**Lu3**] are constructed by a numerical matching of parameters, which gives little information about the underlying characters. However, Moeglin and Waldspurger [**MW3**], [**W2**] have shown that in the case of unipotent representations of a split group $SO(2n+1)$, they do satisfy the required endoscopic character relations, and are therefore the same as the packets implicit in Theorem 1.5.1. The packets of [**DR**] are also constructed by a matching of parameters. At the same time, the authors establish some endoscopic character relations (the required assertions of stability), using character formulas [**AS**] that go back to Murnaghan-Kirillov theory [**Mur**], [**KM**]. Work in progress by Kaletha [**Kal4**] is aimed at the more general endoscopic character relations, and in particular, the problem of comparing the (depth-0, supercuspidal) packets of [**DR**] with those implicit in Theorem 1.5.1.

Theorem 1.5.1 gives the local Langlands correspondence for p-adic F in case $\widetilde{\mathrm{Out}}_N(G) = 1$, and something slightly weaker in case $\widetilde{\mathrm{Out}}_N(G) \neq 1$. In all cases, the theorem yields local Rankin-Selberg L-functions and ε-factors for representations of general quasisplit orthogonal and symplectic groups. For any pair of representations

$$\pi_i \in \widetilde{\Pi}_{\mathrm{temp}}(G_i), \qquad G_i \in \widetilde{\mathcal{E}}_{\mathrm{sim}}(N_i),\ i = 1, 2,$$

we have only to take the corresponding pair of parameters

$$\{\phi_i \in \widetilde{\Phi}_{\mathrm{bdd}}(G_i) : \pi_i \in \widetilde{\Pi}_{\phi_i},\ i = 1, 2\}$$

for the associated local packets. These objects can of course also be regarded as local parameters for general linear groups. They allow us to define the local L-functions

$$L(s, \pi_1 \times \pi_2) = L(s, \phi_1 \times \phi_2)$$

and ε-factors

$$\varepsilon(s, \pi_1 \times \pi_2, \psi_F) = \varepsilon(s, \phi_1 \times \phi_2, \psi_F)$$

in terms of what is already known for general linear groups. We note that similar definitions apply if one of the groups G_i is replaced by a general linear group. In this case, they have been studied in terms of the Langlands-Shahidi method. (See [**Sha5**].)

The remaining two theorems are global. However, their statement requires an extension of the local construction. We could easily formulate the objects of Theorem 1.5.1 for parameters in the larger set $\widetilde{\Psi}^+(G)$. However, it is instructive to introduce an intermediate set

$$\widetilde{\Psi}(G) \subset \widetilde{\Psi}^+_{\mathrm{unit}}(G) \subset \widetilde{\Psi}^+(G),$$

which serves the purpose and is perhaps more natural.

We first introduce an analogous intermediate set

$$\Psi(N) \subset \Psi^+_{\mathrm{unit}}(N) \subset \Psi^+(N)$$

for $GL(N)$. Recall that the local set $\Psi^+(N)$ consists of N-dimensional representations

$$\psi = \ell_1\psi_1 \oplus \cdots \oplus \ell_r\psi_r, \qquad \psi_i \in \Psi^+_{\mathrm{sim}}(N_i),$$

of the locally compact group $L_F \times SU(2)$. Any such ψ corresponds to an irreducible representation of the Levi subgroup

$$GL(N_1, F)^{\ell_1} \times \cdots \times GL(N_r, F)^{\ell_r}$$

of $GL(N, F)$, which is unitary modulo the centre. We write $\Psi^+_{\mathrm{unit}}(N)$ for the set of $\psi \in \Psi^+(N)$ such that the associated induced representation of $GL(N, F)$ is irreducible and unitary. This set can be described explicitly in terms of the parametrization of unitary representations in [**V2**] and [**Tad1**]. It is the natural domain for the set of local factors of automorphic representations in the discrete spectrum of $GL(N)$. The associated set for G is then defined as the intersection

$$\widetilde{\Psi}^+_{\mathrm{unit}}(G) = \widetilde{\Psi}^+(G) \cap \Psi^+_{\mathrm{unit}}(N).$$

Suppose that $\psi \in \widetilde{\Psi}^+_{\mathrm{unit}}(G)$. Any such parameter gives rise to an $\widetilde{\mathrm{Out}}_N(G)$-orbit of standard parabolic subgroups $P = MN$ of G, a parameter $\psi_M \in \widetilde{\Psi}(M)$ and a point λ in the associated open chamber of P in the real vector space

$$\mathfrak{a}^*_M = X^*(M)_F \otimes \mathbb{R}.$$

These objects are characterized by the property that ψ equals the composition of the twist $\psi_{M,\lambda} \in \widetilde{\Psi}^+(M)$ of ψ_M by λ with the L-embedding ${}^LM \subset {}^LG$. Recall that there is a canonical homomorphism

$$H_M : M(F) \longrightarrow \mathfrak{a}_M = \mathrm{Hom}\big(X^*(M)_F, \mathbb{R}\big).$$

The twist $\psi_{M,\lambda}$ is the product of ψ_M with the central Langlands parameter that is dual to the unramified quasicharacter

$$\chi_\lambda : m \longrightarrow e^{\lambda(H_M(m))}, \qquad m \in M(F).$$

The Levi subgroup $M \subset G$ attached to ψ is a product of several general linear groups with a group $G_- \in \widetilde{\mathcal{E}}_{\mathrm{sim}}(N_-)$, for some $N_- \leqslant N$. The stabilizer $\widetilde{\mathrm{Out}}_N(P)$ of P in $\widetilde{\mathrm{Out}}_N(G)$ is isomorphic to $\widetilde{\mathrm{Out}}_{N_-}(G_-)$. The obvious variant of Theorem 1.5.1(a) attaches objects $\widetilde{\Pi}_{\psi_M}$ and $\langle \cdot, \pi_M \rangle$ to the parameter ψ_M. An element $\pi_M \in \widetilde{\Pi}_{\psi_M}$ is then an $\widetilde{\mathrm{Out}}_N(P)$-orbit of representations $\sigma \in \Pi_{\mathrm{unit}}(M)$. Let $\pi_{M,\lambda}$ be the corresponding orbit of representations $\pi_M \otimes \chi_\lambda$, and let $\mathcal{I}_P(\pi_{M,\lambda})$ be the associated $\widetilde{\mathrm{Out}}_N(G)$-orbit of induced representations. With this understanding, we define the packet of ψ as the family

$$\widetilde{\Pi}_\psi = \big\{\pi = \mathcal{I}_P(\pi_{M,\lambda}) : \pi_M \in \widetilde{\Pi}_{\psi_M}\big\}, \tag{1.5.1}$$

a finite set that is bijective with $\widetilde{\Pi}_{\psi_M}$. We are not assuming at this point that the induced representations in question are irreducible or unitary. In other words, an element $\pi \in \widetilde{\Pi}_\psi$ is an $\widetilde{\mathrm{Out}}_N(G)$-orbit of (possibly reducible, possibly nonunitary) representations of $G(F)$. However, the numbers $\operatorname{tr}\pi(f)$ and the operators $\pi(f)$ are nonetheless well defined for functions $f \in \widetilde{\mathcal{H}}(G)$. As functions of λ, these objects extend analytically to entire functions on the complexification $\mathfrak{a}^*_{M,\mathbb{C}}$ of $\mathfrak{a}^*_M$. If $\lambda \in i\mathfrak{a}^*_M$ is purely imaginary and in general position, the induced representations in the corresponding packet are unitary and irreducible, and are therefore among the packets treated in Theorem 1.5.1.

The G-centralizer $S_\psi = S_\psi(G)$ of the given ψ is contained in $\widehat{M}$. It follows that $S_\psi = S_{\psi_M}$, from which we deduce that $\mathcal{S}_\psi = \mathcal{S}_{\psi_M}$. We can therefore define the character

$$s \longrightarrow \langle s, \pi\rangle = \langle s_M, \pi_M\rangle, \qquad \pi \in \widetilde{\Pi}_\psi, \tag{1.5.2}$$

for any element $s = s_M$ in the group $\mathcal{S}_\psi = \mathcal{S}_{\psi_M}$. We thus obtain the local objects of Theorem 1.5.1(a) for any parameter $\psi \in \widetilde{\Psi}^+_{\mathrm{unit}}(G)$. The assertion (a) for these objects, which is to say the canonical construction by endoscopic transfer that we will make precise in §2.2, will follow immediately by analytic continuation.

Suppose now that the field F is global. Our group $G \in \widetilde{\mathcal{E}}_{\mathrm{sim}}(N)$ then represents a global endoscopic datum. The first global theorem is the central result. It gives a decomposition of the automorphic discrete spectrum of G in terms of global parameters $\psi \in \widetilde{\Psi}_2(G)$ and the local objects of Theorem 1.5.1(a) (or rather their extensions (1.5.1) and (1.5.2)). It is best stated in terms of the global Hecke algebra $\mathcal{H}(G)$ on $G(\mathbb{A})$ with respect to the maximal compact subgroup

$$K = \prod_v K_v.$$

Recall that $\mathcal{H}(G)$ is the space of linear combinations of products

$$\prod_v f_v, \qquad f_v \in \mathcal{H}(G_v),$$

such that f_v is the characteristic function of K_v for almost all v. In other words, $\mathcal{H}(G)$ is the restricted tensor product of the local Hecke algebras $\mathcal{H}(G_v)$. Let us write $\widetilde{\mathcal{H}}(G)$ for the restricted tensor product of the local symmetric Hecke algebras $\widetilde{\mathcal{H}}(G_v)$. This is the subspace of functions in $\mathcal{H}(G)$ that are invariant under each of the groups $\widetilde{\mathrm{Out}}_N(G_v)$. For any function f in $\widetilde{\mathcal{H}}(G)$, and any admissible representation of $G(\mathbb{A})$ of the form

$$\pi = \bigotimes_v \pi_v,$$

the operator $\pi(f)$ depends only on the $\widetilde{\mathrm{Out}}_N(G_v)$-orbit of any component π_v. The point is to describe the discrete spectrum as an $\widetilde{\mathcal{H}}(G)$-module.

We are assuming the seed Theorems 1.4.1 and 1.4.2. The second of these implies that the localization mapping $\psi \to \psi_v$ takes the global set $\widetilde{\Psi}(G)$ to the subset $\widetilde{\Psi}^+_{\mathrm{unit}}(G_v)$ of $\widetilde{\Psi}^+_v(N)$. The local theorem we have just stated then attaches a local packet $\widetilde{\Pi}_{\psi_v}$ to ψ and v. We can thus attach a global packet

$$\widetilde{\Pi}_\psi = \Big\{ \bigotimes_v \pi_v : \ \pi_v \in \widetilde{\Pi}_{\psi_v}, \ \langle \cdot, \pi_v \rangle = 1 \text{ for almost all } v \Big\} \tag{1.5.3}$$

of (orbits of) representations of $G(\mathbb{A})$ to any $\psi \in \widetilde{\Psi}(G)$. Any representation $\pi = \bigotimes \pi_v$ in the global packet determines a character

$$\langle x, \pi \rangle = \prod_v \langle x_v, \pi_v \rangle, \qquad x \in \mathcal{S}_\psi, \tag{1.5.4}$$

on the global quotient $\mathcal{S}_\psi$.

THEOREM 1.5.2. *Assume that F is global and that $G \in \widetilde{\mathcal{E}}_{\mathrm{sim}}(N)$. Then there is an $\widetilde{\mathcal{H}}(G)$-module isomorphism*

$$L^2_{\mathrm{disc}}\big(G(F)\backslash G(\mathbb{A})\big) \cong \bigoplus_{\psi \in \widetilde{\Psi}_2(G)} m_\psi \Big(\bigoplus_{\pi \in \widetilde{\Pi}_\psi(\varepsilon_\psi)} \pi \Big), \tag{1.5.5}$$

where m_ψ equals 1 or 2, while

$$\varepsilon_\psi : \ \mathcal{S}_\psi \longrightarrow \{\pm 1\}$$

is a linear character defined explicitly in terms of symplectic ε-factors, and $\widetilde{\Pi}_\psi(\varepsilon_\psi)$ is the subset of representations π in the global packet $\widetilde{\Pi}_\psi$ such that the character $\langle \cdot, \pi \rangle$ on $\mathcal{S}_\psi$ equals ε_ψ.

We need to supplement the statement of Theorem 1.5.2 with a description of the integer m_ψ and the character ε_ψ. The first of these is easy enough. The parameter ψ comes with the L-embedding

$$\widetilde{\psi}_G : \ \mathcal{L}_\psi \times SL(2, \mathbb{C}) \longrightarrow {}^L G,$$

determined up to the action of the group $\widetilde{\mathrm{Aut}}_N(G)$ by conjugation on $\widehat{G}$. We define m_ψ for any $\psi \in \widetilde{\Psi}(G)$ to be the number of $\widehat{G}$-orbits in the $\widetilde{\mathrm{Aut}}_N(G)$-orbit of $\widetilde{\psi}_G$. The theorem applies to a parameter

$$\psi = \psi_1 \boxplus \cdots \boxplus \psi_r$$

in $\widetilde{\Psi}_2(G)$. In this case, it is easy to check that m_ψ equals 1, unless N is even, $\widehat{G}$ equals $SO(N, \mathbb{C})$, and the rank N_i of each of the components ψ_i of ψ is also even, in which case m_ψ equals 2.

The sign character ε_ψ is more interesting. But first we make an observation on general L-functions. Suppose that $\psi \in \widetilde{\Psi}(N)$ is an arbitrary global parameter, and that r is an arbitrary finite dimensional representation of $\mathcal{L}_\psi$, subject only to the condition that its equivalence class is stable under the group $\widetilde{\mathrm{Aut}}(\mathcal{L}_\psi)$. Then r pulls back to a well defined representation r_v of L_{F_v}, for any v. We can therefore define the global L-function

$$L(s, r) = \prod_v L(s, r_v)$$

by an Euler product that converges for the real part of s large. Of course we do not know in general that it has analytic continuation and functional equation. We can still define the global ε-factor as a finite product

$$\varepsilon(s, r, \psi_F) = \prod_v \varepsilon(s, r_v, \psi_{F_v}),$$

where ψ_F is a nontrivial additive character on $\mathbb{A}/F$. We cannot say that this function is independent of ψ_F in general. But if r is symplectic, by which we mean that it takes values in the symplectic subgroup of the underlying general linear group, the value of the local factor $\varepsilon(s, r_v, \psi_{F_v})$ at $s = \frac{1}{2}$ is known to equal $+1$ or -1, and to be independent of ψ_{F_v}. We therefore have a global sign

$$\varepsilon\big(\tfrac{1}{2}, r\big) = \varepsilon\big(\tfrac{1}{2}, r, \psi_F\big) = \pm 1$$

in this case, which is independent of ψ_F.

We shall define ε_ψ if ψ is a general parameter in $\widetilde{\Psi}(G)$. We first define a representation

$$\tau_\psi : \ S_\psi \times \mathcal{L}_\psi \times SL(2, \mathbb{C}) \longrightarrow GL(\hat{\mathfrak{g}})$$

on the Lie algebra $\hat{\mathfrak{g}}$ of $\widehat{G}$ by setting

$$\tau_\psi(s, g, h) = \mathrm{Ad}\big(s \cdot \widetilde{\psi}_G(g, h)\big), \qquad s \in S_\psi, \ (g, h) \in \mathcal{L}_\psi \times SL(2, \mathbb{C}),$$

where $\mathrm{Ad} = \mathrm{Ad}_G$ is the adjoint representation of ${}^L G$. Notice that since it is invariant under the Killing form on $\hat{\mathfrak{g}}$, this representation is orthogonal, and hence self-dual. Let

$$\tau_\psi = \bigoplus_\alpha \tau_\alpha = \bigoplus_\alpha (\lambda_\alpha \otimes \mu_\alpha \otimes \nu_\alpha)$$

be its decomposition into irreducible representations λ_α, μ_α and ν_α of the respective groups S_ψ, $\mathcal{L}_\psi$ and $SL(2, \mathbb{C})$. We then define

$$\varepsilon_\psi(x) = \prod_\alpha{}' \det\big(\lambda_\alpha(s)\big), \qquad s \in S_\psi, \tag{1.5.6}$$

where x is the image of s in $\mathcal{S}_\psi$, and $\prod'$ denotes the product over those indices α with μ_α symplectic and

$$\varepsilon\big(\tfrac{1}{2}, \mu_\alpha\big) = -1. \tag{1.5.7}$$

(The definition (1.5.6) is equivalent to [**A18**, (30.16)], and is conjecturally equivalent to the slightly different formulations [**A8**, (8.4)] and [**A9**, (4.5)] in the earlier papers in which the general multiplicity formula (1.5.5) was postulated.) Theorem 1.5.2 thus asserts that there is an intimate relationship between the automorphic discrete spectrum of G and symplectic root numbers.

Before we state the second global theorem, we recall a property of certain Rankin-Selberg L-functions. Suppose that $\pi \in \mathcal{A}_{\mathrm{cusp}}(N)$ is a cuspidal automorphic representation of $GL(N)$. Then there are two formal products

$$L(s, \pi \times \pi) = L(s, \pi, S^2)\, L(s, \pi, \Lambda^2) \tag{1.5.8}$$

and

$$(1.5.9) \qquad \varepsilon(s, \pi \times \pi, \psi_F) = \varepsilon(s, \pi, S^2, \psi_F)\,\varepsilon(s, \pi, \Lambda^2, \psi_F),$$

where S^2 (resp. Λ^2) is the representation of $GL(N, \mathbb{C})$ on the space of symmetric (resp. skew-symmetric) $(N \times N)$-complex matrices. Each of S^2 and Λ^2 is associated with a Siegel maximal parabolic subgroup $\widehat{P}_+ \subset \widehat{G}_+$, for a split, simple endoscopic group $G_+ \in \widetilde{\mathcal{E}}_{\text{sim}}(N_+)$ with $N_+ = 2N$. The dual group $\widehat{G}_+$ equals $Sp(N_+, \mathbb{C})$ or $SO(N_+, \mathbb{C})$, according to which of the two representations S^2 are Λ^2 we are considering. In each case, the representation is given by the adjoint action of the Levi subgroup $\widehat{M}_+ \cong GL(N, \mathbb{C})$ on the Lie algebra of the unipotent radical. Viewed in this way, one sees that the L-functions on the right hand side of (1.5.8) are among the cases of the Langlands-Shahidi method treated in [**Sha4**]. In both cases, the local L-functions and ε-factors have been constructed so that the formal products (1.5.8) and (1.5.9) become actual products, and so that the global L-functions have analytic continuation with functional equation (1.3.5).

Suppose now that $\phi \in \widetilde{\Phi}_{\text{sim}}(N)$ is cuspidal generic. In other words, ϕ is given by a representation $\pi \in \mathcal{A}_{\text{cusp}}(N)$ that is self-dual. Theorem 1.4.1 asserts that ϕ belongs to the subset

$$\widetilde{\Phi}_{\text{sim}}(G) = \widetilde{\Phi}_{\text{sim}}(N) \cap \widetilde{\Psi}(G),$$

for a unique $G \in \widetilde{\mathcal{E}}_{\text{sim}}(N)$. We will need to understand how G is related to L-functions and ε-factors.

The Rankin-Selberg L-function

$$L(s, \phi \times \phi) = L(s, \pi \times \pi) = L(s, \pi \times \pi^\vee)$$

has a pole of order 1 at $s = 1$. It is known that neither of the corresponding factors $L(s, \phi, S^2)$ and $L(s, \phi, \Lambda^2)$ on the right hand side of (1.5.8) has a zero at $s = 1$. It follows that exactly one of them has a pole at $s = 1$ (which will be of order 1). This motivates the first assertion (a) of the second global theorem, which we can now state.

THEOREM 1.5.3. *Assume that F is global.*

(a) *Suppose that $G \in \widetilde{\mathcal{E}}_{\text{sim}}(N)$ and that ϕ belongs to $\widetilde{\Phi}_{\text{sim}}(G)$. Then $\widehat{G}$ is orthogonal if and only if the symmetric square L-function $L(s, \phi, S^2)$ has a pole at $s = 1$, while $\widehat{G}$ is symplectic if and only if the skew-symmetric square L-function $L(s, \phi, \Lambda^2)$ has a pole at $s = 1$.*

(b) *Suppose that for $i = 1, 2$, ϕ_i belongs to $\widetilde{\Phi}_{\text{sim}}(G_i)$, for simple endoscopic data $G_i = \widetilde{\mathcal{E}}_{\text{sim}}(N_i)$. Then the corresponding Rankin-Selberg ε-factor satisfies*

$$\varepsilon\big(\tfrac{1}{2}, \phi_1 \times \phi_2\big) = 1,$$

if $\widehat{G}_1$ and $\widehat{G}_2$ are either both orthogonal or both symplectic.

The assertion (a) of this theorem was suggested by the corresponding property for L-functions of representations of Galois groups and Weil groups.

It seems to have been first conjectured for automorphic representations by Jacquet and Shalika, at the early stages of their work on Rankin-Selberg L-functions. The assertion (b) is suggested by the corresponding property for ε-factors of orthogonal representations of Galois groups [**FQ**] and Weil groups [**D2**]. It was conjectured for automorphic representations in [**A8**, §8] and [**A9**, §4].

Observe that (a) gives an independent characterization of the group G_ϕ of Theorem 1.4.1. It has been established for representations with Whittaker models by Cogdell, Kim, Piatetskii-Shapiro and Shahidi [**CKPS1**], [**CKPS2**]. In fact, apart from the uniqueness assertion of the first seed Theorem 1.4.1, the statements of Theorem 1.4.1 and 1.4.2, together with that of (a) above, follow from the combined results of [**CKPS2**] and [**GRS2**]. However, we will not be able to use these results. From our standpoint, the essence of the two seed theorems will be in two further statements that we have temporarily suppressed in the interests of simplicity. These give further characterizations of G_ϕ and its completions G_{ϕ_v} in terms of harmonic analysis, and will be what really drives the general argument. They will be included in the assertions of Theorems 2.2.1 and 4.1.2, which we will come to in due course.

The broader significance of (b) will become clear later. This assertion has been proved in special cases by Lapid [**Lap**]. Its role for us will be rather similar to that of (a). Both assertions will be established in the course of proving the other theorems. In fact, they are both an inextricable part of the general induction argument by which all of the theorems will eventually be proved.

CHAPTER 2

Local Transfer

2.1. Langlands-Shelstad-Kottwitz transfer

We now take the field F to be local, a condition that will remain in place throughout Chapter 2. In the first section, we shall review the endoscopic transfer of functions. This is based on the transfer factors of Langlands and Shelstad, and the twisted transfer factors of Kottwitz and Shelstad. It will serve as the foundation for the local classification we stated in §1.5.

Consider our $GL(N)$-coset $\widetilde{G}(N)$. The twisted transfer factors of Kottwitz and Shelstad for $\widetilde{G}(N)$ depend on a choice of F-rational automorphism of $GL(N)$ within the given inner class. For this object, we choose the automorphism $\widetilde{\theta}(N)$ introduced in §1.2 that fixes the standard splitting of $GL(N)$. We then allow $\widetilde{G}(N)$ to stand for the pair $\big(\widetilde{G}^0(N), \widetilde{\theta}(N)\big)$, as well as the coset

$$\widetilde{G}^0(N) \rtimes \widetilde{\theta}(N) = GL(N) \rtimes \theta(N)$$

in $\widetilde{G}^+(N)$ obtained from the pair. We recall that $\widetilde{\theta}(N)$ lies in the inner class of the original automorphism $\theta = \theta(N)$, and therefore gives the same coset in the semidirect product $\widetilde{G}^+(N)$.

The datum

$$\widetilde{G}(N) = \big(\widetilde{G}^0(N), \widetilde{\theta}(N)\big) = \big(\widetilde{G}^0(N), \widetilde{\theta}(N), 1\big)$$

is a special case of a general triplet (G^0, θ, ω), where G^0 is a connected reductive group over F, θ is an automorphism of G over F, and ω is a character on $G^0(F)$. The group generated by θ need not be finite. Its semidirect product with G^0 therefore need not be an algebraic group. However, we can still form the connected variety

$$G = G^0 \rtimes \theta$$

over F, equipped with the two-sided G^0-action

$$x_1(y \rtimes \theta)x_2 = \big(x_1 y \theta(x_2)\big) \rtimes \theta, \qquad x \in G^0,$$

of G^0 on G. The restriction of this action to the diagonal image of G^0 can evidently be identified with the familiar action of G^0 on itself by θ-conjugation. As in the case of $\widetilde{G}(N)$, we shall also write

$$G = (G^0, \theta, \omega).$$

In other words, we allow G to represent the underlying triplet as well as the associate G^0-bitorsor over F.

For the moment, let us take G to be a general triplet over F. We can then form the general local Hecke module $\mathcal{H}(G)$ of G. It consists of the smooth, compactly supported functions on $G(F)$ that are left and right K-finite, relative to the two-sided action on the coset $G(F)$ of a suitable (fixed) maximal compact subgroup K of $G^0(F)$. A semisimple element $\gamma \in G(F)$ will be called *strongly regular* if its G^0-centralizer

$$G_\gamma = (G^0)_\gamma = \{x \in G^0 : x^{-1}\gamma x = \gamma\}$$

is an abelian group, whose group $G_\gamma(F)$ of rational points lies in the kernel of the character ω on $G^0(F)$. For any such γ, and any function $f \in \mathcal{H}(G)$, we form the invariant orbital integral

$$f_G(\gamma) = |D(\gamma)|^{\frac{1}{2}} \int_{G_\gamma(F)\backslash G^0(F)} f(x^{-1}\gamma x)\omega(x)dx.$$

It is normalized here by the usual Weyl discriminant

$$D(\gamma) = \det\big((1 - \mathrm{Ad}(\gamma)_{\mathfrak{g}^0/\mathfrak{g}^0_\gamma}\big),$$

where $\mathfrak{g}^0$ and $\mathfrak{g}^0_\gamma$ denote the Lie algebras of G^0 and G^0_γ respectively. We regard f_G as a function on the set of strongly regular points, and write

$$\mathcal{I}(G) = \{f_G : f \in \mathcal{H}(G)\}$$

for the G^0-invariant Hecke space of such functions.

The functions in $\mathcal{I}(G)$ also have a spectral interpretation. Suppose that π is a unitary extension to $G(F)$ of an irreducible unitary representation (π^0, V) of $G^0(F)$. In other words, π is a function from $G(F)$ to the space of unitary operators on V such that

$$\pi(x_1 x x_2) = \pi^0(x_1)\pi(x)\pi^0(x_2)\omega(x_2), \qquad x_1, x_2 \in G^0(F). \tag{2.1.1}$$

The set of such extensions of π^0 is a $U(1)$-torsor, on which the character

$$\mathrm{tr}\big(\pi(f)\big) = \mathrm{tr}\Big(\int_{G(F)} f(x)\pi(x)dx\Big), \qquad f \in \mathcal{H}(G),$$

is equivariant. We set

$$f_G(\pi) = \mathrm{tr}\big(\pi(f)\big).$$

It can then be shown that either of the two functions $\{f_G(\gamma)\}$ or $\{f_G(\pi)\}$ attached to f determines the other. We can therefore regard f_G as a function of either γ or π. The spectral interpretation is in some ways preferable. This is because the trace Paley-Wiener theorem ([**BDK**], [**CD**], [**Ro1**], [**DM**]) leads to a simple description of $\mathcal{I}(G)$ as a space of functions of π.

In general, an endoscopic datum G' for G represents a 4-tuple $(G', \mathcal{G}', s', \xi')$. According to Langlands, Shelstad and Kottwitz [**LS1**], [**KS**], one can attach a transfer factor Δ to G'. Recall that any Δ comes with some auxiliary data, namely a suitable central extension $\widetilde{G}' \to G'$ over F and an admissible L-embedding $\widetilde{\xi}'\colon \mathcal{G}' \to {}^L\widetilde{G}'$, and is then determined up to a complex multiplicative constant of absolute value 1. (See [**KS**, §2.2]. We shall review these notions later in the global context of §3.2.) The transfer factor is a function $\Delta(\delta', \gamma)$, where δ' is a strongly G-regular stable conjugacy class in $\widetilde{G}'(F)$, and γ is a strongly regular orbit in $G(F)$ under the action of $G^0(F)$ by conjugation. It serves as the kernel function for the transfer mapping, which sends functions $f \in \mathcal{H}(G)$ to functions

$$f'(\delta') = f_\Delta^{\widetilde{G}'}(\delta') = \sum_\gamma \Delta(\delta', \gamma) f_G(\gamma)$$

of δ'. We recall that f' has an equivariance property

$$f'(z'\delta') = \widetilde{\eta}'(z')^{-1} f'(\delta'), \qquad z' \in \widetilde{C}'(F),$$

where $\widetilde{C}'$ is the kernel of the projection $\widetilde{G}' \to G'$, and $\widetilde{\eta}'$ is a character on $\widetilde{C}'(F)$ that depends on the choice of $\widetilde{\xi}'$.

Suppose for example that $G = G^0$ is just a quasisplit group (with trivial character ω), and that $G' = G$. If we set $\widetilde{G}'$ equal to G and $\widetilde{\xi}'$ equal to the identity embedding of $\mathcal{G}' = {}^L G$ into ${}^L\widetilde{G}' = {}^L G$, Δ becomes a constant, which we can take to be 1. In this case, the transfer

$$f^G(\delta) = \sum_{\gamma \to \delta} f_G(\gamma), \qquad f \in \mathcal{H}(G),$$

of f is the stable orbital integral over a strongly regular, stable conjugacy class δ. It is a sum of invariant orbital integrals over the finite set of $G(F)$-conjugacy classes γ in δ. We write $\mathcal{S}(G)$ for the space of functions of δ obtained in this way. More generally, if $\mathfrak{X}_G$ is a closed central subgroup of $G(F)$, and χ is a character on $\mathfrak{X}_G$, we write $\mathcal{S}(G, \chi)$ for the associated space of χ^{-1}-equivariant functions

$$f_\chi^G(\delta) = \int_{\mathfrak{X}_G} f^G(z\delta)\chi(z)dz, \qquad f \in \mathcal{H}(G),$$

of δ. A χ-equivariant linear form S on $\mathcal{H}(G)$ is called *stable* if its value at f depends only on f_χ^G. If this is so, we can identify S with the linear form

$$\widehat{S}(f_\chi^G) = S(f), \qquad f \in \mathcal{H}(G), \tag{2.1.2}$$

on $\mathcal{S}(G, \chi)$. These conventions can then be applied with $\big(\widetilde{G}', \widetilde{C}'(F), \widetilde{\eta}'\big)$ in place of $(G, \mathfrak{X}_G, \chi)$.

The LSK (Langlands-Shelstad-Kottwitz) conjecture asserts that for any G and G', any associated transfer factor Δ, and any $f \in \mathcal{H}(G)$, the function f' belongs to $\mathcal{S}(\widetilde{G}', \widetilde{\eta}')$. The fundamental lemma is a variant of this, which applies to the case that G is unramified. (Among other things [**W6**, 4.4], *unramified* means that F is p-adic, and that G^0 is quasisplit and split over an unramified extension of F.) The fundamental lemma asserts that if f is the characteristic function of a product $K \rtimes \theta$, where K is a θ-stable, hyperspecial maximal compact subgroup of $G^0(F)$, then f' can be taken to be the image in $\mathcal{S}(\widetilde{G}', \widetilde{\eta}')$ of the characteristic function of a hyperspecial maximal compact subgroup of $\widetilde{G}'(F)$. Both of these longstanding conjectures have now been resolved. We include a couple of brief remarks about the proofs, some of which are quite recent.

If F is archimedean, the results are due to Shelstad. In the case $G = G^0$, she established the transfer theorem in terms of the somewhat ad hoc transfer factors she introduced for the purpose [**S3**]. These earlier transfer factors for real groups were the precursors of the systematic transfer factors for both real and p-adic groups in [**LS1**]. In a second paper, Langlands and Shelstad showed that the two transfer factors were in fact the same for real groups, by an indirect argument that used Shelstad's original proof of the transfer theorem [**LS2**, Theorem 2.6.A]. Shelstad has recently expanded her earlier results [**S4**]–[**S6**], establishing the transfer conjecture (again for $G = G^0$) directly in terms of the transfer factors of [**LS1**]. She has recently completed a proof of the general archimedean transfer [**S7**], using the explicit specialization to real groups of the twisted transfer factors of [**KS**]. (See also [**Re**].)

For nonarchimedean F, the fundamental lemma has long been a serious obstacle. Its recent proof by Ngo [**N**] was a breakthrough, which has now opened the way for progress on several fronts. The general proof follows the special case of $G = U(n)$ treated earlier by Laumon and Ngo [**LN**], and is based on the new geometric ideas that were introduced by Goresky, Kottwitz and MacPherson [**GKM1**], [**GKM2**]. The methods in all of these papers exploit the algebraic geometry over fields of positive characteristic. However, by the results of Waldspurger [**W3**] on independence of characteristic (which have also been established by the methods of motivic integration [**CHL**], following [**CH**]), they apply also to our field F of characteristic 0. The paper of Ngo treats both the fundamental lemma for $G = G^0$ and the variant to which Waldspurger had reduced the general case [**W6**]. It therefore resolves the fundamental lemma for any G. As for the LSK conjecture, Waldspurger established some time ago that the case $G = G^0$ would follow from the fundamental lemma [**W1**]. His recent papers [**W6**], [**W7**] extend this implication to the general case. The general results of Waldspurger therefore yield the p-adic LSK conjecture in all cases.

Our main focus in this volume will be on the three cases introduced in §1.2. In the first, we take $G = (G^0, \theta, \omega)$ to be the pair $\widetilde{G}(N) = \big(\widetilde{G}^0(N), \widetilde{\theta}(N)\big)$ (with ω trivial) we were just discussing. In this case, G' represents a twisted endoscopic datum G in the set $\widetilde{\mathcal{E}}(N) = \mathcal{E}\big(\widetilde{G}(N)\big)$. We have already fixed a homomorphism of ${}^L G$ into the dual group $\widehat{G}^0(N) = GL(N, \mathbb{C})$, which of course determines an L-embedding of ${}^L G$ into ${}^L G^0(N)$. This amounts to a canonical choice of auxiliary data. In the second case, G and G' are simply two objects $G \in \widetilde{\mathcal{E}}_{\mathrm{sim}}(N)$ and $G' \in \mathcal{E}(G)$. Here we have also fixed an L-embedding of ${}^L G'$ into ${}^L G$, so there is again no need of further auxiliary data. In the third case, we take $G = (G^0, \theta, \omega)$ to be the pair $\widetilde{G} = (\widetilde{G}^0, \widetilde{\theta})$ (with ω trivial) attached to an even orthogonal group G in $\widetilde{\mathcal{E}}_{\mathrm{sim}}(N)$, and G' to be a twisted endoscopic datum $\widetilde{G}'$ in $\mathcal{E}(\widetilde{G})$. Once again, we have an L-embedding of ${}^L\widetilde{G}'$ into ${}^L\widetilde{G}^0$, and no need for further auxiliary data. This case will generally be easier to manage once we have taken care of the others, and will often be treated as an addendum. The first two cases on the other hand will have a central role in almost all of our arguments. In fact, we shall sometimes find it convenient to treat the corresponding pairs $\big(\widetilde{G}(N), G\big)$ and (G, G') in tandem.

Since the groups we are considering here are quasisplit, we can fix canonical transfer factors. We shall use the Whittaker normalization of [**KS**, §5.3]. In general, a Whittaker datum consists of a rational Borel subgroup and a nondegenerate character on the group of rational points in its unipotent radical. In present circumstances, these objects can be chosen canonically, up to the choice of a fixed nontrivial additive character ψ_F of F.

Consider the $GL(N)$-bitorsor $\widetilde{G}(N)$. We fix a $\widetilde{\theta}(N)$-stable Whittaker datum $\big(B(N), \chi(N)\big)$ for $\widetilde{G}^0(N)$ by taking $B(N)$ to be the standard Borel subgroup of $\widetilde{G}^0(N) = GL(N)$, and $\chi(N)$ to be the nondegenerate character on the unipotent radical $N_{B(N)}(F)$ of $B(N, F)$ attached to ψ_F. That is,

$$\chi(N, x) = \psi_F(x_{1,2} + \cdots + x_{n-1,n}), \qquad x = (x_{i,j}) \in N_{B(N)}(F).$$

Given a twisted endoscopic datum $G \in \widetilde{\mathcal{E}}(N)$, we take

$$\widetilde{\Delta}_N = \widetilde{\Delta}_{\chi(N)} = \widetilde{\Delta}_{S(N)} \cdot \varepsilon\big(\tfrac{1}{2}, \tau_F, \psi_F\big)^{-1}$$

to be the transfer factor assigned to $\big(B(N), \chi(N)\big)$ in [**KS**]. Here $\widetilde{\Delta}_{S(N)}$ is the transfer factor attached in [**KS**, p. 63] to the standard splitting $S(N)$ of $\widetilde{G}^0(N) = GL(N)$, ψ_F is the additive character used to define $\chi(N)$, and τ_G is the representation of the Galois group Γ_F on the space $X^*(T) \otimes \mathbb{Q}$, where T is the standard (diagonal) maximal torus in G. The local ε-factor $\varepsilon\big(\frac{1}{2}, \tau_G, \psi_F\big)$ is trivial unless G has a quasisplit group $SO(2n, \eta_G)$ as a factor, in which case it equals $\varepsilon\big(\frac{1}{2}, \eta_G, \psi_F\big)$. (The general definition in [**KS**] also contains a factor $\varepsilon\big(\frac{1}{2}, \widetilde{\tau}_N, \psi_F\big)$, where $\widetilde{\tau}_N$ is the representation of Γ_F on the space $X^*\big(T(N)\big)^{\widetilde{\theta}(N)} \otimes \mathbb{Q}$ attached to $\widetilde{G}(N)$, but since the associated maximal torus $T(N) \subset B(N)$ is split, this factor equals 1).

A general group $G \in \widetilde{\mathcal{E}}(N)$ is a product of (at most two) quasisplit orthogonal and symplectic groups, together with a number of general linear groups. The discussion of §1.2 leads to a standard splitting for each factor, and hence a splitting S for the product G. We use this splitting to form a Whittaker datum (B, χ) for G. We then have a corresponding transfer factor

$$\Delta = \Delta_\chi = \Delta_S \cdot \varepsilon\big(\tfrac{1}{2}, \tau_{G'}, \psi_F\big)^{-1} \cdot \varepsilon\big(\tfrac{1}{2}, \tau_G, \psi_F\big)$$

for any $G' \in \mathcal{E}(G)$. Similar considerations give rise to a canonical transfer factor

$$\widetilde{\Delta} = \widetilde{\Delta}_\chi = \widetilde{\Delta}_S \cdot \varepsilon\big(\tfrac{1}{2}, \widetilde{\tau}_{G'}, \psi_F\big)^{-1} \cdot \varepsilon\big(\tfrac{1}{2}, \widetilde{\tau}_G, \psi_F\big)$$

for the twisted component $\widetilde{G}$ of an even orthogonal group $G \in \widetilde{\mathcal{E}}_{\mathrm{sim}}(N)$, and a twisted endoscopic datum $\widetilde{G}' \in \mathcal{E}(\widetilde{G})$.

Given the normalized transfer factors $\widetilde{\Delta}_N$, and with the knowledge that the LSK-conjecture holds in general, we have a canonical transfer mapping $\widetilde{f} \to \widetilde{f}^G$ from $\widetilde{\mathcal{H}}(N)$ to $\mathcal{S}(G)$, for any $G \in \widetilde{\mathcal{E}}(N)$. The function $\widetilde{f}^G$ actually lies in the subspace of $\widetilde{\mathrm{O}}\mathrm{ut}_N(G)$-invariant functions in $\mathcal{S}(G)$, since the function $\widetilde{\Delta}_N$ is invariant under the action of this group on the first variable. We also have a canonical transfer mapping $f \to f'$ from $\mathcal{H}(G)$ to $\mathcal{S}(G')$ for any $G' \in \mathcal{E}(G)$, whose image has a similar symmetry property. Finally, in the supplementary case of a twisted (even) orthogonal group $\widetilde{G}$, we have a canonical transfer mapping $\widetilde{f} \to \widetilde{f}'$ from $\mathcal{H}(\widetilde{G})$ to $\mathcal{S}(\widetilde{G}')$ attached to any $\widetilde{G}' \in \mathcal{E}(\widetilde{G})$.

If $G \in \widetilde{\mathcal{E}}_{\mathrm{sim}}(N)$ is simple, we follow §1.5 by defining $\widetilde{\mathcal{H}}(G)$, $\widetilde{\mathcal{I}}(G)$ and $\widetilde{\mathcal{S}}(G)$ as the subspaces of $\widetilde{\mathrm{O}}\mathrm{ut}_N(G)$-invariant functions in $\mathcal{H}(G)$, $\mathcal{I}(G)$ and $\mathcal{S}(G)$. We extend this definition in the natural way to any $G \in \widetilde{\mathcal{E}}(N)$, or indeed to any connected, quasisplit group that is a product of orthogonal, symplectic and general linear factors, by imposing the appropriate symmetry condition at the orthogonal factors of G. For example, if

$$G' = G'_1 \times G'_2, \qquad G'_i \in \widetilde{\mathcal{E}}_{\mathrm{sim}}(N_i),\ i = 1, 2,$$

then

$$\widetilde{\mathcal{S}}(G') = \widetilde{\mathcal{S}}(G'_1) \otimes \widetilde{\mathcal{S}}(G'_2).$$

The mappings $\widetilde{f} \to \widetilde{f}^G$ and $f \to f'$ then take $\widetilde{\mathcal{H}}(N)$ into $\widetilde{\mathcal{S}}(G)$ and $\widetilde{\mathcal{H}}(G)$ into $\widetilde{\mathcal{S}}(G')$. Observe, however, that if $G \in \widetilde{\mathcal{E}}(N)$ is not elliptic, $\widetilde{\mathrm{O}}\mathrm{ut}_N(G)$ is larger than the symmetry group of the subspace $\widetilde{\mathcal{S}}(G)$. In particular, the transfer mapping $\widetilde{f} \to \widetilde{f}^G$ from $\widetilde{\mathcal{H}}(N)$ to $\widetilde{\mathcal{S}}(G)$ is not surjective in this case.

Twisted transfer for $GL(N)$ has a special role to play. It is what leads us to the essential objects for orthogonal and symplectic groups, in their guise as simple endoscopic data $G \in \widetilde{\mathcal{E}}_{\mathrm{sim}}(N)$, on which we will be able to base the local classification. To this end, we will need to know that the mapping $\widetilde{f} \to \widetilde{f}^G$ does take $\widetilde{\mathcal{H}}(N)$ onto $\widetilde{\mathcal{S}}(G)$. The property is part of a broader characterization of the image of the collective transfer mapping

$$\widetilde{f} \longrightarrow \bigoplus_G \widetilde{f}^G, \qquad \widetilde{f} \in \widetilde{\mathcal{H}}(N),\ G \in \widetilde{\mathcal{E}}(N),$$

which we will later also have to know.

We sometimes denote general elements in $\widetilde{\mathcal{E}}(N)$ by M, since they can be regarded as Levi subgroups of elliptic endoscopic groups $G \in \widetilde{\mathcal{E}}_{\mathrm{ell}}(N)$. Recall that any such M comes with an L-embedding ξ_M of $\mathcal{M} = {}^L M$ into ${}^L GL(N)$, which we can treat as an equivalence class of N-dimensional representations of ${}^L M$. We may as well take G also to be a general element in $\widetilde{\mathcal{E}}(N)$. Suppose that its N-dimensional representation $\xi = \xi_G$ can be chosen so that it contains the image of (a representative) of ξ_M. In other words, ξ_M is the composition of an L-embedding ${}^L M \subset {}^L G$ with ξ. This L-embedding identifies $\widehat{M}$ with a Levi subgroup of $\widehat{G}$, which is dual to an

F-rational embedding

$$\lambda_M : M \hookrightarrow G$$

as a Levi subgroup of G. We shall call λ_M a *Levi embedding* of M into G. It is uniquely determined up to composition

$$\alpha \circ \lambda_M \circ \alpha_M$$

by elements $\alpha \in \widetilde{\mathrm{Aut}}_N(G)$ and $\alpha_M \in \widetilde{\mathrm{Aut}}_N(M)$. We note that for any λ_M, the stable orbital integrals of a function $f \in \mathcal{H}(G)$ over G-regular classes in $M(F)$ give a function $f^M = (f_M)^M$ in $\mathcal{S}(M)$.

Consider a family of functions

$$\mathcal{F} = \{f \in \widetilde{\mathcal{H}}(G) : \ G \in \widetilde{\mathcal{E}}(N)\}, \tag{2.1.3}$$

parametrized by the endoscopic data for $\widetilde{G}(N)$. We shall say that $\mathcal{F}$ is a *compatible family* if for any data G and M in $\widetilde{\mathcal{E}}(N)$, and any Levi embedding λ_M of M into G, the associated functions $f \in \widetilde{\mathcal{H}}(G)$ and $h \in \widetilde{\mathcal{H}}(M)$ in $\mathcal{F}$ satisfy

$$f^M = h^M.$$

In other words, the function in $\widetilde{\mathcal{S}}(M)$ attached to M equals the restriction of the corresponding function for G. In particular, the family of stable functions $\{f^G\}$ attached to $\mathcal{F}$ is determined by the subset of functions $f \in \widetilde{\mathcal{H}}(G)$ parametrized by the elliptic elements $G \in \widetilde{\mathcal{E}}_{\mathrm{ell}}(N)$, since any M has a Levi embedding into some elliptic G. We could therefore have formulated $\mathcal{F}$ as a family of functions parametrized by $\widetilde{\mathcal{E}}_{\mathrm{ell}}(N)$.

PROPOSITION 2.1.1. *Suppose that $\mathcal{F}$ is any family of functions* (2.1.3). *Then $\mathcal{F}$ is a compatible family if and only if there is a function $\widetilde{f} \in \widetilde{\mathcal{H}}(N)$ such that*

$$f^G = \widetilde{f}^G, \qquad G \in \widetilde{\mathcal{E}}(N).$$

PROOF. This is the twisted analogue (for $GL(N)$) of a property that plays a significant role in the stabilization of the trace formula. For the general untwisted case, the proof is implicit in [**S3**] and [**A11**], and will be given explicitly in [**A24**]. (See [**A11**, p. 329].) For the twisted group $\widetilde{G}(N)$, the proof follows similar lines, based on work in progress by Shelstad and Mezo on twisted characters for real groups and Waldspurger's p-adic results [**W4**], [**W5**] on twisted endoscopy, which include the twisted analogue of a theorem from [**W1**]. However, since the lemma represents part of the stabilization of the twisted trace formula, which we will be adopting as a general hypothesis in the next chapter, we will not give a formal proof. We will be content simply to outline the main arguments.

Suppose first that F is p-adic. This is the more difficult case. Its proof has three steps, which we discuss briefly in turn. It will be easier for us to refer to the untwisted case in [**A11**], even though the proof for $\widetilde{G}(N)$ is at least partly implicit in [**W4**, V–VI].

The first step is to establish twisted analogues for $\widetilde{G}(N)$ of the adjoint relations [**A11**, (2.4), (2.5)] for geometric transfer factors. Let

$$\widetilde{\Gamma}_{\text{reg,ell}}(N) = \Gamma_{\text{reg,ell}}(\widetilde{G}(N))$$

be the set of $G^0(N,F)$-conjugacy orbits in $G(N,F)$ that are strongly regular and elliptic. (We are using the natural twisted forms of these notions, for which the reader can consult [**KS**, p. 28, 74].) If G belongs to $\widetilde{\mathcal{E}}_{\text{ell}}(N)$, let

$$\widetilde{\Delta}_{N\text{-reg,ell}}(G) = \Delta_{\widetilde{G}(N)\text{-reg,ell}}(G)$$

be the set of stable conjugacy classes in $G(F)$ that are $\widetilde{G}(N)$-strongly regular and elliptic. We then set

$$\widetilde{\Gamma}^{\mathcal{E}}_{\text{reg,ell}}(N) = \Gamma^{\mathcal{E}}_{\text{reg,ell}}(\widetilde{G}(N)) = \coprod_{G\in\widetilde{\mathcal{E}}_{\text{ell}}(N)} (\widetilde{\Delta}_{N\text{-reg,ell}}(G)/\widetilde{\text{O}}\text{ut}_N(G)),$$

where the right hand quotient denotes the set of orbits under the finite group $\widetilde{\text{O}}\text{ut}_N(G)$. The twisted transfer factor for $\widetilde{G}(N)$ restricts to a function

$$\Delta(\delta',\gamma) = \Delta(\delta,\gamma)$$

of $\gamma \in \widetilde{\Gamma}_{\text{reg,ell}}(N)$ and $\delta' \in \widetilde{\Delta}_{N\text{-reg,ell}}(N)$, which depends only on the image δ of δ' in $\widetilde{\Gamma}^{\mathcal{E}}_{\text{reg,ell}}(N)$. The adjoint relations take the form

$$\sum_{\delta\in\widetilde{\Gamma}^{\mathcal{E}}_{\text{reg,ell}}(N)} \Delta(\gamma,\delta)\Delta(\delta,\gamma_1) = \delta(\gamma,\gamma_1)$$

and

$$\sum_{\gamma\in\widetilde{\Gamma}_{\text{reg,ell}}(N)} \Delta(\delta,\gamma)\Delta(\gamma,\delta_1) = \delta(\delta,\delta_1),$$

for elements $\gamma,\gamma_1 \in \widetilde{\Gamma}_{\text{reg,ell}}(N)$ and $\delta,\delta_1 \in \widetilde{\Gamma}^{\mathcal{E}}_{\text{reg,ell}}(N)$, where

$$\Delta(\gamma,\delta) = \overline{\Delta(\delta,\gamma)}$$

is the adjoint transfer factor, and $\delta(\cdot,\cdot)$ is the Kronecker delta. Their proof is similar to that of [**A11**, Lemma 2.2]. In the twisted case here, we use the equivariance relation

$$\Delta(\delta,\gamma) = \langle \text{inv}(\gamma,\gamma_1),\kappa_\gamma\rangle \Delta(\delta,\gamma_1)$$

of [**KS**, Theorem 5.1.D], and the local form of the bijection established in [**KS**, Lemma 7.2.A].

The second step is to establish that the transfer mapping

$$\mathcal{T}^{\mathcal{E}} = \widetilde{\mathcal{T}}^{\mathcal{E}}_N = \bigoplus_G \mathcal{T}^G : \ a \longrightarrow a^{\mathcal{E}} = \bigoplus_G a^G, \qquad a \in \widetilde{\mathcal{I}}_{\text{cusp}}(N),$$

from the invariant space $\widetilde{\mathcal{I}}_{\text{cusp}}(N)$ of cuspidal functions to its endoscopic counterpart

$$\widetilde{\mathcal{I}}^{\mathcal{E}}_{\text{cusp}}(N) = \bigoplus_{G\in\widetilde{\mathcal{E}}_{\text{ell}}(N)} (\mathcal{S}_{\text{cusp}}(G))^{\widetilde{\text{O}}\text{ut}_N(G)}, \tag{2.1.4}$$

is an isomorphism. (Recall that a function a in the space $\widetilde{\mathcal{I}}(N) = \mathcal{I}(\widetilde{G}(N))$ is *cuspidal* if $a(\gamma)$ vanishes for every γ in the complement of $\widetilde{\Gamma}_{\mathrm{reg,ell}}(N)$ in the set $\widetilde{\Gamma}_{\mathrm{reg}}(N)$ of all regular classes.) The proof of this fact is similar to that of its untwisted analogue [**A11**, Proposition 3.5]. One uses the adjoint transfer factor $\Delta(\gamma, \delta)$ to define a candidate

$$\mathcal{T}_{\mathcal{E}} = \widetilde{\mathcal{T}}_{\mathcal{E},N} : \ a^{\mathcal{E}} \longrightarrow a$$

for the inverse mapping. The adjoint relations then tell us that the mapping $\mathcal{T}_{\mathcal{E}}$ is indeed the inverse of $\mathcal{T}^{\mathcal{E}}$, once we know that it takes $\widetilde{\mathcal{I}}^{\mathcal{E}}_{\mathrm{cusp}}(N)$ to $\widetilde{\mathcal{I}}_{\mathrm{cusp}}(N)$. This last fact relies on deeper results of Waldspurger. One has to use his general methods of twisted descent [**W6**], for twisted transfer factors and the corresponding transfer mappings, in place of those of Langlands and Shelstad [**LS2**]. This reduces the problem to a local assertion for (untwisted) p-adic Lie algebras, which one verifies by inverting Waldspurger's kernel formula [**W1**, (1.2)], as in the proof of [**A11**, Proposition 3.5]. (See [**W4**, V.1].)

The third step is to extend the isomorphism $\mathcal{T}^{\mathcal{E}}$ to the full space $\widetilde{\mathcal{I}}(N) = \mathcal{I}(\widetilde{G}(N))$. This is where we are taking the most for granted, namely the extension to the twisted group $\widetilde{G}(N)$ of Theorem 6.2 of [**A11**]. We shall be brief.

Following [**A11**, §1], we form the graded vector space

$$\widetilde{\mathcal{I}}_{\mathrm{gr}}(N) = \bigoplus_{\{\widetilde{M}\}} \big(\mathcal{I}_{\mathrm{cusp}}(\widetilde{M})\big)^{W(\widetilde{M})}, \tag{2.1.5}$$

where $\{\widetilde{M}\}$ ranges over orbits of (semistandard) "Levi subsets" of $\widetilde{G}(N)$ under the Weyl group

$$\widetilde{W}_0^N = W_0^{G^0(N)} \cong S_N,$$

and

$$W(\widetilde{M}) = W^N(\widetilde{M}) = \mathrm{Norm}\big(A_{\widetilde{M}}, \widetilde{G}^0(N)\big)/\widetilde{M}^0$$

is the relative Weyl group for $\widetilde{M}$. (Recall that $\widetilde{M} \subset \widetilde{G}(N)$ is defined [**A4**, §1] as a Levi component of a "parabolic subset" of $\widetilde{G}(N)$, with split component $A_{\widetilde{M}}$.) The twisted tempered characters for $\widetilde{G}(N)$ (which we will introduce formally at the beginning of the next section) then provide a canonical isomorphism from $\widetilde{\mathcal{I}}(N)$ onto $\widetilde{\mathcal{I}}_{\mathrm{gr}}(N)$, as in the untwisted case [**A11**, §4]. If G belongs to $\widetilde{\mathcal{E}}_{\mathrm{ell}}(N)$, we can also form the graded vector space

$$\mathcal{S}_{\mathrm{gr}}(G) = \bigoplus_{\{M\}} \big(\mathcal{S}_{\mathrm{cusp}}(M)\big)^{W(M)}, \tag{2.1.6}$$

where $\{M\}$ ranges over W_0^G-orbits of Levi subgroups of G, and $W(M)$ is the relative Weyl group for M. The spectral construction from [**A11**, §5] then provides a canonical isomorphism from $\mathcal{S}(G)$ onto $\mathcal{S}_{\mathrm{gr}}(G)$. This assertion includes a nontrivial property of stability, and requires proof, unlike the statement for $\widetilde{G}(N)$ above. The necessary justification is furnished by [**A11**, Theorem 6.1], the first main theorem of [**A11**]. It is the second main theorem [**A11**, Theorem 6.2] whose extension to $\widetilde{G}(N)$ we are taking for granted. The extended assertion is that the two isomorphisms are compatible with transfer. In other words, the diagram

$$\begin{array}{ccc} \widetilde{\mathcal{I}}(N) & \xrightarrow{\sim} & \widetilde{\mathcal{I}}_{\mathrm{gr}}(N) \\ {\scriptstyle \mathcal{T}^G}\downarrow & & \downarrow \\ \mathcal{S}(G) & \xrightarrow{\sim} & \mathcal{S}_{\mathrm{gr}}(G)\,, \end{array} \tag{2.1.7}$$

in which the left hand vertical arrow is the (twisted) transfer mapping, and the right hand vertical arrow is built out of restrictions of transfer mappings to cuspidal functions, is commutative.

The assertion of the proposition for our p-adic field F is a consequence of these constructions. We introduce an invariant endoscopic space

$$\widetilde{\mathcal{I}}^{\mathcal{E}}(N) = \Big\{ \bigoplus_{G \in \widetilde{\mathcal{E}}_{\mathrm{ell}}(N)} f^G : \ f \in \mathcal{F} \Big\} \tag{2.1.8}$$

as the subspace of stable functions in

$$\bigoplus_{G \in \widetilde{\mathcal{E}}_{\mathrm{ell}}(N)} \big(\mathcal{S}(G)\big)$$

defined by the stable images of compatible families $\mathcal{F}$. We also form the graded endoscopic space

$$\widetilde{\mathcal{I}}^{\mathcal{E}}_{\mathrm{gr}}(N) = \bigoplus_{\{\widetilde{M}\}} \big(\mathcal{I}^{\mathcal{E}}_{\mathrm{cusp}}(\widetilde{M})\big)^{W(\widetilde{M})}. \tag{2.1.9}$$

It is then not hard to construct a commutative diagram of isomorphisms

$$\begin{array}{ccc} \widetilde{\mathcal{I}}(N) & \xrightarrow{\sim} & \widetilde{\mathcal{I}}_{\mathrm{gr}}(N) \\ {\scriptstyle \mathcal{T}^{\mathcal{E}}}\downarrow & & \downarrow \\ \widetilde{\mathcal{I}}^{\mathcal{E}}(N) & \xrightarrow{\sim} & \widetilde{\mathcal{I}}^{\mathcal{E}}_{\mathrm{gr}}(N) \end{array} . \tag{2.1.10}$$

The upper horizontal isomorphism is as above. The right hand vertical isomorphism is a consequence of the isomorphisms for spaces of cuspidal functions we have described. The lower horizontal isomorphism follows from the definition of a compatible family, and the existence of the corresponding isomorphism in (2.1.7). The three isomorphisms together then tell us that the left hand vertical arrow is also an isomorphism, which from the corresponding arrow in (2.1.7) we see is the endoscopic transfer mapping

$$\mathcal{T}^{\mathcal{E}} = \bigoplus_{G \in \widetilde{\mathcal{E}}_{\mathrm{ell}}(N)} \mathcal{T}^G.$$

The assertion of the proposition follows, for elements $G \in \widetilde{\mathcal{E}}_{\mathrm{ell}}(N)$ and hence for any $G \in \widetilde{\mathcal{E}}(N)$.

Assume now that F is archimedean. In this case, we take for granted the extension of Shelstad's theory of endoscopy, both geometric and spectral, to the twisted group $\widetilde{G}(N)$. (See [**S7**], [**Me**].) Then the archimedean forms of the spaces defined for p-adic F above all have spectral interpretations. They can each be regarded as a Paley-Wiener space of functions in an appropriate space of bounded, self-dual Langlands parameters

$$\phi:\ L_F \longrightarrow {}^L\widetilde{G}^0(N) = GL(N,\mathbb{C}) \rtimes \Gamma_F.$$

Such parameters can of course be identified with self-dual, unitary, N-dimensional representations of W_F. They can be classified by the analysis of §1.2, or rather the extension of this analysis to parameters that are not elliptic. It is then not hard to deduce the archimedean form of the last commutative diagram of isomorphisms, and hence the assertion of the proposition. □

COROLLARY 2.1.2. *Suppose that $G \in \widetilde{\mathcal{E}}_{\mathrm{sim}}(N)$ is simple. Then the (twisted) transfer mapping*

$$\widetilde{f} \longrightarrow \widetilde{f}^G, \qquad \widetilde{f} \in \widetilde{\mathcal{H}}(N),$$

takes $\widetilde{\mathcal{H}}(N)$ onto the subspace

$$\widetilde{\mathcal{S}}(G) = \big(\mathcal{S}(G)\big)^{\widetilde{\mathrm{Out}}_N(G)}$$

of $\mathcal{S}(G)$.

PROOF. We need to check that the transfer mapping $\mathcal{T}^G$ in (2.1.7) takes $\widetilde{\mathcal{I}}(N)$ onto $\widetilde{\mathcal{S}}(G)$. We can either show directly that any function in $\widetilde{\mathcal{S}}(G)$ is the image of a compatible family, or we can examine the lower horizontal mappings in (2.1.7) and (2.1.10). Taking the latter (less direct) course, we compose the inverse of the mapping in (2.1.10) with the projection of $\widetilde{\mathcal{I}}^{\mathcal{E}}(N)$ onto $\mathcal{S}(G)$ and the mapping in (2.1.7). This gives us a mapping

$$\widetilde{\mathcal{I}}^{\mathcal{E}}_{\mathrm{gr}}(N) \longrightarrow \mathcal{S}_{\mathrm{gr}}(G). \tag{2.1.11}$$

We must check that its image is the space of invariants in $\mathcal{S}_{\mathrm{gr}}(G)$ under the natural action of $\widetilde{\mathrm{Out}}_N(G)$.

If we restrict the mapping (2.1.11) to a summand of $\widetilde{\mathcal{I}}^{\mathcal{E}}_{\mathrm{gr}}(N)$ from the decompositions (2.1.9) and (2.1.4), which is then mapped to the corresponding summand of $\mathcal{S}_{\mathrm{gr}}(G)$ in (2.1.6), we obtain an injection from a subspace of $\mathcal{S}_{\mathrm{cusp}}(M)$ into a larger subspace. The problem is simply to characterize its image. It amounts to establishing a isomorphism of groups

$$\mathrm{Out}_{\widetilde{M}}(M)W(\widetilde{M},M) \cong W(M)\widetilde{\mathrm{Out}}_N(G,M), \qquad M \in \mathcal{E}_{\mathrm{ell}}(\widetilde{M}),$$

where $W(\widetilde{M},M)$ is the stabilizer of M in $W(\widetilde{M})$, and $\widetilde{\mathrm{Out}}_N(G,M)$ is the stabilizer of M in $\widetilde{\mathrm{Out}}_N(G)$. This is a straightforward exercise, which we leave to the reader. For example, in the special case that M is the subgroup of diagonal matrices

$$T = \{t : t^\vee = t\}$$

in G, $\mathrm{Out}_{\widetilde{M}}(M)$ is trivial, and the product of groups on the right is indeed equal to the stabilizer of $M = T$ in $W(\widetilde{M}) \cong S_N$. We note here that this property fails if G is composite, and therefore that the condition that G be simple is in fact necessary. □

REMARKS. 1. One direction in the statement of the proposition is more or less obvious. The fact that for any $\widetilde{f} \in \widetilde{\mathcal{H}}(N)$, the set $\{\widetilde{f}^G\}$ is the image of a compatible family $\mathcal{F}$ follows directly from the basic properties of transfer factors. It is the converse, the existence of some $\widetilde{f}$ for any $\mathcal{F}$, that is the main point. The argument we have sketched establishes both parts together.

2. The discussion from the proof of the corollary applies also to parameters in the set $\widetilde{\Psi}(G)$. It gives us another way to see that for any $G \in \widetilde{\mathcal{E}}_{\mathrm{sim}}(N)$, the mapping from $\widetilde{\Psi}(G)$ to $\widetilde{\Psi}(N)$ is injective. We have of course always been treating $\widetilde{\Psi}(G)$ as a subset of $\widetilde{\Psi}(N)$, a property that was already implicit in the discussion of §1.2.

2.2. Characterization of the local classification

The first of the three theorems stated in §1.5 applies to the local field F. It includes a local classification of representations of a quasisplit orthogonal or symplectic group G. However, the description is purely qualitative. In this section, we shall make the assertion precise. We shall state a theorem that characterizes the representations in terms of the LSK-transfer of functions.

The classification is based on the transfer of self-dual representations of $G(N) = GL(N)$, or rather the extensions of such representations to $\widetilde{G}(N)$. Having normalized the LSK-transfer factors, we shall now normalize the extensions of these representations. We need to show that any irreducible self-dual representation π of $\widetilde{G}^0(N,F) = GL(N,F)$ has a canonical extension to the group $\widetilde{G}^+(N,F)$. As was implicit for the transfer factors, we will use the theory of Whittaker models.

Suppose first that π is tempered. Then π has a $\big(B(N),\chi(N)\big)$-Whittaker functional ω, where $\big(B(N),\chi(N)\big)$ is the standard Whittaker datum fixed in the last section. By definition, ω is a nonzero linear form on the underlying space V_∞ of smooth vectors for π such that

$$\omega\big(\pi(n)v\big) = \chi(N,n)\omega(v), \qquad n \in N_{B(N)}(F),\ v \in V_\infty.$$

It is unique up to a scalar multiple. We are assuming that π is self-dual, which is to say that the representation $\pi \circ \widetilde{\theta}(N)$ is equivalent to π. We can therefore choose a nontrivial intertwining operator $\widetilde{I}$ from π to $\pi \circ \widetilde{\theta}(N)$, which is unique up to a nonzero scalar multiple. Since the Whittaker datum $\big(B(N),\chi(N)\big)$ is $\widetilde{\theta}(N)$-stable, the linear form $\omega \circ \widetilde{I}$ on V is also a

$(B(N), \chi(N))$-Whittaker model for π. It therefore equals $c\omega$, for some $c \in \mathbb{C}^*$. We set $\widetilde{\pi}(N) = \pi(\widetilde{\theta}(N))$ equal to the operator $c^{-1}\widetilde{I}$. Then $\widetilde{\pi}(N)$ is the unique intertwining operator from π to $\pi \circ \widetilde{\theta}(N)$ such that

$$\omega = \omega \circ \widetilde{\pi}(N).$$

It provides a unitary extension $\widetilde{\pi}$ of π to the group $\widetilde{G}^+(N, F)$ generated by $\widetilde{G}(N, F)$. In particular, it gives an extension of π to the bitorsor $\widetilde{G}(N, F)$ that satisfies (2.1.1).

Suppose next that π is replaced by a self-dual standard representation ρ. Recall that ρ is a (possibly reducible) representation, induced from the twist $\sigma = \pi_{M,\lambda}$ of an irreducible tempered representation of a Levi subgroup $M(F)$ by a regular, positive real valued character. We can identify M with a block diagonal subgroup

$$GL(m_1) \times \cdots \times GL(m_k)$$

of $GL(N)$, and σ with a representation

$$\pi_1(x_1)|\det x_1|^{\lambda_1} \otimes \cdots \otimes \pi_k(x_k)|\det x_k|^{\lambda_k}, \qquad x \in M(F),$$

for irreducible tempered representations $\pi_1, \dots, \pi_k$ and real numbers $\lambda_1 > \cdots > \lambda_k$. It follows easily from the self-duality of ρ, and the form we have chosen for σ, that

$$\pi_i^\vee \cong \pi_{k+1-i}, \qquad 1 \leqslant i \leqslant k.$$

In particular, M is $\widetilde{\theta}(N)$-stable, and σ is isomorphic to $\sigma \circ \widetilde{\theta}(N)$. By a minor variation of our discussion of the tempered representation π above, we see that σ has a canonical extension to the group

$$\widetilde{M}^+(F) = M(F) \rtimes \langle \widetilde{\theta}(N) \rangle.$$

Notice that the standard block upper triangular subgroup P in $GL(N)$ of type $(m_1, \dots, m_k)$ is also $\widetilde{\theta}(N)$-stable. Its normalizer $\widetilde{P}^+$ in $\widetilde{G}^+(N)$ can be regarded as a parabolic subgroup of $G^+(N)$. The pullback from $\widetilde{M}^+(F)$ to $\widetilde{P}^+(F)$ of the extended representation σ can then be induced to $\widetilde{G}^+(N, F)$. We thus obtain an extension $\widetilde{\rho}$ of ρ to $\widetilde{G}^+(N, F)$.

We note that in the case of $GL(N)$ here, there is a bijection $\phi \to \rho$ from the set $\widetilde{\Phi}(N)$ of self-dual Langlands parameters ϕ for $GL(N)$ and the set of self-dual standard representations $\rho = \rho_\phi$ of $GL(N, F)$. Given $\phi \in \widetilde{\Phi}(N)$, we shall write

$$\widetilde{f}_N(\phi) = \operatorname{tr}(\widetilde{\rho}(\widetilde{f})), \qquad \widetilde{f} \in \widetilde{\mathcal{H}}(N),$$

for the extension $\widetilde{\rho} = \widetilde{\rho}_\phi$ of ρ_ϕ we have just described. Observe that for ρ of the form above, we can replace the vector $\lambda = (\lambda_1, \dots, \lambda_k)$ in $\mathbb{R}^k$ by a complex multiple

$$z\lambda, \qquad z \in \mathbb{C}.$$

This gives us another self-dual standard representation ρ_z, with corresponding Langlands parameter ϕ_z. The function $\widetilde{f}_N(\phi_z)$, defined for z near 1 in terms of the Whittaker model of M, extends to an entire function. Now if z is purely imaginary, ρ_z is tempered, and $\widetilde{f}_N(\phi_z)$ is also defined in terms of the Whittaker model of $GL(N)$. However, it follows easily from the general structure of induced Whittaker models [**Shal**], [**CS**] that the two functions of $z \in i\mathbb{R}$ are equal. In other words, $\widetilde{f}_N(\phi)$ can be obtained for general ϕ by analytic continuation from the case of tempered ϕ.

Suppose finally that π is a general irreducible, self-dual representation of $GL(N, F)$. Then π is the Langlands quotient of a uniquely determined standard representation ρ. The bijective correspondence between irreducible and standard representations implies that ρ is also self-dual. The canonical extension $\widetilde{\rho}$ of ρ to $\widetilde{G}^+(N, F)$ then provides a canonical extension $\widetilde{\pi}$ of π to $\widetilde{G}^+(N, F)$. This is what we needed to construct. In the case that π is unitary, the restriction of this extension to the bitorsor $\widetilde{G}(N, F)$ represents a canonical extension from the general class (2.1.1).

Suppose now that ψ is a parameter in $\widetilde{\Psi}(N)$. As in the global remarks following the statement of Theorem 1.3.3, we define the local Langlands parameter $\phi_\psi \in \widetilde{\Phi}(N)$ by setting

$$\phi_\psi(u) = \psi\left(u, \begin{pmatrix} |u|^{\frac{1}{2}} & 0 \\ 0 & |u|^{-\frac{1}{2}} \end{pmatrix}\right), \qquad u \in L_F.$$

We then have the standard representation $\rho_\psi = \rho_{\phi_\psi}$ of $GL(N, F)$ attached to ϕ_ψ, and its Langlands quotient $\pi_\psi = \pi_{\phi_\psi}$. The irreducible representation π_ψ is unitary and self-dual. It therefore has a canonical extension $\widetilde{\pi}_\psi$ to the bitorsor $\widetilde{G}(N, F)$. We write

$$\widetilde{f}_N(\psi) = \mathrm{tr}\big(\widetilde{\pi}_\psi(\widetilde{f})\big), \qquad \widetilde{f} \in \widetilde{\mathcal{H}}(N). \tag{2.2.1}$$

Our interest will be in the transfer of this linear form to a twisted endoscopic group.

Suppose that ψ belongs to the subset $\widetilde{\Psi}(G)$ of $\widetilde{\Psi}(N)$ attached to a simple datum $G \in \widetilde{\mathcal{E}}_{\mathrm{sim}}(N)$. Then ψ can be identified with an L-homomorphism

$$\widetilde{\psi}_G : \ L_F \times SU(2) \longrightarrow {}^L G.$$

On the other hand, we have just seen that ψ determines a linear form (2.2.1) on the space $\widetilde{\mathcal{H}}(N) = \mathcal{H}\big(\widetilde{G}(N)\big)$. The next theorem uses this linear form to make the local correspondence of Theorem 1.5.1(a) precise. Specifically, it gives an explicit construction of the packet $\widetilde{\Pi}_\psi$ and the pairing $x \to \langle \cdot, \pi \rangle$ of Theorem 1.5.1(a) in terms of (2.2.1), and the endoscopic transfer of functions.

THEOREM 2.2.1. (a) *Suppose that $G \in \widetilde{\mathcal{E}}_{\mathrm{ell}}(N)$, and that ψ belongs to $\widetilde{\Psi}(G)$. Then there is a unique stable linear form*

$$f \longrightarrow f^G(\psi), \qquad f \in \widetilde{\mathcal{H}}(G), \tag{2.2.2}$$

on $\widetilde{\mathcal{H}}(G)$ with the general property

$$\widetilde{f}^G(\psi) = \widetilde{f}_N(\psi), \qquad \widetilde{f} \in \widetilde{\mathcal{H}}(N), \tag{2.2.3}$$

together with a secondary property

$$f^G(\psi) = f^S(\psi_S)f^O(\psi_O), \qquad f \in \widetilde{\mathcal{H}}(G), \tag{2.2.4}$$

in case

$$\begin{aligned} G &= G_S \times G_O, &\quad G_\varepsilon &\in \widetilde{\mathcal{E}}_{\mathrm{sim}}(N_\varepsilon), \\ \psi &= \psi_S \times \psi_O, &\quad \psi_\varepsilon &\in \widetilde{\Psi}(G_\varepsilon), \end{aligned}$$

and

$$f^G = f^S \times f^O, \qquad f^\varepsilon \in \widetilde{\mathcal{S}}(G_\varepsilon), \qquad \varepsilon = O, S,$$

are composite.

(b) *Suppose that $G \in \widetilde{\mathcal{E}}_{\mathrm{sim}}(N)$. Then for every $\psi \in \widetilde{\Psi}(G)$, there is a finite set $\widetilde{\Pi}_\psi$ over $\widetilde{\Pi}_{\mathrm{unit}}(G)$, together with a mapping*

$$\pi \longrightarrow \langle \cdot, \pi \rangle, \qquad \pi \in \widetilde{\Pi}_\psi, \tag{2.2.5}$$

from $\widetilde{\Pi}_\psi$ to the group $\widehat{\mathcal{S}}_\psi$ of irreducible characters on $\mathcal{S}_\psi$, with the following property. If s is a semisimple element in the centralizer $S_\psi = S_\psi(G)$ and (G', ψ') is the preimage of (ψ, s) under the local version of the correspondence (1.4.11) in §1.4, then

$$f'(\psi') = \sum_{\pi \in \widetilde{\Pi}_\psi} \langle s_\psi x, \pi \rangle f_G(\pi), \qquad f \in \widetilde{\mathcal{H}}(G), \tag{2.2.6}$$

where x is the image of s in $\mathcal{S}_\psi$.

REMARKS. 1. If $G \in \widetilde{\mathcal{E}}_{\mathrm{sim}}(N)$ is simple, the uniqueness of the linear form in (a) follows from (2.2.3), since the mapping $\widetilde{f} \to \widetilde{f}^G$ takes $\widetilde{\mathcal{H}}(N)$ onto $\widetilde{\mathcal{S}}(G)$. If $G \notin \widetilde{\mathcal{E}}_{\mathrm{sim}}(N)$ is composite, the uniqueness follows from the product formula (2.2.4).

2. On the right hand side of (2.2.6), we have the point

$$s_\psi = \psi\left(1, \begin{pmatrix} -1 & 0 \\ 0 & -1 \end{pmatrix}\right)$$

in S_ψ, introduced globally in §1.4 and identified with its image in $\mathcal{S}_\psi$. The linear form on the left hand side is a product of linear forms (2.2.6) on the factors of the endoscopic group G' for G. It is then clear that (2.2.6) characterizes the packet $\widetilde{\Pi}_\psi$ and the pairing $\langle x, \pi \rangle$ in (b). It therefore does provide a canonical formulation of the local correspondence of Theorem 1.5.1.

3. If G is composite, the symbol ψ on the right hand of (2.2.3) is understood to be the image of the given parameter under (the not necessarily injective) mapping of $\widetilde{\Psi}(G)$ to $\widetilde{\Psi}(N)$. In this case, the identity (2.2.3) is not

needed to characterize either the linear form in (a) or the local correspondence in (b). However, it will still have an important role in our proofs.

4. Theorem 2.2.1 will be proved at the same time as the theorems stated in Chapter 1, together also with further theorems we will state in due course. The argument, which will take up much of the rest of the volume, will rely on long term induction hypothesis based on the integer N.

5. If F is archimedean and $\psi = \phi$ lies in the subset $\widetilde{\Phi}_{\mathrm{bdd}}(N)$ of generic parameters in $\widetilde{\Psi}(G)$, the assertions of the theorem are included in the general results of Shelstad [**S3**], [**S4**]–[**S7**], and their twisted analogues for $GL(N)$ [**Me**], [**S8**], as we noted in §1.5.

Theorem 2.2.1 includes parameters ψ that are not generic, in the sense that they are nontrivial on the second factor of $L_F \times SU(2)$. One ultimately wants to reduce their study to the case that $\psi = \phi$ is generic. With this in mind, we shall describe how the twisted character on the right hand side of (2.2.3) can be expanded in terms of standard twisted characters.

Let us write $\widetilde{P}(N)$ for the set of "standard representations" of $\widetilde{G}(N, F)$. Elements in this set are induced objects

$$\widetilde{\rho} = \mathcal{I}_{\widetilde{P}}(\widetilde{\pi}_{M,\lambda}), \qquad \lambda \in (\mathfrak{a}^*_{\widetilde{P}})^+,$$

where $\widetilde{P} = \widetilde{M}\widetilde{N}$ is a standard "parabolic subset" [**A4**, §1] of $\widetilde{G}(N)$, and $\widetilde{\pi}_M$ is an extension (2.1.1) to $\widetilde{M}(F)$ of an irreducible tempered representation $\widetilde{\pi}^0_M$ of $\widetilde{M}^0(F)$. (The sets $\widetilde{P}$ and $\widetilde{P}(N)$ are of course different. The first symbol P is an upper case p, while the second is supposed to be an upper case ρ.) By definition, $\widetilde{P}$ is the normalizer in $\widetilde{G}(N)$ of a $\widetilde{\theta}$-stable, standard parabolic subgroup $\widetilde{P}^0 = \widetilde{M}^0\widetilde{N}^0$ of $\widetilde{G}^0(N) = GL(N)$. Its Levi subset $\widetilde{M}$ has a split component $A_{\widetilde{M}} \subset A_{\widetilde{M}^0}$, and an associated real vector space $\mathfrak{a}_{\widetilde{M}} \subset \mathfrak{a}_{\widetilde{M}^0}$ with an open chamber $\mathfrak{a}^+_{\widetilde{P}} \subset \widetilde{\mathfrak{a}}^+_{\widetilde{P}^0}$ and corresponding dual chamber $(\mathfrak{a}^*_{\widetilde{P}})^+ \subset (\mathfrak{a}^*_{\widetilde{P}^0})^+$. As in (2.1.1), there is a simply transitive action

$$u : \widetilde{\pi}_M \longrightarrow \widetilde{\pi}_{M,u}, \qquad u \in U(1),$$

of the group $U(1)$ on the set of extensions $\widetilde{\pi}_M$ of $\widetilde{\pi}^0_M$, and therefore a corresponding action $\widetilde{\rho} \to \widetilde{\rho}_u$ on the set of $\widetilde{\rho} \in \widetilde{P}(N)$ attached to $\widetilde{\pi}^0_M$ and λ. We note that any $U(1)$-orbit in $\widetilde{P}(N)$ contains a simply transitive $(\mathbb{Z}/2\mathbb{Z})$-orbit. It consists of those $\widetilde{\rho}$ such that $\widetilde{\pi}_M$ generates an irreducible representation of the group $\widetilde{M}^+(F)$ generated by $\widetilde{M}(F)$. Within this $(\mathbb{Z}/2\mathbb{Z})$-orbit, we have the element $\widetilde{\rho}$ defined by the Whittaker normalization above.

Any $\widetilde{\rho} \in \widetilde{P}(N)$ has a canonical Langlands quotient $\widetilde{\pi} = \pi_{\widetilde{\rho}}$. It is an extension (2.1.1) to $\widetilde{G}(N, F)$ of an irreducible representation $\widetilde{\pi}^0 \in \Pi(\widetilde{G}^0(N))$ (except for the fact that $\widetilde{\pi}^0$ does not satisfy the unitary condition of (2.1.1)). As in (2.1.1), the set of extensions $\widetilde{\pi}$ attached to $\widetilde{\pi}^0$ in this way is a transitive $U(1)$-orbit. The mapping $\widetilde{\rho} \to \pi_{\widetilde{\rho}}$ is then a $U(1)$-bijection from $\widetilde{P}(N)$ onto the corresponding set $\widetilde{\Pi}(N)$ of extensions to $\widetilde{G}(N, F)$ of irreducible representations of $\widetilde{G}^0(N,F)$. We write $\{\widetilde{P}(N)\}$ and $\{\widetilde{\Pi}(N)\}$ for the set of $U(1)$-orbits in $\widetilde{P}(N)$ and $\widetilde{\Pi}(N)$ respectively.

A standard parabolic subset $\widetilde{P}$ comes with a canonical embedding of its dual chamber $(\mathfrak{a}^*_{\widetilde{P}})^+$ into the closure $\overline{(\mathfrak{a}^*_B)^+}$ of the dual chamber of the standard Borel subgroup B. For any standard representation $\widetilde{\rho} \in \widetilde{P}(N)$ with Langlands quotient $\widetilde{\pi} = \pi_{\widetilde{\rho}}$ in $\widetilde{\Pi}(N)$ as above, we shall write

$$\Lambda_\rho = \Lambda_\pi, \qquad \pi = \widetilde{\pi}^0,\ \rho = \widetilde{\rho}^0,$$

for the corresponding image of the point λ. In general, one writes

$$\Lambda' \leqslant \Lambda, \qquad \Lambda', \Lambda \in \overline{(\mathfrak{a}^*_B)^+},$$

if $\Lambda - \Lambda'$ is a nonnegative integral combination of simple roots of (B, A_B). This determines a partial order on each of the sets $\{\widetilde{P}(N)\}$ and $\{\widetilde{\Pi}(N)\}$.

We thus have two families

$$\widetilde{f}_N(\widetilde{\rho}) = \mathrm{tr}\big(\widetilde{\rho}(\widetilde{f})\big), \qquad \widetilde{f} \in \widetilde{\mathcal{H}}(N),\ \widetilde{\rho} \in \widetilde{P}(N),$$

and

$$\widetilde{f}_N(\widetilde{\pi}) = \mathrm{tr}\big(\widetilde{\pi}(\widetilde{f})\big), \qquad \widetilde{f} \in \widetilde{\mathcal{H}}(N),\ \widetilde{\pi} \in \widetilde{\Pi}(N),$$

of $\widetilde{G}^0(N,F)$-invariant linear forms on $\widetilde{\mathcal{H}}(N)$. Taken up to the action of $U(1)$, they represent two bases of the same vector space. More precisely, there are uniquely determined complex numbers

$$\big\{m(\widetilde{\rho},\widetilde{\pi}), n(\widetilde{\pi},\widetilde{\rho}) : \ \widetilde{\rho} \in \widetilde{P}(N),\ \widetilde{\pi} \in \widetilde{\Pi}(N)\big\},$$

with

$$m(\widetilde{\rho}_u, \widetilde{\pi}_v) = u\, m(\widetilde{\rho},\widetilde{\pi}) v^{-1}, \qquad u, v \in U(1),$$

and

$$n(\widetilde{\pi}_v, \widetilde{\rho}_u) = v\, n(\widetilde{\pi},\widetilde{\rho}) u^{-1}, \qquad u, v \in U(1),$$

such that

$$\widetilde{f}_N(\widetilde{\rho}) = \sum_{\widetilde{\pi} \in \{\widetilde{\Pi}(N)\}} m(\widetilde{\rho},\widetilde{\pi})\, \widetilde{f}_N(\widetilde{\pi}), \qquad \widetilde{\rho} \in \widetilde{P}(N), \tag{2.2.7}$$

and

$$\widetilde{f}_N(\widetilde{\pi}) = \sum_{\widetilde{\rho} \in \{\widetilde{P}(N)\}} n(\widetilde{\pi},\widetilde{\rho})\, f_N(\widetilde{\rho}), \qquad \widetilde{\pi} \in \widetilde{\Pi}(N). \tag{2.2.8}$$

This is a special case of a general result, which can be formulated in the same manner for any triplet $G = (G^0, \theta, \omega)$. (See [**A7**, Proposition 5.1], for example.) The first expansion (2.2.7) follows from the decomposition of the standard representation $\widetilde{\rho}^0$ of $G^0(N,F)$ attached to $\widetilde{\rho}$ into irreducible representations. The second expansion (2.2.8) follows by an inversion of the first, in which the coefficient function $m(\widetilde{\rho},\widetilde{\pi})$ is treated as a unipotent matrix.

Let us be more specific about the matrix $\{m(\widetilde{\rho},\widetilde{\pi})\}$. We recall that $m(\widetilde{\rho},\widetilde{\pi}) = 1$ if $\widetilde{\pi} = \pi_{\widetilde{\rho}}$. In general, if $m(\widetilde{\rho},\widetilde{\pi}) \neq 0$ for a given $\widetilde{\rho}$, the linear form Λ_π satisfies $\Lambda_\pi \leqslant \Lambda_\rho$, with equality holding only when $\widetilde{\pi} = \pi_{\widetilde{\rho}}$. This is the property that is responsible for the matrix being unipotent. In addition, $m(\widetilde{\rho}, \widetilde{\pi})$ vanishes unless $\widetilde{\rho}$ and $\widetilde{\pi}$ have the same central character

$$\eta : F^* \longrightarrow \mathbb{C}^*,$$

and more significantly, the same infinitesimal character

$$\mu : \mathcal{Z}(N) \longrightarrow \mathbb{C}.$$

Here, $\mathcal{Z}(N)$ denotes the center of the universal enveloping algebra of the complexified Lie algebra of $\widetilde{G}^0(N, F)$ if F is archimedean, and the Bernstein center of $\widetilde{G}^0(N, F)$ if F is p-adic. It is an abelian $\mathbb{C}$-algebra, with a two-sided action on $\widetilde{\mathcal{H}}(N)$ that is compatible with the associated two-sided action of the Hecke algebra $\mathcal{H}(N) = \widetilde{\mathcal{H}}^0(N)$. We can therefore regard $\{m(\widetilde{\rho}, \widetilde{\pi})\}$ as a block diagonal matrix, whose blocks are parametrized by θ-stable pairs (μ, η). Since there are only finitely many irreducible or standard representations of $\widetilde{G}^0(N, F)$ with a given infinitesimal character μ, the blocks represent unipotent matrices of finite rank.

We obtain equivalent expansions if we fix representatives in the two sets of $U(1)$-orbits. Let us do so by choosing the Whittaker extensions introduced above. The sets $\{\widetilde{P}(N)\}$ and $\{\widetilde{\Pi}(N)\}$ are then each bijective with the family of Langlands parameters $\widetilde{\Phi}(N)$. For any $\phi \in \widetilde{\Phi}(N)$, we are thus taking $\widetilde{\rho}$ to be the associated standard representation $\widetilde{\rho}_\phi$, and $\widetilde{\pi}$ to be the associated irreducible Langlands quotient $\widetilde{\pi}_\phi$. For any $\psi \in \widetilde{\Psi}(N)$, we take $\widetilde{\pi}$ to be the associated representation $\widetilde{\pi}_\psi$. This gives a set of representatives for the subset of $\{\widetilde{\Pi}(N)\}$ indexed by $\widetilde{\Psi}(N)$. The expansion (2.2.8) for elements in this subset can then be written

$$\widetilde{f}_N(\psi) = \sum_{\phi \in \widetilde{\Phi}(N)} \widetilde{n}(\psi, \phi)\, \widetilde{f}_N(\phi), \qquad \psi \in \widetilde{\Psi}(N), \tag{2.2.9}$$

where

$$\widetilde{n}(\psi, \phi) = n(\widetilde{\pi}_\psi, \widetilde{\rho}_\phi).$$

Before we proceed to the initial study of Theorem 2.2.1, we shall describe a second well known feature of general Langlands parameters $\phi \in \Phi(N)$ for $GL(N)$, their infinitesimal characters. These are by definition the infinitesimal characters μ_ϕ of the standard representations ρ_ϕ of $GL(N, F)$ (or the corresponding Langlands quotients π_ϕ, or indeed, any of the constituents of ρ_ϕ). Our immediate motivation is the fact that the terms on the right hand side of (2.2.9) all have the same infinitesimal character. Our remarks will lead to information about the set $\widetilde{\Phi}(G)$, which we will need for the analysis of (2.2.9). They will also be useful later on.

Suppose first that F is archimedean, and that

$$\phi : W_F \longrightarrow GL(N, \mathbb{C})$$

is a general Langlands parameter for $GL(N)$ over F. The infinitesimal character of ϕ is given by a semisimple adjoint orbit in the Lie algebra of the dual group $GL(N,\mathbb{C})$. To describe it, we first choose a representative of ϕ whose restriction to the subgroup $\mathbb{C}^*$ of W_F takes values in the group of diagonal matrices. Following Langlands' general notation [**L11**, p. 128], we write

$$\lambda^\vee\big(\phi(z)\big) = z^{\langle\mu_\phi,\lambda^\vee\rangle}\overline{z}^{\langle\nu_\phi,\lambda^\vee\rangle}, \qquad z\in\mathbb{C}^*,\ \lambda^\vee\in\mathbb{Z}^N, \tag{2.2.10}$$

for complex diagonal matrices μ_ϕ and ν_ϕ with $\mu_\phi-\nu_\phi$ integral. The infinitesimal character of ϕ is then equal to (the orbit of) μ_ϕ, which we can regard as a multiset of N complex numbers.

Suppose for example that ϕ belongs to the subset $\widetilde{\Phi}_{\mathrm{ell}}(N)$ of $\Phi(N)$. The condition that ϕ be self-dual is equivalent to the relation ${}_t\mu_\phi=-\mu_\phi$. The ellipticity condition implies further that ϕ is a direct sum of distinct, self-dual two-dimensional representations of W_F (together with a self-dual one-dimensional character of F^* in case N is odd), from which one sees that $\nu_\phi=-\mu_\phi$. It follows that

$$\lambda^\vee\big(\phi(z)\big) = (z\overline{z}^{-1})^{\langle\mu_\phi,\lambda^\vee\rangle}, \qquad z\in\mathbb{C}^*,\ \lambda^\vee\in\mathbb{Z}^N,$$

where $\mu_\phi\in\frac{1}{2}\mathbb{Z}^N$ is now a diagonal half integral matrix with ${}_t\mu_\phi=-\mu_\phi$. We know that ϕ belongs to $\widetilde{\Phi}_2(G)$, for a unique $G\in\widetilde{\mathcal{E}}_{\mathrm{ell}}(N)$. The dual group of G is given by

$$\widehat{G}=\widehat{G}_S\times\widehat{G}_O=Sp(N_S,\mathbb{C})\times SO(N_O,\mathbb{C}),$$

where N_O and N_S are defined as the number of components of μ_ϕ, a priori half integers, which are respectively integral and nonintegral. This is an immediate consequence of the explicit description of Langlands parameters for orthogonal and symplectic groups, which we will review in §6.1. The group G is then determined by $\widehat{G}$ and the fact that its quadratic character η_G equals the determinant η_ϕ of ϕ.

Conversely, suppose that $\phi\in\Phi(N)$ satisfies (2.2.10), for a diagonal matrix $\mu_\phi\in\frac{1}{2}\mathbb{Z}^N$ with ${}_t\mu_\phi=-\mu_\phi$. This does not imply that ϕ lies in $\widetilde{\Phi}_{\mathrm{ell}}(N)$, even if the components of μ_ϕ are all distinct. However, we can still attach a canonical element $G\in\widetilde{\mathcal{E}}_{\mathrm{ell}}(N)$ to μ_ϕ and η_ϕ, according to the prescription above. We can also choose a general endoscopic datum $M\in\widetilde{\mathcal{E}}(N)$ such that ϕ is the image of a local Langlands parameter ϕ_M in $\widetilde{\Phi}_2(M)$. It follows from the various definitions that M can be identified with a Levi subgroup of G, and hence that ϕ is the image in $\Phi(N)$ of a canonical element in $\widetilde{\Phi}(G)$. If $G\in\mathcal{E}_{\mathrm{sim}}(N)$ is simple, for example, $\widetilde{\Phi}(G)$ embeds into $\Phi(N)$, so that ϕ itself represents a canonical element in the subset $\widetilde{\Phi}(G)$ of $\Phi(N)$.

Suppose next that F is p-adic. The infinitesimal character of a Langlands parameter

$$\phi:\ L_F=W_F\times SU(2)\longrightarrow GL(N,\mathbb{C})$$

for $GL(N)$ over F is often called the *cuspidal support* of π_ϕ or of ρ_ϕ. It can be regarded as a multiset

$$\mu_\phi = \coprod_{k=1}^{r} \Delta_k,$$

built from "segments" Δ_k with

$$\sum_{k=1}^{r} |\Delta_k| = N,$$

in the teminology of [**Z**]. Let us be more precise.

We first recall [**Z**, §3] that a segment Δ of norm

$$|\Delta| = m = m_s m_u$$

is a set

$$\Delta = [\sigma\alpha^i, \sigma\alpha^j] = \{\sigma\alpha^i, \sigma\alpha^{i-1}, \dots, \sigma\alpha^j\},$$

where $\sigma \in \Pi_{\mathrm{scusp}}(GL(m_s))$ is a (not necessarily unitary) supercuspidal representation of $GL(m_s, F)$, while $m_u - 1 = i - j$ is an integer that is a difference of two given half integers $i \geqslant j$, and $\sigma\alpha^\ell \in \Pi_{\mathrm{scusp}}(GL(m_s))$ is defined by

$$(\sigma\alpha^\ell)(x) = \sigma(x)\,|\det x|^\ell, \qquad x \in GL(m_s, F).$$

(By a half integer, we again mean a number whose product with 2 is an integer!) The associated induced representation of $GL(m, F)$ then has a unique subrepresentation π_Δ. (See [**Z**, §3]. By ordering Δ by decreasing half integers, as above, we have implicitly adopted the convention of Langlands rather than that of Zelevinsky. In other words, the special, essentially square integrable representation π_Δ is a subrepresentation of the induced representation rather than a quotient.) Suppose that a is a multiset consisting of segments $\{\Delta_k\}$ of total norm N. We can then form the associated induced representation

$$\rho_a = \mathcal{I}_P(\pi_{\Delta_1} \otimes \cdots \otimes \pi_{\Delta_r})$$

of $GL(N, F)$, and its corresponding Langlands quotient π_a. The segments Δ are bijective with simple Langlands parameters $\phi_\Delta \in \Phi_{\mathrm{sim}}(m)$. The associated mapping $\phi_\Delta \to \pi_\Delta$ is the Langlands correspondence for the relative discrete series of $GL(m)$. The multisets a are bijective with general Langlands parameters $\phi_a \in \Phi(N)$. The associated two mappings $\phi_a \to \pi_a$ and $\phi_a \to \rho_a$ in this case represent the general Langlands correspondence for irreducible and standard representations of $GL(N)$.

Our description of the infinitesimal character μ_ϕ of a p-adic Langlands parameter $\phi \in \Phi(N)$ will now be clear. We write $\phi = \phi_a$, for a multiset a whose elements are segments $\{\Delta_k\}$. The infinitesimal character is then the multiset obtained by taking the disjoint union over k of the sets Δ_k. For another description, which is a little closer to that of the archimedean case, we write μ_ϕ as at the beginning of §1.3 for the restriction of ϕ to the subgroup

$$\left\{\left(w, \begin{pmatrix} |w|^{\frac{1}{2}} & 0 \\ 0 & |w|^{-\frac{1}{2}} \end{pmatrix}\right) : w \in W_F\right\}$$

of $W_F \times SL(2,\mathbb{C})$. As an N-dimensional representation of W_F, $\mu_\phi \in \Phi(N)$ can be regarded as a Langlands parameter that is trivial on the second factor of L_F. The infinitesimal character is then equal to the multiset μ_a that corresponds to μ_ϕ, which can in turn be identified with μ_ϕ. As we observe from either description, μ_ϕ is a multiset of supercuspidal representations $\sigma\alpha^\ell$, while a is a coarser multiset of segments Δ_k.

Suppose for example that ϕ belongs to the subset $\widetilde{\Phi}_{\mathrm{ell}}(N)$ of $\Phi(N)$. Then ϕ is a direct sum of distinct, irreducible, self-dual representations of L_F, which in turn correspond to distinct, self-dual segments. Since the dual of a general segment $\Delta = [\sigma\alpha^i, \sigma\alpha^j]$ equals $\Delta^\vee = [\sigma^\vee\alpha^{-j}, \sigma^\vee\alpha^{-i}]$, a self-dual segment may be written in the form

$$\Delta = [\sigma\alpha^i, \sigma\alpha^{-i}] = \{\sigma\alpha^i, \sigma\alpha^{i-1}, \dots, \sigma\alpha^{-i}\}, \tag{2.2.11}$$

for a self-dual (unitary) supercuspidal representation $\sigma \in \Pi_{\mathrm{scusp}}(m)$, and a nonnegative half integer i. This segment is symplectic if σ is symplectic and i is an integer, or if σ is orthogonal and i is not an integer. It is orthogonal otherwise, that is, if σ is orthogonal and i is an integer, or if σ is symplectic and i is not an integer. (Recall that the self-dual representation σ was designated symplectic or orthogonal in §1.4 according to whether its corresponding Langlands parameter factors through the associated complex subgroup of $GL(m,\mathbb{C})$.) If G is the unique datum in $\widetilde{\mathcal{E}}_{\mathrm{ell}}(N)$ with $\phi \in \widetilde{\Phi}_2(G)$, its dual group is again a product

$$\widehat{G} = \widehat{G}_S \times \widehat{G}_O = Sp(N_S,\mathbb{C}) \times SO(N_O,\mathbb{C}),$$

given in this case by the partition of the segments of ϕ into those of orthogonal and symplectic type. The group G is then determined by the dual group $\widehat{G}$ and the character $\eta_G = \eta_\phi$.

Conversely, suppose that $\phi \in \Phi(N)$ is a parameter such that μ_ϕ is a disjoint union of self-dual sets (2.2.11). This does not imply that ϕ lies in $\widetilde{\Phi}_{\mathrm{ell}}(N)$. However, as in the archimedean case, we can still attach a canonical element $G \in \widetilde{\mathcal{E}}_{\mathrm{ell}}(N)$ to μ_ϕ and η_ϕ, with the property that ϕ is the image in $\Phi(N)$ of a canonical parameter in $\widetilde{\Phi}(G)$. In the p-adic case, η_ϕ is just the determinant of μ_ϕ, and is therefore superfluous here. However, we will continue to carry it in order to accommodate the Archimedean case.

With these notions in mind, we consider again the expansion (2.2.9). Each parameter $\phi \in \widetilde{\Phi}(N)$ on the right has an infinitesimal character μ_ϕ and determinant η_ϕ. The parameter $\psi \in \widetilde{\Psi}(N)$ on the left is likewise equipped with data $\mu_\psi = \mu_{\phi_\psi}$ and $\eta_\psi = \eta_{\phi_\psi}$. These parameters also come with the linear functionals

$$\Lambda_\phi = \Lambda_{\pi_\phi} = \Lambda_{\rho_\phi}$$

and

$$\Lambda_\psi = \Lambda_{\phi_\psi}.$$

Given ψ, we define

$$\widetilde{\Phi}(N,\psi) = \{\phi \in \widetilde{\Phi}(N) : \ (\mu_\phi, \eta_\phi) = (\mu_\psi, \eta_\psi), \ \Lambda_\phi \leqslant \Lambda_\psi\}.$$

The expansion (2.2.9) then remains valid if the sum is restricted to the subset $\widetilde{\Phi}(N,\psi)$ of $\widetilde{\Phi}(N)$.

LEMMA 2.2.2. *Suppose that the assertion (a) of Theorem* 2.2.1 *holds in the special case that $\psi = \phi$ is generic. It is then valid for any ψ.*

PROOF. The assertion is assumed to hold if $\psi = \phi$ lies in the subset $\widetilde{\Phi}_{\mathrm{bdd}}(N)$ of generic parameters in $\widetilde{\Psi}(N)$. Suppose that ϕ belongs to the larger set of generic parameters in the general family $\widetilde{\Psi}^+(N)$. In other words, $\phi \in \widetilde{\Phi}(N)$ is a general self-dual Langlands parameter. We have observed that the associated (twisted) standard character

$$\widetilde{f}_N(\phi) = \widetilde{f}_N(\widetilde{\rho}_\phi), \qquad \widetilde{f} \in \widetilde{\mathcal{H}}(N),$$

can be obtained by analytic continuation from the tempered case. Since the left hand side of (2.2.3) is also an analytic function in the relevant complex variable, our assumption implies that the obvious variant of assertion (a) of the theorem holds if ψ is replaced by any $\phi \in \Phi(N)$.

Suppose first that G is simple. In this case, we have only the definition (2.2.3) to contend with. We must show that for the given element $\psi \in \widetilde{\Psi}(G)$, the linear form $\widetilde{f}_N(\psi)$ in (2.2.3) depends only on the image of the mapping $\widetilde{f} \to \widetilde{f}^G$. We shall use the expansion (2.2.9) for ψ as an element in $\widetilde{\Psi}(N)$. Given the supposition of the lemma, and its extension above to general parameters $\phi \in \widetilde{\Phi}(G)$, we need only show that if the coefficient $\widetilde{n}(\psi,\phi)$ in (2.2.9) is nonzero for some $\phi \in \widetilde{\Phi}(N)$, then ϕ lies in the subset $\widetilde{\Phi}(G)$ of $\widetilde{\Phi}(N)$. We first note that ψ belongs to the complement of $\widetilde{\Psi}_{\mathrm{ell}}(N)$ in $\widetilde{\Psi}(N)$ if and only if it is the image of a parameter $\widetilde{\psi}_M \in \Psi(\widetilde{M})$ attached to a proper standard Levi subset $\widetilde{M}$ of $\widetilde{G}(N)$. In this case, the terms in (2.2.9) are each induced from analogous terms for $\widetilde{M}$. The required assertion then follows inductively from its analogue for the proper Levi subgroup M of G attached to $\widetilde{M}$. We may therefore assume that ψ lies in the intersection of $\widetilde{\Psi}(G)$ with $\widetilde{\Psi}_{\mathrm{ell}}(N)$, which is just the subset $\widetilde{\Psi}_2(G)$ of $\widetilde{\Psi}(G)$. Another induction argument, combined with the knowledge we now have of the elliptic endoscopic data for G, reduces the problem further to the case that ψ is simple.

We are thus taking both $G \in \widetilde{\mathcal{E}}_{\mathrm{sim}}(N)$ and $\psi \in \widetilde{\Psi}_{\mathrm{sim}}(G)$ to be simple. Then

$$\psi = \mu \otimes \nu, \qquad N = mn,$$

for irreducible unitary representations μ and ν of L_F and $SU(2)$ of respective dimensions m and n. The infinitesimal character μ_ψ of ψ is by definition the infinitesimal character μ_{ϕ_ψ} of the irreducible representation $\pi_\psi = \pi_{\phi_\psi}$ in (2.2.1). It equals the tensor product of the infinitesimal characters of μ and ν, defined in the obvious way, and is naturally compatible with the tensor product of the representations themselves. (The overlapping notation in μ and μ_ψ is unfortunate, but should not cause confusion.) If we compare the remarks for real and p-adic infinitesimal characters above with the local version of the discussion following (1.4.7), we see that the datum in $\widetilde{\mathcal{E}}_{\text{ell}}(N)$ attached to $\mu_\psi = \mu_{\phi_\psi}$ and $\eta_\psi = \eta_{\phi_\psi}$ equals G. But we have agreed that the sum in (2.2.9) can be taken over ϕ in the subset $\widetilde{\Phi}(N,\psi)$ of $\widetilde{\Phi}(N)$. For any such ϕ, the datum attached to $\mu_\phi = \mu_\psi$ and $\eta_\phi = \eta_\psi$ is therefore also equal to G. In particular, ϕ belongs to the subset $\widetilde{\Phi}(G)$ of $\widetilde{\Phi}(N)$. This is what we had to show.

We have established that for simple $G \in \widetilde{\mathcal{E}}_{\text{sim}}(N)$, and any $\psi \in \widetilde{\Psi}(G)$, the set

$$\widetilde{\Phi}(G,\psi) = \widetilde{\Phi}(N,\psi)$$

is contained in the subset $\widetilde{\Phi}(G)$ of $\widetilde{\Phi}(N)$. The definition (2.2.3) is thus valid for G. It takes the form

$$f^G(\psi) = \sum_{\phi\in\widetilde{\Phi}(G,\psi)} \tilde{n}(\psi,\phi)\, f^G(\phi), \qquad f \in \widetilde{\mathcal{H}}(G), \tag{2.2.12}$$

which we obtain from (2.2.9).

Suppose now that $G \in \widetilde{\mathcal{E}}_{\text{ell}}(N)$ and $\psi \in \widetilde{\Psi}(G)$ are composite, as in (2.2.4). In this case, the linear form (2.2.2) is defined by the product (2.2.4) and what we have established for the simple data G_S and G_O. The problem in this case is to show that (2.2.3) is valid.

It is again enough to assume that ψ belongs to the subset $\widetilde{\Psi}_{\text{ell}}(N)$ of $\widetilde{\Psi}(N)$. In other words, ψ lies in the subset $\widetilde{\Psi}_2(G)$ of $\widetilde{\Psi}(G)$. We are now dealing with decompositions $G = G_S \times G_O$, $\psi = \psi_S \times \psi_O$ and $N = N_S + N_O$ that are nontrivial. However, we shall treat them in the same way, by applying the expansion (2.2.9) with ϕ summed over the subset $\widetilde{\Phi}(N,\psi)$ of $\widetilde{\Phi}(N)$. Any $\phi \in \widetilde{\Phi}(N,\psi)$ has the same infinitesimal character and determinant as the original parameter $\psi \in \widetilde{\Psi}_{\text{ell}}(N)$. It follows from the earlier remarks for real and p-adic infinitesimal characters that ϕ is the image of a canonical parameter in $\widetilde{\Phi}(G)$. In particular, we can identify ϕ with an element in $\widetilde{\Phi}(G)$, even though the mapping from $\widetilde{\Phi}(G)$ to $\widetilde{\Phi}(N)$ is not injective. With this interpretation, we obtain a canonical decomposition

$$\phi = \phi_S \oplus \phi_O, \qquad \phi_S \in \widetilde{\Phi}(N_S,\psi_S),\ \phi_O \in \widetilde{\Phi}(N_O,\psi_O), \tag{2.2.13}$$

for any $\phi \in \widetilde{\Phi}(N,\psi)$.

It is convenient to write $\widetilde{\mathcal{I}}(N,\psi)$ for the finite-dimensional space of complex valued functions $\tilde{f}_N(\phi)$ on the finite set $\widetilde{\Phi}(N,\psi)$. This space represents a quotient of the invariant Hecke space $\widetilde{\mathcal{I}}(N)$ of $\widetilde{G}(N)$, since the twisted trace Paley-Wiener theorem for $GL(N)$ implies that the restriction mapping from $\widetilde{\mathcal{I}}(N)$ to $\widetilde{\mathcal{I}}(N,\psi)$ is surjective. We can identify it with the space of functions on the set

$$\widetilde{P}(N,\psi) = \{\widetilde{\rho}_\phi : \ \phi \in \widetilde{\Phi}(N,\psi)\},$$

since $\widetilde{f}_N(\widetilde{\rho}_\phi) = \widetilde{f}_N(\phi)$ for any $\phi \in \widetilde{\Phi}(N,\psi)$. It is also canonically isomorphic to the space of functions on the set

$$\widetilde{\Pi}(N,\psi) = \{\widetilde{\pi}_\phi : \ \phi \in \widetilde{\Phi}(N,\psi)\},$$

where we recall that like $\widetilde{\rho}_\phi$, $\widetilde{\pi}_\phi$ represents the canonical Whittaker extension to $\widetilde{G}(N,F)$ of the underlying representation of $GL(N,F)$. Indeed, the general expansions (2.2.7) and (2.2.8) remain valid with the sets $\widetilde{P}(N,\psi)$ and $\widetilde{\Pi}(N,\psi)$ in place of $\{\widetilde{P}(M)\}$ and $\{\widetilde{\Pi}(N)\}$. For any function $\widetilde{f}_N \in \widetilde{\mathcal{I}}(N,\psi)$, the corresponding function on $\widetilde{\Pi}(N,\psi)$ is defined by this modification

$$\widetilde{f}_N(\widetilde{\pi}_{\phi_1}) = \sum_{\phi\in\widetilde{\Phi}(N,\psi)} n(\widetilde{\pi}_{\phi_1},\widetilde{\rho}_\phi)\,\widetilde{f}_N(\widetilde{\rho}_\phi), \qquad \phi_1 \in \widetilde{\Phi}(N,\psi),$$

of (2.2.8).

We define a linear transformation

$$\widetilde{\mathcal{I}}(N,\psi) \longrightarrow \widetilde{\mathcal{I}}(N_S,\psi_S)\otimes\widetilde{\mathcal{I}}(N_O,\psi_O) \tag{2.2.14}$$

by mapping $\widetilde{f}_N \in \widetilde{\mathcal{I}}(N,\psi)$ to the function

$$\widetilde{f}_{N_S,N_O}(\phi_S \times \phi_O) = \widetilde{f}_N(\phi), \qquad \phi = \phi_S \oplus \phi_O,$$

in the tensor product. The mapping is actually an isomorphism, by virtue of the canonical decomposition (2.2.13) of any parameter $\phi \in \widetilde{\Phi}(N,\psi)$. It is closely related to the twisted transfer mapping $\widetilde{f} \to \widetilde{f}^G$. More precisely, it is part of a commutative diagram of isomorphisms

$$\begin{array}{ccc}
\widetilde{\mathcal{I}}(N,\psi) & \longrightarrow & \widetilde{\mathcal{I}}(N_S,\psi_S)\otimes\widetilde{\mathcal{I}}(N_O,\psi_O) \\
\downarrow & & \downarrow \\
\widetilde{\mathcal{I}}(G,\psi) & \longrightarrow & \widetilde{\mathcal{I}}(G_S,\psi_S)\otimes\widetilde{\mathcal{I}}(G_O,\psi_O),
\end{array}$$

in which $\widetilde{\mathcal{I}}(G,\psi)$ is the quotient of $\widetilde{\mathcal{I}}(G)$ defined by restriction of functions to the subset $\widetilde{\Phi}(G,\psi) = \widetilde{\Phi}(N,\psi)$ of $\widetilde{\Phi}(G)$. The left hand vertical arrow is the corresponding lift of the transfer mapping $\widetilde{f}_N \to \widetilde{f}^G$, while the right hand vertical arrow is its analogue for $\psi_S \times \psi_O$. This diagram is a consequence of the supposition of the lemma, or rather its extension to nontempered generic parameters ϕ described at the beginning of the proof.

We shall show that the isomorphism dual to (2.2.14) maps the product $\widetilde{\pi}_{\psi_S} \times \widetilde{\pi}_{\psi_O}$, regarded as a linear form on the right hand tensor product, to the linear form $\widetilde{\pi}_\psi$ on $\widetilde{\mathcal{I}}(N,\psi)$. At the same time, we shall show that the coefficients in (2.2.9) satisfy the natural formula

$$\widetilde{n}(\psi,\phi) = \widetilde{n}(\psi_S,\phi_S)\,\widetilde{n}(\psi_O,\phi_O), \qquad \phi \in \widetilde{\Phi}(N,\psi). \tag{2.2.15}$$

It follows from the definition of (2.2.14) that the dual isomorphism takes the linear form $\widetilde{\rho}_S \times \widetilde{\rho}_O = \widetilde{\rho}_{\phi_S} \times \widetilde{\rho}_{\phi_O}$ attached to any $\phi \in \widetilde{\Phi}(N,\psi)$ to the linear form $\widetilde{\rho} = \widetilde{\rho}_\phi$. Viewed directly in terms of standard representations, this correspondence is a composition

$$\widetilde{\rho}_S \times \widetilde{\rho}_O \longrightarrow \rho_S \times \rho_O \longrightarrow \mathcal{I}_{P_\psi}(\rho_S \times \rho_O) = \rho \longrightarrow \widetilde{\rho}, \tag{2.2.16}$$

where P_ψ is the standard parabolic subgroup of $GL(N)$ of type (N_S,N_O). The expansion (2.2.7) for $\widetilde{\rho}$ is obtained from the decomposition of $\widetilde{\rho}$ (as a standard representation of $G^+(N,F)$) into irreducible constituents. Its analogues for $\widetilde{\rho}_S$ and $\widetilde{\rho}_O$ are of course obtained in the same way. The representation $\widetilde{\pi}_\psi$ is the Langlands quotient of $\widetilde{\rho}_\psi = \widetilde{\rho}_{\phi_\psi}$, or more precisely, the extension to $\widetilde{G}^+(N,F)$ of the Langlands quotient π_ψ of ρ_ψ that is compatible with the Whittaker extension $\widetilde{\rho}_\psi$ of ρ_ψ, while $\widetilde{\pi}_{\psi_S} \times \widetilde{\pi}_{\psi_O}$ is the Langlands quotient of $\widetilde{\rho}_{\psi_S} \times \widetilde{\rho}_{\psi_O}$. Using a local form of the global notation (1.3.6), we could write $\widetilde{\pi}_{\psi_S} \boxplus \widetilde{\pi}_{\psi_O}$ for the image of $\widetilde{\pi}_{\psi_S} \times \widetilde{\pi}_{\psi_O}$ under the composition (2.2.16). But since the representations π_{ψ_S} and π_{ψ_O} are unitary, the induced representation $\mathcal{I}_{P_\psi}(\pi_{\psi_S} \times \pi_{\psi_O})$ is irreducible [**Be**]. It follows that $\widetilde{\pi}_{\psi_S} \boxplus \widetilde{\pi}_{\psi_O}$, a priori a quotient of $\widetilde{\rho}_\psi$, is actually the Langlands quotient $\widetilde{\pi}_\psi$. This is what we wanted to show. Moreover, if we apply the dual of (2.2.14) to the terms in the product of the expansions (2.2.9) for ψ_S and ψ_O, we see that

$$\widetilde{f}_N(\widetilde{\pi}_{\psi_S} \boxplus \widetilde{\pi}_{\psi_O}) = \sum_{\phi \in \widetilde{\Phi}(N,\psi)} \widetilde{n}(\psi_S,\phi_S)\,\widetilde{n}(\psi_O,\phi_O)\,\widetilde{f}_N(\phi),$$

in the obvious notation. The expansion (2.2.9) itself can be written

$$\widetilde{f}_N(\widetilde{\pi}_\psi) = \sum_{\phi \in \widetilde{\Phi}(N,\psi)} \widetilde{n}(\psi,\phi)\,\widetilde{f}_N(\phi).$$

Since the left hand sides are equal, the required decomposition (2.2.15) of the coefficients follows as well.

We can now complete our proof of the second half of Lemma 2.2.2. Suppose that $\widetilde{f}$ is any function in $\widetilde{\mathcal{H}}(N)$. According to the definition (2.2.12) of $\widetilde{f}^G(\psi)$, we can write

$$\begin{aligned}\widetilde{f}^G(\psi) &= \widetilde{f}^G(\psi_S \times \psi_O)\\ &= \sum_{\phi_S,\phi_O} \widetilde{n}(\psi_S,\phi_S)\,\widetilde{n}(\psi_O,\phi_O)\,\widetilde{f}^G(\phi_S \times \phi_O),\end{aligned}$$

for a double sum over $\phi_S \in \widetilde{\Phi}(N_S, \psi_S)$ and $\phi_O \in \widetilde{\Phi}(N_O, \psi_O)$. This in turn equals

$$\sum_{\phi_S, \phi_O} \widetilde{n}(\psi_S, \phi_S)\, \widetilde{n}(\psi_O, \phi_O)\, \widetilde{f}_{N_S,N_O}(\phi_S \times \phi_O)$$

$$= \sum_{\phi \in \widetilde{\Phi}(N,\psi)} \widetilde{n}(\psi, \phi)\, \widetilde{f}_N(\phi),$$

by the definition of $\widetilde{f}_{N_S,N_O}$, the commutative diagram above, and the coefficient formula (2.2.15). The last sum is then equal $\widetilde{f}_N(\psi)$, according to (2.2.9). We have established that $\widetilde{f}^G(\psi)$ equals $\widetilde{f}_N(\psi)$, which is the required identity (2.2.3). □

REMARKS. 1. My original proof was incorrectly based on the analogue of (2.2.9) for the connected group $GL(N)$. I thank the referee for pointing this out, and for suggestions that motivated the revised proof above.

2. The analogue of the decomposition (2.2.15) holds more generally for the coefficients

$$n(\widetilde{\pi}, \widetilde{\rho}), \qquad \widetilde{\pi} \in \widetilde{\Pi}(N, \psi),\ \rho \in \widetilde{P}(N, \psi),$$

in (2.2.8). This is because each induced representation

$$\mathcal{I}_{P_\psi}(\pi_S \times \pi_O), \qquad \widetilde{\pi}_S \in \widetilde{\Pi}(N_S, \psi_S),\ \widetilde{\pi}_O \in \widetilde{\Pi}(N_O, \psi_O),$$

is irreducible, an observation I owe to the referee. The point is that there is no interaction between the infinitesimal characters of π_S and π_O. To be precise, we can attach an "affine $\mathbb{Z}$-module" to the infinitesimal character μ_π of any $\pi \in \Pi(N)$ by writing

$$\mathbb{Z}[\mu_\pi] = \{\sigma\alpha^\ell :\ \sigma \in \mu_\pi,\ \ell \in \mathbb{Z}\}$$

in the earlier notation for p-adic F, and

$$\mathbb{Z}[\mu_\pi] = \{s + \ell :\ s \in \mu_\pi,\ \ell \in \mathbb{Z}\}$$

if F is archimedean and μ_π is treated as a multiset of order N. The sets $\mathbb{Z}[\mu_{\pi_S}]$ and $\mathbb{Z}[\mu_{\pi_O}]$ are then disjoint, from which it follows by standard methods that the induced representation is irreducible. This presumably leads to a generalization of (2.2.3) that is tied to (2.2.8) (and a general representation $\widetilde{\pi}_\phi$) rather than (2.2.9) (with the representation $\widetilde{\pi}_\psi$).

3. The generic case of Theorem 2.2.1(a), which was our hypothesis for the lemma, will be established later by global means (Lemmas 5.4.2 and 6.6.3). It is possible that the general case of Theorem 2.2.1(a), the actual assertion of the lemma, could also be proved globally as a nongeneric supplement of Lemma 5.4.2. (See Lemmas 7.3.1 and 7.3.2, which represent nongeneric supplements of the other results in §5.4.) However, the local proof we have given here is perhaps more instructive.

The following lemma gives another reduction of Theorem 2.2.1. This one is much easier, and has to some extent been implicit in our earlier discussion.

LEMMA 2.2.3. *Suppose that Theorem* 2.2.1 *is valid if N is replaced by any integer $N_* < N$. Then it also holds for any parameter $\psi \in \widetilde{\Psi}(G)$, for $G \in \widetilde{\mathcal{E}}_{\rm sim}(N)$, such that the group S_ψ has infinite center.*

PROOF. Suppose that $\psi \in \widetilde{\Psi}(G)$ is as given. The centralizer in ${}^L G$ of any nontrivial central torus in S_ψ is then the L-group ${}^L M$ of a proper Levi subgroup M of G. We fix M, and choose an L-homomorphism ψ_M from $L_F \times SU(2)$ to ${}^L M$ whose image in $\widetilde{\Phi}(G)$ equals ψ. The centralizer S_{ψ_M} in $\widehat{M}$ of the image of ψ_M is then equal to the original centralizer S_ψ in $\widehat{G}$ of the image of ψ.

The Levi subgroup M is a product of general linear groups with a group $G_- \in \widetilde{\mathcal{E}}_{\rm sim}(N_-)$, for some $N_- < N$. (See (2.3.4), for example.) It follows from the supposition of the lemma that the natural analogue of Theorem 2.2.1 holds for M. This gives us a stable linear form $h^M(\psi_M)$ on $\widetilde{\mathcal{S}}(M)$, a packet $\widetilde{\Pi}_{\psi_M}$ over $\widetilde{\Pi}_{\rm unit}(M)$, and a mapping $\pi_M \longrightarrow \langle \cdot, \pi_M\rangle$ from $\widetilde{\Pi}_{\psi_M}$ to $\widehat{\mathcal{S}}_{\psi_M}$. We define the corresponding objects for G by setting

$$f^G(\psi) = f^M(\psi_M), \qquad f \in \widetilde{\mathcal{H}}(G),$$

$$\widetilde{\Pi}_\psi = \{\pi \subset \mathcal{I}_P(\pi_M) : \ \pi_M \in \widetilde{\Pi}_{\pi_M}\}, \qquad P \in \mathcal{P}(M),$$

and

$$\langle \cdot, \pi\rangle = \langle \cdot, \mathcal{I}_P(\pi_M)\rangle = \langle \cdot, \pi_M\rangle.$$

The induced representations $\mathcal{I}_P(\pi_M)$ are presumably irreducible, which of course would mean that π equals $\mathcal{I}_P(\pi_M)$. But in any case, the required character identity (2.2.6) then follows from the familiar descent formula

$$\sum_{\pi \subset \mathcal{I}_P(\pi_M)} f_G(\pi) = \mathrm{tr}\big(\mathcal{I}_P(\pi_M, f)\big) = f_M(\pi_M), \qquad f \in \widetilde{\mathcal{H}}(G). \qquad \square$$

Theorem 2.2.1 characterizes the local classification in terms of two kinds of local endoscopy, twisted endoscopy for $GL(N)$ and ordinary endoscopy for the group $G \in \widetilde{\mathcal{E}}_{\rm sim}(N)$. Recall that we have also been thinking of a third case, that of twisted endoscopy for an even orthogonal group. This case is not needed for the statement of the classification. However, it has a role in the proof, and it also provides supplementary information about characters. We shall state what is needed as a theorem, to be established along with everything else.

Even orthogonal groups $G \in \widetilde{\mathcal{E}}_{\rm sim}(N)$ of course represent the case that the group $\widetilde{\rm Out}_N(G)$ is nontrivial. They are responsible for our general notation $\widetilde{\Psi}(G)$ and $\widetilde{\Pi}_\psi$ (rather than $\Psi(G)$ and Π_ψ), since it is in this case that an element in $\widetilde{\Pi}_\psi$ can be a nontrivial $\widetilde{\rm Out}_N(G)$-orbit of irreducible representations π (rather than a singleton). Twisted endoscopy is defined by the nontrivial automorphism $\widetilde{\theta} = \mathrm{Int}\big(\widetilde{w}(N)\big)$ of order 2 in $\widetilde{\rm Out}_N(G)$. It applies to the subset $\Psi(\widetilde{G})$ of $\widetilde{\rm Out}_N(G)$-fixed parameters ψ in $\widetilde{\Psi}(G)$, or more accurately, those $\psi \in \widetilde{\Psi}(G)$ that can be represented by an $\widetilde{\mathrm{Out}}_N(G)$-fixed L-homomorphism from $L_F \times SU(2)$ to ${}^L G$. For any such ψ, we would expect the elements in $\widetilde{\Pi}_\psi$ to be singletons. In other words, they should consist of irreducible representations π of $G(F)$ that extend to the group $\widetilde{G}^+(F)$ generated by $\widetilde{G}(F)$. For it is clear that the corresponding linear characters (2.2.5) on $\mathcal{S}_\psi$ extend to characters on the group $\widetilde{\mathcal{S}}_\psi^+$ generated by the $\mathcal{S}_\psi$-bitorsor $\widetilde{\mathcal{S}}_\psi = \mathcal{S}_\psi(\widetilde{G})$. (Here we are following the obvious local analogue of global notation near the end of §1.4.) However, we shall establish this only in case ψ is generic.

THEOREM 2.2.4. *Suppose that N is even, that $G \in \widetilde{\mathcal{E}}_{\mathrm{sim}}(N)$ is orthogonal in the sense that $\widehat{G} = SO(N,\mathbb{C})$, and that ψ lies in the subset $\Psi(\widetilde{G})$ of $\widetilde{\theta}$-stable elements in $\widetilde{\Psi}(G)$.*

(a) *Suppose that $\widetilde{s}$ is a semisimple element in the $\widetilde{G}$-twisted centralizer $\widetilde{S}_\psi$, and that*

$$(\widetilde{G}', \widetilde{\psi}'), \qquad \widetilde{G}' \in \mathcal{E}(\widetilde{G}),\ \widetilde{\psi}' \in \Psi(\widetilde{G}'),$$

is the preimage of $(\psi, \widetilde{s})$ under the analogue of the correspondence (1.4.11). Then we have an identity

$$\widetilde{f}'(\widetilde{\psi}') = \sum_{\pi \in \widetilde{\Pi}_\psi} \langle s_\psi \widetilde{x}, \widetilde{\pi}\rangle \widetilde{f}_{\widetilde{G}}(\widetilde{\pi}), \qquad \widetilde{f} \in \mathcal{H}(\widetilde{G}), \tag{2.2.17}$$

where $\widetilde{x}$ is the image of $\widetilde{s}$ in $\widetilde{\mathcal{S}}_\psi$, $\widetilde{\pi}$ is any extension of π to $\widetilde{G}^+(F)$, and $\langle \cdot, \widetilde{\pi}\rangle$ is a corresponding extension of the linear character (2.2.5) to $\widetilde{\mathcal{S}}_\psi^+$ such that the product

$$\langle s_\psi \widetilde{x}, \widetilde{\pi}\rangle \widetilde{f}_{\widetilde{G}}(\widetilde{\pi}), \qquad \widetilde{x} \in \widetilde{\mathcal{S}}_\psi,$$

in (2.2.17) depends only on π (as an element in $\Pi_{\mathrm{unit}}(G)$).

(b) *If $\phi = \psi$ is generic, each $\pi \in \widetilde{\Pi}_\phi$ is an $\widetilde{\mathrm{Out}}_N(G)$-stable representation of $G(F)$, which consequently does have an extension $\widetilde{\pi}$ to $\widetilde{G}^+(F)$. (In general, if $\pi \in \widetilde{\Pi}_\psi$ does not have an extension, we agree simply to set $\widetilde{\pi}$ equal to 0).*

REMARKS. 1. The theorem relates twisted characters on $\widetilde{G}(F)$ with stable characters on the twisted endoscopic group $\widetilde{G}'(F)$. Recall that $\widetilde{G}'$ is determined by a pair of odd positive integers $(\widetilde{N}_1', \widetilde{N}_2')$ whose sum equals N, and a pair of quadratic characters $(\widetilde{\eta}_1', \widetilde{\eta}_2')$ on Γ_F whose product equals the quadratic character η that defines G as a quasisplit group. The left hand side of (2.2.17) is the value of $\widetilde{f}'$, the transfer of $\widetilde{f}$ to $\mathcal{S}(\widetilde{G}')$, at a product of stable linear forms

$$\psi' = \widetilde{\psi}_1' \times \widetilde{\psi}_2', \qquad \widetilde{\psi}_i' \in \widetilde{\Psi}(\widetilde{N}_i'),$$

each defined by Theorem 2.2.1(a). On the right hand side, the sum is taken over those $\pi \in \widetilde{\Pi}_\psi$ that as irreducible representations of $G(F)$ have extensions to $\widetilde{G}^+(F)$. If π does have an extension, there is in general no canonical choice (in contrast to the case of twisted $GL(N)$). However, the theorem asserts that the summand in (2.2.17) is independent of the choice, which is all we need.

2. The extension $\langle\,\cdot,\widetilde{\pi}\rangle$ of $\langle\,\cdot,\pi\rangle$ is uniquely determined by the extension $\widetilde{\pi}$ of π. Given the condition that the product of $\langle\,\cdot,\widetilde{\pi}\rangle$ with $\widetilde{f}_{\widetilde{G}}(\widetilde{\pi})$ depends only on π as a representation of $G(F)$ (which allows for the possibility that $\pi \in \widetilde{\Pi}_\psi$ occurs several times with the same linear character $\langle\cdot,\pi\rangle$), this is easily established from (2.2.17). Indeed, if (2.2.17) is valid for two different extensions $\langle\,\cdot,\widetilde{\pi}_1\rangle$ and $\langle\,\cdot,\widetilde{\pi}_1\rangle'$ of (2.2.5) to $\widetilde{\mathcal{S}}_\psi^+$, for a given $\pi_1 \in \Pi_\psi$ with extension $\widetilde{\pi}_1$ to $\widetilde{G}^+(F)$, we can identify the corresponding summands on the right hand side. This implies that $\widetilde{f}_{\widetilde{G}}(\widetilde{\pi}_1)$ vanishes for any $\widetilde{f} \in \mathcal{H}(\widetilde{G})$, which is contrary to the basic properties of irreducible characters.

3. We could have denoted the packet of ψ in this case by Π_ψ rather than $\widetilde{\Pi}_\psi$, since we expect it to consist of singletons, and therefore be a subset of $\Pi_{\mathrm{unit}}(G)$ (composed of $\widetilde{\mathrm{Out}}_N(G)$-stable, irreducible, unitary representations). We shall sometimes do so in the future.

4. Suppose that ψ lies in $\Psi_2(\widetilde{G})$, for example, and that $\pi \in \widetilde{\Pi}_\psi$ does happen to map to an $\widetilde{\mathrm{Out}}_N(G)$-orbit

$$\{\pi_1, \pi_2 : \pi_i \in \Pi_{\mathrm{unit}}(G)\}$$

of order 2 in $\Pi_{\mathrm{unit}}(G)$. It then turns out that the multiplicity $\widetilde{m}_\psi(\pi) = \widetilde{m}_\psi(\pi, \widetilde{\mathcal{S}}_\psi)$ of π over $\widetilde{\mathcal{S}}_\psi$ is *even*. This property can be established by global means, as we shall observe at the end of §7.4. It implies that there is an $\widetilde{\mathrm{Out}}(G)$-stable representation

$$\tfrac{1}{2}\widetilde{m}_\psi(\pi)(\langle x,\pi\rangle(\pi_1(y)\oplus\pi_2(y))), \qquad s \in \mathcal{S}_\psi,\ y \in G(F), \tag{2.2.18}$$

of $\mathcal{S}_\psi \times G(F)$ whose restriction from the group algebra $\mathcal{H}(G)$ to its symmetric subalgebra $\widetilde{\mathcal{H}}(G)$ equals the contribution of π (as an element in a multiset over $\widehat{\mathcal{S}}_\psi$) to the right hand side of (2.26).

2.3. Normalized intertwining operators

Intertwining operators play an essential role in the proof of our theorems. Global intertwining operators are the main terms in the discrete part of the trace formula. It is important to understand their local factors in order to interpret the stabilization of this global object. Local intertwining operators also play a role in the local classification. They lead to a partial construction of the packets $\widetilde{\Pi}_\psi$ and pairings $\langle\cdot,\pi\rangle$ of Theorem 2.2.1, which reduces their study to the case that π belongs to $\widetilde{\Psi}_2(G)$.

We shall devote the remaining three sections of the chapter to this topic. One of the main problems is to normalize local intertwining operators. This has been done quite generally by Shahidi [**Sha4**], for inducing representations with Whittaker models. However, our inducing representations will frequently not have Whittaker models, even when they are tempered. We will apply global methods to their study, both in the initial stage here and in arguments from later chapters.

The problem is not simply to normalize intertwining operators, but to do so in a way that is compatible with endoscopic transfer. To focus on the different questions that arise, it will be best to separate the problem into three distinct steps. We shall deal with these in succession, the first two in this section, and the third in the next.

The first step is to normalize intertwining operators between representations induced from different parabolic subgroups. These objects are familiar from Harish-Chandra's later work [**Ha4**], [**Ha5**] on the Plancherel formula. We begin by introducing a slightly nonstandard family of normalizing factors.

Recall that if ϕ is an N-dimensional representation of the local Langlands group, we can form the local L-function $L(s, \phi)$ and ε-factor $\varepsilon(s, \phi, \psi_F)$. The ε-factor follows the notation [**T2**, (3.6.4)] of Langlands, and depends on a choice of nontrivial additive character ψ_F of F. For p-adic F, it has the general form

$$\varepsilon(s, \phi, \psi_F) = \varepsilon(\phi, \psi_F)\, q_F^{-n(s-\frac{1}{2})},$$

for a nonzero complex number

$$\varepsilon(\phi, \psi_F) = \varepsilon\big(\tfrac{1}{2}, \phi, \psi_F\big),$$

and an integer $n = n(\phi, \psi_F)$. The integer n is the simpler of the two constants, since it can be expressed explicitly in terms of the Artin conductor of ϕ and the conductor of ψ_F [**T2**, (3.6.4), (3.6.5)]. In particular, the quotient

$$\delta(\phi, \psi_F) = \varepsilon(0, \phi, \psi_F)\, \varepsilon\big(\tfrac{1}{2}, \phi, \psi_F\big)^{-1} = (q_F)^{\frac{n}{2}} \tag{2.3.1}$$

is a more elementary object than the ε-factor.

Suppose now that G represents an element in $\widetilde{\mathcal{E}}_{\mathrm{sim}}(N)$, and that M is a fixed Levi subgroup of G. Intertwining operators will be attached to parameters and representations for M. For the time being, therefore, we shall denote such objects by symbols ϕ, ψ, π, etc., that have hitherto been reserved for G.

Suppose then that $\phi \in \widetilde{\Phi}(M)$ is a Langlands parameter for M. The normalizing factors of ϕ include special values of local L-functions, which are only defined for parameters in general position. To take care of this, we let λ be a point in general position in the vector space

$$\mathfrak{a}^*_{M,\mathbb{C}} = X^*(M)_F \otimes \mathbb{C}.$$

The associated twist

$$\phi_\lambda(w) = \phi(w)|w|^\lambda, \qquad w \in L_F,$$

of ϕ is then in general position. Suppose that P and P' belong to $\mathcal{P}(M)$, the set of parabolic subgroups of G over F with Levi component M. We write $\rho_{P'|P}$ for the adjoint representation of ${}^LM$ on the quotient

$$\widehat{\mathfrak{n}}_{P'}/\widehat{\mathfrak{n}}_{P'} \cap \widehat{\mathfrak{n}}_P,$$

where $\widehat{\mathfrak{n}}_{P'}$ denotes the Lie algebra of the unipotent radical of $\widehat{P}'$. The composition

$$\rho^\vee_{P'|P} \circ \phi_\lambda \tag{2.3.2}$$

of ϕ_λ with the contragredient of $\rho_{P'|P}$ is a finite dimensional representation of L_F. We define a corresponding local normalizing factor

$$r_{P'|P}(\phi_\lambda) = r_{P'|P}(\phi_\lambda, \psi_F)$$

as the quotient

$$L(0, \rho^\vee_{P'|P} \circ \phi_\lambda)\, \delta(\rho^\vee_{P'|P} \circ \phi_\lambda, \psi_F)^{-1} L(1, \rho^\vee_{P'|P} \circ \phi_\lambda)^{-1}. \tag{2.3.3}$$

Both M and ϕ have parallel decompositions

$$M \cong GL(N_1') \times \cdots \times GL(N_{r'}') \times G_-, \qquad G_- \in \widetilde{\mathcal{E}}_{\mathrm{sim}}(N_-), \tag{2.3.4}$$

and

$$\phi \cong \phi_1' \times \cdots \times \phi_{r'}' \times \phi_-, \qquad \phi_- \in \widetilde{\Psi}(G_-),$$

for positive integers $N_1', \ldots, N_{r'}'$ and N_- such that

$$2N_1' + \cdots + 2N_{r'}' + N_- = N.$$

(Keep in mind that each general linear factor of M has a diagonal embedding $g \to g \times g^\vee$ into G.) We shall ultimately argue by induction on N. This means that if M is proper in G, we will be able to assume that the local Langlands correspondence of Theorem 1.5.1 holds for the factor G_- of M. In particular, we will be able to assume that any $\pi \in \widetilde{\Pi}(M)$ belongs to a unique L-packet $\widetilde{\Pi}_\phi$, and thereby write

$$r_{P'|P}(\pi_\lambda) = r_{P'|P}(\phi_\lambda). \tag{2.3.5}$$

In case F is archimedean, the notation here matches that of [**A7**], provided that ψ_F is taken to be the standard additive character [**T2**, (3.2.4), (3.2.5)]. For in this case, the general ε-factors are independent of s. The general quotient (2.3.1) is then equal to 1, and can be removed from the definition (2.3.3).

The formula (2.3.5) requires a word of explanation. It is not a definition, since the right hand side is defined (2.3.3) in terms of local Artin L-functions and δ-factors, while the left hand side is the corresponding quotient

$$L(0, \pi_\lambda, \rho^\vee_{P'|P})\, \delta(\pi_\lambda, \rho^\vee_{P'|P})^{-1} L(1, \pi_\lambda, \rho^\vee_{P'|P})^{-1}$$

of representation theoretic objects. This in turn is defined as the quotient

$$L(0, \pi_{\phi,\lambda}, \rho^\vee_{P'|P})\, \delta(\pi_{\phi,\lambda}, \rho^\vee_{P'|P})^{-1} L(1, \pi_{\phi,\lambda}, \rho^\vee_{P'|P})^{-1}$$

attached to the product

$$\widetilde{M}^0(N) = GL(N_1') \times \cdots \times GL(N_{r'}') \times GL(N_-)$$

of general linear groups, where π_ϕ is the representation of $\widetilde{M}^0(N, F)$ given by ϕ, and where the finite dimensional representation $\rho_{P'|P}$ is identified with its transfer from ${}^L M$ to the dual group of $\widetilde{M}^0(N)$. The Rankin-Selberg constituents of each side of (2.3.5) are equal, thanks to the local classification (Theorem 1.3.1) for $GL(N)$. The symmetric and exterior square constituents are more subtle. However, it is known that their L-functions and δ-factors are equal [**He2**]. It also seems likely that the work [**CST**] in progress will resolve the deeper question on ε-factors. In any case, the identity is valid.

We need a more general formulation, which applies to nongeneric parameters ψ. Suppose first ψ is an n-dimensional representation of the product of L_F with $SU(2)$. Then ψ extends to a representation of the product of L_F with the complex group $SL(2,\mathbb{C})$. Its pullback

$$\phi_\psi(u) = \psi\left(u, \begin{pmatrix} |u|^{\frac{1}{2}} & 0 \\ 0 & |u|^{-\frac{1}{2}} \end{pmatrix}\right), \qquad u \in L_F,$$

becomes an N-dimensional representation of L_F. We define

$$L(s,\psi) = L(s,\phi_\psi),$$
$$\varepsilon(s,\psi,\psi_F) = \varepsilon(s,\phi_\psi,\psi_F),$$

and

$$\delta(\psi,\psi_F) = \delta(\phi_\psi,\psi_F).$$

In following standard notation here, we have had to use the symbol ψ in two different roles. To minimize confusion, we shall always include the subscript F whenever ψ denotes an additive character.

Suppose now that $\psi \in \widetilde{\Psi}(M)$ is a general parameter for the Levi subgroup M. For $P, P' \in \mathcal{P}(M)$ as above, the meromorphic function

$$r_{P'|P}(\psi_\lambda) = r_{P'|P}(\phi_{\psi_\lambda}) = r_{P'|P}(\phi_{\psi,\lambda}) \tag{2.3.6}$$

of $\lambda \in \mathfrak{a}^*_{M,\mathbb{C}}$ depends implicitly on the additive character ψ_F. It will serve as our general normalizing factor.

The basic unnormalized intertwining operators are defined as for example in [**A7**, (1.1)]. We take π to be an irreducible unitary representation of $M(F)$, with twist

$$\pi_\lambda(m) = \pi(m)e^{\lambda(H_M(m))}, \qquad m \in M(F),$$

by $\lambda \in \mathfrak{a}^*_{M,\mathbb{C}}$. We then have the familiar operator

$$J_{P'|P}(\pi_\lambda) : \mathcal{H}_P(\pi) \longrightarrow \mathcal{H}_{P'}(\pi)$$

that intertwines the induced representations $\mathcal{I}_P(\pi_\lambda)$ and $\mathcal{I}_{P'}(\pi_\lambda)$. It is defined as a meromorphic function of λ by analytic continuation of an integral, taken over the space

$$N_{P'}(F) \cap N_P(F)\backslash N_{P'}(F), \tag{2.3.7}$$

which converges absolutely for $\mathrm{Re}(\lambda)$ in an affine chamber in $\mathfrak{a}^*_M$. (See [**A7**, §1].)

We do have to specify the underlying invariant measure on (2.3.7). The standard splitting on the quasisplit group G determines a Chevalley basis, and hence an invariant F-valued differential form of highest degree on the quotient (2.3.7). Its absolute value, together with the Haar measure on F that is self-dual with respect to ψ_F, then determines an invariant measure on (2.3.7). This is the measure that we take in the definition of $J_{P'|P}(\pi_\lambda)$. It follows easily from [**T2**, (3.6.5)] that the dependence of $J_{P'|P}(\pi_\lambda)$ on ψ_F is parallel to that of the normalizing factor $r_{P'|P}(\psi_\lambda)$. We leave the reader to check that the measure is compatible with the convention in [**A7**] for real groups. In other words, if F is archimedean and ψ_F is the standard additive character, the canonical measure defined on p. 31 of [**A7**] in terms of a certain bilinear form coincides with the measure on (2.3.7) chosen here. We will therefore be able to apply the results of [**A7**] without modification.

Our version of the normalized intertwining operators will require an assumption of the kind mentioned above. We assume that M is proper in G, and that the local Theorem 1.5.1 holds if G is replaced by M. There is of course nothing new to prove for the general linear factors of M. The assumption refers specifically to the factor G_- of M. It will later be treated as an induction hypothesis, applied to the integer $N_- < N$. We can then assume the existence of the packet $\widetilde{\Pi}_\psi$ of ($\mathrm{Out}_{N_-}(G_-)$-orbits) of representations of $M(F)$ for every $\psi \in \widetilde{\Psi}(M)$. We define the normalized intertwining operator attached to any $\psi \in \widetilde{\Psi}(M)$ and $\pi \in \widetilde{\Pi}_\psi$ by

$$R_{P'|P}(\pi_\lambda, \psi_\lambda) = r_{P'|P}(\psi_\lambda)^{-1} J_{P'|P}(\pi_\lambda). \tag{2.3.8}$$

It is independent of the additive character ψ_F.

We note that (2.3.8) differs in some respects from the standard normalized intertwining operator, conjectured in general in [**L5**, p. 281–282], and established for archimedean F in [**A7**]. For example, the δ-factor in (2.3.3) differs from the ε-factor of [**L5**]. This is essentially because the remarks of [**L5**] apply to elements w in the relative Weyl group

$$W(M) = W^G(M) = \mathrm{Norm}(A_M, G)/M,$$

or more precisely, representatives of such elements in $G(F)$, rather than to pairs (P, P') of parabolic subgroups. (We have already noted that the δ-factor reduces to 1 in the archimedean case of [**A7**].) Another point is that π here represents an orbit of irreducible representations of $M(F)$ under the group

$$\widetilde{\mathrm{O}}\mathrm{ut}_{N_-}(M) = \widetilde{\mathrm{O}}\mathrm{ut}_P(M) \cong \widetilde{\mathrm{O}}\mathrm{ut}_{N_-}(G_-).$$

However, any use we make of (2.3.8) will be a context that is independent of the representative of the orbit. Finally, the normalizing factor (2.3.6) is defined in terms of the L-function and δ-factor of ψ, rather than π (which is to say, the Langlands parameter ϕ such that π lies in Π_ϕ). If ψ is not generic, the two possible normalizing factors can differ. The nonstandard form (2.3.6) turns out to be appropriate for the comparison of trace formulas.

PROPOSITION 2.3.1. *Assume that the local and global theorems are valid if N is replaced by any integer $N_- < N$. Then the operators (2.3.8) satisfy* *the multiplicative property*

$$R_{P''|P}(\pi_\lambda, \psi_\lambda) = R_{P''|P'}(\pi_\lambda, \psi_\lambda) R_{P'|P}(\pi_\lambda, \psi_\lambda), \tag{2.3.9}$$

for any P, P' and P'' in $\mathcal{P}(M)$, as well as the adjoint condition

$$R_{P'|P}(\pi_\lambda, \psi_\lambda)^* = R_{P|P'}(\pi_{-\overline{\lambda}}, \psi_{-\overline{\lambda}}). \tag{2.3.10}$$

In particular, $R_{P'|P}(\pi_\lambda, \psi_\lambda)$ is unitary and hence analytic if λ is purely imaginary, and the operator

$$R_{P'|P}(\pi, \psi) = R_{P'|P}(\pi_0, \psi_0)$$

is therefore defined.

PROOF. Suppose first that $\phi = \psi$ is generic, which we recall means that it is trivial on the extra factor $SU(2)$. In this case, $\widetilde{\Pi}_\phi = \widetilde{\Pi}_\psi$ is an L-packet, and (2.3.8) is the normalized intertwining operator

$$R_{P'|P}(\pi_\lambda) = r_{P'|P}(\pi_\lambda)^{-1} J_{P'|P}(\pi_\lambda), \qquad \pi \in \widetilde{\Pi}_\psi,$$

studied for real groups in [**A7**]. Since the required assertions were established for real groups in [**A7**], we can assume that F is p-adic here. Moreover, by standard reductions [**A7**, p. 29], it suffices to consider the case that π represents an element in the set $\Pi_2(M)$ of representations of $M(F)$ that are square integrable modulo the center. In fact, we shall see that it suffices to consider the case that the general linear components of π are supercuspidal. Imposing these conditions on π, we write

$$\pi = \pi_1' \times \cdots \times \pi_{r'}' \times \pi_-, \qquad \pi_i = \Pi_{\mathrm{scusp}}\big(GL(N_i')\big),\ \pi_- \in \widetilde{\Pi}_2(G_-), \tag{2.3.11}$$

for its decomposition relative to (2.3.4).

To handle this basic case, we shall use global means. Having reserved the symbol F in this chapter for our local field, we write $\dot{F}$ for a global field. If u is any valuation of $\dot{F}$, we write $S_\infty(u) = S_\infty \cup \{u\}$ and $S_\infty^u = S_\infty - \{u\}$, where S_∞ is the set of archimedean valuations of $\dot{F}$. We shall apply the following two lemmas to the components π_- and π_i' of π.

LEMMA 2.3.2. *Given the local objects $F, G_- \in \widetilde{\mathcal{E}}_{\mathrm{sim}}(N_-)$ and $\pi_- \in \Pi_2(G_-)$, we can find corresponding global objects $\dot{F}$, $\dot{G}_- \in \widetilde{\mathcal{E}}_{\mathrm{sim}}(N_-)$ and $\dot{\pi}_- \in \Pi_2(\dot{G}_-)$, together with a valuation u of $\dot{F}$, with the following properties.*

(i) $(F, G_-, \pi_-) = (\dot{F}_u, \dot{G}_{-,u}, \dot{\pi}_{-,u})$.

(ii) *For any $v \notin S_\infty(u)$, $\dot{\pi}_{-,v}$ has a vector fixed by a special maximal compact subgroup of $\dot{G}_-(\dot{F}_v)$.*

We shall establish a stronger version of Lemma 2.3.2 in §6.2, as an application of the simple invariant trace formula. In the meantime, we shall take the lemma for granted.

LEMMA 2.3.3 (Henniart, Shahidi, Vignéras). *We can choose the objects $\dot{F}$ and u of the last lemma so that the other local factors $\pi_i \in \Pi_{\mathrm{scusp}}(GL(N_i'))$ of π also have global analogues. More precisely, for each i we can find a cuspidal automorphic representation $\dot{\pi}_i$ of $GL(N_i')$ over $\dot{F}$ such that $\pi_i = \dot{\pi}_{i,u}$, and such that for any $v \notin S_\infty(u)$, $\dot{\pi}_{i,v}$ has a vector fixed by a maximal compact subgroup of $GL(N_i', \dot{F}_v)$.*

This lemma is a consequence of the fact that π_i has a Whittaker model. In the form stated here, it is Proposition 5.1 of [**Sha4**], which applies to a generic representation of any quasisplit group over F. We note, however, that π_- need not be generic, which is why we are using Lemma 2.3.2 in place of the analogue of Lemma 2.3.3 for G_-. □

Applying the lemmas to the components of the representation (2.3.11), we obtain an automorphic representation

$$\dot{\pi} = \dot{\pi}_1' \times \cdots \times \dot{\pi}_{r'}' \times \dot{\pi}_-$$

of the group

$$\dot{M} = \dot{M}_1' \times \cdots \times \dot{M}_{r'}' \times \dot{G}_-, \qquad \dot{M}_i' = GL(N_i'),$$

such that $\dot{\pi}_u = \pi$. We can identify $\dot{M}$ with a Levi subgroup of a group $\dot{G}$ over $\dot{F}$, which represents a simple endoscopic datum in $\widetilde{\mathcal{E}}_{\mathrm{sim}}(N)$ over $\dot{F}$, and has the property that $\dot{G}_u = G$. We then have the global induced representation

$$\mathcal{I}_P(\dot{\pi}), \qquad P \in \mathcal{P}(\dot{M}),$$

and the global intertwining operator, defined by analytic continuation in λ of the product

$$M_{P'|P}(\dot{\pi}_\lambda) = \bigotimes_v J_{P_v'|P_v}(\dot{\pi}_{v,\lambda}),$$

from $\mathcal{I}_P(\dot{\pi}_\lambda)$ to $\mathcal{I}_{P'}(\dot{\pi}_\lambda)$.

The operator $M_{P'|P}(\dot{\pi}_\lambda)$ here is a minor modification of the global operator that is at the heart of Langlands' theory of Eisenstein series. In this setting Langlands' functional equation takes the form

$$M_{P''|P}(\dot{\pi}_\lambda) = M_{P''|P'}(\dot{\pi}_\lambda) M_{P'|P}(\dot{\pi}_\lambda). \tag{2.3.12}$$

(See for example [**A3**, §1].) We shall combine it with what can be obtained from the intertwining operators at places $v \neq u$.

We can write

$$M_{P'|P}(\dot{\pi}_\lambda) = r_{P'|P}(\dot{\pi}_\lambda)\, R_{P'|P}(\dot{\pi}_\lambda), \tag{2.3.13}$$

for the normalized global intertwining operator

$$R_{P'|P}(\dot{\pi}_\lambda) = \bigotimes_v R_{P'|P}(\dot{\pi}_{v,\lambda}), \tag{2.3.14}$$

and the global normalizing factor defined by analytic continuation in λ of a product

$$r_{P'|P}(\dot{\pi}_\lambda) = \prod_v r_{P'|P}(\dot{\pi}_{v,\lambda}).$$

The factor of v in this last product is our canonical normalizing factor for the generic parameter $\dot{\phi}_v$. It is defined by the analogue for $\dot{\pi}_{v,\lambda}$ of (2.3.3) and (2.3.5). The existence of the corresponding L-function and δ-factor is guaranteed by the implicit assumption that Theorem 1.5.1 holds for the localization $\dot{G}_{-,v}$ of $\dot{G}_-$. We note that the measures and additive characters used to define the local factors in (2.3.14) can be chosen to be compatible with corresponding global objects. This follows from the well known fact [**T1**, (3.3)] that if each $\psi_{\dot{F}_v}$ is obtained from a fixed, nontrivial additive character $\psi_{\dot{F}}$ on $\dot{\mathbb{A}}/\dot{F}$, the associated product measure on $\dot{\mathbb{A}}$ is self-dual (relative to $\psi_{\dot{F}}$) and assigns volume 1 to the quotient $\dot{\mathbb{A}}/\dot{F}$.

We claim that

$$R_{P''|P}(\dot{\pi}_{v,\lambda}) = R_{P''|P'}(\dot{\pi}_{v,\lambda})R_{P'|P}(\dot{\pi}_{v,\lambda}), \tag{2.3.15}$$

for any valuation $v \neq u$ of $\dot{F}$. For archimedean v, this is the concrete form of [**A7**, Theorem 2.1(R_2)], established as Proposition 3.1 in the Appendix of [**A7**]. For p-adic v, we use the property (ii) of $\dot{\pi}_{-,v}$ in Lemma 2.3.2. Combined with the theory of p-adic spherical functions [**Ma**] and Whittaker models [**C**], it implies that $\dot{\pi}_{-,v}$ has a Whittaker model. Since for any i, the generic representation $\dot{\pi}_{i,v}$ of $GL(N_i, \dot{F}_v)$ also has a Whittaker model, the same is true of the representation $\dot{\pi}_v$ of $\dot{M}(\dot{F}_v)$. We can therefore apply Shahidi's results [**Sha4**] to the induced representation $\mathcal{I}_P(\dot{\pi}_{v,\lambda})$. They imply that Langlands' conjectural normalizing factors have the desired properties. In particular, the formula (2.3.15) does hold for v, as claimed. (We will discuss Shahidi's results further in §2.5.)

The global normalizing factors have an expression

$$r_{P'|P}(\dot{\pi}_\lambda) = L(0, \dot{\pi}_\lambda, \rho^\vee_{P'|P})\, \delta(\dot{\pi}_\lambda, \rho^\vee_{P'|P})^{-1} L(1, \dot{\pi}_\lambda, \rho^\vee_{P'|P})^{-1}$$

in terms of global L-functions and δ-factors

$$\delta(\dot{\pi}_\lambda, \rho^\vee_{P'|P}) = \varepsilon(0, \dot{\pi}_\lambda, \rho^\vee_{P'|P})\varepsilon\big(\tfrac{1}{2}, \pi_\lambda, \rho^\vee_{P'|P}\big)^{-1}.$$

We claim that they satisfy the identity

$$r_{P''|P}(\dot{\pi}_\lambda) = r_{P''|P'}(\dot{\pi}_\lambda)r_{P'|P}(\dot{\pi}_\lambda). \tag{2.3.16}$$

By standard reductions, it suffices to consider (2.3.16) when the Levi subgroup $\dot{M}$ of $\dot{G}$ is maximal. (See [**A7**, §2] for example.) In this case there are only two groups P and $\overline{P}$ in $\mathcal{P}(\dot{M})$, and it suffices to take $P' = \overline{P}$ and $P'' = P$, since $r_{P|P}(\dot{\pi}_\lambda) = 1$. But $\rho_{\overline{P}|P}$ is just the direct sum of the standard (tensor product) representation of $GL(N'_1, \mathbb{C}) \times {}^L\dot{G}_-$ with either the symmetric square or skew symmetric square representation of $GL(N'_1, \mathbb{C})$. We are assuming that the global theorems hold for $\dot{G}_-$. This allows us to express $L(s, \dot{\pi}_\lambda, \rho^\vee_{\overline{P}|P})$ as the product of a Rankin-Selberg L-function with either a symmetric square or skew symmetric square L-function. In particular, this

L-function has analytic continuation and functional equation

$$L(s, \dot{\pi}_\lambda, \rho^\vee_{\bar{P}|P}) = \varepsilon(s, \dot{\pi}_\lambda, \rho^\vee_{\bar{P}|P}) L(1-s, \dot{\pi}_\lambda, \rho_{\bar{P}|P}).$$

It follows that

$$r_{\bar{P}|P}(\dot{\pi}_\lambda) = L(1, \dot{\pi}_\lambda, \rho_{\bar{P}|P}) L(1, \dot{\pi}_\lambda, \rho^\vee_{\bar{P}|P})^{-1} \varepsilon\big(\tfrac{1}{2}, \dot{\pi}_\lambda, \rho^\vee_{\bar{P}|P}\big).$$

Similarly, we have

$$r_{P|\bar{P}}(\dot{\pi}_\lambda) = L(1, \dot{\pi}_\lambda, \rho_{P|\bar{P}}) L(1, \dot{\pi}_\lambda, \rho^\vee_{P|\bar{P}})^{-1} \varepsilon\big(\tfrac{1}{2}, \dot{\pi}_\lambda, \rho^\vee_{P|\bar{P}}\big).$$

Since $\rho^\vee_{P|\bar{P}} = \rho_{\bar{P}|P}$, the product of the four L-functions equals 1. Another application of the functional equation tells us that the product of the two ε-factors also equals 1. The product of the two normalizing factors thus equals 1. In other words, the identity (2.3.16) holds in the special case at hand, and hence for any M, P' and P, as claimed.

Consider the product expression for the operator $M_{P'|P}(\dot{\pi}_\lambda)$ given by (2.3.13) and (2.3.14). We need only combine the multiplicative property (2.3.12) of this operator with its analogues (2.3.16) and (2.3.15) for the scalar factor in (2.3.13) and the factors with $v \neq u$ in (2.3.14). It follows that the same property holds for the remaining factor, that corresponding to $v = u$ in (2.3.13). This is the required specialization

$$R_{P''|P}(\pi_\lambda) = R_{P''|P'}(\pi_\lambda) R_{P'|P}(\pi_\lambda)$$

of (2.3.9) to the case that $\phi = \psi$ is generic, and π is of the form (2.3.11).

It is easy to relax the conditions on π. From the remarks leading to [**A7**, (2.2)], we see that the last product formula remains valid for representations (2.3.11) whose general linear components π_i are *induced* supercuspidal. Since any representation in $\Pi_2\big(GL(N_i)\big)$ is a subrepresentation of such a π_i, and the corresponding normalizing factors can be related by [**A7**, Proposition 5.2], the product formula holds for any $\pi \in \widetilde{\Pi}_2(M)$. Similar remarks then allow us to extend the formula to any representation $\pi \in \widetilde{\Pi}(M)$. (See [**A7**, §2], for example.) The required formula (2.3.9) is thus valid if $\psi = \phi$ is any parameter in $\widetilde{\Phi}(M)$.

Suppose now that the local parameter ψ for M is arbitrary. The normalizing factor in (2.3.8) is attached to ψ rather than π. This is the distinction noted prior to the statement of the proposition. It is pertinent because the set $\widetilde{\Pi}_\psi$ to which π belongs will not in general be an L-packet. In other words, the local Langlands parameter ϕ of π may be different from the local Langlands parameter ϕ_ψ of ψ. By our assumption on G_-, we are free to regard both ϕ_ψ and ϕ as local Langlands parameters for a product of general linear groups. The normalizing factor (2.3.6) to which the proposition applies is taken relative to ϕ_ψ. The extension of the required identity (2.3.9) from tempered representations to arbitrary Langlands quotients, which was established for any F on p. 30 of [**A7**], is implicitly based on the parameter ϕ. We must show that if (2.3.9) holds for the one set of normalizing factors, it also holds for the other.

As Langlands parameters attached to a product of general linear groups, ϕ_ψ and ϕ correspond to irreducible representations π_ψ and π_ϕ of the corresponding products of groups over F. The two representations are related by the algorithm of Zelevinsky, used in the proof of Lemma 2.2.2, (and its analogue [**AH**] for $F = \mathbb{R}$.) In other words, π_ψ and π_ϕ are *block equivalent*, in the sense of D. Vogan. (See [**A7**, p. 42].) It then follows from [**A7**, Proposition 5.2] that the quotient

$$r_{P'|P}(\phi_\lambda, \psi_\lambda) = r_{P'|P}(\phi_\lambda)^{-1} r_{P'|P}(\psi_\lambda)$$

of the two normalizing factors satisfies the analogue

$$r_{P''|P}(\phi_\lambda, \psi_\lambda) = r_{P''|P'}(\phi_\lambda, \psi_\lambda) r_{P'|P}(\phi_\lambda, \psi_\lambda)$$

of (2.3.9). This implies that the two formulations of the identity (2.3.9), the one here and the original version in [**A7**], are equivalent. The identity therefore holds as stated.

We still have the adjoint condition (2.3.10) to verify. It is a straightforward matter to establish a general identity

$$J_{P'|P}(\pi_\lambda)^* = J_{P|P'}(\pi^*_{-\overline{\lambda}}),$$

for any irreducible representation π of $M(F)$ with adjoint π^*. One first takes λ to lie in the domain of absolute convergence of the integral that defines $J_{P'|P}(\pi_\lambda)$. One obtains the formula in this case by a suitable change of variables in the quasi-invariant measure on $P'(F)\backslash G(F)$ determined by the inner product on $\mathcal{H}_{P'}(\pi)$. The formula for arbitrary λ then follows by analytic continuation. In (2.3.10), π lies in the packet $\widetilde{\Pi}_\psi$, and is hence unitary by our assumption that Theorem 1.5.1 holds for G_-. (We of course know that the unitary condition holds also for the remaining general linear factors of M.) In other words, $\pi = \pi^*$.

It is also not difficult to see that

$$\overline{r_{P'|P}(\psi_\lambda)} = r_{P|P'}(\psi_{-\overline{\lambda}}),$$

for $\psi \in \widetilde{\Psi}(M)$ as in (2.3.10). In the special case that $\psi = \phi$ is generic, this follows from the explicit formulas in [**T2**] for the terms in (2.3.3). For a general parameter $\psi \in \widetilde{\Psi}(M)$, we write the parameter $\phi_\psi \in \widetilde{\Phi}(M)$ as a twist ϕ_μ, where $\phi \in \widetilde{\Phi}_{\mathrm{temp}}(M)$ is the restriction of ψ to the subgroup L_F of $L_F \times SU(2)$, and μ is a point in $\mathfrak{a}^*_M$. Since the Langlands parameter $\phi_{-\mu}$ is $\widehat{M}$-conjugate to ϕ_μ, we obtain

$$\begin{aligned}\overline{r_{P'|P}(\psi_\lambda)} = \overline{r_{P'|P}(\phi_{\mu+\lambda})} &= r_{P|P'}(\phi_{-\mu-\overline{\lambda}})\\ &= r_{P|P'}(\phi_{\mu-\overline{\lambda}}) = r_{P|P'}(\psi_{-\overline{\lambda}}),\end{aligned}$$

as claimed. The adjoint condition (2.3.10) then follows from the definition (2.3.8), since π is assumed to be unitary.

The identity (2.3.9) reduces to

$$R_{P|P'}(\pi_\lambda) = R_{P'|P}(\pi_\lambda)^{-1}.$$

when $P'' = P$. If we take λ to be purely imaginary in (2.3.10), we then see that

$$\begin{aligned} R_{P'|P}(\pi_\lambda, \psi_\lambda)^* &= R_{P|P'}(\pi_{-\bar{\lambda}}, \psi_{-\bar{\lambda}}) \\ &= R_{P|P'}(\pi_\lambda, \psi_\lambda) = R_{P'|P}(\pi_\lambda, \psi_\lambda)^{-1}. \end{aligned}$$

In other words $R_{P'|P}(\pi_\lambda)$ is unitary, as noted in the last assertion of the proposition. The proof of Proposition 2.3.1 is complete. □

We have constructed normalized intertwining operators $R_{P'|P}(\pi, \psi)$ between induced representations $\mathcal{I}_P(\pi)$ and $\mathcal{I}_{P'}(\pi)$. This is just the first step. It remains to convert these objects to self-intertwining operators of $\mathcal{I}_P(\pi)$, which will be attached to elements $w \in W(M)$ that stabilize π.

Suppose that $P \in \mathcal{P}(M)$, $\pi \in \Pi_{\text{unit}}(M)$ and $w \in W(M)$ are fixed. We can certainly choose a representative $\widetilde{w}$ of w in the normalizer $N(M)$ of M in $G(F)$. This gives us another representation

$$(w\pi)(m) = \pi(\widetilde{w}^{-1} m \widetilde{w}), \qquad m \in M(F),$$

of $M(F)$ on the underlying space V_π of π, which depends only on the image of $\widetilde{w}$ in the quotient of $N(M)$ by the center of $M(F)$. It also gives us an intertwining isomorphism

$$\ell(\widetilde{w}, \pi) : \mathcal{H}_{P'}(\pi) \longrightarrow \mathcal{H}_P(w\pi)$$

from $\mathcal{I}_{P'}(\pi)$ to $\mathcal{I}_P(w\pi)$ by left translation

$$(\ell(\widetilde{w}, \pi)\phi')(x) = \phi'(\widetilde{w}^{-1}x), \qquad \phi' \in \mathcal{H}_{P'}(\pi),\ x \in G(F), \tag{2.3.17}$$

where

$$P' = w^{-1}P = \widetilde{w}^{-1} P \widetilde{w}^{-1}.$$

If $w\pi$ is equivalent to π, we can in addition choose an intertwining isomorphism $\widetilde{\pi}(w)$ of V_π from $w\pi$ to π. Its pointwise action on $\mathcal{H}_P(w\pi)$ intertwines $\mathcal{I}_P(w\pi)$ with $\mathcal{I}_P(\pi)$. The composition

$$\widetilde{\pi}(w) \circ \ell(\widetilde{w}, \pi) \circ R_{P'|P}(\pi, \psi), \qquad \pi \in \widetilde{\Pi}_\psi,$$

of intertwining maps

$$\mathcal{H}_P(\pi) \longrightarrow \mathcal{H}_{P'}(\pi) \longrightarrow \mathcal{H}_P(w\pi) \longrightarrow \mathcal{H}_P(\pi)$$

will then be a self intertwining operator of $\mathcal{I}_P(\pi)$. It is in the choices of $\widetilde{w}$ and $\widetilde{\pi}(w)$ that we will need to be aware of the implications for endocopic transfer. We will deal with $\widetilde{w}$ at the end of this section, and $\widetilde{\pi}(w)$ in the beginning of the next. In each of these remaining two steps, we will also have to adjust the relevant intertwining map by its own scalar normalizing factor.

There is a natural way to choose the representative $\widetilde{w}$ of w. The group G comes with a splitting

$$S = \big(T, B, \{X_\alpha\}\big).$$

We assume that M and P are standard, in the sense that they contain T and B respectively. Let w_T be the representative of w in the Weyl group $W_F(G,T)$ that stabilizes the simple roots of $(B \cap M, T)$. We then take $\widetilde{w} = \widetilde{w}_S$ to be the representative of w_T in $G(F)$ attached to the splitting S, as on p. 228 of [**LS1**]. Thus,

$$\widetilde{w} = \widetilde{w}_{\alpha_1} \cdots \widetilde{w}_{\alpha_r},$$

where $w_T = w_{\alpha_1} \cdots w_{\alpha_r}$ is a reduced decomposition of w_T relative to the simple roots of (G,T), and

$$\widetilde{w}_\alpha = \exp(X_\alpha)\exp(-X_{-\alpha})\exp(X_\alpha),$$

in the notation of [**LS1**].

However, our representative $\widetilde{w}$ introduces another difficulty. For it is well known that the mapping $w \rightarrow \widetilde{w}$ from $W(M)$ to $N(M)$ is not multiplicative in w. The same is therefore true of the intertwining operator $\ell(\widetilde{w},\pi)$. The obstruction is the co-cycle of [**LS1**, Lemma 2.1.A], which plays a critical role in the construction of transfer factors. To compensate for it, we must introduce the factor $\varepsilon\big(\frac{1}{2},\cdot\big)$ that was taken out of the original definition (2.3.1). For global reasons, we will use the representation theoretic ε-factor, given by the representation π_ψ of the product $\widetilde{M}^0(N,F)$ of general linear groups. We are assuming now that π belongs to the packet $\widetilde{\Pi}_\psi$ attached to a given $\psi \in \widetilde{\Psi}(M)$. Since M is proper, we can assume as before that the packet exists, and that π is unitary. The choice of P allows us to identify $W(M)$ with the Weyl group $W(\widehat{M})$ of $\widehat{M}$. (We are regarding $\widehat{M}$ here as a complex group with an L-action of the Galois group $\Gamma = \Gamma_F$, so that $W(\widehat{M})$ is the group of Γ-invariant elements in the quotient $\mathrm{Norm}_{\widehat{G}}(\widehat{M})/\widehat{M}$.) We define the ε-factor

$$\varepsilon_P(w,\psi) = \varepsilon\big(\tfrac{1}{2},\pi_\psi,\rho^\vee_{w^{-1}P|P},\psi_F\big), \qquad w \in W(M). \tag{2.3.18}$$

We know from the general results of Shahidi that we must also include a supplementary term. It is the λ-factor

$$\lambda(w) = \lambda(w,\psi_F) = \prod_a \lambda_a(\psi_F), \qquad w \in W(M), \tag{2.3.19}$$

of [**KeS**, (4.1)] and [**Sha4**, (3.1)], in which a ranges over the reduced roots of (B, A_B) such that $w_T a < 0$, and A_B is the split component of B (or T) over F. This term is built out of the complex numbers $\lambda(E/F,\psi_F)$ attached to finite extensions E/F by Langlands [**L3**] to account for the behaviour of ε-factors under induction. In the case at hand,

$$\lambda_a(\psi_F) = \lambda(F_a/F,\psi_F),$$

where F_a is either a quadratic extension of F or F itself. This field is given by

$$G_{a,\mathrm{sc}} = \mathrm{Res}_{F_a/F}\big(SL(2)\big),$$

where $G_{a,\mathrm{sc}}$ is the simply connected cover of the derived group of G_a, the Levi subgroup of G of semisimple rank one attached to a.

It is known that neither $\varepsilon_P(w,\psi)$ nor $\lambda(w)$ is multiplicative in w. However, the next lemma tells us that their combined obstruction matches that of $\ell_P(\widetilde{w},\pi)$.

LEMMA 2.3.4. *The product*

$$\ell(w,\pi,\psi) = \lambda(w)^{-1}\varepsilon_P(w,\psi)\ell(\widetilde{w},\pi), \qquad w \in W(M), \tag{2.3.20}$$

satisfies the condition

$$\ell(w'w,\pi,\psi) = \ell(w',w\pi,w\psi)\ell(w,\pi,\psi), \qquad w',w \in W(M).$$

PROOF. We consider general elements w' and w in $W(M)$. Since $\widetilde{w}'\widetilde{w}$ and $\widetilde{w'w}$ are two representatives in $G(F)$ of $w'w$, and since they preserve the same splitting of M, they satisfy

$$\widetilde{w}'\widetilde{w} = \widetilde{w'w}\, z(w',w),$$

for a point $z(w',w)$ in the center of $M(F)$. It follows that

$$\ell(\widetilde{w'w},\pi) = \eta_\pi\big(z(w',w)\big)\ell(\widetilde{w}',w\pi)\ell(\widetilde{w},\pi),$$

where η_π is the central character of π.

Suppose for a moment that M equals the minimal Levi subgroup T. The elements w' and w then belong to the rational Weyl group $W(T) = W_F(T)$ of T. We write $\Sigma_B(w',w)$ for the subset of roots α in Σ_B, the set of all roots of (B,T), such that $w\alpha \notin \Sigma_B$ and $w'w\alpha \in \Sigma_B$. The corresponding sum

$$\lambda(w',w) = \sum_\alpha \alpha^\vee, \qquad \alpha \in \Sigma_B(w',w),$$

is a co-character in $X_*(T)$, which is fixed by the Galois group $\Gamma = \Gamma_F$, since both w' and w are defined over F. It then follows from [**LS1**, Lemma 2.1.A] that

$$z(w',w) = (-1)^{\lambda(w',w)},$$

where the right hand side is understood as a point in

$$T(F) = \big(X_*(T)\otimes \overline{F}^*\big)^\Gamma.$$

Suppose now that M is arbitrary, subject only to our requirement that $P \in \mathcal{P}(M)$ contains B. For any $w \in W(M)$, we have constructed the rational Weyl element $w_T \in W(T)$. The definitions tell us that

$$z(w',w) = (-1)^{\lambda(w',w)}, \qquad w',w \in W(M),$$

where

$$\lambda(w',w) = \sum_\alpha \alpha^\vee, \qquad \alpha \in \Sigma_B(w'_T,w_T).$$

In other words, $z(w',w) = z(w'_T,w_T)$ and $\lambda(w',w) = \lambda(w'_T,w_T)$. We claim that $\lambda(w',w)$ belongs to the subgroup $X_*(A_M)$ of co-characters of the split component A_M of the center of M. To see this, consider an element u in the rational Weyl group W_0^M of (M,T). If α is any root in $\Sigma_B(w'_T,w_T)$,

$u\alpha$ obviously belongs to Σ_B. Moreover, we see that $w_T(u\alpha) \notin \Sigma_B$ and $w'_T(w_T(u\alpha)) \in \Sigma_B$, since w_T and $w'_T w_T$ each conjugate u to another element in W_0^M. In other words, $u\alpha$ also lies in $\Sigma_B(w'_T, w_T)$. The Weyl group W_0^M thus acts by permutation on $\Sigma_B(w'_T, w_T)$, and consequently fixes the sum $\lambda(w', w)$. The claim then follows from the fact that $\lambda(w', w)$ is also fixed by Γ. In particular, $z(w', w)$ lies in the group $A_M(F) = X_*(A_M) \otimes F^*$. The split torus A_M is the product of the centers of the general linear factors of M in (2.3.4). We can therefore write

$$\eta_\pi\big(z(w', w)\big) = \eta_\psi\big(z(w', w)\big),$$

where η_ψ is the central character of the representation π_ψ, since π and π_ψ are equal on these general linear factors.

We have established an obstruction

$$\ell(\widetilde{w'w}, \pi)\, \ell(\widetilde{w}, \pi)^{-1}\, \ell(\widetilde{w}', w\pi)^{-1} = \eta_\psi\big((-1)^{\lambda^\vee(w',w)}\big)$$

for the operator $\ell(\widetilde{w}, \pi)$ to be multiplicative in w. Before we compare it with the corresponding obstructions for the other two factors in (2.3.20), let us first take note of its L-group interpretation.

The complex dual torus $\widehat{A}_M$ of A_M can be identified with the "split part" of the cocenter of $\widehat{M}$, namely the largest connected quotient of $\widehat{M}$ on which the action of ${}^L M$ by conjugation is trivial. This provides the second of the two canonical isomorphisms

$$X_*(A_M) \cong X^*(\widehat{A}_M) \cong X^*(\widehat{M})_F.$$

Moreover, we can extend any character on $\widehat{M}$ in the set $X^*(\widehat{M})_F$ to a character on ${}^L M$ that is trivial on the semidirect factor W_F of ${}^L M$. If $\phi \in \widetilde{\Phi}(M)$ is a Langlands parameter with central character η_ϕ on $A_M(F)$, it is then easy to see that

$$\eta_\phi(x^{\lambda^\vee}) = (\lambda^\vee \circ \phi)(u) = \lambda^\vee\big(\phi(u)\big), \qquad x \in F^*,\ \lambda^\vee \in X_*(A_M), \tag{2.3.21}$$

where $\lambda^\vee$ is identified with a character on ${}^L M$, and u is any element in W_F whose image in the abelian quotient $F^* \cong W_F^{\mathrm{ab}}$ equals x. Let

$$\lambda^\vee(w', w) = \sum_\beta \lambda_\beta^\vee$$

be the natural decomposition of the character $\lambda^\vee = \lambda^\vee(w', w)$ into a sum over roots β in the subset

$$\widehat{\Sigma}_P^r(w', w) = \big\{\beta \in \widehat{\Sigma}_P^r :\ w\beta < 0,\ w'w\beta > 0\big\}$$

of the reduced roots $\widehat{\Sigma}_P^r$ of $(\widehat{P}, A_{\widehat{M}})$. The summand of β thus equals

$$\lambda_\beta^\vee = \sum_\alpha \alpha^\vee, \qquad \alpha \in \Sigma_\beta,$$

where Σ_β is the subset of roots α of (G, T) whose coroot $\alpha^\vee$ restricts to a positive multiple of β on $A_{\widehat{M}}$. The components $\lambda_\beta^\vee$ of $\lambda^\vee(w', w)$ each belong to the group $A_{\widehat{M}}$, by virtue of the argument we applied above to $\lambda^\vee(w',w)$. It follows that the obstruction for $\ell(\widetilde{w},\pi)$ can be written as a product

$$\eta_\psi\big((-1)^{\lambda^\vee(w',w)}\big) = \prod_\beta (\lambda_\beta^\vee \circ \phi_\psi)(-1) \tag{2.3.22}$$

over $\beta \in \widehat{\Sigma}_P^r(w',w)$.

For any $\beta \in \widehat{\Sigma}_P^r$, let $\rho_\beta = \rho_{-\beta}^\vee$ be the adjoint representation of ${}^LM$ on the subspace

$$\widehat{\mathfrak{n}}_\beta = \bigoplus_\alpha \widehat{\mathfrak{n}}_{\alpha^\vee}, \qquad \alpha \in \Sigma_\beta,$$

of $\widehat{\mathfrak{n}}_P$ attached to β. Th ε-term in (2.3.20) then has a decomposition

$$\varepsilon_P(w,\psi) = \prod_\beta \varepsilon\Big(\tfrac{1}{2}, \pi_\psi, \rho_\beta, \psi_F\Big), \qquad \beta \in \widehat{\Sigma}_P^r,\ w\beta < 0.$$

This is similar to the decomposition we have come to expect of normalizing factors. For example, it tells us that the obstruction for $\varepsilon_P(w,\psi)$ to be multiplicative in w, which in general we write in the inverse form

$$e(w',w) = \varepsilon_P(w',w\psi)\,\varepsilon_P(w,\psi)\,\varepsilon_P(w'w,\psi)^{-1},$$

is trivial if the length of $w'w$ equals the sum of the lengths of w' and w. For arbitrary elements w' and w in $W(M)$, we observe from this decomposition that

$$e(w',w) = \prod_\beta \varepsilon\Big(\tfrac{1}{2}, \pi_\psi, \rho_\beta, \psi_P\Big)\,\varepsilon\Big(\tfrac{1}{2}, \pi_\psi, \rho_{-\beta}, \psi_F\Big), \qquad \beta \in \widehat{\Sigma}_P^r(w',w).$$

For each β, the product of the two representation theoretic ε-factors here equals the corresponding product of Artin ε-factors. It then follows from the general property [**T2**, (3.6.8)] of Artin ε-factors that

$$\varepsilon\Big(\tfrac{1}{2}, \pi_\psi, \rho_\beta, \psi_F\Big)\,\varepsilon\Big(\tfrac{1}{2}, \pi_\psi, \rho_{-\beta}, \psi_F\Big) = (\det\,\rho_\beta \circ \phi_\psi)(-1),$$

where the right hand side is understood as the value of the determinant of $\rho_\beta \circ \phi_\psi$ at any $u \in W_F$ whose image in $F^* \cong W_F^{\mathrm{ab}}$ equals (-1). We can therefore write the obstruction for $\varepsilon_P(w,\psi)$ as a product

$$e(w',w) = \prod_\beta (\det \rho_\beta \circ \phi_\psi)(-1) \tag{2.3.23}$$

over $\beta \in \Sigma_P^r(w',w)$.

Suppose for example that G is split. Then the supplementary term $\lambda(w)$ in (2.3.20) is trivial. Moreover, for any $\beta \in \widehat{\Sigma}_P^r$, we can regard ρ_β as a representation of the complex connected group $\widehat{M} = {}^LM$ on $\widehat{\mathfrak{n}}_\beta$. Its weights on $\widehat{T}$ are the characters $\alpha^\vee$ parametrized by $\alpha \in \Sigma_\beta$, each of which occurs with multiplicity 1. It follows that the restriction of ρ_β to $\widehat{T}$ has determinant equal to the character $\lambda_\beta^\vee$. Since every element in $\widehat{M}$ is conjugate to an element in $\widehat{T}$, the characters $\lambda_\beta^\vee$ and $\det \rho_\beta$ on $\widehat{M}$ are equal. It follows that

$$\prod_\beta (\lambda_\beta^\vee \circ \phi_\psi)(-1) = \prod_\beta (\det \rho_\beta \circ \phi_\psi)(-1), \qquad \beta \in \widehat{\Sigma}_P^r(w', w),$$

so that the obstructions for the two primary terms in (2.3.20) match. The lemma is therefore valid in case G is split.

Assume now that $G \in \widetilde{\mathcal{E}}_{\mathrm{sim}}(N)$ is a general element in $\widetilde{\mathcal{E}}_{\mathrm{sim}}(N)$. Our remaining concern is the nonsplit group $G = SO(2n, \eta_{E/F})$, where $\eta = \eta_G = \eta_{E/F}$ is the quadratic character on F^* attached to a fixed quadratic extension E/F. This is the case in which the supplementary term $\lambda(w)$ in (2.3.20) is nontrivial. We recall that $\lambda(w)$ is given by a product (2.3.19) of λ-factors $\lambda(F_a/F, \psi_F)$, taken over the reduced roots a of (B, A_B) with $w_T a < 0$. Each a determines an orbit $\{\alpha\}$ of roots of (G, T) under the Galois group $\Gamma = \Gamma_F$. The field F_a is the extension of F that corresponds to the stabilizer of a in Γ. It equals either F, in which case the associated factor $\lambda(F/F, \psi_F)$ is trivial, or the quadratic extension E, in which case the factor satisfies

$$\lambda(E/F, \psi_F)^2 = \eta_{E/F}(-1).$$

(See [**KeS**, (2.7)].) It follows that the obstruction for $\lambda(w)$ to be a character in w equals the product

$$\lambda(w'w)\,\lambda(w)^{-1}\lambda(w')^{-1} = \prod_\beta \eta_{E/F}(-1)^{\mu_\beta} \tag{2.3.24}$$

over $\beta \in \Sigma_P^r(w', w)$, where μ_β is the number of roots a with $F_a = E$ such that the associated orbit $\{\alpha\}$ is contained in Σ_β. We need to combine this with the other two obstructions (2.3.23) and (2.3.22).

For the general group $G \in \widetilde{\mathcal{E}}_{\mathrm{sim}}(N)$, the characters $\lambda_\beta^\vee$ and $\det \rho_\beta$ on $\widehat{M}$ attached to any $\beta \in \widehat{\Sigma}_P$ remain equal. However, we must also consider their extensions to the nonconnected group ${}^L M = \widehat{M} \rtimes \Gamma_{E/F}$. As an element $\lambda^\vee$ in $X_*(A_M)$, following the general identity (2.3.21), $\lambda_\beta^\vee$ has a canonical extension that is trivial on the factor $\Gamma_{E/F}$ of ${}^L M$. The character $\det \rho_\beta$ is the determinant of a representation ρ_β that is defined a priori on ${}^L M$. It is generally not trivial on $\Gamma_{E/F}$.

Fix the root $\beta \in \widehat{\Sigma}_P^r$. The quotient $\Gamma_{E/F}$ of W_F acts by permutation on the coroot spaces $\widehat{\mathfrak{n}}_{\alpha^\vee}$ in $\widehat{\mathfrak{n}}_P$, in a way that is compatible with the natural action of $\Gamma_{E/F}$ on the roots $\alpha \in \Sigma_\beta$. To describe the extension of $\det \rho_\beta$ to ${}^L M$, it suffices to consider the case that P is a maximal parabolic subgroup of the group G, which we are implicitly assuming equals $SO(2n, \eta_{E/F})$. The Levi component M of P then takes the form

$$M \cong GL(m) \times SO(2n - 2m, \eta_{E/F}), \qquad m < n.$$

The set of roots α can be written as

$$\Sigma_\beta = \{\alpha_{ij}^{\pm}(t) = t_i(t_j)^{\pm 1} : 1 \leqslant i \leqslant m-1,\ m \leqslant j \leqslant n\},$$

where t represents a diagonal matrix (t_i) in the standard maximal torus T of G. The group $\Gamma_{E/F}$ acts simply transitively on any pair

$$p_i = (\alpha_{in}^{+}, \alpha_{in}^{-}), \qquad i \leqslant m-1,$$

and acts trivially on each of the remaining roots of Σ_β. Consequently, the determinant of the restriction of ρ_β to the factor $\Gamma_{E/F}$ of ${}^L M$ equals a product of copies of the nontrivial character of $\Gamma_{E/F}$, taken over the set of pairs p_i. It follows that

$$\det \rho_\beta(1 \rtimes \sigma) = (-1)^{m-1},$$

for the nontrivial element $\sigma \in \Gamma_{E/F}$. This will allow us to compare the characters $\lambda_\beta^\vee$ and $\det \rho_\beta$ on the factor $\Gamma_{E/F}$ of ${}^L M$. Moreover, it is clear from the definitions that $m-1$ equals the integer μ_β assigned to β in (2.3.24).

In comparing the two primary obstructions (2.3.22) and (2.3.23), we need to choose an element u in the Weil group W_F whose image in $F^* \cong W_F^{\mathrm{ab}}$ equals (-1). We write

$$\phi_\psi(u) = m_\psi(u) \rtimes \sigma(u)$$

where $m_\psi(u)$ belongs to $\widehat{M}$, and $\sigma(u)$ lies in the quotient $\Gamma_{E/F}$ of W_F. We are assuming here that G is not split, so that E is defined as a quadratic extension of F. We then observe that the element $\sigma(u) \in \{\pm 1\}$ equals $\eta_{E/F}(-1)$, a property that follows from the fact that $\eta_{E/F}(-1)$ equals 1 if and only if the element $(-1) \in F^*$ lies in the image of the norm mapping from E^* to F^*. For any $\beta \in \widehat{\Sigma}_P^r(w', w)$, we then have

$$\begin{aligned}
(\det \rho_\beta \circ \phi_\psi)(-1) &= \det \rho_\beta\big(\phi_\psi(u)\big) \\
&= \det \rho_\beta\big(m_\psi(u)\big)\ \det \rho_\beta\big(1 \rtimes \sigma(u)\big) \\
&= \lambda_\beta^\vee\big(m_\psi(u)\big)\, \eta_{E/F}(-1)^{\mu_\beta} \\
&= \lambda_\beta^\vee\big(\phi_\psi(u)\big)\, \eta_{E/F}(-1)^{\mu_\beta} \\
&= (\lambda_\beta^\vee \circ \phi_\psi)(-1)\, \eta_{E/F}(-1)^{\mu_\beta}.
\end{aligned}$$

It remains only to take the product over β of the numbers in each side of the identity we have obtained. We conclude that the obstruction (2.3.23) equals the product of the original obstruction (2.3.22) with the supplementary obstruction (2.3.24). It then follows from the definitions that the threefold product (2.3.20) is indeed multiplicative in w, as required. □

REMARKS. 1. The operator $\ell(w, \pi, \phi)$ makes sense for an arbitrary pair of groups (G, M), under the appropriate hypothesis on the local Langlands correspondence for M. The proof of the lemma should carry over to the general case, although I have not checked the details. I thank the referee for pointing out that $\lambda(w)$ is generally not a character in w.

2. It is interesting to see the local root numbers (2.3.18) appearing as explicit factors of the operators (2.3.20). They remind us of the global root numbers (1.5.7) used to construct the sign character (1.5.6). As orthogonal root numbers, however, they are less deep, and better understood [**D2**] than the symplectic root numbers (1.5.7) and their local analogues.

We can think of Lemma 2.3.4 as a normalization of the "miniature" intertwining operator $\ell(\widetilde{w}, \pi)$. It is reminiscent of the main property (2.3.9) of Proposition 2.3.1, if somewhat simpler. We combine the two normalized operators by writing

$$R_P(w, \pi, \psi) = \ell(w, \pi, \psi) R_{w^{-1}P|P}(\pi, \psi), \qquad w \in W(M). \tag{2.3.25}$$

This operator then has a decomposition

$$R_P(w, \pi_\lambda, \psi_\lambda) = r_P(w, \psi_\lambda)^{-1} J_P(\widetilde{w}, \pi_\lambda), \qquad \pi \in \widetilde{\Pi}_\psi, \tag{2.3.26}$$

for the unnormalized intertwining operator

$$J_P(\widetilde{w}, \pi_\lambda) = \ell(\widetilde{w}, \pi_\lambda) J_{w^{-1}P|P}(\pi_\lambda)$$

from $\mathcal{H}_P(\pi_\lambda)$ to $\mathcal{H}_P(w\pi_\lambda)$, and a corresponding normalizing factor

$$r_P(w, \psi_\lambda) = \lambda(w) \varepsilon_P(w, \psi_\lambda)^{-1} r_{w^{-1}P|P}(\psi_\lambda)$$

that reduces to the product of $\lambda(w)$ with the more familiar quotient

$$L(0, \pi_{\psi,\lambda}, \rho^\vee_{w^{-1}P|P}) \varepsilon(0, \pi_{\psi,\lambda}, \rho^\vee_{w^{-1}P|P}, \psi_F)^{-1} L(1, \pi_{\psi,\lambda}, \rho^\vee_{w^{-1}P|P})^{-1}. \tag{2.3.27}$$

We are using the ψ-variant of (2.3.5) here, while continuing to suppress the implicit dependence on the additive character ψ_F from the notation. It follows from Proposition 2.3.1 and Lemma 2.3.4 that the normalized intertwining operator

$$R_P(w, \pi, \psi) : \mathcal{H}_P(\pi) \longrightarrow \mathcal{H}_P(w\pi)$$

satisfies the usual co-cycle relation

$$R_P(w'w, \pi, \psi) = R_P(w', w\pi, w\psi) R_P(w, \pi, \psi), \qquad w', w \in W(M). \tag{2.3.28}$$

This is essentially what has been conjectured in general by Langlands [**L5**, Appendix II].

2.4. Statement of the local intertwining relation

The purpose of this section is to state a theorem that relates local intertwining operators with endoscopic transfer. The relation will have global implications for our later analysis of trace formulas. It will also be closely tied to Theorem 2.2.1, both as a supplement and as a part of the proof.

We are assuming that G belongs to $\widetilde{\mathcal{E}}_{\mathrm{sim}}(N)$, and that M is a proper Levi subgroup of G. There is still some unfinished business from the last section. It is to convert the intertwining operator $R_P(w, \pi, \psi)$ from $\mathcal{I}_P(\pi)$ to $\mathcal{I}_P(w\pi)$ to a self-intertwining operator for $\mathcal{I}_P(\pi)$. This is the last step in the normalization process, and the one that is most closely tied to endoscopy.

Recall that ψ lies in $\widetilde{\Psi}(M)$, and that π belongs to the packet $\widetilde{\Pi}_\psi$. We will usually take w to be in the subgroup

$$W_\psi(\pi) = W(\pi) \cap W_\psi(M) = \{w \in W(M) : \ w\pi \cong \pi, \ w\psi \cong \psi\} \tag{2.4.1}$$

of elements w in $W(M) \cong W(\widehat{M})$ that stabilize the equivalence classes of both π and ψ. We will then be able to introduce an intertwining operator $\widetilde{\pi}(w)$ from $w\pi$ to π.

Suppose that w lies in $W_\psi(\pi)$, and that $\widetilde{\pi}(w)$ is an intertwining operator from $w\pi$ to π. We can think of $\widetilde{\pi}$ as an extension of π to the $M(F)$-bitorsor

$$\widetilde{M}_w(F) = M(F) \rtimes \widetilde{w},$$

in the sense of (2.1.1), or even an extension of π to the group $\widetilde{M}_w^+(F)$ generated by $\widetilde{M}_w(F)$, if $\widetilde{\pi}(w)$ is chosen so that its order equals that of w. The product

$$R_P(w, \widetilde{\pi}, \psi) = \widetilde{\pi}(w) \circ R_P(w, \pi, \psi) \tag{2.4.2}$$

is then a self-intertwining operator of $\mathcal{I}_P(\pi)$, which of course depends on the choice of $\widetilde{\pi}(w)$. Suppose that the operators $\widetilde{\pi}(w)$ on V_π could be chosen to be multiplicative in w. It would then follow from (2.3.28) and the definition of $R_P(w, \pi, \psi)$ that the mapping

$$w \longrightarrow R_P(w, \widetilde{\pi}, \psi), \qquad w \in W_\psi(\pi),$$

is a homomorphism from $W_\psi(\pi)$ to the space of intertwining operators of $\mathcal{I}_P(\pi)$. However, there is no canonical choice for $\widetilde{\pi}(w)$ in general. While this does not necessarily preclude an ad hoc construction, it represents a problem that left unattended would soon lead to trouble. The solution is provided by an application of Theorem 2.2.4 to M.

We first recall the various finite groups that can be attached to the parameter ψ. We are regarding ψ as an element in $\widetilde{\Psi}(M)$. However, its composition with the standard embedding of ${}^L M$ into ${}^L G$ is an element in $\widetilde{\Psi}(G)$, which we shall also denote by ψ. From these two objects, we obtain the two complex reductive groups

$$S_\psi(M) \subset S_\psi(G) = S_\psi,$$

and their two finite quotients

$$\mathcal{S}_\psi(M) \subset \mathcal{S}_\psi(G) = \mathcal{S}_\psi.$$

Recall that $\mathcal{S}_\psi$ is the group of connected components in the quotient $\overline{S}_\psi = S_\psi / Z(\widehat{G})^\Gamma$. It is clear that $S_\psi(M)$ is the subgroup of elements in S_ψ that leave the complex torus

$$A_{\widehat{M}} = \big(Z(\widehat{M})^{\Gamma_F}\big)^0$$

pointwise fixed, and it is easily seen that $\mathcal{S}_\psi(M)$ is the image of $S_\psi(M)$ in $\mathcal{S}_\psi$.

There are other interesting finite groups as well. We write $N_\psi(G,M)$ for the normalizer of the complex torus $A_{\widehat{M}}$ in S_ψ and $\mathfrak{N}_\psi(G,M)$ for the group of components in the quotient

$$\overline{N}_\psi(G,M) = N_\psi(G,M)/Z(\widehat{G})^\Gamma.$$

Then $\mathcal{S}_\psi(M)$ is a normal subgroup of $\mathfrak{N}_\psi(G,M)$. Its quotient is isomorphic to the Weyl group

$$W_\psi(G,M) = W(S_\psi, A_{\widehat{M}}),$$

which is to say, the group of automorphisms of $A_{\widehat{M}}$ induced from $\overline{S}_\psi$. We write $W^0_\psi(G,M)$ for the normal subgroup of automorphisms in $W_\psi(G,M)$ that are induced from the connected component $\overline{S}^0_\psi$, and

$$R_\psi(G,M) = W_\psi(G,M)/W^0_\psi(G,M)$$

for its quotient. On the other hand, there is a canonical injection from $W^0_\psi(G,M)$ into $\mathfrak{N}_\psi(G,M)$. This map is well defined, since the centralizer of $A_{\widehat{M}}$ in S^0_ψ is connected, and is therefore contained in $S^0_\psi(M)$. In other words, $W^0_\psi(G,M)$ can also be regarded as a normal subgroup of $\mathfrak{N}_\psi(G,M)$. The quotient $\mathcal{S}_\psi(G,M)$ of $\mathfrak{N}_\psi(G,M)$ by $W^0_\psi(G,M)$ is a subgroup of $\mathcal{S}_\psi$. We can summarize the relations among these finite groups in a commutative diagram

(2.4.3)

$$\begin{array}{ccccccccc}
 & & & & 1 & & 1 & & \\
 & & & & \downarrow & & \downarrow & & \\
 & & & & W^0_\psi(G,M) & = & W^0_\psi(G,M) & & \\
 & & & & \downarrow & & \downarrow & & \\
1 & \longrightarrow & \mathcal{S}_\psi(M) & \longrightarrow & \mathfrak{N}_\psi(G,M) & \longrightarrow & W_\psi(G,M) & \longrightarrow & 1 \\
 & & \Vert & & \downarrow\uparrow & & \downarrow\uparrow & & \\
1 & \longrightarrow & \mathcal{S}_\psi(M) & \longrightarrow & \mathcal{S}_\psi(G,M) & \longrightarrow & R_\psi(G,M) & \longrightarrow & 1 \\
 & & & & \downarrow & & \downarrow & & \\
 & & & & 1 & & 1 & &
\end{array}$$

of short exact sequences. The dotted arrows stand for splittings determined by the parabolic subgroup

$$\overline{P}_\psi = (\widehat{P} \cap S^0_\psi)/Z(\widehat{G})^{\Gamma_F}$$

of $\overline{S}^0_\psi$, or rather the chamber of $\overline{P}_\psi$ in the Lie algebra of $A_{\widehat{M}}$.

Suppose that u belongs to $\mathfrak{N}_\psi(G,M)$. We write w_u for the image of u in $W_\psi(G,M)$ given by the horizontal short exact sequence at the center of (2.4.3). Since w_u stabilizes $A_{\widehat{M}}$, it normalizes $\widehat{M}$, and can therefore be treated as an element in $W(\widehat{M})$. The choice of P then allows us to identify

w_u with an element in $W(M)$. Recall that M is a product of groups (2.3.4), all of which all have standard splittings. We set

$$\widetilde{M}_u = \widetilde{M}_{w_u} = (M, \widetilde{w}_u),$$

where we recall that $\widetilde{w}_u$ is the automorphism in the inner class of w_u that preserves the corresponding splitting of M. Then $\widetilde{M}_u$ is a twisted group, and ψ belongs to the corresponding subset $\Psi(\widetilde{M}_u)$ of $\widetilde{\Psi}(M)$. As earlier, the centralizer quotient

$$\widetilde{\mathcal{S}}_{\psi,u} = \mathcal{S}_\psi(\widetilde{M}_u)$$

is an $\mathcal{S}_\psi(M)$-bitorsor. We shall write $\widetilde{u}$ for the element in this set defined by u.

Assume that π represents a $\widetilde{w}_u$-stable element in the packet $\widetilde{\Pi}_\psi = \widetilde{\Pi}_\psi(M)$. We can then choose an intertwining operator $\widetilde{\pi}(w_u)$ from $w_u\pi$ to π, from which we obtain the intertwining operator $R_P(w_u, \widetilde{\pi}, \psi)$ of $\mathcal{I}_P(\pi)$ as in (2.4.2). As we have said, there is generally no canonical choice for $\widetilde{\pi}(w_u)$. However, M is a product (2.3.4) of groups, which include a factor G_- to which either Theorem 2.2.1 or Theorem 2.2.4 can be applied. In particular, the products

$$\langle \widetilde{u}, \widetilde{\pi} \rangle\, \widetilde{\pi}(w_u)$$

and

$$\langle \widetilde{u}, \widetilde{\pi} \rangle\, R_P(w_u, \widetilde{\pi}, \psi) \tag{2.4.4}$$

can be defined, and are independent of the choice of $\widetilde{\pi}(w_u)$. (The automorphism $\widetilde{w}_u$ of M can also include a permutation of general linear factors in (2.3.4), but the extension of Theorem 2.2.1 to this slightly more general situation is clear. We have taken the liberty of expressing the assertion (2.2.6) of this theorem in the same form as its counterpart (2.2.17) in Theorem 2.2.4, with the extension $\langle \cdot, \cdot \rangle$ attached to the Whittaker extension of the general linear components of π being trivial.) We therefore obtain a canonical linear form

$$f_G(\psi, u) = \sum_{\pi \in \widetilde{\Pi}_\psi} \langle \widetilde{u}, \widetilde{\pi} \rangle \operatorname{tr}\big(R_P(w_u, \widetilde{\pi}, \psi)\mathcal{I}_P(\pi, f)\big) \tag{2.4.5}$$

in $f \in \mathcal{H}(G)$. The convention here is similar to that of Theorem 2.2.4, in that the summand on the right is zero if π happens not to be $\widetilde{w}_u$-stable.

We will later have to consider the role of u in the vertical split exact sequence at the center of (2.4.3). We will then write x_u and w_u^0 for the images of u in the respective groups $\mathcal{S}_\psi(G, M)$ and $W_\psi^0(G, M)$.

Suppose now that s is any semisimple element in the complex reductive group S_ψ. We write

$$f'_G(\psi, s) = f'(\psi') \tag{2.4.6}$$

for the linear form in $f \in \widetilde{\mathcal{H}}(G)$ attached to the preimage (G', ψ') of (ψ, s) under the local form of the correspondence (1.4.11). Any such s projects to an element in the quotient $\mathcal{S}_\psi = \mathcal{S}_\psi(G)$, which may or may not lie in the subgroup $\mathcal{S}_\psi(G, M)$.

THEOREM 2.4.1 (Local intertwining relation for G). *For any u in the local normalizer $\mathfrak{N}_\psi(G, M)$, the identity*

$$f'_G(\psi, s_\psi s) = f_G(\psi, u), \qquad f \in \widetilde{\mathcal{H}}(G), \tag{2.4.7}$$

holds for any semisimple element $s \in S_\psi$ that projects onto the image x_u of u in $\mathcal{S}_\psi(G, M)$.

The most important case is when ψ lies in the subset $\widetilde{\Psi}_2(M)$ of $\widetilde{\Psi}(M)$. Then the group

$$T_\psi = A_{\widehat{M}}$$

is a maximal torus in S_ψ, and $\mathcal{S}_\psi(G, M)$ equals the full group $\mathcal{S}_\psi$. In this case, we write

$$\mathfrak{N}_\psi = \mathfrak{N}_\psi(G, M),$$
$$W_\psi = W_\psi(G, M),$$
$$W^0_\psi = W^0_\psi(G, M),$$

and

$$R_\psi = R_\psi(G, M).$$

The last group R_ψ can be regarded as the R-group of the parameter ψ, since it is closely related to the reducibility of the induced representations

$$\mathcal{I}_P(\pi), \qquad P \in \mathcal{P}(M),\ \pi \in \widetilde{\Pi}_\psi(M).$$

We shall often treat ψ strictly as an element in $\widetilde{\Psi}(G)$ in this case, and we shall write

$$\mathcal{S}^1_\psi = \mathcal{S}_\psi(M),$$

since M is uniquely determined up to the action of the group $\widetilde{\mathrm{Aut}}_N(G)$.

LEMMA 2.4.2. *Suppose that for any M, Theorem 2.4.1 is valid whenever ψ belongs to $\widetilde{\Psi}_2(M)$. Then it holds for any ψ in the general set $\widetilde{\Psi}(M)$.*

The proof of the lemma will be straightforward, but it does depend on some further observations. We shall postpone it until the end of this section. In the meantime, we shall discuss an important implication of the local intertwining relation postulated by the theorem.

We first observe that the groups in (2.4.3) can be described in concrete terms. For any $\psi \in \widetilde{\Psi}(G)$, we write

$$\psi = \ell_1\psi_1 \oplus \cdots \oplus \ell_r\psi_r,$$

following the local form of the notation (1.4.1). As in the global case of §1.4, the subscripts $k \in K_\psi$ range over disjoint indexing sets $I_\psi = I^+_\psi(G) \coprod I^-_\psi(G)$, J_ψ and $J^\vee_\psi$. We recall that if $i \in I_\psi$, $\psi_i \in \widetilde{\Psi}_{\mathrm{sim}}(G_i)$ is a simple parameter for an endoscopic datum $G_i \in \widetilde{\mathcal{E}}_{\text{sim}}(N_i)$, while if $j \in J_\psi$, ψ_j belongs to the complement of $\widetilde{\Psi}_{\text{sim}}(N_i)$ in $\Psi_{\text{sim}}(N_i)$. The subset $I_\psi^+(G)$ consists of those i such that $\widehat{G}$ and $\widehat{G}_i$ are of the same type (both orthogonal or both symplectic), while $I_\psi^-(G)$ consists of those i such that $\widehat{G}$ and $\widehat{G}_i$ are of opposite type. We then have

$$S_\psi = \Big(\prod_{i\in I_\psi^+(G)} O(\ell_i,\mathbb{C})\Big)^+ \times \Big(\prod_{i\in I_\psi^-(G)} Sp(\ell_i,\mathbb{C})\Big) \times \Big(\prod_{j\in J_\psi} GL(\ell_j,\mathbb{C})\Big) \tag{2.4.8}$$

as in (1.4.8), where $(\cdot)^+ = (\cdot)_\psi^+$ denotes the kernel of the character

$$\xi_\psi^+ : \prod_i g_i \longrightarrow \prod_i (\det g_i)^{N_i}, \qquad g_i \in O(\ell_i,\mathbb{C}),\ i \in I_\psi^+(G).$$

The identity component of this group obviously equals

$$S_\psi^0 = \Big(\prod_{i\in I_\psi^+(G)} SO(\ell_i,\mathbb{C})\Big) \times \Big(\prod_{i\in I_\psi^-(G)} Sp(\ell_i,\mathbb{C})\Big) \times \Big(\prod_{j\in J_\psi} GL(\ell_j,\mathbb{C})\Big).$$

Finally $\overline{S}_\psi$ is the quotient of S_ψ by a central subgroup, of order 2 if N is even and order 1 if N is odd, which may or may not lie in S_ψ^0.

Consider for simplicity the basic case that ψ is the image of a parameter in $\widetilde{\Psi}_2(M)$, for the given Levi subgroup M. We can then identify the complex torus $T_\psi = A_{\widehat{M}}$ with the product of standard maximal tori in the simple factors of S_ψ^0. It is clear that $\pi_0(S_\psi)$ is the group of connected components in the left hand factor $(\,\cdot\,)^+$ of S_ψ, and that its image in $\pi_0(\overline{S}_\psi)$ is the group $\mathcal{S}_\psi$. The subgroup $\mathcal{S}_\psi^1$ is given by the contribution of the subfactors $O(\ell_i,\mathbb{C})$ with ℓ_i odd, while the quotient R_ψ is given by the subfactors $O(\ell_i,\mathbb{C})$ with ℓ_i even. The group $\mathfrak{N}_\psi$ is an extension of $\mathcal{S}_\psi$ by the product W_ψ^0 of standard Weyl groups of the simple factors of S_ψ^0.

It is worth looking at one example in greater detail. Consider the lower horizontal short exact sequence

$$1 \longrightarrow \mathcal{S}_\psi^1 \longrightarrow \mathcal{S}_\psi \longrightarrow R_\psi \longrightarrow 1 \tag{2.4.9}$$

from (2.4.3), in case N is even and $\widehat{G}$ is orthogonal. We write

$$I^+ = I_\psi^+(G)$$

for simplicity, and identify the group of connected components in the product

$$\prod_{i\in I^+} O(\ell_i,\mathbb{C})$$

from (2.4.8) with the group Σ of functions

$$\sigma : I^+ \longrightarrow \mathbb{Z}/2\mathbb{Z}.$$

The point is to account for the indices $i \in I^+$ in terms of the parity of both associated integers ℓ_i and N_i. For example, the group $\pi_0(S_\psi)$ corresponds to the subgroup

$$\Sigma' = \Big\{\sigma \in \Sigma : \prod_i{}' \sigma(i) = 1\Big\},$$

where the product is over the set I_o^+ of indices i with N_i odd. The group $\mathcal{S}_\psi = \pi_0(\overline{S}_\psi)$ corresponds to the quotient

$$\overline{\Sigma}' = \Sigma' / \langle\overline{\sigma}\rangle,$$

where $\overline{\sigma}$ is the element in Σ' (of order 1 or 2) such that $\overline{\sigma}(i) = -1$ whenever i belongs to the set $I^{+,o}$ of indices with ℓ_i odd. (The set $I^{+,o} \cap I_o^+$ of i with *both* ℓ_i and N_i odd is even, so $\overline{\sigma}$ does lie in Σ'.) The group $\mathcal{S}_\psi^1$ corresponds to the subgroup $(\overline{\Sigma}')^1$ of $\sigma \in \overline{\Sigma}'$ such that $\sigma(i) = 1$ whenever i belongs to the subset $I^{+,e}$ of indices with ℓ_i even. The R-group R_ψ then corresponds to the associated quotient of $\overline{\Sigma}'$, which is canonically isomorphic to the group R' of functions

$$\rho : I^{+,e} \longrightarrow \mathbb{Z}/2\mathbb{Z},$$

with the supplementary requirement that

$$\prod_i{}' \rho(i) = 1,$$

in the exceptional case that $I^{+,o} \cap I_o^+$ is empty. The short exact sequence (2.4.9) can thus be identified with the more concrete exact sequence

$$1 \longrightarrow (\overline{\Sigma}')^1 \longrightarrow \overline{\Sigma}' \longrightarrow R' \longrightarrow 1. \tag{2.4.10}$$

The other two cases (with N odd or $\widehat{G}$ symplectic) are similar, but simpler. In all three cases, the short exact sequence (2.4.9) (and its analogue (2.4.10)) splits. However, if N is even and $\widehat{G}$ is orthogonal there is generally no canonical splitting. The sign character

$$\varepsilon(\rho) = \prod_i{}'' \rho(i), \qquad \rho \in R', \tag{2.4.11}$$

where the product is over the intersection $I^{+,e} \cap I_o^+$, is a distinguishing feature of this case. It represents the obstruction to a canonical splitting of (2.4.10). Observe also that as an element in R_ψ, ρ can act by outer automorphism on the even orthogonal factor $G_- \in \widetilde{\mathcal{E}}_{\mathrm{ell}}(N_-)$, as well as the general linear factors. It is not hard to infer from our discussion that the action on G_- is nontrivial if and only if $\varepsilon(\rho) = -1$. In this case, the subparameter ψ_- of ψ for G_- lies in the subset of the $\Psi(\widetilde{G}_-)$ of $\widetilde{\Psi}(G_-)$. The character (2.4.11) thus governs the ambiguity of both factors in the product (2.4.4). In fact, it is fair to say that this character is what ultimately forces us to deal with twisted endoscopy for even orthogonal groups, as stated in Theorem 2.2.4, if only to resolve the ambiguity in the definition (2.4.2).

We now continue with our discussion of the local intertwining relation. The last lemma tells us that its general proof can be reduced to the case that $\psi \in \widetilde{\Psi}_2(M)$. In dealing with this basic case, we shall make a slight change of notation. Following an earlier suggestion, we will generally take ψ to be simply a parameter in the set $\widetilde{\Psi}(G)$. We can then choose a pair (M, ψ_M), where ψ_M lies in the set $\widetilde{\Psi}_2(M, \psi)$ of parameters in $\widetilde{\Psi}_2(M)$ whose image equals ψ. As we have noted, the Levi subgroup M is uniquely determined as an $\widetilde{\mathrm{Out}}_N(G)$-orbit of G-conjugacy classes by ψ.

We are assuming that M is proper in G, which is to say that ψ does not belong to $\widetilde{\Psi}_2(G)$. We can therefore assume as usual that we have constructed the packet $\widetilde{\Pi}_{\psi_M}$ of ($\widetilde{\mathrm{Out}}_{N_-}(M)$-orbits of) representations π_M of $M(F)$. We shall combine this with the local intertwining relation to construct the packet $\widetilde{\Pi}_\psi$ of ($\widetilde{\mathrm{Out}}_N(G)$-orbits) of representations π of $G(F)$. The construction provides an important reduction of the local classification. It also gives us an independent way to view the packet $\widetilde{\Pi}_\psi$, in the more concrete terms of induced representations and intertwining operators. Its proof amounts to a reordering of some of the discussion of this section. We shall try to present it so as to be suggestive of more general situations.

PROPOSITION 2.4.3. *Assume that for any proper Levi subgroup M of G, Theorem* 2.4.1 *holds for parameters in $\widetilde{\Psi}_2(M)$. Then if ψ is any parameter in the complement of $\widetilde{\Psi}_2(G)$ in $\widetilde{\Psi}(G)$, the packet $\widetilde{\Pi}_\psi$ and pairing $\langle x, \pi\rangle$ of Theorem* 2.2.1 *exist, and satisfy* (2.2.6)*.*

PROOF. The packet $\widetilde{\Pi}_{\psi_M}$ provides a representation

$$\Pi_{\psi_M} = \bigoplus_{(\xi_M, \pi_M)} (\xi_M \otimes \pi_M) \qquad (2.4.12)$$

of the group $\mathcal{S}^1_\psi \times M(F)$. We let π_M range here over a suitable set of representatives of the $\widetilde{\mathrm{Out}}_{N_-}(M)$-orbits that comprise the packet $\widetilde{\Pi}_{\psi_M}$ and ξ_M range over the associated (1-dimensional) representations

$$\xi_M(x_M) = \langle x_M, \pi_M\rangle$$

of $\mathcal{S}^1_\psi$. We are of course free to identify Π_{ψ_M} with a representation of the group algebra $C(\mathcal{S}^1_\psi) \otimes \mathcal{H}(M)$. Its restriction $\widetilde{\Pi}_{\psi_M}$ to the subalgebra $C(\mathcal{S}^1_\psi) \otimes \widetilde{\mathcal{H}}(M)$ is then independent of the choice of representatives. The overlapping notation is deliberate.

The group $\mathfrak{N}_\psi$ acts on $M(F)$ by

$$u: \; m \longrightarrow \widetilde{w}_u m \widetilde{w}_u^{-1}, \qquad m \in M(F),\ u \in \mathfrak{N}_\psi,$$

where the image w_u of u in W_ψ is regarded as an element in $W(M)$. The associated semidirect product

$$M(F) \rtimes \mathfrak{N}_\psi$$

is then an extension of the group W_ψ by the direct product

$$M(F) \times \mathcal{S}^1_\psi \cong \mathcal{S}^1_\psi \times M(F).$$

We claim that there is a canonical extension of the representation of Π_{ψ_M} to $M(F) \rtimes \mathfrak{N}_\psi$.

To see this, we have first to expand slightly on the discussion of the short exact sequence (2.4.10) above. The action of $\mathfrak{N}_\psi$ on $M(F)$ obviously stabilizes the orthogonal or symplectic factor G_- of M. Let $\mathfrak{N}_{\psi,-}$ be the subgroup of elements that act on G_- by inner automorphism. If $\widehat{G}_-$ is symplectic or N is odd, for example, $\mathfrak{N}_{\psi,-}$ equals $\mathfrak{N}_\psi$. In the other case that $\widehat{G}_-$ is orthogonal and N is even, $\mathfrak{N}_{\psi,-}$ is the kernel of the pullback to $\mathfrak{N}_\psi$ of the sign character (2.4.11) on R_ψ. This implies that $\mathfrak{N}_{\psi,-}$ is again equal to $\mathfrak{N}_\psi$, unless the factor ψ_- of ψ_M lies in the subset of $\Psi_2(\widetilde{G}_-)$ of $\widetilde{\Psi}_2(G_-)$. In all cases, there is a canonical splitting

$$\mathfrak{N}_{\psi,-} = \mathcal{S}_\psi^1 \oplus W_{\psi,-},$$

where $W_{\psi,-}$ is the image of $\mathfrak{N}_{\psi,-}$ in W_ψ. It follows from the definitions that any constituent π_M of (2.4.12) has a canonical extension to $M(F) \rtimes W_{\psi,-}$. The representation Π_{ψ_M} therefore has a canonical extension to the subgroup $M(F) \rtimes \mathfrak{N}_{\psi,-}$ (of index 1 or 2) of $M(F) \rtimes \mathfrak{N}_\psi$.

The claim is that Π_{ψ_M} can be extended canonically to the larger group $M(F) \rtimes \mathfrak{N}_\psi$. Suppose that π_M is W_ψ-stable. Applying Theorems 2.2.4 and 2.2.1 separately to the components of π_M relative to the decomposition (2.3.4), as in the definition (2.4.4), we see that $\xi_M \otimes \pi_M$ has a canonical extension to $M(F) \rtimes \mathfrak{N}_\psi$. We expect this condition to hold for all π_M, but we cannot rule out the possibility that some representation π_M in (2.4.12) not be $\widetilde{\mathrm{O}}\mathrm{ut}_{N_-}(G_-)$-stable. Then $\mathfrak{N}_{\psi,-}$ is proper in $\mathfrak{N}_\psi$, and ψ_- lies in $\Psi_2(\widetilde{G}_-)$. In this case, we appeal to Remark 4, following the statement of Theorem 2.2.4, to choose the corresponding terms on the right hand side of (2.4.12). That is, we choose the contribution to (2.4.12) of the $\widetilde{\mathrm{O}}\mathrm{ut}_{N_-}(G_-)$-orbit of π_M (a set of order 2 that can also be regarded as the $\mathfrak{N}_\psi$-orbit or the W_ψ-orbit of π_M) according to the analogue for π_M of the representation (2.2.18). The group $\mathfrak{N}_\psi$ (or W_ψ) then acts by permutation on the associated summands of (2.4.12). We conclude that the full sum (2.4.12) does have a canonical extension to $M(F) \rtimes \mathfrak{N}_\psi$, as claimed. We shall denote it by $\widetilde{\Pi}_{\psi_M}$, despite the further ambiguity in the notation.

Let

$$\Pi_\psi = \mathcal{I}_P(\Pi_{\psi_M}) = \bigoplus_{(\xi_M,\pi_M)} \big(\xi_M \otimes \mathcal{I}_P(\pi_M)\big) \tag{2.4.13}$$

be the representation of $\mathcal{S}_\psi^1 \times G(F)$ induced parabolically from the representation Π_{ψ_M} of $\mathcal{S}_\psi^1 \times M(F)$. Our canonical extension of Π_{ψ_M} to $M(F) \rtimes \mathfrak{N}_\psi$ can then be combined with the intertwining operators $R_P(w, \pi_M, \psi_M)$ from the last section. A modest generalization of (2.4.2), in which the irreducible representation π of $M(F)$ is replaced by the (reducible) restriction of Π_{ψ_M} to $M(F)$, and the extension $\widetilde{\pi}$ of π to $\widetilde{M}_{w_u}(F)$ is replaced by our canonical extension $\widetilde{\Pi}_{\psi_M}$ of Π_{ψ_M} to $M(F) \rtimes \mathfrak{N}_\psi$, allows us to attach an intertwining operator

$$R_P(u, \widetilde{\Pi}_{\psi_M}, \psi_M) = \widetilde{\Pi}_{\psi_M}(u) \circ R_P(w_u, \Pi_{\psi_M}, \psi_M) \tag{2.4.14}$$

to any $u \in \mathfrak{N}_\psi$. This operator then commutes with the restriction of Π_ψ to $G(F)$. The product

$$\Pi_\psi(u, g) = R_P(u, \widetilde{\Pi}_{\psi_M}, \psi_M)\Pi_\psi(g), \qquad u \in \mathfrak{N}_\psi,\ g \in G(F), \tag{2.4.15}$$

therefore gives a canonical extension of the representation Π_ψ to the product $\mathfrak{N}_\psi \times G(F)$.

We consider the character

$$\operatorname{tr}\big(\widetilde{\Pi}_\psi(u, f)\big), \qquad u \in \mathfrak{N}_\psi,\ f \in \widetilde{\mathcal{H}}(G),$$

of the restriction $\widetilde{\Pi}_\psi$ of Π_ψ to the product of $\mathfrak{N}_\psi$ with the symmetric Hecke algebra $\widetilde{\mathcal{H}}(G)$. Suppose that $\mathfrak{o}$ is an $\mathfrak{N}_\psi$-orbit of representations π_M. If the orbit contains one element, the restriction of the operator $\widetilde{\Pi}_{\psi_M}(u)$ in (2.4.14) to the subspace of $(\xi_M \otimes \pi_M)$ in (2.4.12) equals

$$\widetilde{\xi}_M(u) \otimes \widetilde{\pi}_M(w_u),$$

where $\widetilde{\xi}_M(u) = \langle \widetilde{u}, \widetilde{\pi}_M \rangle$. It follows from (2.4.2) that the restriction of the intertwining operator (2.4.14) to the corresponding subspace in (2.4.13) equals

$$\widetilde{\xi}_M(u) R_P(w_u, \widetilde{\pi}_M, \psi_M).$$

The contribution of this subspace to the character is therefore

$$\langle \widetilde{u}, \widetilde{\pi}_M \rangle \operatorname{tr}\big(R_P(w_u, \widetilde{\pi}_M, \psi_M)\mathcal{I}_P(\pi_M, f)\big).$$

On the other hand, if $\mathfrak{o}$ contains two elements, the operator $\Pi_{\psi_M}(u)$ interchanges the subspaces of the two representations $(\xi_M \otimes \pi_M)$. The restriction of (2.4.14) therefore interchanges the corresponding two subspaces of (2.4.13), so the contribution of these subspaces to the character vanishes. It follows from the definition (2.4.5) (with our notation now dictating a sum over $\pi_M \in \widetilde{\Pi}_{\psi_M}$ in place of the sum over $\pi \in \widetilde{\Pi}_\psi$ in (2.4.5)) that

$$\operatorname{tr}\big(\widetilde{\Pi}_\psi(u, f)\big) = f_G(\psi, u), \qquad u \in \mathfrak{N}_\psi,\ f \in \widetilde{\mathcal{H}}(G). \tag{2.4.16}$$

We will be concerned with the restriction of Π_ψ to the subgroup $\mathcal{S}_\psi \times G(F)$ of $\mathfrak{N}_\psi \times G(F)$, relative to the splitting of the left hand vertical exact sequence in (2.4.3). The local intertwining relation suggests that the representation is trivial on the complementary subgroup W^0_ψ. We will indeed establish this fact by global means later, so there will be no loss of information in the restriction. As a representation of $\mathcal{S}_\psi \times G(F)$, Π_ψ has a decomposition

$$\Pi_\psi = \bigoplus_{(\xi,\pi)} (\xi \otimes \pi),$$

for irreducible representations ξ of $\mathcal{S}_\psi$ and π of $G(F)$. The sum is of course finite, but the constituents can have multiplicities. We define the packet $\widetilde{\Pi}_\psi$ to be the disjoint union of the set of π, each identified with its associated

$\widetilde{\mathrm{Out}}_N(G)$-orbit, that occur. Any $\pi \in \widetilde{\Pi}_\psi$ then comes with a corresponding character

$$\xi(x) = \langle x, \pi \rangle, \qquad x \in \mathcal{S}_\psi,$$

of the abelian group $\mathcal{S}_\psi$. It follows that

$$\operatorname{tr}\big(\widetilde{\Pi}_\psi(x, f)\big) = \sum_{\pi \in \widetilde{\Pi}_\psi} \langle x, \pi \rangle f_G(\pi), \qquad x \in \mathcal{S}_\psi,\ f \in \widetilde{\mathcal{H}}(G). \tag{2.4.17}$$

We have constructed the packet $\widetilde{\Pi}_\psi$ and pairing $\langle x, \pi \rangle$. To show that they satisfy (2.2.6), we apply our assumption that Theorem 2.4.1 is valid. Combined with (2.4.16) and (2.4.17), it tells us that

$$\begin{aligned} f'_G(\psi, s_\psi s) &= f_G(\psi, u) \\ &= \operatorname{tr}\big(\widetilde{\Pi}_\psi(u, f)\big) = \operatorname{tr}\big(\widetilde{\Pi}_\psi(x, f)\big) \\ &= \sum_{\pi \in \widetilde{\Pi}_\psi} \langle x, \pi \rangle\, f_G(\pi), \end{aligned}$$

for elements s and u that project to the same point x in $\mathcal{S}_\psi$. A similar identity holds for $f'_G(\psi, s)$, provided that we replace x by the point $s_\psi^{-1} x = s_\psi x$. Since

$$f'_G(\psi, s) = f'(\psi'),$$

we obtain

$$f'(\psi') = \sum_{\pi \in \widetilde{\Pi}_\psi} \langle s_\psi x, \pi \rangle\, f_G(\pi), \qquad f \in \widetilde{\mathcal{H}}(G).$$

This is the required identity (2.2.6). □

REMARKS. 1. It is only at the end of the proof that we appeal to our hypothesis that Theorem 2.4.1 holds for G. In particular, the actual construction (2.4.17) of the packet and pairing of Theorem 2.2.1 depends only on the application of Theorems 2.2.1 and 2.2.4 to the smaller group G_-.

2. The structure of the proof appears to be quite general. We shall add several comments on what might be expected if G is a general connected group.

(i) There are certainly examples of quasisplit groups G for which the finite groups $\mathcal{S}_\psi^1$, $\mathcal{S}_\psi$ and R_ψ are nonabelian. This will make the associated representations Π_{ψ_M} and Π_ψ more interesting. (In general, Π_ψ would continue to denote both a packet and a representation.) It seems likely that there will again be a canonical extension of the representation Π_{ψ_M} of $\mathcal{S}_\psi^1 \times M(F)$ to the semidirect product $M(F) \rtimes \mathfrak{N}_\psi$. However, there will also be new phenomena.

(ii) The stabilizer $W_\psi(\pi_M)$ of a representation $\pi_M \in \Pi_{\psi_M}$ could be proper in W_ψ. However, we would expect, at least if $\psi = \phi$ is generic (and in contrast to the hypothetical possibility treated in the proof of the proposition) that $W_\psi(\pi_M)$ is also equal to the stabilizer of the corresponding irreducible representation ξ_M of $\mathcal{S}_\psi^1$ attached to π_M. It is conceivable that the representation π_M will not have an extension to $M(F) \rtimes W_\psi(\pi_M)$. The obstruction would presumably reduce to a complex valued 2-cocycle on the image $R_\psi(\pi_M)$ of $W_\psi(\pi_M)$ in R_ψ, of the kind considered in [**A10**, §3]. However, we could expect the same obstruction (or rather its inverse) to account for the failure of ξ_M to have an extension to the preimage $\mathfrak{N}_\psi(\pi_M)$ of $W_\psi(\pi_M)$ in $\mathfrak{N}_\psi$. In fact, there ought to be a canonical extension of the representation $\xi_M \otimes \pi_M$ of $\mathcal{S}_\psi^1 \times M(F)$ to the semidirect product $M(F) \rtimes \mathfrak{N}_\psi(\pi_M)$. One could then induce this extended representation to the full group $M(F) \rtimes \mathfrak{N}_\psi$. The direct sum of the representations so obtained, taken over the W_ψ-orbits $\{\pi_M\}$ in the packet Π_{ψ_M}, would be the canonical extension of the representation Π_{ψ_M} from $\mathcal{S}_\psi^1 \times M(F)$ to $M(F) \rtimes \mathfrak{N}_\psi$. With this object in hand, one would then be able to construct the representation Π_ψ of $\mathcal{S}_\psi \times G(F)$ from induced representations $\mathcal{I}_P(\pi_M)$ and intertwining operators $R_P(w, \widetilde{\pi}_M, \psi)$, as in the proof of the proposition.

(iii) If G is not quasisplit, the situation is more complicated. In this case, one would need to replace the groups $\mathcal{S}_\psi^1$ and $\mathcal{S}_\psi$ by extensions $\mathcal{S}_{\psi,\mathrm{sc}}^1$ and $\mathcal{S}_{\psi,\mathrm{sc}}$, defined as in [**A19**, §3]. With this proviso, the arguments for quasisplit groups would presumably carry over. We shall discuss inner twists of orthogonal and symplectic groups in Chapter 9.

We return to our quasisplit classical group $G \in \widetilde{\mathcal{E}}_{\mathrm{sim}}(N)$. We have still to establish the reduction asserted in Lemma 2.4.2. However, we shall first state a variant of the local intertwining relation, which applies to a twisted orthogonal group $\widetilde{G}$.

Recall that $\widetilde{G} = (\widetilde{G}^0, \widetilde{\theta})$, where $G = \widetilde{G}^0$ is an even orthogonal group in $\widetilde{\mathcal{E}}_{\mathrm{sim}}(N)$. The statement will be almost the same as that of Theorem 2.4.1. In particular, M is a proper Levi subgroup of G, and $P \in \mathcal{P}(M)$ is a parabolic subgroup of G. (We do not assume that P is normalized by any element in $\widetilde{G}$.) If π lies in the packet $\widetilde{\Pi}_\psi$, $\mathcal{I}_P(\pi)$ denotes the corresponding representation induced from $P(F)$ to the group $\widetilde{G}^+(F)$ generated by $\widetilde{G}(F)$. The normalized intertwining operator (2.4.2) is then the earlier operator attached to G and M, but with one difference. The element w must be taken here from the rational Weyl set $W(\widetilde{G}, M)$ induced from $\widetilde{G}$. This implies that the four groups in the lower right hand block in (2.4.3) have also to be formulated as sets, with $\widetilde{G}$ in place of G. For any u in the set $\mathfrak{N}_\psi(\widetilde{G}, M)$, we construct the objects w_u, $\widetilde{w}_u$, $\widetilde{M_u}$ and $\widetilde{x}_u$ as above, and we define the linear form $\widetilde{f}_{\widetilde{G}}(\psi, u)$ on $\mathcal{H}(\widetilde{G})$ as in (2.4.5). We also define the linear form $\widetilde{f}'_{\widetilde{G}}(\psi, \widetilde{s})$ on $\mathcal{H}(\widetilde{G})$ as in (2.4.6), for any semisimple element $\widetilde{s}$ in the centralizer $\widetilde{S}_\psi = S_\psi(\widetilde{G})$.

THEOREM 2.4.4 (Local intertwining relation for $\widetilde{G}$). *For any $u \in \mathfrak{N}_\psi(\widetilde{G}, M)$, the identity*

$$\widetilde{f}'_{\widetilde{G}}(\psi, s_\psi \widetilde{s}) = \widetilde{f}_{\widetilde{G}}(\psi, u), \qquad \widetilde{f} \in \mathcal{H}(\widetilde{G}), \tag{2.4.18}$$

holds for any semisimple element $\widetilde{s} \in \widetilde{S}_\psi$ that projects to the image $\widetilde{x}_u$ of u in $\mathcal{S}_\psi(\widetilde{G}, M)$.

This theorem joins our growing body of unproved assertions that will be established later by interlocking induction arguments. In the meantime, we assume as necessary that it holds if G is replaced by an even orthogonal group $G_- \in \widetilde{\mathcal{E}}_{\mathrm{sim}}(N_-)$, for any $N_- < N$.

PROOF OF LEMMA 2.4.2. We are given a local parameter $\psi \in \widetilde{\Psi}(M)$, where M is a Levi subgroup of the simple group $G \in \widetilde{\mathcal{E}}_{\mathrm{sim}}(N)$. We are also given a point $u \in \mathfrak{N}_\psi(G, M)$, and a point $s \in S_\psi$ whose image in the quotient $\mathcal{S}_\psi$ equals the image x_u of u in the subgroup $\mathcal{S}_\psi(G, M)$ of $\mathcal{S}_\psi$. We have to show that the linear form

$$f_G(\psi, u) = \sum_{\pi \in \widetilde{\Pi}_\psi} \langle \widetilde{u}, \widetilde{\pi} \rangle \operatorname{tr}\big(R_P(w_u, \widetilde{\pi}, \psi) \mathcal{I}_P(\pi, f)\big)$$

equals its endoscopic counterpart $f'_G(\psi, s_\psi s)$.

The parameter ψ is the image of a "square integrable" parameter $\psi_1 \in \widetilde{\Psi}_2(M_1)$, for a Levi subgroup M_1 of M. The point u is a coset in the complex group S_ψ that normalizes the torus $A_{\widehat{M}}$ in S^0_ψ. It has a representative that also normalizes the maximal torus T_ψ in S^0_ψ. In other words, we can choose a representative u_1 of u in the group $\mathfrak{N}_{\psi_1}$, which one sees is determined up to translation by the subgroup $W^0_{\psi_1}(M, M_1)$ of $\mathfrak{N}_{\psi_1}$ in the chain

$$W^0_{\psi_1}(M, M_1) \subset W^0_{\psi_1}(G, M_1) \subset \mathfrak{N}_{\psi_1}(G, M_1) = \mathfrak{N}_{\psi_1}.$$

The point s can of course be identified with an element $s_1 \in S_{\psi_1}(G)$, since the groups $S_{\psi_1}(G)$ and $S_\psi(G)$ are equal. Moreover, s_1 projects to the image x_{u_1} of u_1 in the group $\mathcal{S}_{\psi_1}(G) = \mathcal{S}_{\psi_1}(G, M_1)$, since x_{u_1} equals x_u. The information we are given from the lemma therefore tells us that the linear form

$$f_G(\psi_1, u_1) = \sum_{\pi_1 \in \widetilde{\Pi}_{\psi_1}} \langle \widetilde{u}_1, \widetilde{\pi}_1 \rangle \operatorname{tr}\big(R_{P_1}(w_{u_1}, \widetilde{\pi}_1, \psi_1) \mathcal{I}_{P_1}(\pi_1, f)\big) \tag{2.4.19}$$

equals its endoscopic counterpart $f'_G(\psi_1, s_{\psi_1} s_1)$. Since $s_{\psi_1} s_1$ equals $s_\psi s$, the linear forms $f'_G(\psi_1, s_{\psi_1} s_1)$ and $f'_G(\psi, s_\psi s)$ are equal. We have therefore to verify that $f_G(\psi_1, u_1)$ equals $f_G(\psi, u)$.

The problem is simply to interpret the terms in the expression (2.4.19) for $f_G(\psi_1, u_1)$. Assuming an appropriate choice of M_1, we can arrange that the parabolic subgroup $P_1 \in \mathcal{P}(M_1)$ is contained in P. The Weyl element in (2.4.19) then satisfies

$$w_{u_1} = w^M_{u_1} w_u,$$

where $w_{u_1}^M \in W^M(M_1)$ is a Weyl element for M, and the point $w_u \in W(M)$ is identified with the Weyl element in $W(M_1)$ that stabilizes the parabolic subgroup $R_1 = P_1 \cap M$ of M. The induced representation in (2.4.19) satisfies

$$\mathcal{I}_{P_1}(\pi_1) = \mathcal{I}_P\big(\mathcal{I}_{R_1}^M(\pi_1)\big),$$

by induction in stages, while the intertwining operator has a corresponding decomposition

$$R_{P_1}(w_{u_1}, \widetilde{\pi}_1, \psi_1) = R_{R_1}^M(w_{u_1}^M, \widetilde{\pi}_1, \psi_1) R_P(w_u, \widetilde{\pi}, \psi).$$

We assume implicitly that the analogues of Theorems 2.2.1 and 2.2.4, and of Theorems 2.4.1 and 2.4.4, hold for the proper Levi subgroup M in place of G. They tell us that for any function $h \in \mathcal{H}(\widetilde{M}_u)$, the sum

$$\sum_{\pi_1 \in \widetilde{\Pi}_{\psi_1}} \langle \widetilde{u}_1, \widetilde{\pi}_1 \rangle \operatorname{tr}\big(R_{R_1}^M(w_{u_1}^M, \widetilde{\pi}_1, \psi_1) \mathcal{I}_{R_1}^M(\pi_1, h)\big)$$

equals the linear form

$$h'_M(\psi_1, s_{\psi_1} s_1) = \sum_{\pi \in \widetilde{\Pi}_\psi} \langle \widetilde{u}, \widetilde{\pi} \rangle\, h_M(\widetilde{\pi}).$$

Finally, from general principles, we know that the trace in (2.4.19) fibres as a product over the Hilbert space

$$\mathcal{H}_{P_1}(\pi_1) = \mathcal{H}_P\big(\mathcal{H}_{R_1}^M(\pi_1)\big).$$

It follows that the expression for $f_G(\psi_1, u_1)$ reduces to the earlier expression for $f_G(\psi, u)$, as required. □

We have now established preliminary reductions of our local theorems. Lemma 2.4.2 reduces the local intertwining relation of Theorem 2.4.1 to the case of parameters in the set $\widetilde{\Psi}_2(M)$. Proposition 2.4.3 interprets the local, intertwining relation itself as an explicit construction of some of the packets $\widetilde{\Pi}_\psi$ of Theorem 2.2.1, and hence reduces this theorem to parameters in the set $\widetilde{\Psi}_2(G)$. Similar reductions, which we shall leave to the reader, apply to the assertions of Theorems 2.2.4 and 2.4.4. For the basic cases that remain, we will need global methods. We shall establish them, along with the supplementary local assertions of Theorem 1.5.1(b), in Chapters 6 and 7.

2.5. Relations with Whittaker models

The theory of Whittaker models has been an important part of representation theory for many years. For example, it is at the heart of the theorem of multiplicity 1 for $GL(N)$ [**Shal**], and its generalization Theorem 1.3.2 by Jacquet and Shalika. In the hands of Shahidi, Whittaker models have yielded a broader understanding of intertwining operators. We shall review a couple of his main results, for comparison with our earlier discussion.

Suppose for the moment that G is an arbitrary quasisplit, connected reductive group over the local field F. Suppose also that $(B, T, \{X_k\})$ is a Γ_F-stable splitting of G over F, where

$$X_k = X_{\alpha_k}, \qquad 1 \leqslant k \leqslant n,$$

are fixed root vectors for the simple roots of (B, T). If ψ_F is a fixed nontrivial additive character on F, the function

$$\chi(u) = \psi_F(u_1 + \cdots + u_n), \qquad u \in N_B(F),$$

is a nondegenerate character on the unipotent radical $N_B(F)$ of $B(F)$. As usual, $\{u_k\}$ are the coordinates of $\log(u)$ relative to the simple root vectors $\{X_k\}$, or more correctly, any basis of root vectors of $\mathfrak{n}_B(F)$ that includes $\{X_k\}$. Then χ represents a Whittaker datum (B, χ) for $G(F)$.

We are interested in irreducible tempered representations $\pi \in \Pi_{\text{temp}}(G)$ of $G(F)$ that have a (B, χ)-Whittaker model. As in the special case of $G = GL(N)$ from §2.2, a (B, χ)-*Whittaker functional* for π is a nonzero linear form ω on the underlying space of smooth vector V_∞ for π such that

$$\omega\big(\pi(u)v\big) = \chi(u)\,\omega(v), \qquad u \in N_B(F),\ v \in V_\infty.$$

We recall that the vector space spanned by the (B, χ)-Whittaker functionals for π has dimension at most 1 [**Shal**], and that π is said to be *generic* if the space is actually nonzero. For any such ω, the function

$$W(x, v) = \omega\big(\pi(x)v\big), \qquad x \in G(F),\ v \in V,$$

satisfies

$$W\big(x, \pi(y)v\big) = W(xy, v), \qquad y \in G(F),$$

and therefore represents an intertwining operator from π to the representation $\mathcal{I}_{N_B}(\chi)$ of $G(F)$ induced from the character χ of $N_B(F)$. The *Whittaker model* of ω is the corresponding space of functions

$$W(\pi, \chi) = \big\{W(x) = W(x, v) : v \in V_\infty\big\}$$

of x, equipped with the representation of $G(F)$ by right translation. Conversely, suppose that $W(\pi, \chi)$ is a subrepresentation of $\mathcal{I}_{N_B}(\chi)$ that is equivalent to π. If

$$v \longrightarrow W(x, v), \qquad v \in V,\ x \in G(F),$$

is a nontrivial intertwining operator from π to $W(\pi, \chi)$, the linear form

$$\omega(v) = W(1, v), \qquad v \in V_\infty,$$

is a (B, χ)-Whittaker functional for π. Whittaker models and Whittaker functionals are thus essentially the same. Our focus will be on the functionals.

Suppose that M is standard Levi subgroup of G over F. Then M is the Levi component of a parabolic subgroup $P = MN_P$ of G that contains B. The Whittaker datum (B, χ) for G restricts to a Whittaker datum (B_M, χ_M) for $M(F)$, which can be expressed as above in terms of the splitting of M attached to that of G. Assume now that π and ω are attached to M instead of G. That is, π is an irreducible representation of $M(F)$, with a (B_M, χ_M)-Whittaker functional ω. There is then a canonical Whittaker functional $\Omega_{\chi,\omega}(\pi)$ for the induced representation $\mathcal{I}_P(\pi)$. It is defined in terms of the Whittaker integral

$$W_{\chi,\omega}(x, h, \pi_\lambda) = \int_{N_*(F)} \omega\big(h_{\pi,\lambda}(w_*^{-1} n_* x)\big)\, \chi(n_*)^{-1}\, dn_*, \tag{2.5.1}$$

whose ingredients we recall.

The vector h lies in the space $\mathcal{H}_{P,\infty}(\pi)$ of smooth functions in $\mathcal{H}_P(\pi)$. We are following the convention that $\mathcal{H}_P(\pi)$ is a Hilbert space of functions from K to V, which does not change if π is replaced with its twist π_λ by a point $\lambda \in \mathfrak{a}^*_{M,\mathbb{C}}$. We have therefore to write

$$h_{\pi,\lambda}(x) = \pi\big(M_P(x)\big)\, h\big(K_P(x)\big)\, e^{(\lambda+\rho_P)(H_P(x))}, \qquad x \in G(F),$$

in the notation of [**A7**, p. 26], to obtain a vector in the usual space on which $\mathcal{I}_P(\pi_\lambda)$ acts. The group $N_* = N_{P_*}$ is the unipotent radical of the standard parabolic subgroup $P_* = M_* N_*$ that is "adjoint" to P, in the sense that

$$M_* = w_*\, M\, w_*^{-1}, \qquad w_* = w_\ell w_\ell^M,$$

where w_ℓ and w_ℓ^M are the longest elements in the restricted Weyl groups of G and M respectively. The Whittaker integral (2.5.1) converges absolutely for $\mathrm{Re}(\lambda)$ in a certain chamber, and has analytic continuation as an entire function of λ. (See [**CS**, Proposition 2.1] and [**Sha1**, Proposition 3.1].) Its value $W_{\chi,\omega}(x, h, \pi)$ is therefore a well defined intertwining operator from $\mathcal{I}_P(\pi)$ to $\mathcal{I}_{N_B}(\chi)$. We are interested in the corresponding (B, χ)-Whittaker functional

$$\Omega_{\chi,\omega}(\pi) :\ h \longrightarrow \Omega_{\chi,\omega}(h, \pi) = W_{\chi,\omega}(1, h, \pi), \qquad h \in \mathcal{H}_{P,\infty}(\pi),$$

for $\mathcal{I}_P(\pi)$.

The constructions from the earlier sections of this chapter pertain to the special case of $G \in \widetilde{\mathcal{E}}_{\mathrm{sim}}(N)$. To compare them with the results of Shahidi, we take $\pi \in \Pi_{\mathrm{temp}}(M)$ to be tempered. We assume that Theorem 1.5.1 holds for the Levi subgroup M, which we take to be proper. Then π belongs to the packet $\widetilde{\Pi}_\phi$ of a parameter $\phi \in \widetilde{\Phi}_{\mathrm{bdd}}(M)$. We can denote the normalizing factor from the end of §2.3 by

$$r_P(w, \pi_\lambda) = r_P(w, \phi_\lambda), \qquad w \in W(M),$$

since ϕ is uniquely determined by π. For the same reason, we write

$$R_P(w, \pi_\lambda) = R_P(w, \pi_\lambda, \phi_\lambda)$$

for the normalized intertwining operator in (2.3.26). Suppose in addition that π is generic, and is equipped with a (B_M, χ_M)-Whittaker functional ω. If π and $w\pi$ are equivalent, there is then a canonical choice for the intertwining operator $\widetilde{\pi}(w)$ from $w\pi$ to π. To see this, recall that the representative

$\widetilde{w}$ of w in $G(F)$ is defined by the splitting of G. It follows from [**Sp**, Proposition 11.2.11] that $\widetilde{w}$ preserves the splitting of M, and therefore stabilizes the Whittaker functional ω. We can therefore choose $\widetilde{\pi}(w)$ uniquely so that

$$\omega = \omega \circ \widetilde{\pi}(w). \tag{2.5.2}$$

The definition (2.4.2) thus gives us a *canonical* self-intertwining operator

$$R_P(w, \widetilde{\pi}) = R_P(w, \widetilde{\pi}, \phi)$$

of $\mathcal{I}_P(\pi)$, in the case $\pi \in \Pi_{\mathrm{temp}}(M)$ is both generic and equivalent to $w\pi$.

In this section, we have taken G to be an arbitrary quasisplit group. With the assumption that π is generic, Shahidi was nonetheless able to construct the local L and ε-functions that appear in the normalizing factors $r_P(w, \pi_\lambda)$. This was a local ingredient of the paper [**Sha4**] that became known as the Langlands-Shahidi method, in which Langlands's original results [**L4**] were extended, and the analytic continuation and functional equation for a broad class of automorphic L-functions were established. In particular, the local definitions of §2.3 and §2.4 all carry over to the general group G, if $\pi \in \Pi_{\mathrm{temp}}(M)$ is a representation that is generic. We therefore have the normalizing factor $r_P(w, \pi_\lambda)$, and the normalized intertwining operator

$$R_P(w, \pi_\lambda) = r_P(w, \pi_\lambda)^{-1}\, J_P(\widetilde{w}, \pi_\lambda), \qquad w \in W(M),$$

in (2.3.26), to go with the basic unnormalized intertwining operator $J_P(\widetilde{w}, \pi_\lambda)$. By the general results of Harish-Chandra [**Ha4**] and the properties of $r_P(w, \pi_\lambda)$ established by Shahidi, these functions all have analytic continuation as meromorphic functions of $\lambda \in \mathfrak{a}^*_{M,\mathbb{C}}$. Moreover, if w belongs to the stabilizer $W(\pi)$ of π in $W(M)$, we obtain the canonical self-intertwining operator

$$R_P(w, \widetilde{\pi}) = \widetilde{\pi}(w) \circ R_P(w, \pi)$$

of $\mathcal{I}_P(\pi)$ from the natural generalization of (2.5.2).

THEOREM 2.5.1 (Shahidi). *Suppose that G is a quasisplit group over F with Levi subgroup M, and that $\pi \in \Pi_{\mathrm{temp}}(M)$ is generic.*

(a) *The normalized intertwining operators*

$$R(w, \pi) = R_P(w, \pi), \qquad w \in W(M),$$

are unitary, and satisfy the relation.

$$R(w'w, \pi) = R(w', w\pi)\, R(w, \pi), \qquad w', w \in W(M). \tag{2.5.3}$$

(b) *Suppose that w belongs to the subgroup $W(\pi)$ of $W(M)$. Then the canonical self intertwining operator $R_P(w, \widetilde{\pi})$ satisfies the relation*

$$\Omega_{\chi,\omega}(\pi) = \Omega_{\chi,\omega}(\pi) \circ R_P(w, \widetilde{\pi}), \tag{2.5.4}$$

if ω is a (B_M, χ_M)-Whittaker functional for π.

The property (a) is [**Sha4**, Theorem 7.9]. Its proof is contained in several papers, which include [**Sha1**] and [**Sha2**] as well as [**Sha4**]. Since the argument is by induction on the length of w, the assertion has to be proved in slightly greater generality. Namely, w must be taken from the more general set $W(M, M')$ of Weyl elements that conjugate M to a second standard Levi subgroup M', while w' is taken from a second set $W(M', M'')$. The operators in (2.5.3) are otherwise defined exactly as above, and the required identity takes the same form.

Shahidi's starting point is the meromorphic scalar valued function $C_{\chi,\omega}(w, \pi_\lambda)$, defined by the relation

$$\Omega_{\chi,\omega}(\pi_\lambda) = C_{\chi,\omega}(w, \pi_\lambda)\,\big(\Omega_{\chi,\omega}(w\pi_\lambda) \circ J(\widetilde{w}, \pi_\lambda)\big)$$

given by the unnormalized intertwining operator

$$J(\widetilde{w}, \pi_\lambda) = J_P(\widetilde{w}, \pi_\lambda).$$

The existence of this function follows from the multiplicity 1 of Whittaker functionals, and the fact that $J(\widetilde{w}, \pi_\lambda)$ is meromorphic in λ. Since

$$J(\widetilde{w'w}, \pi_\lambda) = r(w', w, \pi_\lambda)\, J(\widetilde{w}', w\pi_\lambda)\, J(\widetilde{w}, \pi_\lambda),$$

for a meromorphic scalar valued function $r(w', w, \pi_\lambda)$, one sees that

$$r_P(w', w, \pi_\lambda) = C_{\chi,\omega}(w'w, \pi_\lambda)^{-1} C_{\chi,\omega}(w', w\pi_\lambda)\, C_{\chi,\omega}(w, \pi_\lambda).$$

To split the 2-cycle $r(w', w, \pi_\lambda)$ in terms of suitable normalizing factors, it then suffices to compute the local coefficients $C_{\chi,\omega}(w, \pi_\lambda)$. Therein lies the problem. Shahidi's construction of these objects eventually leads to the construction of L and ε-functions, which he uses to express $C_{\chi,\omega}(w, \pi_\lambda)$ and to define normalizing factors $r_P(w, \pi_\lambda)$ such that the corresponding normalized operators satisfy (2.5.3).

The general version of (2.5.4) must be written differently. It takes the form

$$\Omega_{\chi,\omega}(\pi) = \Omega_{\chi,\omega}(w\pi) \circ R(w, \pi), \tag{2.5.5}$$

for a general element $w \in W(M, M')$. In the special case that $M' = M$ and w belongs to the subgroup $W(\pi)$ of $W(M) = W(M, M)$, we have the intertwining operator $\widetilde{\pi}(w)$ from $w\pi$ to π that stabilizes ω. It follows from the definitions that the right hand side of (2.5.5) equals

$$\Omega_{\chi,\omega}(w\pi) \circ \widetilde{\pi}(w)^{-1} \circ R_P(w, \widetilde{\pi}) = \Omega_{\chi,\omega}(\pi) \circ R_P(w, \widetilde{\pi}).$$

Therefore (2.5.5) does reduce to (2.5.4) in this case.

The proof of (2.5.5) is only implicit in [**Sha4**]. The right hand side of (2.5.5) equals the value at $\lambda = 0$ of the operator

$$\begin{aligned}&\Omega_{\chi,\omega}(w\pi_\lambda) \circ \big(r_P(w, \pi_\lambda)^{-1}\, J(\widetilde{w}, \pi_\lambda)\big)\\ =&r_P(w, \pi_\lambda)^{-1}\, C_{\chi,\omega}(w, \pi_\lambda)^{-1}\, \Omega_{\chi,\omega}(\pi_\lambda).\end{aligned}$$

The problem is to show that the product

$$r_P(w, \pi_\lambda)\, C_{\chi,\omega}(w, \pi_\lambda)$$

is analytic at $\lambda = 0$, with value at $\lambda = 0$ equal to 1. We shall have to leave the reader to extract this fact from the formulas of [**Sha4**] and [**Sha2**], following the calculation from the special case in [**KeS**, §4]. We shall try to review the calculation in the paper [**A27**] in preparation. □

We will have a special interest in the following five cases:

(i) F arbitrary, $G = GL(N)$, M arbitrary;
(ii) $F = \mathbb{C}$, $G \in \widetilde{\mathcal{E}}_{\text{sim}}(N)$, M arbitrary;
(iii) F arbitrary, $G \in \widetilde{\mathcal{E}}_{\text{sim}}(N)$, $M \cong GL([N/2])$;
(iv) F arbitrary, $G \in \widetilde{\mathcal{E}}_{\text{sim}}(N)$, $M = T$ minimal;
(v) F arbitrary, $G \in \widetilde{\mathcal{E}}_{\text{sim}}(N)$, $M_{\text{der}} \cong Sp(2)$.

In cases (i)–(iv), it is known that any $\pi \in \Pi_{\text{temp}}(M)$ is generic. In the last case (v), G is of the form $Sp(2n)$, while M is an product of several copies of $GL(1)$ with the group $SL(2) = Sp(2)$. From the general relation between representations of $SL(2)$ and $GL(2)$ described in the introduction of [**LL**], we see that any $\pi_- \in \Pi_{\text{temp}}\big(SL(2)\big)$ is generic, for *some* choice of Whittaker datum. The same is therefore true for any $\pi \in \Pi_{\text{temp}}(M)$. This case will not be needed in any of our future arguments, unlike the other four. We have included it here for general perspective, and because it will arise in a natural setting later in Lemma 6.4.1. It follows that in all five cases, the canonical self intertwining operator $R_P(w, \widetilde{\pi})$ is defined for any $w \in W(\pi)$, and satisfies (2.5.4).

COROLLARY 2.5.2. *Suppose that* (F, G, M) *is as in one of the cases* (i)–(v), *that* $\pi \in \Pi_{\text{temp}}(M)$, *and that* $w \in W(\pi)$. *Then if* $(\Pi, \mathcal{V})$ *is the unique irreducible* (B, χ)*-generic subrepresentation of* $\mathcal{I}_P(\pi)$, *the operator* $R_P(w, \widetilde{\pi})$ *satisfies*

$$R_P(w, \widetilde{\pi})\phi = \phi, \qquad \phi \in \mathcal{V}_\infty.$$

In particular, if $\mathcal{I}_P(\pi)$ *is irreducible, we have*

$$R_P(w, \widetilde{\pi}) = 1.$$

PROOF. The restriction of $R_P(w, \widetilde{\pi})$ to the irreducible subspace $\mathcal{V}$ of $\mathcal{H}_P(\pi)$ is a nonzero scalar. Since π is generic, $R_P(w, \widetilde{\pi})$ satisfies (2.5.4). Since the Whittaker functional $\Omega_{\chi,\omega}(\pi)$ on $\mathcal{H}_{P,\infty}(\pi)$ is supported on the subspace $\mathcal{V}$, the scalar in question equals 1. □

We note that in the cases (i) and (ii), the induced representation $\mathcal{I}_P(\pi)$ is always irreducible. In (iii), suppose that π corresponds to the Langlands parameter $\phi \in \Phi_2(M)$, and that $w \in W(\pi)$ is nontrivial. The centralizer S_ϕ is then isomorphic to either $Sp(2, \mathbb{C})$ or $O(2, \mathbb{C})$. We will apply the corollary later to the case that the centralizer is $Sp(2, \mathbb{C})$, in the proof of Lemma 5.4.6, after first showing that the operator $R_P(w, \widetilde{\pi})$ is a scalar. In (iv), our interest will be in the case that π is trivial on the maximal compact subgroup of $T(F)$. The representation $\mathcal{I}_P(\pi)$ need not be irreducible in this case. However, the generic subrepresentation Π is easily identified. For it follows from (2.5.1) (together with an approximation argument in case F is archimedean) that Π is the irreducible constituent of $\mathcal{I}_P(\pi)$ that contains the trivial representation of K. It is only in the last case (v) that the L-packets for M can be nontrivial.

In the same paper, Shahidi conjectured that any tempered L-packet for the quasisplit group G has a (B, χ)-generic constituent [**Sha4**, Conjecture 9.4], motivated by the proof of the property for archimedean F that followed from the results of [**Kos**] and [**V1**]. Shelstad [**S6**] has recently established a strong form of this conjecture for real groups. Namely, if the transfer factors are normalized as in [**KS**, (5.3)], the (B, χ)-generic representation in a tempered packet is the unique representation for which the linear character (2.2.5) is trivial. For nonarchimedean F and $G \in \widetilde{\mathcal{E}}_{\text{sim}}(N)$, Konno [**Kon**] showed that the conjecture would follow from local twisted endoscopy for G, which is to say, the assertion of Theorem 2.2.1. In §8.3, we will show that the strong form of the conjecture (apart from the uniqueness condition) is also valid in this case, after proving Theorem 2.2.1 in Chapter 6. In retrospect, we could have assumed Shahidi's conjecture inductively for M, and then appealed to his conditional proof of Proposition 2.3.1 in [**Sha4**, §9]. I have retained the proof of the proposition in §2.3 for its different perspective, and its role in the transition to the self intertwining operators (2.4.4).

We need to expand on the case (i) of $GL(N)$. It will be important to establish an analogue for the component $\widetilde{G}(N)$ of the identity of Corollary 2.5.2. Of equal importance, but harder, is the task of proving an identity for general parameters $\psi \in \widetilde{\Psi}(N)$.

Suppose that M is a standard Levi subgroup of $GL(N)$, and that

$$\psi : L_F \times SU(2) \longrightarrow {}^L M$$

is a general parameter in $\Psi(M)$. Then ψ corresponds to an irreducible unitary representation π_ψ of $M(F)$, which is the Langlands quotient of a standard representation ρ_ψ. We write $\widetilde{W}(M)$ for the Weyl set of outer automorphisms of M induced from the component $\widetilde{G}(N)$, and $\widetilde{W}_\psi(M)$ for the stabilizer of ψ in $\widetilde{W}(M)$. Any element in $\widetilde{W}_\psi(M)$ then stabilizes π_ψ and ρ_ψ as well as ψ. We assume that the set $\widetilde{W}_\psi(M)$ is nonempty. This implies that the image of ψ in $\Psi(N)$, which we again denote by ψ, lies in the subset $\widetilde{\Psi}(N)$.

Suppose that w belongs to $\widetilde{W}_\psi(M)$. Then

$$w = \widetilde{\theta}(N) \circ w^0,$$

where $\widetilde{\theta}(N)$ is the standard outer automorphism of $GL(N)$ defined in §1.3, and w^0 belongs to a set $W(M, M')$. Bearing in mind that $\widetilde{\theta}(N)$ acts as an involution on the set of standard parabolic subgroups of $GL(N)$, we note that M' is the standard Levi subgroup that is paired with M. The representative

$$\widetilde{w} = \widetilde{\theta}(N) \circ \widetilde{w}^0$$

of w in $\widetilde{G}(N,F)$ preserves the standard Whittaker datum (B_M,χ_M) for M. We would like to define twisted intertwining operators by some variant of (2.4.2) and (2.3.25).

The operator

$$R_P(w,\pi_\psi) = \ell(w,\pi_\psi)\, R_{w^{-1}P|P}(\pi_\psi)$$

in (2.3.25) has to be slightly modified. For we cannot define the operator

$$\ell(\widetilde{w},\pi_\psi) = \mathcal{H}_{w^{-1}P}(\pi_\psi) \longrightarrow \mathcal{H}_P(w\pi_\psi)$$

in the product

$$\ell(w,\pi_\psi) = \varepsilon_P(w,\psi)\,\ell(\widetilde{w},\pi_\psi)$$

by a simple left translation by $\widetilde{w}^{-1}$, since the operation would not take values in $\mathcal{H}_P(w\pi_\psi)$. We instead write

$$(2.5.6)\qquad \big(\widetilde{\ell}(\widetilde{w},\pi_\psi)\,\phi\big)(x) = \phi\big(\widetilde{w}^{-1}\,x\,\widetilde{\theta}(N)\big), \qquad \phi\in\mathcal{H}_{w^{-1}P}(\pi_\psi),$$

in order that the right hand side make sense, as well as

$$\widetilde{\ell}(w,\pi_\psi) = \varepsilon_P(w,\psi)\,\widetilde{\ell}(\widetilde{w},\pi_\psi).$$

The product

$$\widetilde{R}_P(w,\pi_\psi) = \widetilde{\ell}(w,\pi_\psi)\,R_{w^{-1}P|P}(\pi_\psi)$$

is then an operator

$$\widetilde{R}_P(w,\pi_\psi):\ \mathcal{H}_P(\pi_\psi) \longrightarrow \mathcal{H}_P(w\pi_\psi).$$

If ω is a (B_M,χ_M)-Whittaker functional for ρ_ψ, we define the intertwining operator $\widetilde{\rho}_\psi(w)$ from $w\rho_\psi$ to ρ_ψ uniquely by analytic continuation, and the property

$$\omega\circ\widetilde{\rho}_\psi(w) = \omega.$$

This lifts to an intertwining operator $\widetilde{\pi}_\psi(w)$ from $w\pi_\psi$ to π_ψ. The product

$$\widetilde{R}_P(w,\widetilde{\pi}_\psi) = \widetilde{\pi}_\psi(w)\circ\widetilde{R}_P(w,\pi_\psi)$$

is then a canonical operator on $\mathcal{H}_P(\pi_\psi)$. We note that its construction is a variant of both the definition at the beginning of §2.2, and the definition of the operator in Theorem 2.5.1(b).

The twisted operator $\widetilde{R}_P(w,\widetilde{\pi}_\psi)$ does not intertwine $\mathcal{I}_P(\pi_\psi)$ with itself. Indeed, it follows from (2.5.6) and the other definitions that

$$\widetilde{R}_P(w,\widetilde{\pi}_\psi):\ \mathcal{I}_P(\pi_\psi) \longrightarrow \mathcal{I}_P(\pi_\psi)\circ\widetilde{\theta}(N).$$

In other words, $\widetilde{R}_P(\omega,\widetilde{\pi}_\psi)$ intertwines $\mathcal{I}_P(\pi_\psi)$ with the representation

$$\widetilde{\theta}(N)^{-1}\,\mathcal{I}_P(\pi_\psi) = \mathcal{I}_P(\pi_\psi)\circ\widetilde{\theta}(N).$$

Recall that we have already introduced an operator with the same intertwining property. In §2.2, we defined an operator

$$\widetilde{\mathcal{I}}_P(\pi_\psi,N) = \widetilde{\Pi}(N), \qquad \Pi = \mathcal{I}_P(\pi_\psi),$$

with

$$\widetilde{\mathcal{I}}_P(\pi_\psi,N):\ \mathcal{I}_P(\pi_\psi) \longrightarrow \mathcal{I}_P(\pi_\psi)\circ\widetilde{\theta}(N),$$

in terms of a Whittaker functional for $\mathcal{I}_P(\pi_\psi)$. How are the two objects related?

THEOREM 2.5.3. *The intertwining operator attached to* $\psi \in \Psi(M)$ *and* $w \in \widetilde{W}_\psi(M)$ *satisfies*

$$\widetilde{R}_P(w, \widetilde{\pi}_\psi) = \widetilde{\mathcal{I}}_P(\pi_\psi, N). \tag{2.5.7}$$

In case $\phi = \psi$ is generic, one can apply the formula (2.5.5) to the element $w^0 \in W(M, M')$. The required identity (2.5.7) then follows without much difficulty from the various definitions. The general case is more difficult, and as far as I know, has not been investigated. It requires further techniques, based on some version of minimal K-types. Rather than introduce them here, we shall leave the general proof of Theorem 2.5.3 for a separate paper [**A26**].

There is another way to interpret the identity (2.5.7). We begin by writing $\widetilde{S}_\psi(N)$ as in §1.4 for the centralizer in $\widehat{\widetilde{G}}(N)$ of the image of ψ. Then $\widetilde{S}_\psi(N)$ is a bitorsor under the centralizer $\widetilde{S}^0_\psi(N)$ of the image of ψ in $\widehat{\widetilde{G}}{}^0(N)$. The analogue of the diagram (2.4.3) for the $G^0 = \widetilde{G}^0(N)$ torsor $G = \widetilde{G}(N)$ certainly makes sense. If we define its objects in the natural way in terms of the bitorsor

$$S_\psi = \widetilde{S}_\psi(N),$$

we see that

$$\mathfrak{N}_\psi(G, M) = W_\psi(G, M) = \widetilde{W}_\psi(M),$$

and that this last set is a bitorsor under the group

$$W^0_\psi(G, M) = W_\psi(M).$$

All the other sets in the diagram are trivial. We shall use the diagram to formulate the analogue for $\widetilde{G}(N)$ of the local intertwining relation of Theorem 2.4.1. The formalities of the process will be helpful in the next section in understanding the spectral terms of the twisted trace formula for $GL(N)$. The actual intertwining relation for $\widetilde{G}(N)$ will be an essential ingredient of the global comparison in Chapter 4.

The twisted intertwining relation can be most easily stated if we inflate $\mathcal{I}_P(\pi_\psi)$ to an induced representation of $\widetilde{G}^+(N, F)$. With this interpretation, $\mathcal{I}_P(\pi_\psi)$ acts on the larger Hilbert space

$$\widetilde{\mathcal{H}}^+_P(\pi_\psi) = \mathcal{H}_P(\pi_\psi) \oplus \widetilde{\mathcal{H}}_P(\pi_\psi),$$

where $\widetilde{\mathcal{H}}_P(\pi_\psi)$ is the space of functions supported on the component $\widetilde{G}(N, F)$. For any $w \in \widetilde{W}_\psi(M)$, we then obtain a linear transformation

$$R_P(w, \widetilde{\pi}_\psi) : \mathcal{H}_P(\pi_\psi) \longrightarrow \widetilde{\mathcal{H}}_P(\pi_\psi)$$

by setting

$$\big(R_P(w, \widetilde{\pi}_\psi)\phi\big)(x) = \big(\widetilde{R}_P(w, \widetilde{\pi}_\psi)\phi\big)\big(x\widetilde{\theta}(N)^{-1}\big),$$

for $\phi \in \mathcal{H}_P(\pi_\psi)$ and $x \in \widetilde{G}(N,F)$. In other words, $R_P(w, \widetilde{\pi}_\psi)$ is defined *exactly* as in (2.4.2) and (2.3.25), namely, with the translation operation $\ell(\widetilde{w}, \pi_\psi)$ defined by excluding the right translation by $\widetilde{\theta}(N)$ from (2.5.6). If $\widetilde{f}$ belongs to the Hecke space $\widetilde{\mathcal{H}}(N)$, its integration against $\mathcal{I}_P(\pi_\psi)$ gives a linear transformation

$$\mathcal{I}_P(\pi_\psi, \widetilde{f}) : \widetilde{\mathcal{H}}_P(\pi_\psi) \longrightarrow \mathcal{H}_P(\pi_\psi)$$

that we can compose with $R_P(w, \widetilde{\pi}_\psi)$. Following (2.4.5), we set

$$\widetilde{f}_N(\psi, u) = \mathrm{tr}\big(R_P(w, \widetilde{\pi}_\psi)\mathcal{I}_P(\pi_\psi, \widetilde{f})\big), \qquad \widetilde{f} \in \widetilde{\mathcal{H}}(N), \tag{2.5.8}$$

for any point $u = w$ in the set $\mathfrak{N}_\psi(G, M) = \widetilde{W}_\psi(M)$.

The representation $\mathcal{I}_P(\pi_\psi)$ of $\widetilde{G}^0(N,F)$ on $\mathcal{H}_P(\pi_\psi)$ also has an *extension* to the bitorsor $\widetilde{G}(N,F)$. This is provided by the operator

$$\widetilde{\mathcal{I}}_P(\pi_\psi, N) = \widetilde{\mathcal{I}}_P\big(\pi_\psi, \widetilde{\theta}(N)\big),$$

which then yields the linear form

$$\widetilde{f}_N(\psi) = \mathrm{tr}\big(\widetilde{\mathcal{I}}_P(\pi_\psi, \widetilde{f})\big), \qquad \widetilde{f} \in \widetilde{\mathcal{H}}(N),$$

of (2.2.1). Observe that if

$$\big(\mathcal{I}_P(\pi_\psi, N)\phi\big)(x) = \big(\widetilde{\mathcal{I}}_P(\pi_\psi, N)\phi\big)\big(x\widetilde{\theta}(N)^{-1}\big),$$

for $\phi \in \mathcal{H}_P(\pi_\psi)$ and $x \in \widetilde{G}(N,F)$, then

$$\widetilde{f}_N(\psi) = \mathrm{tr}\big(\mathcal{I}_P(\pi_\psi, \widetilde{f})\mathcal{I}_P(\pi_\psi, N)\big) = \mathrm{tr}\big(\mathcal{I}_P(\pi_\psi, N)\mathcal{I}_P(\pi_\psi, \widetilde{f})\big).$$

It follows from Theorem 2.5.3 that

$$\widetilde{f}_N(\psi) = \widetilde{f}_N(\psi, u), \qquad u \in \widetilde{W}_\psi(M). \tag{2.5.9}$$

Suppose that s is a semisimple element in the torsor $\widetilde{S}_\psi(N)$. The pair (ψ, s) then has a preimage under the general correspondence (1.4.11), which consists of a datum $G \in \widetilde{\mathcal{E}}(N)$, and a parameter in $\widetilde{\Psi}(G)$ that we continue to denote by ψ. If the assertion (a) of Theorem 2.2.1 is valid, we have the linear form

$$\widetilde{f}^G(\psi) = \widetilde{f}_N(\psi), \qquad \widetilde{f} \in \widetilde{\mathcal{H}}(N),$$

on $\widetilde{\mathcal{S}}(G)$. Following (2.4.6), we then set

$$\widetilde{f}_N^G(\psi, s) = \widetilde{f}^G(\psi), \qquad \widetilde{f} \in \widetilde{\mathcal{H}}(N). \tag{2.5.10}$$

COROLLARY 2.5.4 (Local intertwining relation for $\widetilde{G}(N)$). *Assume that the assertion (a) of Theorem* 2.2.1 *is valid for any pair*

$$(G, \psi), \qquad G \in \widetilde{\mathcal{E}}(N),\ \psi \in \widetilde{\Psi}(G).$$

Then if $\psi \in \Psi(M)$ is as in Theorem 2.5.3, *the identity*

$$\widetilde{f}_N^G(\psi, s_\psi s) = \widetilde{f}_N(\psi, u), \qquad \widetilde{f} \in \widetilde{\mathcal{H}}(N), \tag{2.5.11}$$

holds for any $u \in \widetilde{W}_\psi(M)$ and any semisimple element $s \in \widetilde{S}_\psi(N)$.

PROOF. It follows from (2.5.10) and (2.5.9) that $\widetilde{f}_N^G(\psi, s)$ equals $\widetilde{f}_N(\psi, u)$, for any s and u. The two linear forms are therefore independent of the points s and u. This is something we could have expected from general principles, given that the set $\widetilde{S}_\psi(N)$ is connected. Its proof, together with that of the equality of the two linear forms, is thus an immediate consequence of Theorem 2.5.3. The required identity (2.5.11) obviously follows if we replace s by $s_\psi s$. □

We return briefly to the case that $G \in \widetilde{\mathcal{E}}_{\text{sim}}(N)$ is a quasisplit orthogonal or symplectic group. A central theme of Chapter 2 has been that the general self-intertwining operator $R_P(w, \widetilde{\pi}, \psi)$ in (2.4.2) has no canonical definition. Only when the operator is balanced with the pairing $\langle \widetilde{x}_u, \widetilde{\pi} \rangle$ in (2.4.4) do we obtain a canonical object in general. We have seen in this section that the definition (2.4.2) can be made canonical in the special case that the parameter $\phi = \psi$ and the representation π are generic. Before we move on, we remind ourselves of another case that leads to a canonical definition. To describe it, we take ψ to be a general parameter in $\widetilde{\Psi}_2(M)$, with $\pi \in \widetilde{\Pi}(M)$ being a general element in the corresponding packet $\widetilde{\Pi}_\psi$.

We are assuming that w lies in the stabilizer $W_\psi(\pi)$ of ψ and π. As an element in $W(M)$, w stabilizes the orthogonal or symplectic part G_- of M. Suppose that it acts on G_- by inner automorphism. This is always so unless N is even and $\widehat{G}$ is orthogonal, in which case it is equivalent to the condition that the value corresponding to w of the explicit sign character (2.4.11) be 1. The representative $\widetilde{w}$ of w then commutes with G_-. We can therefore define the intertwining operator $\widetilde{\pi}(w)$ from $w\pi$ to π by asking that it stabilize the relevant Whittaker functional on each of the general linear factors of M. The product (2.4.2) is then a *canonical* self intertwining operator $R_P(w, \widetilde{\pi}, \psi)$ of $\mathcal{I}_P(\pi)$. It owes its existence to the fact that the extension $\langle \cdot, \widetilde{\pi} \rangle$ of the character $\langle \cdot, \pi \rangle$ from $\mathcal{S}_\phi(M)$ to $\mathcal{S}_\phi(\widetilde{M})$ in (2.4.5) is canonical.

Suppose for example that w lies in the subgroup W_ψ^0 of elements in W_ψ induced by points in the connected group $S_\psi^0(G)$. The corresponding value of (2.4.11) is then 1, for trivial reasons, and we obtain a canonical self intertwining operator $R_P(w, \widetilde{\pi}, \psi)$. The local intertwining relation for G suggests that $R_P(w, \widetilde{\pi}, \psi)$ equals 1. We shall establish this fact later, in the course of proving the intertwining relation.

There is one other application of the theory of Whittaker models that we will need. It is a kind of converse of the theorem of Konno, for nonarchimedean F, which we will need at one point in the proof of the local intertwining relation. Since the results in [**Kon**] rely on the exponential map, we do have to assume that the residual characteristic char(F) of F is not equal to 2.

LEMMA 2.5.5. *Suppose that F is nonarchimedean, and that the objects $G \in \widetilde{\mathcal{E}}_{\mathrm{sim}}(N)$, $\phi \in \widetilde{\Phi}(G)$, $\phi_M \in \widetilde{\Phi}_2(M, \phi)$, $M \neq G$ and (B_M, χ_M) over F have the following three properties.*

(i) $p = \mathrm{char}(F) \neq 2$.

(ii) *The element $\pi_M \in \widetilde{\Pi}_{\phi_M}$ with $\langle \cdot, \pi_M \rangle = 1$ is the unique (B_M, χ_M)-generic representation in the packet $\widetilde{\Pi}_{\phi_M}$.*

(iii) *For any pair (s, u) as in the statement of Theorem 2.4.1, there is a constant $e(s, u)$ such that*

$$f'_G(\phi, s) = e(s, u)\, f_G(\phi, u), \qquad f \in \widetilde{\mathcal{H}}(G).$$

Then $e(s, u) = 1$ for every (s, u). In other words, the local intertwining relation is valid for ϕ and ϕ_M.

Results of this nature are probably known. The proof in any case is not difficult, given Theorem 2.5.1(b) and the papers [**Kon**], [**MW1**] and [**Rod2**]. Since we need to move on, we shall leave the details for [**A27**]. We shall use Lemma 2.5.5 only in very special cases, which will come up in the proofs of Lemma 6.4.1 and Lemma 6.6.2.

CHAPTER 3

Global Stabilization

3.1. The discrete part of the trace formula

We can now begin to set in place the means for proving the three main theorems. The methods are based on the trace formula. In recent years, it has been possible to compare many terms in the trace formula directly with their stable analogues for endoscopic groups. The resulting cancellation has led to a corresponding identity for the remaining terms. These are the essential spectral terms, the ones which carry the desired information about automorphic representations. They comprise what is called the discrete part of the trace formula.

We shall review the discrete part of the trace formula in this section. The theorems we are trying to prove concern quasisplit orthogonal and symplectic groups. In the long term, we are of course interested in more general groups so it would be sensible to think as broadly as possible in discussing the general theory. We shall do so whenever practical, with the understanding that the results will ultimately be specialized to orthogonal and symplectic groups, and the corresponding global parameters defined in §1.4.

We assume until further notice that the field F is global. Suppose for the moment that G is a connected, reductive algebraic group over F. The discrete part of the trace formula for G is any linear form on the Hecke algebra $\mathcal{H}(G)$. We recall that $\mathcal{H}(G)$ is the space of smooth, compactly supported, complex-valued functions on $G(\mathbb{A})$ that are K-finite relative to the left and right actions of a suitable maximal compact subgroup $K \subset G(\mathbb{A})$. It will be convenient to have this linear form depend on two other quantities, which we will use to account for two minor technical complications.

The first point concerns the nominal failure for G to have any discrete spectrum. This is governed by the split component A_G of the center of G, or equivalently, the quotient of $G(\mathbb{A})$ by $G(\mathbb{A})^1$. The subgroup $G(\mathbb{A})^1$ of $G(\mathbb{A})$ equals the kernel of the homomorphism

$$H_G : \, G(\mathbb{A}) \longrightarrow \mathfrak{a}_G$$

from $G(\mathbb{A})$ onto the real vector space

$$\mathfrak{a}_G = \mathrm{Hom}_{\mathbb{Z}}\big(X^*(G)_F, \mathbb{R}\big),$$

which is defined by the familiar prescription

$$e^{\langle H_G(x), \chi\rangle} = |\chi(x)|, \qquad x \in G(\mathbb{A}),\ \chi \in X^*(G)_F.$$

It has a group theoretic complement

$$A^+_{G,\infty} = A_{G_{\mathbb{Q}}}(\mathbb{R})^0, \qquad G_{\mathbb{Q}} = R_{F/\mathbb{Q}}(G),$$

in $G(\mathbb{A})$, where $R_{F/\mathbb{Q}}(\cdot)$ denotes the restriction of scalars, and $(\,\cdot\,)^0$ stands for the connected component of 1. In particular, the mapping H_G restricts to a group theoretic isomorphism from $A^+_{G,\infty}$ onto $\mathfrak{a}_G$. It gives rise to an isometric isomorphism

$$L^2\big(G(F)A^+_{G,\infty}\backslash G(\mathbb{A})\big) \xrightarrow{\sim} L^2\big(G(F)\backslash G(\mathbb{A})^1\big).$$

We are therefore free to work with the discrete spectrum of the left hand space in place of $L^2_{\mathrm{disc}}\big(G(F)\backslash G(\mathbb{A})^1\big)$.

To allow flexibility for induction arguments, it is useful to treat a slightly more general situation. We choose a closed subgroup $\mathfrak{X}_G$ of the full center $Z\big(G(\mathbb{A})\big)$ of $G(\mathbb{A})$ such that the product $\mathfrak{X}_G Z\big(G(F)\big)$ is closed and cocompact in $Z\big(G(\mathbb{A})\big)$. (In general, we write $Z(H)$ for the center of any group H.) The quotient

$$G(F)\mathfrak{X}_G\backslash G(\mathbb{A}) = G(F)\backslash G(\mathbb{A})/\mathfrak{X}_G$$

then has finite invariant volume, as in the special case that $\mathfrak{X}_G = A^+_{G,\infty}$. We also fix a character χ on the quotient

$$Z\big(G(F)\big) \cap \mathfrak{X}_G\backslash\mathfrak{X}_G = Z\big(G(F)\big)\backslash Z\big(G(F)\big)\mathfrak{X}_G,$$

and write

$$L^2_{\mathrm{disc}}\big(G(F)\backslash G(\mathbb{A}),\chi\big) \subset L^2\big(G(F)\backslash G(\mathbb{A}),\chi\big)$$

for the space of χ-equivariant functions on $G(F)\backslash G(\mathbb{A})$ that are square-integrable modulo $\mathfrak{X}_G$, and decompose discretely under the action of $G(\mathbb{A})$. The central character datum $(\mathfrak{X}_G,\chi)$ for G gives rise to corresponding data for Levi subgroups M of G. Given M, we write $A^{+,G}_{M,\infty}$ for the kernel in $A^+_{M,\infty}$ of the composition

$$A^+_{M,\infty} \xrightarrow{H_M} \mathfrak{a}_M \longrightarrow \mathfrak{a}_G.$$

The product

$$\mathfrak{X}_M = (A^{+,G}_{M,\infty})\mathfrak{X}_G$$

is then an extension of $\mathfrak{X}_G$, to which we can pull back the character χ. We obtain a triplet $(M,\mathfrak{X}_M,\chi)$ that satisfies the same conditions as $(G,\mathfrak{X}_G,\chi)$. In particular, we obtain a representation of $M(\mathbb{A})$ on $L^2_{\mathrm{disc}}\big(M(F)\backslash M(\mathbb{A}),\chi\big)$ that decomposes discretely. If $P = N_P M$ belongs to the set $\mathcal{P}(M)$ of parabolic subgroups of G with Levi component M, we write

$$\mathcal{I}_P(\chi) = \mathcal{I}^G_P(\chi)$$

for the corresponding parabolically induced representation. It acts on the Hilbert space $\mathcal{H}_P(\chi)$ of left $N_P(\mathbb{A})$-invariant functions ϕ on $G(\mathbb{A})$ such that the function $\phi(mk)$ on $M(\mathbb{A}) \times K$ belongs to the space

$$L^2_{\mathrm{disc}}\big(M(F)\backslash M(\mathbb{A}),\chi\big) \otimes L^2(K).$$

This discussion can of course be applied to the special case above, namely where $\mathfrak{X}_G = A^+_{G,\infty}$, $\chi = 1$, and $\mathfrak{X}_M = A^+_{M,\infty}$.

The second point has to do with analytic estimates. W. Müller has established [**Mu**] that the restriction of any operator

$$\mathcal{I}_P(\chi, f) = \int_{G(\mathbb{A})} f(x)\mathcal{I}_P(\chi, x)dx, \qquad f \in \mathcal{H}(G),$$

to the discrete spectrum is of trace class. Rather than work with the estimates of Müller, however, we shall rely on the consequences in [**A5**] of what was later called the multiplier convergence estimate [**A14**, §3]. These allow us to work with small subrepresentations of $\mathcal{I}_P(x)$, defined by restricted Archimedean infinitesimal characters.

We fix a minimal Levi subgroup M_0 of G, which we can assume is in good position relative to K. We can then construct the real vector space

$$\mathfrak{h} = i\mathfrak{h}_K \oplus \mathfrak{h}_0$$

in terms of the group $M_{0,\mathbb{Q}} = R_{F/\mathbb{Q}}(M_0)$, as in [**A2**, §3]. Thus, $\mathfrak{h}_0$ is the Lie algebra of a maximal real split torus in the real group $M_{0,\mathbb{Q}}(\mathbb{R}) = M_0(F_\infty)$, and $\mathfrak{h}_K$ is a Cartan subalgebra of the Lie algebra of the compact real group

$$K \cap M_{0,\mathbb{Q}}(\mathbb{R}) = K \cap M_0(F_\infty).$$

The complexification $\mathfrak{h}_\mathbb{C}$ of $\mathfrak{h}$ is a Cartan subalgebra of the Lie algebra of the complex group $G_\mathbb{Q}(\mathbb{C})$, whose real form $\mathfrak{h}$ is invariant under the complex Weyl group W. The role of the real space $\mathfrak{h} = \mathfrak{h}_G$ is simply to control infinitesimal characters. We represent the Archimedean infinitesimal character of an irreducible representation π of $G(\mathbb{A})$ by a complex valued linear form $\mu_\pi = \mu_{\pi,\mathbb{R}} + i\mu_{\pi,I}$ on $\mathfrak{h}$. The imaginary part $\mu_{\pi,I}$, regarded as a W-orbit in $\mathfrak{h}^*$, is the object of interest. It has a well defined norm $\|\mu_{\pi,I}\|$, relative to a fixed, W-invariant Hermitian metric on $\mathfrak{h}^*_\mathbb{C}$. There is then a decomposition

$$\mathcal{I}_P(\chi) = \bigoplus_{t \geqslant 0} \mathcal{I}_{P,t}(\chi)$$

of $\mathcal{I}_P(\chi)$, where $\mathcal{I}_{P,t}(\chi)$ is the subrepresentation of $\mathcal{I}_P(\chi)$ composed of those irreducible constituents π with $\|\mu_{\pi,I}\| = t$. The endoscopic comparison of trace formulas reduces ultimately to a reciprocity law among the representations $\mathcal{I}_{P,t}(\chi)$ attached to different groups.

The discrete part of the trace formula will be a linear form that depends on χ and t. We may as well build the dependence on χ into the test function f. Let $\mathcal{H}(G, \chi)$ be the equivariant Hecke algebra of functions f on $G(\mathbb{A})$ that satisfy

$$f(xz) = f(x)\chi(z)^{-1}, \qquad z \in \mathfrak{X}_G,$$

and are compactly supported modulo $\mathfrak{X}_G$. The operator

$$\mathcal{I}_{P,t}(\chi, f) = \int_{G(\mathbb{A})/\mathfrak{X}_G} f(x)\mathcal{I}_{P,t}(\chi, x)dx$$

on $\mathcal{H}_P(\chi)$ is then defined for f in $\mathcal{H}(G,\chi)$. In this setting, the discrete part of the trace formula is the linear form

$$I_{\mathrm{disc},t}(f) = I^G_{\mathrm{disc},t}(f), \qquad t \geqslant 0,\ f \in \mathcal{H}(G,\chi),$$

defined by

(3.1.1)
$$\sum_{M\in\mathcal{L}} |W_0^M||W_0^G|^{-1} \sum_{w\in W(M)_{\mathrm{reg}}} |\det(w-1)_{\mathfrak{a}_M^G}|^{-1}\mathrm{tr}\big(M_{P,t}(w,\chi)\mathcal{I}_{P,t}(\chi,f)\big).$$

The outer sum in (3.1.1) is over the finite set $\mathcal{L} = \mathcal{L}(M_0)$ of Levi subgroups of G that contain M_0. The inner sum is over the set of regular elements

$$W(M)_{\mathrm{reg}} = \big\{w \in W(M) :\ \det(w-1)_{\mathfrak{a}_M^G} \neq 0\big\}$$

in the relative Weyl group

$$W(M) = W^G(M) = \mathrm{Norm}(A_M, G)/M$$

for G and M. Here $\mathfrak{a}_M^G$ is the canonical complement of $\mathfrak{a}_G$ in $\mathfrak{a}_M$, so $W(M)_{\mathrm{reg}}$ is the set of w whose fixed subspace in $\mathfrak{a}_M$ equals the minimal space $\mathfrak{a}_G$. Following standard practice, we have written $W_0^M = W^M(M_0)$ and $W_0^G = W^G(M_0)$ for the Weyl groups with respect to M_0.

The remaining ingredient of (3.1.1) comes from the standard global intertwining operator $M_P(w,\chi)$ that is at the heart of Langlands' theory of Eisenstein series. We can define $M_P(w,\chi)$ as the value at $\lambda = 0$ of a meromorphic composition

$$M_P(w,\chi_\lambda) = \ell(w)\circ M_{P'|P}(\chi_\lambda), \qquad P' = w^{-1}P,\ \lambda \in (\mathfrak{a}_M^G)^*_{\mathbb{C}}, \tag{3.1.2}$$

of operator valued functions on $\mathcal{H}_P(\chi)$. The factor $\ell(w)$ stands for the mapping from $\mathcal{H}_{P'}(\chi)$ to $\mathcal{H}_P(\chi)$ defined by left translation by any representative $\widetilde{w}^{-1}$ of w^{-1} in $G(F)$, while $P' = w^{-1}P$ is the parabolic subgroup $\widetilde{w}^{-1}P\widetilde{w}$ in $\mathcal{P}(M)$. The other factor is the operator

$$M_{P'|P}(\chi_\lambda) :\ \mathcal{H}_P(\chi) \longrightarrow \mathcal{H}_{P'}(\chi), \qquad P, P' \in \mathcal{P}(M),$$

whose value

$$\big(M_{P'|P}(\chi_\lambda)\phi\big)(x), \qquad \phi \in \mathcal{H}_P(\chi),\ x \in G(\mathbb{A}),$$

is defined for the real part of λ in a certain cone in $(\mathfrak{a}_M^G)^*$ (and for ϕ in the dense subspace of smooth functions in $\mathcal{H}_P$) by the familiar intertwining integral

$$\int_{N_{P'}(\mathbb{A})\cap N_P(\mathbb{A})\backslash N_{P'}(\mathbb{A})} \phi(nx)e^{(\lambda+\rho_P)(H_P(nx))}dn \cdot e^{-(\lambda+\rho_{P'})(H_{P'}(x))}.$$

We recall that H_P is the mapping from $G(\mathbb{A})$ to $\mathfrak{a}_M$ defined by

$$H_P(nmk) = H_M(m), \qquad n \in N_P(\mathbb{A}),\ m \in M(\mathbb{A}),\ k \in K,$$

that ρ_P is the usual linear form on $\mathfrak{a}_M$ defined by half the sum of the roots (with multiplicity) of (P, A_M), and that

$$\chi_\lambda(u) = \chi(u)\, e^{\lambda(H_M(u))}, \qquad u \in \mathfrak{X}_M,$$

is the twist of χ by λ.

The operator $M_P(w, \chi_\lambda)$ leaves the subspace $\mathcal{H}_{P,t}(\chi)$ of $\mathcal{H}_P(\chi)$ invariant, and intertwines the two induced representations $\mathcal{I}_{P,t}(\chi_\lambda)$ and $\mathcal{I}_{P,t}(\chi_{w\lambda})$ on this subspace. It has analytic continuation as a meromorphic function of $\lambda \in (\mathfrak{a}_M^G)^*_{\mathbb{C}}$, whose values at $i(\mathfrak{a}_M^G)^*$ are analytic and unitary. Thus, $M_P(w, \chi_\lambda)$ restricts to a unitary operator $M_{P,t}(w, \chi)$ on the Hilbert space $\mathcal{H}_{P,t}(\chi)$, which intertwines the representation $\mathcal{I}_{P,t}(\chi)$. This operator is the last of the terms in (3.1.1). It is also the most interesting. Its analysis includes the local intertwining relation stated in the last chapter, and will be an important aspect of future chapters.

More generally, suppose that

$$G = (G^0, \theta, \omega) \tag{3.1.3}$$

is an arbitrary triplet over F. Then G^0 is a connected reductive group over F, θ is a semisimple automorphism of G over F, and in the global context here, ω is a character on $G^0(\mathbb{A})$ that is trivial on the subgroup $G^0(F)$. As in the local case of §2.1, we also write

$$G = G^0 \rtimes \theta$$

more narrowly for the associated G^0-bitorsor over F, with distinguished point

$$\theta = 1 \rtimes \theta.$$

The Hecke module $\mathcal{H}(G, \chi)$ is then a space of functions on $G(\mathbb{A})$ defined exactly as in the case $G = G^0$ above. We note that there are two morphisms from $G \times G$ to G^0. They are defined in the obvious way as formal algebraic operations on the bitorsor by

$$\begin{cases} y_1^{-1} y_2 = \theta^{-1}(x_1^{-1} x_2), \\ y_1 y_2^{-1} = x_1 x_2^{-1}, \end{cases} \tag{3.1.4}$$

for two points $y_i = x_i \rtimes \theta$ in G. Our general convention of letting G stand for both a G^0-torsor and the underlying triplet should not lead to confusion. It is similar to our convention for endoscopy, in which a symbol G' represents both an endoscopic group and the underlying endoscopic datum.

The discrete part of the trace formula for G again takes the form (3.1.1), provided that the terms are properly interpreted. We recall how these terms can be understood in the general twisted case.

For a start, the central character datum $(\mathfrak{X}_G, \chi)$ has to be adapted to G (rather than G^0). We write

$$\mathfrak{a}_G = \mathfrak{a}_{G^0}^\theta = \{H \in \mathfrak{a}_{G^0} : \ \mathrm{Ad}(\theta)H = H\} \subset \mathfrak{a}_{G^0}$$

for the subspace of θ-fixed vectors in the real vector space $\mathfrak{a}_{G^0}$, and

$$A_G = (A_{G^0}^\theta)^0 \subset A_{G^0}$$

for the maximal F-split torus in the centralizer of G in G^0. We assume implicitly that ω is trivial on the subgroup

$$A^+_{G,\infty} = A_{G_{\mathbb{Q}}}(\mathbb{R})^0, \qquad G_{\mathbb{Q}} = R_{F/\mathbb{Q}}(G),$$

of $G^0(\mathbb{A})$, since the twisted trace formula for G otherwise becomes trivial. If A is any ring that contains F, we write

$$Z\big(G(A)\big) = Z\big(G^0(A)\big)^\theta \subset Z\big(G^0(A)\big)$$

for the centralizer of $G(A)$ in $G^0(A)$. We then take

$$\mathfrak{X}_G \subset Z\big(G(\mathbb{A})\big)$$

to be any closed subgroup satisfying the conditions above, with the further requirement that it lie in the kernel of ω, and χ to be any character on the quotient of $\mathfrak{X}_G$ by its intersection with $Z\big(G(F)\big) \cap \mathfrak{X}_G$. The standard example is of course the pair

$$(\mathfrak{X}_G, \chi) = (A^+_{G,\infty}, 1).$$

For any such choice, we can form the equivariant Hecke module $\mathcal{H}(G,\chi)$ of functions in $G(\mathbb{A})$, relative to $(\mathfrak{X}_G, \chi)$ and a suitably fixed maximal compact subgroup $K \subset G^0(\mathbb{A})$. The general version of (3.1.1) will be a linear form in a function f in $\mathcal{H}(G,\chi)$.

The general version of (3.1.1) is again based on a fixed minimal Levi subgroup $M_0 \subset G^0$. Suppose that M belongs to the set $\mathcal{L} = \mathcal{L}(M_0)$ of Levi subgroups of G^0 that contain M_0. The composition

$$\mathfrak{a}_M \longrightarrow \mathfrak{a}_{G^0} \longrightarrow \mathfrak{a}_{G^0,\theta},$$

together with the natural isomorphism between $\mathfrak{a}_G$ and the space of θ-covariants

$$\mathfrak{a}_{G^0,\theta} = \mathfrak{a}_{G^0}/\big\{X - \mathrm{Ad}(\theta)X : \ X \in \mathfrak{a}_{G^0}\big\}$$

in $\mathfrak{a}_{G^0}$, gives a canonical linear projection from $\mathfrak{a}_M$ onto $\mathfrak{a}_G$. We write $\mathfrak{a}^G_M$ for its null space in $\mathfrak{a}_M$. We then have the regular set

$$W_{\mathrm{reg}}(M) = W^G_{\mathrm{reg}}(M) = \big\{w \in W(M) : \ \det(w-1)_{\mathfrak{a}^G_M} \neq 0\big\},$$

in the Weyl set

$$W(M) = W^G(M) = \mathrm{Norm}(A_M, G)/M$$

of outer automorphisms of M induced by the conjugation action of G on G^0. The set $W_{\mathrm{reg}}(M)$ is generally quite distinct from its untwisted analogue $W^0_{\mathrm{reg}}(M) = W^{G^0}_{\mathrm{reg}}(M)$, defined for G^0 above. These conventions account for the general indices of summation M and w, and the corresponding coefficients

$$|\det(w-1)_{\mathfrak{a}^G_M}|^{-1},$$

in the twisted interpretation of (3.1.1).

We observe that $(\mathfrak{X}_G, \chi)$ again provides a central character datum $(\mathfrak{X}_M, \chi)$ for any $M \in \mathcal{L}$. The first component is the extension

$$\mathfrak{X}_M = (A^{+,G}_{M,\infty})\mathfrak{X}_G$$

of $\mathfrak{X}_G$, and the second component is the character on $\mathfrak{X}_M$ obtained by pulling back χ. The only difference from the case $G = G^0$ above is that $A^{+,G}_{M,\infty}$ is now the kernel of the projection of $A^+_{M,\infty}$ onto the subspace $\mathfrak{a}_G$ attached to G (rather than G^0).

The remaining term is the trace in (3.1.1). Suppose that P belongs to the set $\mathcal{P}(M)$ of parabolic subgroups of G^0 with Levi component M. We write $\mathcal{H}^0_{P,t}(\chi)$ for the Hilbert space of the induced representation defined for G^0 above. We reserve the symbol $\mathcal{H}_{P,t}(\chi)$ for its analogue for G, namely the space of complex-valued functions ϕ on $G(\mathbb{A})$ such that for any $y \in G(\mathbb{A})$, the function

$$\phi(xy), \qquad x \in G^0(\mathbb{A}),$$

belongs to $\mathcal{H}^0_{P,t}(\chi)$. The operators in (3.1.1) attached to $f \in \mathcal{H}(G, \chi)$ and $w \in W(M)$ make sense in this more general context. However, they have now to be interpreted as linear transformations

$$\mathcal{I}_{P,t}(\chi, f) : \mathcal{H}_{P,t}(\chi) \longrightarrow \mathcal{H}^0_{P,t}(\chi) \tag{3.1.5}$$

and

$$M_{P,t}(w, \chi) : \mathcal{H}^0_{P,t}(\chi) \longrightarrow \mathcal{H}_{P,t}(\chi). \tag{3.1.6}$$

The first transformation equals

$$\mathcal{I}_{P,t}(\chi, f) = \mathcal{I}^G_{P,t}(\chi, f) = \int_{G(\mathbb{A})/\mathfrak{X}_G} f(y)\mathcal{I}^G_{P,t}(\chi, y)dy,$$

for the induced mapping

$$\big(\mathcal{I}^G_P(\chi, y)\phi\big)(x) = \phi(xy)\omega(\theta^{-1}xy), \qquad \phi \in \mathcal{H}_{P,t}(\chi),\ x \in G^0(\mathbb{A}),$$

from $\mathcal{H}_{P,t}(\chi)$ to $\mathcal{H}^0_{P,t}(\chi)$. (There is no need to include ω in the notation, as it is implicit in superscript G.) The second transformation is given by (3.1.2), with the translation $\ell(w)$ defined now as a map from $\mathcal{H}^0_{P'}(\chi)$ to $\mathcal{H}_P(\chi)$ by (3.1.4). The product

$$M_{P,t}(w, \chi)\mathcal{I}_{P,t}(\chi, f)$$

is then an operator on $\mathcal{H}_{P,t}(\chi)$, for which the trace in (3.1.1) is defined.

With this notation, all of the terms in the expression (3.1.1) makes sense. The discrete part of the trace formula for $G = (G^0, \theta, \omega)$ is the linear form

$$I_{\mathrm{disc},t}(f) = I^G_{\mathrm{disc},t}(f), \qquad t \geqslant 0,\ f \in \mathcal{H}(G, \chi),$$

defined by the general form of this expression. It is the main spectral component of a general formula usually known as the twisted trace formula. The invariant (untwisted) trace formula was established in [**A5**]. Its general (twisted) analogue has recently been proved in full detail [**LW**], following notes from the seminar [**CLL**]. Our notation here generally follows that of paper [**A5**], which included some of the formal aspects of the twisted trace formula, but with three minor differences that might be worth pointing out explicitly.

In [**A5**], G represents a component of a (nonconnected) reductive algebraic group, and ω is trivial. This is slightly less general than the triplet (G^0, θ, ω) here. However, the earlier discussion extends without difficulty. The given datum of [**A5**] is also less precise, in that it specifies the underlying automorphism of G^0 only up to inner automorphism. The point of view here is more suitable to stablization, which we will discuss in the next section.

Secondly, the induced Hilbert spaces $\mathcal{H}^0_{P,t}$ and $\mathcal{H}_{P,t}$ do not take this form in [**A5**]. They reflect the fact here that the G^0-torsor G need not be attached to a component of a reductive algebraic group. They also streamline the notation in all cases.

Finally, we reiterate that the elements $M \in \mathcal{L}$ here are Levi subgroups of G^0. The complementary geometric terms in the twisted trace formula require another notion, that of a Levi *subset* M of G, which we can recall from the local context of §2.2. To keep the two kinds of objects straight, we wrote $\mathcal{L}^0 = \mathcal{L}^{G^0}(M_0)$ in [**A5**] for the set of Levi subgroups of G^0 containing M_0 and $\mathcal{L} = \mathcal{L}^G(M_0)$ for the collection of Levi subsets of G, noting at the same time that there was an injective mapping $M \to M^0$ from $\mathcal{L}$ to $\mathcal{L}^0$. In the present volume, the discrete part $I_{\mathrm{disc},t}$ is the only term we will consider. Since it is based only on the Levi subgroups of G^0, we are free here to denote these objects by M, and to write $\mathcal{L}$ for the set of M that contain M_0. It is with this notation that the expression from $I_{\mathrm{disc},t}(f)$ takes exactly the same form (3.1.1) in general as it does in the original case of $G = G^0$.

The starting point for the trace formula is the kernel of an integral operator. In the general case at hand, the operator is a composition

$$\mathcal{I}_G(\chi, f) \circ M_G(\theta, \chi), \qquad f \in \mathcal{H}(G, \chi),$$

of what amount to specializations to $M = G^0$ of the transformations (3.1.5) and (3.1.6), but with the χ-equivariant Hilbert space $L^2\big(G^0(F)\backslash G^0(\mathbb{A}), \chi\big)$ in place of

$$\mathcal{H}^0_{G^0,t}(\chi) = L^2_{\mathrm{disc},t}\big(G^0(F)\backslash G^0(\mathbb{A}), \chi\big).$$

The reader can check that the kernel of this operator equals

$$K(x, y) = \sum_{\gamma \in G(F)} f(x^{-1}\gamma y)\omega(y), \qquad x, y \in G^0(F)\backslash G^0(\mathbb{A}).$$

As a general rule, we have not tried to be explicit about measures. In the global setting here, we note simply that the operator $\mathcal{I}_{P,t}(\chi, f)$ in the formula (3.1.1) for $I_{\mathrm{disc},t}(f)$ depends on a choice of $G^0(\mathbb{A})$-invariant measure on $G(\mathbb{A})/\mathfrak{X}_G$. This amounts to a choice of Haar measure on $G^0(\mathbb{A})/\mathfrak{X}_G$. Observe that a Haar measure on the group $\mathfrak{X}_G$ determines a Haar measure on $A^+_{G,\infty}$ such that

$$\mathrm{vol}\big(\mathfrak{X}_G Z(G(F))\backslash Z(G(\mathbb{A}))\big) = \mathrm{vol}\big(A^+_{G,\infty} Z(G(F))\backslash Z(G(\mathbb{A}))\big),$$

for any Haar measure on $Z\big(G(\mathbb{A})\big)$. The linear form $I_{\mathrm{disc},t}(f)$ thus depends on a choice of $G^0(\mathbb{A})$-invariant measure on $G(\mathbb{A})/A^+_{G,\infty}$, or equivalently, a Haar measure on $G^0(\mathbb{A})/A^+_{G,\infty}$.

Finally, we observe that the linear form $I_{\mathrm{disc},t}(f)$ is "admissible" in f. In other words, it can be written as a linear combination of twisted characters

$$\mathrm{tr}\big(\pi(f)\big), \qquad \pi^0 \in \Pi^0,$$

where Π^0 is a *finite* set of irreducible unitary representations of $G^0(\mathbb{A})$, and π is an extension of π^0 to $G(\mathbb{A})$, in the sense of the global analogue of (2.1.1). The set $\Pi^0 = \Pi^0_t(f)$ depends on f, but only through its K-type.

To be more precise, let us agree that a Hecke type for G means a pair $(\tau_\infty, \kappa^\infty)$, where κ^∞ is an open compact subgroup of $G^0(\mathbb{A}^\infty)$ and τ_∞ is a finite set of irreducible representations of a maximal compact subgroup K_∞ of $G(\mathbb{A}_\infty)$. We have written

$$\mathbb{A}_\infty = \{a \in \mathbb{A} : \ a_v = 0, \ \text{if } v \notin S_\infty\} = \prod_{v \in S_\infty} F_v = F_\infty$$

and

$$\mathbb{A}^\infty = \{a \in \mathbb{A} : \ a_v = 0, \ \text{if } v \in S_\infty\}$$

here for the respective subrings of archimedean and finite adeles in $\mathbb{A}$. We also write

$$\mathbb{A}^S = \{a \in \mathbb{A} : \ a_v = 0, \ \text{if } v \in S\}$$

and

$$A^\infty_S = \{a \in \mathbb{A}^\infty : \ a_v = 0, \ \text{if } v \notin S\},$$

if S is any set of valuations of F that contains the set S_∞ of archimedean valuations. We can then decompose κ^∞ as a product $\kappa^\infty_S K^S$, where $S \supset S_\infty$ is finite, κ^∞_S is an open compact subgroup of $G^0(\mathbb{A}^\infty_S)$, and K^S is a product over $v \notin S$ of hyperspecial maximal compact subgroups $K_v \subset G^0(F_v)$. Assume that the product $\kappa = K_\infty \kappa^\infty$ is contained in the maximal compact subgroup K (and is in particular a subgroup of finite index in K). We shall say that $(\tau_\infty, \kappa^\infty)$ is a *Hecke type for* f if f is bi-invariant under translation by κ^∞, and transforms under left and right translation by K_∞ according to representations in the set τ_∞. Any $f \in \mathcal{H}(G,\chi)$ of course has a Hecke type.

The assertion above is that the finite set Π^0 depends only on a choice of Hecke type for f. This follows from the definition (3.1.1) of $I_{\mathrm{disc},t}(f)$ and the theory of Eisenstein series, specifically Langlands' decomposition of $L^2\big(G^0(F)\backslash G^0(\mathbb{A})\big)$ in terms of residues of cuspidal Eisenstein series [**L5**, Chapter 7]. (See [**A5**, Lemma 4.1]. Once κ^∞ has been fixed, the archimedean infinitesimal character of π is controlled by the K_∞-type τ_∞ and the norm

$t = \|\mu_{\pi,I}\|$ of its imaginary part.) Notice that for the extension π of π^0, $\mu_{\pi,I} = \mu_{\pi^0,I}$ is a linear form on the quotient

$$\mathfrak{h} = \mathfrak{h}^0_\theta = \mathfrak{h}^0/\{H - \mathrm{Ad}(\theta)H : H \in \mathfrak{h}^0\}$$

of the real vector space $\mathfrak{h}^0 = \mathfrak{h}_{G^0}$ attached to G^0.

3.2. Stabilization

In this section we shall formally state the condition on which our theorems depend. We have just described a linear form $I^G_{\mathrm{disc},t}(f)$ on $\mathcal{H}(G,\chi)$. The condition is that it can be stabilized. There has been considerable progress on this problem in recent years, and a general solution in case $G = G^0$ is now in place. One can hope that its extension to twisted groups might soon be within reach. For our part here, we require a solution only for the two twisted groups $G = \widetilde{G}(N)$ and $G = \widetilde{G}$, initially introduced in §1.2.

We continue to suppose for the time being that G represents a general triplet (G^0, θ, ω) over the global field F. We need to discuss what it means to stabilize the linear form $I^G_{\mathrm{disc},t}(f)$. Let us first review a few of the background notions from the beginning of [**KS**], which underlie both the local discussion of §2.1 and the global discussion we are about to undertake here.

General endoscopic transfer has to be formulated in terms of several supplementary data. These include a connected quasisplit group G^* over F, equipped with an inner class of inner twistings ψ: $G^0 \to G^*$ of the connected group G^0. Given G^*, one fixes an F-automorphism θ^* of G^* that preserves some F-splitting, and is of the form

$$\theta^* = \mathrm{Int}(g_\theta)\, \psi\, \theta\, \psi^{-1},$$

for some element $g_\theta \in G^*_{\mathrm{sc}}$. (As usual, G^*_{sc} denotes the simply connected cover of derived group of G^*.) The automorphism θ of G^0 also induces an automorphism $\widehat{\theta}$ of the dual group $\widehat{G}^0$ of G^0 that preserves some Γ_F-splitting, and is determined up to conjugation. One fixes a 1-cocycle a_ω from the Weil group W_F of F to the center $Z(\widehat{G}^0)$ of $\widehat{G}^0$ that is the Langlands dual of the automorphic character ω on $G^0(\mathbb{A})$. This then gives rise to an L-automorphism

$${}^L\theta = {}^L\theta_\omega : \; g \times w \longrightarrow \widehat{\theta}(g) a_\omega(w)^{-1} \times w, \qquad g \times w \in {}^L G^0,$$

of the L-group ${}^L G^0$ of G^0. We use it to define the dual $\widehat{G}^0$-bitorsor

$$\widehat{G} = \widehat{G}_\omega = \widehat{G}^0 \rtimes {}^L\theta_\omega,$$

on which the L-group ${}^L G^0$ acts by conjugation.

These matters are discussed (with slightly different notation) in the early stages (2.1) of [**KS**]. So is the general notion of endoscopic datum G' for G.

As in the local case mentioned briefly in §2.1, a general endoscopic datum G' represents a 4-tuple $(G', \mathcal{G}', s', \xi')$. We recall that the first component, denoted also by G', is a connected quasisplit group over F. The other components are a split extension $\mathcal{G}'$ of W_F by $\widehat{G}'$, a semisimple element

s' in $\widehat{G}$, and an L-embedding ξ' of $\mathcal{G}'$ into ${}^LG^0$. These four components are required to satisfy the conditions (2.1.1)–(2.1.4) of [**KS**], the most basic being the assertion (2.1.4b) that $\xi'(\widehat{G}')$ equals the connected centralizer of s' in $\widehat{G}^0$. Similar definitions apply to the local completions F_v of F. In particular, an endoscopic datum G' for G over F localizes to an endoscopic datum G'_v for G_v over F_v. The general notion of *isomorphism* between endoscopic data is defined on p. 18 of [**KS**]. As in our earlier special cases, we write

$$\mathrm{Out}_G(G') = \mathrm{Aut}_G(G')/\mathrm{Int}_G(G')$$

for the group of outer automorphisms of G' over F obtained from automorphisms of G' as an endoscopic datum. We also write $\mathcal{E}(G)$ for the set of isomorphism classes of endoscopic data G' that are locally relevant to G, by which we mean that for every v, $G'(F_v) = G'_v(F_v)$ contains elements that are norms from $G(F_v)$ ([**KS**, p. 29]). Endoscopic data that are not locally relevant play no role in transfer, and can be ignored.

For a given endoscopic datum G', the group $\mathcal{G}'$ need not be L-isomorphic to ${}^LG'$. Even if it is, there often is not a canonical isomorphism. To remedy this defect, one attaches an auxiliary datum $(\widetilde{G}', \widetilde{\xi}')$ to G', as in the local discussion of §2.1. We recall that $\widetilde{G}'$ represents a central extension of G' by a torus $\widetilde{C}'$ over F that is induced, in the sense that it is a product of tori of the form $R_{E/F}(\mathbb{G}_m)$, and $\widetilde{\xi}'$ is an L-embedding of $\mathcal{G}'$ into ${}^L\widetilde{G}'$. For example, one can always take $\widetilde{G}'$ to be a z-extension of G' (See [**KS**, (2.2)].) The L-embedding $\widetilde{\xi}'$ gives rise to a cohomology class in $H^1(F, \widehat{\widetilde{C}}')$, which in turn yields an automorphic character $\widetilde{\eta}'$ on $\widetilde{C}'(\mathbb{A})/\widetilde{C}'(F)$, by the global Langlands correspondence for tori. (With a suitable choice of $\widetilde{\xi}'$, we can assume that $\widetilde{\eta}'$ is unitary.)

The global transfer factor for G and G' (and the auxiliary datum $(\widetilde{G}', \widetilde{\xi}')$) is a canonical function $\Delta(\delta', \gamma)$ of two adèlic variables δ' and γ. These lie in the adèlized varieties of strongly G-regular, stable conjugacy classes in $\widetilde{G}'$ and G respectively. It again serves as the kernel of a transfer mapping, which takes functions $f \in \mathcal{H}(G)$ to functions

$$f'(\delta') = \sum_{\gamma} \Delta(\delta', \gamma) f_G(\gamma)$$

of δ'. The global transfer factor can be written as a (noncanonical) product

$$\Delta(\delta', \gamma) = \prod_v \Delta_v(\delta'_v, \gamma_v) \tag{3.2.1}$$

of local transfer factors. The transfer mapping therefore takes a decomposable function

$$f = \prod_v f_v$$

in $\mathcal{H}(G)$ to the decomposable function

$$f'(\delta') = \prod_v f'_v(\delta'_v)$$

of δ'. The LSK transfer conjecture, now a theorem, tells us that for any v, f'_v belongs to the space $\mathcal{S}(\widetilde{G}'_v, \widetilde{\eta}'_v)$. The fundamental lemma tells us that for almost all v, f'_v is the image in $\mathcal{S}(\widetilde{G}'_v, \widetilde{\eta}'_v)$ of the characteristic function of a hyperspecial maximal compact subgroup of $\widetilde{G}'(F_v)$. It follows that f' belongs to the restricted tensor product

$$\mathcal{S}(\widetilde{G}', \widetilde{\eta}') = \widetilde{\bigotimes_v} \mathcal{S}(\widetilde{G}'_v, \widetilde{\eta}'_v)$$

of local stable Hecke algebras.

The general situation is actually slightly more complicated than we have indicated. It comes with a cohomology class $z' \in H^1\big(F, Z(G^*_{\mathrm{sc}})_{\theta^*}\big)$ with values in the group of θ^*-covariants

$$Z(G^*_{\mathrm{sc}})_{\theta^*} = Z(G^*_{\mathrm{sc}})/\{z\theta^*(z)^{-1} : z \in Z(G^*_{\mathrm{sc}})\}$$

in the center of G^*_{sc} [**KS**, Lemma 3.1.A]. For much of [**KS**], the authors assume that this class (or rather its local analogue) is trivial. They explain how to take care of the more general situation in (5.4), at the expense of replacing $\mathcal{S}(\widetilde{G}', \widetilde{\eta}')$ by a twisted stable Hecke algebra relative to an inner automorphism $\widetilde{\theta}'$ of $\widetilde{G}'$ over F. However, it seems likely that by choosing $(\widetilde{G}', \widetilde{\xi}')$ suitably, for example so that $\widetilde{G}'$ is a z-extension of G' with connected center, one could arrange that $\widetilde{\theta}'$ be trivial. It ought to be easy to check this point. I have not done so, since the most general case will not be our main focus here. Instead, let us simply assume as a condition on G that $(\widetilde{G}', \widetilde{\xi}')$ can be chosen for each G' so that $\widetilde{\theta}'$ is trivial, and therefore that any transfer mapping does indeed take values in the untwisted Hecke algebra $\mathcal{S}(\widetilde{G}', \widetilde{\eta}')$.

There is also another point. The auxiliary datum $(\widetilde{G}', \widetilde{\xi}')$ gives rise to a surjective, affine linear mapping

$$\widetilde{\mathfrak{h}}' \longrightarrow \mathfrak{h} = \mathfrak{h}^0_\theta$$

between the real vector spaces attached to $\widetilde{G}'$ and G. This is dual to a mapping

$$\mu_{\pi,\mathbb{R}} + i\mu_{\pi,I} \longrightarrow \mu_{\pi',\mathbb{R}} + i\mu_{\pi',I}$$

of archimedean infinitesimal characters of corresponding irreducible tempered representations π_∞ and π'_∞. (The representation π'_∞ of $\widetilde{G}'(F_\infty)$ is required to have central character $\widetilde{\eta}'_\infty$ on $\widetilde{C}'(F_\infty)$, while π_∞ is understood to be an extension of a representation to $G(F_\infty)$ of the form (2.1.1). They can be any corresponding representations in Shelstad's endoscopic transfer of archimedean L-packets.) We can always modify the L-homomorphism $\widetilde{\xi}'$ by tensoring it with an automorphic character on $\widetilde{G}'(\mathbb{A})$. For simplicity, we assume that this has been done so that the affine mapping from $\widetilde{\mathfrak{h}}'$ to $\mathfrak{h}$ descends to a linear isomorphism from $\widetilde{\mathfrak{h}}'$ to $\mathfrak{h}$. We can then assume that

$$\|\mu_{\pi,I}\| = \|\mu_{\pi',I}\|',$$

for the dual of a suitable Hermitian norm $\|\cdot\|'$ on $\mathfrak{h}^*_{\mathbb{C}}$. (The alternative would be to take $\|\cdot\|$ to be an affine norm, in the sense that it is a translate in $\mathfrak{h}^*_{\mathbb{C}}$ of a Hermitian norm, and $\|\cdot\|'$ to be an affine norm on $\widetilde{\mathfrak{h}}'_{\mathbb{C}}$ that is related to $\|\cdot\|$ in a way that depends on $(\widetilde{G}', \widetilde{\xi}')$.)

In the last section, we worked with a central character datum $(\mathfrak{X}_G, \chi)$ for G. To relate it to global transfer, we recall that for any G', there is a canonical injection of the group of covariants

$$Z(G^0)_\theta = Z(G^0)/\{z\theta(z)^{-1} : \ z \in Z(G^0)\}$$

of $Z(G^0)$ into $Z(G')$. (See [**KS**, p. 53].) We assume here that χ is trivial on the kernel of the restriction of this mapping to $\mathfrak{X}_G$. Then χ pulls back to a character on the image of $\mathfrak{X}_G$ in $Z\big(G'(\mathbb{A})\big)$, and hence also to a character (which we denote again by χ) on its preimage

$$\widetilde{\mathfrak{X}}' = \mathfrak{X}_{\widetilde{G}'}$$

in $Z\big(\widetilde{G}'(\mathbb{A})\big)$. Recall also that the automorphic character $\widetilde{\eta}'$ on $\widetilde{C}'(\mathbb{A})$ attached to $(\widetilde{G}', \widetilde{\xi}')$ is actually defined on the preimage of $Z\big(G^0(\mathbb{A})\big)$ in $Z\big(\widetilde{G}'(\mathbb{A})\big)$ [**KS**, p. 53 and p. 112]. It therefore restricts to a character on $\widetilde{\mathfrak{X}}'$. The product

$$\widetilde{\chi}' = \chi\widetilde{\eta}'$$

then represents a third character on $\widetilde{\mathfrak{X}}'$. The pair $(\widetilde{\mathfrak{X}}', \widetilde{\chi}')$ will serve as our central character datum for $\widetilde{G}'$. It follows from the discussion on p. 112 of [**KS**] that the global transfer of functions $f \to f'$ can be defined if f belongs to $\mathcal{H}(G,\chi)$, and descends to a mapping from $\mathcal{H}(G,\chi)$ to $\mathcal{S}(\widetilde{G}', \widetilde{\chi}')$. By definition, $\mathcal{S}(\widetilde{G}', \widetilde{\chi}')$ is the space of strongly regular, stable orbital integrals of functions in the equivariant global Hecke algebra $\mathcal{H}(\widetilde{G}', \widetilde{\chi}')$. We can therefore identify any stable linear form S' on $\mathcal{H}(\widetilde{G}', \widetilde{\chi}')$ with a linear form $\widehat{S}'$ on $\mathcal{S}(\widetilde{G}', \widetilde{\chi}')$ by the global analogue of the local prescription (2.1.2).

For any G', we write

$$\overline{Z}(\widehat{G}')^\Gamma = Z(\widehat{G}')^\Gamma/\big(Z(\widehat{G}')^\Gamma \cap Z(\widehat{G})^\Gamma\big) \cong Z(\widehat{G}')^\Gamma Z(\widehat{G})/Z(\widehat{G}), \tag{3.2.2}$$

where

$$Z(\widehat{G}) = Z(\widehat{G}^0)^{\widehat{\theta}}$$

is the subgroup of $\widehat{\theta}$-fixed points in the center of $\widehat{G}^0$. The general endoscopic datum G' is called *elliptic* if $\overline{Z}(\widehat{G}')^\Gamma$ is finite. We write $\mathcal{E}_{\text{ell}}(G)$ as usual for the subset of elliptic isomorphism classes in $\mathcal{E}(G)$. The elliptic classes represent the global endoscopic data that are used to stabilize the trace formula.

In their stabilization of the regular elliptic terms in the twisted trace formula, Kottwitz and Shelstad introduced global coefficients

$$\iota(G, G'), \qquad G' \in \mathcal{E}_{\text{ell}}(G).$$

These objects are defined by an explicit formula [**KS**, p. 115], which generalizes the formula of Kottwitz [**K3**, Theorem 8.3.1] for the original coefficients introduced by Langlands [**L10**] in case $G = G^0$. They are also essentially the same as the provisional coefficients defined in [**A9**, (3.3)]. These were introduced to describe a conjectural stabilization of the discrete part of the trace formula [**A9**, Hypothesis 3.1].

As an aside, we note that the discrete part of the trace formula is the spectral analogue of what was called the elliptic part of the trace formula, in the case $G = G^0$ studied in [**A15**]. The elliptic part of the trace formula is a sum of invariant orbital integrals, geometric terms that include semisimple orbital integrals. The semisimple elliptic part was stabilized for $G = G^0$ in [**K5**]. It is the regular (semisimple) elliptic part that was stabilized for any G in [**KS**], generalizing the first results of Langlands [**L10**] for $G = G^0$. We note that all of these partial stabilizations were originally dependent on the LSK-transfer conjecture and the fundamental lemma. They are now unconditional.

What is meant by a stabilization of some part of the trace formula? For the discrete part, it is a decomposition

$$I_{\text{disc},t}(f) = \sum_{G' \in \mathcal{E}_{\text{ell}}(G)} \iota(G, G') \widehat{S}'_{\text{disc},t}(f'), \qquad f \in \mathcal{H}(G, \chi), \tag{3.2.3}$$

of the given linear form in terms of stable linear forms

$$S'_{\text{disc},t} = S^{G'}_{\text{disc},t} : \mathcal{H}(\widetilde{G}', \widetilde{\chi}') \longrightarrow \mathbb{C}$$

attached to elliptic endoscopic data. The correspondence

$$f \longrightarrow f' = f^{G'}$$

is the global transfer mapping from $\mathcal{H}(G, \chi)$ to $\mathcal{S}(\widetilde{G}', \widetilde{\chi}')$ we have just described. The stable linear form $S'_{\text{disc},t}$ is universal, in that it depends only on $\widetilde{G}'$ (and not G). The coefficient $\iota(G, G')$ depends on G as well as G', but not on which part of the trace formula is being stabilized. The set of summation $\mathcal{E}_{\text{ell}}(G)$ is generally infinite in the global case here. However, the sum can be taken over a finite subset of $\mathcal{E}_{\text{ell}}(G)$, which depends only on a choice of finite set $S \supset S_\infty$ of valuations outside of which f is unramified.

We note that if $G = G^0$ is quasisplit, the stable linear form $S_{\text{disc},t} = S^G_{\text{disc},t}$ on $G(\mathbb{A})$ is defined inductively by (3.2.3). The assertion in this case is simply that the difference

$$I_{\text{disc},t}(f) - \sum_{G' \neq G} \iota(G, G') \widehat{S}'_{\text{disc},t}(f'), \qquad f \in \mathcal{H}(G, \chi),$$

is stable. For any other G, there are no further definitions that can be made. The assertion then becomes an identity, which ties the original linear form

$I_{\mathrm{disc},t}(f)$ to the seemingly unrelated forms

$$f \longrightarrow \widehat{S}'_{\mathrm{disc},t}(f').$$

In general, the summand of G' in (3.2.3) is assumed to be independent of the choice of $(\widetilde{G}', \widetilde{\xi}')$, while the linear form $S'_{\mathrm{disc},t}$ on $\widetilde{G}'(\mathbb{A})$ is to be independent of $\widetilde{\xi}'$. But the transfer mapping $f \to f'$ does depend on $\widetilde{\xi}'$. This apparent contradiction can be resolved directly by induction, and an appeal to [**A5**, Lemma 4.3].

If $G = G^0$ is a connected group, the decomposition (3.2.3) was established in [**A16**], subject to a general condition on the fundamental lemma. As we have noted, the standard fundamental lemma has now been established. The condition in [**A16**] was actually a generalization of the fundamental lemma, which was conjectured [**A12**, Conjecture 5.1] for weighted orbital integrals. It was used in the stabilization of the general geometric terms, its role being roughly similar to that of the standard fundamental lemma in the analysis of the elliptic geometric terms. The weighted fundamental lemma has now been established by Chaudouard and Laumon [**CL1**], [**CL2**], using the global methods introduced by Ngo [**N**], and a reduction by Waldspurger [**W7**], [**W8**] to fields of positive characteristic. This removes the condition that had qualified the results for $G = G^0$ in [**A16**]. In particular, the decomposition (3.2.3) is now valid if G is any connected group.

If G is a general triplet (3.1.3), the decomposition (3.2.3) remains conjectural. The problem is to extend the results of the papers [**A14**]–[**A16**] to the twisted case, or in other words, to stabilize the twisted trace formula. The linear form $I_{\mathrm{disc},t}(f)$ whose stabilization (3.2.3) we seek is given by the explicit formula (3.1.1). However, there is no direct way to stabilize the terms in this formula without first assuming the theorems we are trying to prove in this volume (or rather their general analogues for G). One has instead to stabilize all the complementary terms in the twisted trace formula. Once this is done, the decomposition will follow from the twisted trace formula itself, and the stable trace formula for each $\widetilde{G}'$ (which was established for connected quasisplit groups in [**A16**]). Some of the techniques will certainly carry over to the twisted case without much change. Others will call for refinement, and no doubt new ideas. Still, there is reason to be hopeful that the general stabilization can be established in the not too far distant future [**W9**], [**W10**]. In any case, it is not our intention in this volume to discuss the complementary terms in the (twisted) trace formula. We shall instead simply take the decomposition (3.2.3) as a hypothesis in the cases under consideration.

We note that the setting for the papers [**A14**]–[**A16**] was actually slightly different from that of a connected reductive group $G = G^0$. It pertains to what we called a K-group, a somewhat artificial object consisting of a finite disjoint union of connected groups over F with extra structure. The point is that one can arrange for G to be one of these components. The identity (3.2.3) for $G = G^0$ follows from the results of [**A16**] (with minor adjustments for the slightly different kind of central character datum $(\mathfrak{X}_G, \chi)$) by extending a given function f on $G(\mathbb{A})$ to be zero on the complementary adelic components of the K-group.

Before we leave the general case, we recall the general formula

$$\iota(G, G') = |\pi_0(\kappa_G)|^{-1} k(G, G') |\overline{Z}(\widehat{G}')^\Gamma|^{-1} |\mathrm{Out}_G(G')|^{-1} \tag{3.2.4}$$

for the global coefficients. Here, $\pi_0(\kappa_G)$ is the group of connected components in the complex group

$$\kappa_G = Z(\widehat{G})^\Gamma \cap \big(Z(\widehat{G}^0)^\Gamma\big)^0,$$

and

$$k(G, G') = \big|\ker^1\big(F, Z(\widehat{G}^0)\big)\big|^{-1} \big|\ker^1\big(F, Z(\widehat{G}')\big)\big|,$$

where $\ker^1(F, \cdot)$ denotes the subgroup of locally trivial classes in the global cohomology group $H^1(F, \cdot)$. The group $\overline{Z}(\widehat{G}')^\Gamma$ given by (3.2.2) is finite, since G' is assumed to be elliptic. Notice that $\pi_0(\kappa_G)$ equals the group of fixed points of the automorphism $\widehat{\theta}$, acting on the complex torus

$$\big(Z(\widehat{G}^0)^\Gamma\big)^0 / \big(Z(\widehat{G})^\Gamma\big)^0.$$

On the other hand, the dual automorphism θ acts as a linear isomorphism $\mathrm{Ad}(\theta)$ on the real vector space $\mathfrak{a}_{G^0}$, for which $\mathfrak{a}_G$ is the subspace of pointwise fixed vectors. The order of the group of fixed points of $\widehat{\theta}$ equals the absolute value of the determinant of $\big(\mathrm{Ad}(\theta) - 1\big)$ on the subspace $\mathfrak{a}_{G^0}^G$ of $\mathfrak{a}_{G^0}$. (See [**SpS**, II.17].) It follows that

$$\big|\pi_0(\kappa_G)\big|^{-1} = \big|\det\big(\mathrm{Ad}(\theta) - 1\big)_{\mathfrak{a}_{G^0}^G}\big|^{-1}. \tag{3.2.5}$$

This factor in (3.2.4) does not occur in the corresponding formula on p. 115 of [**KS**]. (It was taken from the formula [**A9**, (3.5)], which was used in a provisional stabilization of general spectral terms.) The discrepancy can be traced to the choice of Haar measure on $G(\mathbb{A})/G(F)A_{G,\infty}^+$ on p. 74 of [**KS**]. With the conventions of [**KS**], the determinant of $(w - 1)$ in (3.1.1) would have to be taken on the space $\mathfrak{a}_M^{G^0}$ rather than $\mathfrak{a}_M^G$.

We return to our special cases of study. For the two principal cases, either G equals the triplet $\widetilde{G}(N) = \big(\widetilde{G}(N)^0, \widetilde{\theta}(N), 1\big)$, or $G = G^0$ represents an element in $\widetilde{\mathcal{E}}_{\mathrm{sim}}(N)$. It is pretty clear how to specialize the general notions above. In the first case, we take the inner twist to be the identity morphism from $\widetilde{G}(N)^0$ onto $\widetilde{G}(N)^* = GL(N)$, and we set $\widetilde{\theta}(N)^* = \widetilde{\theta}(N)$. We also set $\widehat{\widetilde{\theta}}(N) = {}^L\widetilde{\theta}(N) = \widetilde{\theta}(N)$, regarded as an automorphism of either $\widehat{\widetilde{G}}(N)^0 = GL(N, \mathbb{C})$ or ${}^L\widetilde{G}(N)^0 = GL(N, \mathbb{C}) \times \Gamma_F$. In both cases, an endoscopic datum $G' \in \mathcal{E}_{\mathrm{ell}}(G)$ (or rather its isomorphism class) takes the simplified form from Chapter 1. That is, $\mathcal{G}'$ is just the L-group ${}^LG'$ and ξ' is the canonical L-embedding of ${}^LG'$ into ${}^LG^0$ that accompanies G' as an endoscopic datum. Since there is no need for the supplementary datum $(\widetilde{G}', \widetilde{\xi}')$, we simply set $\widetilde{G}' = G'$ and $\widetilde{\xi}' = 1$. There is also no need of central character data. We can therefore take $\mathfrak{X}_G = \widetilde{\mathfrak{X}}' = 1_G$ in all cases.

We also have the supplementary case, in which G is the triplet $\widetilde{G} = (\widetilde{G}^0, \widetilde{\theta}, 1)$ attached to an even orthogonal group $\widetilde{G}^0 \in \widetilde{\mathcal{E}}_{\mathrm{sim}}(N)$. Similar remarks apply here. In particular, we take $\psi = 1$ and $\widetilde{\theta}^* = \widetilde{\theta}$. There is again no call for auxiliary datum, which leaves us free to have $\widetilde{G}'$ and $\widetilde{\xi}'$ stand for components in an endoscopic datum $(\widetilde{G}', \widetilde{\mathcal{G}}', \widetilde{s}', \widetilde{\xi}')$ for $\widetilde{G}$.

The general formula (3.2.4) for $\iota(G, G')$ contains a quotient $k(G, G')$ of orders of groups of locally trivial cohomology classes. In the three cases of interest, the action of $\Gamma = \Gamma_F$ on either $Z(\widehat{G}^0)$ or $Z(\widehat{G}')$ factors through an abelian quotient of Γ. It follows that these cohomology groups are trivial and that $k(G, G') = 1$. The general formula therefore reduces to

$$\iota(G, G') = \left|\pi_0(\kappa_G)\right|^{-1} \left|\overline{Z}(\widehat{G}')^{\Gamma}\right|^{-1} \left|\mathrm{Out}_G(G')\right|^{-1}. \tag{3.2.6}$$

For example, in the twisted general linear group $\widetilde{G}(N)$, we can write

$$\widetilde{\iota}(N, G) = \iota\big(\widetilde{G}(N), G\big) = \tfrac{1}{2}\left|\overline{Z}(\widehat{G})^{\Gamma}\right|^{-1} \left|\widetilde{\mathrm{O}}\mathrm{ut}_N(G)\right|^{-1},$$

for any $G \in \mathcal{E}_{\mathrm{ell}}(N)$, since $\kappa_{\widetilde{G}(N)} = \mathbb{Z}/2\mathbb{Z}$ in this case.

The global transfer factor $\Delta(\delta', \gamma)$ is canonical, but the local factors in (3.2.1) are generally not. However, in the quasisplit case where we are now working, the local transfer factors can be specified uniquely. In fact, as we saw in the explicit discussion of our three special cases in §2.1, there are two ways to normalize the local transfer factors. One depends on a choice of splitting over F_v, while the other depends on a choice of Whittaker datum over F_v. They differ by the ε-factor

$$\varepsilon\big(\tfrac{1}{2}, r_v, \psi_{F_v}\big)$$

of a virtual representation r_v of Γ_v [**KS**, §5.3]. It is a consequence of the global definitions ([**LS1**, §6.2–6.3], [**KS**, §7.3]) that the product formula (3.2.1) is valid for quasisplit G if the local transfer factors are all normalized in terms of a common splitting over F. If on the other hand, they are all normalized in terms of a common global Whittaker datum, as has been the understanding for our three special cases here, the product changes by the global ε-factor $\varepsilon\big(\frac{1}{2}, r\big)$ of a virtual representation r of Γ. But it is a consequence of the general construction in §5.3 of [**KS**] that the virtual representation r is orthogonal. It then follows from [**FQ**] that the global ε-factor $\varepsilon\big(\frac{1}{2}, r\big)$ equals 1. The product formula (3.2.1) is therefore valid for the Whittaker normalizations of the local transfer factors we have adopted.

We shall state formally the hypothesis on which the results of this volume rest. Keep in mind that (3.2.3) has now been established in case $G = G^0$, and in particular, if G is any element in $\widetilde{\mathcal{E}}_{\mathrm{sim}}(N)$. The question therefore concerns our other two basic cases.

HYPOTHESIS 3.2.1. *Suppose that G equals either $\widetilde{G}(N)$ or $\widetilde{G}$, in the notation introduced initially in §1.2. Then the stabilization (3.2.3) of $I^G_{\mathrm{disc},t}(f)$ holds for any $t \geqslant 0$ and $f \in \mathcal{H}(G)$.* □

3.3. Contribution of a parameter ψ

The decomposition (3.2.3) is what will drive the proofs of our theorems. In the end, our results will come from the interplay of formulas obtained by specializing (3.2.3) to the cases $G = \widetilde{G}(N)$ and $G \in \widetilde{\mathcal{E}}_{\mathrm{sim}}(N)$. We have first to bring the global parameters $\psi \in \widetilde{\Psi}(N)$ back into the discussion. In this section, we shall describe the contribution of ψ to the terms in the specializations of (3.2.3).

We may as well continue to work more broadly while this is still feasible. We do not have global parameters ψ for general G, but we are free to replace them by Hecke eigenfamilies c for G. Some of this discussion will be rather formal. The reader might prefer to pass directly to Corollary 3.3.2, which contains its application to the parameters ψ.

For simplicity, suppose first that G is a connected reductive group over the global field F. In §1.3, we introduced the set $\mathcal{C}_{\mathrm{aut}}(G)$ of (equivalence classes of) families c of semisimple classes in ${}^LG$. We may as well work with the subset $\mathcal{C}_{\mathrm{aut}}(G,\chi)$ of classes in $\mathcal{C}_{\mathrm{aut}}(G)$ that are compatible with the central character datum $(\mathfrak{X}_G, \chi)$. Any class $c \in \mathcal{C}_{\mathrm{aut}}(G)$ determines a character ζ_v on $Z\big(G(F_v)\big)$ for almost all v, namely the central character of the unramified representation of $G(F_v)$ attached to c_v. Recall that χ is a character on the closed subgroup $\mathfrak{X}_G$ of $Z\big(G(\mathbb{A})\big)$. We define $\mathcal{C}_{\mathrm{aut}}(G,\chi)$ somewhat artificially as the set of c in $\mathcal{C}_{\mathrm{aut}}(G)$ such that χ extends to a character ζ on $Z\big(G(F)\big)\backslash Z\big(G(\mathbb{A})\big)$ whose restriction to $Z\big(G(F_v)\big)$ equals ζ_v for almost any v.

We need to see how the set $\mathcal{C}_{\mathrm{aut}}(G,\chi)$ is related to the terms in the explicit expression (3.1.1) for the discrete part

$$I_{\mathrm{disc},t}(f), \qquad f \in \mathcal{H}(G,\chi),$$

of the trace formula. Consider the operator $\mathcal{I}_{P,t}(\chi, f)$ in this expression. It is isomorphic to a direct sum of induced representations of the form

$$\pi = \mathcal{I}_P(\pi_M), \qquad \|\mu_{\pi,I}\| = t,$$

in which π_M is taken from the set of irreducible subrepresentations of $L^2_{\mathrm{disc}}\big(M(F)\backslash M(\mathbb{A}),\chi\big)$. For any such π, the class $c(\pi)$ belongs to $\mathcal{C}_{\mathrm{aut}}(G,\chi)$. If c is an arbitrary class in $\mathcal{C}_{\mathrm{aut}}(G,\chi)$, we write

$$\mathcal{I}_{P,t,c}(\chi, f) = \bigoplus_{\{\pi:\, c(\pi)=c\}} \mathcal{I}_{P,\pi}(\chi, f),$$

where $\pi = \mathcal{I}_P(\pi_M)$ as above, and $\mathcal{I}_{P,\pi}(\chi)$ is the subrepresentation of $\mathcal{I}_{P,t}(\chi)$ corresponding to π. We also write $M_{P,t,c}(w,\chi)$ for the restriction of the operator $M_{P,t}(w,\chi)$ in (3.1.1) to the invariant subspace $\mathcal{H}_{P,t,c}(\chi)$ on which $\mathcal{I}_{P,t,c}(\chi)$ acts. Then

$$\operatorname{tr}\big(M_{P,t}(w,\chi)\mathcal{I}_{P,t}(\chi,f)\big) = \sum_{c\in\mathcal{C}_{\mathrm{aut}}(G,\chi)} \operatorname{tr}\big(M_{P,t,c}(w,\chi)\mathcal{I}_{P,t,c}(\chi,f)\big).$$

It follows that

$$I_{\mathrm{disc},t}(f) = \sum_{c\in\mathcal{C}_{\mathrm{aut}}(G,\chi)} I_{\mathrm{disc},t,c}(f), \tag{3.3.1}$$

where $I_{\mathrm{disc},t,c}(f)$ is the c-variant of $I_{\mathrm{disc},t}(f)$, obtained by replacing the trace in (3.1.1) by the summand of c in its decomposition above.

Suppose now that G represents an arbitrary triplet (G^0,θ,ω). To extend the decomposition to this case, we need only agree on the meaning of the associated set $\mathcal{C}_{\mathrm{aut}}(G,\chi)$. We define it to be the subset of classes $c\in\mathcal{C}(G^0,\chi)$ that are compatible with θ and ω, in the sense that for almost all valuations v, the associated conjugacy classes c_v satisfy

$$\widehat{\theta}_v(c_v) = c_v\, c(\omega_v).$$

With this interpretation of $\mathcal{C}_{\mathrm{aut}}(G,\chi)$, the decomposition (3.3.1) remains valid as stated. We note that in general, the sum in (3.3.1) can be taken over a finite set that depends on f only through a choice of Hecke type.

Recall that $\mathcal{H}(G,\chi)$ is the χ^{-1}-equivariant Hecke module of $G(\mathbb{A})$, relative to a suitably chosen maximal compact subgroup

$$K = \prod_v K_v$$

of $G^0(\mathbb{A})$. It is a direct limit

$$\mathcal{H}(G,\chi) = \varinjlim_S \mathcal{H}(G,K^S,\chi),$$

where $\mathcal{H}(G,K^S,\chi)$ is the space of functions in $\mathcal{H}(G,\chi)$ that are bi-invariant under the product

$$K^S = \prod_{v\notin S} K_v, \qquad K_v\subset G^0(F_v),$$

of hyperspecial maximal compact subgroups. The set $\mathcal{C}_{\mathrm{aut}}(G,\chi)$ is a direct limit

$$\mathcal{C}_{\mathrm{aut}}(G,\chi) = \varinjlim_S \mathcal{C}^S_{\mathrm{aut}}(G,\chi),$$

where $\mathcal{C}^S_{\mathrm{aut}}(G,\chi)$ is the relevant variant of the set defined in §1.3, composed of families of semisimple conjugacy classes

$$c^S = \{c_v:\ v\notin S\}.$$

In both limits, $S\supset S_\infty$ represents a large finite set of valuations outside of which G is unramified. The two are related through the unramified Hecke algebra

$$\mathcal{H}^S_{\mathrm{un}} = \mathcal{H}^S_{\mathrm{un}}(G^0) = C^\infty_c\big(K^S\backslash G^0(\mathbb{A}^S)/K^S\big)$$

on $G^0(\mathbb{A}^S)$. On the one hand, an element $c^S \in \mathcal{C}^S_{\mathrm{aut}}(G,\chi)$ determines a complex-valued character

$$h \longrightarrow \widehat{h}(c^S), \qquad h \in \mathcal{H}^S_{\mathrm{un}}, \tag{3.3.2}$$

on $\mathcal{H}^S_{\mathrm{un}}$. On the other, there is an action

$$f \longrightarrow f_h, \qquad f \in \mathcal{H}(G, K^S, \chi),\ h \in \mathcal{H}^S_{\mathrm{un}},$$

of $\mathcal{H}^S_{\mathrm{un}}$ on $\mathcal{H}(G, K^S, \chi)$ that is characterized by multipliers. Namely, if π is an extension to $G(\mathbb{A})$ of an irreducible unitary representation π^0 of $G^0(\mathbb{A})$, which is unramified outside of S and satisfies the global analogue of (2.1.1), then

$$\mathrm{tr}\big(\pi(f_h)\big) = \widehat{h}\big(c^S(\pi)\big)\mathrm{tr}\big(\pi(f)\big).$$

It is easy to describe the decomposition (3.3.1) in terms of eigenvalues of $\mathcal{H}^S_{\mathrm{un}}$. Any function $f \in \mathcal{H}(G,\chi)$ belongs to $\mathcal{H}(G, K^S, \chi)$, for some S. We fix S and f, and consider the linear form

$$I_{\mathrm{disc},t}(f_h), \qquad h \in \mathcal{H}^S_{\mathrm{un}},$$

on $\mathcal{H}^S_{\mathrm{un}}$. It follows from the expression (3.1.1) for $I_{\mathrm{disc},t}(f)$ that this linear form is a finite sum of eigenforms. More precisely, we can write

$$I_{\mathrm{disc},t}(f) = \sum_{c^S} I_{\mathrm{disc},t,c^S}(f),$$

where I_{disc,t,c^S} is a linear form on $\mathcal{H}(G, K^S, \chi)$ such that

$$I_{\mathrm{disc},t,c^S}(f_h) = \widehat{h}(c^S) I_{\mathrm{disc},t,c^S}(f). \tag{3.3.3}$$

The sum is over a finite subset of $\mathcal{C}^S_{\mathrm{aut}}(G,\chi)$, which again depends on f only through a choice of Hecke type (necessarily of the form $(\tau_\infty, \kappa^\infty_S K^S)$ in this case). If c belongs to $\mathcal{C}_{\mathrm{aut}}(G,\chi)$, we can then write

$$I_{\mathrm{disc},t,c}(f) = \sum_{c^S \to c} I_{\mathrm{disc},t,c^S}(f), \tag{3.3.4}$$

the sum being over the preimage of c in $\mathcal{C}^S_{\mathrm{aut}}(G,\chi)$. This is the summand of c on the right hand side of (3.3.1). For a given f, it is independent of the choice of S.

Suppose that $G' \in \mathcal{E}_{\mathrm{ell}}(G)$ is an elliptic endoscopic datum, with G again being a general triplet (3.1.3). If $(\widetilde{G}', \widetilde{\xi}')$ is an auxiliary datum for G', we have the subset $\mathcal{C}_{\mathrm{aut}}(\widetilde{G}')$ of $\mathcal{C}(\widetilde{G}')$. The properties of $(\widetilde{G}', \widetilde{\xi}')$ are such that an element $c' \in \mathcal{C}(\widetilde{G}', \widetilde{\chi}')$ transfers to a family c of conjugacy classes for G. However, because we do not know the principle of functoriality for G and $\widetilde{G}'$, we cannot say that c lies in $\mathcal{C}_{\mathrm{aut}}(G,\chi)$. To sidestep this formal difficulty, let us simply write

$$\mathcal{C}_{\mathbb{A}}(G,\chi) = \varinjlim_{S} \mathcal{C}^S_{\mathbb{A}}(G,\chi)$$

for the set of *all* equivalence classes of families c^S of semisimple conjugacy classes in ${}^LG$ that are compatible with χ, and such that for each v, c_v

projects a Frobenius class in W_{F_v} at v. It follows from the various constructions that there is a canonical mapping $c' \to c$ from $\mathcal{C}_{\mathbb{A}}(\widetilde{G}', \widetilde{\chi}')$ to $\mathcal{C}_{\mathbb{A}}(G, \chi)$. We can of course take $\mathcal{C}_{\mathbb{A}}(G, \chi)$ to be the domain of summation in (3.3.1) if we set $I_{\mathrm{disc},t,c}(f) = 0$ for c in the complement of $\mathcal{C}_{\mathrm{aut}}(G, \chi)$ in $\mathcal{C}_{\mathbb{A}}(G, \chi)$.

We would like to convert the decomposition (3.2.3) for $I_{\mathrm{disc},t}(f)$ into a parallel decomposition for $I_{\mathrm{disc},t,c}(f)$. We need to be clear about our assumptions, since (3.2.3) remains conjectural in the twisted case. However, the decomposition is valid in case $G = G^0$, and in particular if $G = G^*$ is quasisplit. This means that the stable linear forms on the right hand side of (3.2.3) are defined unconditionally for any G. As an assumption on G, (3.2.3) can therefore be regarded as a conjectural identity, both sides of which are well defined.

LEMMA 3.3.1. *Assume that G satisfies (3.2.3).*

(a) *Suppose that G equals G^0 and is quasisplit. Then there is a decomposition*

$$S_{\mathrm{disc},t}(f) = \sum_{c \in \mathcal{C}_{\mathbb{A}}(G,\chi)} S_{\mathrm{disc},t,c}(f), \qquad f \in \mathcal{H}(G, \chi), \tag{3.3.5}$$

for stable linear forms $S_{\mathrm{disc},t,c} = S^G_{\mathrm{disc},t,c}$ that satisfy the analogues of (3.3.3) and (3.3.4), and vanish for all c outside a finite subset of $\mathcal{C}_{\mathbb{A}}(G, \chi)$ that depends on f only through a choice of Hecke type.

(b) *Suppose that G is arbitrary. Then for any $c \in \mathcal{C}_{\mathbb{A}}(G, \chi)$, there is a decomposition*

$$I_{\mathrm{disc},t,c}(f) = \sum_{G' \in \mathcal{E}_{\mathrm{ell}}(G)} \iota(G, G') \sum_{c' \to c} \widehat{S}'_{\mathrm{disc},t,c'}(f'), \qquad f \in \mathcal{H}(G, \chi), \tag{3.3.6}$$

where c' is summed over the set of classes in $\mathcal{C}_{\mathbb{A}}(\widetilde{G}', \widetilde{\chi}')$ that map to c.

PROOF. We prove that the stable linear form $S'_{\mathrm{disc},t,c}$ exists for any $G' \in \mathcal{E}_{\mathrm{ell}}(G)$. Assume inductively that if $G' \neq G$, this form satisfies the analogue for G' of the assertion (a).

Suppose first that $G = G^0$ is quasisplit, as in (a). If $f \in \mathcal{H}(G, K^S, \chi)$ and $c^S \in \mathcal{C}^S_{\mathbb{A}}(G, \chi)$, we set

$$S_{\mathrm{disc},t,c^S}(f) = I_{\mathrm{disc},t,c^S}(f) - \sum_{G' \neq G} \iota(G, G') \sum_{c',^S \to c^S} \widehat{S}'_{\mathrm{disc},t,c',^S}(f'),$$

the inner sum being over elements $c',^S$ in $\mathcal{C}^S_{\mathbb{A}}(\widetilde{G}', \widetilde{\chi}')$ that map to c^S. If h belongs to the unramified Hecke algebra $\mathcal{H}^S_{\mathrm{un}}$ for G, the fundamental lemma for spherical functions [**Hal**] implies that

$$(f_h)' = f'_{h'}.$$

We can then write

$$\begin{aligned}
&S_{\mathrm{disc},t,c^S}(f_h)\\
=&I_{\mathrm{disc},t,c^S}(f_h) - \sum_{G'\neq G} \iota(G,G') \sum_{c',^S\to c^S} \widehat{S}'_{\mathrm{disc},t,c',^S}((f_h)')\\
=&I_{\mathrm{disc},t,c^S}(f_h) - \sum_{G'\neq G} \iota(G,G') \sum_{c',^S\to c^S} \widehat{S}'_{\mathrm{disc},t,c',^S}(f'_{h'})\\
=&\widehat{h}(c^S)I_{\mathrm{disc},t,c^S}(f) - \widehat{h}(c^S)\sum_{G'}\iota(G,G')\sum_{c',^S}\widehat{S}'_{\mathrm{disc},t,c',^S}(f')\\
=&\widehat{h}(c^S)S_{\mathrm{disc},t,c^S}(f),
\end{aligned}$$

by the stable analogue for G' of (3.3.3), and the identity

$$\widehat{h}'(c',^S) = \widehat{h}(c^S).$$

This is the stable analogue for G of (3.3.3)

To complete the proof of (a), we sum over c^S in $\mathcal{C}^S_{\mathbb{A}}(G,\chi)$. We obtain

$$\begin{aligned}
&\sum_{c^S} S_{\mathrm{disc},t,c^S}(f)\\
=&\sum_{c^S} I_{\mathrm{disc},t,c^S}(f) - \sum_{c^S}\sum_{G'\neq G}\iota(G,G')\sum_{c',^S\to c^S}\widehat{S}'_{\mathrm{disc},t,c',^S}(f')\\
=&I_{\mathrm{disc},t}(f) - \sum_{G'\neq G}\iota(G,G')\sum_{c'}\widehat{S}'_{\mathrm{disc},t,c'}(f')\\
=&I_{\mathrm{disc},t}(f) - \sum_{G'\neq G}\iota(G,G')\widehat{S}'_{\mathrm{disc},t}(f')\\
=&S_{\mathrm{disc},t}(f),
\end{aligned}$$

by the stable analogues for G' of (3.3.4) and (3.3.5) (which we assume inductively), and the fact that the double sum over $c',^S$ and c^S reduces to a simple sum over $c' \in \mathcal{C}_{\mathbb{A}}(\widetilde{G}',\widetilde{\chi}')$. The triple sum over c^S, G' and $c',^S$ above can be taken over a finite set that depends on f only through a choice of Hecke type. This follows from the corresponding property for $I_{\mathrm{disc},t}(f)$, the finiteness of the sum over G', the uniformity properties of the transfer mapping $f \to f'$ that follow from the two theorems in [**A11**, §6], and our induction hypothesis above for G'. We set

$$S_{\mathrm{disc},t,c}(f) = \sum_{c^S\to c} S_{\mathrm{disc},t,c^S}(f).$$

This definition gives the stable analogue for G of (3.3.4). It thus completes our proof of the assertion (a) for G, as well as the accompanying induction argument.

Suppose that G is arbitrary, as in (b), and that f again belongs to $\mathcal{H}(G, K^S, \chi)$. To establish (3.3.6), we consider the expression

$$\sum_{G' \in \mathcal{E}_{\mathrm{ell}}(G)} \iota(G, G') \sum_{c'^{,S} \to c^S} \widehat{S}'_{\mathrm{disc},t,c'^{,S}}(f') \tag{3.3.7}$$

attached to any $c^S \in \mathcal{C}^S_{\mathbb{A}}(G, \chi)$. If f is replaced by its transform f_h under a general element $h \in \mathcal{H}^S_{\mathrm{un}}$, the expression is multiplied by the factor $\widehat{h}(c^S)$. Recall that the linear form $I_{\mathrm{disc},t,c^S}(f)$ transforms the same way under the action of $\mathcal{H}^S_{\mathrm{un}}$, and that its sum over c^S in $\mathcal{C}^S_{\mathbb{A}}(G, \chi)$ equals $I_{\mathrm{disc},t}(f)$. But the sum over c^S of (3.3.7) equals

$$\sum_{G'} \iota(G, G') \widehat{S}'_{\mathrm{disc},t}(f') = I_{\mathrm{disc},t}(f)$$

as well, by (3.2.3) and the stable analogues of (3.3.4) and (3.3.5) for G'. Thus, both $I_{\mathrm{disc},t,c^S}(f)$ and (3.3.7) represent the c^S-component of the linear form $I_{\mathrm{disc},t}(f)$, relative to its decomposition into eigenfunctions under the action of $\mathcal{H}^S_{\mathrm{un}}$. Therefore (3.3.7) equals $I_{\mathrm{disc},t,c^S}(f)$. The required decomposition (3.3.6) is then given by the sum of each side of this identity over those c^S that map to the given class $c \in \mathcal{C}_{\mathbb{A}}(G, \chi)$. □

REMARK. The spherical fundamental lemma [**Hal**] was needed to justify the first steps in the proof above. As we noted in the introduction, it is closely related to the the two theorems of [**A11**] that we used in the proof of Proposition 2.1.1 (and will use again in Corollary 6.7.4). In fact, in our case of an orthgonal or symplectic group $G \in \widetilde{\mathcal{E}}_{\mathrm{sim}}(N)$, it is not hard to see that the specialization of these theorems to unramified spherical functions, combined with the spherical fundamental lemma for $SL(2)$, imply the spherical fundamental lemma for G. This is perhaps not surprising, given that the results of [**Hal**] and [**A11**] are proved by a similar global argument, based on the simple trace formula and the fundamental lemma for units. (The paper [**Hal**] predates [**A11**], and was itself motivated by an earlier result [**Clo2**] of Clozel.) It is perhaps worth noting that the results of [**Hal**] and [**A11**] differ in a way that is suggestive (though very much simpler) of a difference between two general ways of trying to classify representations of p-adic groups, by types and by characters.

More generally, suppose that $\widetilde{\mathcal{C}}_{\mathbb{A}}(G, \chi)$ is the family of equivalence classes relative to some given equivalence relation on $\mathcal{C}_{\mathbb{A}}(G, \chi)$. If c now belongs to $\widetilde{\mathcal{C}}_{\mathbb{A}}(G, \chi)$, we write $I_{\mathrm{disc},t,c}(f)$ for the sum of the corresponding linear forms attached to elements of $\mathcal{C}_{\mathbb{A}}(G, \chi)$ in this equivalence class. We obtain the obvious variant

$$I_{\mathrm{disc},t}(f) = \sum_{c \in \widetilde{\mathcal{C}}_{\mathbb{A}}(G,\chi)} I_{\mathrm{disc},t,c}(f)$$

of the decomposition (3.3.1). Similarly, we have the stable variant

$$S_{\mathrm{disc},t}(f) = \sum_{c \in \widetilde{\mathcal{C}}_{\mathbb{A}}(G,\chi)} S_{\mathrm{disc},t,c}(f)$$

of (3.3.5), in case $G = G^0$ is quasisplit. If G' belongs to $\mathcal{E}_{\mathrm{ell}}(G)$, we can pull the equivalence relation back to $\mathcal{C}_{\mathbb{A}}(\widetilde{G}', \widetilde{\chi}')$. We write

$$I'_{\mathrm{disc},t,c}(f') = \sum_{c' \to c} I'_{\mathrm{disc},t,c'}(f') \tag{3.3.8}$$

and

$$S'_{\mathrm{disc},t,c}(f') = \sum_{c' \to c} S'_{\mathrm{disc},t,c'}(f') \tag{3.3.9}$$

for any $c \in \widetilde{\mathcal{C}}_{\mathbb{A}}(G,\chi)$ and $f' \in \mathcal{H}(\widetilde{G}', \widetilde{\chi}')$, the sum being over the elements in $\mathcal{C}_{\mathbb{A}}(\widetilde{G}', \widetilde{\chi}')$ that map to c. This is compatible with the definition above for G and $\mathcal{C}_{\mathbb{A}}(G,\chi)$, in case c lies in the image of the map and is consequently identified with a class in $\widetilde{\mathcal{C}}_{\mathbb{A}}(\widetilde{G}', \widetilde{\chi}')$. It is also compatible with the summation over c' in the decomposition (3.3.6). We can therefore rewrite (3.3.6) slightly more simply as

$$I_{\mathrm{disc},t,c}(f) = \sum_{G' \in \mathcal{E}_{\mathrm{ell}}(G)} \iota(G,G') \widehat{S}'_{\mathrm{disc},t,c}(f'), \tag{3.3.10}$$

for any c in $\widetilde{\mathcal{C}}_{\mathbb{A}}(G,\chi)$. This is the form of the general decomposition we will specialize.

Our discussion so far has been quite formal and considerably more general than we need. It simplifies in the two cases of $G = \widetilde{G}(N)$ and $G \in \widetilde{\mathcal{E}}_{\mathrm{sim}}(N)$ that are our primary concern. We recall that in both of these cases, we take $\chi = 1$, $\widetilde{G}' = G'$ and $\widetilde{\eta}' = 1$.

Suppose first that G equals $\widetilde{G}(N)$. It follows from Theorems 1.3.2 and 1.3.3 that the mapping $\psi \to c(\psi)$ from the set $\widetilde{\Psi}(N) = \Psi(\widetilde{G}(N))$ to the subset $\widetilde{\mathcal{C}}(N) = \mathcal{C}(\widetilde{G}(N))$ of $\mathcal{C}_{\mathbb{A}}(\widetilde{G}(N))$ is a bijection. These theorems tell us in addition that the mapping $c^S \to c$ is injective on the preimage of $\widetilde{\mathcal{C}}(N)$ in the set $\widetilde{\mathcal{C}}_{\mathbb{A}}(N) = \mathcal{C}_{\mathbb{A}}(\widetilde{G}(N))$, and hence that the sums over c^S above are all superfluous in this case. Any element $\psi \in \widetilde{\Psi}(N)$ also gives rise to an archimedean infinitesimal character, the norm of whose imaginary part we denote by $t(\psi)$. With this notation, we write

$$I_{\mathrm{disc},\psi}(\widetilde{f}) = I_{\mathrm{disc},t(\psi),c(\psi)}(\widetilde{f}), \qquad \widetilde{f} \in \widetilde{\mathcal{H}}(N). \tag{3.3.11}$$

Suppose next that G represents an element in $\widetilde{\mathcal{E}}_{\mathrm{sim}}(N)$. Following the general notation (3.3.8) and (3.3.9), we set

$$I_{\mathrm{disc},\psi}(f) = I_{\mathrm{disc},t(\psi),c(\psi)}(f) \tag{3.3.12}$$

and

$$S_{\mathrm{disc},\psi}(f) = S_{\mathrm{disc},t(\psi),c(\psi)}(f), \tag{3.3.13}$$

for any $\psi \in \widetilde{\Psi}(N)$ and $f \in \mathcal{H}(G)$. We will obviously have a special interest in the case that ψ lies in the subset $\widetilde{\Psi}(G)$ of parameters in $\widetilde{\Psi}(N)$ attached to G. However, the existence of this subset will require a running induction hypothesis based on Theorem 1.4.1, so we will be restricted to the coarser set $\Psi(N)$ for some time. Observe that the notation (3.3.12) and (3.3.13) holds more generally if G belongs to the larger set $\widetilde{\mathcal{E}}_{\mathrm{ell}}(N)$.

We shall state the specialization of (3.3.10) separately as a corollary of Lemma 3.3.1. It is of course Hypothesis 3.2.1 that tells us that the condition of the lemma holds if $G = \widetilde{G}(N)$.

COROLLARY 3.3.2. *Suppose that ψ belongs to $\widetilde{\Psi}(N)$.*

(a) *If $G = \widetilde{G}(N)$, we have*

$$I_{\mathrm{disc},\psi}(\widetilde{f}) = \sum_{G \in \widetilde{\mathcal{E}}_{\mathrm{ell}}(N)} \widetilde{\iota}(N,G)\widehat{S}^{G}_{\mathrm{disc},\psi}(\widetilde{f}^{G}), \qquad \widetilde{f} \in \widetilde{\mathcal{H}}(N). \tag{3.3.14}$$

(b) *If $G \in \widetilde{\mathcal{E}}_{\mathrm{sim}}(N)$, we have*

$$I_{\mathrm{disc},\psi}(f) = \sum_{G' \in \mathcal{E}_{\mathrm{ell}}(G)} \iota(G,G')\widehat{S}'_{\mathrm{disc},\psi}(f'), \qquad f \in \mathcal{H}(G). \tag{3.3.15}$$

□

Suppose that G and ψ are as in (b). The linear form $I_{\mathrm{disc},\psi}(f)$ on the left hand side of (3.3.15) then has a spectral expansion

$$\sum_{M \in \mathcal{L}} |W_0^M||W_0^G|^{-1} \sum_{w \in W(M)_{\mathrm{reg}}} \big|\det(w-1)_{\mathfrak{a}_M^G}\big|^{-1} \mathrm{tr}\big(M_{P,\psi}(w)\mathcal{I}_{P,\psi}(f)\big),$$

where

$$M_{P,\psi}(w)\mathcal{I}_{P,\psi}(f) = M_{P,t(\psi),c(\psi)}(w,1_M)\mathcal{I}_{P,t(\psi),c(\psi)}(1_M,f).$$

It is understood here that the last expression vanishes if $c(\psi)$ lies in the the complement of $\widetilde{\mathcal{C}}_{\mathrm{aut}}(G)$, and that 1_M stands for the trivial character on $A^{+}_{M,\infty}$. The same formula obviously applies to the left hand side of (3.3.14) if we set $G = \widetilde{G}(N)$ and $f = \widetilde{f}$.

We have come at length to the decompositions (3.3.14) and (3.3.15) that for any $\psi \in \widetilde{\Psi}(N)$ will serve as the foundation for our proofs. They are the culmination of the general discussion of the last three sections. The discussion has been broader than necessary, in the hope that it might offer some general perspective, and in addition, to serve as a foundation for future investigations. We turn now to the business at hand. We shall begin a study of the implications of these decompositions for the proofs of our theorems.

3.4. A preliminary comparison

Our interest will be focused on the decompositions (3.3.14) and (3.3.15) attached to a parameter $\psi \in \widetilde{\Psi}(N)$. One of our goals will be to establish transparent expansions of the two sides of each identity that we can compare.

This will be the topic of Chapter 4. In the remaining two sections of this chapter, we shall establish a complementary result that is more modest.

Suppose first that G equals the twisted general linear group $\widetilde{G}(N)$. The classification of Theorems 1.3.2 and 1.3.3 then characterizes the automorphic spectrum in terms of self-dual families c. For we have observed that the mapping $\psi \to c(\psi)$ in this case is a bijection from $\widetilde{\Psi}(N)$ onto the subset $\widetilde{\mathcal{C}}(N)$ of $\mathcal{C}_{\mathbb{A}}(\widetilde{G}(N))$. We recall that $\widetilde{\mathcal{C}}(N)$ represents the set of self-dual representations in the automorphic spectrum of $GL(N)$. In particular, it represents all the representations that occur in the formula (3.1.1) for $\widetilde{G}(N)$. The summand of c in (3.3.1) therefore vanishes in this case unless $c = c(\psi)$ for some $\psi \in \widetilde{\Psi}(N)$. The decomposition (3.3.1) thus reduces to

$$(3.4.1) \qquad I_{\mathrm{disc},t}(\widetilde{f}) = \sum_{\{\psi\in\widetilde{\Psi}(N):t(\psi)=t\}} I_{\mathrm{disc},\psi}(\widetilde{f}), \qquad \widetilde{f} \in \widetilde{\mathcal{H}}(N).$$

We would like to establish a similar formula if G is any element in $\widetilde{\mathcal{E}}_{\mathrm{sim}}(N)$. Following the general convention in §3.3, we write $\widetilde{\mathcal{C}}_{\mathbb{A}}(G) = \widetilde{\mathcal{C}}_{\mathbb{A}}(G, 1_G)$ for the set of fibres of the mapping

$$\mathcal{C}_{\mathbb{A}}(G) \longrightarrow \widetilde{\mathcal{C}}_{\mathbb{A}}(N) = \mathcal{C}_{\mathbb{A}}(\widetilde{G}(N)).$$

The mapping then descends to an injection from $\widetilde{\mathcal{C}}_{\mathbb{A}}(G)$ to $\widetilde{\mathcal{C}}_{\mathbb{A}}(N)$. In particular, we can identify $\widetilde{\mathcal{C}}_{\mathbb{A}}(G)$ with a subset of $\widetilde{\mathcal{C}}_{\mathbb{A}}(N)$. Of course $\widetilde{\mathcal{C}}_{\mathbb{A}}(N)$ also contains $\widetilde{\mathcal{C}}(N)$, a subset we can always identify with $\widetilde{\Psi}(N)$. The set $\widetilde{\mathcal{C}}_{\mathbb{A}}(G)$ is designed for the symmetric subalgebra $\widetilde{\mathcal{H}}(G)$ of $\mathcal{H}(G)$. Recall from §1.5 that $\widetilde{\mathcal{H}}(G)$ equals $\mathcal{H}(G)$ unless $\widehat{G} = SO(2m, \mathbb{C})$, in which case it consists of functions that at each v are symmetric under the automorphism of G we have fixed. The fibres in $\mathcal{C}_{\mathbb{A}}(G)$ of points in $\widetilde{\mathcal{C}}_{\mathbb{A}}(G)$ have a similar description. However, so that there be no misunderstanding, we note that even if f belongs to the symmetric Hecke algebra $\widetilde{\mathcal{H}}(G)$, the function $I_{\mathrm{disc},t,c}(f)$ of $c \in \mathcal{C}_{\mathbb{A}}(G)$ is not constant on the fibres. Indeed, a fibre can be infinite, while this function has finite support. However, if c is taken to be a class in $\widetilde{\mathcal{C}}_{\mathbb{A}}(G)$ instead of $\mathcal{C}_{\mathbb{A}}(G)$, $I_{\mathrm{disc},t,c}(f)$ is the sum of the function over the fibre of c in $\mathcal{C}_{\mathbb{A}}(G)$.

The following proposition will reduce the study of automorphic spectra of our groups $G \in \widetilde{\mathcal{E}}_{\mathrm{sim}}(N)$ to the subsets attached to parameters $\psi \in \widetilde{\Psi}(N)$. Its proof over the next two sections will be a model for the more elaborate global comparisons that will occupy us later.

PROPOSITION 3.4.1. *Suppose that $G \in \widetilde{\mathcal{E}}_{\mathrm{sim}}(N)$, $f \in \widetilde{\mathcal{H}}(G)$, $t \geqslant 0$ and $c \in \widetilde{\mathcal{C}}_{\mathbb{A}}(G)$. Then*

$$I_{\mathrm{disc},t,c}(f) = 0 = S_{\mathrm{disc},t,c}(f),$$

unless

$$(t, c) = \big(t(\psi), c(\psi)\big),$$

for some $\psi \in \widetilde{\Psi}(N)$.

We shall begin the proof later in this section, and complete it in the next. Before doing so, we shall establish two important corollaries of the proposition, and comment on some related matters.

The first corollary follows immediately from the decompositions (3.3.1) and (3.3.5) (stated in terms of $\widetilde{\mathcal{C}}_{\mathbb{A}}(G)$ rather than $\mathcal{C}_{\mathbb{A}}(G)$), and the definitions (3.3.12) and (3.3.13).

COROLLARY 3.4.2. *For any $t \geqslant 0$ and $f \in \widetilde{\mathcal{H}}(G)$, we have*

$$I_{\mathrm{disc},t}(f) = \sum_{\{\psi \in \widetilde{\Psi}(N)\colon t(\psi)=t\}} I^G_{\mathrm{disc},\psi}(f) \tag{3.4.2}$$

and

$$S^G_{\mathrm{disc},t}(f) = \sum_{\{\psi \in \widetilde{\Psi}(N)\colon t(\psi)=t\}} S^G_{\mathrm{disc},\psi}(f). \tag{3.4.3}$$

□

The other corollary bears directly on our main global theorem. For a group $G \in \widetilde{\mathcal{E}}_{\mathrm{sim}}(N)$, we define an invariant subspace

$$L^2_{\mathrm{disc},t,c}\big(G(F)\backslash G(\mathbb{A})\big), \qquad t \geqslant 0,\ c \in \widetilde{\mathcal{C}}_{\mathbb{A}}(G),$$

of $L^2_{\mathrm{disc}}\big(G(F)\backslash G(\mathbb{A})\big)$ in the obvious way. It is the direct sum

$$\bigoplus_{\{\pi\colon \|\mu_{\pi,I}\|=t,\, c(\pi)=c\}} m(\pi)\pi,$$

of the irreducible representations π of $G(\mathbb{A})$ attached to t and c that occur in $L^2_{\mathrm{disc}}\big(G(F)\backslash G(\mathbb{A})\big)$ with positive multiplicity $m(\pi)$. Then

$$L^2_{\mathrm{disc}}\big(G(F)\backslash G(\mathbb{A})\big) = \bigoplus_{t,c} L^2_{\mathrm{disc},t,c}\big(G(F)\backslash G(\mathbb{A})\big).$$

If ψ belongs to $\widetilde{\Psi}(N)$, we set

$$L^2_{\mathrm{disc},\psi}\big(G(F)\backslash G(\mathbb{A})\big) = L^2_{\mathrm{disc},t(\psi),c(\psi)}\big(G(F)\backslash G(\mathbb{A})\big).$$

COROLLARY 3.4.3. *If $t \geqslant 0$ and $c \in \widetilde{\mathcal{C}}_{\mathbb{A}}(G)$, then*

$$L^2_{\mathrm{disc},t,c}\big(G(F)\backslash G(\mathbb{A})\big) = 0,$$

unless

$$(t,c) = \big(t(\psi), c(\psi)\big),$$

for some $\psi \in \widetilde{\Psi}(N)$. In particular, we have a decomposition

$$L^2_{\mathrm{disc}}\big(G(F)\backslash G(\mathbb{A})\big) = \bigoplus_{\psi \in \widetilde{\Psi}(N)} L^2_{\mathrm{disc},\psi}\big(G(F)\backslash G(\mathbb{A})\big). \tag{3.4.4}$$

PROOF OF COROLLARY 3.4.3. Assume that (t,c) is not of the form $\big(t(\psi), c(\psi)\big)$. The proposition then tells us $I_{\mathrm{disc},t,c}(f)$ vanishes. The formula for $I_{\mathrm{disc},t,c}(f)$ given by the c-analogue of (3.1.1) is a sum over Levi subgroups M of G. If M is proper in G, it is a product of general linear groups with a group $G_- \in \widetilde{\mathcal{E}}_{\mathrm{sim}}(N_-)$, for some integer $N_- < N$. We can assume inductively that the corollary holds if N is replaced by N_-. Since the same property holds for the general linear factors of M, the analogue of the corollary holds for M itself. The operator $\mathcal{I}_{P,t,c}(1_M, f)$ in the summand of M is induced from the (t, c)-component of the automorphic discrete spectrum of M, or more correctly, the component of the discrete spectrum given by some partition of (t, c) among the factors of M. This operator then vanishes, by our assumption on (t, ψ), and so therefore does the summand of M.

The remaining term in the formula for $I_{\mathrm{disc},t,c}(f)$ is the summand

$$\mathrm{tr}\big(\mathcal{I}_{G,t,c}(1_G, f)\big)$$

corresponding to $M = G$. We have established that it vanishes. But

$$\mathcal{I}_{G,t,c}(f) = \mathcal{I}_{G,t,c}(1_G, f)$$

is the operator on the space $L^2_{\mathrm{disc},t,c}\big(G(F)\backslash G(\mathbb{A})\big)$ by right convolution of f. Its trace equals the sum

$$\sum_{\pi} m(\pi)\mathrm{tr}\big(\pi(f)\big), \qquad \|\mu_{\pi,I}\| = t, \; c(\pi) = c,$$

of the irreducible subrepresentations of this space. The function f belongs only to the (locally) symmetric Hecke algebra $\widetilde{\mathcal{H}}(G)$. However, since the multiplicities $m(\pi)$ are all positive, it is clear that the sum cannot vanish for all $f \in \widetilde{\mathcal{H}}(G)$ unless it also vanishes for all $f \in \mathcal{H}(G)$. It follows that $m(\pi) = 0$ for each π. In other words, the sum is over an empty set, and the assertions of the corollary hold. □

It will be convenient to write

(3.4.5) $$R^G_{\mathrm{disc},\psi}(f) = \mathcal{I}_{G,t(\psi),c(\psi)}(f), \qquad f \in \mathcal{H}(G),$$

for any $G \in \widetilde{\mathcal{E}}_{\mathrm{sim}}(N)$ and $\psi \in \widetilde{\Psi}(N)$. Then $R^G_{\mathrm{disc},\psi}$ stands for the regular representation of $G(\mathbb{A})$ on $L^2_{\mathrm{disc},\psi}\big(G(F)\backslash G(\mathbb{A})\big)$. The second corollary of (the yet unproven) Proposition 3.4.1 reduces the study of the automorphic discrete spectrum of G to that of the subrepresentations $R^G_{\mathrm{disc},\psi}$. Given the statement of Theorem 1.5.2, we will obviously need to reduce the problem further to parameters ψ that lie in the subset $\widetilde{\Psi}(G)$ of $\widetilde{\Psi}(N)$. This question turns out to be surprisingly difficult. We will be able to give a partial answer below, after some elementary remarks on central characters. However, its full resolution will not come before the end of our general induction argument in Chapter 8.

Recall that any $G \in \widetilde{\mathcal{E}}_{\mathrm{sim}}(N)$ comes with a character η_G of Γ_F of order 1 or 2. If this character is nontrivial, it determines G uniquely. The same is true if $\eta_G = 1$ and N is odd. However, if $\eta_G = 1$ and N is even, there is a second element $G^\vee \in \widetilde{\mathcal{E}}_{\mathrm{sim}}(N)$ with $\eta_{G^\vee} = 1$. Its dual group $\widehat{G}^\vee$ equals $SO(N, \mathbb{C})$ if $\widehat{G} = Sp(N, \mathbb{C})$ and $Sp(N, \mathbb{C})$ if $\widehat{G} = SO(N, \mathbb{C})$. In other words, $G^\vee$ is the split group $SO(N)$ if G equals $SO(N + 1)$, while $G^\vee$ equals $SO(N + 1)$ if G is the split group $SO(N)$.

Any parameter $\psi \in \widetilde{\Psi}(N)$ also comes with a character η_ψ on Γ_F of order 1 or 2. It is the central character n_{π_ψ}, regarded as an idèle class character of order 1 or 2, of the self-dual automorphic representation π_ψ of $GL(N)$ attached to ψ. If $\eta_\psi \neq \eta_G$, the Tchebotarev density theorem tells us that we can find a valuation v, at which ψ, G, η_ψ and η_G are unramified, such that the local characters $\eta_{\psi,v} = \eta_{\psi_v}$ and $\eta_{G,v} = \eta_{G_v}$ are distinct. It follows from the definitions that the conjugacy class $c(\psi_v)$ in $GL(N,\mathbb{C})$ does not meet the image of ${}^LG_v$. This implies that

$$R^G_{\mathrm{disc},\psi} = 0, \qquad \eta_\psi \neq \eta_G,\ G \in \widetilde{\mathcal{E}}_{\mathrm{sim}}(N). \tag{3.4.6}$$

On the other hand, it follows from the construction of the subset $\widetilde{\Psi}(G)$ of $\widetilde{\Psi}(N)$ in §1.4, which we will presently formalize with a running induction hypothesis, that $\eta_\psi = \eta_G$ if ψ belongs to $\widetilde{\Psi}(G)$. This gives a partial characterization of the complement of $\widetilde{\Psi}(G)$ in $\widetilde{\Psi}(N)$, for our simple endoscopic datum $G \in \widetilde{\mathcal{E}}_{\mathrm{sim}}(N)$. Our partial answer to the question posed above is that $R^G_{\mathrm{disc},\psi}$ vanishes for any ψ in the complement of $\widetilde{\Psi}(G)$ in $\widetilde{\Psi}(N)$, except possibly in the case that $\eta_\psi = 1$, N is even and G is split (which is to say that $\eta_G = \eta_\psi = 1$). It is this last case that contains major unresolved difficulties.

Suppose again that $\eta_\psi \neq \eta_G$, for some $G \in \widetilde{\mathcal{E}}_{\mathrm{sim}}(N)$, and that $\psi \in \widetilde{\Psi}(N)$. If v is a valuation such that $\eta_{\psi,v} \neq \eta_{G,v}$ as above, the conjugacy class $c(\psi_v)$ in $GL(N,\mathbb{C})$ does not meet the image of ${}^LM_v$ in $GL(N,\mathbb{C})$, for any Levi subgroup M of G. This implies that

$$\mathcal{I}_{P,t(\psi),c(\psi)}(f) = 0, \qquad P \in \mathcal{P}(M),\ f \in \widetilde{\mathcal{H}}(G).$$

It follows from the analogue of (3.1.1) for ψ that $I_{\mathrm{disc},\psi}(f) = 0$. Applying this assertion to the left hand side of (3.3.15), we see inductively that a similar assertion holds for $S^G_{\mathrm{disc},\psi}(f)$. In other words,

$$I^G_{\mathrm{disc},\psi}(f) = 0 = S^G_{\mathrm{disc},\psi}(f), \qquad f \in \widetilde{\mathcal{H}}(G),\ \eta_\psi \neq \eta_G,\ G \in \widetilde{\mathcal{E}}_{\mathrm{sim}}(N). \tag{3.4.7}$$

There is one other remark to be made before we start the proof of the proposition. It concerns a minor point of possible confusion. Our convention of letting the same symbol denote both a group and the endoscopic datum it represents has usually been harmless. For we agreed in the general case (3.2.3) that the linear form $S'_{\mathrm{disc},t}$ depends on G' only as a group (or more generally, on the pair $(\widetilde{G}', \widetilde{\chi}')$), and not on the other components $\mathcal{G}'$, s' and ξ' of the endoscopic datum. However, the c-component (3.3.9) of $S'_{\mathrm{disc},t}$ (as well as its analogue (3.3.8) for $I'_{\mathrm{disc},t}$) depends by definition on G' as an endoscopic datum. In the case $G \in \widetilde{\mathcal{E}}_{\mathrm{sim}}(N)$ we are considering here, this means that the ψ-component (3.3.13) of the stable linear form $S_{\mathrm{disc},t} = S^G_{\mathrm{disc},t}$ (as well as its analogue (3.3.12) for $I_{\mathrm{disc},t} = I^G_{\mathrm{disc},t}$) depends nominally on G as an endoscopic datum. The question arises when N is odd. In this case, $G = Sp(N-1)$ is the unique group with $\widehat{G} = SO(N,\mathbb{C})$, while the endoscopic datum is further parametrized by the quadratic character η_G.

However, this distinction is still innocuous. For the ψ-component (3.3.13) vanishes if $\eta_G \neq \eta_\psi$, and in the case that $\eta_G = \eta_\psi$, it remains unchanged if ψ is replaced with a translate $\psi \otimes \eta$ by a quadratic character η, and G is adjusted accordingly.

PROOF OF PROPOSITION 3.4.1. (First step): We assume inductively that the proposition holds if N is replaced by any integer $N_- < N$. We will let $G \in \widetilde{\mathcal{E}}_{\mathrm{sim}}(N)$ vary, so we need to fix a general pair

$$(t, c), \qquad t \geqslant 0,\ c \in \widetilde{\mathcal{C}}_{\mathbb{A}}(N),$$

that is independent of G. We assume that (t, c) is not of the form $\big(t(\psi), c(\psi)\big)$ for any $\psi \in \widetilde{\Psi}(N)$. We must establish that the linear forms $I_{\mathrm{disc},t,c}(f)$ and $S_{\mathrm{disc},t,c}(f)$ both vanish for any $G \in \widetilde{\mathcal{E}}_{\mathrm{sim}}(N)$ and $f \in \widetilde{\mathcal{H}}(G)$.

We fix G for the moment, and begin as in the proof of Corollary 3.4.3. Our induction hypothesis implies that the corollary also holds if N is replaced by $N_- < N$. It follows that the (t, c)-component of the discrete spectrum of any proper Levi subgroup M of G vanishes. The operator $\mathcal{I}_{P,t,c}(f)$ in the summand of M in the c-analogue of (3.1.1) therefore vanishes, and so then does the summand itself. Since the only remaining summand corresponds to $M = G$, we see that

$$I_{\mathrm{disc},t,c}(f) = \mathrm{tr}\big(\mathcal{I}_{G,t,c}(f)\big), \tag{3.4.8}$$

for any $f \in \widetilde{\mathcal{H}}(G)$.

We can apply similar arguments to the decomposition (3.3.10) given by the c-analogue of (3.2.3). Consider an index of summation $G' \in \mathcal{E}_{\mathrm{ell}}(G)$ in this decomposition with $G' \neq G$. Then G' is a proper product

$$G' = G'_1 \times G'_2, \qquad G'_i \in \mathcal{E}_{\mathrm{sim}}(N'_i),\ N'_i < N.$$

The corresponding linear form $\widehat{S}'_{\mathrm{disc},t,c}(f')$ in (3.3.10) is defined in turn by a sum (3.3.9). It follows from our induction hypothesis and our condition on (t, c) that each of the summands in (3.3.9) vanish. The linear form in (3.3.10) thus vanishes, and so therefore does the summand of G'. The remaining term in the decomposition (3.3.10) of $I_{\mathrm{disc},t,c}(f)$ is the stable linear form $S_{\mathrm{disc},t,c}(f)$. We have shown that

$$I_{\mathrm{disc},t,c}(f) = S_{\mathrm{disc},t,c}(f). \tag{3.4.9}$$

Next, we apply (3.3.10) again, but this time with $\widetilde{G}(N)$ in place of G. We have then to replace f by a function $\widetilde{f}$ in $\widetilde{\mathcal{H}}(N)$. The datum G plays the role of G' in (3.3.10), and is summed over the general indexing set $\widetilde{\mathcal{E}}_{\mathrm{ell}}(N) = \mathcal{E}_{\mathrm{ell}}\big(\widetilde{G}(N)\big)$. If G lies in the complement of $\widetilde{\mathcal{E}}_{\mathrm{sim}}(N)$ in $\widetilde{\mathcal{E}}_{\mathrm{ell}}(N)$, it equals a proper product

$$G = G_S \times G_O, \qquad G_\varepsilon \in \widetilde{\mathcal{E}}_{\mathrm{sim}}(N_\varepsilon),\ N_\varepsilon < N.$$

Arguing as above, we see from our induction hypothesis, the condition on (t, c), and the definition (3.3.9), that

$$\widehat{S}^G_{\mathrm{disc},t,c}(\widetilde{f}^G) = 0$$

in this case. The sum on the right hand side of (3.3.10) can therefore be taken over G in the subset $\widetilde{\mathcal{E}}_{\mathrm{sim}}(N)$ of $\widetilde{\mathcal{E}}_{\mathrm{ell}}(N)$. The left hand side of the identity is the linear form $I_{\mathrm{disc},t,c}(\widetilde{f})$. It vanishes by the remarks preceding (3.4.1) and our condition on (t, c). The identity becomes

$$\sum_{G \in \widetilde{\mathcal{E}}_{\mathrm{sim}}(N)} \widetilde{\imath}(N, G)\widehat{S}^G_{\mathrm{disc},t,c}(\widetilde{f}^G) = 0. \tag{3.4.10}$$

This completes the first step.

To state what we have established so far, we shall introduce a global form of the local object (2.1.3) from §2.1. We shall formulate it as a family of functions

$$\mathcal{F} = \{f \in \widetilde{\mathcal{H}}(G) : \ G \in \widetilde{\mathcal{E}}_{\mathrm{ell}}(N)\} \tag{3.4.11}$$

parametrized by global endoscopic data that are elliptic, with the condition that $f = 0$ for all but finitely many G. Let us say that $\mathcal{F}$ is a *decomposable compatible family* if for each v, there is a local compatible family of functions

$$\mathcal{F}_v = \{f_v \in \widetilde{\mathcal{H}}(G_v) : \ G_v \in \widetilde{\mathcal{E}}_v(N)\}$$

parametrized by the local endoscopic groups $G_v(F_v)$ such that for any $G \in \widetilde{\mathcal{E}}_{\mathrm{ell}}(N)$ with completions $G_v \in \widetilde{\mathcal{E}}_v(N)$, the corresponding function satisfies

$$f = \prod_v f_v.$$

We shall then say simply that $\mathcal{F}$ is a *compatible family* if there is a finite set of decomposable compatible families

$$\mathcal{F}_i = \{f_i \in \widetilde{\mathcal{H}}(G) : \ G \in \widetilde{\mathcal{E}}_{\mathrm{ell}}(N)\},$$

such that the function attached to any G satisfies

$$f = \sum f_i.$$

This is a natural analogue of the local definition. We observe from Proposition 2.1.1 that $\mathcal{F}$ is a compatible family if and only if there is a function $\widetilde{f}$ in $\widetilde{\mathcal{H}}(N)$ such that

$$f^G = \widetilde{f}^G, \qquad G \in \widetilde{\mathcal{E}}_{\mathrm{ell}}(N).$$

In other words, the function $f \in \mathcal{F}$ attached to any $G \in \widetilde{\mathcal{E}}_{\mathrm{ell}}(N)$ has the same image in $\widetilde{\mathcal{S}}(G)$ as $\widetilde{f}$.

The notion of Hecke type can obviously be formulated here. By a *Hecke type for* $\widetilde{\mathcal{E}}_{\mathrm{ell}}(N)$, we shall mean a pair

$$\big(S, \{(\tau_\infty, \kappa^\infty)\}\big), \tag{3.4.12}$$

where $S \supset S_\infty$ is a finite set of valuations, and $\{(\tau_\infty, \kappa^\infty)\}$ is a finite set of Hecke types, one for each $G \in \widetilde{\mathcal{E}}_{\mathrm{ell}}(N)$ that is unramified outside of S, such that

$$\kappa^\infty = \kappa_S^\infty K^S, \qquad \kappa_S^\infty \subset G(\mathbb{A}_S^\infty),$$

for a hyperspecial maximal compact subgroup K^S of $G(\mathbb{A}^S)$. Then (3.4.12) is a *Hecke type for* $\mathcal{F}$ if the function f attached to G vanishes whenever G ramifies outside of S, and has Hecke type $(\tau_\infty, \kappa^\infty)$ if G is unramified outside of S. Any compatible family $\mathcal{F}$ clearly has a Hecke type.

Suppose that $\mathcal{F}$ is any compatible family (3.4.11), and that $\widetilde{f} \in \widetilde{\mathcal{H}}(N)$ is chosen so that $\widetilde{f}^G = f^G$ for any G. We can then write

$$\begin{aligned}\sum_{G \in \widetilde{\mathcal{E}}_{\mathrm{sim}}(N)} \widetilde{\iota}(N,G)\mathrm{tr}\big(\mathcal{I}_{G,t,c}(f)\big) &= \sum_G \widetilde{\iota}(N,G) S_{\mathrm{disc},t,c}(f)\\ &= \sum_G \widetilde{\iota}(N,G)\widehat{S}_{\mathrm{disc},t,c}(f^G)\\ &= \sum_G \widetilde{\iota}(N,G)\widehat{S}_{\mathrm{disc},t,c}(\widetilde{f}^G),\end{aligned}$$

by (3.4.8) and (3.4.9). According to (3.4.10), this last sum vanishes. We conclude that

$$\sum_{G \in \widetilde{\mathcal{E}}_{\mathrm{sim}}(N)} \widetilde{\iota}(N,G)\mathrm{tr}\big(\mathcal{I}_{G,t,c}(f)\big) = 0, \qquad f \in \mathcal{F}, \tag{3.4.13}$$

if $\mathcal{F}$ is any compatible family (3.4.11). Since $\widetilde{\iota}(N,G) > 0$, we can write the summand of G as a linear combination

$$\sum_\pi c_G(\pi)\mathrm{tr}\big(\pi(f)\big), \qquad \pi \in \Pi_{\mathrm{unit}}(G),$$

of irreducible unitary characters on $G(\mathbb{A})$ with nonnegative coefficients $c_G(\pi)$. We can then write (3.4.13) in the form

$$\sum_{G \in \widetilde{\mathcal{E}}_{\mathrm{ell}}(N)} \sum_{\pi \in \Pi_{\mathrm{unit}}(G)} c_G(\pi)\mathrm{tr}\big(\pi(f)\big) = 0, \qquad f \in \mathcal{F}, \tag{3.4.14}$$

for nonnegative coefficients $c_G(\pi)$, which actually vanish if G lies in the complement of $\widetilde{\mathcal{E}}_{\mathrm{sim}}(N)$. The identity (3.4.14) is valid for any compatible family $\mathcal{F}$, and the double sum can be taken over a finite set that depends on $\mathcal{F}$ only through the choice of a Hecke type.

In the next section we will use the identity (3.4.14) to show that all of the coefficients $c_G(\pi)$ vanish. In fact, we will establish general vanishing properties, which will be a foundation for more subtle comparisons later on. The coefficients are defined by the decomposition of the (t,c)-component $\mathcal{I}_{G,t,c}$ of the representation of $G(\mathbb{A})$ on the discrete spectrum. If they vanish, we see that

$$\mathrm{tr}\big(\mathcal{I}_{G,t,c}(f)\big) = \widetilde{\iota}(N,G)^{-1}\sum_\pi c_G(\pi)\mathrm{tr}\big(\pi(f)\big) = 0,$$

for any $G \in \widetilde{\mathcal{E}}_{\mathrm{sim}}(N)$ and any $f \in \mathcal{H}(G)$. It then follows from (3.4.8) and (3.4.9) that

$$I_{\mathrm{disc},t,c}(f) = S_{\mathrm{disc},t,c}(f) = 0,$$

as required. In other words, we will have a proof of Proposition 3.4.1 once we have shown that the coefficients $c_G(\pi)$ vanish.

3.5. On the vanishing of coefficients

We have to complete the proof of Proposition 3.4.1. In the last section, we reduced the problem to proving that the coefficients $c_G(\pi)$ in the identity (3.4.14) all vanish. We shall now deduce that in any such identity, which we recall is associated to the global field F and the positive integer N, the coefficients automatically vanish.

PROPOSITION 3.5.1. *Suppose that there are nonnegative coefficients*

$$c_G(\pi), \qquad G \in \widetilde{\mathcal{E}}_{\mathrm{ell}}(N),\ \pi \in \Pi(G),$$

such that for every global compatible family $\mathcal{F}$ (3.4.11), *the function*

$$(G, \pi) \longrightarrow c_G(\pi) f_G(\pi), \qquad f \in \mathcal{F},$$

is supported on a finite set that depends only on a choice of Hecke type for $\mathcal{F}$, *and such that the double sum*

$$\sum_G \sum_\pi c_G(\pi) f_G(\pi), \qquad f \in \mathcal{F}, \tag{3.5.1}$$

vanishes. Then $c_G(\pi) = 0$, *for every* G *and* π.

The proof of Proposition 3.5.1 is primarily local. In order to focus on the essential ideas, we shall revert briefly to the local notation of Chapter 2. Thus, in contrast to the convention that has prevailed to this point in Chapter 3, and until further notice later in the section, we take F to be a local field. The sets $\widetilde{\mathcal{E}}(N)$, $\widetilde{\Pi}(G)$, $\widetilde{\mathcal{H}}(G)$, and so on, are then to be understood as local objects over F.

We shall take advantage of this interlude to recall some notions from local harmonic analysis, both for the proof and for later use. We shall review the theory of the representation theoretic R-group [**A10**, §1–3], which is founded on work of Harish-Chandra, Knapp, Stein and Zuckerman on intertwining operators (see [**Ha4**], [**Kn**], [**KnS**], [**KnZ1**] and [**Si2**]). For these remarks, we may as well allow G to be an arbitrary connected reductive group over F.

As in the global notation of §3.1, we let

$$\mathcal{L} = \mathcal{L}^G = \mathcal{L}^G(M_0),$$

be the finite set of Levi subgroups of G that contain a fixed minimal Levi subgroup M_0, and we write W_0^G for the Weyl group $W(M_0)$ of M_0. The set of $G(F)$-conjugacy classes of Levi subgroups of G is bijective with the set of W_0^G-orbits in $\mathcal{L}$. We then write $T(G)$ for the set of W_0^G-orbits of triplets

$$\tau = \tau_r = (M, \sigma, r), \qquad M \in \mathcal{L},\ \sigma \in \Pi_2(M),\ r \in R(\sigma), \tag{3.5.2}$$

where $R(\sigma)$ is the R-group of σ in G. We recall that $R(\sigma)$ is given by a short exact sequence

$$1 \longrightarrow W^0(\sigma) \longrightarrow W(\sigma) \longrightarrow R(\sigma) \longrightarrow 1,$$

where $W(\sigma)$ is the stabilizer of σ in the Weyl group $W(M)$ of M, and $W^0(\sigma)$ is the Weyl group of a root system defined by the vanishing of Plancherel densities. Any choice of positive chamber for the root system determines a splitting of the sequence, and thereby allows us to identity $R(\sigma)$ with a subgroup of $W(\sigma)$.

It is known how to construct general normalized intertwining operators

$$R_P(w, \sigma) : \mathcal{I}_P(\sigma) \longrightarrow \mathcal{I}_P(\sigma), \qquad w \in W(\sigma),$$

from the basic intertwining integrals, which satisfy analogues of (2.3.9) and (2.3.10). (The focus of the discussion of §2.2.3 and §2.3.4 was more specific, namely to establish particular normalizations that would be compatible with endoscopy.) The group $W^0(\sigma)$ becomes the subgroup of elements $w \in W(\sigma)$ such that $R_P(w, \sigma)$ is a scalar.

What is not known in general is whether these normalizations can be chosen to be multiplicative in w. The question really pertains to the R-group $R(\sigma)$ [**A10**, p. 91, Remark 1], and is equivalent to whether the irreducible representation σ has an extension to a semidirect product $M(F) \rtimes R(\sigma)$. One sidesteps the problem by introducing a finite central extension

$$1 \longrightarrow Z_\sigma \longrightarrow \widetilde{R}(\sigma) \longrightarrow R(\sigma) \longrightarrow 1$$

of $R(\sigma)$. It then becomes possible to attach normalized intertwining operators $R_P(r, \sigma)$ to elements r in $\widetilde{R}(\sigma)$ such that

$$R_P(zr, \sigma) = \chi_\sigma(z)^{-1} R_P(r, \sigma), \qquad z \in Z_\sigma,\ r \in \widetilde{R}(\sigma),$$

for a fixed character χ_σ of Z_σ, and so that the mapping

$$r \longrightarrow R_P(r, \sigma), \qquad r \in \widetilde{R}(\sigma),$$

is a homomorphism from $\widetilde{R}(\sigma)$ to the space of intertwining operators of $\mathcal{I}_P(\sigma)$. We write $\widetilde{T}(G)$ for the set of W_0^G-orbits of triplets of the form (3.5.2), but with r in the extension $\widetilde{R}(\sigma)$ of $R(\sigma)$. We then set

$$f_G(\tau) = f_G(\tau_r) = \operatorname{tr}\big(R_P(r, \sigma)\mathcal{I}_P(\sigma, f)\big), \qquad f \in \mathcal{H}(G), \tag{3.5.3}$$

for any element $\tau = \tau_r$ in $\widetilde{T}(G)$.

The groups $R(\sigma)$ and $\widetilde{R}(\sigma)$ do not have to be abelian. However, we can still work with the set $\Pi\big(\widetilde{R}(\sigma), \chi_\sigma\big)$ of irreducible representations ξ of $\widetilde{R}(\sigma)$

whose Z_σ-central character equals χ_σ. For any such ξ, the linear form

$$f_G(\rho_\xi) = |R(\sigma)|^{-1} \sum_{r \in R(\sigma)} \operatorname{tr}\big(\xi(r)\big) f_G(\tau_r), \qquad f \in \mathcal{H}(G), \tag{3.5.4}$$

is the character of a subrepresentation of $\mathcal{I}_P(\sigma)$, which turns out to be irreducible if σ is tempered.

We write $T_{\text{temp}}(G)$ and $\widetilde{T}_{\text{temp}}(G)$ for the subsets of triplets τ in the respective sets $T(G)$ and $\widetilde{T}(G)$ for which the second component σ of (3.5.2) belongs to $\Pi_{2,\text{temp}}(M)$, the subset of tempered representations in $\Pi_2(M)$. For any such σ, we write $\Pi_\sigma(G)$ for the set of irreducible constituents of the induced tempered representations $\mathcal{I}_P(\sigma)$ of $G(F)$. It is well known that the set $\Pi_{\text{temp}}(G)$ of all irreducible tempered representations of $G(F)$ is a disjoint union, over W_0^G-orbits of pairs (M, σ), of the sets $\Pi_\sigma(G)$. (See for example [**A10**, Proposition 1.1].) Moreover, the basis theorem of Harish-Chandra [**Ha4**] tells us that if σ belongs to $\Pi_{2,\text{temp}}(M)$, and $\{r\}$ is a set of representatives in $\widetilde{R}(\sigma)$ of the conjugacy classes in $R(\sigma)$, the corresponding set of linear forms

$$r \longrightarrow f_G(\tau_r), \qquad f \in \mathcal{H}(G),$$

is a basis of the space spanned by the characters of representations in $\Pi_\sigma(G)$. It follows that the mapping

$$\xi \longrightarrow \pi_\xi = \rho_\xi, \qquad \xi \in \Pi\big(\widetilde{R}(\sigma), \chi_\sigma\big),$$

is a bijection from $\Pi\big(\widetilde{R}(\sigma), \chi_\sigma\big)$ to $\Pi_\sigma(G)$. The full set $\Pi_{\text{temp}}(G)$ can therefore be identified with the family of W_0^G-orbits of triplets

$$\pi = \pi_\xi \cong (M, \sigma, \xi), \qquad M \in \mathcal{L},\ \sigma \in \Pi_{2,\text{temp}}(M),\ \xi \in \Pi\big(\widetilde{R}(\sigma), \chi_\sigma\big).$$

(See [**A10**, §3], where the set $T_{\text{temp}}(G)$ was denoted by $T(G)$. We need to reserve the symbol $T(G)$ here for nontempered triplets.)

One thus obtains a rough classification of $\Pi_{\text{temp}}(G)$ from local harmonic analysis. It is independent of the finer endoscopic classification, which for simple endoscopic data $G \in \widetilde{\mathcal{E}}_{\text{sim}}(N)$ is contained in the assertions of the local theorems. We will eventually relate the two classifications in Chapter 6, as part of the proof of the local theorems. This will obviously entail letting σ represent an element $\pi_M \in \widetilde{\Pi}_{\phi_M}$, with $\phi_M \in \widetilde{\Phi}_2(M, \phi)$ being a preimage of a generic parameter $\phi \in \widetilde{\Phi}(G)$. The resulting relations will then confirm that $W(\sigma) = W_\phi$, $W^0(\sigma) = W_\phi^0$, $R(\sigma) = R_\phi$, and therefore that the two kinds of R-groups are the same.

In the general nontempered case, we have to distinguish between irreducible representations $\pi \in \Pi(G)$ and standard representations $\rho \in P(G)$. The general relations between the two kinds of representations are parallel to those between extensions $\widetilde{\pi} \in \widetilde{\Pi}(N)$ and $\widetilde{\rho} \in \widetilde{P}(N)$ discussed in §2.2. Let us review them.

There is a bijection

$$\pi \longrightarrow \rho_\pi, \qquad \pi \in \Pi(G),$$

between $\Pi(G)$ and $P(G)$ such that π is the Langlands quotient of ρ. The character of any π has a decomposition

$$f_G(\pi) = \sum_{\rho \in P(G)} n(\pi, \rho)\, f_G(\rho), \qquad f \in \mathcal{H}(G),$$

into standard characters. The coefficients $n(\pi, \rho)$ are uniquely determined integers, with

$$n(\pi, \rho_\pi) = 1,$$

which have finite support in ρ for any π. Both π and ρ determine real linear forms Λ_π and Λ_ρ. These objects lie in the dual closed chamber $\overline{(\mathfrak{a}_0^*)^+}$ in $\mathfrak{a}_0 = \mathfrak{a}_{M_0}$ attached to a preassigned minimal parabolic subgroup $P_0 \in \mathcal{P}(M_0)$ of G, and are a measure of the failure of the representations to be tempered. In particular, $\Lambda_\pi = \Lambda_{\rho_\pi}$ is the point in the relatively open cone

$$(\mathfrak{a}_P^*)^+ \subset \overline{(\mathfrak{a}_0^*)^+}, \qquad P \supset P_0$$

that represents the nonunitary part of ρ_π, as a representation induced from $P(F)$. If $n(\pi, \rho) \neq 0$, then $\Lambda_\rho \leqslant \Lambda_\pi$, in the usual sense that $\Lambda_\pi - \Lambda_\rho$ is a nonnegative integral combination of simple roots of (P_0, A_0), with equality $\Lambda_\rho = \Lambda_\pi$ holding if and only if ρ equals ρ_π. (See [**A7**, Proposition 5.1], which is based on results in [**BoW**] and underlying ideas of Vogan.)

Standard characters can in turn be decomposed into the virtual characters parametrized by $\widetilde{T}(G)$. For it follows from the definition of a standard representation, and the general classification of $\Pi_{\mathrm{temp}}(G)$ described above, that the set $P(G)$ of standard representations of $G(F)$ can be identified with the family of W_0^G-orbits of triplets

$$\rho = \rho_\xi \cong (M, \sigma, \xi), \qquad M \in \mathcal{L},\ \sigma \in \Pi_2(M),\ \xi \in \Pi\big(\widetilde{R}(\sigma), \chi_\sigma\big). \tag{3.5.5}$$

We define

$$\langle \rho, \tau \rangle = |R(\sigma)|^{-1} \mathrm{tr}\big(\xi(r)\big),$$

for $\rho = \rho_\xi$ as in (3.5.5) and $\tau = \tau_r$ as in the analogue for $\widetilde{T}(G)$ of (3.5.2). We also set $\langle \rho, \tau \rangle = 0$ for elements $\rho \in P(G)$ and $\tau \in \widetilde{T}(G)$ that are complementary, in the sense that their corresponding triplets do not share the same W_0^G-orbit of pairs (M, σ). It then follows that

$$f_G(\pi) = \sum_{\tau \in T(G)} n(\pi, \tau) f_G(\tau), \qquad \pi \in \Pi(G),\ f \in \mathcal{H}(G), \tag{3.5.6}$$

where

$$n(\pi, \tau) = \sum_{\rho \in P(G)} n(\pi, \rho) \langle \rho, \tau \rangle, \qquad \tau \in \widetilde{T}(G).$$

If τ belongs to $\widetilde{T}(G)$, we set $\Lambda_\tau = \Lambda_\rho$, for any element $\rho \in P(G)$ with $\langle \rho, \tau \rangle \neq 0$. Then Λ_τ is a linear form in $\overline{(\mathfrak{a}_0^*)^+}$, with the property that $\Lambda_\tau \leqslant \Lambda_\pi$ if $n(\pi, \tau) \neq 0$.

Elliptic elements in $T(G)$ have a special role. We recall that $T_{\mathrm{ell}}(G)$ is the subset of (W_0^G-orbits of) triplets τ_r in (3.5.2) such that r is regular, in the sense that

$$\mathfrak{a}_G = \{H \in \mathfrak{a}_M : rH = H\}.$$

In general, we will write

$$G_{\mathrm{reg}} = G_{\text{str-reg}}$$

for the open connected subset of strongly regular points in G, and

$$G_{\mathrm{reg,ell}}(F) = G_{\text{str-reg,ell}}(F)$$

for the subset of elements in $G_{\mathrm{reg}}(F)$ that are elliptic over F. Thus, $G_{\mathrm{reg,ell}}(F)$ is the subset of elements in $G(F)$ whose centralizer is a torus T such that $T(F)/A_G(F)$ is compact. (As in Chapter 2, we will generally use the simpler notation on the left, since we will not need to consider regular points that are not strongly regular.) Elements in $T_{\mathrm{ell}}(G)$ have the important property that they are uniquely determined by the restriction of their virtual characters to $G_{\mathrm{reg,ell}}(F)$.

To be more precise, let $\{\tau\}$ be a set of representatives of the set $T_{\mathrm{ell}}(G)$ in its preimage $\widetilde{T}_{\mathrm{ell}}(G)$ in $\widetilde{T}(G)$. Then the corresponding set of distributions

$$\tau \longrightarrow f_G(\tau), \qquad f \in C_c^\infty\big(G_{\mathrm{reg,ell}}(F)\big),$$

on $G_{\mathrm{reg,ell}}(F)$ is a basis for the space spanned by the restriction to $G_{\mathrm{reg,ell}}(F)$ of characters of representations in either of the sets $\Pi(G)$ or $P(G)$. (See [**Ka**] and [**A11**]. This is also a consequence of the orthogonality relations in [**A11**, §6].) It follows from the definitions that $T(G)$ is a disjoint union, over W_0^G-orbits of Levi subgroups M, of images in $T(G)$ of the corresponding elliptic sets $T_{\mathrm{ell}}(M)$. For any $\tau \in T(G)$, we write M_τ for a Levi subgroup such that τ lies in the image of $T_{\mathrm{ell}}(M_\tau)$.

Suppose that $\pi \in \Pi(G)$. If the associated standard representation $\rho_\pi \in P(G)$ corresponds to the triplet $(M_\pi, \sigma_\pi, \xi_\pi)$ under (3.5.5), we define an element τ_π in either $T(G)$ or $\widetilde{T}(G)$ by the triplet $(M_\pi, \sigma_\pi, 1)$. The two elements ρ_π and τ_π are such that both $n(\pi, \rho_\pi)$ and $\langle \rho_\pi, \tau_\pi \rangle$ are nonzero, and they satisfy

$$\Lambda_{\tau_\pi} = \Lambda_{\rho_\pi} = \Lambda_\pi.$$

Conversely, suppose that $\rho \in P(G)$ and $\tau \in T(G)$ are elements such that both $n(\pi, \rho)$ and $\langle \rho, \tau \rangle$ are nonzero (τ being identified here with a representative in $\widetilde{T}(G)$). Then

$$\Lambda_\tau = \Lambda_\rho \leqslant \Lambda_\pi.$$

If these linear forms are all equal, M_τ contains M_π (with M_τ and M_π taken to be suitable representatives of the associated W_0^G-orbits in $\mathcal{L}$). Moreover, with the assumption that $\Lambda_\tau = \Lambda_\pi$, the group M_τ equals M_π if and only if $\tau = \tau_\pi$ and $\rho = \rho_\pi$. We observe that in this case, we have

$$n(\pi, \tau_\pi) = n(\pi, \rho_\pi)\langle \rho_\pi, \tau_\pi \rangle = 1 \cdot |R(\sigma_\pi)|^{-1} \dim(\xi_\pi) > 0.$$

We return to the case that G represents a twisted endoscopic datum for $GL(N)$. We will now have no need of the extension $\widetilde{R}(\sigma)$ of $R(\sigma)$ from the general case. For it is clear from the structure of the Levi subgroup M (2.3.4) and the corresponding Weyl group $W(M)$, that any representation of $\sigma \in \Pi_2(M)$ extends to the semidirect product $M(F) \rtimes W(\sigma)$, and in particular, to its subgroup $M(F) \rtimes R(\sigma)$. This leaves us free to take $\widetilde{R}(\sigma) = R(\sigma)$. We can therefore make a change in the notation above, letting $\widetilde{T}(G)$ denote the set of $\widetilde{\mathrm{O}}\mathrm{ut}_N(G)$-orbits in $T(G)$ as usual, rather than triplets (3.5.2) with $r \in \widetilde{R}(\sigma)$. A similar understanding applies to the sets $\widetilde{T}_{\mathrm{ell}}(G)$, $\widetilde{T}(M)$, etc.

The statement of Proposition 3.5.1 makes sense over the local field F. The set $\mathcal{F}$ has of course to be understood as a local compatible family (2.1.3), with the implication that the global set $\widetilde{\mathcal{E}}_{\mathrm{ell}}(N)$ is replaced by the general local set $\widetilde{\mathcal{E}}(N)$. A Hecke type for $\mathcal{F}$ will be family of objects, parametrized by the finite set of $G \in \widetilde{\mathcal{E}}(N)$, consisting of open compact subgroups $K_0 \subset G(F)$ if F is p-adic, and finite sets $\tau_{\mathbb{R}}$ of irreducible representations of maximal compact subgroups $K_{\mathbb{R}} \subset G(\mathbb{R})$ if $F = \mathbb{R}$.

LEMMA 3.5.2. *Proposition* 3.5.1 *is valid, stated with $\widetilde{\mathcal{E}}(N)$ in place of $\widetilde{\mathcal{E}}_{\mathrm{ell}}(N)$, over the local field F.*

PROOF. We are given that the sum

$$\sum_G \sum_\pi c_G(\pi) f_G(\pi), \tag{3.5.7}$$

over $G \in \widetilde{\mathcal{E}}(N)$ and $\pi \in \Pi(G)$, vanishes for any compatible family $\mathcal{F}$ over the local field F. The function f of course stands for the element of $\mathcal{F}$ indexed by G. We shall let $\mathcal{F}$ vary over the compatible families attached to a fixed local Hecke type.

By the general discussion above, we can write the sum (3.5.7) as

$$\begin{aligned} &\sum_{G \in \widetilde{\mathcal{E}}(N)} \sum_{\pi \in \Pi(G)} \sum_{\tau \in T(G)} c_G(\pi) n(\pi, \tau) f_G(\tau) \\ &= \sum_G \sum_\pi \sum_{M \in \mathcal{L}^G} \sum_{\tau \in T_{\mathrm{ell}}(M)} |W_0^G(\tau)||W_0^G|^{-1} c_G(\pi) n(\pi,\tau) f_M(\tau), \end{aligned}$$

where $W_0^G(\tau)$ is the stabilizer of (M, τ) in W_0^G. The coefficient $n(\pi, \tau)$ in the last expression is understood as a W_0^G-invariant function of $\tau \in \widetilde{T}_{\mathrm{ell}}(M)$, which is to say, a function whose product with $f_M(\tau)$ depends only on the image of τ in $T(G)$. The multiple sum can be taken over a finite set of indices (G, π, M, τ), which depends only on the given local Hecke type.

A Levi subgroup M of G can be treated in its own right as an element in $\widetilde{\mathcal{E}}(N)$. Conversely, for any $M \in \widetilde{\mathcal{E}}(N)$, there will generally be several $G \in \widetilde{\mathcal{E}}(N)$ in which M embeds as a Levi subgroup. The best way to account for these groups is as elements in the twisted analogue $\widetilde{\mathcal{E}}_M(N) = \mathcal{E}_M(\widetilde{G}(N))$ of the set $\mathcal{E}_{M'}(G)$ defined (for any connected reductive group G and any endoscopic datum M' for G) on p. 227 of [**A12**]. (See [**Wh**, Appendix, p. 434–436].) We leave the reader to formulate the definition of $\widetilde{\mathcal{E}}_M(N)$ as a set of twisted endoscopic data (G, s, ξ) for $\widetilde{G}(N)$, which contain M as a Levi subgroup, and are taken up to translation of s by elements in the product

$$\big(Z(\widehat{G})^{\Gamma}\big)^0 Z\big(\widehat{\widetilde{G}}(N)\big).$$

(The inclusion of the connected component $\big(Z(\widehat{G})^{\Gamma}\big)^0$ is a departure from the convention of [**A12**], and is designed to make the set $\widetilde{\mathcal{E}}_M(N)$ finite.) We can then replace the last double sum over $G \in \widetilde{\mathcal{E}}(N)$ and $M \in \mathcal{L}^G$ by a double sum over $M \in \widetilde{\mathcal{E}}(N)$ and $G \in \widetilde{\mathcal{E}}_M(N)$, provided that we adjust the coefficients to compensate for the fact that several elements in $\widetilde{\mathcal{E}}_M(N)$ could represent the same isomorphism class in $\widetilde{\mathcal{E}}(N)$. It follows that (3.5.7) equals

$$\sum_{M \in \widetilde{\mathcal{E}}(N)} \sum_{\tau \in T_{\mathrm{ell}}(M)} \sum_{G \in \widetilde{\mathcal{E}}_M(N)} \sum_{\pi \in \Pi(G)} c_G(\pi) n(\pi, \tau) \gamma_G(\tau) f_M(\tau),$$

for positive constants $\gamma_G(\tau)$.

For any $M \in \widetilde{\mathcal{E}}(N)$, the function $f_M(\tau)$ in the last sum depends on G, since it comes from the function $f \in \mathcal{F}$ attached to G. The family

$$f_M(\tau), \qquad M \in \widetilde{\mathcal{E}}(N),\ G \in \widetilde{\mathcal{E}}_M(N),\ \tau \in \widetilde{T}_{\mathrm{ell}}(M),$$

of such functions is indexed by $\mathcal{F}$. As $\mathcal{F}$ varies, the corresponding set of families is a natural vector space. It is not hard to see from the definition of a compatible family (2.1.3), together with the trace Paley-Wiener theorem [**CD**] [**DM**], that this vector space is a direct sum over $M \in \widetilde{\mathcal{E}}(N)$ of the subspaces of (families of) functions that are supported on $\widetilde{T}_{\mathrm{ell}}(M)$. It follows that the sum

$$\sum_{\tau \in T_{\mathrm{ell}}(M)} \sum_{G \in \widetilde{\mathcal{E}}_M(N)} \sum_{\pi \in \Pi(G)} c_G(\pi)\, n(\pi, \tau)\, \gamma_G(\tau)\, f_M(\tau)$$

vanishes for any M, and any compatible family $\mathcal{F}$.

We fix $M \in \widetilde{\mathcal{E}}(N)$, and consider the function $f_M(\tau)$ attached to $G \in \widetilde{\mathcal{E}}_M(G)$. By the definition (2.1.3), f_M lies in the subspace of $\widetilde{\mathrm{Out}}_N(M)$-invariant functions in $\mathcal{I}(M)$. We will assume that f_M also lies in the subspace $\mathcal{I}_{\mathrm{cusp}}(M)$ of cuspidal functions in $\mathcal{I}(M)$, but this will be the only other constraint. It then follows from the trace Paley-Wiener theorem for M, and the linear independence of the set of distributions on $M_{\text{str-reg,ell}}(F)$ attached to $T_{\mathrm{ell}}(M)$, that we can take $f_M(\tau)$ to be any $\widetilde{\mathrm{Out}}_N(M)$-invariant function in the natural Paley-Wiener space on $\widetilde{T}_{\mathrm{ell}}(M)$. Since we are working with a preassigned Hecke type, we are actually appealing here to an implicit consequence of the trace Paley-Weiner theorem. Namely, if Γ is a finite set of irreducible representations of the maximal compact subgroup K_M of $M(F)$, and $\mathcal{I}(M)_\Gamma$ and $\mathcal{H}(M)_\Gamma$ are the corresponding subspaces of

$\mathcal{I}(M)$ and $\mathcal{H}(M)$, there is a suitable section

$$\mathcal{I}(M)_\Gamma \longrightarrow \mathcal{H}(M)_\Gamma.$$

(See [**A5**, Lemma A.1], for example.) The only other constraint imposed by the definition of $\mathcal{F}$ is that as G varies, the images f^M in $\widetilde{\mathcal{S}}(M)$ of the functions $f \in \widetilde{\mathcal{H}}(G)$ are all equal. It follows from this that we can take the functions $f_M(\tau)$ parametrized by G to be equal. In particular, we can choose the common function $f_M(\tau)$ so that it isolates any $\widetilde{\mathrm{Out}}_N(M)$-orbit in $T_{\mathrm{ell}}(M)$. It follows that the sum

$$c_M(\tilde{\tau}) = \sum_{G \in \widetilde{\mathcal{E}}_M(N)} \sum_{\pi \in \Pi(G)} \sum_{\tau \in \tilde{\tau}} c_G(\pi)\, n(\pi,\tau)\, \gamma_G(\tau) \tag{3.5.8}$$

vanishes for any $M \in \widetilde{\mathcal{E}}(N)$ and any $\widetilde{\mathrm{Out}}_N(M)$-orbit $\tilde{\tau}$ in $\widetilde{T}_{\mathrm{ell}}(M)$.

It is instructive to introduce an equivalence relation on the set of pairs

$$(G,\pi), \qquad G \in \widetilde{\mathcal{E}}(N),\ \pi \in \Pi(G).$$

We write $(G,\pi) \sim (G_1,\pi_1)$ if there is a pair

$$(M,\tau), \qquad \tau \in \widetilde{T}_{\mathrm{ell}}(M),$$

such that both $n(\pi,\tau)$ and $n(\pi_1,\tau)$ are nonzero. We can then define *τ-equivalence* to be the equivalence relation generated by this basic relation. If P is any τ-equivalence class, we write P^* for the adjoint equivalence class in the set of pairs (M,τ). It consists of those (M,τ) such that $n(\pi,\tau)$ is nonzero for some (G,π) in P. One sees without difficulty that the classes P and P^* are both finite. These two equivalence relations are natural variants of the familiar relations of block equivalence, which are defined on the two sets $\Pi(G)$ and $P(G)$ attached to any connected reductive group G over F. We shall use them in the analysis of (3.5.8).

Assume that the assertion of the lemma is false. We can then fix a τ-equivalence class P such that the subset

$$P' = \big\{(G,\pi) \in P :\ c_G(\pi) \neq 0\big\}$$

is not empty. We define

$$\lambda' = \max\big\{\|\Lambda_\pi\| :\ (G,\pi) \in P'\big\}$$

and

$$\mu' = \min\big\{\dim(M_\pi) :\ (G,\pi) \in P',\ \|\Lambda_\pi\| = \lambda'\big\}.$$

(As in the general setting of §3.2, the Hermitian norm $\|\cdot\|$ on the space $\mathfrak{a}_0^*$ attached to G is defined in terms of a suitable, preassigned norm on the space $\widetilde{\mathfrak{a}}_0^*$ attached to $\widetilde{G}(N)$.) Let (G',π') be a fixed pair in P' such that

$$\big(\|\Lambda_{\pi'}\|,\ \dim(M_{\pi'})\big) = (\lambda',\mu').$$

The pair

$$(M',\tau') = (M_{\pi'},\tau_{\pi'})$$

belongs to P^*, since we can treat

$$\tau' = (M', \pi', 1)$$

as an element in $\widetilde{T}_{\mathrm{ell}}(M')$. It also has the property that

$$\big(\|\Lambda_{\tau'}\|, \dim(M_{\tau'})\big) = (\lambda', \mu').$$

Moreover, G' is represented by a datum in $\widetilde{\mathcal{E}}_{M'}(N)$. We shall use the fact that $c_{M'}(\widetilde{\tau}') = 0$ to derive a contradiction.

Suppose that (G, π) is any pair in P' such that G is represented by a datum in $\widetilde{\mathcal{E}}_{M'}(N)$, and that τ is any element in the $\widetilde{\mathrm{O}}\mathrm{ut}_N(M')$-orbit $\widetilde{\tau}'$ such that $n(\pi, \tau) \neq 0$. The latter condition implies that $\Lambda_\tau \leqslant \Lambda_\pi$, that $M_\tau \supset M_\pi$ in case $\Lambda_\tau = \Lambda_\pi$, and that $\tau = \tau_\pi$ in case both $\Lambda_\tau = \Lambda_\pi$ and $M_\tau = M_\pi$ hold. But it is clear that

$$\big(\|\Lambda_\tau\|, \dim(M_\tau)\big) = \big(\|\Lambda_{\tau'}\|, \dim(M_{\tau'})\big) = \big(\|\Lambda_{\pi'}\|, \dim(M_{\pi'})\big),$$

all three pairs being equal to the extremal pair (λ', μ'). It then follows from the definitions that the two relations $\Lambda_\tau = \Lambda_\pi$ and $M_\tau = M_\pi$ do indeed hold, so that $\tau = \tau_\pi$. The coefficient $n(\pi, \tau)$ is therefore strictly positive. The summand of G, π and τ in the expression (3.5.8) for $c_{M'}(\widetilde{\tau}')$ is consequently positive.

The first two sums in (3.5.8) reduce to a sum over pairs (G, π) in P, according to the definition of τ-equivalence. We have just seen that a summand of (G, π) and τ is either positive or zero. Since the summand of (G', π') and τ' is positive by assumption, the coefficient $c_{M'}(\widetilde{\tau}')$ itself is positive. This gives the desired contradiction, and completes the proof of Lemma 3.5.2. □

We now go back to the global setting. For the rest of the chapter (and indeed until the end of Chapter 5) we assume that the field F is global. We need to establish Proposition 3.5.1. We shall do so by applying the local proof above to the completions F_v of F.

PROOF OF PROPOSITION 3.5.1. Let $S \supset S_\infty$ be a finite set of valuations of the global field F. We shall write $\widetilde{\mathcal{E}}_S(N)$ for the finite set of products

$$G_S = \prod_{v \in S} G_v, \qquad G_v \in \widetilde{\mathcal{E}}_v(N),$$

where $\widetilde{\mathcal{E}}_v(N)$ denotes the set $\mathcal{E}\big(\widetilde{G}_v(N)\big)$ of twisted endoscopic data for $GL(N)$ over F_v, as earlier. We can then form the set

$$\Pi(G_S) = \Big\{\pi_S = \bigotimes_{v \in S} \pi_v : \ \pi_v \in \Pi(G_v)\Big\}$$

of irreducible representations of G_S. Let us also write $\widetilde{\mathcal{E}}_{\mathrm{ell}}(N, S)$ for the finite set of global endoscopic data $G \in \widetilde{\mathcal{E}}_{\mathrm{ell}}(N)$ that are unramified outside of S, and $\Pi(G, S)$ for the set of irreducible representations of $G(\mathbb{A})$ that are unramified outside of S. For a given pair

$$(G_S, \pi_S), \qquad G_S \in \widetilde{\mathcal{E}}_S(N), \ \pi_S \in \Pi(G_S),$$

we then define

$$c_{G_S}(\pi_S) = \sum_{(G,\pi)} c_G(\pi), \qquad G \in \widetilde{\mathcal{E}}_{\mathrm{ell}}(N,S),\ \pi \in \Pi(G,S).$$

The sum here is taken over the preimage of (G_S, π_S) under the localization mapping

$$(G,\pi) \longrightarrow (G_S, \pi_S),$$

and can be restricted to a finite set, in view of the given condition on the Hecke type. We are also given that each coefficient $c_G(\pi)$ is nonnegative, so the same is true of $c_{G_S}(\pi_S)$. It therefore suffices to show that $c_{G_S}(\pi_S)$ vanishes for any S and any (G_S, π_S).

Suppose that

$$\mathcal{F}_S = \prod_{v \in S} \mathcal{F}_v = \Big\{ f_S = \prod_v f_v :\ f_v \in \widetilde{\mathcal{H}}(G_v),\ G_v \in \widetilde{\mathcal{E}}_v(N) \Big\}$$

is a product of local compatible families $\mathcal{F}_v$. Then $\mathcal{F}_S$ determines a decomposable global compatible family

$$\mathcal{F} = \big\{ f \in \mathcal{H}(G) :\ G \in \widetilde{\mathcal{E}}_{\mathrm{ell}}(N) \big\}$$

in the natural way. That is, f vanishes unless G lies in $\widetilde{\mathcal{E}}_{\mathrm{ell}}(N,S)$, in which case f is the product of the function $f_S \in \widetilde{\mathcal{H}}(G_S)$ with the characteristic function of a hyperspecial maximal compact subgroup K^S of $G(\mathbb{A}^S)$. The expression (3.5.1) becomes a double sum

$$\sum_{G_S} \sum_{\pi_S} c_{G_S}(\pi_S) f_{S,G_S}(\pi_S), \qquad f_S \in \mathcal{F}_S, \tag{3.5.9}$$

which can be taken over a finite set that depends only on a choice of Hecke type for $\mathcal{F}_S$. We are told that (3.5.9) vanishes. We must use this to deduce that each $c_{G_S}(\pi_S)$ vanishes.

We have simply to duplicate the proof of Lemma 3.5.2, amplifying the notation in the appropriate way (generally without comment). We find that (3.5.9) equals

$$\sum_{M_S} \sum_{\tau_S} \sum_{G_S} \sum_{\pi_S} c_{G_S}(\pi_S) n(\pi_S, \tau_S) \gamma_{G_S}(\tau_S) f_{S,M_S}(\tau_S), \tag{3.5.10}$$

for sums over $M_S \in \widetilde{\mathcal{E}}_S(N)$, $\tau_S \in T_{\mathrm{ell}}(M_S)$, $G_S \in \widetilde{\mathcal{E}}_{M_S}(N)$ and $\pi_S \in \Pi(G_S)$, and for positive constants $\gamma_{G_S}(\tau_S)$. It is understood here that $\widetilde{\mathcal{E}}_{M_S}(N)$ stands for the product of the local sets $\widetilde{\mathcal{E}}_{M_v}(N)$ (rather than the global object defined on [**A12**, p. 241]). We can then take each $f_{S,M_S}(\tau_S)$ to be any general, cuspidal, $\widetilde{\mathrm{Out}}_N(M_S)$-invariant Paley-Wiener function that is independent of G_S. Varying f_S, we conclude that for any $M_S \in \widetilde{\mathcal{E}}_S(N)$, and any orbit $\widetilde{\tau}_S$ of $\widetilde{\mathrm{Out}}_N(M_S)$, the sum

$$c_{M_S}(\widetilde{\tau}_S) = \sum_{G_S \in \widetilde{\mathcal{E}}_{M_S}(N)} \ \sum_{\pi_S \in \Pi(G_S)} \ \sum_{\tau_S \in \widetilde{\tau}_S} c_{G_S}(\pi_S) n(\pi_S, \tau_S) \gamma_{G_S}(\tau_S) \tag{3.5.11}$$

vanishes.

Following the proof of the lemma further, we fix a τ_S-equivalence class

$$P_S = \{(G_S, \pi_S) : \ G_S \in \widetilde{\mathcal{E}}_S(N), \ \ \tau_S \in \Pi(G_S)\}.$$

To apply this part of the local argument, we enumerate the elements v in S as $v_1, \dots, v_n$, and introduce a lexicographic order on the associated local objects $\|\Lambda_{\pi_v}\|$ and $\dim(M_{\pi_v})$. We take the "alphabet" A to be the set of pairs $x = (\lambda, \mu)$, for λ a nonnegative real number and μ a nonnegative integer. It has a linear order defined by

$$x = (\lambda, \mu) > x' = (\lambda', \mu'),$$

if $\lambda > \lambda'$, or if $\lambda = \lambda'$ and $\mu < \mu'$. We take the "dictionary" D_S to be the set $A^S = A^n$, with the associated lexicographic order. There is then a mapping

$$(G_S, \pi_S) \longrightarrow w(G_S, \pi_S) = \prod_{v \in S} \big(\|\Lambda_{\pi_v}\|, \ \dim(M_{\pi_v})\big)$$

from P_S to D_S, as well as a mapping

$$(M_S, \tau_S) \longrightarrow w(M_S, \tau_S) = \prod_{v \in S} \big(\|\Lambda_{\tau_v}\|, \ \dim(M_{\tau_v})\big)$$

from the adjoint class P_S^* to D_S.

We need to show that the set

$$P'_S = \{(G_S, \pi_S) \in P_S : \ c_{G_S}(\pi_S) \neq 0\}$$

is empty. Suppose that it is not. We can then take the largest element

$$w' = w(G'_S, \pi'_S), \qquad (G'_S, \pi'_S) \in P'_S,$$

in the image of P'_S in D_S. The choice of (G'_S, π'_S) in the preimage also gives us an associated pair

$$(M'_S, \tau'_S) = (M_{\pi'_S}, \tau_{\pi'_S})$$

in P_S^*. Suppose that (G_S, π_S) is any pair in P'_S, and that τ_S is any element in the orbit $\widetilde{\tau}'_S$ of τ'_S such that $n(\pi_S, \tau_S) \neq 0$. It follows from the lexicographic order that

$$w(M_S, \tau_S) \leqslant w(G_S, \pi_S),$$

with equality if and only if $\tau_S = \tau_{\pi_S}$. By definition, we have

$$w(M_S, \tau_S) = w(M'_S, \tau'_S) = w(G'_S, \pi'_S) = w'.$$

It then follows from the definition of w' that $\tau_S = \tau_{\pi_S}$, and hence that $n(\pi_S, \tau_S)$ is strictly positive. The summand of G_S, π_S and τ_S in (3.5.11) is therefore positive.

The inner two sums in (3.5.11) reduce to a double sum over pairs (G_S, π_S) in P_S. Any summand in (3.5.11) is consequently either positive or zero. Since the summand of (G'_S, π'_S) and τ'_S is positive, the sum itself is positive, contradicting the fact that $c_{M'_S}(\widetilde{\tau}'_S)$ vanishes. This completes the proof of Proposition 3.5.1. □

The proof we have just completed is largely formal, though this may not be apparent in the general framework we have had to adopt. It is essentially a consequence of the decomposition of irreducible representations into standard representations. We will also need a more technical generalization of the proposition. We shall state it as a corollary, since its proof is the same.

COROLLARY 3.5.3. *Suppose that we have coefficients $c_G(\pi)$ that are as in the proposition, except that for a fixed valuation v and a simple datum $G_1 \in \widetilde{\mathcal{E}}_{\mathrm{sim}}(N)$, the given sum (3.5.1) equals a separate expression*

$$\sum_{\tau_v \in T(G_{1,v})} d_1(\tau_v, f_1^v) f_{v,G_1}(\tau_v), \tag{3.5.12}$$

in which the function f_1 corresponding to G_1 in the compatible family $\mathcal{F}$ is taken to be a product

$$f_1 = f_v f_1^v, \qquad f_v \in \widetilde{\mathcal{H}}(G_{1,v}),\ f_1^v \in \widetilde{\mathcal{H}}\big(G(\mathbb{A}^v)\big),$$

and where the coefficient

$$d_1(\tau_v, f_1^v), \qquad \tau_v \in T(G_{1,v}),$$

is an $\widetilde{\mathrm{Out}}_N(G_{1,v})$-invariant function of τ_v, which is supported on a finite set that depends only on a choice of Hecke type for f_v, and which equals 0 for any τ_v of the form $(M_v, \sigma_v, 1)$. Then the coefficients $c_G(\pi)$ and $d_1(\tau_v, f_1^v)$ all vanish.

PROOF. Suppose that the coefficients $c_G(\pi)$ all vanish. Then the sum (3.5.1) vanishes for any $\mathcal{F}$, and so therefore does the expression (3.5.12). Since G_1 is simple, the mapping $f_v \to f_v^{G_1}$ from $\widetilde{\mathcal{H}}(G_{1,v})$ to $\widetilde{\mathcal{S}}(G_{1,v})$ is surjective by Corollary 2.1.2. We can therefore choose the compatible family $\mathcal{F}$ so that the function $f_{v,G_1}(\tau_v)$ takes preassigned values on any finite set of $\widetilde{\mathrm{Out}}_N(G_{1,v})$-orbits τ_v. It follows that the coefficients $d_1(\tau_v, f_1^v)$ also vanish. We have thus only to show that each $c_G(\pi)$ vanishes.

As in the proof of the proposition, we fix a finite set of valuations $S \supset S_\infty$, which we can assume contains v. We then see that it is enough to show that the coefficients $c_{G_S}(\pi_S)$ all vanish. Following the next step in the proof of the proposition, we see that the original sum (3.5.1) takes the form (3.5.10), an expression that is therefore equal to (3.5.12). We can again choose the function $f_{S,M_S}(\tau_S)$ in (3.5.10) as we wish, and in particular, so that it isolates a given $\widetilde{\mathrm{Out}}_N(M_S)$-orbit $\widetilde{\tau}_S$. If the v component of any representative of this orbit is of the form $(M_v, \sigma_v, 1)$, the expression (3.5.12) vanishes, by the given condition on the coefficient $d_1(\pi_v, f_1^v)$. It follows that the coefficient $c_{M_S}(\widetilde{\tau}_S)$ defined by (3.5.11) vanishes in this case.

This is all we need to complete the argument in the proof of the proposition. For it was the vanishing of $c_{M'_S}(\widetilde{\tau}'_S)$, in the case of the chosen pair

$$(M'_S, \tau'_S) = (M_{\pi'_S}, \tau_{\pi'_S}),$$

from which we deduced that the coefficients $c_{M_S}(\pi_S)$ all vanish. But τ'_S is of the form

$$\tau'_S = \tau_{\pi'_S} = (M_{\pi'_S}, \sigma_{\pi'_S}, 1),$$

by our definition of maximality. The assertions of the corollary follow. □

With the proof of Proposition 3.5.1, we have established that the coefficients $c_G(\pi)$ in the earlier identity (3.4.14) vanish. This completes the proof of Proposition 3.4.1, the problem we started with. In particular, the expansions of Corollaries 3.4.2 and 3.4.3 are valid. They tell us that for any $G \in \widetilde{\mathcal{E}}_{\mathrm{sim}}(N)$, the discrete spectrum and the discrete part of the trace formula are both delimited to the original global parameters $\psi \in \widetilde{\Psi}(N)$ of Chapter 1. We can therefore concentrate on the contributions of these parameters. This leads to a finer analysis, which will occupy the rest of the volume, and in which Proposition 3.5.1 and its corollary will have a critical role.

CHAPTER 4

The Standard Model

4.1. Statement of the stable multiplicity formula

We assume that G is as in the theorems stated in Chapter 1. It is therefore a connected, quasisplit orthogonal or symplectic group, which is to say that it represents an element in the set

$$\widetilde{\mathcal{E}}_{\mathrm{sim}}(N) = \mathcal{E}_{\mathrm{sim}}\big(\widetilde{G}(N)\big), \qquad N \geqslant 1, \ \widetilde{G}(N) = GL(N) \rtimes \widetilde{\theta},$$

of isomorphism classes of simple, twisted endoscopic data for $GL(N)$ over F. We will later need to broaden the discussion somewhat, in order to include at least the case $G = \widetilde{G}(N)$. It will also be useful to add more general comments from time to time for added perspective, and to lay foundations for future study. However, we shall always try to be explicit whenever G represents something other than the basic case of an element in $\widetilde{\mathcal{E}}_{\mathrm{sim}}(N)$.

We also assume that the underlying field F is global. We have seen that the only families c that contribute to the discrete part of the trace formula for G come from parameters $\psi \in \widetilde{\Psi}(N)$. We expect that ψ must actually belong to the subset $\widetilde{\Psi}(G)$ of $\widetilde{\Psi}(N)$. But as we have noted, this fact is deep, and will be established only after a sustained analysis of other questions. We must therefore work with a general parameter $\psi \in \widetilde{\Psi}(N)$ for the time being.

Observe that in even mentioning the set $\widetilde{\Psi}(G)$, we are making the implicit assumption that Theorem 1.4.1 is valid. This was the seed theorem, which when applied to the simple generic constituents ϕ_i of ψ led us in §1.4 to a definition of $\widetilde{\Psi}(G)$. In this sense, Chapter 4 resembles Chapter 2 in relying on some cases of our stated theorems for its definitions and arguments. We recall that Chapter 3 was more direct, since it was independent of any part of the local and global theorems. At the beginning of §4.3, we shall take on some formal induction hypotheses, which replace these informal implicit assumptions, and which will carry the argument into later chapters.

In §3.3, we described the spectral expansion

(4.1.1)
$$I_{\mathrm{disc},\psi}(f) = \sum_{M} |W_0^M||W_0^G|^{-1} \sum_{w} |\det(w-1)_{\mathfrak{a}_M^G}|^{-1} \operatorname{tr}\big(M_{P,\psi}(w)\mathcal{I}_{P,\psi}(f)\big)$$

for the contribution of our parameter $\psi \in \widetilde{\Psi}(N)$ at a test function $f \in \widetilde{\mathcal{H}}(G)$. We also established an endoscopic expansion

$$I_{\mathrm{disc},\psi}(f) = \sum_{G' \in \mathcal{E}_{\mathrm{ell}}(G)} \iota(G, G')\, \widehat{S}'_{\mathrm{disc},\psi}(f') \tag{4.1.2}$$

for this contribution. Our long term goal is to extract as much information as possible from the identity of these two expansions. In this section, we shall state a formula for the linear form $\widehat{S}'_{\mathrm{disc},\psi}(f')$ on the right hand side of (4.1.2).

It will suffice to describe the case that $G' = G$. We have therefore to state a putative formula for the stable linear form

$$S_{\mathrm{disc},\psi}(f) = S^G_{\mathrm{disc},\psi}(f), \qquad f \in \widetilde{\mathcal{H}}(G).$$

The formula depends on local information yet to be established. Specifically, it is based on the stable linear form postulated in Theorem 2.2.1(a), or rather a global product

$$f^G(\psi) = \prod_v f_v^G(\psi_v), \qquad f = \prod_v f_v, \tag{4.1.3}$$

of these objects. Theorem 1.4.2 asserts that this product makes sense if ψ lies in the subset $\widetilde{\Psi}(G)$ of $\widetilde{\Psi}(N)$. However, the formula also has a global component, which will add to the complexity of its eventual proof.

The stable linear form $S_{\mathrm{disc},\psi}(f)$ is uniquely determined by the expansion (4.1.2). The stable linear form $f^G(\psi)$ is uniquely determined by the conditions of Theorem 2.2.1(a). Our formula asserts that the first linear form equals a multiple of the second by an explicitly given constant, which vanishes unless ψ lies in $\widetilde{\Psi}(G)$. It is this constant that contains the global information. We will call it the *stable multiplicity* of ψ.

The global constant has three factors that will be familiar from the definitions of §1.5. They are the integer m_ψ, the inverse of the order of the finite group $\mathcal{S}_\psi$, and a special value

$$\varepsilon^G(\psi) = \varepsilon_\psi(s_\psi) = \varepsilon^G_\psi(s_\psi) \tag{4.1.4}$$

of the sign character ε^G_ψ. The point $s_\psi \in \mathcal{S}_\psi$ was defined in §1.4 as the image of the nontrivial central element of $SL(2,\mathbb{C})$. The integer $m_\psi = m_\psi(G)$ of course depends on G, and vanishes (by definition) unless ψ belongs to $\widetilde{\Psi}(G)$, in which case the other objects S_ψ and ε^G_ψ make sense. The product of the three factors is therefore well defined, and vanishes unless ψ lies in $\widetilde{\Psi}(G)$. There will also be a fourth factor. This is the number $\sigma(\overline{S}^0_\psi)$ constructed from ψ in [**A9**]. We shall review its definition.

The factor $\sigma(\overline{S}^0_\psi)$ is part of a general construction that applies to any complex (not necessarily connected) reductive group S. To allow for induction, we take S more generally to be any union of connected components in some complex reductive group. We write S^0 for the connected component of 1, and

$$Z(S) = \mathrm{Cent}(S, S^0)$$

for the centralizer of S in S^0. Let T be a fixed maximal torus in S^0. We can then form the Weyl set

$$W(S) = \mathrm{Norm}(T, S)/T$$

of automorphisms of T induced from S. The Weyl group $W^0 = W(S^0)$ of S^0 is of course an obvious special case. We write $W_{\mathrm{reg}}(S)$ for the subset of elements $w \in W(S)$ that are regular, in the sense that the fixed point set of w in T is finite. This property holds for w if and only if the determinant $\det(w-1)$ is nonzero, where $(w-1)$ is regarded as a linear transformation on the real vector space

$$\mathfrak{a}_T = \mathrm{Hom}\big(X^*(T), \mathbb{R}\big).$$

Finally, we define the sign

$$s^0(w) = \mathrm{sgn}^0(w) = \pm 1$$

of an element $w \in W$ to be the parity of the number of positive roots of (S^0, T) mapped by w to negative roots. Given these various objects, we attach a real number

$$i(S) = |W(S)|^{-1} \sum_{w \in W_{\mathrm{reg}}(S)} s^0(w) |\det(w-1)|^{-1} \tag{4.1.5}$$

to S.

The number $i(S)$ bears a formal resemblance to the spectral expansion (4.1.1) of $I_{\mathrm{disc},\psi}(f)$. The rest of the construction amounts to the definition of a number $e(S)$ that bears a similar resemblance to the endoscopic expansion (4.1.2). This second number is defined inductively in terms of factors $\sigma(S_1)$ attached to *connected* complex reductive groups S_1.

Let us write S_{ss} for the set of semisimple elements in S. For any $s \in S_{\mathrm{ss}}$, we set

$$S_s = \mathrm{Cent}(s, S^0), \tag{4.1.6}$$

the centralizer of s in the connected group S^0. Then S_s is also a complex reductive group, with identity component

$$S_s^0 = (S_s)^0 = \mathrm{Cent}(s, S^0)^0.$$

(The notation here differs from the convention used in some places, in which the symbol S_s is reserved for the identity component of the centralizer. We must also take care not to confuse elements $s \in S_{\mathrm{ss}}$ with the sign character in (4.1.5), which we will always denote with the superscript 0.) If Σ is any subset of S that is invariant under conjugation by S^0, we shall write $\mathcal{E}(\Sigma)$ for the set of equivalence classes in

$$\Sigma_{\mathrm{ss}} = \Sigma \cap S_{\mathrm{ss}},$$

with the equivalence relation defined by setting

$$s' \sim s$$

if

$$s' = s^0 z s (s^0)^{-1}, \qquad s^0 \in S^0, \ z \in Z(S_s^0)^0.$$

The essential case is the subset

$$S_{\text{ell}} = \{ s \in S_{\text{ss}} : \ |Z(S_s^0)| < \infty \} \tag{4.1.7}$$

of S. For among other things, the equivalence relation in S_{ell} that defines the quotient

$$\mathcal{E}_{\text{ell}}(S) = \mathcal{E}(S_{\text{ell}})$$

is simply S^0-conjugacy. The resemblance here with the (more complicated) definition of endoscopic data [**KS**] is not accidental.

The following proposition is a restatement of Theorem 8.1 of [**A9**].

PROPOSITION 4.1.1. *There are unique constants $\sigma(S_1)$, defined whenever S_1 is a connected complex reductive group, such that for any S, the number*

$$e(S) = \sum_{s \in \mathcal{E}_{\text{ell}}(S)} |\pi_0(S_s)|^{-1} \sigma(S_s^0) \tag{4.1.8}$$

equals $i(S)$, and such that

$$\sigma(S_1) = \sigma(S_1/Z_1)|Z_1|^{-1}, \tag{4.1.9}$$

for any central subgroup Z_1 of S_1.

To see the uniqueness of the constants $\sigma(S_1)$, one notes that (4.1.9) implies that $\sigma(S_1)$ equals 0 if the center $Z(S_1)$ is infinite. If $Z(S_1)$ is finite, we define $\sigma(S_1)$ inductively from (4.1.8) by setting

$$\sigma(S_1)|Z(S_1)| = i(S_1) - \sum_s |\pi_0(S_{1,s})|^{-1} \sigma(S_{1,s}^0),$$

where s is summed over the complement of $\mathcal{E}_{\text{ell}}(Z(S_1))$ in $\mathcal{E}_{\text{ell}}(S_1)$. We refer the reader to [**A9**, §8] for the proof of the general identities (4.1.8) and (4.1.9). □

REMARK. If S is a single connected component, it is a bitorsor under the actions of S^0 by left and right translation. The constructions actually make sense if S is taken to be any bitorsor, relative to a complex connected reductive group S^0. Proposition 4.1.1 remains valid in this setting, since the group multiplication on S plays no role in the proof. More generally one could take S to any finite union of S^0-torsors.

We return to our parameter $\psi \in \widetilde{\Psi}(N)$. Recall that S_ψ stands for the centralizer in $\widehat{G}$ of the image of (some representative of) ψ. As in the local discussion from §2.4, we shall often work directly with the quotient

$$\overline{S}_\psi = S_\psi / Z(\widehat{G})^\Gamma, \qquad \Gamma = \Gamma_F. \tag{4.1.10}$$

It is of course the group

$$\mathcal{S}_\psi = \pi_0(\overline{S}_\psi) = \overline{S}_s/\overline{S}_s^0$$

of connected components of $\overline{S}_\psi$ that governs the main assertion of our basic global Theorem 1.5.2. The fourth global factor is the number $\sigma(\overline{S}_\psi^0)$, defined for the connected reductive group $S_1 = \overline{S}_\psi^0$ by the proposition.

We can now state the formula for $S_{\mathrm{disc},\psi}(f)$.

THEOREM 4.1.2 (Stable multiplicity formula). *Given $\psi \in \widetilde{\Psi}(N)$ and $G \in \widetilde{\mathcal{E}}_{\mathrm{sim}}(N)$, we have*

$$S_{\mathrm{disc},\psi}(f) = m_\psi \, |\mathcal{S}_\psi|^{-1} \sigma(\overline{S}_\psi^0) \, \varepsilon^G(\psi) \, f^G(\psi), \tag{4.1.11}$$

for any $f \in \widetilde{\mathcal{H}}(G)$. In particular, $S_{\mathrm{disc},\psi}(f)$ vanishes unless ψ belongs to the subset $\widetilde{\Psi}(G)$ of $\widetilde{\Psi}(N)$.

This represents one of the main results of the volume. It will have to be proved at the same time as the theorems stated in Chapter 1, and their local refinements from Chapter 2. However, we will be able to apply it inductively to endoscopic groups $G' \in \mathcal{E}_{\mathrm{ell}}(G)$ that are proper. We write the specialization of the formula to G' in slightly different terms.

Recall that elements $\psi \in \widetilde{\Psi}(G)$ can be identified with $\widetilde{\mathrm{Aut}}_N(G)$-orbits of L-homomorphisms

$$\widetilde{\psi}_G : \mathcal{L}_\psi \times SL(2,\mathbb{C}) \longrightarrow {}^L G.$$

Following a suggestion from §1.4, we write $\Psi(G)$ for the corresponding set of orbits under the subgroup $\widehat{G}$ of $\widetilde{\mathrm{Aut}}_N(G)$. There is then a surjective mapping $\psi_G \to \psi$ from $\Psi(G)$ to $\widetilde{\Psi}(G)$, for which the order of the fibre of ψ equals the integer m_ψ. If f lies in $\widetilde{\mathcal{H}}(G)$ and ψ_G maps to ψ, we write

$$f^G(\psi_G) = f^G(\psi).$$

The corresponding set $\Psi(G')$ for G' consists of $\widehat{G}'$-orbits of L-homomorphisms

$$\psi' : \mathcal{L}_\psi \times SL(2,\mathbb{C}) \longrightarrow {}^L G'.$$

The composition of any such ψ' with the endoscopic L-embedding ξ' of ${}^L G'$ into ${}^L G$ is an element ψ_G in $\Psi(G)$, which in turn maps into $\widetilde{\Psi}(N)$. This gives a mapping from $\Psi(G')$ to $\widetilde{\Psi}(N)$, whose image is contained in the subset $\widetilde{\Psi}(G)$ of $\widetilde{\Psi}(N)$. We write

$$\Psi(G',\psi) = \Psi(G',G,\psi)$$

for the fibre in $\Psi(G')$ of our given element $\psi \in \widetilde{\Psi}(N)$, a set that is empty unless ψ belongs to $\widetilde{\Psi}(G)$.

We want to specialize the formula of Theorem 4.1.2 to the linear form $\widehat{S}'_{\mathrm{disc},\psi}(f')$ in (4.1.2). It is a consequence of the definitions (3.3.9) and (3.3.13) that

$$\widehat{S}'_{\mathrm{disc},\psi}(f') = \sum_{\psi'} \widehat{S}'_{\mathrm{disc},\psi'}(f'),$$

where ψ' is summed over those elements in $\widetilde{\Psi}(G')$ that map to ψ. Since this sum is over the image of $\Psi(G',\psi)$ in $\widetilde{\Psi}(G')$, the analogue for G' of the integer m_ψ leads to a sum over $\Psi(G',\psi)$. The specialization of the theorem to G' then yields the following corollary.

COROLLARY 4.1.3. *Given* $G \in \widetilde{\mathcal{E}}_{\mathrm{sim}}(N)$, $G' \in \mathcal{E}_{\mathrm{ell}}(G)$, $\psi \in \widetilde{\Psi}(N)$ *and* $f \in \widetilde{\mathcal{H}}(G)$, *we have an expansion*

$$\widehat{S}'_{\mathrm{disc},\psi}(f') = \sum_{\psi' \in \Psi(G',\psi)} |\mathcal{S}_{\psi'}|^{-1} \sigma(\overline{S}^0_{\psi'})\, \varepsilon'(\psi')\, f'(\psi'), \tag{4.1.12}$$

for the global sign

$$\varepsilon'(\psi') = \varepsilon^{G'}(\psi') = \varepsilon^{G'}_{\psi'}(s_{\psi'}).$$

□

REMARK. Suppose that the endoscopic group G' is proper, in the sense that it is distinct from G. In other words, it is a proper product

$$G' = G'_1 \times G'_2, \qquad G'_i \in \widetilde{\mathcal{E}}_{\mathrm{sim}}(N'_i),\ N'_1 + N'_2 = N,$$

of simple endoscopic groups. We note that the informal assumptions under which the corollary makes sense depend only on the positive integer N. They could be replaced by a formal induction hypothesis that the theorems we have stated so far, including Theorem 4.1.2, are valid if N is replaced by a smaller integer N_-. Corollary 4.1.3 would then hold for G'.

The group $\overline{S}_\psi$ given by (4.1.10) is of course insensitive to whether we treat ψ as an element in $\Psi(G)$ or $\widetilde{\Psi}(G)$. With either interpretation of ψ, the group will be at the heart of our analysis of the linear form $I_{\mathrm{disc},\psi}(f)$. As ψ varies, the properties of $\overline{S}_\psi$ characterize an important chain of subsets of $\Psi(G)$. We write

$$\Psi_{\mathrm{sim}}(G) \subset \Psi_2(G) \subset \Psi_{\mathrm{ell}}(G) \subset \Psi_{\mathrm{disc}}(G) \subset \Psi(G), \tag{4.1.13}$$

where

$$\Psi_{\mathrm{sim}}(G) = \{\psi \in \Psi(G) : \ |\overline{S}_\psi| = 1\},$$

$$\Psi_2(G) = \{\psi \in \Psi(G) : \ \overline{S}_\psi \text{ is finite}\},$$

$$\Psi_{\mathrm{ell}}(G) = \{\psi \in \Psi(G) : \ \overline{S}_{\psi,s} \text{ is finite for some } s \in \overline{S}_{\psi,\mathrm{ss}}\},$$

and

$$\Psi_{\mathrm{disc}}(G) = \{\psi \in \Psi(G) : \ Z(\overline{S}_\psi) \text{ is finite}\}.$$

These sets are stable under the action of the group $\widetilde{\mathrm{Out}}_N(G)$ (of order 1 or 2). The associated sets of orbits give a corresponding chain of subsets of $\widetilde{\Psi}(G)$, which can be regarded as a refinement of the earlier chain (1.4.7) from §1.4. It is also clear that we would have similar chains if F were taken to be a local field.

The subsets (4.1.13) of $\Psi(G)$ (or rather their analogues for $\widetilde{\Psi}(G)$) have a direct bearing on how we interpret the corresponding linear forms $I_{\mathrm{disc},\psi}(f)$. For example, $\widetilde{\Psi}_{\mathrm{disc}}(G)$ should be the subset of parameters $\psi \in \widetilde{\Psi}(G)$ such that $I_{\mathrm{disc},\psi}(f)$ is nonzero. Elements ψ in the smaller set $\widetilde{\Psi}_{\mathrm{ell}}(G)$ are characterized by the property that for some $G' \in \mathcal{E}_{\mathrm{ell}}(G)$, the subset

$$\Psi_2(G',\psi) = \Psi_2(G') \cap \Psi(G',\psi)$$

of $\Psi(G')$ is nonempty. In other words, ψ should contribute to the discrete spectrum of G'. The subset $\widetilde{\Psi}_2(G)$ of $\widetilde{\Psi}_{\mathrm{ell}}(G)$ consists of those ψ for which we can take $G' = G$. This is of course to say that ψ should contribute to the discrete spectrum of G, or equivalently, that the term with $M = G$ in the expansion (4.1.1) of $I_{\mathrm{disc},\psi}(f)$ is nonzero. The smallest set $\widetilde{\Psi}_{\mathrm{sim}}(G)$ consists simply of the parameters $\psi \in \widetilde{\Psi}_2(G)$ such that for any $G' \in \mathcal{E}_{\mathrm{ell}}(G)$ distinct from G, the set $\Psi_2(G',\psi)$ (or for that matter the set $\Psi(G',\psi)$) is empty.

These last remarks are intended as motivation for what follows. They offer some guidance for our efforts to compare the two expansions (4.1.1) and (4.1.2). Notice that the properties that define the subsets in (4.1.13) are given entirely in terms of the group $\overline{S}_\psi$. They make sense if $\overline{S}_\psi$ is replaced by any complex reductive group, or even a disjoint union S of bitorsors under a connected group S^0. For example, one could take S to be bitorsor under a complex reductive (but not necessarily connected) group S^*, a set that can obviously be regarded as a finite disjoint union of bitorsors under the connected group $S^0 = (S^*)^0$. It is necessary to formulate the definitions in this setting if one wants to replace G by an arbitrary triplet (3.1.2).

4.2. On the global intertwining relation

We continue to work with a fixed group $G \in \widetilde{\mathcal{E}}_{\mathrm{sim}}(N)$ and a parameter $\psi \in \widetilde{\Psi}(N)$, both taken over the global field F. Our long term aim will be to compare the spectral expansion (4.1.1) of $I_{\mathrm{disc},\psi}(f)$ with its endoscopic expansion (4.1.2). In the last section, we stated a formula for the linear form

$$\widehat{S}'_{\mathrm{disc},\psi}(f'), \qquad f \in \widetilde{\mathcal{H}}(G),$$

in the endoscopic expansion. In this section, we shall establish an expansion for the analogous term

$$\mathrm{tr}\big(M_{P,\psi}(w)\mathcal{I}_{P,\psi}(f)\big), \qquad f \in \widetilde{\mathcal{H}}(G), \tag{4.2.1}$$

in the spectral expansion. We shall also reformulate the local intertwining relation stated in Chapter 2 as a global relation. This will ultimately serve as a link between the two formulas.

The trace (4.2.1) depends on fixed elements $M \in \mathcal{L}$, $P \in \mathcal{P}(M)$ and $w \in W(M)_{\mathrm{reg}}$. The Levi subgroup M is a product of several general linear groups with an orthogonal or symplectic group of the same form as G, as in the local case (2.3.4). We can therefore construct the sets

$$\widetilde{\Psi}_2(M) \subset \widetilde{\Psi}(M)$$

of parameters as products of sets of the sort we have already defined. We write $\widetilde{\Psi}_M(G)$ for the image of $\widetilde{\Psi}_2(M)$ in $\widetilde{\Psi}(G)$ under the natural mapping. We can also form the set $\Psi_2(M)$. For later reference, we write

$$\Psi_2(M,\psi) = \Psi_2(M,G,\psi)$$

for the fibre of our given element $\psi \in \widetilde{\Psi}(N)$ in $\Psi_2(M)$, a set that is empty unless ψ belongs to $\widetilde{\Psi}_M(G)$. We may as well also write $\widetilde{\Psi}_2(M,\psi)$ for the fibre of ψ in $\widetilde{\Psi}_2(M)$.

Once again, the set $\widetilde{\Psi}(G)$ (or for that matter, any of corresponding sets attached to M) depends on the implicit assumption that Theorem 1.4.1 is valid. The rules here are the same as in the last section. Each object we introduce is predicated on the validity of any cases of the stated theorems required for its existence. As we have noted, we will replace this implicit assumption with formal induction hypotheses at the beginning of §4.3.

Our informal assumptions will include Theorem 1.5.2, since among other things, we will need to introduce a global analogue of the expression (2.4.5). In particular, we can assume that the relative discrete spectrum of M decomposes into subspaces indexed by parameters in $\widetilde{\Psi}_2(M)$. The trace (4.2.1) comes from the contribution of ψ to a representation induced from the discrete spectrum of M. It thus vanishes if ψ does not lie in $\widetilde{\Psi}_M(G)$. We shall therefore assume that ψ belongs to $\widetilde{\Psi}_M(G)$, and in particular, that it lies in the subset $\widetilde{\Psi}(G)$ of $\widetilde{\Psi}(N)$.

Our central algebraic object will continue to be the complex reductive group $\overline{S}_\psi$. We digress to consider some of its global implications, including our formulation of the global intertwining relation.

The group $\overline{S}_\psi$ gives rise to several finite groups, which are global analogues of the finite groups introduced in §2.4. For example, the basic component group $\mathcal{S}_\psi = \pi_0(\overline{S}_\psi)$ has a normal subgroup $\mathcal{S}^1_\psi$. This is isomorphic to the component group $\mathcal{S}_{\psi_M}$ attached to any ψ_M in the subset $\widetilde{\Psi}_2(M,\psi)$ of elements in $\widetilde{\Psi}_2(M)$ that map to ψ. Our assumption that ψ_M lies in $\widetilde{\Psi}_2(M)$ means that the quotient

$$\overline{T}_\psi = Z(\widehat{M})^\Gamma/Z(\hat{G})^\Gamma = A_{\widehat{M}}/(A_{\widehat{M}} \cap Z(\hat{G})^\Gamma), \tag{4.2.2}$$

where

$$A_{\widehat{M}} = \big(Z(\widehat{M})^\Gamma\big)^0,$$

is a maximal torus in $\overline{S}_\psi$. The associated finite groups W^0_ψ and W_ψ of automorphisms can then be identified with the full Weyl groups of $\overline{S}^0_\psi$ and

$\overline{S}_\psi$. They take their places in the global version

(4.2.3)

$$\begin{array}{ccccccccc}
 & & & & 1 & & 1 & & \\
 & & & & \downarrow & & \downarrow & & \\
 & & & & W_\psi^0 & = & W_\psi^0 & & \\
 & & & & \downarrow & & \downarrow & & \\
1 & \longrightarrow & \mathcal{S}_\psi^1 & \longrightarrow & \mathfrak{N}_\psi & \longrightarrow & W_\psi & \longrightarrow & 1 \\
 & & \Vert & & \downarrow \uparrow & & \downarrow \uparrow & & \\
1 & \longrightarrow & \mathcal{S}_\psi^1 & \longrightarrow & \mathcal{S}_\psi & \longrightarrow & R_\psi & \longrightarrow & 1 \\
 & & & & \downarrow & & \downarrow & & \\
 & & & & 1 & & 1 & &
\end{array}$$

of the commutative diagram (2.4.3) (or rather the special case of (2.4.3) in which ψ is represented by an element in $\widetilde{\Psi}_2(M)$).

It is interesting to compare the general diagram (4.2.3) with its specialization to parameters in the subsets (4.1.13). The set $\Psi_{\mathrm{ell}}(G)$ in (4.1.13), for example, is closely related to the R-group R_ψ in (4.2.3). One sees easily that ψ belongs to $\widetilde{\Psi}_{\mathrm{ell}}(G)$ if and only if there is an element $w \in R_\psi$ whose fixed point set in $\overline{T}_\psi$ (relative to the embedding of R_ψ into W_ψ defined by a choice of Borel subgroup $\overline{B}_\psi \subset \overline{S}_\psi$) is finite. This condition in turn implies that $\overline{S}_\psi^0 = \overline{T}_\psi$. Such elements will be an important part of our future study.

Consider a point u from the group

$$\mathfrak{N}_\psi = \overline{N}_\psi(G, \overline{T}_\psi)/\overline{T}_\psi$$

at the center of the diagram. Following the local notation from §2.4, we shall write w_u and x_u for the images of u in W_ψ and $\mathcal{S}_\psi$ respectively. We can identify the first point w_u with an element in the group $W(M)$. This gives a global twisted group

$$\widetilde{M}_u = \widetilde{M}_{w_u} = (M, \widetilde{w}_u).$$

We have also been following the later convention from §2.4, in which ψ_M denotes an element in $\widetilde{\Psi}_2(M)$ that maps to ψ. Then ψ_M lies in the subset $\Psi_2(\widetilde{M}_u)$ of $\widetilde{\Psi}_2(M)$. We again write $\widetilde{u}$ (somewhat superfluously) for the image of u in the associated $\mathcal{S}_{\psi_M}$-bitorsor

$$\widetilde{\mathcal{S}}_{\psi_M, u} = \mathcal{S}_{\psi_M}(\widetilde{M}_u).$$

This notation will be useful to describe the normalization of the global intertwining operator in (4.2.1).

We are assuming that the analogue of Theorem 1.4.2 is valid for M. This implies that the localization $\psi_{M,v}$ of ψ_M at any v belongs to the corresponding set $\widetilde{\Psi}^+(M_v, \psi_v)$. The localization ψ_v of ψ therefore belongs to $\widetilde{\Psi}^+(G_v)$. It follows easily that $\overline{S}_\psi$ embeds into $\overline{S}_{\psi_v}$, and that there is a morphism of the diagram (4.2.3) into its local counterpart (2.4.3). In particular, any element u in $\mathfrak{N}_\psi$ has a local image u_v in the group $\mathfrak{N}_{\psi_v}(G_v, M_v)$, for any valuation v. We can therefore define the local linear form

$$f_{v,G}(\psi_v, u_v), \qquad f_v \in \widetilde{\mathcal{H}}(G_v),$$

by (2.4.5) (with ψ_{M_v}, π_{M_v} and u_v taking the place here of the local symbols ψ, π and u in (2.4.5)). We obtain a global linear form

$$f_G(\psi, u), \qquad f \in \widetilde{\mathcal{H}}(G),$$

by setting

$$f_G(\psi, u) = \prod_v f_{v,G}(\psi_v, u_v), \qquad f = \prod_v f_v, \tag{4.2.4}$$

for a product that by the discussion of §2.5 can be taken over a finite set. It is clear from §2.4 that the definition has an equivalent global formulation

$$f_G(\psi, u) = \sum_{\pi_M \in \widetilde{\Pi}_{\psi_M}} \langle \widetilde{u}, \widetilde{\pi}_M \rangle \operatorname{tr}\big(R_P(w_u, \widetilde{\pi}_M, \psi_M)\mathcal{I}_P(\pi_M, f)\big), \tag{4.2.4$'$}$$

where $R_P(w_u, \widetilde{\pi}_M, \psi_M)$ is a normalized global intertwining operator. This operator is defined as a product over v of its local analogues, constructed in Chapter 2 by (2.3.26), (2.4.2) and (2.4.4). In particular, it depends implicitly on an extension $\widetilde{\pi}_M$ of π_M to the torsor $\widetilde{M}_u(\mathbb{A})$. The pairing $\langle \widetilde{u}, \widetilde{\pi}_M \rangle$ represents the corresponding extension of the character $\langle \cdot, \pi_M \rangle$ on the group $\mathcal{S}_{\psi_M} \cong \mathcal{S}^1_\psi$ to the bitorsor $\widetilde{\mathcal{S}}_{\psi_M,u}$, which is again defined as a product of its local analogues.

Consider also a semisimple element s in the original group $\overline{S}_\psi$. It has a local image s_v in the group $\overline{S}_{\psi_v}$, for any valuation v. We can therefore define the local linear form

$$f'_{v,G}(\psi_v, s_v), \qquad f_v \in \widetilde{\mathcal{H}}(G_v),$$

by (2.4.6) (with ψ_v and s_v in place of ψ and s). We obtain a global linear form

$$f'_G(\psi, s), \qquad f \in \widetilde{\mathcal{H}}(G),$$

by setting

$$f'_G(\psi, s) = \prod_v f'_{v,G}(\psi_v, s_v), \qquad f = \prod_v f_v, \tag{4.2.5}$$

for a product that can again be taken over a finite set. The global formulation of this definition is of course

$$f'_G(\psi, s) = f'(\psi'), \tag{4.2.5$'$}$$

where (G', ψ') is the preimage of (ψ, s) under the global correspondence (1.4.11).

The local intertwining relation stated as Theorem 2.4.1 applies to the local factors in (4.2.4) and (4.2.5). We can state the corresponding global identity for the products as a corollary of this theorem.

COROLLARY 4.2.1 (Global intertwining relation for G). *For any u in the global normalizer $\mathfrak{N}_\psi$, the identity*

$$f'_G(\psi, s_\psi s) = f_G(\psi, u), \qquad f \in \widetilde{\mathcal{H}}(G), \tag{4.2.6}$$

holds for any semisimple element $s \in \overline{S}_\psi$ that projects onto the image x_u of u in $\mathcal{S}_\psi$. □

We remind ourselves that Theorem 2.4.1 has yet to be established. Its proof in fact will be one of our major concerns. The same therefore goes for the corollary. We have formulated it here to give our discussion some sense of direction, and in particular, to see how we will eventually relate the summands of (4.1.1) and (4.1.2).

We note that the diagram (4.2.3) has an obvious analogue if $\overline{S}_\psi$ is replaced by any complex reductive group S. More generally, it makes sense if we take S to be a bitorsor under a complex reductive group S^*. (We write S^* here because we need to reserve the symbol $S^0 = (S^*)^0$ for the connected component of 1 in S^*.) The corresponding objects $\mathfrak{N}$, W, $\mathcal{S}$ and R in the diagram are then bitorsors under their analogues $\mathfrak{N}^*$, W^*, $\mathcal{S}^*$ and R^* for the group S^*. On the other hand, the objects W^0 and $\mathcal{S}^1$ are then equal to the associated groups for S^*. In particular, $\mathcal{S}^1$ is the projection onto the quotient $\mathcal{S}^* = \pi_0(S^*)$ of a subgroup S^1 of S^* with

$$S^0 \subset S^1 \subset S^*.$$

For example, we can take

$$S = \overline{\widetilde{S}}_\psi = S_\psi(\widetilde{G})/Z(\widehat{\widetilde{G}})^\Gamma,$$

where $\widetilde{G}$ is a twisted orthogonal group over F, and ψ belongs to the subset $\Psi(\widetilde{G})$ of $\widetilde{\Psi}(G)$. It would be easy to formulate a global intertwining relation for $\widetilde{G}$, as a corollary of Theorem 2.4.4 that is parallel to Corollary 4.2.1. We will not do so, since we do not need it. However, we will formulate a global supplement of Theorem 1.5.2 for $\widetilde{G}$. We shall state it here for application to the fixed group M above, even though its proof will only come later.

We are supposing that $\widetilde{G} = (\widetilde{G}^0, \widetilde{\theta})$, where $G = \widetilde{G}^0$ is an even orthogonal group in $\widetilde{\mathcal{E}}_{\text{sim}}(N)$ over the global field F. Consider a parameter $\psi \in \widetilde{\Psi}_2(G)$. For any function $\phi \in L^2_{\text{disc},\psi}\big(G(F)\backslash G(\mathbb{A})\big)$, and any $y \in \widetilde{G}(\mathbb{A})$, the function

$$\big(R^{\widetilde{G}}_{\text{disc},\psi}(y)\phi\big)(x) = \phi(\widetilde{\theta}^{-1}xy), \qquad x \in G(\mathbb{A}), \tag{4.2.7}$$

also belongs to $L^2_{\text{disc},\psi}\big(G(F)\backslash G(\mathbb{A})\big)$. We thus obtain a canonical extension of the representation $R^G_{\text{disc},\psi}$ of $G(\mathbb{A})$ to the group $\widetilde{G}(\mathbb{A})^+$ generated by $\widetilde{G}(\mathbb{A})$. Theorem 1.5.2, together with Theorem 2.2.1, describes the character of the original representation in terms of the transfer of characters from $GL(N)$.

The theorem we are about to state, in combination with Theorem 2.2.4, plays a similar role in its extension.

Suppose that ψ lies in the subset

$$\Psi_2(\widetilde{G}) = \Psi(\widetilde{G}) \cap \widetilde{\Psi}_2(G)$$

of $\widetilde{\Psi}_2(G)$. Then the set

$$\widetilde{\mathcal{S}}_\psi = \pi_0(\overline{\widetilde{S}}_\psi)$$

is a (nonempty) $\mathcal{S}_\psi$-torsor. Any localization ψ_v of ψ lies in the subset $\Psi^+(\widetilde{G}_v)$ of $\widetilde{\Psi}^+(G_v)$, and determines a mapping $\widetilde{x} \to \widetilde{x}_v$ from $\widetilde{\mathcal{S}}_\psi$ to $\widetilde{\mathcal{S}}_{\psi_v}$, according to Theorem 1.4.2. Suppose that π is a representation in the global packet $\widetilde{\Pi}_\psi$, equipped with an extension to the group $\widetilde{G}(\mathbb{A})^+$ (or equivalently, an intertwining operator $\pi(\widetilde{\theta})$ of π of order 2). We can then choose extensions of the local components π_v of π to the groups $\widetilde{G}(F_v)^+$ whose tensor product is compatible with the extension of π. Theorem 2.2.4 asserts that there is a corresponding product

$$\langle \widetilde{x}, \widetilde{\pi} \rangle = \prod_v \langle \widetilde{x}_v, \widetilde{\pi}_v \rangle, \qquad \widetilde{x} \in \mathcal{S}_\psi,$$

of extensions of linear characters $\langle \cdot, \pi_v \rangle$, which determines an extension of the linear character $\langle \cdot, \pi \rangle$ on $\mathcal{S}_\psi$ to the group $\widetilde{\mathcal{S}}_\psi^+$ generated by $\widetilde{\mathcal{S}}_\psi$. Assume that $\langle \cdot, \pi \rangle$ equals the sign character ε_ψ On the one hand, ε_ψ has a canonical extension to $\widetilde{\mathcal{S}}_\psi^+$. This follows from the definition (1.5.6) and the fact that the adjoint action of the complex group $\widehat{G} = SO(N, \mathbb{C})$ on its Lie algebra has a canonical extension to the group $O(N, \mathbb{C})$. On the other, Theorem 1.5.2 asserts that π occurs in the decomposition of $R^G_{\mathrm{disc},\psi}$. Moreover, we have just seen that $R^G_{\mathrm{disc},\psi}$ has a canonical extension to $\widetilde{G}(\mathbb{A})^+$. The theorem stated below asserts that the restriction of this extension to the subspace of π coincides with the extension of ε_ψ under the local correspondence of extensions postulated by Theorem 2.2.4.

THEOREM 4.2.2. *Suppose that $\widetilde{G}$, with $G \in \widetilde{\mathcal{E}}_{\mathrm{sim}}(N)$, is a twisted orthogonal group over F, and that ψ belongs to $\widetilde{\Psi}_2(G)$.*

(a) *If ψ lies in the complement of $\Psi_2(\widetilde{G})$ in $\widetilde{\Psi}_2(G)$, we have*

$$\mathrm{tr}\big(R^{\widetilde{G}}_{\mathrm{disc},\psi}(\widetilde{f})\big) = 0, \qquad \widetilde{f} \in \mathcal{H}(\widetilde{G}).$$

(b) *Assume that ψ lies in $\Psi_2(\widetilde{G})$, and that π is a representation in the global packet $\widetilde{\Pi}_\psi$ such that the linear character $\langle \cdot, \pi \rangle$ on $\mathcal{S}_\psi$ is equal to $\varepsilon_\psi = \varepsilon_\psi^{-1}$. Then the restriction to π of the canonical extension of $R^G_{\mathrm{disc},\psi}$ to $\widetilde{G}(\mathbb{A})^+$ corresponds to the canonical extension of ε_ψ to $\widetilde{\mathcal{S}}_\psi^+$ under the product of local correspondences of extensions given by Theorem* 2.2.4.

We will now take up the expansion of the global trace (4.2.1). As an aid to the reader, we shall try to formulate it so that it is roughly parallel to its endoscopic counterpart, Corollary 4.1.3. In particular, we shall state it as a corollary of Theorem 4.2.2 and Theorem 1.5.2, just as its endoscopic analogue was stated as a corollary of Theorem 4.1.2. We recall that Theorem 4.1.2 was the stable multiplicity formula, while Theorem 1.5.2 is the ordinary multiplicity formula (with its twisted supplement Theorem 4.2.2). We will want to apply the two corollaries in the next two sections, long before any of these three theorems have been proved. We will then have to apply them under a common induction hypothesis.

Our proof of the expansion of (4.2.1) in the remainder of this section will be more complicated than that of its endoscopic counterpart. In addition to the two theorems on which it depends as a corollary, the proof also requires an analogue of Theorem 4.2.2 for the original $GL(N)$-torsor $\widetilde{G}(N)$. However, this is elementary, unlike Theorem 4.2.2, in the sense that we will be able to establish it on the spot.

Consider a parameter ψ in the global set $\widetilde{\Psi}_{\mathrm{sim}}(N) = \Psi_{\mathrm{sim}}\big(\widetilde{G}(N)\big)$. The analogue

$$(4.2.8)\quad \big(R^N_{\mathrm{disc},\psi}(y)\phi\big)(x) = \phi\big(\widetilde{\theta}(N)^{-1}xy\big), \qquad y \in \widetilde{G}(N,\mathbb{A}),\ x \in GL(N,\mathbb{A}),$$

of (4.2.7), defined for $\phi \in L^2_{\mathrm{disc},\psi}\big(GL(N,F)\backslash GL(N,\mathbb{A})\big)$, is a canonical extension of the representation $R^N_{\mathrm{disc},\psi}$ of $GL(N,\mathbb{A})$ to the group $\widetilde{G}(N,\mathbb{A})^+$. It corresponds to a canonical extension of the automorphic representation π_ψ of $GL(N,\mathbb{A})$ attached to ψ. On the other hand, for any completion ψ_v of ψ, we also have the extension of the representation π_{ψ_v} of $GL(N,F_v)$ described in §2.1.

LEMMA 4.2.3. *Assume that $\psi \in \widetilde{\Psi}_{\mathrm{sim}}(N)$. Then the extension of the automorphic representation π_ψ of $GL(N,\mathbb{A})$ defined by (4.2.8) equals the tensor product of local extensions $\widetilde{\pi}_{\psi_v}(N)$ of representations π_{ψ_v} defined in terms of Whittaker functionals in §2.1.*

PROOF. The definition (4.2.8) makes sense if $R^N_{\mathrm{disc},\psi}$ is replaced by the right regular representation R^N of $G(N,\mathbb{A})$ on any invariant space of functions on the quotient of $GL(N,\mathbb{A})$ by $GL(N,F)$. In particular, the operator

$$\widetilde{R}(N) = R^N\big(\widetilde{\theta}(N)\big)$$

intertwines R^N with $R^N \circ \widetilde{\theta}(N)$. It suffices to show that the intertwining operator $\widetilde{R}_{\mathrm{disc},\psi}(N)$ of $R^N_{\mathrm{disc},\psi}$ corresponds to the intertwining operator

$$\widetilde{\pi}_\psi(N) = \bigotimes_v \widetilde{\pi}_{\psi_v}(N)$$

of

$$\pi_\psi = \bigotimes_v \pi_{\psi_v}.$$

We write $\psi = \mu \boxtimes \nu$ and $N = mn$, where μ is a unitary cuspidal automorphic representation of $GL(m)$, and $\nu = \nu^n$ is the irreducible representation of $SL(2,\mathbb{C})$ of dimension n.

Let $\mathcal{I}_P(\sigma_{\mu,\lambda})$ be the induced representation of $GL(N, \mathbb{A})$ defined as in the statement of Theorem 1.3.3, but with a general vector $\lambda \in \mathbb{C}^n$ in place of the fixed vector

$$\rho_P = \left(\tfrac{n-1}{2}, \tfrac{n-2}{2}, \cdots, -\tfrac{(n-1)}{2}\right)$$

of exponents. It acts on a Hilbert space $\mathcal{H}_P(\sigma_\mu)$ that is independent of λ. Given that μ is self-dual, we see that there are isomorphisms

$$\mathcal{I}_P(\sigma_{\mu,\lambda}) \cong \mathcal{I}_P(\sigma_{\mu^\vee,\lambda^\vee}) \cong \mathcal{I}_P(\sigma_{\mu,\lambda^\vee})^\vee,$$

in which

$$\lambda^\vee = (-\lambda_n, \ldots, -\lambda_1).$$

It follows that there is a canonical intertwining isomorphism

$$\widetilde{\mathcal{I}}_P(\sigma_{\mu,\lambda}, N) : \mathcal{I}_P(\sigma_{\mu,\lambda}) \longrightarrow \mathcal{I}_P(\sigma_{\mu,\lambda^\vee})^\vee = \mathcal{I}_P(\sigma_{\mu,\lambda^\vee}) \circ \widetilde{\theta}(N)$$

that is compatible with the corresponding two standard global Whittaker functionals (namely, the global analogues of (2.5.1) for λ and $\lambda^\vee$) on the space $\mathcal{H}_P(\sigma_\mu)$. Suppose now that λ is in general position. Then for any ϕ in the subspace $\mathcal{H}_P^0(\sigma_\mu)$ of κ-finite vectors in $\mathcal{H}_P(\sigma_\mu)$, the Eisenstein series

$$E(\phi, \lambda) : x \longrightarrow E(x, \phi, \lambda)$$

is a well defined automorphic form on $GL(N, \mathbb{A})$. Let $R^N(\mu, \lambda)$ be the restriction of R^N, regarded as a representation of the Hecke algebra $\widetilde{\mathcal{H}}(N)$, to the space

$$\mathcal{A}(\mu, \lambda) = \{E(\phi, \lambda) : \phi \in \mathcal{H}_P^0(\sigma_\mu)\}$$

The mapping

$$E(\lambda) : \phi \longrightarrow E(\phi, \lambda)$$

is then an intertwining isomorphism between the irreducible representations $\mathcal{I}_P(\sigma_{\mu,\lambda})$ and $R^N(\mu, \lambda)$ of $\widetilde{\mathcal{H}}(N)$. Our third ingredient is the intertwining mapping $\widetilde{R}(N)$ above. It takes $\mathcal{A}(\mu, \lambda)$ onto the space $\mathcal{A}(\mu^\vee, \lambda^\vee) = \mathcal{A}(\mu, \lambda^\vee)$, as one sees readily from an examination of the relevant constant terms, and therefore represents an intertwining isomorphism

$$\widetilde{R}(N) : R^N(\mu, \lambda) \longrightarrow R^N(\mu, \lambda^\vee)^\vee = R^N(\mu, \lambda^\vee) \circ \widetilde{\theta}(N).$$

The main point is to show that

$$\widetilde{R}(N) E(\phi, \lambda) = E\big(\widetilde{\mathcal{I}}_P(\sigma_{\mu,\lambda}, N)\phi, \lambda^\vee\big). \tag{4.2.9}$$

The putative identity (4.2.9) asserts that the diagram

$$\begin{array}{ccc} \mathcal{I}_P(\sigma_{\mu,\lambda}) & \xrightarrow{\widetilde{\mathcal{I}}_P(\sigma_{\mu,\lambda},N)} & \mathcal{I}_P(\sigma_{\mu,\lambda^\vee}) \circ \widetilde{\theta}(N) \\ {\scriptstyle E(\lambda)}\downarrow & & \downarrow{\scriptstyle E(\lambda^\vee)} \\ R^N(\mu, \lambda) & \xrightarrow{\widetilde{R}(N)} & R^N(\mu, \lambda^\vee) \circ \widetilde{\theta}(N), \end{array}$$

is commutative. Since the four arrows each represent intertwining isomorphisms between irreducible representations, we can write

$$E(\lambda^\vee)\widetilde{\mathcal{I}}_P(\sigma_{\mu,\lambda}, N) = c_\lambda \widetilde{R}(N)E(\lambda),$$

for a multiplicative complex number $c_\lambda \in \mathbb{C}^*$. We must prove that $c_\lambda = 1$.

The group $GL(N)$ has a standard global Whittaker datum $(B, \chi) = (B(N), \chi(n))$. It is defined as in the local discussion of §2.5, except that χ is a left $N_B(F)$-invariant, nondegenerate character on $N_B(\mathbb{A})$. The Eisenstein series in (4.2.9) has a canonical automorphic χ-Whittaker functional

$$(\omega E)(\phi, \lambda) = \int_{N_B(F)\backslash N_B(\mathbb{A})} E(n, \phi, \lambda)\, \overline{\chi(n)}\, dn.$$

A change of variables in this integral, combined with the definitions of χ and $\widetilde{\theta}(N)$ in terms of the standard splitting of $GL(N)$, tells us that

$$\omega\big(\widetilde{R}(N)E(\phi, \lambda)\big) = \omega\big(E(\phi, \lambda)\big).$$

Let us write $\Omega(\phi, \lambda)$ for the χ-Whittaker functional for $\mathcal{I}_P(\sigma_{\mu,\lambda})$ with respect to which we have normalized the isomorphism $\widetilde{\mathcal{I}}_P(\sigma_{\mu,\lambda}, N)$. It is induced from the automorphic χ_{M_P}-Whittaker functional on the space of cuspidal functions on $M_P(\mathbb{A})$, and is defined by the global analogues of the integral (2.5.1). We then have

$$\Omega\big(\widetilde{\mathcal{I}}_P(\sigma_{\mu,\lambda}, N)\phi, \lambda^\vee\big) = \Omega(\phi, \lambda), \qquad \phi \in \mathcal{H}_P^0(\sigma_\mu),$$

by definition. Moreover, one can show from the basic definition of $E(\phi, \lambda)$ as an absolutely convergent series on an open set of λ that

$$\omega\big(E(\lambda)\phi\big) = \Omega(\phi, \lambda), \qquad \phi \in \mathcal{H}_P^0(\sigma_\mu).$$

(See [**Sha1**, p. 351].) Combining these identities, we see that

$$\begin{aligned}\omega\big(E(\lambda^\vee)\widetilde{\mathcal{I}}_P(\sigma_{\mu,\lambda}, N)\phi\big) &= \Omega\big(\widetilde{\mathcal{I}}_P(\sigma_{\mu,\lambda}, N)\phi, \lambda^\vee\big)\\ &= \Omega(\phi, \lambda) = \omega\big(E(\lambda)\phi\big)\\ &= \omega\big(\widetilde{R}(N)E(\lambda)\phi\big).\end{aligned}$$

It follows that $c_\lambda = 1$, as required, and therefore that the identity (4.2.9) is valid.

The embedding of the global Langlands quotient π_ψ into the automorphic discrete spectrum is provided by residues of Eisenstein series. More precisely, it is the sum of iterated residues of $E(x, \phi, \lambda)$ at the point $\lambda = \rho_P = \rho_P^\vee$ given by the general scheme in [**L5**, Chapter 7], and its specialization to $GL(N)$ in [**MW2**], that yields the intertwining isomorphism from π_ψ to $R^N_{\mathrm{disc},\psi}$. We recall that the residue scheme is noncanonical. In particular, it could be replaced by its "adjoint", in which the iterated residues are taken in $\lambda^\vee$ rather than λ. We are trying to relate the operators $\widetilde{R}(N)$ and $\widetilde{\pi}_\psi(N)$. But $\widetilde{\pi}_\psi(N)$ is the Langlands quotient of the value at $\lambda = \rho_P$ of the operator $\widetilde{\mathcal{I}}_P(\sigma_{\mu,\lambda}, N)$. The lemma then follows from an application of the residue operation to the left hand side of (4.2.9), and its adjoint to the right hand side. □

We now consider the global trace (4.2.1). We shall expand it into an expression that is roughly parallel to the formula (4.1.12) of Corollary 4.1.3, and which, as we noted above, is a corollary of Theorems 4.2.2 and 1.5.2.

Any parameter $\psi_M \in \widetilde{\Psi}_2(M)$ determines a subspace

$$L^2_{\mathrm{disc},\psi_M}\big(M(F)\backslash M(\mathbb{A})\big) \subset L^2_{\mathrm{disc}}\big(A_M(\mathbb{R})^0 M(F)\backslash M(\mathbb{A})\big)$$

of the discrete spectrum. This subspace in turn has an $M(\mathbb{A})$-equivariant decomposition

$$L^2_{\mathrm{disc},\psi_M}\big(M(F)\backslash M(\mathbb{A})\big) \cong \bigoplus_{\pi_M} m_{\psi_M}(\pi_M)\pi_M,$$

for irreducible representations π_M of $M(\mathbb{A})$ such that the Hecke eigenfamily $c(\pi_M)$ for π_M maps to its analogue $c(\psi_M)$ for π_M, and nonnegative multiplicities $m_{\psi_M}(\pi_M)$. The induced representation in (4.2.1) therefore has a decomposition

$$\mathcal{I}_{P,\psi}(f) \cong \bigoplus_{\psi_M} \bigoplus_{\pi_M} m_{\psi_M}(\pi_M)\,\mathcal{I}_P(\pi_M, f),$$

where the sum over ψ_M can be restricted to the subset $\widetilde{\Psi}_2(M,\psi)$ of $\widetilde{\Psi}_2(M)$. Let us write $M_P(w,\pi_M)$ for the transfer of the intertwining operator $M_{P,\psi}(w)$ in (4.2.1) to the space on which $\mathcal{I}_P(\pi_M, f)$ acts. We can then express the global trace (4.2.1) as a double sum

$$\sum_{\psi_M}\sum_{\pi_M} m_{\psi_M}(\pi_M)\,\mathrm{tr}\big(M_P(w,\pi_M)\,\mathcal{I}_P(\pi_M, f)\big),$$

with the understanding that the trace vanishes if $w\pi_M$ is not equivalent to π_M.

The original intertwining operator $M_{P,\psi}(w)$ in (4.2.1) is attached to a pair $G \in \mathcal{E}_{\mathrm{sim}}(G)$ and $M \in \mathcal{L}$. It acts on the Hilbert space $\mathcal{H}_{P,\psi}$ of the induced representation $\mathcal{I}_{P,\psi}$. It is defined by analytic continuation of Langlands' original integral over

$$N_{w^{-1}P}(\mathbb{A}) \cap N_P(\mathbb{A})\backslash N_{w^{-1}P}(\mathbb{A}).$$

Since the definition also includes a left translation by w^{-1}, the transfer

$$M_P(w,\pi_M) = J_P(\widetilde{w},\pi_M)$$

of $M_{P,\psi}(w)$ to the Hilbert space of $\mathcal{I}_P(\pi_M, f)$ is the global analogue of the unnormalized operator $J_P(\widetilde{w},\pi_\lambda)$ (with π_M in place of π_λ) in (2.3.26). The global form of the entire expression (2.3.26) is the product

$$R_P(w,\pi_M,\psi_M) = r_P(w,\psi_M)^{-1}M_P(w,\pi_M),$$

where $r_P(w,\psi_M)$ is the product of the global λ-factor $\lambda(w)$ with the global normalizing factor

$$L(0,\pi_{\psi_M},\rho^\vee_{w^{-1}P|P})\,\varepsilon(0,\pi_{\psi_M},\rho^\vee_{w^{-1}P|P})^{-1}\,L(1,\pi_{\psi_M},\rho^\vee_{w^{-1}P|P})^{-1} \qquad (4.2.10)$$

defined by analytic continuation of a product of local normalizing factors (2.3.27). Since the global λ-factor equals 1, the global normalizing factor is in fact just equal to (4.2.10).

In the discussion of Chapter 2, the next step was to transform the local object (2.3.26), which occurs on the right hand side of (2.4.2), into the self-intertwining operator on the left hand side of (2.4.2). In the global setting here, $R_P(w, \pi_M, \psi_M)$ is already a self-intertwining operator, under the condition $w \in W_\psi(\pi_M)$ of (2.4.2). In fact, a global version of the intertwining operator $\pi(\widetilde{w})$ of (2.4.2) is built into the definition of $R_P(w, \pi_M, \psi_M)$. It comes from operators (4.2.7) and (4.2.8), applied to the relevant factors of M, and taken at values $y = \widetilde{w}$ and $x = 1$ at which they become trivial. For the general linear factors of M, we must use Lemma 4.2.3 to verify that the analogue of (4.2.8) equals the tensor product of the local Whittaker extensions that are implicit in the global self-intertwining operator in (4.2.4)$'$. We can therefore write

$$M_P(w, \pi_M) = r_P(w, \psi_M)\, R_P(w, \widetilde{\pi}_w, \psi_M), \qquad w \in W_\psi(\pi_M),$$

in the global form of the notation (2.4.2). It remains to apply Theorems 1.5.2 and 4.2.2 to the other factor G_- of M.

We have expressed the global trace (4.2.1) as a double sum over ψ_M and π_M. We first apply Theorem 4.2.2(a) to the G_--component of the global multiplicity $m_{\psi_M}(\pi)$ that occurs in any summand. We see that the summand vanishes unless ψ_M belongs to the subset

$$\Psi_2(\widetilde{M}_w, \psi) = \Psi(\widetilde{M}_w) \cap \widetilde{\Psi}_2(M, \psi)$$

of w-fixed elements in $\widetilde{\Psi}_2(M, \psi)$. We can also restrict the second sum to elements π_M in the packet $\widetilde{\Pi}_{\psi_M}$. It follows that (4.2.1) equals

$$\sum_{\psi_M \in \Psi_2(\widetilde{M}_w, \psi)} \sum_{\pi_M \in \widetilde{\Pi}_{\psi_M}} m_{\psi_M}(\pi_M)\, r_P(w, \psi_M)\, \mathrm{tr}\big(R_P(w, \widetilde{\pi}_M, \psi_M)\, \mathcal{I}_P(\pi_M, f)\big).$$

Next, we apply Theorem 1.5.2 to G_-. This gives us the multiplicity formula

$$m_{\psi_M}(\pi_M) = m_{\psi_M} |\mathcal{S}_{\psi_M}|^{-1} \sum_{x_M \in \mathcal{S}_{\psi_M}} \varepsilon_{\psi_M}(x_M) \langle x_M, \pi_M \rangle,$$

by Fourier inversion on the abelian group $\mathcal{S}_{\psi_M}$. The last step will be to apply Theorem 4.2.2(b) to G_-.

The condition that ψ_M lies in $\Psi_2(\widetilde{M}_w, \psi)$ means that w lies in the subgroup W_ψ of $W(M)$. (Like the other groups in the diagram (4.2.3), W_ψ is in fact defined in terms of an implicit representative ψ_M of ψ.) Let us write

$$\mathfrak{N}_\psi(w) \;\cong\; \widetilde{\mathcal{S}}_{\psi_M, u}, \qquad w_u = w,$$

for the fibre of w in $\mathfrak{N}_\psi$, relative to the horizontal exact sequence at the center of (4.2.3). We can then write

$$\sum_{x_M \in \mathcal{S}_{\psi_M}} \varepsilon_{\psi_M}(x_M) \langle x_M, \pi_M \rangle = \sum_{u \in \mathfrak{N}_\psi(w)} \varepsilon_{\psi_M}(\widetilde{u}) \langle \widetilde{u}, \widetilde{\pi}_M \rangle,$$

where the summand on the right is an extension to the group $\widetilde{\mathcal{S}}^{+}_{\psi_M,u}$ of the character on $\mathcal{S}_{\psi_M}$ defined by the summand on the left. The extension must be trivial if the character on the left is trivial, but can otherwise be arbitrary, since both sums then automatically vanish. We take $\varepsilon_{\psi_M}(\widetilde{u})$ to be the canonical extension of $\varepsilon_{\psi_M}(x_M)$. If the linear character $\langle x_M, \pi_M\rangle$ on $\mathcal{S}_{\psi_M}$ equals $\varepsilon^{-1}_{\psi_M}$, the pairing $\langle \widetilde{u}, \widetilde{\pi}_M\rangle$ will then have to be the canonical extension of $\varepsilon^{-1}_{\psi_M}$. Applying Theorem 4.2.2(b) to M in this case, we conclude that $R_P(w, \widetilde{\pi}_M, \psi_M)$ is defined in terms of an implicit extension of π_M to $\widetilde{M}_w(\mathbb{A})^+$ that is compatible with the product of local extensions associated by Theorem 2.2.4 to $\langle \widetilde{u}, \widetilde{\pi}_M\rangle$. In case $\langle x_M, \pi_M\rangle$ is distinct from $\varepsilon^{-1}_{\psi_M}$, which is to say that π_M does not occur in the discrete spectrum, we have only to define $R_P(w, \widetilde{\pi}_M, \psi_M)$ by any product of local intertwining operators (2.4.2) that is compatible with the extension of $\langle \widetilde{u}, \widetilde{\pi}_M\rangle$ we have chosen.

Having agreed upon these conventions, we take the sum over $u \in \mathfrak{N}_\psi(w)$ outside the sum over $\pi_M \in \widetilde{\Pi}_{\psi_M}$. It then follows from the definition (4.2.4) that the resulting sum

$$\sum_{\pi_M \in \widetilde{\Pi}_{\psi_M}} \langle \widetilde{u}, \widetilde{\pi}_M\rangle \operatorname{tr}\big(R_P(w, \widetilde{\pi}_M, \psi_M)\, \mathcal{I}_P(\pi_M, f)\big)$$

equals $f_G(\psi, u)$. Combining this with the other terms discussed above, we obtain the expansion of (4.2.1) that represents our common corollary of Theorems 4.2.2 and 1.5.2.

COROLLARY 4.2.4. *Given $G \in \widetilde{\mathcal{E}}_{\mathrm{sim}}(N)$, $M \in \mathcal{L}$, $\psi \in \widetilde{\Psi}_M(G)$ and $f \in \widetilde{\mathcal{H}}(G)$, we have an expansion*

$$\sum_{\psi_M \in \Psi_2(\widetilde{M}_w, \psi)} |\mathcal{S}_{\psi_M}|^{-1} \sum_{u \in \mathfrak{N}_\psi(w)} r_P(w, \psi_M)\, \varepsilon_{\psi_M}(\widetilde{u})\, f_G(\psi, u) \tag{4.2.11}$$

for the global trace

$$\operatorname{tr}\big(M_{P,\psi}(w)\, \mathcal{I}_{P,\psi}(f)\big), \qquad w \in W(M). \qquad \square$$

REMARKS. 1. Suppose that the Levi subgroup M is proper in G. In other words, it is a product of general linear groups with a simple endoscopic group $G_- \in \widetilde{\mathcal{E}}_{\mathrm{sim}}(N_-)$ that is a proper subgroup of G. As in Corollary 4.1.3, the informal assumptions under which this corollary makes sense depend only on the positive integer N. They could be replaced by a formal induction hypothesis that the theorems stated so far, including Theorems 1.5.2 and 4.2.2, are valid if N is replaced by a smaller integer N_-. Corollary 4.2.4 would then hold for M.

2. Corollary 4.2.4 bears a formal resemblance to Corollary 4.1.3 on several accounts. This is no accident. We shall see in the next two sections that their roles in our comparison of the two expansions (4.1.1) and (4.1.2) are entirely parallel.

The global intertwining relation (4.2.6) and the formula (4.2.11) for the global trace (4.2.1) are obviously designed to be used together. They have important implications for the two expansions (4.1.1) and (4.1.2), which we shall analyse over the next few sections. However, we should first describe twisted versions of the formulas for $GL(N)$, since they will also be needed.

Suppose then that G equals the component $\widetilde{G}(N)$, rather than an element in $\widetilde{\mathcal{E}}_{\mathrm{sim}}(N)$. Then G represents the triplet

$$\big(\widetilde{G}(N)^0, \widetilde{\theta}(N), 1\big) = \big(GL(N), \widetilde{\theta}(N), 1\big),$$

as in the general discussion of Chapter 3. We take ψ to be a parameter in the set

$$\widetilde{\Psi}(G) = \widetilde{\Psi}(N) = \Psi\big(\widetilde{G}(N)\big),$$

and f to be a function in the space

$$\widetilde{\mathcal{H}}(G) = \widetilde{\mathcal{H}}(N) = \mathcal{H}\big(\widetilde{G}(N)\big).$$

We then have the spectral expansion (4.1.1) (which has been established in general), and the endoscopic expansion (4.1.2) (provided by Hypothesis 3.2.1).

One of our aims has been to arrange matters so that the twisted case $G = \widetilde{G}(N)$ does not require separate notation. It is for this reason that we write

$$S_\psi = S_\psi(G) = \mathrm{Cent}\big(\mathrm{Im}(\psi), \widehat{G}\big)$$

and

$$\overline{S}_\psi = S_\psi / Z(\widehat{G}^0)^\Gamma = S_\psi / Z(\widehat{G}^0),$$

where $\widehat{G}$ is the $\widehat{G}^0$-torsor $\widehat{G}^0 \rtimes \widehat{\theta}$. These two sets are bitorsors under the corresponding complex reductive groups

$$S_\psi^* = S_\psi(G^0) = \mathrm{Cent}(\mathrm{Im}(\psi), \widehat{G}^0)$$

and

$$\overline{S}_\psi^* = S_\psi^* / Z(\widehat{G}^0)^\Gamma = S_\psi^* / Z(\widehat{G}^0).$$

Similarly, the quotient

$$\mathcal{S}_\psi = \pi_0(\overline{S}_\psi)$$

is a bitorsor under the finite group

$$\mathcal{S}_\psi^* = \pi_0(\overline{S}_\psi^*).$$

Observe that S_ψ^* is a product of general linear groups, and that $Z(\widehat{G}^0) = Z(\widehat{G}^0)^\Gamma$ equals the group $\mathbb{C}^*$, embedded diagonally in S_ψ^*. In particular, the groups S_ψ^* and $\overline{S}_\psi^*$ are both connected. Therefore $\overline{S}_\psi^*$ equals $\overline{S}_\psi^0$, so the sets $\mathcal{S}_\psi^*$ and $\mathcal{S}_\psi$ are both trivial in this case.

Despite the fact that $\mathcal{S}_\psi$ is trivial, the constructions of §4.1 are still interesting when applied to the bitorsor $S = \overline{S}_\psi$. Proposition 4.1.1 gives a noteworthy relationship among the associated numbers $i(S) = i(\overline{S}_\psi)$ and $\sigma(S_s^0) = \sigma(\overline{S}_{\psi,s}^0)$. The properties formulated in terms of $\overline{S}_\psi$ at the end of §4.1 still make sense, and provide a chain of subsets (4.1.13) of the set $\Psi(G) = \widetilde{\Psi}(G)$. In this case, the subsets $\Psi_{\mathrm{sim}}(G)$ and $\Psi_2(G)$ are equal. The formula of Corollary 4.1.3 also makes sense as stated, that is, with G' representing any datum in the set $\mathcal{E}_{\mathrm{ell}}(G) = \widetilde{\mathcal{E}}_{\mathrm{ell}}(N)$. It is an immediate consequence of Theorem 4.1.2 (with G being as originally understood, an element in $\widetilde{\mathcal{E}}_{\mathrm{sim}}(N)$.)

The earlier discussion of this section also extends to the twisted case. The diagram (4.2.3) becomes simpler, since the sets in the lower horizontal exact sequence are now all trivial. It follows that the sets $\mathfrak{N}_\psi$ and W_ψ are equal, and are both bitorsors under the finite group W^0_ψ. The families $\Psi_2(M,\psi)$, $\Psi(M,\psi)$, $\Psi(G',\psi)$, etc., all have obvious meaning. It is understood here that M is as in the twisted form of (4.1.1), a Levi subgroup of $G^0 = GL(N)$, which we assume is proper. Furthermore, the notation (4.2.4) and (4.2.5) continues to make sense. In the case of (4.2.4), we note the packet $\widetilde{\Pi}_{\psi_M} = \Pi_{\psi_M}$ attached to any $\psi_M \in \Psi_2(M,\psi)$ contains one representation π_M, and that the corresponding global factor $\langle \widetilde{u}, \widetilde{\pi}_M\rangle$ is trivial. The other definition (4.2.5) is formulated in terms of the correspondence

$$(G',\psi') \longrightarrow (\psi, s), \qquad G' \in \mathcal{E}(G),\ s \in \overline{S}_{\psi,\mathrm{ss}},$$

which is given by the obvious construction.

The statements of Corollaries 4.2.1 and 4.2.4 also extend without change in notation. In the formula (4.2.11) of Corollary 4.2.4, the factors $|\mathcal{S}_{\psi_M}|$ and $\varepsilon_{\psi_M}(\tilde{u})$ are both equal to 1, while $r_P(w,\psi_M)$ represents a global normalizing factor for $G^0 = GL(N)$. The derivation of this formula is similar. (The formula cannot be described as a corollary in this case, since Theorem 4.2.2 no longer has a role.) The proof of the twisted global intertwining relation (4.2.6), or rather its reduction to the local relation Corollary 2.5.4, also remains the same. We recall that Corollary 2.5.4 depends on the definitions provided by Theorem 2.2.1 (which of course still has to be proved).

The discussion of this section thus holds uniformly if G is either an element in $\widetilde{\mathcal{E}}_{\mathrm{sim}}(N)$ or if G equals $\widetilde{G}(N)$ itself. We shall apply it in this generality over the next few sections.

4.3. Spectral terms

We are now at the stage where we can begin the proof of our results in earnest. There are essentially nine theorems. They are the original three theorems of classification, Theorems 1.5.1, 1.5.2 and 1.5.3 (which for their statements implicitly include Theorems 1.4.1 and 1.4.2), their local refinements Theorems 2.2.1, 2.2.4, 2.4.1 and 2.4.4 from Chapter 2, and the global refinements Theorems 4.1.2 and 4.2.2. We shall prove all of the theorems together. The argument will be a multilayered induction, which will take up much of the rest of the volume. We will start with the spectral expansion (4.1.1) in this section, and the endoscopic expansion (4.1.2) in the next.

We need to be clear about our induction assumptions. For a start, they are to be distinguished from our structural condition (Hypothesis 3.2.1) on the stabilization of the twisted trace formula. Our primary induction hypothesis has already been applied informally (and often implicitly) a number of times. It concerns the rank N.

Recall that the nine theorems all apply to a group G in the set $\widetilde{\mathcal{E}}_{\mathrm{sim}}(N) = \mathcal{E}_{\mathrm{sim}}\big(\widetilde{G}(N)\big)$, where

$$\widetilde{G}(N) = GL(N) \rtimes \widetilde{\theta}(N),$$

for some positive integer N. We fix N. We then assume inductively that the theorems all hold if G is replaced by any group $G_- \in \widetilde{\mathcal{E}}_{\mathrm{sim}}(N_-)$ (and hence also for any G_- in one of the larger sets $\widetilde{\mathcal{E}}_{\mathrm{ell}}(N_-)$ or $\widetilde{\mathcal{E}}(N_-)$), for any positive integer N_- with $N_- < N$. We will later have occasion to take on more delicate induction hypotheses.

We assume for the rest of this chapter that F is global. For any $G \in \widetilde{\mathcal{E}}_{\mathrm{sim}}(N)$ and $\psi \in \widetilde{\Psi}(N)$, we have been writing $R^G_{\mathrm{disc},\psi}$ for the representation of the symmetric global Hecke algebra $\widetilde{\mathcal{H}}(G)$ on the ψ-component $L^2_{\mathrm{disc},\psi}\big(G(F)\backslash G(\mathbb{A})\big)$ of the discrete spectrum. It follows from Corollary 3.4.3 (one of the results we have established, rather than just stated) that the representation of $\widetilde{\mathcal{H}}(G)$ on the full discrete spectrum is a direct sum over $\psi \in \widetilde{\Psi}(N)$ of the subrepresentations $R^G_{\mathrm{disc},\psi}$. Theorem 1.5.2 (one of our stated, but yet unproven theorems) describes the decomposition of $R^G_{\mathrm{disc},\psi}$ into irreducible representations, at least up to an understanding of the local constituents of the packet $\widetilde{\Pi}_\psi$. It asserts that $R^G_{\mathrm{disc},\psi}$ is zero if ψ does not belong to the subset $\widetilde{\Psi}_2(G)$ of $\widetilde{\Psi}(N)$, and that if ψ does lie in $\widetilde{\Psi}_2(G)$, then

$$R^G_{\mathrm{disc},\psi} = \bigoplus_{\pi \in \widetilde{\Pi}_\psi} m_\psi(\pi)\, \pi,$$

where

$$m_\psi(\pi) = \begin{cases} m_\psi, & \text{if } \langle \cdot, \pi \rangle = \varepsilon_\psi^{-1} = \varepsilon_\psi, \\ 0 & \text{otherwise.} \end{cases}$$

This is of course one of the assertions we can apply inductively (having already done so implicitly in the last section). Its validity for G, however, will be among the most difficult things we have to prove.

Suppose for example that ψ belongs to the subset $\widetilde{\Psi}(G)$ of $\widetilde{\Psi}(N)$, but lies in the complement of $\widetilde{\Psi}_2(G)$. Then ψ factors through a parameter $\psi_M \in \Psi_2(M)$, for a proper Levi subgroup M of G. By our induction assumption, applied to the factors of M as a product of groups, the image of ψ_M in $\widetilde{\Psi}_2(M, \psi)$ contributes to the automorphic discrete spectrum of M (taken modulo the center). The theory of Eisenstein series then tells us that ψ contributes to the continuous spectrum of G. Theorem 1.5.2 includes the assertion that ψ does not contribute to the discrete spectrum of G. In other words, the automorphic spectrum of G has no *embedded eigenvalues*, in the sense of unramified Hecke eigenfamilies in $\mathcal{C}_{\mathrm{aut}}(G)$. This question is obviously an important part of the theorem. It will be a focal point of the long term study we are about to undertake.

For the next few sections, we shall be working from the slightly broader perspective reached at the end of the last section. That is, we assume that G is either an element in $\widetilde{\mathcal{E}}_{\mathrm{sim}}(N)$ or equal to $\widetilde{G}(N)$ itself. The basic expansions (4.1.1) and (4.1.2), which are our starting points, were originally formulated in this generality. The rest of the discussion of §4.1 and §4.2 (apart from Theorem 4.1.2, which is expressly limited to the case that G is a connected quasisplit group) also applies.

We note that the operator

$$R^G_{\mathrm{disc},\psi}(f), \qquad f \in \widetilde{\mathcal{H}}(G),$$

makes sense if $G = \widetilde{G}(N)$. It is defined by the extension (4.2.8). Equivalently, it is simply the composition of the two factors in (4.2.1), under the condition that $M = P = G^0$. In particular, its trace gives the corresponding contribution to the term with $M = G^0$ in (4.1.1). The other contribution to this term comes from the sole element $w = \theta$ in $W_{\mathrm{reg}}(M)$. We observe that

$$|\det(w-1)_{\mathfrak{a}^G_M}|^{-1} = |\det(\theta-1)_{\mathfrak{a}^G_{G^0}}|^{-1} = |\kappa_G|^{-1},$$

by (3.2.5), recalling at the same time that

$$|\kappa_G|^{-1} = |\pi_0(\kappa_G)|^{-1} = \tfrac{1}{2},$$

in the case $G = \widetilde{G}(N)$ at hand. The operator $R^G_{\mathrm{disc},\psi}(f)$ is straightforward in this case, since we know from Theorem 1.3.2 that $GL(N)$ has no embedded eigenvalues. In particular, the operator equals 0 unless ψ belongs to $\widetilde{\Psi}_{\mathrm{sim}}(G)$. We include it in the discussion in order to have uniform statements of the results.

We fix $\psi \in \widetilde{\Psi}(N)$ and $f \in \widetilde{\mathcal{H}}(G)$. We are looking at the expansion (4.1.1), with G now allowed to be either an element in $\widetilde{\mathcal{E}}_{\mathrm{sim}}(N)$ or $\widetilde{G}(N)$ itself. The term with $M = G^0$ equals the product

$$r^G_{\mathrm{disc},\psi}(f) = |\kappa_G|^{-1}\mathrm{tr}\big(R^G_{\mathrm{disc},\psi}(f)\big).$$

It follows that the difference

$$I_{\mathrm{disc},\psi}(f) - r^G_{\mathrm{disc},\psi}(f) \tag{4.3.1}$$

equals the sum of those terms in (4.1.1) with $M \neq G$. The summand of M depends only on the W^G_0-orbit of M. We can therefore sum over the set $\{M\}$ of W^G_0-orbits in $\mathcal{L}$ distinct from G, provided that we multiply by the number

$$|W^G_0||W^M_0|^{-1}|W(M)|^{-1}$$

of elements in a given orbit. The difference (4.3.1) therefore equals

$$\sum_{\{M\}\neq\{G\}} |W(M)|^{-1} \sum_{w\in W_{\mathrm{reg}}(M)} |\det(w-1)_{\mathfrak{a}^G_M}|^{-1}\,\mathrm{tr}\big(M_{P,\psi}(w)\,\mathcal{I}_{P,\psi}(f)\big).$$

We can apply Corollary 4.2.4, which is founded on our underlying induction hypothesis, to the factor

$$\mathrm{tr}\big(M_{P,\psi}(w)\,\mathcal{I}_{P,\psi}(f)\big).$$

The difference (4.3.1) becomes a fourfold sum, over $\{M\} \neq \{G\}$, $w \in W_{\mathrm{reg}}(M)$, $\psi_M \in \Psi_2(\widetilde{M}_w, \psi)$ and $u \in \mathfrak{N}_\psi(w)$, of the product of

$$|W(M)|^{-1}|\det(w-1)_{\mathfrak{a}_M^G}|^{-1}$$

with

$$|\mathcal{S}_{\psi_M}|^{-1} r_P(w, \psi_M)\, \varepsilon_{\psi_M}(\widetilde{u})\, f_G(\psi, u).$$

Despite its notation, the second factor is independent of the representative ψ_M of ψ. We will therefore be able to remove the sum over the set of ψ_M, so long as we multiply the summand by its order. The result will be clearer if we first rearrange the double sum over w and ψ_M.

We start by writing the sum over $\psi_M \in \Psi_2(\widetilde{M}_w, \psi)$ as a double sum over ψ_G in $\Psi(G, \psi)$ and ψ_M in $\Psi_2(\widetilde{M}_w, \psi_G)$, the subset of parameters in $\Psi_2(\widetilde{M}_w, \psi)$ that map to the $\widehat{G}$-orbit ψ_G. We then interchange the sums over w and ψ_G. This allows us to combine the resulting sums over w and ψ_M into a double sum over the set

$$V_\psi = \big\{(\psi_M, w) \in \Psi_2(M, \psi_G) \times W_{\mathrm{reg}}(M) :\ \mathrm{Int}(w) \circ \psi_M = \psi_M\big\}.$$

Now, the projection $(\psi_M, w) \to w$ gives a canonical fibration

$$V_\psi \longrightarrow \{W_{\psi,\mathrm{reg}}\}$$

of V_ψ over the set of orbits of W_ψ by conjugation on $W_{\psi,\mathrm{reg}}$. We claim that the group $W(M)$ acts transitively on the fibres. Since ψ_G is a $\widehat{G}$-orbit, and since $A_{\widehat{M}}$ is a maximal torus in $S^0_{\psi_M}$, for any $\psi_M \in \Psi_2(M, \psi)$, any two parameters in $\Psi_2(M, \psi_G)$ are conjugate by an element in $\widehat{G}$ that stabilizes $A_{\widehat{M}}$. Therefore $W(M)$ acts transitively on $\Psi_2(M, \psi_G)$. The claim follows from the fact that the stabilizer in $W(M)$ of any parameter in $\Psi_2(M, \psi_G)$ is isomorphic to W_ψ. The summand is constant on the fibres. We can therefore replace the double sum over V_ψ with a constant multiple of a simple sum over $\{W_{\psi,\mathrm{reg}}\}$. The scaling constant is slightly simpler if we take the simple sum over the set $W_{\psi,\mathrm{reg}}$, rather than its quotient $\{W_{\psi,\mathrm{reg}}\}$. With this change, we have then only to multiply the summand by the integer

$$|\Psi_2(M, \psi_G)| = |W(M)||W_\psi|^{-1}.$$

For any ψ_G, there is at most one orbit $\{M\}$ such that the set $\Psi_2(M, \psi_G)$ is not empty. Different elements $\psi_G \in \Psi(G, \psi)$ can give different orbits $\{M\}$, but the corresponding summands remain equal. We can therefore remove the exterior sum over ψ_G, provided that we multiply the summand by the second integer

$$|\Psi(G, \psi)| = m_\psi.$$

We conclude that the difference (4.3.1) equals

(4.3.2)

$$m_\psi |W_\psi|^{-1} \sum_w |\det(w-1)_{\mathfrak{a}_M^G}|^{-1} |\mathcal{S}_{\psi_M}|^{-1} \sum_u r_P(w,\psi_M)\, \varepsilon_{\psi_M}(\widetilde{u})\, f_G(\psi, u),$$

where w and u are summed over $W_{\psi,\mathrm{reg}}$ and $\mathfrak{N}_\psi(w)$ respectively, and (M, ψ_M) is any pair with $M \neq G^0$ and $\psi_M \in \Psi_2(\widetilde{M}_w, \psi)$. Of course, if there is no such (M, ψ_M), the expression is understood to be 0. This is the case if ψ lies in either $\widetilde{\Psi}_2(G)$ or in the complement of the subset $\widetilde{\Psi}(G)$ of $\widetilde{\Psi}(N)$.

The case that $G = \widetilde{G}(N)$, which is now part of our analysis, is generally the simpler of the two. However, it does have one minor complication. This is because $\mathfrak{a}_G$, the $(+1)$-eigenspace of the operator θ on $\mathfrak{a}_{G^0}$, is a proper subspace of $\mathfrak{a}_{G^0}$. (In fact, it is clear that $\dim \mathfrak{a}_{G^0} = 1$ and $\dim \mathfrak{a}_G = 0$.) We obtain a decomposition

$$|\det(w-1)_{\mathfrak{a}_M^G}| = |\det(w-1)_{\mathfrak{a}_M^{G^0}}|\, |\det(w-1)_{\mathfrak{a}_{G^0}^G}|,$$

for any $w \in W_\psi$. Since the restriction of w to $\mathfrak{a}_{G^0}$ equals θ, we use (3.2.5) to write

$$|\det(w-1)_{\mathfrak{a}_{G^0}^G}| = |\det(\theta-1)_{\mathfrak{a}_{G^0}^G}| = |\kappa_G| \tag{4.3.3}$$

for the second factor on the right. The first factor may be written as

$$|\det(w-1)_{\mathfrak{a}_M^{G^0}}| = |\det(w-1)|,$$

where the operator $(w-1)$ on the right is to be regarded as an endomorphism of the real vector space $\mathfrak{a}_{\overline{T}_\psi}$, as in §4.1. Indeed, $\mathfrak{a}_M^{G^0}$ is isomorphic to the dual of $\mathfrak{a}_{\overline{T}_\psi}$, since as a maximal torus in the complex connected group $\overline{S}_\psi^0 = \overline{S}_\psi^*$, the group $\overline{T}_\psi$ is isomorphic to $A_{\widehat{M}}/Z(\widehat{G}^0)^\Gamma$. We substitute the resulting product into our expression (4.3.2) for the difference (4.3.1).

We also add a couple of housekeeping changes to the notation in (4.3.2). The group $\mathcal{S}_{\psi_M}$ is isomorphic to the group $\mathcal{S}_\psi^1$ in the diagram (4.2.3). It then follows from the relevant two short exact sequences in the diagram that

$$|W_\psi|^{-1}|\mathcal{S}_{\psi_M}|^{-1} = |W_\psi|^{-1}|\mathcal{S}_\psi^1|^{-1} = |\mathfrak{N}_\psi|^{-1} = |\mathcal{S}_\psi|^{-1}|W_\psi^0|^{-1}.$$

If u is any element in the coset $\mathfrak{N}_\psi(w)$, we set

$$\varepsilon_\psi^1(u) = \varepsilon_{\psi_M}(\widetilde{u})$$

and

$$r_\psi^G(w) = r_P(w, \psi_M),$$

since these functions depend only on ψ. With these changes, we have now left ourselves with no reference to M in our expression for (4.3.1). This is not a problem, since the orbit of M is still implicit in the torsor $\overline{S}_\psi$. However, we will need a marker to rule out the case that $M = G^0$. We take care of this by setting

$$W'_{\psi,\mathrm{reg}} = \{w \in W_{\psi,\mathrm{reg}} :\ w \neq 1\},$$

since it is clear that

$$W'_{\psi,\mathrm{reg}} = \begin{cases} \varnothing, & \text{if } \overline{S}_\psi \text{ is finite,} \\ W_{\psi,\mathrm{reg}}, & \text{otherwise.} \end{cases}$$

Collecting the various terms, we see that the difference (4.3.1) equals

$$C_\psi |W_\psi^0|^{-1} \sum_{w \in W'_{\psi,\mathrm{reg}}} \sum_{u \in \mathfrak{N}_\psi(w)} |\det(w-1)|^{-1} r_\psi^G(w)\, \varepsilon_\psi^1(u)\, f_G(\psi, u), \tag{4.3.4}$$

where

$$C_\psi = m_\psi |\kappa_G|^{-1} |\mathcal{S}_\psi|^{-1}. \tag{4.3.5}$$

Keep in mind that if ψ does not belong to the subset $\widetilde{\Psi}(G)$ of $\widetilde{\Psi}$, then m_ψ and C_ψ both vanish by definition, and so therefore does the expression. The expression consequently makes sense, even though $\mathcal{S}_\psi$ is not defined in this case.

There is a critical lemma, which gives a different way to express the product of the two signs $r_\psi^G(w)$ and $\varepsilon_\psi^1(u)$. To state it, we need to recall the original sign function $s^0(w) = s_\psi^0(w)$ on W_ψ that came up in (4.1.5). If only for reasons of symmetry, we shall write

$$s_\psi^0(u) = s_\psi^0(w_u),$$

for its pullback to $\mathfrak{N}_\psi$. Then $s_\psi^0(u)$ equals (-1) raised to a power given by the number of roots of $(\overline{B}_\psi, \overline{T}_\psi)$ mapped by w_u to negative roots. This object depends on ψ only through the group W_ψ. It it not to be confused with the deeper sign character $\varepsilon_\psi(x) = \varepsilon_\psi^G(x)$ on $\mathcal{S}_\psi$, defined by (1.5.6) in terms of symplectic root numbers attached to ψ, or its twisted analogue $\varepsilon_\psi^1(u)$ for M. It is of course also unrelated to the fixed point s_ψ in $\mathcal{S}_\psi$.

LEMMA 4.3.1. *The sign characters satisfy the relation*

$$r_\psi^G(w_u)\, \varepsilon_\psi^1(u) = \varepsilon_\psi^G(x_u)\, s_\psi^0(u), \qquad u \in \mathfrak{N}_\psi.$$

We shall give the proof of this lemma in §4.6. Assuming it for now, we make the appropriate substitution in the last expression (4.3.4) for the difference (4.3.1). The next step is to rearrange the double sum in (4.3.4) by using the vertical exact sequence of $\mathfrak{N}_\psi$ in the diagram (4.2.3) in place of the horizontal exact sequence. Accordingly, we write

$$\mathfrak{N}_\psi(x), \qquad x \in \mathcal{S}_\psi,$$

for the fibre of x under the mapping from $\mathfrak{N}_\psi$ to $\mathcal{S}_\psi$. We also write $W_\psi(x)$ for the bijective image of $\mathfrak{N}_\psi(x)$ in W_ψ. We then let $\mathfrak{N}_{\psi,\mathrm{reg}}(x)$ and $\mathfrak{N}'_{\psi,\mathrm{reg}}(x)$ denote the preimages in $\mathfrak{N}_\psi(x)$ of the respective subsets

$$W_{\psi,\mathrm{reg}}(x) = W_\psi(x) \cap W_{\psi,\mathrm{reg}}$$

and

$$W'_{\psi,\mathrm{reg}}(x) = W_\psi(x) \cap W'_{\psi,\mathrm{reg}}$$

of $W_\psi(x)$. With this notation, we obtain the following result.

LEMMA 4.3.2. *The difference*

$$I_{\mathrm{disc},\psi}(f) - r^G_{\mathrm{disc},\psi}(f)$$

equals

$$C_\psi \sum_{x \in \mathcal{S}_\psi} \varepsilon^G_\psi(x)\, |W^0_\psi|^{-1} \sum_{u \in \mathfrak{N}'_{\psi,\mathrm{reg}}(x)} s^0_\psi(u)\, |\det(w_u - 1)|^{-1} f_G(\psi, u), \tag{4.3.6}$$

for any $\psi \in \widetilde{\Psi}(N)$ *and* $f \in \widetilde{\mathcal{H}}(G)$. □

The existence of the pairing $\langle x_M, \pi_M \rangle$ for $M \neq G^0$, which is implicit in (4.2.4), is part of our basic induction hypothesis. Suppose that we happen to know also that it is defined when $M = G^0$. Since ψ is assumed to belong to $\widetilde{\Psi}_M(G)$ in (4.2.4), this is only relevant to the case that $\psi \in \widetilde{\Psi}_2(G)$. It is essentially a local hypothesis, namely that for any $\psi_G \in \Psi_2(G, \psi)$, the localization $\psi_{G,v}$ belongs to $\Psi^+(G_v, \psi_v)$, and the local pairing $\langle x_{G,v}, \pi_{G,v} \rangle$ is defined. In particular, it is weaker than asking that the global multiplicity formula of Theorem 1.5.2 hold. However, it does allow us to form the difference

$${}^0 r^G_{\mathrm{disc},\psi}(f) = r^G_{\mathrm{disc},\psi}(f) - |\kappa_G|^{-1} m_\psi\, |\mathcal{S}_\psi|^{-1} \sum_{x \in \mathcal{S}_\psi} \varepsilon^G_\psi(x) \sum_{\pi \in \widetilde{\Pi}_\psi} \langle x, \pi \rangle\, f_G(\pi),$$

between the (normalized) trace $r^G_{\mathrm{disc},\psi}(f)$ and its expected value.

It follows easily from the definitions (4.2.4) and (4.3.5) that

$${}^0 r^G_{\mathrm{disc},\psi}(f) = r^G_{\mathrm{disc},\psi}(f) - C_\psi \sum_{x \in \mathcal{S}_\psi} \varepsilon^G_\psi(x)\, f_G(\psi, x), \tag{4.3.7}$$

in case $\psi \in \widetilde{\Psi}_2(G)$. In particular, the existence of the pairing $\langle x_M, \pi_M \rangle$ for $M = G^0$ amounts to the existence of the function $f_G(\psi, x)$ for $\psi \in \widetilde{\Psi}_2(G)$. In this case, the second term in the difference (4.3.7) is equal to the expression given by Lemma 4.3.2, but with the index of summation $\mathfrak{N}_{\psi,\mathrm{reg}}(x)$ in place of $\mathfrak{N}'_{\psi,\mathrm{reg}}(x)$. This is of course because $\mathfrak{N}_{\psi,\mathrm{reg}}(x)$ equals the point $\{x\}$ when ψ belongs to $\widetilde{\Psi}_2(G)$. (The original indexing set $\mathfrak{N}'_{\psi,\mathrm{reg}}(x)$ is empty in this case.) If ψ does not lie in $\widetilde{\Psi}_2(G)$, on the other hand, $\mathfrak{N}'_{\psi,\mathrm{reg}}(x)$ equals $\mathfrak{N}_{\psi,\mathrm{reg}}(x)$. In this case, we simply set

$${}^0 r^G_{\mathrm{disc},\psi}(f) = r^G_{\mathrm{disc},\psi}(f), \tag{4.3.8}$$

since the expected value of the trace $r^G_{\mathrm{disc},\psi}(f)$ is then equal to 0. It follows that the formula of Lemma 4.3.2 remains valid with ${}^0 r^G_{\psi,\mathrm{disc}}(f)$ in place of $r^G_{\psi,\mathrm{disc}}(f)$, and $\mathfrak{N}_{\psi,\mathrm{reg}}(x)$ in place of $\mathfrak{N}'_{\psi,\mathrm{reg}}(x)$.

Suppose that in addition to being well defined, the function $f_G(\psi, u)$ depends only on the image x of u in $\mathcal{S}_\psi$. This is implied by the global intertwining identity, but is obviously weaker. The function can then be taken outside the sum over u in $\mathfrak{N}_{\psi,\mathrm{reg}}(x)$. Lemma 4.3.2, or rather its minor extension above, can then be formulated as follows.

Corollary 4.3.3. *Suppose that the function*

$$f_G(\psi, x) = f_G(\psi, u), \qquad x \in \mathcal{S}_\psi, \ u \in \mathfrak{N}_\psi(x),$$

is defined , and depends only on x. Then the difference

$$I_{\mathrm{disc},\psi}(f) - {}^0 r^G_{\mathrm{disc},\psi}(f)$$

equals

$$C_\psi \sum_{x \in \mathcal{S}_\psi} i_\psi(x)\, \varepsilon^G_\psi(x)\, f_G(\psi, x), \tag{4.3.9}$$

where

$$i_\psi(x) = |W^0_\psi|^{-1} \sum_{w \in W_{\psi,\mathrm{reg}}(x)} s^0_\psi(w) |\det(w-1)|^{-1}.$$

□

There is perhaps more notation in this section than is strictly necessary. As in other sections, it has been designed to suggest how various results fit together, and thus offer some guidance to the course of the overall argument. In the case here, it is meant to emphasize the intrinsic symmetry between formulas in this section and those of the next.

4.4. Endoscopic terms

This section is parallel to the last one. It is aimed at the terms in the endoscopic expansion (4.1.2) of $I_{\mathrm{disc},\psi}(f)$. We shall derive a finer expansion of (4.1.2) that can be compared with the one we have established for (4.1.1).

We are working in the setting adopted in the last section. Then $\widetilde{G}(N)$ is fixed, and G is allowed to denote either an element in $\widetilde{\mathcal{E}}_{\mathrm{sim}}(N)$ or $\widetilde{G}(N)$ itself. The parameter $\psi \in \widetilde{\Psi}(N)$ is also fixed, as is the function of $f \in \widetilde{\mathcal{H}}(G)$.

The terms in (4.1.2) are parametrized by elliptic endoscopic data $G' \in \mathcal{E}_{\mathrm{ell}}(G)$. The subset $\mathcal{E}_{\mathrm{sim}}(G)$ of $\mathcal{E}_{\mathrm{ell}}(G)$ is of particular interest, because its terms are not amenable to induction. In recognition of this, we set

$$s^G_{\mathrm{disc},\psi}(f) = \sum_{G' \in \mathcal{E}_{\mathrm{sim}}(G)} \iota(G, G')\, \widehat{S}'_{\mathrm{disc},\psi}(f').$$

If G' belongs to the complement of $\mathcal{E}_{\mathrm{sim}}(G)$ in $\mathcal{E}_{\mathrm{ell}}(G)$, it is represented by a product of groups to which we can apply our induction hypothesis. In particular, we can apply the stable multiplicity identity of Corollary 4.1.3 to the corresponding term $\widehat{S}'_{\mathrm{disc},\psi}(f')$ in (4.1.2). We can of course also apply the specialization at the end of §3.2 of the general formula (3.2.4) to the coefficient $\iota(G, G')$ in (4.1.2). We see that the difference

$$I_{\mathrm{disc},\psi}(f) - s^G_{\mathrm{disc},\psi}(f) \tag{4.4.1}$$

equals

$$\sum_{G'} \iota(G, G') \sum_{\psi'} \widehat{S}'_{\mathrm{disc},\psi'}(f'), \tag{4.4.2}$$

where G' is summed over the complement of $\mathcal{E}_{\mathrm{sim}}(G)$ in $\mathcal{E}_{\mathrm{ell}}(G)$ and ψ' is summed over $\Psi(G', \psi)$, while

$$\widehat{S}'_{\mathrm{disc},\psi'}(f') = |\mathcal{S}_{\psi'}|^{-1}\sigma(\overline{S}^0_{\psi'})\,\varepsilon'(\psi')\,f'(\psi') \tag{4.4.3}$$

and

$$\iota(G,G') = |\kappa_G|^{-1}|\overline{Z}(\widehat{G}')^{\Gamma}|^{-1}|\mathrm{Out}_G(G')|^{-1}. \tag{4.4.4}$$

We recall that $\Psi(G', \psi)$ is the set of $\widehat{G}'$-orbits of L-embeddings of $\mathcal{L}_\psi \times SL(2,\mathbb{C})$ to ${}^LG^0$ that map to ψ, and that

$$\varepsilon'(\psi') = \varepsilon^{G'}_{\psi'}(s_{\psi'}).$$

If ψ does not lie in $\widetilde{\Psi}(G)$, a subset of $\widetilde{\Psi}(N)$ that is proper if $G \in \widetilde{\mathcal{E}}_{\mathrm{sim}}(N)$ and is equal to $\widetilde{\Psi}(N)$ if $G = \widetilde{G}(N)$, each set $\Psi(G',\psi)$ is empty. The difference (4.4.1) thus vanishes in this case. We can therefore assume that ψ does lie in $\widetilde{\Psi}(G)$.

The computations of the present section are again founded on the centralizer

$$\overline{S}_\psi = S_\psi / Z(\widehat{G}^0)^{\Gamma}.$$

This object is well defined (as a $\widehat{G}^0$-orbit) under our assumption that $\psi \in \widetilde{\Psi}(G)$. It is represented by the original complex reductive group (4.1.10) if $G \in \widetilde{\mathcal{E}}_{\mathrm{sim}}(N)$, and a bitorsor under the complex connected reductive group

$$\overline{S}^*_\psi = S_\psi(G^0)/Z(\widehat{G}^0)^{\Gamma},$$

in case $G = \widetilde{G}(N)$.

Any semisimple element $s \in \overline{S}_{\psi,\mathrm{ss}}$ in $\overline{S}_\psi$ gives rise to a pair

$$(G',\psi'), \qquad G' \in \mathcal{E}(G),\ \psi' \in \Psi(G',\psi),$$

which maps to (ψ, s) under the correspondence (1.4.11) of §1.4. The subset

$$\overline{S}_{\psi,\mathrm{ell}} = \{s \in \overline{S}_{\psi,\mathrm{ss}} :\ |Z(\overline{S}^0_{\psi,s})| < \infty\}$$

of $\overline{S}_{\psi,\mathrm{ss}}$ consists of those elements s such that the endoscopic datum $G' = G'_s$ belongs to $\mathcal{E}_{\mathrm{ell}}(G)$. Our concern will be the subset

$$\overline{S}'_{\psi,\mathrm{ell}} = \{s \in \overline{S}_{\psi,\mathrm{ell}} :\ G'_s \notin \mathcal{E}_{\mathrm{sim}}(G)\}$$

of $\overline{S}_{\psi,\mathrm{ell}}$. In the case that $G \in \widetilde{\mathcal{E}}_{\mathrm{sim}}(N)$, for example, $\overline{S}'_{\psi,\mathrm{ell}}$ is just the subset of elements $s \in \overline{S}_{\psi,\mathrm{ell}}$ with $s \neq 1$. Following notation from §4.1, we write

$$\mathcal{E}_{\psi,\mathrm{ell}} = \mathcal{E}(\overline{S}_{\psi,\mathrm{ell}})$$

and

$$\mathcal{E}'_{\psi,\mathrm{ell}} = \mathcal{E}(\overline{S}'_{\psi,\mathrm{ell}})$$

for the set of orbits in $\overline{S}_{\psi,\mathrm{ell}}$ and $\overline{S}'_{\psi,\mathrm{ell}}$ respectively under the action of $\overline{S}^0_\psi$ by conjugation. These are of course contained in the finite set

$$\mathcal{E}_\psi = \mathcal{E}(\overline{S}_\psi) = \mathcal{E}(\overline{S}_{\psi,\mathrm{ss}})$$

defined in §4.1. Our aim is to replace the double sum over G' and ψ' in our expression for (4.4.1) by a simple sum over $\mathcal{E}'_{\psi,\mathrm{ell}}$. This leads to some rescaling constants, which we need to compute.

Consider the set of pairs

$$Y'_{\mathrm{disc},\psi}(G) = \{y = (\psi_G, s_G)\},$$

where ψ_G is an actual L-homomorphism from $\mathcal{L}_\psi \times SL(2,\mathbb{C})$ to ${}^L G^0$ that maps to ψ, and s_G belongs to $\overline{S}'_{\psi_G,\mathrm{ell}}$. The centralizer set $\overline{S}'_{\psi_G,\mathrm{ell}}$ here is obviously the analogue of $\overline{S}'_{\psi,\mathrm{ell}}$, which is to say that it is defined directly for the L-homomorphism ψ_G rather than some unspecified representative of the equivalence class of L-homomorphisms $\widetilde{\psi}_G$ attached to ψ in §1.4. There is a natural left action of the group $\widehat{G}^0$ on $Y'_{\mathrm{ell}}(G,\psi)$, defined by conjugation on the two components of a pair y. We would like to replace the double sum over G' and ψ' in (4.4.2) by a sum over the corresponding set

$$\widehat{G}^0 \backslash\backslash Y'_{\mathrm{disc},\psi}(G)$$

of $\widehat{G}^0$-orbits in $Y'_{\mathrm{disc},\psi}(G)$. (We will denote general orbit spaces by a double bar, so we can reserve a single bar for the special case of cosets.)

Any pair $y = (\psi_G, s_G)$ in $Y'_{\mathrm{disc},\psi}(G)$ is the image of a unique pair $x = (G', \psi')$, under the correspondence (1.4.11). It is understood here that G' is simply an elliptic (nonsimple) endoscopic datum for G, taken up to translation of the associated semisimple element $s' \in \widehat{G}$ by $Z(\widehat{G}^0)^\Gamma$, and that ψ' is an actual L-homomorphism from $\mathcal{L}_\psi \times SL(2,\mathbb{C})$ to ${}^L G'$ that maps to ψ. However, it is the isomorphism class of (G', ψ'), as an element in $\mathcal{E}_{\mathrm{ell}}(G) \times \Psi(G',\psi)$, that indexes the double sum (4.2.2). This class maps to a subset of the $\widehat{G}^0$-orbit of (ψ_G, s_G). The summand of (G',ψ') actually depends only on the $\widehat{G}^0$-orbit, a property that follows easily from the fact that f belongs to the subspace $\widetilde{\mathcal{H}}(G)$ of $\mathcal{H}(G)$. We have therefore to count the number of isomorphism classes (G',ψ') that map to the $\widehat{G}^0$-orbit of (ψ_G, s_G).

To describe the $\widehat{G}^0$-orbit of $y = (\psi_G, s_G)$, we first note that the stabilizer of ψ_G in $\widehat{G}^0$ is the group

$$S^*_{\psi_G} = S_{\psi_G}(G^0) = \mathrm{Cent}\big(\mathrm{Im}(\psi_G), \widehat{G}^0\big).$$

It then follows that the stabilizer of y in $\widehat{G}^0$ is the stabilizer

$$S^+_y = S^+_{\psi_G, s_G} = \{g \in S^*_{\psi_G} : \ g s_G g^{-1} = s_G\}$$

of the $Z(\widehat{G}^0)^\Gamma$-coset s_G in $S^*_{\psi_G}$. The orbit of y in $\widehat{G}^0$ is therefore bijective with the quotient coset space $\widehat{G}^0/S^+_y$.

To describe the isomorphism classes (G',ψ') in the $\widehat{G}^0$-orbit of y, we note that the class of G' in $\mathcal{E}_{\mathrm{ell}}(G)$ can also be treated as a $\widehat{G}^0$-orbit. Its stabilizer in $\widehat{G}^0$ equals the subgroup $\mathrm{Aut}_G(G')$. (These assertions follow directly from the definitions in [**KS**, p. 18–19], and the fact that we are regarding G' as a

$Z(\widehat{G}^0)^\Gamma$-orbit of endoscopic data.) The natural mapping from the orbit of y to the class of G' can then be identified with the projection

$$\widehat{G}^0/S_y^+ \longrightarrow \widehat{G}^0/\mathrm{Aut}_G(G').$$

In particular, the stabilizer of ψ' in $\mathrm{Aut}_G(G')$ can be identified with S_y^+. However, the class of ψ' in $\Psi(G',\psi)$ is not its orbit under $\mathrm{Aut}_G(G')$, but rather its orbit (by conjugation) under the subgroup $\widehat{G}'$ of $\mathrm{Aut}_G(G')$. This is the reason that there are several isomorphism classes in the $\widehat{G}^0$-orbit of y. Now the $\widehat{G}'$-orbit of ψ' is the same as its orbit under the product

$$\mathrm{Int}_G(G') = \widehat{G}'\, Z(\widehat{G}^0)^\Gamma,$$

since $Z(\widehat{G}^0)^\Gamma$ commutes with the image of ψ'. We recall that $\mathrm{Int}_G(G')$ is a normal subgroup of $\mathrm{Aut}_G(G')$, whose quotient is the finite group $\mathrm{Out}_G(G')$. It follows from these remarks that the set of pairs (G',ψ') in the $\widehat{G}^0$-orbit of y is bijective with the set

$$\mathrm{Aut}_G(G')/S_y^+\,\mathrm{Int}_G(G') \cong \mathrm{Out}_G(G')/\big(S_y^+\,\mathrm{Int}_G(G')/\mathrm{Int}_G(G')\big).$$

Writing

$$S_y^+\,\mathrm{Int}_G(G')/\mathrm{Int}_G(G') \cong S_y^+/S_y^+ \cap \widehat{G}'\, Z(\widehat{G}^0)^\Gamma,$$

we see that the number of such pairs is equal to the quotient

$$|\mathrm{Out}_G(G')|\,|S_y^+/S_y^+ \cap \widehat{G}'\, Z(\widehat{G}^0)^\Gamma|^{-1}. \tag{4.4.5}$$

We have established that the double sum over G' and ψ' in (4.4.2) can be replaced by a sum over $\widehat{G}^0$-orbits of pairs $y = (\psi_G, s_G)$, provided that the summand is multiplied by (4.4.5). The next step is to replace the new sum by an iterated sum over ψ_G and s_G. The outer sum will be over the set $\Psi(G,\psi)$ of $\widehat{G}^0$-orbits of L-homomorphisms ψ_G that map to the given element $\psi \in \widetilde{\Psi}(G)$. Since the function lies in $\widetilde{\mathcal{H}}(G)$, the corresponding summand is independent of ψ_G, and the outer sum collapes. In other words, we can identify ψ_G with ψ, provided that we multiply by the order

$$m_\psi = |\Psi(G,\psi)|. \tag{4.4.6}$$

The stabilizer of ψ_G in $\widehat{G}^0$ is by definition the group

$$S_{\psi_G}(G^0) \cong S_\psi(G^0) = S_\psi^*.$$

The remaining inner sum can therefore be taken over the finite set

$$\begin{aligned} S_{\psi_G}(G^0)\backslash\backslash \overline{S}'_{G,\mathrm{ell}} &\cong \overline{S}_\psi^*\backslash\backslash S'_{\psi,\mathrm{ell}} \\ &\cong \overline{S}_\psi^*\backslash\backslash \overline{S}'_{\psi,\mathrm{ell}} \end{aligned}$$

of orbits $s_G = s$ in $\overline{S}'_{\psi,\mathrm{ell}}$, under action by conjugation of either of the groups S_ψ^* or $\overline{S}_\psi^* = S_\psi^*/Z(\widehat{G}^0)^\Gamma$.

We prefer to take orbits of the connected component $\overline{S}^0_\psi = (\overline{S}^*_\psi)^0$ of $\overline{S}^*_\psi$ rather than $\overline{S}^*_\psi$. The stabilizer in $\overline{S}^*_\psi$ of an element $s \in \overline{S}'_{\psi,\mathrm{ell}}$ is the centralizer

$$\overline{S}^+_{\psi,s} = S^+_{\psi,s}/Z(\widehat{G}^0)^\Gamma = \mathrm{Cent}(s, \overline{S}^*_\psi)$$

of s in $\overline{S}^*_\psi$. The $\overline{S}^*_\psi$-orbit of s is therefore bijective with $\overline{S}^*_\psi/\overline{S}^+_{\psi,s}$. The $\overline{S}^0_\psi$-orbit of s is bijective with $\overline{S}^0_\psi/\overline{S}_{\psi,s}$, where we recall that $\overline{S}_{\psi,s}$ is the centralizer of s in $\overline{S}^0_\psi$. We can therefore sum s over the set

$$\overline{S}^0_\psi \backslash\backslash \overline{S}'_{\psi,\mathrm{ell}} = \mathcal{E}(\overline{S}'_{\psi,\mathrm{ell}}) = \mathcal{E}'_{\psi,\mathrm{ell}}$$

of $\overline{S}^0_\psi$-orbits in $\overline{S}'_{\psi,\mathrm{ell}}$, rather than the set $\overline{S}^*_\psi \backslash\backslash \overline{S}'_{\psi,\mathrm{ell}}$ above, provided that we multiply the summand by the quotient

$$|\overline{S}^+_{\psi,s}/\overline{S}_{\psi,s}|\,|\overline{S}^*_\psi/\overline{S}^0_\psi|^{-1}. \tag{4.4.7}$$

The last few observations have been directed at the sum over G' and ψ' in (4.4.2). We have now established that this double sum can be replaced by a simple sum over the set $\mathcal{E}'_{\psi,\mathrm{ell}}$, provided that the summand is multiplied by the product of the factors (4.4.5), (4.4.6), and (4.4.7). The summand itself becomes the product of these three factors with the right hand sides of (4.4.4) and (4.4.3). We thus obtain an expanded formula for the difference (4.4.1) with which we started. Multiplying the various factors together, we see that (4.4.1) can be written as the sum over s in $\mathcal{E}'_{\psi,\mathrm{ell}}$ of the product of two expressions

$$|S^+_{\psi,s}/S^+_{\psi,s} \cap \widehat{G}'Z(\widehat{G}^0)^\Gamma|^{-1}\,|\mathcal{S}_{\psi'}|^{-1}\,|\overline{Z}(\widehat{G}')|^{-1}\,|\overline{S}^+_{\psi,s}/\overline{S}_{\psi,s}|$$

and

$$|\kappa_G|^{-1}\, m_\psi\, |\overline{S}^*_\psi/\overline{S}^0_\psi|^{-1}\, \sigma(\overline{S}^0_{\psi'})\, \varepsilon'(\psi')\, f'(\psi'),$$

in which (G', ψ') maps to the pair

$$y = (\psi, s) = (\psi_G, s_G).$$

We consider each of the two expressions in turn.

To simplify the first expression, we note that

$$\overline{S}^0_{\psi,s} = (\overline{S}^+_{\psi,s})^0 = \mathrm{Cent}(s, \overline{S}^0_\psi)^0,$$

and hence that

$$|\mathcal{S}_{\psi'}| = |\pi_0(\overline{S}_{\psi'})| = |\overline{S}^+_{\psi,s} \cap \overline{\widehat{G}}'/(\overline{S}^+_{\psi,s})^0\,\overline{Z}(\widehat{G}')|,$$

where $\overline{\widehat{G}}'$ denotes the quotient

$$\widehat{G}'Z(\widehat{G}^0)/Z(\widehat{G}^0) \cong \widehat{G}'/Z(\widehat{G}').$$

Consequently,

$$\begin{aligned}
&|S^+_{\psi,s}/S^+_{\psi,s}\cap \widehat{G}'Z(\widehat{G}^0)|^{-1}|\mathcal{S}_{\psi'}|^{-1}\\
=&|\overline{S}^+_{\psi,s}/\,\overline{S}^+_{\psi,s}\,\cap\,\overline{\widehat{G}}'|^{-1}|\overline{S}^+_{\psi,s}\,\cap\,\overline{\widehat{G}}'/(\overline{S}^+_{\psi,s})^0\,\overline{Z}(\widehat{G}')|^{-1}\\
=&|\overline{S}^+_{\psi,s}/\overline{S}^0_{\psi,s}\,\overline{Z}(\widehat{G}')|^{-1}.
\end{aligned}$$

The first expression therefore equals

$$|\overline{S}_{\psi,s}/\overline{S}^0_{\psi,s}\,\overline{Z}(\widehat{G}')|^{-1}|\overline{Z}(\widehat{G}')|^{-1},$$

a product that also equals

$$|\pi_0(\overline{S}_{\psi,s}|^{-1}|\overline{S}^0_{\psi,s}\cap\overline{Z}(\widehat{G}')|^{-1}.$$

The product of the first three factors in the second expression equals the constant C_ψ of (4.3.5), since

$$|\overline{S}^*_\psi/\overline{S}^0_\psi| = |\overline{S}_\psi/\overline{S}^0_\psi| = |\mathcal{S}_\psi|.$$

The fourth factor $\sigma(\overline{S}^0_{\psi'})$ in the expression has been defined (4.1.8) if $\overline{S}^0_{\psi'}$ is replaced by any complex, connected reductive group S_1. We recall the property (4.1.9) from §4.1, which asserts that

$$\sigma(S_1) = \sigma(S_1/Z_1)\,|Z_1|^{-1},$$

for any central subgroup Z_1 of S_1. We can therefore write

$$\begin{aligned}
\sigma(\overline{S}^0_{\psi'}) &= \sigma\big(\overline{S}^0_{\psi,s}/\overline{S}^0_{\psi,s}\cap\overline{Z}(\widehat{G}')\big)\\
&= \sigma(\overline{S}^0_{\psi,s})\,|\overline{S}^0_{\psi,s}\cap\overline{Z}(\widehat{G}')|.
\end{aligned}$$

Taking the product of the resulting two expressions, we conclude that the difference (4.4.1) equals

$$C_\psi \sum_{s\in\mathcal{E}'_{\psi,\mathrm{ell}}} |\pi_0(\overline{S}_{\psi,s})|^{-1}\,\sigma(\overline{S}^0_{\psi,s})\,\varepsilon'(\psi')\,f'(\psi'), \tag{4.4.8}$$

where (G',ψ') maps to the pair (ψ,s).

As at this point in the discussion of the last section, we will need a lemma on signs. Its purpose is to express the sign

$$\varepsilon'(\psi') = \varepsilon^{G'}_{\psi'}(s_{\psi'})$$

explicitly in terms of (ψ,s). To state it, we write x_s for the image of s in $\mathcal{S}_\psi$.

LEMMA 4.4.1. *The sign characters for G and G' satisfy the relation*

$$\varepsilon^{G'}_{\psi'}(s_{\psi'}) = \varepsilon^G_\psi(s_\psi x_s),$$

for any elements G', ψ' and s such that (G',ψ') maps to (ψ,s).

We shall give the proof of this lemma in §4.6, along with that of its predecessor Lemma 4.3.1 from the last section. Assuming it for now, we make the appropriate substitution in the expression (4.4.8) we have obtained for (4.4.1). We also write $f'(\psi')$ as $f'_G(\psi, s)$, according to the definition (4.2.5). We thus find that (4.4.1) equals

$$C_\psi \sum_{s \in \mathcal{E}'_{\psi,\mathrm{ell}}} |\pi_0(\overline{S}_{\psi,s})|^{-1}\, \sigma(\overline{S}^0_{\psi,s}) \varepsilon^G_\psi(s_\psi x_s)\, f'_G(\psi, s).$$

The last expression more closely resembles that of Lemma 4.3.2 if we fibre the sum over points $x \in \mathcal{S}_\psi$. Accordingly, we write

$$\mathcal{E}_\psi(x), \qquad x \in \mathcal{S}_\psi,$$

for the fibre of x under the canonical mapping from $\mathcal{E}_\psi$ to $\mathcal{S}_\psi$. We then set

$$\mathcal{E}_{\psi,\mathrm{ell}}(x) = \mathcal{E}_\psi(x) \cap \mathcal{E}_{\psi,\mathrm{ell}}$$

and

$$\mathcal{E}'_{\psi,\mathrm{ell}}(x) = \mathcal{E}_\psi(x) \cap \mathcal{E}'_{\psi,\mathrm{ell}}.$$

With this notation, we obtain the following result.

LEMMA 4.4.2. *The difference*

$$I_{\mathrm{disc},\psi}(f) - s^G_{\mathrm{disc},\psi}(f)$$

equals

$$C_\psi \sum_{x \in \mathcal{S}_\psi} \varepsilon^G_\psi(s_\psi x) \sum_{s \in \mathcal{E}'_{\psi,\mathrm{ell}}(x)} |\pi_0(\overline{S}_{\psi,s})|^{-1}\, \sigma(\overline{S}^0_{\psi,s})\, f'_G(\psi, s),$$

for any $\psi \in \widetilde{\Psi}(N)$ and $f \in \widetilde{\mathcal{H}}(G)$. □

The existence of the linear form $f'(\psi')$, for any G' in the complement of $\mathcal{E}_{\mathrm{sim}}(G)$, is part of our basic induction hypothesis. Suppose that we happen to know also that $f'(\psi')$ is defined for any $G' \in \mathcal{E}_{\mathrm{sim}}(G)$ and $\psi' \in \Psi(G', \psi)$. This is primarily a local assumption, namely that for any valuation v, the localization ψ'_v belongs to $\Psi^+(G'_v, \psi_v)$, and Theorem 2.2.1(a) holds for (G'_v, ψ'_v). It is weaker than asking that the stable multiplicity formula of Theorem 4.1.2 hold for G' and ψ'. However, it does allow us to define the difference

$$(4.4.9) \qquad {}^0\widehat{S}'_{\mathrm{disc},\psi}(f') = \widehat{S}'_{\mathrm{disc},\psi}(f') - \sum_{\psi' \in \Psi(G',\psi)} |\mathcal{S}_{\psi'}|^{-1} \sigma(\overline{S}^0_{\psi'}) \varepsilon'(\psi')\, f'(\psi')$$

between $\widehat{S}'_{\mathrm{disc},\psi}(f')$ and its expected value, as well as the sum

$$(4.4.10) \qquad {}^0s^G_{\mathrm{disc},\psi}(f) = \sum_{G' \in \mathcal{E}_{\mathrm{sim}}(G)} \iota(G, G')\, {}^0\widehat{S}'_{\mathrm{disc},\psi}(f'),$$

which represents the difference between $s^G_{\mathrm{disc},\psi}(f)$ and its expected value.

The difference

$$s^G_{\mathrm{disc},\psi}(f) - {}^0s^G_{\mathrm{disc},\psi}(f)$$

equals the sum over $G' \in \mathcal{E}_{\mathrm{sim}}(G)$ and $\psi' \in \Psi(G', \psi)$ of the product of the right hand sides of (4.4.3) and (4.4.4). Using (4.2.5), (4.3.5) and other definitions above, we can write it as a sum similar to that of Lemma 4.4.2. Indeed, we have only to apply the discussion above to summands indexed by pairs (G', ψ') with $G' \in \mathcal{E}_{\mathrm{sim}}(G)$. These indices are attached to pairs (ψ, s) in which s belongs to the complement of $\overline{S}'_{\psi,\mathrm{ell}}$ in $\overline{S}_{\psi,\mathrm{ell}}$. In particular, the existence of the linear forms $f'(\psi')$ for $G' \in \mathcal{E}_{\mathrm{sim}}(G)$ amounts to the existence of $f'_G(\psi, s)$, as a function defined for s in the entire domain $\overline{S}_{\psi,\mathrm{ell}}$. The arguments above carry over to these summands without change. It follows that the formula of Lemma 4.4.2 remains valid with ${}^0 s^G_{\psi,\mathrm{disc}}(f)$ in place of $s^G_{\psi,\mathrm{disc}}(f)$, and $\mathcal{E}_{\psi,\mathrm{ell}}(x)$ in place of $\mathcal{E}'_{\psi,\mathrm{ell}}(x)$.

It is convenient to change variables in the resulting sum over s. The set of $\overline{S}^0_\psi$-orbits in $\overline{S}_\psi(x)_{\mathrm{ell}}$ is invariant under translation by the point s_ψ. Since s_ψ lies in the center of $\overline{S}^*_\psi$, the centralizer $\overline{S}_{\psi,s}$ equals $\overline{S}_{\psi,s_\psi s}$. A change of variables from s to $s_\psi s$ therefore yields the expression

$$C_\psi \sum_{x \in \mathcal{S}_\psi} \sum_{s \in \mathcal{E}_{\psi,\mathrm{ell}}(x)} |\pi_0(\overline{S}_{\psi,s})|^{-1} \sigma(\overline{S}^0_{\psi,s}) \, \varepsilon^G_\psi(x) \, f'_G(\psi, s_\psi s).$$

This is equal to the difference between $I_{\mathrm{disc},\psi}(f)$ and ${}^0 s^G_{\psi,\mathrm{disc}}(f)$.

Suppose that in addition to being well defined, the function $f'_G(\psi, s)$ depends only on the image x of s in $\mathcal{S}_\psi$. This is implied by the global intertwining identity, but is again weaker. In fact, it is easily derived from basic definitions, as we shall see in the next section. The function $f'_G(\psi, s_\psi s)$ above can then be taken outside the last sum over s. Lemma 4.4.2, or rather its minor extension above, can then be formulated as follows.

COROLLARY 4.4.3. *Suppose that the function*

$$f'_G(\psi, x) = f'_G(\psi, s), \qquad x \in \mathcal{S}_\psi, \ s \in \mathcal{E}_\psi(x),$$

is defined, and depends only on x. Then the difference

$$I_{\mathrm{disc},\psi}(f) - {}^0 s^G_{\mathrm{disc},\psi}(f)$$

equals

$$C_\psi \sum_{x \in \mathcal{S}_\psi} e_\psi(x) \, \varepsilon^G_\psi(x) \, f'_G(\psi, s_\psi x), \tag{4.4.11}$$

where

$$e_\psi(x) = \sum_{s \in \mathcal{E}_{\psi,\mathrm{ell}}(x)} |\pi_0(\overline{S}_{\psi,s})|^{-1} \sigma(\overline{S}^0_{\psi,s}). \qquad \square$$

In Corollaries 4.3.3 and 4.4.3, we now have two parallel expansions for the discrete part of the trace formula, or rather its ψ-component $I_{\mathrm{disc},\psi}(f)$. The comparison of these expansions will be a central preoccupation from this point on. As an elementary example of what is to come, let us consider the simplest of cases.

Suppose that $\psi \in \widetilde{\Psi}(N)$ and $G \in \widetilde{\mathcal{E}}_{\mathrm{sim}}(N)$ are such that ψ does *not* belong to $\widetilde{\Psi}(G)$. Then

$$^0 r^G_{\mathrm{disc},\psi}(f) = r^G_{\mathrm{disc},\psi}(f) = \mathrm{tr}\big(R^G_{\mathrm{disc},\psi}(f)\big)$$

and

$$^0 s^G_{\mathrm{disc},\psi}(f) = s^G_{\mathrm{disc},\psi}(f) = S^G_{\mathrm{disc},\psi}(f) = {}^0 S^G_{\mathrm{disc},\psi}(f),$$

since the expected values of $r^G_{\mathrm{disc},\psi}(f)$ and $s^G_{\mathrm{disc},\psi}(f)$ are by definition equal to 0. Strictly speaking, the premises of the two corollaries are not valid, since the relevant functions are not defined in the case $\psi \notin \widetilde{\Psi}(G)$. However, the expansions of Lemmas 4.3.2 and 4.4.2 hold. The coefficient C_ψ in these expansions vanishes, since the factor m_ψ in (4.3.5) equals 0. It follows that

(4.4.12)
$$\mathrm{tr}\big(R^G_{\mathrm{disc},\psi}(f)\big) = S^G_{\mathrm{disc},\psi}(f) = {}^0 S^G_{\mathrm{disc},\psi}(f), \qquad \psi \notin \widetilde{\Psi}(G),\ f \in \widetilde{\mathcal{H}}(G).$$

In particular, the trace of $R^G_{\mathrm{disc},\psi}(f)$ is a stable linear form on $\widetilde{\mathcal{H}}(G)$ in this case. We of course expect that each term in the identity vanishes, but we are not yet ready to prove this. Observe that we could also have derived (4.4.12) directly from our induction hypotheses and the original expansions (4.1.1) and (4.1.2). In particular, the identity does not depend on the sign formulas of Lemmas 4.3.1 and 4.4.1.

4.5. The comparison

We shall now compare the two expansions we have established. In particular, we shall try to show that for many ψ, the conditions of the parallel expansions of Corollaries 4.3.3 and 4.4.3 are valid. We recall that ψ belongs to the set $\widetilde{\Psi}(N) = \Psi\big(\widetilde{G}(N)\big)$ of parameters for the twisted general linear group $\widetilde{G}(N) = GL(N) \rtimes \widetilde{\theta}(N)$ over the global field F. The most difficult case is when ψ lies in $\widetilde{\Psi}_{\mathrm{ell}}(N)$, which would mean that ψ belongs to $\widetilde{\Psi}_2(G)$, for some datum G (a priori unique) in $\widetilde{\mathcal{E}}_{\mathrm{ell}}(N)$. Its analysis will be taken up in later chapters. In this section, we assume that ψ lies in the complement of $\widetilde{\Psi}_{\mathrm{ell}}(N)$.

The two corollaries were based on the further assumption that two functions are defined, and depend only on a point $x \in \mathcal{S}_\psi$. Suppose that ψ belongs to $\widetilde{\Psi}(G)$, where as in §4.3 and §4.4, G is either a given datum in $\widetilde{\mathcal{E}}_{\mathrm{sim}}(N)$ or equal to $\widetilde{G}(N)$. Our condition that ψ lies in the complement of $\widetilde{\Psi}_{\mathrm{ell}}(N)$ implies that the functions are at least defined. For the function $f_G(\psi, u)$ of Corollary 4.3.3, this follows from our induction hypothesis that the pairing $\langle \widetilde{u}, \widetilde{\pi}_M \rangle$ in the definition $(4.2.4)'$ exists for the Levi subgroup $M \neq G^0$ attached to ψ. The function $f'_G(\psi, s)$ of Corollary 4.4.3 is also amenable to induction. Its existence reduces to the analogue of Theorem 2.2.1(a) for the group $M \neq G^0$, as the reader can readily verify. Alternatively, the existence of $f'_G(\psi, s)$ can be regarded as part of an analysis of the linear form $f'(\psi')$ in the definition $(4.2.5)'$ that we will undertake presently (following the statements of formulas (4.5.7) and (4.5.8) below).

The main part of the assumption underlying the two corollaries is that each of the two functions depends only on the image x of u or s in $\mathcal{S}_\psi$. To see how the comparison works, we temporarily take on a broader assumption, which includes the premise of Corollaries 4.3.3 and 4.4.3. We suppose for the moment that the global intertwining relation holds for any G in either of the corollaries. In other words, we suppose that

$$f'_G(\psi, s_\psi x) = f_G(\psi, x), \qquad f \in \widetilde{\mathcal{H}}(G),\ x \in \mathcal{S}_\psi, \tag{4.5.1}$$

where G is either an element in $\widetilde{\mathcal{E}}_{\mathrm{sim}}(N)$ with $\psi \in \widetilde{\Psi}(G)$, or is equal to $\widetilde{G}(N)$ itself.

According to Corollary 4.4.3, the difference

$$I_{\mathrm{disc},\psi}(f) - {}^0 s^G_{\mathrm{disc},\psi}(f), \qquad f \in \widetilde{\mathcal{H}}(G), \tag{4.5.2}$$

equals

$$C_\psi \sum_{x \in \mathcal{S}_\psi} e_\psi(x)\, \varepsilon^G_\psi(x)\, f'_G(\psi, s_\psi x).$$

Corollary 4.3.3 tells us that the difference

$$I_{\mathrm{disc},\psi}(f) - {}^0 r^G_{\mathrm{disc},\psi}(f), \qquad f \in \widetilde{\mathcal{H}}(G), \tag{4.5.3}$$

equals

$$C_\psi \sum_{x \in \mathcal{S}_\psi} i_\psi(x)\, \varepsilon^G_\psi(x)\, f_G(\psi, x).$$

We have now reached the point where we appeal to Proposition 4.1.1. This result, which we recall is a kind of miniature replica of the basic identity between (4.1.2) and (4.1.1), tells us that the coefficients $e_\psi(x)$ and $i_\psi(x)$ in the two expressions are equal. It follows from (4.5.1) that the expressions themselves are equal. The two differences (4.5.2) and (4.5.3) are therefore equal. We conclude from the definitions (4.4.10) and (4.3.8) that

$$|\kappa_G|^{-1} \operatorname{tr}\big(R^G_{\mathrm{disc},\psi}(f)\big) = \sum_{G' \in \mathcal{E}_{\mathrm{sim}}(G)} \iota(G, G')\, {}^0\widehat{S}'_{\mathrm{disc},\psi}(f), \tag{4.5.4}$$

for any G and any $f \in \widetilde{\mathcal{H}}(G)$.

Suppose first that G is one of the groups in $\widetilde{\mathcal{E}}_{\mathrm{sim}}(N)$. Then $G' = G$ is the only element in the set $\mathcal{E}_{\mathrm{sim}}(G)$. The equation (4.5.4) reduces to

$$\operatorname{tr}\big(R^G_{\mathrm{disc},\psi}(f)\big) = {}^0 S^G_{\mathrm{disc},\psi}(f), \tag{4.5.5}$$

since $|\kappa_G| = 1$ in this case. In particular, the trace of $R^G_{\mathrm{disc},\psi}(f)$ is a stable linear form on $\widetilde{\mathcal{H}}(G)$. Notice that (4.5.5) has the same form as the identity (4.4.12) established at the end of the last section, in case ψ does not lie in $\widetilde{\Psi}(G)$. Consequently, (4.5.5) is valid if G is replaced by any datum in $\widetilde{\mathcal{E}}_{\mathrm{sim}}(N)$.

Next, suppose that G equals $\widetilde{G}(N)$. The character

$$\operatorname{tr}\bigl(R^G_{\mathrm{disc},\psi}(f)\bigr), \qquad f \in \mathcal{H}(G),$$

then vanishes, since our condition that $\psi \notin \widetilde{\Psi}_{\mathrm{ell}}(N)$ means that ψ cannot contribute to the discrete spectrum of $\widetilde{G}(N)^0 = GL(N)$. The general identity (4.5.4) reduces to

$$\sum_{G' \in \mathcal{E}_{\mathrm{sim}}(G)} \iota(G,G')\, {}^0\widehat{S}'_{\mathrm{disc},\psi'}(f') = 0 \tag{4.5.6}$$

in this case. The groups G' here represent the data in $\widetilde{\mathcal{E}}_{\mathrm{sim}}(N)$ we have just treated, and for which we have established (under the condition (4.5.1)) that

$${}^0\widehat{S}'_{\mathrm{disc},\psi}(f') = \operatorname{tr}\bigl(R^{G'}_{\mathrm{disc},\psi}(h')\bigr),$$

for any function $h' \in \widetilde{\mathcal{H}}(G')$ whose image in $\widetilde{\mathcal{S}}(G')$ equals f'.

Continuing to suppose that the general condition (4.5.1) is valid, we combine (4.5.5) and (4.5.6), and then let G range over all of the elements in $\widetilde{\mathcal{E}}_{\mathrm{sim}}(N)$. Appealing to Proposition 2.1.1, or rather its global extension described in §3.4, we see that

$$\sum_{G \in \widetilde{\mathcal{E}}_{\mathrm{sim}}(N)} \widetilde{\iota}(N,G)\operatorname{tr}\bigl(R^G_{\mathrm{disc},\psi}(f)\bigr) = 0,$$

for any compatible family of functions

$$\bigl\{f \in \widetilde{\mathcal{H}}(G):\ G \in \widetilde{\mathcal{E}}_{\mathrm{ell}}(N)\bigr\}.$$

(The groups G that represent elements in the complement of $\widetilde{\mathcal{E}}_{\mathrm{sim}}(N)$ in $\widetilde{\mathcal{E}}_{\mathrm{ell}}(N)$ are part of the original definition (3.4.11), but they play no role here. For since they are composite, we can assume inductively that the corresponding representations $R^G_{\mathrm{disc},\psi}$ are zero.) The coefficients

$$\widetilde{\iota}(N,G) = \iota\bigl(\widetilde{G}(N), G\bigr)$$

are positive. The coefficients of the decompositon of $R^G_{\mathrm{disc},\psi}(f)$ into irreducible representations are of course also positive. We can therefore apply Proposition 3.5.1, the basic vanishing property we have established for compatible families. It tells us that

$$R^G_{\mathrm{disc},\psi}(f) = 0, \qquad f \in \widetilde{\mathcal{H}}(G), \tag{4.5.7}$$

for any $G \in \widetilde{\mathcal{E}}_{\mathrm{sim}}(N)$. In other words, ψ does not contribute to the discrete spectrum of G. This in turn implies that

$${}^0S^G_{\mathrm{disc},\psi}(f) = 0, \qquad f \in \widetilde{\mathcal{H}}(G). \tag{4.5.8}$$

In other words, the formula of Theorem 4.1.2 for $S^G_{\mathrm{disc},\psi}(f)$ is valid for any $G \in \widetilde{\mathcal{E}}_{\mathrm{sim}}(N)$.

We shall see that the comparison just completed has broad application. In particular, we shall show that for many parameters $\psi \in \widetilde{\Psi}(N)$, the condition (4.5.1) is indeed valid for every G with $\psi \in \widetilde{\Psi}(G)$. This will allow us to conclude that the global theorems hold for the given parameter $\psi \notin \widetilde{\Psi}_{\mathrm{ell}}(N)$.

The putative identity (4.5.1) includes the condition that each side is a well defined function of $x \in \mathcal{S}_\psi$. In other words, each side depends only on the image in $\mathcal{S}_\psi$ of the datum in terms of which it is defined. This is of course the premise on which the two Corollaries 4.3.3 and 4.4.3 were based. It is not difficult to verify much of it directly. We shall do so, as a necessary prelude to being able to establish cases of (4.5.1) by induction.

The left hand side $f'_G(\psi, s_\psi x)$ of (4.5.1) can be treated uniformly for $G \in \widetilde{\mathcal{E}}_{\mathrm{sim}}(N)$ or $G = \widetilde{G}(N)$. It is defined $(4.2.5)'$ in terms of a semisimple element $s \in \overline{S}_\psi$ whose projection onto $\mathcal{S}_\psi$ equals x. It equals

$$^*f'(^*\psi') = f^{*G'}(^*\psi'), \qquad f \in \widetilde{\mathcal{H}}(G),$$

where $(^*G', {}^*\psi')$ is the preimage of the pair

$$(\psi, {}^*s), \qquad {}^*s = s_\psi s,$$

under the general correspondence

$$(G', \psi') \longrightarrow (\psi, s).$$

We shall establish for any G that $f'(\psi')$ depends only on the image x of s.

Note first that if s is replaced by an $\overline{S}{}^0_\psi$-conjugate s_1, (G', ψ') is replaced by its image (G'_1, ψ'_1) under an isomorphism of endoscopic data. It follows from the definitions that

$$f'_1(\psi'_1) = f'(\psi'), \qquad f'_1 = f^{G'_1}.$$

Since any automorphism of the complex reductive group $\overline{S}{}^0_\psi$ stablilizes a conjugate of $(\overline{T}_\psi, \overline{B}_\psi)$, we can choose s_1 so that $\mathrm{Int}(s_1)$ stablilizes $(\overline{T}_\psi, \overline{B}_\psi)$ itself. With this property, the representative

$$s_x = s_1$$

of x in $\overline{S}_\psi$ is determined up to a $\overline{T}_\psi$-translate. In particular, the complex torus

$$\overline{T}_{\psi,x} = \mathrm{Cent}(s_x, \overline{T}_\psi)^0$$

in $\overline{T}_\psi$ is uniquely determined by x. Now any point in $\overline{T}_\psi s_x$ can be written in the form

$$ts_xt^{-1}t_x = ts_xt_xt^{-1}, \qquad t \in \overline{T}_\psi,\ t_x \in \overline{T}_{\psi,x}.$$

It will therefore be enough to show that the linear form

$$f'_x(\psi'_x) = f'_1(\psi'_1)$$

remains unchanged under translation of s_x by any element $t_x \in \overline{T}_{\psi,x}$.

The centralizer $\widehat{M}{}^0_x$ of $\overline{T}_{\psi,x}$ in $\widehat{G}{}^0$ is a Γ_F-stable Levi subgroup of $\widehat{G}{}^0$ that contains the dual $\widehat{M}$ of the underlying Levi subgroup M. Since the automorphism $\mathrm{Int}(s_x)$ centralizes $\overline{T}_{\psi,x}$, it stabilizes not only $\widehat{M}_x^0$, but also some parabolic subgroup $\widehat{P}_x^0 \in \mathcal{P}(\widehat{M}_x^0)$. The $\widehat{M}_x^0$-torsor

$$\widehat{M}_x = \widehat{M}_x^0 \rtimes \mathrm{Int}(s_x)$$

can therefore be treated as a Levi subset of $\widehat{G}$, which is in turn dual to a Levi subset M_x of G. (Remember that we are including the case $G = \widetilde{G}(N)$ in our analysis. As in §2.2, we are using terminology here that originated in [**A5**, §1].) Now (ψ, s) is the image of a pair

$$(\psi_{M_x}, s_{M_x}), \qquad \psi_{M_x} \in \widetilde{\Psi}(M_x),\ s_{M_x} \in S_{\psi_{M_x}},$$

attached to M_x under the L-embeddings ${}^L M_x^0 \subset {}^L G^0$ and ${}^L M_x \subset {}^L G$. This pair is in turn the image of an endoscopic pair (M_x', ψ_{M_x}') for M_x. In particular, we obtain an elliptic endoscopic datum M_x' for M_x, which can be identified with a Levi subgroup of G'. It follows that

$$f'(\psi') = f'_{M_x}(\psi'_{M_x}),$$

where

$$f'_{M_x} = f^{M_x'} = (f')_{M_x'}$$

is the transfer of f to M_x'. Since $f'_{M_x}(\psi'_{M_x})$ does not change under translation of s_x by $\overline{T}_{\psi,x}$, the same is true of $f_x'(\psi_x')$. We conclude that the function

$$f'_G(\psi, x) = f'(\psi') = f'_x(\psi'_x)$$

does indeed depend only on x. So therefore does the left hand side $f'_G(\psi, s_\psi x)$ of (4.5.1).

The right hand side $f_G(\psi, x)$ of (4.5.1) is less clear-cut. Recall that it is defined (4.2.4)$'$ in terms of a point u in $\mathfrak{N}_\psi$ whose projection x_u onto $\mathcal{S}_\psi$ equals x. The fibre of x under this projection is a torsor under the action of the subgroup W_ψ^0 of $\mathfrak{N}_\psi^*$. (We have to include the subscript $*$, since the case $G = \widetilde{G}(N)$ remains a part of the discussion.)

The dependence of $f_G(\psi, x)$ on u is twofold. It depends on its projection $w = w_u$ onto W_ψ, through the normalized intertwining operator in the trace

$$\mathrm{tr}\big(R_P(w, \widetilde{\pi}_M, \psi_M)\mathcal{I}_P(\pi_M, f)\big),$$

and on the value at $\widetilde{u}$ of $\widetilde{\xi} = \langle\,\cdot, \widetilde{\pi}_M\rangle$, an extension of the character $\langle\,\cdot, \pi_M\rangle$ from $\mathcal{S}_\psi^1$ to $\mathcal{S}_\psi^1 u$ that in turn determines the associated extension $\widetilde{\pi}_M$ of the representation π_M. Since $\mathcal{S}_\psi^1 \times W_\psi^0$ is a normal subgroup of $\mathfrak{N}_\psi^*$, we can assume that each $\widetilde{\xi}$ is constant on the W_ψ^0-orbit of u. That is, we can arrange that

$$\widetilde{\xi}(\widetilde{w^0 u}) = \widetilde{\xi}(\widetilde{w}^0\, \widetilde{u}) = \widetilde{\xi}(\widetilde{u}), \qquad w^0 \in W_\psi^0.$$

In particular, in the case that u lies in $\mathcal{S}_\psi^1 \times W_\psi^0$, we take $\widetilde{\xi}$ to be the product of the character $\langle\,\cdot, \pi_M\rangle$ on $\mathcal{S}_\psi^1$ with the trivial character on W_ψ^0.

We then conclude, from the earlier stages of the proof of Proposition 2.4.3 for example, that

$$\widetilde{\pi}_M(w^0 w) = \widetilde{\pi}_M(w),$$

and hence that the corresponding intertwining operators for G satisfy

$$R_P(w^0 w, \widetilde{\pi}_M, \psi_M) = R_P(w^0, \widetilde{\pi}_M, \psi_M)\, R_P(w, \widetilde{\pi}_M, \psi_M), \qquad w^0 \in W^0_\psi.$$

The condition that $f_G(\psi, x)$ depend only on x can at this point be seen to be equivalent to the requirement that the operators $R_P(w^0, \widetilde{\pi}_M, \psi_M)$ equal 1 for all $w^0 \in W^0_\psi$. If the operators are all trivial, the condition on $f_G(\psi, x)$ follows immediately from (4.2.4)$'$ and the conventions above. The other direction is perhaps less clear, since f ranges only over the subspace $\widetilde{\mathcal{H}}(G)$ of $\mathcal{H}(G)$, but it can be deduced from the fact that $R_P(w^0, \widetilde{\pi}_M, \psi_M)$ is unitary. Since we will ultimately establish the condition on $f_G(\psi, x)$ by showing directly that the operators $R_P(w^0, \widetilde{\pi}_M, \psi_M)$ are all trivial, there will be no harm in assuming henceforth that the two conditions are indeed equivalent. We will not be in a position, though, to establish either condition in general for some time.

Suppose however that $G = \widetilde{G}(N)$. The global diagram (4.2.3) for $\widetilde{G}(N)$ maps into the local diagram (2.4.3) attached to any completion ψ_v. In particular, we have an injection $w^0 \longrightarrow w^0_v$ from W^0_ψ to $W^0_{\psi_v}$, and a decomposition

$$R_P(w^0, \widetilde{\pi}_M, \psi) = \bigotimes_v R_P(w^0_v, \widetilde{\pi}_{M,v}, \psi_v).$$

The factors on the right are local normalized intertwining operators for $G^0 = GL(N)$. It follows from Theorem 2.5.3 that they are all equal to the identity. The global operator $R_P(w^0, \widetilde{\pi}_M, \psi)$ is therefore also equal to 1. The right hand side $f_G(\psi, x)$ of (4.5.1) therefore does depend only on x, in the case $G = \widetilde{G}(N)$.

Consider the other case that $G \in \widetilde{\mathcal{E}}_{\mathrm{sim}}(N)$. As the Weyl group of $(\overline{S}^0_\psi, \overline{T}_\psi)$, W^0_ψ is generated by simple reflections $\{w^0_\alpha\}$. Suppose that

$$\dim(\overline{T}_\psi) \geqslant 2.$$

Then any w^0_α centralizes a torus of positive dimension in $\overline{T}_\psi$. The centralizer of this torus in $\widehat{G}$ is a proper Levi subgroup $\widehat{M}_\alpha$ of $\widehat{G}$ that contains $\widehat{M}$, and is dual to a Levi subgroup M_α of G that contains M. If ψ_α is the image in $\widetilde{\Psi}(M_\alpha)$ of the chosen parameter $\psi_M \in \Psi_2(M, \psi)$, w^0_α belongs to the Weyl group of $(\overline{S}^0_{\psi_\alpha}, \overline{T}_\psi)$. We can assume inductively that

$$R_P(w^0_\alpha, \widetilde{\pi}_M, \psi_M) = R_{P \cap M_\alpha}(w^0_\alpha, \widetilde{\pi}_M, \psi_\alpha) = 1.$$

It follows that $R_P(w^0, \widetilde{\pi}_M, \psi_M)$ equals 1 for any $w^0 \in W^0_\psi$. Thus, the right hand side $f_G(\psi, x)$ of (4.5.1) again depends only on x, in case $G \in \widetilde{\mathcal{E}}_{\mathrm{sim}}(N)$ and $\dim(\overline{T}_\psi) \geqslant 2$. It is the complementary case that $\dim(\overline{T}_\psi) = 1$ that causes difficulty. We shall treat some (though not all) of it by a more sophisticated comparison argument later in the section.

How do the groups $S_\psi = S_\psi(G)$ vary, as G ranges over elements $G \in \widetilde{\mathcal{E}}_{\mathrm{sim}}(N)$ with $\psi \in \widetilde{\Psi}(G)$? Suppose for a moment that G is fixed. The centralizer $S_\psi = S_\psi(G)$ is then equal to a product

$$\Big(\prod_{i \in I_\psi^+(G)} O(\ell_i, \mathbb{C})\Big)_\psi^+ \times \Big(\prod_{i \in I_\psi^-(G)} Sp(\ell_i, \mathbb{C})\Big) \times \Big(\prod_{j \in J_\psi} GL(\ell_j, \mathbb{C})\Big), \tag{4.5.9}$$

in the notation of §1.4. As we observed in §3.4, the condition that $\psi \in \widetilde{\Psi}(G)$ implies that the character η_G equals η_ψ. If η_ψ is nontrivial or N is odd, G is the only datum in $\widetilde{\mathcal{E}}_{\mathrm{sim}}(N)$ with $\eta_G = \eta_\psi$. If η_ψ is trivial and N is even, however, there is a second datum $G^\vee \in \widetilde{\mathcal{E}}_{\mathrm{sim}}(N)$ with $\eta_{G^\vee} = \eta_\psi$. The parameter ψ then belongs to $\widetilde{\Psi}(G^\vee)$ if and only if ℓ_i is even for each $i \in I_\psi^+(G)$. With this condition, we have $I_\psi^+(G^\vee) = I_\psi^-(G)$ and $I_\psi^-(G^\vee) = I_\psi^+(G)$, and the centralizer $S_\psi^\vee = S_\psi^\vee(G^\vee)$ equals

$$\Big(\prod_{i \in I_\psi^+(G)} Sp(\ell_i, \mathbb{C})\Big) \times \Big(\prod_{i \in I_\psi^-(G)} O(\ell_i, \mathbb{C})\Big)_\psi^+ \times \Big(\prod_{j \in J_\psi} GL(\ell_j, \mathbb{C})\Big). \tag{4.5.10}$$

PROPOSITION 4.5.1. *Suppose that ψ lies in the complement of $\widetilde{\Psi}_{\mathrm{ell}}(N)$ in $\widetilde{\Psi}(N)$, and in addition, that it lies in the complement of $\widetilde{\Psi}_{\mathrm{ell}}(G)$ for every $G \in \widetilde{\mathcal{E}}_{\mathrm{sim}}(N)$. Then*

$$\mathrm{tr}\big(R^G_{\mathrm{disc},\psi}(f)\big) = 0 = {}^0S^G_{\mathrm{disc},\psi}(f), \qquad f \in \widetilde{\mathcal{H}}(G),$$

for any $G \in \widetilde{\mathcal{E}}_{\mathrm{sim}}(N)$. In other words, the global Theorems 1.5.2 *and* 4.1.2 *hold for G and ψ.*

REMARKS. 1. The first condition on ψ has been a de facto hypothesis for the last three sections. The second condition is new, and has been imposed in order to avoid subtleties that will have to be treated later. It does not imply the first, since it does not preclude parameters $\psi \in \widetilde{\Psi}_2(G)$, for elements G in the complement of $\widetilde{\mathcal{E}}_{\mathrm{sim}}(N)$ in $\widetilde{\mathcal{E}}_{\mathrm{ell}}(N)$.

2. There are other global theorems besides the two mentioned in the statement of the proposition that must be established in general. However, Theorem 1.5.3 applies to parameters that are simple (and generic), and therefore lie in $\widetilde{\Psi}_{\mathrm{ell}}(N)$. It is irrelevant to the case at hand. The same is true of the global Theorem 4.2.2. The remaining global assertion is the intertwining relation of Corollary 4.2.1, for both the case $G \in \widetilde{\mathcal{E}}_{\mathrm{sim}}(N)$ and $G = \widetilde{G}(N)$. This is just the putative identity (4.5.1). If we are able to establish it directly, we will thus obtain not just the statement of the proposition, but in fact all of the global results that pertain to ψ.

PROOF. We shall try to establish (4.5.1) directly by induction, and thereby obtain the required assertion from the comparison above. We shall find that this is possible for many, but not all of the given parameters ψ. There will be some cases that require a more delicate comparison. To see where the problems arise, let us begin with any parameter ψ in the complement of $\widetilde{\Psi}_{\mathrm{ell}}(N)$, and see how far we can go.

We have to verify (4.5.1) in both the case $G \in \widetilde{\mathcal{E}}_{\mathrm{sim}}(N)$ and $G = \widetilde{G}(N)$. The identity makes sense only if the set $\mathcal{S}_\psi = \mathcal{S}_\psi(G)$ is defined, which is to say that ψ belongs to $\widetilde{\Psi}(G)$. This condition always holds if $G = \widetilde{G}(N)$. However, it is quite possible that ψ not lie in $\widetilde{\Psi}(G)$ for any $G \in \widetilde{\mathcal{E}}_{\mathrm{sim}}(N)$. If this is the case, (4.5.5) is still valid for each G, as we saw in (4.4.12), and the comparison that yields the required identities (4.5.7) and (4.5.8) can be carried out. We shall therefore assume implicitly that $\psi \in \widetilde{\Psi}(G)$, for any of the data $G \in \mathcal{E}_{\mathrm{sim}}(N)$ we consider below.

Assume for the present that

(i) $$\dim(\overline{T}_\psi) \geqslant 2, \qquad G \in \widetilde{\mathcal{E}}_{\mathrm{sim}}(N),$$

as in the remarks prior to the statement of the proposition. Then as we have seen in all cases, the two sides of (4.5.1) each depend only on x. Moreover, the function on the left hand side satisfies the relation

$$f'_G(\psi, x) = f'(\psi') = f'_x(\psi'_x) = f'_{M_x}(\psi_{M_x}, x),$$

for $M_x \subset G$ and $\psi_{M_x} \in \widetilde{\Psi}(M_x)$ as in the earlier remarks. It follows easily from the definitions that $M_{s_\psi x} = M_x$ and $\psi_{s_\psi x} = \psi_x$. The left hand side itself therefore satisfies

$$f'_G(\psi, s_\psi x) = f'_{M_x}(\psi_{M_x}, s_\psi x).$$

On the other hand, the earlier representative s_x of x in $\overline{S}_\psi$ lies in the subset $\overline{N}_\psi$ of $\overline{S}_\psi$. We take its $\overline{T}_\psi$-coset u_x to represent x in $\mathfrak{N}_\psi$, and on the right hand side of (4.5.1). The coset u_x centralizes the complex torus

$$\overline{T}_{\psi,x} = \mathrm{Cent}(s_x, \overline{T}_\psi)^0$$

in $\overline{T}_\psi$. Since $\overline{T}_{\psi,x}$ can be identitied with the Γ-split part of the centralizer of $\widehat{M}_x$, the right hand side satisfies the relation

$$f_G(\psi, x) = f_{M_x}(\psi_{M_x}, x).$$

Assume also that

(ii) $$\dim(\overline{T}_{\psi,x}) \geqslant 1, \qquad G \in \widetilde{\mathcal{E}}_{\mathrm{sim}}(N), \ x \in \mathcal{S}_\psi.$$

Since the center of any $G \in \widetilde{\mathcal{E}}_{\mathrm{sim}}(N)$ is finite, M_x is a proper Levi subgroup of G in this case. Consider the other case that $G = \widetilde{G}(N)$. The given condition that ψ not lie in $\widetilde{\Psi}_{\mathrm{ell}}(N)$ means that either the multiplicity ℓ_k of some component ψ_k of ψ is larger than 1, or the indexing set $J = J_\psi$ is not empty. In the first instance, the subgroup S^*_ψ of $\widehat{G}^0$ has an $\mathrm{Int}(s_x)$-stable factor $GL(\ell_k, \mathbb{C})$ of positive semisimple rank. In the second, S^*_ψ has a subgroup isomorphic to $\mathbb{C}^* \times \mathbb{C}^*$, whose factors are interchanged by $\mathrm{Int}(s_x)$. It follows easily that the condition (ii) holds also in case $G = \widetilde{G}(N)$, and that M_x is again proper in G. We can therefore argue by induction. More precisely, we can assume inductively that the analogue of (4.5.1) for M_x holds in all cases. Since the conditions of the proposition imply that (i) holds also in the case $\widetilde{G} = \widetilde{G}(N)$, we see in all cases that

$$f'_G(\psi, s_\psi x) = f'_{M_x}(\psi_{M_x}, s_\psi x) = f_{M_x}(\psi_{M_x}, x) = f_G(\psi, x).$$

The identity (4.5.1) therefore holds for all G. We note for future reference that in case $G = \widetilde{G}(N)$, (4.5.1) in fact holds for *any* ψ in the complement of $\widetilde{\Psi}_{\mathrm{ell}}(N)$.

We have established the required identity (4.5.1) for ψ, provided that the conditions (i) and (ii) on ψ hold for any G in $\widetilde{\mathcal{E}}_{\mathrm{sim}}(N)$ and any x in $\mathcal{S}_\psi = \mathcal{S}_\psi(G)$. For which ψ do the two conditions not hold?

Consider the product formula (4.5.9) for the group S_ψ attached to a given $G \in \widetilde{\mathcal{E}}_{\mathrm{sim}}(N)$ with $\psi \in \widetilde{\Psi}(G)$. The second and third components of the product are connected complex groups. We can therefore assume that the representative s_x of x described above lies in the first component, the nonconnected group defined by the product over $I_\psi^+(G)$. To check the conditions (i) and (ii), we need only describe the rank of the complex group (4.5.9), and the rank of the centralizer of s_x in (4.5.9).

The two conditions clearly hold unless the third component, the product over J_ψ, is either trivial or consists of just one factor $GL(1, \mathbb{C}) \cong \mathbb{C}^*$. In the second case we have an infinite central subgroup of S_ψ, whose centralizer is a proper Levi subgroup of $\widehat{G}$. The identity (4.5.1) then follows by induction. (We note that ψ lies in the complement of $\widetilde{\Psi}_{\mathrm{disc}}(G)$ in this case, and that the global theorems also follow directly from the expansions for (4.5.2) and (4.5.3).) We may therefore assume that the indexing set J_ψ for ψ is empty. The two conditions remain valid unless the second component, the product over $I_\psi^-(G)$, is either trivial or consists of one factor $Sp(2, \mathbb{C})$. Putting aside the second possibility for the moment, we suppose that the set $I_\psi^-(G)$ is also empty. The conditions (i) and (ii) are then still valid unless $\ell_i \leqslant 2$ for every $i \in I_\psi^+(G)$, or $\ell_i = 3$ for one $i \in I_\psi^+(G)$ and $\ell_i = 1$ for the remaining indices i. Putting aside the second possibility again, we suppose that

$$S_\psi = \Big(\prod_{i \in I_\psi^+(G)} O(\ell_i, \mathbb{C}) \Big)_\psi^+, \qquad \ell_i \in \{1, 2\}.$$

Recall that $(\,\cdot\,)_\psi^+$ stands for the kernel of a character $\xi_\psi^+ = \xi_\psi^+(G)$ of order 1 or 2. If there were an element

$$s = \prod_i s_i, \qquad s_i \in O(\ell_i, \mathbb{C}),$$

in the kernel, with $\det(s_i) = -1$ whenever $\ell_i = 2$, ψ would belong to the subset $\widetilde{\Psi}_{\mathrm{ell}}(G)$ of $\widetilde{\Psi}(G)$, since the centralizer of s in S_ψ would then be finite. This outcome is expressly ruled out by the hypotheses of the proposition. On the other hand, if there is no such s, S_ψ has a central torus $\mathbb{C}^*$. In this case, (4.5.1) again follows inductively from its reduction to a proper Levi subgroup of G.

We have thus seen that (4.5.1) holds for any given ψ that does not fall into one of the two cases set aside above. In the first of these cases, the index set J_ψ is empty, while the product over $I_\psi^-(G)$ in (4.5.9) reduces simply to $Sp(2,\mathbb{C})$. Observe that the conditions (i) and (ii) remain valid in this case unless the factors in the product over $I_\psi^+(G)$ are all equal to $O(1,\mathbb{C}) \cong \mathbb{Z}/2\mathbb{Z}$. In the second case, the index sets J_ψ and $I_\psi^-(G)$ are both empty, while the product over $I_\psi^+(G)$ reduces to a product of $O(3,\mathbb{C})$ with several copies of the group $O(1,\mathbb{C}) \cong \mathbb{Z}/2\mathbb{Z}$. The two outstanding cases thus take the form

$$\begin{cases} \psi = 2\psi_1 \boxplus \psi_2 \boxplus \cdots \boxplus \psi_r, \\ S_\psi \cong Sp(2,\mathbb{C}) \times (\mathbb{Z}/2\mathbb{Z})^{r'-1}, \qquad r \geqslant 1, \end{cases} \tag{4.5.11}$$

and

$$\begin{cases} \psi = 3\psi_1 \boxplus \psi_2 \boxplus \cdots \boxplus \psi_r, \\ S_\psi \cong SO(3,\mathbb{C}) \times (\mathbb{Z}/2\mathbb{Z})^{r'}, \qquad r \geqslant 1, \end{cases} \tag{4.5.12}$$

where $S_\psi = S_\psi(G)$ for some $G \in \widetilde{\mathcal{E}}_{\mathrm{sim}}(N)$, and

$$r' = \begin{cases} r, & \text{if } \xi_\psi^+ = 1, \\ r-1, & \text{otherwise.} \end{cases}$$

In these remaining cases, it is understood that $G \in \widetilde{\mathcal{E}}_{\mathrm{sim}}(N)$ is some datum with $\psi \in \widetilde{\Psi}(G)$. According to the discussion leading up to (4.5.10), G is uniquely determined by this condition, *except* for the case of (4.5.11) with $r = 1$. In this case, however, there is a second element $G^\vee \in \widetilde{\mathcal{E}}_{\mathrm{sim}}(N)$ with $\psi \in \widetilde{\Psi}(G^\vee)$. It then follows from (4.5.11) that the group $S_\psi^\vee = S_\psi(G^\vee)$ equals $\big(O(2,\mathbb{C})\big)_\psi^+$. There are two possibilities. If the character $\xi_\psi^{+,\vee} = \xi_\psi^+(G^\vee)$ is 1, $S_\psi^\vee$ equals $O(2,\mathbb{C})$, and ψ lies in $\widetilde{\Psi}_{\mathrm{ell}}(G^\vee)$, a condition on ψ that is ruled out by the proposition. We can therefore assume that $\xi_\psi^{+,\vee} \neq 1$ if $r = 1$ in (4.5.11). In this case, $S_\psi^\vee \cong \mathbb{C}^*$, and the analogue of (4.5.1) for $G^\vee$ follows inductively from its reduction to a proper Levi subgroup of $G^\vee$.

Let us summarize our progress to this point. We have established (4.5.1) for any ψ that satisfies the conditions of the proposition, and that does not belong to one of the two exceptional cases (4.5.11) and (4.5.12). The required global theorems for such ψ then follow from the general comparison at the beginning of the section. It remains to treat the exceptional cases by a more subtle comparison.

Suppose that ψ and G are as in (4.5.11). We cannot use the earlier comparison, because we do not know that the global identity (4.5.1) holds. However, we can still apply Corollary 4.4.3 to the difference (4.5.2), since the required condition was established in general prior to the statement of the proposition. We can also apply Corollary 4.3.3 to the difference (4.5.3). Here, we do not know that the function $f_G(\psi, x)$ in the resulting expression depends only on the image x of the element $u \in \mathfrak{N}_\psi(x)$ used to define it. However, in the case at hand, there is only one element $w = w_x$ in the subset $W_{\psi,\mathrm{reg}}(x)$ of $W_\psi(x)$ that indexes the internal sum in the expression (4.3.9) of Corollary 4.3.3, and hence only one element in $u = u_x$ in the corresponding subset $\mathfrak{N}_{\psi,\mathrm{reg}}(x)$ of $\mathfrak{N}_\psi(x)$. We take the liberty of setting

$$f_G(\psi, x) = f_G(x, u_x)$$

in order to use the notation of Corollary 4.3.3. Substituting the identity $e_\psi(x) = i_\psi(x)$, we see that the difference

$$(4.5.13) \qquad {}^0 s^G_{\mathrm{disc},\psi}(f) - {}^0 r^G_{\mathrm{disc},\psi}(f) = {}^0 S^G_{\mathrm{disc},\psi}(f) - \mathrm{tr}\big(R^G_{\mathrm{disc},\psi}(f)\big)$$

equals

$$C_\psi \sum_{x \in \mathcal{S}_\psi} i_\psi(x)\, \varepsilon^G_\psi(x)\big(f_G(\psi, x) - f'_G(\psi, s_\psi x)\big).$$

The constant C_ψ is defined in (4.3.5) as a product

$$C_\psi = m_\psi\, |\kappa_G|^{-1} |\mathcal{S}_\psi|^{-1} = m_\psi\, |\mathcal{S}_\psi|^{-1}$$

of positive factors. The coefficient $i_\psi(x)$ satisfies

$$\begin{aligned} i_\psi(x) &= |W^0_\psi|^{-1} \sum_{w \in W_{\psi,\mathrm{reg}}(x)} s^0_\psi(w)\, |\det(w-1)|^{-1} \\ &= |W^0_\psi|^{-1}\, s^0_\psi(w_x)\, |\det(w_x - 1)|^{-1} = -\tfrac{1}{4}, \end{aligned}$$

since $|W^0_\psi| = 2$, $|\det(w_x - 1)| = 2$ and $s^0_\psi(w_x) = -1$, in this case. It is the minus sign here that will be the critical factor. From the form (4.5.11) of $\mathcal{S}_\psi$, we observe that the R-group R_ψ is trivial. It follows that there is an isomorphism $x \to x_M$ from $\mathcal{S}_\psi$ onto $\mathcal{S}_{\psi_M}$, where ψ_M represents a fixed element in $\Psi_2(M, \psi)$. The coefficient $\varepsilon^G_\psi(x)$ then satisfies

$$\varepsilon^G_\psi(x) = \varepsilon^M_{\psi_M}(x_M) = \varepsilon_{\psi_M}(x_M),$$

a reduction one can infer either directly from the definition (1.5.6), or as a special case of Lemma 4.3.1. We can therefore write

$$C_\psi\, i_\psi(x)\, \varepsilon^G_\psi(x) = -n_\psi\, m_{\psi_M}\, |\mathcal{S}_{\psi_M}|^{-1}\, \varepsilon_{\psi_M}(x_M),$$

for the positive constant

$$n_\psi = \tfrac{1}{4}\, m_\psi\, m^{-1}_{\psi_M}.$$

The remaining terms in our expression for (4.5.13) are the linear forms $f'_G(\psi, s_\psi x)$ and $f_G(\psi, x)$. It follows from the definition $(4.2.5)'$ and the application of (2.2.6) inductively to M that

$$\begin{aligned} f'_G(\psi, x) &= f'(\psi') = f'(\psi'_M) \\ &= \sum_{\pi_M \in \widetilde{\Pi}_{\psi_M}} \langle s_\psi x_M, \pi_M \rangle\, f_M(\pi_M), \end{aligned}$$

where (M', ψ'_M) is the endoscopic pair attached to (ψ_M, x_M). A similar formula holds with $s_\psi x$ in place of x and x_M in place of $s_\psi x_M$, since

$$s_\psi(s_\psi x)_M = s_\psi(s_\psi x_M) = (s_\psi)^2 x_M = x_M.$$

We can therefore write

$$\begin{aligned} - C_\psi \sum_{x \in \mathcal{S}_\psi} i_\psi(x)\, \varepsilon_\psi^G(x)\, f'_G(\psi, s_\psi x) &= n_\psi\, m_{\psi_M}\, |\mathcal{S}_{\psi_M}|^{-1} \sum_{x_M \in \mathcal{S}_{\psi_M}} \sum_{\pi_M \in \widetilde{\Pi}_{\psi_M}} \varepsilon_{\psi_M}(x_M) \langle x_M, \pi_M \rangle\, f_M(\pi_M) \\ &= n_\psi \sum_{\pi_M \in \widetilde{\Pi}_{\psi_M}} m_{\psi_M}(\pi_M)\, \mathrm{tr}\big(\mathcal{I}_P(\pi_M, f)\big), \end{aligned}$$

by applying the multiplicity formula

$$m_{\psi_M}(\pi_M) = m_{\psi_M}\, |\mathcal{S}_{\psi_M}|^{-1} \sum_{x_M \in \mathcal{S}_{\psi_M}} \varepsilon_{\psi_M}(x_M) \langle x_M, \pi_M \rangle$$

of Theorem 1.5.2 inductively to the proper Levi subgroup M of G, and substituting the usual adjoint relation

$$f_M(\pi_M) = \mathrm{tr}\big(\mathcal{I}_P(\pi_M, f)\big)$$

of f_M. The other linear form

$$f_G(\psi, x) = f_G(\psi, u_x)$$

is given by (4.2.4)′. It can be written as

$$\sum_{\pi_M \in \widetilde{\Pi}_{\psi_M}} \langle x_M, \pi_M \rangle\, \mathrm{tr}\big(R_P(w_x, \widetilde{\pi}_M, \psi_M)\, \mathcal{I}_P(\pi_M, f)\big),$$

since $w_x = w_u$, and $\langle x_M, \pi_M \rangle = \langle \widetilde{u}, \widetilde{\pi}_M \rangle$ if $u = u_x$. We can therefore write

$$\begin{aligned} C_\psi \sum_{x \in \mathcal{S}_\psi} i_\psi(x)\, \varepsilon_\psi^G(x)\, f_G(\psi, x) &= - n_\psi \sum_{\pi_M \in \widetilde{\Pi}_{\psi_M}} m_{\psi_M}(\pi_M)\, \mathrm{tr}\big(R_P(w_x, \widetilde{\pi}_M, \psi_M)\, \mathcal{I}_P(\pi_M, f)\big) \end{aligned}$$

by again applying the multiplicity formula.

We have expanded (4.5.13) into a (signed) sum of two rather similar expressions. The difference (4.5.13) thus equals

$$n_\psi \sum_{\pi_M \in \widetilde{\Pi}_{\psi_M}} m_{\psi_M}(\pi_M)\, \mathrm{tr}\big(\big(1 - R_P(w_x, \widetilde{\pi}_M, \psi_M)\big)\mathcal{I}_P(\pi_M, f)\big). \tag{4.5.14}$$

It follows that the stable linear form ${}^0S^G_{\mathrm{disc},\psi}(f)$ on the right hand side of (4.5.13) equals the sum of $\mathrm{tr}\big(R^G_{\mathrm{disc},\psi}(f)\big)$ with (4.5.14).

We have taken G to be the fixed group in $\widetilde{\mathcal{E}}_{\mathrm{sim}}(N)$ that is implicit in (4.5.11). We therefore write G^* for a general group in $\widetilde{\mathcal{E}}_{\mathrm{sim}}(N)$. If ψ does not lie in $\widetilde{\Psi}(G^*)$, the identity (4.4.12) (with G^* in place of G) is valid. This condition applies to any $G^* \neq G$, if $r \geqslant 2$. If $r = 1$, we have seen that there is one group $G^* \neq G$ with $\psi \in \widetilde{\Psi}(G^*)$, namely $G^* = G^\vee$. But we have also seen that $\xi_\psi^{+,\vee} \neq 1$ in this case (given our assumption that $\psi \notin \widetilde{\Psi}_{\mathrm{ell}}(G^\vee)$), and that the analogue of (4.5.1) for G^* follows by induction. It thus follows from (4.5.5) that

$$ {}^0 S^{G^*}_{\mathrm{disc},\psi}(f^*) = \mathrm{tr}\big(R^{G^*}_{\mathrm{disc},\psi}(f^*)\big), \qquad f^* \in \widetilde{\mathcal{H}}(G^*), $$

for any $G^* \in \widetilde{\mathcal{E}}_{\mathrm{sim}}(N)$ distinct from G.

We saw earlier in the proof that the analogue of (4.5.1) for $\widetilde{G}(N)$ holds for any $\psi \notin \widetilde{\Psi}_{\mathrm{ell}}(N)$, and in particular, for the case (4.5.11) we are considering. It then follows from (4.5.6) that

$$ \sum_{G^* \in \widetilde{\mathcal{E}}_{\mathrm{sim}}(N)} \widetilde{\iota}(N,G^*)\, {}^0\widehat{S}^{G^*}_{\mathrm{disc},\psi}(\widetilde{f}^{G^*}) = 0, $$

for any $\widetilde{f} \in \widetilde{\mathcal{H}}(N)$. Substituting for each of the stable linear forms on the left, we see that the sum of the expression

$$ \sum_{G^* \in \widetilde{\mathcal{E}}_{\mathrm{sim}}(N)} \widetilde{\iota}(N,G^*)\, \mathrm{tr}\big(R^{G^*}_{\mathrm{disc},\psi}(f^*)\big) \tag{4.5.15} $$

with (4.5.14) vanishes, for any compatible family of functions

$$ \{f^* \in \mathcal{H}(G^*) : \ G^* \in \widetilde{\mathcal{E}}_{\mathrm{sim}}(N)\}. $$

The global intertwining operator $R_P(w_x, \widetilde{\pi}_M, \psi_M)$ in (4.5.14) has square equal to 1. Its eigenvalues are therefore equal to 1 or -1. The expression (4.5.14) is in consequence a nonnegative sum of irreducible characters. The same is true of the other expression (4.5.15). We can therefore apply Proposition 3.5.1 to the sum of the two expressions. We conclude that

$$ R_P(w_x, \widetilde{\pi}_M, \psi_M) = 1, $$

for any $\pi_M \in \widetilde{\Pi}_{\psi_M}(M)$ with $m_{\psi_M}(\pi_M) \neq 0$, and that

$$ \mathrm{tr}\big(R^{G^*}_{\mathrm{disc},\psi}(f^*)\big) = {}^0 S^{G^*}_{\mathrm{disc},\psi}(f^*) = 0, \qquad f^* \in \widetilde{\mathcal{H}}(G^*), $$

for every $G^* \in \widetilde{\mathcal{E}}_{\mathrm{sim}}(N)$. This completes the proof of the proposition for the first exceptional case (4.5.11).

The argument for the remaining exceptional case (4.5.12) is similar. It is actually slightly simpler, since in this case G is the only element in $\widetilde{\mathcal{E}}_{\mathrm{sim}}(N)$ with $\psi \in \widetilde{\Psi}(G)$. In particular, there is no distinction to be made between $r = 1$ and $r > 1$. The comparison is otherwise identical, and the required global theorems in this last case follow as above. Proposition 4.5.1 thus holds in all cases. □

COROLLARY 4.5.2. *Suppose that ψ is as in the proposition. Then the global intertwining relation (4.5.1) of Corollary 4.2.1 is valid unless ψ falls* *into one of the two exceptional classes (4.5.11) or (4.5.12), in which case we have the weaker identity*

$$\sum_{x \in \mathcal{S}_\psi} \varepsilon_\psi^G(x)\big(f'_G(\psi, s_\psi x) - f_G(\psi, x)\big) = 0, \qquad f \in \widetilde{\mathcal{H}}(G), \tag{4.5.16}$$

for any $G \in \widetilde{\mathcal{E}}_{\mathrm{sim}}(N)$ with $\psi \in \widetilde{\Psi}(G)$. If $G = \widetilde{G}(N)$, the identity (4.5.1) holds without restriction, and indeed, for any $\psi \notin \widetilde{\Psi}_{\mathrm{ell}}(N)$.

PROOF. The assertions of the corollary were established in the course of proving the proposition. For example, we verified (4.5.1) in the generality above in order to apply the earlier comparison. In the case (4.5.11), the identity (4.5.16) follows from the fact, established at the end of the proof, that the difference (4.5.13) vanishes. The case of (4.5.12) is similar. □

The most interesting aspect of the last proof is the treatment of the exceptional cases (4.5.11) and (4.5.12). For example, if $r > 1$ in (4.5.11), G is the only datum in $\widetilde{\mathcal{E}}_{\mathrm{sim}}(N)$ with $\psi \in \widetilde{\Psi}(G)$. The same goes for (4.5.12), with $r \geqslant 1$. If $G^* \in \widetilde{\mathcal{E}}_{\mathrm{sim}}(N)$ is distinct from G, and satisfies the further condition that $\xi_{G^*} \neq \xi_\psi$, we know a priori that $R_\psi^{G^*} = 0$ and ${}^0S_\psi^{G^*} = 0$. However, this does not account for the case that $G^* = G^\vee$, in which $\eta_G = \eta_\psi = 1$ and N is even. We really do need the argument above to deal with this case, as well as to compensate for not knowing (4.5.1) in case $G^* = G$. Looking back at the argument, the reader can see that we have been lucky with the sign $s_\psi^0(w_x)$, in which w_x was the unique element in $W_{\psi,\mathrm{reg}}(x)$. If $s_\psi^0(w_x)$ had somehow turned out to equal 1 instead of -1, the argument would have failed, perhaps dooming the entire enterprise.

The role of the sign $s_\psi^0(w_x)$ has broader implications. It is representative of a whole collection of arguments, of which our treatment of (4.5.11) and (4.5.12) is among the simplest and most transparent. The more elaborate comparisons will be applied in later chapters to the cases excluded by Propositon 4.5.1. In general, $s_\psi^0(w_x)$ is an essential ingredient of the interlocking sign Lemmas 4.3.1 and 4.4.1, which were instrumental in the general reductions of §4.3 and §4.4, and hence also Proposition 4.5.1, but which have so far been taken for granted. This underlines the significance of the next section, in which we will establish the two lemmas in general.

4.6. The two sign lemmas

Lemmas 4.3.1 and 4.4.1 are parallel. They played critical roles in the corresponding expansions of §4.3 and §4.4, which in turn led to the general results of §4.5. Both lemmas concern the basic sign character $\varepsilon_\psi = \varepsilon_\psi^G$ attached to ψ in §1.5. The first can be regarded as a reduction of the character to its analogue for a Levi subgroup, while the second amounts to a reduction to endoscopic groups. We shall prove them in this section, thereby completing the proof of the general reduction obtained in the last section.

We revert to the setting of §4.3–§4.4. That is, we take G to be either a group in $\widetilde{\mathcal{E}}_{\text{sim}}(N)$ or equal to $\widetilde{G}(N)$ itself. In the second case, ${}^L G$ and $\overline{S}_\psi$ are only bitorsors, under the respective groups ${}^L G^0$ and $\overline{S}_\psi^*$. Following our earlier convention, we shall understand a representation of ${}^L G$ or $\overline{S}_\psi$ to be a morphism that extends to a representation of the group ${}^L G^+$ or $\overline{S}_\psi^+$ generated by ${}^L G$ or $\overline{S}_\psi$. For example, the canonical extension of ε_ψ^G to the quotient $\mathcal{S}_\psi$ of $\overline{S}_\psi$ in the case $G = \widetilde{G}(N)$ is a character in this sense. That is, it extends to a character (of order 2) on the quotient $\mathcal{S}_\psi^+$ of $\overline{S}_\psi^+$.

We will need a certain amount of preliminary discussion before we can prove the two sign lemmas. This will be useful in its own right. For we shall see that it gives some insight into the meaning of the sign characters ε_ψ.

Suppose that $\psi \in \widetilde{\Psi}(G)$ is fixed. We assume that Theorems 1.4.1 and 1.4.2 hold for the self-dual generic constituents of ψ. The group $\mathcal{L}_\psi$ is then defined, and has local embeddings (1.4.14). The sign character is defined (1.5.6) in terms of the irreducible constituents

$$\tau_\alpha = \lambda_\alpha \otimes \mu_\alpha \otimes \nu_\alpha$$

of the representation

$$\tau_\psi(s, g, h) = \mathrm{Ad}_G\big(s \cdot \widetilde{\psi}_G(g, h)\big)$$

of the product

$$\overline{S}_\psi \times \mathcal{L}_\psi \times SL(2, \mathbb{C}), \tag{4.6.1}$$

acting on the Lie algebra $\hat{\mathfrak{g}}$ of $\widehat{G}^0$. What we need here is a decomposition of τ_ψ into (possibly reducible) representations whose $\mathcal{L}_\psi$-constituents have good L-functions.

Recall that $\mathcal{L}_\psi$ is a fibre product over $k \in \{K_\psi\}$ of L-groups ${}^L G_k$. The factor ${}^L G_k$ comes in turn with a complex homomorphism into $GL(m_k, \mathbb{C})$. For any k, we write g_k for the image in $GL(m_k, \mathbb{C})$ of a point $g \in \mathcal{L}_\psi$ under the composition

$$\mathcal{L}_\psi \longrightarrow {}^L G_k \longrightarrow GL(m_k, \mathbb{C}).$$

Recall also that $\{K_\psi\}$ is the set of orbits of the involution $k \to k^\vee$ on the set

$$K_\psi = I_\psi \coprod J_\psi \coprod J_\psi^\vee.$$

We may as well identify $\{K_\psi\}$ with the union of I_ψ and J_ψ, even though choice of the subset J_ψ of K_ψ is not canonical. We will then write

$$g_{k^\vee} = g_k^\vee = {}^t g_k^{-1}, \qquad g \in \mathcal{L}_\psi,$$

for any index $k^\vee \in J_\psi^\vee$.

The decomposition of the representation τ_ψ gives rise to two general families of representations of $\mathcal{L}_\psi$. The first consists of the representations

$$R_{kk'}(g)X = g_k \cdot X \cdot {}^t g_{k'}, \qquad g \in \mathcal{L}_\psi,\ k, k' \in K_\psi,$$

on complex $(m_k \times m_{k'})$-matrix space. They satisfy the natural relations

$$R_{kk'}^{\vee} \cong R_{k^{\vee}(k')^{\vee}} \quad \text{and} \quad R_{kk'} \cong R_{k'k},$$

and they include also the representations

$$R_{k(k')^{\vee}}(g)X = g_k \cdot X \cdot g_{k'}^{-1}, \qquad g \in \mathcal{L}_{\psi},\ k, k' \in K_{\psi}.$$

Their L-functions are the standard Rankin-Selberg L-functions

$$L(s, R_{kk'}) = L(s, \mu_k \times \mu_{k'}),$$

where $\mu_k \times \mu_{k'}$ is the cuspidal automorphic representation of $GL(m_k) \times GL(m_{k'})$ attached to the (k, k')-component of ψ. The other family consists of the symmetric square representations

$$S_k(g) = S^2(g_k), \qquad g \in \mathcal{L}_{\psi},\ k \in K_{\psi},$$

and the skew-symmetric square representations

$$\Lambda_k(g) = \Lambda^2(g_k), \qquad g \in \mathcal{L}_{\psi},\ k \in K_{\psi}.$$

They satisfy the relations

$$S_k^{\vee} \cong S_{k^{\vee}} \quad \text{and} \quad \Lambda_k^{\vee} \cong \Lambda_{k^{\vee}},$$

and also have familiar L-functions. We shall say that a finite dimensional representation σ of $\mathcal{L}_{\psi}$ is *standard* if it belongs to one of these two general families. (This is not to be confused with the notion of a standard representation $\rho \in P(G)$ of $G(F)$, which we discussed at some length in §3.5, §2.2, and other sections.)

The standard representations that are self-dual consist of the following three sets

$$\{R_{kk'} :\ k, k' \in I_{\psi}\},$$

$$\{R_{kk^{\vee}} :\ k \in K_{\psi}\},$$

and

$$\{S_k, \Lambda_k :\ k \in I_{\psi}\}.$$

We recall here that I_{ψ} is the subset of indices $k \in K_{\psi}$ such that $\widehat{G}_k$ equals a classical group $SO(m_k, \mathbb{C})$ or $Sp(m_k, \mathbb{C})$, rather than $GL(m_k, \mathbb{C})$. Among these self-dual representations, the ones in the second and third sets are all orthogonal. Those in the first set are orthogonal if $\widehat{G}_k$ and $\widehat{G}_{k'}$ are of the same type, and symplectic if $\widehat{G}_k$ and $\widehat{G}_{k'}$ are of opposite type. Observe that a standard representation could be reducible. If it is symplectic, however, it must be irreducible.

The global L-function $L(s, \sigma)$ attached to a standard representation σ of $\mathcal{L}_{\psi}$ has analytic continuation, with functional equation

$$L(s, \sigma) = \varepsilon(s, \sigma)\, L(1 - s, \sigma^{\vee}).$$

If σ belongs to the third general family above, it is understood that

$$\varepsilon(s, \sigma) = \varepsilon(s, \mu, \sigma)$$

is the automorphic ε-factor for the cuspidal automorphic representation μ of $GL(m_k)$ attached to σ, since it is not known that the arithmetic ε-factors are the same as the automorphic ε-factors in this case. However, if σ is self-dual and symplectic, it is an irreducible Rankin-Selberg representation, for which the two kinds of ε-factors are known to be equal. In particular, the ε-factor $\varepsilon(s, \sigma)$ is a finite product of local arithmetic ε-factors, as in the definition of ε_ψ^G.

LEMMA 4.6.1. *There is a decomposition*

$$\tau_\psi = \bigoplus_\kappa \tau_\kappa = \bigoplus_\kappa (\lambda_\kappa \otimes \sigma_\kappa \otimes \nu_\kappa), \qquad \kappa \in \mathcal{K}_\psi, \tag{4.6.2}$$

of τ_ψ relative to the product (4.6.1), for standard representations σ_κ, irreducible representations λ_κ and ν_κ, and an indexing set $\mathcal{K}_\psi$ with an involution

$$\kappa \leftrightarrow \kappa^\vee$$

such that

$$\tau_\kappa^\vee \cong \tau_{\kappa^\vee}.$$

PROOF. It is clear that we can introduce the notion of a standard representation ρ for the product

$$\mathcal{A}_\psi = \mathcal{L}_\psi \times SL(2, \mathbb{C})$$

by copying the definition above for $\mathcal{L}_\psi$. One can then establish a decomposition

$$\tau_\psi = \bigoplus_\iota (\lambda_\iota \otimes \rho_\iota), \qquad \iota \in \mathcal{I}_\psi,$$

that satisfies the required properties of (4.6.2), but with standard representations ρ_ι of $\mathcal{A}_\psi$ in place of the tensor products $\sigma_\kappa \otimes \nu_\kappa$. This is a straightforward consequence of the definition (1.4.1) of ψ and the structure of the Lie algebra $\hat{\mathfrak{g}}$, which we leave to the reader. The exercise becomes a little more transparent if we take $g_k^\vee$ to be the "second transpose" ${}_tg^{-1}$, which we can do because the assertion depends only on the inner class of the automorphism.

The second step is to observe that any standard representation of $\mathcal{A}_\psi$ has a decomposition

$$\rho_\iota = \bigoplus_{\kappa \in \mathcal{K}_{\psi,\iota}} (\sigma_\kappa \otimes \nu_\kappa),$$

for standard representation σ_κ of $\mathcal{L}_\psi$ and irreducible representations ν_κ of $SL(2, \mathbb{C})$. If ρ_ι is a Rankin-Selberg representation $R_{kk'}$ for $\mathcal{A}_\psi$, σ_κ is the same Rankin-Selberg representation for $\mathcal{L}_\psi$, while ν_κ ranges over the irreducible constituents of the tensor product

$$\nu_k \otimes \nu_{k'} \tag{4.6.3}$$

representation of $SL(2, \mathbb{C})$. If ρ_ι is a standard representation S_k or Λ_k for $\mathcal{A}_\psi$, σ_κ alternates between the corresponding two kinds of standard representation of $\mathcal{L}_\psi$, and ν_κ ranges over appropriate constituents of (4.6.3) (with $k' = k$). The lemma follows. □

COROLLARY 4.6.2. *If a constituent σ_κ of (4.6.2) is diagonal, in the sense that it is of the form R_{kk}, $R_{kk^\vee}$, S_k or Λ_k, the corresponding representation ν_κ of $SL(2,\mathbb{C})$ is odd dimensional.*

PROOF. The assertion follows from the proof of the lemma, and the fact that the irreducible constituents of the tensor product (4.6.3) are odd dimensional if $k' = k$. □

We can use the lemma to express the sign character ε_ψ in terms of standard representations. We claim that

$$\varepsilon_\psi(x) = \varepsilon_\psi^G(x) = \prod_\kappa \det\big(\lambda_\kappa(s)\big), \qquad s \in \overline{S}_\psi,$$

where $x = x_s$ is the image of s in $\mathcal{S}_\psi$, and the product is over the set of indices $\kappa \in \mathcal{K}_\psi$ such that σ_κ is symplectic, and satisfies $\varepsilon\big(\frac{1}{2},\sigma_\kappa\big) = -1$. The standard representations σ_κ parametrized by this set are irreducible, so the corresponding representations λ_κ of $\overline{S}_\psi$ are among the factors λ_α in the product (1.5.6) that defines ε_ψ^G. There could be other factors in (1.5.6), since the diagonal standard representations σ_κ are reducible, and might have symplectic constituents even though they are not symplectic themselves. However, these supplementary factors occur in pairs λ_α and $\lambda_\alpha^\vee$, whose contributions to (1.5.6) cancel. Indeed, either the corresponding subrepresentation τ_κ of τ_ψ is not self-dual, in which case $\tau_\kappa^\vee$ also contributes to the product, or σ_κ is self-dual but not symplectic, since it is reducible, in which case its symplectic factors occur in pairs. The claim follows.

We can in fact pare the product further by excluding those factors with $\nu_\kappa(s_\psi) = 1$. These are the factors for which the irreducible representation ν_κ of $SL(2,\mathbb{C})$ is odd dimensional, and hence orthogonal. For if the corresponding subrepresentation τ_κ of τ_ψ is not self-dual, or if it occurs in τ_ψ with even multiplicity, we again obtain pairs of factors that cancel. If τ_κ is self-dual, and has odd multiplicity, it is orthogonal (since τ_ψ is orthogonal). This implies that the representation λ_κ of $\overline{S}_\psi$ is symplectic (since σ_κ is symplectic and ν_k is orthogonal). It thus has determinant 1, and does not contribute to the product. We can therefore write

$$\varepsilon_\psi(x_s) = \prod_{\kappa\in\mathcal{K}_\psi^-} \det\big(\lambda_\kappa(s)\big), \qquad s \in \overline{S}_\psi, \tag{4.6.4}$$

where $\mathcal{K}_\psi^-$ is the subset of indices $\kappa \in \mathcal{K}_\psi$ such that

(i) σ_κ is symplectic,

(ii) $\varepsilon\big(\frac{1}{2},\sigma_\kappa\big) = -1$,

and

(iii) $\nu_\kappa(s_\psi) = -1$.

We shall work for a time with the restriction

$$\tau_{\psi,1} = \mathrm{Ad}_G \circ \widetilde{\psi}_G$$

of τ_ψ to the subgroup

$$\mathcal{A}_\psi = \mathcal{L}_\psi \times SL(2,\mathbb{C})$$

of the product (4.6.1). Let $\psi_M \in \Psi_2(M, \psi)$ be a fixed embedding from $\mathcal{L}_\psi$ to ${}^L M$ attached to ψ, for some fixed Levi subgroup $M \subset G^0$ with dual Levi subgroup ${}^L M \subset {}^L G^0$. Then

$$\mathrm{Ad}_G \circ \widetilde{\psi}_G = \mathrm{Ad}_{G,M} \circ \psi_M,$$

where $\mathrm{Ad}_{G,M}$ is the restriction of Ad_G to ${}^L M$. We fix a decomposition

$$\mathrm{Ad}_{G,M} \circ \psi_M = \bigoplus_{h \in H_\psi} (\sigma_h \otimes \nu_h)$$

of this representation, which we assume is compatible with both the decomposition (4.6.2) of τ_ψ, and the root space decomposition

$$\hat{\mathfrak{g}} = \bigoplus_{\xi} \hat{\mathfrak{g}}_\xi = \widehat{\mathfrak{m}} \oplus \Big(\bigoplus_{\alpha \in \widehat{\Sigma}_M} \hat{\mathfrak{g}}_\alpha \Big)$$

of $\hat{\mathfrak{g}}$ with respect to the Γ-split component

$$A_{\widehat{M}} = \big(Z(\widehat{M})^\Gamma\big)^0$$

of the center of $\widehat{M}$. The first condition tells us that we can write

$$\sigma_h \otimes \nu_h = \sigma_\kappa \otimes \nu_\kappa, \qquad h \in H_\psi,$$

for a surjective mapping $h \to \kappa$ from the indexing set H_ψ onto $\mathcal{K}_\psi$. In the second condition, ξ ranges over the roots of $(\widehat{G}^0, A_{\widehat{M}})$ (including 0), while $\widehat{\Sigma}_M$ denotes the usual set of nonzero roots. This condition means that H_ψ is a disjoint union of sets $H_{\psi,\xi}$ such that if ρ_ξ is the restriction of $\mathrm{Ad}_{G,M}$ to the subspace $\hat{\mathfrak{g}}_\xi$ of $\hat{\mathfrak{g}}$, then

$$\rho_\xi \circ \psi_M = \bigoplus_{h \in H_{\psi,\xi}} (\sigma_h \otimes \nu_h). \tag{4.6.5}$$

Let $H^-_{\psi,\xi}$ be the preimage of $\mathcal{K}^-_\psi$ in $H_{\psi,\xi}$, and set

$$\hat{\mathfrak{g}}^-_{\psi,\xi} = \bigoplus_{h \in H^-_{\psi,\xi}} \hat{\mathfrak{g}}_h,$$

where $\hat{\mathfrak{g}}_h$ is the subspace of $\hat{\mathfrak{g}}_\xi$ on which the irreducible representation

$$\mu_h \otimes \nu_h = \sigma_h \otimes \nu_h$$

of $\mathcal{A}_\psi$ acts. The bitorsor $\overline{S}_\psi$ does not generally act on the space $\hat{\mathfrak{g}}^-_{\psi,\xi}$ (or for that matter on the larger root space $\hat{\mathfrak{g}}_\xi$). However, the product $\overline{S}_\psi \times \mathcal{A}_\psi$ does act on the sum

$$\hat{\mathfrak{g}}^-_\psi = \bigoplus_{\xi} \hat{\mathfrak{g}}^-_{\psi,\xi} = \bigoplus_{h \in H^-_\psi} \hat{\mathfrak{g}}_h,$$

where

$$H^-_\psi = \coprod_{\xi} H^-_{\psi,\xi}$$

is the preimage of $\mathcal{K}_\psi^-$ in H_ψ. There is consequently a $\overline{S}_\psi^*$-equivariant morphism from $\overline{S}_\psi$ into the semisimple algebra $\mathrm{End}_{\mathcal{A}_\psi}(\hat{\mathfrak{g}}_\psi^-)$. We can therefore write

$$\varepsilon_\psi(x_s) = \det\big(s, \mathrm{End}_{\mathcal{A}_\psi}(\hat{\mathfrak{g}}_\psi^-)\big), \qquad s \in \overline{S}_\psi. \tag{4.6.6}$$

We shall be concerned also with the subspace

$$\widehat{\mathfrak{m}}_\psi^- = \hat{\mathfrak{g}}_{\psi,0}^-$$

of $\hat{\mathfrak{g}}_\psi^-$, and its quotient

$$\hat{\mathfrak{g}}_\psi^-/\widehat{\mathfrak{m}}_\psi^- = \bigoplus_{\alpha\in\widehat{\Sigma}_M} \hat{\mathfrak{g}}_{\psi,\alpha}^-.$$

The group $\mathcal{A}_\psi$ of course acts on these two spaces. So does the normalizer $\overline{N}_\psi$ in $\overline{S}_\psi$ of the maximal torus

$$\overline{T}_\psi = Z(\widehat{M})^\Gamma/Z(\widehat{G}^0)^\Gamma = A_{\widehat{M}}/\big(A_{\widehat{M}} \cap Z(\widehat{G}^0)\big)^\Gamma$$

in $\overline{S}_\psi^*$. We write

$$\varepsilon_\psi^M(u) = \det\big(n, \mathrm{End}_{\mathcal{A}_\psi}(\widehat{\mathfrak{m}}_\psi^-)\big), \qquad n \in \overline{N}_\psi,$$

and

$$\varepsilon_\psi^{G/M}(u) = \det\big(n, \mathrm{End}_{\mathcal{A}_\psi}(\hat{\mathfrak{g}}_\psi^-/\widehat{\mathfrak{m}}_\psi^-)\big), \qquad n \in \overline{N}_\psi,$$

where u is the projection of n onto the quotient

$$\mathfrak{N}_\psi = \overline{N}_\psi/\overline{T}_\psi.$$

Then

$$\varepsilon_\psi^M(u)\,\varepsilon_\psi^{G/M}(u) = \varepsilon_\psi(x_u), \qquad u \in \mathfrak{N}_\psi,$$

where we recall that x_u is the image of u in $\mathcal{S}_\psi$.

Consider the adjoint action of the product $\overline{N}_\psi \times \mathcal{A}_\psi$ on the block diagonal subalgebra $\widehat{\mathfrak{m}}_\psi^-$ of $\hat{\mathfrak{g}}_\psi^-$. The formula for $\varepsilon_\psi^M(u)$ above matches an earlier definition of the extension $\varepsilon_{\psi_M}^M(\widetilde{u})$ of the character $\varepsilon_{\psi_M}^M$ to $\widetilde{S}_{\psi_M,u}$, which appears for example in (4.2.11). We can therefore write

$$\varepsilon_\psi^M(u) = \varepsilon_{\psi_M}^M(\widetilde{u}) = \varepsilon_\psi^1(u),$$

in the notation introduced prior to the statement of Lemma 4.3.1. On the other hand, the subgroup $\mathcal{S}_\psi^1$ of $\mathfrak{N}_\psi^*$ leaves invariant each of the spaces $\hat{\mathfrak{g}}_{\psi,\alpha}^-$. Since the actions of $\mathcal{S}_\psi^1$ on $\hat{\mathfrak{g}}_{\psi,\alpha}^-$ and $\hat{\mathfrak{g}}_{\psi,-\alpha}^-$ are dual, their contributions to $\varepsilon_\psi^{G/M}(u)$ cancel. It follows that the value

$$\varepsilon_\psi^{G/M}(w_u) = \varepsilon_\psi^{G/M}(u)$$

depends only on the image w_u of u in W_ψ. We have shown that our sign character $\varepsilon_\psi(x) = \varepsilon_\psi^G(x)$ satisfies a decomposition formula

$$\varepsilon_\psi^1(u)\,\varepsilon_\psi^{G/M}(w_u) = \varepsilon_\psi(x_u), \qquad u \in \mathfrak{N}_\psi. \tag{4.6.7}$$

In preparation for the proof of Lemma 4.3.1, we shall also investigate the global normalizing factor

$$r_\psi(w) = r^G_\psi(w) = r_P(w, \psi_M),$$

given by (4.2.10). It equals the value at $\lambda = 0$ of the quotient

$$L(0, \rho_{P,w} \circ \psi_{M,\lambda})\, \varepsilon(0, \rho_{P,w} \circ \psi_{M,\lambda})^{-1}\, L(1, \rho_{P,w} \circ \psi_{M,\lambda})^{-1}, \tag{4.6.8}$$

where $\rho_{P,w} = \rho^\vee_{w^{-1}P|P}$ is contragredient of the adjoint representation of ${}^LM$ on

$$w^{-1}\widehat{\mathfrak{n}}_P w / w^{-1}\widehat{\mathfrak{n}}_P w \cap \widehat{\mathfrak{n}}_P,$$

and $\psi_{M,\lambda}$ represents the formal twist of the L-homomorphism ψ_M by a point λ in the complex vector space

$$\mathfrak{a}^*_{M,\mathbb{C}} = X^*(M)_F \otimes \mathbb{C}.$$

We are now using the Artin notation for L-functions introduced in §1.5, or rather its extension to finite-dimensional representations of the product $\mathcal{A}_\psi$, since the automorphic representation π_{ψ_M} in (4.2.10) will not be an explicit part of the discussion here. It is of course implicit in the parameter ψ_M, and we must indeed remember to give the ε-factor in (4.6.8) its automorphic interpretation.

As a global normalizing factor, we know that (4.6.8) is a meromorphic function of λ, which is analytic at $\lambda = 0$. This is a consequence of the general theory of Eisenstein series, and the normalization of local intertwining operators from Chapter 2. However, we will need to be more explicit. We write

$$\widehat{\Sigma}_{P,w} = \{\alpha \in \widehat{\Sigma}_P :\ w\alpha \notin \widehat{\Sigma}_P\}$$

as usual, where $\widehat{\Sigma}_P \subset \widehat{\Sigma}_M$ denotes the set of roots in $(\widehat{P}, A_{\widehat{M}})$. Then

$$\rho_{P,w} = \bigoplus_{\alpha \in \widehat{\Sigma}_{P,w}} \rho^\vee_{-\alpha} \cong \bigoplus_{\alpha \in \widehat{\Sigma}_{P,w}} \rho_\alpha,$$

since the Killing form on $\hat{\mathfrak{g}}$ determines an isomorphism from $\rho^\vee_{-\alpha}$ to ρ_α. From the decomposition (4.6.5) of $\rho_\alpha \circ \psi_M$ into standard representations, we see that the L-function

$$L(s, \rho_{P,w} \circ \psi_{M,\lambda})$$

in (4.6.5) has analytic continuation with functional equation. It follows that (4.6.8) equals

$$L(1, \rho^\vee_{P,w} \circ \psi_{M,\lambda})\, L(1, \rho_{P,w} \circ \psi_{M,\lambda})^{-1}.$$

We can thus write

$$\begin{aligned} r_\psi(w) &= \lim_{\lambda\to 0} L(1, \rho_{P,w}^\vee \circ \psi_{M,\lambda})\, L(1, \rho_{P,w} \circ \psi_{M,\lambda})^{-1} \\ &= \lim_{\lambda\to 0} \prod_{\alpha\in\widehat{\Sigma}_{P,w}} \big(L(1, \rho_\alpha^\vee \circ \psi_{M,\psi})\, L(1, \rho_\alpha, \psi_{M,\lambda})^{-1}\big) \\ &= \lim_{\lambda\to 0} \prod_{\alpha\in\widehat{\Sigma}_{P,w}} \big(L(1-\alpha(\lambda), \rho_\alpha^\vee \circ \psi_M)\, L(1+\alpha(\lambda), \rho_\alpha \circ \psi_M)^{-1}\big) \\ &= \lim_{\lambda\to 0} \prod_{\alpha\in\widehat{\Sigma}_{P,w}} \big((-\alpha(\lambda))^{a_{-\alpha}} (\alpha(\lambda))^{-a_\alpha}\big), \end{aligned}$$

where a_α denotes the order of pole of $L(s, \rho_\alpha \circ \psi_M)$ at $s = 1$. It follows from (4.6.5) that

$$L(s, \rho_\alpha \circ \psi_M) = \prod_{h\in H_{\psi,\alpha}} L(s, \sigma_h \otimes \nu_h).$$

In particular,

$$a_\alpha = \sum_{h\in H_{\psi,\alpha}} a_h,$$

where

$$a_h = \mathrm{ord}_{s=1}\, L(s, \sigma_h \otimes \nu_h).$$

Moreover, since $L(s, \sigma_h \otimes \nu_h)$ can be regarded as a standard automorphic L-function (of type $R_{kk'}$, S_k or Λ_k, but attached to unitary, not necessarily cuspidal representations), we have the adjoint relation

$$L\big(s, (\sigma_h \otimes \nu_h)^\vee\big) = \overline{L(\overline{s}, \sigma_h \otimes \nu_h)}.$$

It follows that

$$a_h = \mathrm{ord}_{s=1}\, L\big(s, (\sigma_h \otimes \nu_h)^\vee\big).$$

Since $\rho_{-\alpha} \cong \rho_\alpha^\vee$, the orders $a_{-\alpha}$ and a_α are then equal. We conclude that

$$r_\psi(w) = \prod_{\alpha\in\widehat{\Sigma}_{P,w}} \prod_{h\in H_{\psi,\alpha}} (-1)^{a_h}. \tag{4.6.9}$$

Observe that from the last adjoint relation, the contributions to (4.6.9) of distinct representations $(\sigma_h \otimes \nu_h)^\vee$ and $\sigma_h \otimes \nu_h$ cancel. The interior product in (4.6.9) need therefore be taken over only those h with $\sigma_h^\vee = \sigma_h$.

The question then is to determine the parity of a_h, for any $h \in H_{\psi,\alpha}$ with $\sigma_h^\vee \cong \sigma_h$. The irreducible representation ν_h maps the diagonal matrix $diag(|t|^{\frac{1}{2}}, |t|^{-\frac{1}{2}})$ in $SL(2,\mathbb{C})$ to the diagonal matrix

$$diag\big(|t|^{\frac{1}{2}(n_h-1)}, |t|^{\frac{1}{2}(n_h-3)}, \dots, |t|^{-\frac{1}{2}(n_h-1)}\big),$$

in $GL(n_h, \mathbb{C})$, where $n_h = \dim(\nu_h)$. It follows from the definitions that

$$L(s, \sigma_h \otimes \nu_h) = \prod_{i=1}^{n_h} L\big(s + \tfrac{1}{2}(n_h - 2i + 1), \sigma_h\big). \tag{4.6.10}$$

We must therefore describe the order of $L(s, \sigma_h)$ at any real half integer. As a standard (unitary, cuspidal) automorphic L-function of type $R_{kk'}$, S_k

or Λ_k, $L(s, \sigma_h)$ is nonzero whenever $\mathrm{Re}(s) \geqslant 1$ or $\mathrm{Re}(s) \leqslant 0$. Moreover, its only possible poles in this region are at $s = 1$ and $s = 0$. The relevant half integers in (4.6.10) are thus confined to the points 0, $\frac{1}{2}$, and 1. The question separates into two cases. If $\dim(\nu_h)$ is even, we have only to consider the order of $L(s, \sigma_h)$ at $s = \frac{1}{2}$. If $\dim(\nu_h)$ is odd, we need to consider its order at the points $s = 0$ and $s = 1$. In both cases, we shall assume that the assertions of Theorem 1.5.3 can be applied inductively to the simple generic constituents of ψ.

Suppose first that $\dim(\nu_h)$ is even. Then σ_h cannot be diagonal, in the sense of Corollary 4.6.2. It must consequently be an irreducible, self-dual Rankin-Selberg representation of type $R_{kk'}$. Its L-function satisfies the functional equation

$$L(s, \sigma_h) = \varepsilon(s, \sigma_h)\, L(1 - s, \sigma_h). \tag{4.6.11}$$

It follows that $L(s, \sigma_k)$ will have even or odd order at $s = \frac{1}{2}$ according to whether $\varepsilon\big(\frac{1}{2}, \sigma_h\big)$ equals $+1$ or -1. The contribution of h to (4.6.9) is therefore just the sign of $\varepsilon\big(\frac{1}{2}, \sigma_h\big)$. Observe that if σ_h is orthogonal, Theorem 1.5.3(b) asserts that the sign is $+1$, and can be ignored. In other words, we can assume that σ_h is symplectic, and therefore that h lies in the subset $H^-_{\psi,\alpha}$ of $H_{\psi,\alpha}$. We thus obtain the contribution to the formula (4.6.9) for $r_\psi(w)$ of those h with $\dim(\nu_h)$ even. It equals the product

$$r^-_\psi(w) = \prod_{\alpha \in \widehat{\Sigma}_{P,w}} (-1)^{|H^-_{\psi,\alpha}|}. \tag{4.6.12}$$

Suppose next that $\dim(\nu_h)$ is odd. Then σ_h is a self-dual representation of $\mathcal{L}_\psi$, which is typically diagonal, in the sense of Corollary 4.6.2. Its L-function still satisfies the functional equation (4.6.11), but with the understanding that the ε-factor $\varepsilon(s, \sigma_h)$ is given its automorphic interpretation. In particular, the orders of $L(s, \sigma_h)$ at $s = 0$ and $s = 1$ are equal. It follows from (4.6.10) that the contribution of h to (4.6.9) is just 1 if $\dim(\nu_h)$ is greater than 1, and is equal to the order of $L(s, \sigma_h)$ at $s = 1$ if ν_h is the trivial one-dimensional representation of $SL(2, \mathbb{C})$. Theorem 1.5.3(a) tells us this order, in case σ_h is of the form S_k or Λ_k. It amounts to the assertion that $L(s, \sigma_h)$ has a pole at $s = 1$, necessarily of order 1, if and only if the representation σ_h of $\mathcal{L}_\psi$ contains the trivial representation. The same assertion has long been known in the other case that σ_h is of Rankin-Selberg type $R_{kk'}$. This gives us the contribution of h to (4.6.9) in case $\dim(\nu_h)$ is odd. It equals (-1) if the representation $\sigma_h \otimes \nu_h$ of $\mathcal{A}_\psi$ contains the trivial 1-dimensional representation, and is $(+1)$ otherwise.

It is easy to characterize the total contribution to (4.6.9) of those h with $\dim(\nu_h)$ odd. The representation $\sigma_h \otimes \nu_h$ of $\mathcal{A}_\psi$ acts on a subspace $\hat{\mathfrak{g}}_{\alpha,h}$ of the root space $\hat{\mathfrak{g}}_\alpha$ of $\hat{\mathfrak{g}}$. It contains the trivial representation if and only if $\hat{\mathfrak{g}}_{\alpha,h}$ has nontrivial intersection with the Lie algebra of $\overline{S}^*_\psi$. This is to say that α is a root $(\overline{S}^*_\psi, \overline{T}_\psi)$, and that $\hat{\mathfrak{g}}_{\alpha,h}$ contains the corresponding root space. As α and h range over $\widehat{\Sigma}_{P,w}$ and $H_{\psi,\alpha}$, we obtain a factor (-1) in the product formula (4.6.9) for $r_\psi(w)$ for every positive root α of $(\overline{S}_\psi, \overline{T}_\psi)$ that is mapped by w to a negative root. The product of these factors is equal to the ordinary sign character $s^0_\psi(w)$ on W_ψ introduced in §4.1. The character $s^0_\psi(w)$ thus gives the contribution from those h with $\dim(\nu_h)$ odd.

This completes our analysis of the product (4.6.9) in terms of its even and odd parts. We have shown that the global normalizing factor $r_\psi(w) = r^G_\psi(w)$ has a decomposition

$$r_\psi(w) = r^-_\psi(w)\, s^0_\psi(w), \qquad w \in W_\psi. \tag{4.6.13}$$

PROOF OF LEMMA 4.3.1. We must establish an identity

$$r_\psi(w_u)\, \varepsilon^1_\psi(u) = \varepsilon_\psi(x_u)\, s^0_\psi(u), \qquad u \in \mathfrak{N}_\psi,$$

among the four signs, where $s^0_\psi(u) = s^0_\psi(w_u)$. The decompositions (4.6.7) and (4.6.13) reduce the problem to showing that the two characters r^-_ψ and $\varepsilon^{G/M}_\psi$ on W_ψ are equal. We are assuming that ψ and M are as in §4.3. In fact we assume that M is proper in G^0, and therefore subject to our induction hypotheses, since the required assertion is trivial if $M = G^0$.

We fix $w \in W_\psi$. Let $\overline{T}^+_{\psi,w}$ be the group generated in $\overline{S}^+_\psi$ by the subset $\overline{T}_\psi w$ of $\overline{S}_\psi$, and let Λ^-_ψ be the representation of this group on the space

$$V^-_\psi = \mathrm{End}_{\mathcal{A}_\psi}(\hat{\mathfrak{g}}^-_\psi / \widehat{\mathfrak{m}}^-_\psi).$$

We can then write

$$\varepsilon^{G/M}_\psi(w) = \det\big(\Lambda^-_\psi(w)\big),$$

since the determinant is independent of the preimage of w in $\overline{T}^+_{\psi,w}$.

The group $\overline{T}^+_{\psi,w}$ is an extension by $\overline{T}_\psi$ of the cyclic group $\langle w\rangle$ generated by w. Its representation theory is well understood. For example, there is a surjective mapping $\Lambda \to \mathfrak{o}$, from the set of irreducible representations Λ of $\overline{T}^+_{\psi,w}$ onto the set of $\langle w\rangle$-orbits $\mathfrak{o}$ of characters $t \to t^\xi$ on $\overline{T}_\psi$, such that

$$\mathrm{tr}\big(\Lambda(t)\big) = \sum_{\xi \in \mathfrak{o}} t^\xi, \qquad t \in \overline{T}_\psi.$$

It is a consequence of our definitions that

$$\mathrm{tr}\big(\Lambda^-_\psi(t)\big) = \sum_{\alpha \in \widehat{\Sigma}_M} |H^-_{\psi,\alpha}|\, t^\alpha, \qquad t \in \overline{T}_\psi.$$

This leads to a decomposition

$$\Lambda^-_\psi = \bigoplus_{\mathfrak{o}_\alpha} \bigoplus_{h \in H^-_{\psi,\alpha}} \Lambda_h,$$

where $\mathfrak{o}_\alpha$ ranges over the $\langle w\rangle$-orbits of roots $\alpha \in \widehat{\Sigma}_M$, and Λ_h is an irreducible representation of $\overline{T}^+_{\psi,w}$ that maps to $\mathfrak{o}_\alpha$. We can therefore write

$$\varepsilon_\psi^{G/M}(w) = \prod_{\mathfrak{o}_\alpha} \prod_{h \in H^-_{\psi,\alpha}} \det\big(\Lambda_h(w)\big). \tag{4.6.14}$$

An orbit $\mathfrak{o}_\alpha$ is said to be *symmetric* if it contains the root $(-\alpha)$. If α is not symmetric, $\mathfrak{o}_\alpha$ and $\mathfrak{o}_{-\alpha}$ are two distinct orbits of equal order. They give rise to pairwise contragredient sets of representations Λ_h, whose contributions to the formula (4.6.14) cancel. We can therefore restrict the outer product in (4.6.14) to orbits $\mathfrak{o}_\alpha$ that are symmetric.

Other factors in the product (4.6.14) can also pair off. Recall that any index $h \in H^-_{\psi,\alpha}$ gives rise to an irreducible representation $\mu_h \otimes \nu_h$ of $\mathcal{A}_\psi$. The self-dual representations μ_h and ν_h are both symplectic, since $\varepsilon\big(\frac{1}{2}, \mu_h\big) = -1$ and $\nu_h(s_\psi) = -1$, so their tensor product is orthogonal. On the other hand, the representation

$$\bigoplus_h (\Lambda_h \otimes \sigma_h \otimes \nu_h)$$

of $\overline{T}^+_{\psi,w} \times \mathcal{A}_\psi$ on $\hat{\mathfrak{g}}^-_\psi$ is invariant under the Killing form, and hence orthogonal. This does not mean that the individual representations Λ_h need to be orthogonal. However, the representations Λ_h that are not orthogonal will occur in pairs, whose contributions to (4.6.14) cancel. We can therefore take the interior product in (4.6.14) over the subset $H^-_{\psi,\alpha,o}$ of indices $h \in H^-_{\psi,\alpha}$ such that Λ_h is orthogonal.

Consider an orthogonal representation

$$\Lambda_h, \qquad h \in H^-_{\psi,\alpha,o}.$$

Elements of the Weyl group of a complex orthogonal group can be represented by permutation matrices. One sees from this that the image of $\overline{T}^+_{\psi,w}$ under Λ_h is the semidirect product of the image of $\overline{T}_\psi$ with the (cyclic) quotient of $\langle w\rangle$ of order $|\mathfrak{o}_\alpha|$. It then follows that the restriction of Λ_h to $\langle w\rangle$ is isomorphic to the permutation representation of $\langle w\rangle$ on $\mathfrak{o}_\alpha$. In particular, $\Lambda_h(w)$ is a well defined operator, whose eigenvalues are the set of roots of 1 of order $|\mathfrak{o}_\alpha|$. We are assuming that the orbit $\mathfrak{o}_\alpha$ is symmetric. It therefore has even order, from which it follows that

$$\det\big(\Lambda_h(w)\big) = -1.$$

The orthogonal representations thus each contribute a sign to the product (4.6.14). The formula reduces to

$$\varepsilon_\psi^{G/M}(w) = \prod_{\mathfrak{o}_\alpha \text{ symmetric}} (-1)^{|H^-_{\psi,\alpha,o}|}. \tag{4.6.15}$$

The analysis of the product formula (4.6.12) for $r^-_\psi(w)$ is slightly different. Since

$$|H^-_{\psi,w\alpha}| = |H^-_{\psi,\alpha}|,$$

the factor of α in the product depends only on the orbit $\mathfrak{o}_\alpha$. If $\mathfrak{o}_\alpha$ is symmetric, we see after a moment's thought that the integer

$$\left|\mathfrak{o}_\alpha \cap \widehat{\Sigma}_{P,w}\right|$$

equals the number of sign changes in the sequence

$$(\alpha, w\alpha, \ldots, -\alpha)$$

(relative to the order on $\widehat{\Sigma}_M$ defined $\widehat{\Sigma}_P$). Since this number is odd, the contribution to (4.6.12) of all the roots in a symmetric orbit is the same as the contribution of any positive root in the orbit. If $\mathfrak{o}_\alpha$ is not symmetric, $\mathfrak{o}_\alpha$ and $\mathfrak{o}_{-\alpha}$ are disjoint, and the integer

$$\left|\mathfrak{o}_\alpha \cap \widehat{\Sigma}_{P,w}\right| + \left|\mathfrak{o}_{-\alpha} \cap \widehat{\Sigma}_{P,w}\right|$$

equals the number of sign changes in the sequence

$$(\alpha, w\alpha, \ldots, \alpha).$$

Since the number is even in this case, the nonsymmetric orbits contribute nothing to (4.6.12). The product (4.6.12) therefore reduces to

$$\prod_{\mathfrak{o}_\alpha \text{ symmetric}} (-1)^{|H^-_{\psi,\alpha}|}.$$

As we have noted, indices h in the complement of $H^-_{\psi,\alpha,o}$ in $H^-_{\psi,\alpha}$ occur in pairs, and consequently contribute nothing to the product. The right hand side of (4.6.12) therefore reduces to the right hand side of (4.6.15). This establishes the required identity

$$\varepsilon_\psi^{G/M}(w) = r^-_\psi(w),$$

and completes the proof of Lemma 4.3.1. □

We observe that both assertions of Theorem 1.5.3 were needed to prove Lemma 4.3.1. They were each invoked in the preamble that led to the decomposition (4.6.13) of $r_\psi(w)$. For example, we used the assertion (b) on orthogonal ε-factors to obtain an exponent in (4.6.12) that included only indices h with σ_h symplectic. This reduction was essential. Had we been forced also to include orthogonal Rankin-Selberg representations σ_h (with a corresponding enlargement of the exponent $|H^-_{\psi,\alpha}|$ in (4.6.12)), our product (4.6.12) would then not have matched the formula for $\varepsilon_\psi^{G/M}$ we obtained in the proof of the lemma. Our application of both assertions (a) and (b) is of course predicated on our induction hypothesis. In the case of (b), we can assume that the assertion holds for any pair of simple generic parameters

$$\mu_i \in \widetilde{\Phi}_{\text{sim}}(G_i), \qquad G_i \in \widetilde{\mathcal{E}}(m_i),\ i = 1, 2, \tag{4.6.16}$$

with $m_1 + m_2 < N$.

We can actually manage with a weaker assumption on Theorem 1.5.3(b). Our proof of Lemma 4.3.1 for ψ requires only that Theorem 1.5.3(b) apply to any relevant pair

$$\psi_i = \mu_i \boxtimes \nu_i, \qquad i = 1, 2,$$

of distinct simple constituents of ψ. In other words, when Theorem 1.5.3(b) implies that the contribution to (4.6.9) of a pair is 1, we require that the contribution actually be equal to 1. The pair (ψ_1, ψ_2) gives rise to a Rankin-Selberg representation $\psi_1 \times \psi_2$ of $\mathcal{A}_\psi$, and corresponds to a root α in (4.6.9). It is relevant to Theorem 1.5.3(b) only if the Rankin-Selberg representation $\mu_1 \times \mu_2$ of $\mathcal{L}_\psi$ is orthogonal, and the tensor product $\nu_1 \otimes \nu_2$ decomposes into irreducible representations $SL(2, \mathbb{C})$ of *even* dimension. In fact, we see from (4.6.10) and the proof of Lemma 4.6.1 that $\nu_1 \otimes \nu_2$ must decompose into an *odd* number of such representations for the application of Theorem 1.5.3(b) to (4.6.9) to be nontrivial. The assumption on Theorem 1.5.3(b) can then be formulated as in (4.6.16), but with (ψ_1, ψ_2) in place of (μ_1, μ_2). We will adopt this weaker form of the induction assumption at the beginning of the next chapter, and then resolve it in Lemma 5.1.6. The induction hypothesis on Theorem 1.5.3(a), implicit in the proof, will not be treated until much later. It will be resolved only at the end of §8.2, as part of the final step.

PROOF OF LEMMA 4.4.1. This lemma will be easier to prove than the last. We have to show that

$$\varepsilon_{\psi'}^{G'}(s_{\psi'}) = \varepsilon_\psi(s_\psi x_s), \qquad s \in \overline{S}_{\psi,\mathrm{ss}},$$

where (G', ψ') is the preimage of (ψ, s) under the correspondence (1.4.11). We shall follow the general argument set out in [**A9**, p. 51–53].

We begin with the formula

$$\varepsilon_\psi(x_s) = \varepsilon_\psi^G(x_s) = \prod_{\kappa \in \mathcal{K}_\psi^-} \det\big(\lambda_\kappa(s)\big)$$

obtained (4.6.4) from any decomposition

$$\tau_\psi = \bigoplus_{\kappa \in \mathcal{K}_\psi} (\lambda_\kappa \otimes \sigma_\kappa \otimes \nu_\kappa)$$

that satisfies the conditions of Lemma 4.6.1. We could in fact weaken these conditions. For example, a decomposition in which the representations λ_κ of $\overline{S}_\psi$ are allowed to be reducible gives the same formula for $\varepsilon_\psi^G(x_s)$.

Notice that the point s_ψ belongs both the group $SL(2, \mathbb{C})$ and the centralizer S_ψ^*. Therefore

$$\lambda_\kappa(s_\psi) = \nu_\kappa(s_\psi) = -1,$$

for any κ in the subset $\mathcal{K}_\psi^-$ of $\mathcal{K}_\psi$. It follows that

$$\varepsilon_\psi^G(s_\psi) = \prod_{\kappa \in \mathcal{K}_\psi^-} (-1)^{\dim(\lambda_\kappa)}.$$

We will of course be free to apply this formula if G is replaced by the endoscopic group G'.

Suppose that s, G' and ψ' are as in the putative identity. The dual group $\widehat{G}'$ is then the connected centralizer in $\widehat{G}^0$ of the point $s' = s$. Its Lie algebra $\hat{\mathfrak{g}}'$ is just the kernel of $\mathrm{Ad}(s)$ in $\hat{\mathfrak{g}}$. The parameter ψ' for G' is defined by the L-embedding

$$\widetilde{\psi}' : \mathcal{A}_\psi \longrightarrow {}^L G',$$

which in turn is obtained from $\widetilde{\psi}_G$ by restriction of the codomain to the L-subgroup ${}^L G'$ of ${}^L G^0$. It follows that $\tau_{\psi',1}$, the restriction of $\tau_{\psi'}$ to $\mathcal{A}_\psi$, is simply the mapping given by restricting the representation $\tau_{\psi,1}$ to the subspace $\hat{\mathfrak{g}}'$ of $\hat{\mathfrak{g}}$. For any $\kappa \in \mathcal{K}_\psi$, let λ_κ^s be the representation of $\overline{S}_{\psi'}$ on the $(+1)$-eigenspace of $\lambda_\kappa(s)$. If $\mathcal{K}_{\psi'}$ denotes the set of $\kappa \in \mathcal{K}_\psi$ with $\lambda_\kappa^s \neq 0$, we can then write

$$\tau_{\psi'} = \bigoplus_{\kappa \in \mathcal{K}_{\psi'}} (\lambda_\kappa^s \otimes \sigma_\kappa \otimes \nu_\kappa).$$

This decomposition satisfies the conditions of Lemma 4.6.1 (weakened to the extent that λ_κ^s could be reducible). It follows that

$$\varepsilon_{\psi'}^{G'}(s_{\psi'}) = \prod_{\kappa \in \mathcal{K}_{\psi'}^-} (-1)^{\dim(\lambda_\kappa^s)},$$

where $\mathcal{K}_{\psi'}^-$ is the intersection of $\mathcal{K}_{\psi'}$ with $\mathcal{K}_\psi^-$.

The original formula (4.6.4) tells us that $\varepsilon_\psi^G(x_s)$ equals the product of all the eigenvalues, counting multiplicities, of all of the operators

$$\{\lambda_\kappa(s) : \kappa \in \mathcal{K}_\psi^-\}.$$

But the contragredient $\lambda_\kappa \to \lambda_\kappa^\vee$ can be treated as an involution on the set of representations λ_κ with $\kappa \in \mathcal{K}_\psi^-$. It follows that if ξ is an eigenvalue not equal to $(+1)$ or (-1), its inverse ξ^{-1} is another eigenvalue, but with the same total multiplicity. Therefore

$$\varepsilon_\psi^G(x_s) = (-1)^{m(-1)},$$

where $m(-1)$ is the total multiplicity of the eigenvalue (-1). By the same token, the number

$$\Big(\sum_{\kappa \in \mathcal{K}_\psi^-} \big(\dim(\lambda_\kappa) - \dim(\lambda_\kappa^s)\big)\Big) - m(-1),$$

being the sum of multiplicities of eigenvalues distinct from (± 1), is an even integer. It follows that

$$\begin{aligned}\varepsilon_\psi^G(s_\psi x_s) &= \varepsilon_\psi^G(s_\psi)\,\varepsilon_\psi^G(x_s)\\ &= (-1)^{m(-1)} \prod_{\kappa\in\mathcal{K}_\psi^-} (-1)^{\dim(\lambda_\kappa)}\\ &= (-1)^{m(-1)} \prod_{\kappa\in\mathcal{K}_\psi^-} (-1)^{(\dim(\lambda_\kappa)-\dim(\lambda_\kappa^s))}\cdot \varepsilon_{\psi'}^{G'}(s_{\psi'})\\ &= \varepsilon_{\psi'}^{G'}(s_{\psi'}),\end{aligned}$$

as required. This completes the proof of Lemma 4.4.1. □

4.7. On the global theorems

The remaining two sections of the chapter represent a digression. The purpose of this section is to relate the global theorems we have stated with the trace formula. The discussion amounts to a series of informal remarks, which will help us to understand the formal proofs later on. In the next section, we shall sketch how our earlier comparison of spectral and endoscopic terms extends to general groups. A reader wishing to go on with the main argument, which we left off at the end of §4.5, can skip these sections and proceed directly to Chapter 5. We will however include a formal lemma at the end of the section, which we will use later after establishing the conditions on which it is based.

This section is intended as more technical motivation for the global theorems stated in Chapter 1. Among other things, we shall see how the theorems are consequences of the stable multiplicity formula of Theorem 4.1.2. This will be for guidance only, since in the end we will have to establish all of the theorems together. However, it might explain how the multiplicity formula of Theorem 1.5.2, and the properties of poles and signs in Theorem 1.5.3, are in some sense implicit in the trace formula. The remarks of this section can also be regarded as an illustration of the general comparison in the simplest situation – the case of a parameter $\psi \in \widetilde{\Psi}_{\mathrm{ell}}(N)$ that was excluded earlier, and which is ultimately the hardest to prove.

For this exercise, we shall assume whatever we need of the local theorems we have stated. We shall also suppose that the global seed Theorem 1.4.1 and its complement Theorem 1.4.2 are both valid, and that the resulting global parameters for groups $G \in \widetilde{\mathcal{E}}_{\mathrm{sim}}(N)$ have the rough spectral properties that motivated the initial remarks in §1.2. Specifically, we assume that there is no contribution to the discrete spectrum of G from any parameter ψ that does not lie in $\widetilde{\Psi}_2(G)$. We shall treat the fine spectral assertions of Theorem 1.5.2 and parts (a) and (b) of Theorem 1.5.3 in reverse order.

We begin with a general global parameter

$$\psi = \psi_1 \boxplus \cdots \boxplus \psi_r, \qquad \psi_i \in \widetilde{\Psi}_{\mathrm{sim}}(N_i), \tag{4.7.1}$$

in $\widetilde{\Psi}_{\mathrm{ell}}(N)$. This was implicitly excluded from earlier sections, specifically in the comparisons of §4.3 and §4.4 that led to Proposition 4.5.1. However, with our heuristic assumptions above, the comparisons are valid for ψ, and in fact are quite transparent. They might give us a better overall view if we retrace some of the steps. We shall begin with $G = \widetilde{G}(N)$ and f equal to a function $\widetilde{f} \in \widetilde{\mathcal{H}}(N)$.

Suppose for a moment that $r = 1$. Then

$$\psi = \psi_1 = \mu \boxtimes \nu$$

is simple, and has a contribution $R^0_{\mathrm{disc},\psi}$ to the relative discrete spectrum of $G^0 = GL(N)$. We note that the sign character ε^G_ψ is trivial in this case, since as in Corollary 4.6.2, the tensor product of the unipotent part ν with itself is a sum of irreducible odd dimensional representations. We also recall that $R^G_{\mathrm{disc},\psi}$, the canonical extension of $R^{G^0}_{\mathrm{disc},\psi}$, is a product of the corresponding local extensions, according to the theory of local and global Whittaker models for $GL(N)$. Therefore

$$I^G_{\mathrm{disc},\psi}(f) = r^G_{\mathrm{disc},\psi}(f) = |\kappa_G|^{-1}\mathrm{tr}(R_{\mathrm{disc},\psi}(f)) = \tfrac{1}{2}\,\mathrm{tr}\big(R^G_{\mathrm{disc},\psi}(f)\big) = \tfrac{1}{2}\,f_G(\psi).$$

This can be regarded as an elementary analogue for $G = \widetilde{G}(N)$ of either Theorem 1.5.2 or Theorem 4.2.2. We have used it, at least implicitly, in our earlier discussion in §4.3 and §4.4.

If $r > 1$, $I^G_{\mathrm{disc},\psi}(f)$ equals the difference (4.3.1) with which we began the comparison in §4.3, since ψ does not contribute to the discrete spectrum of $G = \widetilde{G}(N)$. Therefore $I^G_{\mathrm{disc},\psi}(f)$ equals the expression (4.3.2) we obtained for the difference. To analyze it, we take M to be the standard Levi subgroup of $G^0 = GL(N)$ corresponding to the partition $(N_1, \dots, N_r)$. Then ψ is the image of a parameter $\psi_M \in \Psi_2(M, \psi)$. There is one element w in the set W_ψ. This element stabilizes M, and induces the standard outer automorphism on each of the factors $GL(N_i)$ of M. It follows that

$$\big|\det(w-1)_{\mathfrak{a}^G_M}\big|^{-1} = \big(\tfrac{1}{2}\big)^r.$$

The other coefficients in (4.3.2) satisfy

$$m_\psi = |W_\psi| = |\mathcal{S}_{\psi_M}| = 1,$$

according to their definitions, and can be ignored. The fibre $\mathfrak{N}_\psi(w)$ of w in $\mathfrak{N}_\psi$ consists of one element u. We obtain

$$f_G(\psi, u) = \mathrm{tr}\big(R_P(w, \widetilde{\pi}_M, \psi_M)\mathcal{I}_P(\pi_M, f)\big) = f_G(\psi),$$

again by the theory of Whittaker models for $GL(N)$, applied this time to both M and G. The expression (4.3.2) therefore reduces to

$$\big(\tfrac{1}{2}\big)^r r_P(w, \psi_M)\, f_G(\psi), \qquad G = \widetilde{G}(N),\ f \in \widetilde{\mathcal{H}}(G).$$

We have shown that

$$I_{\mathrm{disc},\psi}(\widetilde{f}) = I^N_{\mathrm{disc},\psi}(\widetilde{f}) = \big(\tfrac{1}{2}\big)^r\, r_{\widetilde{P}}(w, \psi_{\widetilde{M}})\, \widetilde{f}_N(\psi), \qquad \widetilde{f} \in \widetilde{\mathcal{H}}(N),$$

where ψ is as in (4.7.1), $\widetilde{P}$ is a parabolic subgroup of $\widetilde{G}^0(N)$ with Levi component $\widetilde{M} = M$, w is the unique element in the Weyl set $\widetilde{W}_\psi^N = W_\psi(\widetilde{G}(N), \widetilde{M})$ and $r_{\widetilde{P}}(w, \psi_{\widetilde{M}})$ is the global normalizing factor for $\widetilde{G}(N)$. We also have the stable decomposition

$$I_{\mathrm{disc},\psi}^N(\widetilde{f}) = \sum_{G \in \widetilde{\mathcal{E}}_{\mathrm{ell}}(N)} \widetilde{\imath}(N, G)\, \widehat{S}_{\mathrm{disc},\psi}^G(\widetilde{f}^G).$$

Our heuristic assumptions, together with (4.4.12), tell us that $S_{\mathrm{disc},\psi}^G$ vanishes if ψ does not belong to the subset $\widetilde{\Psi}_2(G)$ of $\widetilde{\Psi}_{\mathrm{ell}}(N)$. We can therefore fix G to be the unique element in $\widetilde{\mathcal{E}}_{\mathrm{ell}}(N)$ such that G does belong to $\widetilde{\Psi}_2(G)$. The local Theorem 2.2.1, which we are also assuming, tells us that $\widetilde{f}_N(\psi)$ equals $\widetilde{f}^G(\psi)$. It follows that

$$\widehat{S}_{\mathrm{disc},\psi}^G(\widetilde{f}^G) = \widetilde{\imath}(N, G)^{-1} \left(\tfrac{1}{2}\right)^r r_{\widetilde{P}}(w, \psi_{\widetilde{M}})\, \widetilde{f}^G(\psi), \qquad \widetilde{f} \in \widetilde{\mathcal{H}}(N). \tag{4.7.2}$$

Consider now the case that $r = 2$ in (4.7.1), and that ψ_1 and ψ_2 are of opposite type. In other words, one is orthogonal and the other is symplectic. The datum G with $\psi \in \widetilde{\Psi}_2(G)$ is then of composite form

$$G = G_1 \times G_2, \qquad G_i \in \widetilde{\mathcal{E}}_{\mathrm{sim}}(N_i).$$

The stable multiplicity formula stated in Theorem 4.1.2 is global, and is not included in our heuristic assumptions here. However, for our simple parameters $\psi_i \in \widetilde{\Psi}_{\mathrm{sim}}(G_i)$, the formula follows inductively from our discussion of the case $r = 1$ above, with another appeal to Corollary 4.6.2 to confirm that the signs $\varepsilon^{G_i}(\psi_i)$ equal 1. The stable multiplicity formula therefore holds for the simple parameter ψ for the product G. It takes the reduced form

$$S_{\mathrm{disc},\psi}^G(f) = m_\psi\, f^G(\psi), \qquad f \in \widetilde{\mathcal{H}}(G),$$

since the coefficients $|\mathcal{S}_\psi|^{-1}$, $\sigma(\overline{S}_\psi^0)$ and $\varepsilon^G(\psi)$ are all trivial in this case. Combining this with (4.7.2), we obtain

$$m_\psi\, \widetilde{f}^G(\psi) = \widetilde{\imath}(N, G)^{-1} \left(\tfrac{1}{2}\right)^2 r_{\widetilde{P}}(w, \psi_{\widetilde{M}})\, \widetilde{f}^G(\psi), \qquad \widetilde{f} \in \widetilde{\mathcal{H}}(N).$$

It is easy to see that

$$\begin{aligned}\widetilde{\imath}(N, G)^{-1}\left(\tfrac{1}{2}\right)^2 &= \left(\tfrac{1}{2}|\overline{Z}(\widehat{G})^\Gamma|^{-1}\, |\widetilde{\mathrm{O}}\mathrm{ut}_N(G)|^{-1}\right)^{-1}\left(\tfrac{1}{2}\right)^2 \\ &= |\widetilde{\mathrm{O}}\mathrm{ut}_N(G)| = m_\psi,\end{aligned}$$

given the formula for $\widetilde{\imath}(N, G)$ at the end of §3.2, the fact that $|\overline{Z}(\widehat{G})^\Gamma| = 2$ for the composite G, and the definition in §1.5 of m_ψ. The resulting cancellation then gives

$$\widetilde{f}^G(\psi) = r_{\widetilde{P}}(w, \psi_{\widetilde{M}})\, \widetilde{f}^G(\psi), \qquad \widetilde{f} \in \widetilde{\mathcal{H}}(N).$$

We thus obtain an identity

$$r_{\widetilde{P}}(w, \psi_{\widetilde{M}}) = 1, \tag{4.7.3}$$

in our case of $r = 2$, and ψ_1 and ψ_2 of opposite type.

Actually, we should be a little careful in the last step of the justification of (4.7.3), since the mapping

$$\widetilde{f} \longrightarrow \widetilde{f}^G = \widetilde{f}^1 \times \widetilde{f}^2, \qquad \widetilde{f}^i \in \widetilde{\mathcal{H}}(G_i),$$

does not take $\widetilde{\mathcal{H}}(N)$ onto $\widetilde{\mathcal{H}}(G)$. However, an inductive application of Theorem 1.5.2 to G_1 and G_2 tells us that the linear form

$$f^G(\psi), \qquad f \in \widetilde{\mathcal{H}}(G),$$

is a sum of characters. The conclusion (4.7.3) then follows as a simple special case of Proposition 3.5.1.

The identity (4.7.3) leads to a proof of Theorem 1.5.3(b). To see this, suppose that

$$\mu_i \in \widetilde{\Phi}_{\mathrm{sim}}(m_i), \qquad i = 1, 2,$$

are simple generic parameters of the same type, either both orthogonal or both symplectic. Set $\psi_1 = \mu_1$ and $\psi_2 = \mu_2 \boxtimes \nu_2$, where ν_2 is the irreducible two-dimensional representation of $SL(2,\mathbb{C})$. The sum ψ of ψ_1 and ψ_2 then satisfies the conditions above, with

$$N = m_1 + 2m_2.$$

We calculated the global normalizing factor $r_{\widetilde{P}}(w, \psi_{\widetilde{M}})$ in §4.6. In the case here, $\widetilde{P}$ is the standard maximal parabolic subgroup of $GL(N)$ of type $(N_1, N_2) = (m_1, 2m_2)$. The representation $\rho_{\widetilde{P},w} \circ \psi_{\widetilde{M}}$ of $\mathcal{A}_\psi$ in (4.6.8) acts on the Lie algebra $\mathfrak{n}_{\widetilde{P}}$, and is the exterior tensor product of the orthogonal Rankin-Selberg representation σ of $\mathcal{L}_\psi$ attached to $\mu_1 \times \mu_2$ with the representation ν_2 of $SL(2,\mathbb{C})$. It follows from a special case of (4.6.9), (4.6.10), and (4.6.11), specifically the description below (4.6.11) of the contribution of the representation $\sigma = \sigma_h$ to (4.6.9), that

$$r_{\widetilde{P}}(w, \psi_{\widetilde{M}}) = \varepsilon\big(\tfrac{1}{2}, \sigma\big) = \varepsilon\big(\tfrac{1}{2}, \mu_1 \times \mu_2\big).$$

It then follows from (4.7.3) that

$$\varepsilon\big(\tfrac{1}{2}, \mu_1 \times \mu_2\big) = 1. \tag{4.7.4}$$

This is the assertion (b) of Theorem 1.5.3, with $\mu_1 \times \mu_2$ and (m_1, m_2) in place of $\phi_1 \times \phi_2$ and (N_1, N_2).

The assertion (a) of Theorem 1.5.3 is harder, even with the heuristic assumptions we have allowed ourselves. Its formal proof will be one of our long term goals. We shall make only a few general remarks here, just to give a sense of the direction of the later proof.

Theorem 1.5.3(a) concerns the case $r = 1$. Suppose that $\psi = \phi$ belongs to the set $\widetilde{\Phi}_{\mathrm{sim}}(G)$ of simple generic parameters, for a simple datum $G \in \widetilde{\mathcal{E}}_{\mathrm{sim}}(N)$, as in the assertion. The statement was of course motivated by the expected properties of global L-functions. If ϕ is orthogonal, according to the criterion of Theorem 1.4.1, the group $\mathcal{L}_\phi = \mathcal{L}_\psi$ is an extension of W_F by the complex orthogonal group $\widehat{G}_\phi$. In this case, the representation $S^2 \circ \widetilde{\phi}_G$ of $\mathcal{L}_\phi$ contains the trivial representation, and the L-function

$L(s,\phi,S^2)$ would be expected to have a pole at $s = 1$. Similarly, if ϕ is symplectic, the L-function $L(s,\phi,\Lambda^2)$ would be expected to have a pole at $s = 1$. We will eventually have to establish these criteria by harmonic analysis and the trace formula. The technique relies on the introduction of a supplementary parameter

$$\psi_+ = \psi \boxplus \psi$$

in $\widetilde{\Psi}(N_+)$, with $N_+ = 2N$.

The global parameter ψ_+ does not lie in the elliptic set $\widetilde{\Psi}_{\mathrm{ell}}(N_+)$, since it has a repeated factor. Let $G_+ \in \widetilde{\mathcal{E}}_{\mathrm{sim}}(N_+)$ be the twisted endoscopic datum such that ψ_+ lies in $\widetilde{\Psi}(G_+)$, and such that $\widehat{G}_+$ and $\widehat{G}$ are of the same type, either both orthogonal or both symplectic. The quadratic characters η_{ψ_+} and η_{G_+} attached to ψ_+ and G_+ are both equal to 1. We therefore have the companion datum $G_+^\vee$ in $\widetilde{\mathcal{E}}_{\mathrm{sim}}(N_+)$, and ψ_+ is also contained in its parameter set $\widetilde{\Psi}(G_+^\vee)$. The groups G_+ and $G_+^\vee$ share a maximal Levi subgroup M_+. It is isomorphic to $GL(N)$, and comes with a canonical element $\psi_{+,M_+} \in \Psi(M_+,\psi_+)$ (which we may as well identify with ψ_+). Let $\rho_+ = \rho(G_+)$ and $\rho_+^\vee = \rho(G_+^\vee)$ be the adjoint representations of $\widehat{M}_+ \cong GL(N,\mathbb{C})$ on the Lie algebras $\mathfrak{n}_+$ and $\mathfrak{n}_+^\vee$ of the unipotent radicals of corresponding maximal parabolic subgroups of $\widehat{G}_+$ and $\widehat{G}_+^\vee$. (We caution ourselves that $\rho_+^\vee$ does not denote the contragredient of ρ_+ here.) Then $\rho_+^\vee$ is of the same type as G, namely S^2 if $\widehat{G}$ is orthogonal and Λ^2 if $\widehat{G}$ is symplectic, while ρ_+ is of type opposite to G. On the other hand, the centralizer groups satisfy

$$S_{\psi_+}^\vee = S_{\psi_+}(G_+^\vee) \cong Sp(2,\mathbb{C})$$

and

$$S_{\psi_+} = S_{\psi_+}(G_+) \cong \begin{cases} O(2,\mathbb{C}), & \text{if } N \text{ is even,} \\ SO(2,\mathbb{C}), & \text{if } N \text{ is odd.} \end{cases}$$

It follows that the Lie algebra of $S_{\psi_+}^\vee$ intersects $\mathfrak{n}_+^\vee$ in a 1-dimensional space, while the Lie algebra of S_{ψ_+} intersects $\mathfrak{n}_+$ only at $\{0\}$.

These properties are direct consequences of the structure of the Lie algebras of $\widehat{G}_+$ and $\widehat{G}_+^\vee$. They reduce the problem to the behaviour of global normalizing factors. To be more specific, we have first to note that the heuristic assumptions of this section preclude a contribution from ψ_+ to the discrete spectrum of either G_+ or $G_+^\vee$. The analogue for either of these groups of the expression (4.3.2) is therefore simply the discrete part of the corresponding trace formula, represented by the first term in the difference (4.3.1).

Consider then the expression for

$$I_{\mathrm{disc},\psi_+}^\vee(f_+^\vee) = I_{\mathrm{disc},\psi_+}^{G_+^\vee}(f_+^\vee), \qquad f_+^\vee \in \widetilde{\mathcal{H}}(G_+^\vee),$$

given by the analogue of (4.3.2). The double sum in (4.3.2) collapses to one simple summand in this case, of which the various terms are easy to describe explicitly. In particular, by (4.6.9) and (4.6.10), the analogue of the global normalizing factor $r_P(w, \psi_M)$ reduces to the sign

$$(-1)^{a_+^\vee}, \qquad a_+^\vee = \mathrm{ord}_{s=1} L(s, \rho_+^\vee \circ \psi_+).$$

This sign is supposed to equal (-1), since it comes from the L-function that is expected to have a simple pole. One can also describe all of the terms in the expression for

$$I^{N_+}_{\mathrm{disc},\psi_+}(\widetilde{f}_+), \qquad \widetilde{f}_+ \in \widetilde{\mathcal{H}}(N_+),$$

given by its analogue of (4.3.2). In this case, the relevant global normalizing factor does in fact equal (-1). We choose $f_+^\vee$ and $\widetilde{f}_+$ so that they transfer to the same function in $\widetilde{\mathcal{S}}(G_+^\vee)$. It then follows easily from the global intertwining relation for $G_+^\vee$ (which reduces to the local intertwining relation we are assuming) and its analogue for $\widetilde{G}(N_+)$ (which we already know) that the analogues for $\widetilde{f}_+$ and $f_+^\vee$ of the linear form $f_G(\psi, u)$ in (4.3.2) are equal. We then see immediately that the two expressions are equal up to a multiple that is positive if and only if $L(s, \rho_+^\vee \circ \psi_+)$ does have a pole (necessarily simple) at $s = 1$. A little more thought reveals that the absolute value of this multiple is actually equal to the coefficient $\widetilde{\iota}(N_+, G_+^\vee)$. The required assertion (a) is therefore equivalent to the undetermined sign in an identity

$$I^{N_+}_{\mathrm{disc},\psi_+}(\widetilde{f}_+) = \pm\, \widetilde{\iota}(N_+, G_+^\vee)\, I^\vee_{\mathrm{disc},\psi_+}(f_+^\vee)$$

being *positive*. How might one use the existence of this identity to determine the sign?

We can write

$$I^\vee_{\mathrm{disc},\psi_+}(f_+^\vee) = S^\vee_{\mathrm{disc},\psi_+}(f_+^\vee),$$

since the only elliptic endoscopic datum for $G_+^\vee$ through which ψ_+ factors in $G_+^\vee$ itself. For similar reasons, we can write

$$I^{N_+}_{\mathrm{disc},\psi_+}(\widetilde{f}_+) = \widetilde{\iota}(N_+, G_+)\, \widehat{S}_{\mathrm{disc},\psi_+}(\widetilde{f}_+^{G_+}) + \widetilde{\iota}(N_+, G_+^\vee)\, \widehat{S}^\vee_{\mathrm{disc},\psi_+}(\widetilde{f}_+^{G_+^\vee}). \tag{4.7.5}$$

Now the factor $\sigma(\overline{S}^0_{\psi_+})$ in the putative stable multiplicity formula for S_{disc,ψ_+} vanishes, since the group

$$\overline{S}^0_{\psi_+} = \overline{S}_{\psi_+}(G_+)^0 \cong GL(1, \mathbb{C})$$

has infinite center. We therefore expect that S_{disc,ψ_+} vanishes (as a linear form on $\widetilde{\mathcal{H}}(G_+)$). If it does, we see that

$$I_{\mathrm{disc},\psi_+}(\widetilde{f}_+) = \widetilde{\iota}(N_+, G_+^\vee)\, I^\vee_{\mathrm{disc},\psi_+}(f_+^\vee),$$

given that $\widetilde{f}_+$ and $f_+^\vee$ transfer to the same function in $\widetilde{\mathcal{S}}(G_+^\vee)$. In other words, the identity above will indeed have the required positive sign.

Our heuristic assumptions do not include the stable multiplicity formula of Theorem 4.1.2. However, it is possible to establish that S_{disc,ψ_+} vanishes from the assumptions we do have, and the expression for

$$I_{\mathrm{disc},\psi_+}(f_+) = I^{G_+}_{\mathrm{disc},\psi_+}(f_+), \qquad f_+ \in \widetilde{\mathcal{H}}(G_+),$$

given by the analogue of (4.3.2).

If N is odd, the argument is easy. In this case, the group $\overline{S}_{\psi_+} = \overline{S}^0_{\psi_+}$ has infinite center, and ψ_+ does not belong to the subset $\widetilde{\Psi}_{\mathrm{disc}}(G_+)$ of $\widetilde{\Psi}(G_+)$ defined in §4.1, from which we see that $I_{\mathrm{disc},\psi_+}(f_+)$ vanishes. Since $G \times G$ is not an endoscopic group for G_+ in this case (it is actually a twisted endoscopic group), ψ_+ does not factor through any proper endoscopic datum for G_+. It follows that

$$S_{\mathrm{disc},\psi_+}(f_+) = I_{\mathrm{disc},\psi_+}(f_+) = 0,$$

as required. If N is even, the heuristic argument is more complicated. In this case, the expression for $I_{\mathrm{disc},\psi_+}(f_+)$ contains a sign

$$(-1)^{a_+}, \qquad a_+ = \mathrm{ord}_{s=1}\, L(s, \rho_+ \circ \psi_+),$$

which equals (-1) if Theorem 1.5.3(a) if false. One combines this information, and the fact that the corresponding sign $(-1)^{a_+^\vee}$ for $I^\vee_{\mathrm{disc},\psi_+}(f_+^\vee)$ then equals $+1$, with the endoscopic decomposition

$$I_{\mathrm{disc},\psi_+}(f_+) = S_{\mathrm{disc},\psi_+}(f_+) + \iota(G_+, G'_+)\, \widehat{S}'_{\mathrm{disc},\psi_+}(f'_+)$$

for G_+ and $G'_+ = G \times G$, and its analogue (4.7.5) for $\widetilde{G}(N_+)$, G_+ and $G_+^\vee$. Since

$$S'_{\mathrm{disc},\psi_+} = S^G_{\mathrm{disc},\psi} \otimes S^G_{\mathrm{disc},\psi},$$

an expression we can evaluate easily given that $\psi \in \widetilde{\Psi}_{\mathrm{sim}}(G)$ is simple, we find that the first decomposition gives a formula for $S_{\mathrm{disc},\psi_+}(f_+)$. An analysis of the three terms in the second decomposition then leads to a contradiction, namely that (4.7.5) cannot hold if the two signs are as above. In other words, Theorem 1.5.3(a) must be valid.

These last remarks have been quite sketchy. They are intended to serve as a heuristic argument in support of assertion (a) of Theorem 1.5.3. There is no point in being more precise now, since we will later have to revisit the argument in more detail and greater generality. As we will see, the formal proof of the assertion will be much more difficult without our heuristic assumption that ψ_+ contributes nothing to the discrete spectra of G_+ and $G_+^\vee$.

Having discussed the two assertions of Theorem 1.5.3, we take ψ as in (4.7.1), and assume that both assertions are valid (where applicable) for the simple constituents ψ_i of ψ. This allows us to apply the two sign lemmas proved in §4.6 to ψ. We shall use the lemmas, both for $G = \widetilde{G}(N)$ and $G \in \widetilde{\mathcal{E}}_{\mathrm{sim}}(N)$, to derive the stable multiplicity formula of Theorem 4.1.2.

With ψ fixed, we take G to be the unique datum in $\widetilde{\mathcal{E}}_{\mathrm{ell}}(N)$ such that ψ lies in the subset $\widetilde{\Psi}_2(G)$ of $\widetilde{\Psi}_{\mathrm{ell}}(N)$. We of course want to assume that G is simple in the case we are studying. The first step is to apply Lemma 4.3.1, with $\widetilde{G}(N)$ in place of G, to the global normalizing factor in (4.7.2).

We obtain

$$\begin{aligned} r_{\widetilde{P}}(w, \psi_{\widetilde{M}}) = \widetilde{r}_{\psi}^{N}(w) &= \widetilde{r}_{\psi}^{N}(w)\, \widetilde{\varepsilon}_{\psi}^{1}(u) \\ &= \widetilde{\varepsilon}_{\psi}^{N}(x_u)\, \widetilde{s}_{\psi}^{0}(u) = \widetilde{\varepsilon}_{\psi}^{N}(x_u), \end{aligned}$$

where u is the unique element in the fibre $\widetilde{\mathfrak{N}}_{\psi}(w)$. We are using the fact that

$$\widetilde{\varepsilon}_{\psi}^{1}(u) = \widetilde{\varepsilon}_{\psi_M}(u) = 1,$$

following an earlier remark in the special case $r = 1$, and that

$$\widetilde{s}_{\psi}^{0}(u) = \widetilde{s}_{\psi}^{0}(w_u) = 1,$$

since the group $\widetilde{S}_{\psi}(N)^0$ is abelian. The point $x = x_u$ is the unique element in the set $\mathcal{S}_{\psi}(N)$. It represents a point $s \in \widetilde{S}_{\psi}(N)$ such that the pair (G, ψ) is the preimage of (ψ, s) under the correspondence (1.4.11). The next step is to apply Lemma 4.4.1, with $\big(\widetilde{G}(N), G\big)$ in place of (G, G'), to the sign $\widetilde{\varepsilon}_{\psi}^{N}(x) = \varepsilon_{\psi}^{\widetilde{G}(N)}(x)$. We obtain

$$\begin{aligned} \widetilde{\varepsilon}_{\psi}^{N}(x_u) = \widetilde{\varepsilon}_{\psi}^{N}(x) &= \widetilde{\varepsilon}_{\psi}^{N}(s_{\psi}x) \\ &= \varepsilon_{\psi}^{G}(s_{\psi}) = \varepsilon^{G}(\psi), \end{aligned}$$

since the point s_{ψ} lies in the connected group $\widetilde{S}_{\psi}^{*}(N) = \widetilde{S}_{\psi}^{0}(N)$. The identity (4.7.2) then takes the form

$$\widehat{S}_{\mathrm{disc},\psi}^{G}(\widetilde{f}^{G}) = \widetilde{\iota}(N, G)^{-1}\big(\tfrac{1}{2}\big)^{r}\, \varepsilon^{G}(\psi)\, \widetilde{f}^{G}(\psi), \qquad \widetilde{f} \in \widetilde{\mathcal{H}}(N).$$

The remaining terms in the identity are easily treated. Given the formula for $\widetilde{\iota}(N, G)$ at the end of §3.2, and the fact that $\overline{Z}(\widehat{G})^{\Gamma} = \{1\}$ in the case here that G is simple, we obtain

$$\widetilde{\iota}(N, G)^{-1} = \big(\tfrac{1}{2}\,|\overline{Z}(\widehat{G})^{\Gamma}|^{-1}\,|\widetilde{\mathrm{O}}\mathrm{ut}_N(G)|^{-1}\big)^{-1} = 2\,|\widetilde{\mathrm{O}}\mathrm{ut}_N(G)|.$$

The resulting power $\big(\frac{1}{2}\big)^{r-1}$ can then be related to the finite group

$$\mathcal{S}_{\psi}(G) = \mathcal{S}_{\psi} = \overline{S}_{\psi} = S_{\psi}/Z(\widehat{G})^{\Gamma}$$

attached to G and ψ. One finds that

$$|\mathcal{S}_{\psi}|^{-1} = |\widetilde{\mathrm{O}}\mathrm{ut}_N(G, \psi)|\,\big(\tfrac{1}{2}\big)^{r-1},$$

with $\widetilde{\mathrm{O}}\mathrm{ut}_N(G, \psi)$ being stabilizer of ψ in $\widetilde{\mathrm{O}}\mathrm{ut}_N(G)$, after first noting that the product

$$|Z(\widehat{G})^{\Gamma}|\,|\widetilde{\mathrm{O}}\mathrm{ut}_N(G)|$$

equals 1, 2 or 4 in the respective cases that $\widehat{G}$ equals $SO(N, \mathbb{C})$ (N odd), $Sp(N, \mathbb{C})$ or $SO(N, \mathbb{C})$ (N even). We then note that the integer m_{ψ} in the putative multiplicity formula satisfies

$$m_{\psi} = |\widetilde{\mathrm{O}}\mathrm{ut}_N(G)|\,|\widetilde{\mathrm{O}}\mathrm{ut}_N(G, \psi)|^{-1},$$

according to the definition in §1.5. It follows that

$$\widehat{S}^G_{\mathrm{disc},\psi}(\widetilde{f}^G) = m_\psi\, |\mathcal{S}_\psi|^{-1}\, \varepsilon^G(\psi)\, \widetilde{f}^G(\psi), \qquad f \in \widetilde{\mathcal{H}}(N).$$

This in turn can be written as

$$S^G_{\mathrm{disc},\psi}(f) = m_\psi\, |\mathcal{S}_\psi|^{-1}\, \varepsilon^G(\psi)\, f^G(\psi), \qquad f \in \widetilde{\mathcal{H}}(G), \tag{4.7.6}$$

since the mapping $\widetilde{f} \to \widetilde{f}^G$ takes $\widetilde{\mathcal{H}}(N)$ onto $\widetilde{\mathcal{S}}(G)$.

The identity (4.7.6) is the stable multiplicity formula of Theorem 4.1.2 for the parameter $\psi \in \widetilde{\Psi}_2(G)$, since the constant $\sigma(\overline{S}^0_\psi)$ equals 1 for the trivial group $\overline{S}^0_\psi$. We can therefore apply its analogues from Corollary 4.1.3 to the data G' in the endoscopic expansion

$$I_{\mathrm{disc},\psi}(f) = \sum_{G' \in \mathcal{E}_{\mathrm{ell}}(G)} \iota(G,G')\, \widehat{S}'_{\mathrm{disc},\psi}(f'), \qquad f \in \widetilde{\mathcal{H}}(G),$$

for G. The right hand side becomes a double sum

$$\sum_{G'} \sum_{\psi'} \iota(G,G')\, |\mathcal{S}_{\psi'}|^{-1}\, \varepsilon'(\psi')\, f'(\psi'), \tag{4.7.7}$$

over $G' \in \mathcal{E}_{\mathrm{ell}}(G)$ and $\psi' \in \Psi(G',\psi)$. The analysis of this expression then follows the general discussion in §4.4 that led to Lemma 4.4.2. It will be useful to revisit briefly the argument in the simpler case here, especially since the case $\psi \in \widetilde{\Psi}_2(G)$ was formally ruled out of the earlier discussion. (See [**K5**, §12] and [**K6**, p. 191].)

The outer sum in (4.7.7) can be taken over $\widehat{G}$-orbits of elliptic endoscopic data G', in which the corresponding element $s' \in \widehat{G}$ is treated as a $Z(\widehat{G})^\Gamma$-coset. The inner sum is over the set of $\widehat{G}'$-orbits of homomorphisms ψ' from $\mathcal{A}_\psi$ to ${}^LG'$ that map to ψ. We can replace it by the coarser set of orbits under the stabilizer $\mathrm{Aut}_G(G')$ of G' in $\widehat{G}$, so long as we multiply the summand by the quotient

$$|\mathrm{Out}_G(G')|\, |\mathrm{Out}_G(G',\psi')|^{-1}.$$

Again, $\mathrm{Out}_G(G',\psi')$ is the stabilizer of ψ' (as a $\widehat{G}'$-orbit) in the finite group

$$\mathrm{Out}_G(G') = \mathrm{Aut}_G(G')/\widehat{G}'.$$

The first coefficient in (4.7.7) equals

$$\iota(G,G') = |\overline{Z}(\widehat{G}')^\Gamma|^{-1}\, |\mathrm{Out}_G(G')|^{-1}.$$

We can therefore write (4.7.7) as a double sum

$$\sum_{(G',\psi')} |\mathcal{S}_{\psi'}|^{-1}\, |\overline{Z}(\widehat{G}')^\Gamma|^{-1}\, |\mathrm{Out}_G(G',\psi')|^{-1}\, \varepsilon'(\psi')\, f'(\psi') \tag{4.7.8}$$

over $\widehat{G}$-orbits of pairs.

The main step is to apply the correspondence (1.4.11) to the indices (G',ψ'). We thereby transform the sum in (4.7.8) to a double sum over $\widehat{G}$-orbits of pairs (ψ_G, s_G), where ψ_G is an L-homomorphism from $\mathcal{A}_\psi$ to

${}^L G$ that maps to ψ, and s_G is an element in the centralizer $\overline{S}_{\psi_G}$. The set of $\widehat{G}$-orbits of ψ_G is the finite set $\Psi(G, \psi)$ of order m_ψ. We can therefore remove the sum over ψ_G, identifying each ψ_G with ψ, if we multiply the summand by m_ψ. The sum over s_G then becomes a sum over elements s in the underlying finite abelian group $\mathcal{S}_\psi = \overline{S}_\psi$. It remains to express the summand in (4.7.8) in terms of ψ and s.

It follows easily from the definitions that

$$\begin{aligned} \mathrm{Out}_G(G', \psi') &\cong S^+_{\psi,s}/S^+_{\psi,s} \cap \widehat{G}' \\ &\cong \overline{S}^+_{\psi,s}/\overline{S}^+_{\psi,s} \cap \overline{\widehat{G}}', \end{aligned}$$

where

$$\begin{aligned} S^+_{\psi,s} &= \{g \in S_\psi : \ gsg^{-1} = s\}, \qquad s \in \overline{S}_\psi, \\ \overline{S}^+_{\psi,s} &= S^+_{\psi,s}/Z(\widehat{G})^\Gamma = \mathrm{Cent}(s, \overline{S}_\psi), \end{aligned}$$

and

$$\overline{\widehat{G}}' = \widehat{G}'/Z(\widehat{G})^\Gamma.$$

Since $\overline{S}_\psi$ is abelian, we also have

$$\overline{S}^+_{\psi,s} = \overline{S}_\psi = \mathcal{S}_\psi,$$

and

$$\overline{S}^+_{\psi,s} \cap \overline{\widehat{G}}' = S_\psi(G')/Z(\widehat{G})^\Gamma.$$

The quotient of this second group by the subgroup

$$\overline{Z}(\widehat{G}')^\Gamma = Z(\widehat{G}')^\Gamma/Z(\widehat{G})^\Gamma$$

equals

$$S_\psi(G')/Z(\widehat{G}')^\Gamma = \overline{S}_\psi(G') = \mathcal{S}_\psi(G') \cong \mathcal{S}_{\psi'}.$$

We thus obtain a reduction

$$|\mathcal{S}_{\psi'}|^{-1} \, |\overline{Z}(\widehat{G}')^\Gamma|^{-1} \, |\mathrm{Out}_G(G', \psi')|^{-1} = |\mathcal{S}_\psi|^{-1}$$

for the product of the first three coefficients in (4.7.8). Applying Lemma 4.4.1 to the fourth coefficient, we obtain

$$\varepsilon'(\psi') = \varepsilon^{G'}_{\psi'}(s_{\psi'}) = \varepsilon^G_\psi(s_\psi s) = \varepsilon_\psi(s_\psi s).$$

Finally, we can write the fifth term as

$$f'(\psi') = \sum_{\pi \in \widetilde{\Pi}_\psi} \langle s_\psi s, \pi \rangle \, f_G(\pi),$$

by applying Theorem 2.2.1 to each completion ψ_v of ψ.

We have now expressed all five factors in the summand of (4.7.8) in terms of ψ and s. We can therefore replace the double sum over G' and s' by the simple sum over $s \in \mathcal{S}_\psi$ described above. We obtain

$$\begin{aligned} I_{\mathrm{disc},\psi}(f) &= m_\psi\, |\mathcal{S}_\psi|^{-1} \sum_{s \in \mathcal{S}_\psi} \sum_{\pi \in \widetilde{\Pi}_\psi} \varepsilon_\psi(s_\psi s) \langle s_\psi s, \pi \rangle f_G(\pi) \\ &= m_\psi\, |\mathcal{S}_\psi|^{-1} \sum_{\pi} \sum_{x \in \mathcal{S}_\psi} \varepsilon_\psi(x) \langle x, \pi \rangle f_G(\pi) \\ &= \sum_{\pi \in \widetilde{\Pi}_\psi} m_\psi(\pi)\, f_G(\pi), \end{aligned}$$

where

$$m_\psi(\pi) = \begin{cases} m_\psi, & \text{if } \langle \cdot, \pi \rangle = \varepsilon_\psi^{-1} = \varepsilon_\psi, \\ 0, & \text{otherwise.} \end{cases} \tag{4.7.9}$$

On the other hand, we can write

$$I_{\mathrm{disc},\psi}(f) = \mathrm{tr}\big(R^G_{\mathrm{disc},\psi}(f)\big), \qquad f \in \widetilde{\mathcal{H}}(G).$$

This is because ψ does not contribute to any of the terms with $M \neq G$ in the expansion (4.1.1) of $I_{\mathrm{disc},\psi}(f)$. We have thus established the multiplicity formula

$$\mathrm{tr}\big(R^G_{\mathrm{disc},\psi}(f)\big) = \sum_{\pi \in \widetilde{\Pi}_\psi} m_\psi(\pi)\, f_G(\pi), \qquad f \in \widetilde{\mathcal{H}}(G), \tag{4.7.10}$$

postulated by Theorem 1.5.2, under the heuristic assumptions we have made. In so doing, we have shown that the sign character $\varepsilon_\psi(x)$ is forced on us by the known dependence of global intertwining operators on L-functions for $GL(N)$.

It has perhaps been difficult to follow the stream of informal remarks of this section. Let us recapitulate what we have done. After taking on some plausible assumptions, we have sketched a proof of four main global assertions – the triviality of orthogonal ε-factors (Theorem 1.5.3(b)), the existence of poles of L-functions (Theorem 1.5.3(a)), the stable multiplicity formula for parameters $\psi \in \widetilde{\Psi}_2(G)$ (Theorem 4.1.2), and the spectral multiplicity formula for parameters $\psi \in \widetilde{\Psi}_2(G)$ (Theorem 1.5.2). The stable multiplicity formula is at the heart of things. Had we taken *it* as an assumption (albeit a less plausible one), we could easily have established the other three assertions. In particular, under this condition, our justification of the spectral multiplicity formula above can be regarded as a formal proof. We shall state this as a lemma for future use.

LEMMA 4.7.1. *Assume that the stable multiplicity formula (4.7.6) holds for any $N \geqslant 1$, $G \in \widetilde{\mathcal{E}}_{\mathrm{sim}}(N)$ and $\psi \in \widetilde{\Psi}_2(G)$. Then the spectral multiplicity formula (4.7.10) also holds for any N, G and ψ.* □

Our proof of the lemma is represented by the elementary argument above, beginning in the paragraph following (4.7.6). The condition on the stable multiplicity formula is to be interpreted broadly. It implicitly includes Theorems 1.4.1, 1.4.2 and 2.2.1 (in the definition of the set $\widetilde{\Psi}_2(G)$ and the linear form $f^G(\psi)$). We could also add the condition that there be no contribution to the discrete spectrum of G from parameters not in $\widetilde{\Psi}_2(G)$ (so that the formula of the lemma accounts for everything). These conditions are among the most difficult we will have to prove. They are why the simple arguments of this section have to be replaced by the rigorous proofs that will occupy the rest of the volume. We will appeal to Lemma 4.7.1 several times later, in each case at the end of an extended argument that establishes the other conditions.

4.8. Remarks on general groups

We shall conclude with a brief description of how some of the earlier discussion of the chapter might be extended. A general heuristic comparison of spectral and endoscopic terms was the topic of [**A9**]. This followed the special case of the "generic" discrete spectrum treated in [**K3**, §10–12]. The article [**A9**] is not so easy to read, for among other reasons, the fact that it was written without the benefit of the volume [**KS**] of Kottwitz and Shelstad.

We are in a stronger position here. In addition to [**KS**], we now have the results of the earlier sections of this chapter on which to base our general remarks. The special cases treated in these sections, namely the groups $G \in \widetilde{\mathcal{E}}_{\mathrm{sim}}(N)$ and the twisted group $G = \widetilde{G}(N)$ that are the main focus of the volume, are clearly simpler. For a start, they come with a well defined substitute $\mathcal{L}_\psi$ for the global Langlands group. In addition, the locally trivial 1-cocycles that complicate the general theory are not present. There is also the further simplification that the groups $\mathcal{S}_\psi$ are all abelian. However, we can still sketch how some of our arguments extend to the general case, referring as necessary to the relevant sections of [**A9**].

For this discussion, we take G to be a general triplet (G^0, θ, ω) over the global field F. Following [**KS**, 2.1] (but with the notation of §3.2), we form the dual set

$$\widehat{G} = \widehat{G}^0 \rtimes \widehat{\theta} = \widehat{G}^0 \rtimes {}^L\theta,$$

which is a bitorsor under the dual group $\widehat{G}^0$. We also form the L-set

$${}^LG = \widehat{G} \rtimes W_F = {}^LG^0 \rtimes {}^L\theta,$$

which is an L-bitorsor under the L-group ${}^LG^0$. Recall that ${}^L\theta = {}^L\theta_\omega$ is an extension of the automorphism $\widehat{\theta}$ from $\widehat{G}^0$ to ${}^LG^0$, which depends on ω. It must be used in $\widehat{G}$ when we form its semidirect product with W_F. Recall also that an endoscopic datum G' for G represents a 4-tuple $(G', \mathcal{G}', s', \xi')$ that satisfies (analogues of) the conditions (2.1.1)–(2.1.4) of [**KS**]. We are taking s' here to be a semisimple element in $\widehat{G}$. We observe that in the notation here, the condition (2.1.4a) of [**KS**] asserts that

$$s'\xi'(g') = a'(w')\,\xi'(g')\,s', \qquad g' \in \mathcal{G}',$$

where w' is the image of g' in W_F, and $a'(w)$ is a locally trivial 1-cocycle from W_F to $Z(\widehat{G}^0)$.

We must of course assume that we have a suitable substitute for the Langlands group L_F. We take it to be an object in the category of locally compact topological groups. We suppose that we have at our disposal an extension

$$L_F^* \longrightarrow W_F,$$

with compact connected kernel K_F^*, that takes the place of L_F. We could of course assume simply that L_F^* is the actual Langlands group L_F, but in practice, L_F^* would be a larger group. For example, we will suggest later in §8.5 how to replace most of our complex groups $\mathcal{L}_\psi$ with one locally compact group.

We suppose that along with L_F^*, we are also given a distinguished family $\Phi^*_{\mathrm{bdd}}(G^0)$ of L-homomorphisms

$$\phi:\ L_F^* \longrightarrow {}^LG^0.$$

These objects should have bounded image in $\widehat{G}^0$, and be determined up to the general equivalence relation introduced in [**K3**]. Namely, two L-homomorphisms ϕ and ϕ' are *equivalent* if

$$\phi'(u) = z(u)\,g\,\phi(u)\,g^{-1}, \qquad u \in L_F^*, \tag{4.8.1}$$

where g belongs to $\widehat{G}^0$, and z is the pullback of a locally trivial 1-cocycle from W_F to $Z(\widehat{G}^0)$. For any $\phi \in \Phi^*_{\mathrm{bdd}}(G^0)$, we set $S_\phi(G^0)$ equal to the group of elements $s \in \widehat{G}^0$ such that

$$s\phi(u) = z(u)\,\phi(u)\,s, \qquad u \in L_F^*, \tag{4.8.2}$$

where z is again the pullback of a locally trivial 1-cocycle. Our guiding intuition is to be based on the following expectation: there should be a canonical L-homomorphism

$$L_F \longrightarrow L_F^*$$

from the actual Langlands group to L_F^*, determined up to conjugacy by the kernel K_F^*, such that the restriction mapping from $\Phi^*_{\mathrm{bdd}}(G)$ to the set $\Phi_{\mathrm{bdd}}(G)$ attached to L_F is a bijection that preserves the centralizers $S_\phi(G^0)$.

Given the set $\Phi^*_{\mathrm{bdd}}(G^0)$, we define $\Psi(G) = \Psi^*(G)$ to be the family of equivalence classes of L-homomorphisms

$$\psi:\ A_F^* \longrightarrow {}^LG^0, \qquad A_F^* = L_F^* \times SU(2),$$

whose restriction to L_F^* lies in $\Phi^*_{\mathrm{bdd}}(G^0)$, and for which the centralizer set

$$S_\psi = S_\psi(G)$$

is not empty. Equivalence is defined here by the evident analogue of (4.8.1). The set $S_\psi(G)$ is defined as in (4.8.2), but with ψ in place of ϕ, and s ranging over elements in $\widehat{G}$ rather than $\widehat{G}^0$. The elements ψ in $\Psi(G)$ are then to be treated as the global parameters for G.

We would actually need to assume more broadly that L_F^* comes also with a distinguished family $\Phi_{\mathrm{bdd}}^*(H)$ of special parameters for any connected group H that arises in the endoscopic study of G. For example, H could be a Levi subgroup M of G^0, or a twisted endoscopic group G' for G, or perhaps an auxiliary extension $\widetilde{G}'$ of G'. In fact, we would typically permit G itself to vary over some suitable family that allows for induction arguments. This of course is exactly what we have done in the special cases we are treating in this volume. In general, we assume that the sets $\Phi_{\mathrm{bdd}}^*(H)$ are functorial with respect to endoscopic embeddings among the associated L-groups. We then define the general families of parameters $\Psi(H)$ and the centralizers $S_\psi(H)$ for H, as we did above for G.

Finally, we suppose that L_F^* comes with local homomorphisms

$$L_{F_v} \longrightarrow L_F^*$$

over the corresponding Weil groups, which are defined as usual up to conjugation. We then obtain localization mappings

$$\psi \longrightarrow \psi_v, \qquad \psi \in \Psi(G),$$

from $\Psi(G)$ to the local parameter sets $\Psi^+(G_v)$. We can of course form the subset $\Psi(G,\chi)$ of global parameters in $\Psi(G)$ that are attached to a given central character datum $(\mathfrak{X}_G,\chi)$ for G. The localizations then provide a mapping

$$\psi \longrightarrow \big(t(\psi), c(\psi)\big), \qquad \psi \in \Psi(G,\chi),$$

as in the special case of §3.3. We recall that $t(\psi)$ is the nonnegative real number given by the norm of the imaginary part of the archimedean infinitesimal character of ψ, and that $c \in \mathcal{C}_{\mathbb{A}}(G,\chi)$ is the equivalence class of families of conjugacy classes in ${}^LG$ attached to unramified places v of F. For any such pair (t,c), we define $\Psi(G,t,c)$ to be the set of $\psi \in \Psi(G,\chi)$ such that

$$\big(t(\psi), c(\psi)\big) = (t,c).$$

If c belongs instead to some quotient $\widetilde{\mathcal{C}}_{\mathbb{A}}(G,\chi)$ of $\mathcal{C}_{\mathbb{A}}(G,\chi)$, as in §3.3, we can of course define $\Psi(G,t,c)$ to be the union of sets associated to the points in the fibre of (t,c) in $\mathcal{C}_{\mathbb{A}}(G,\chi)$. We can also define the set

$$\Psi(G',t,c) = \Psi(\widetilde{G}',\widetilde{\xi}',t,c),$$

for any element $G' \in \mathcal{E}(G)$ with auxiliary datum $(\widetilde{G}',\widetilde{\xi}')$, as the union of subsets of $\Psi(\widetilde{G}',\widetilde{\chi}')$ attached to points in the fibre of (t,c) in $\mathcal{C}_{\mathbb{A}}(\widetilde{G}',\widetilde{\chi}')$.

The general goal is of course to understand the discrete part of the trace formula, and the spectral information it contains. In §3.3, we observed that it has a canonical decomposition

$$I_{\mathrm{disc},t}(f) = \sum_{c \in \widetilde{\mathcal{C}}_{\mathbb{A}}(G,\chi)} I_{\mathrm{disc},t,c}(f), \qquad f \in \mathcal{H}(G),$$

into c-components. It is therefore enough to understand any such component, and specifically, to compare the c-analogues of the spectral and endoscopic expansions (4.1.1) and (4.1.2). In our special case of groups $G \in \widetilde{\mathcal{E}}_{\text{sim}}(N)$, where we can appeal to strong multiplicity 1 for $GL(N)$ and its generalization in Theorem 1.3.2, we indexed c-components in (3.3.12) and (3.3.13) directly in terms of parameters $\psi \in \widetilde{\Psi}(N)$. This is not possible in general. Instead, one must somehow decompose the linear forms

$$\operatorname{tr}\big(M_{P,t,c}(w,\chi)\,\mathcal{I}_{P,t,c}(\chi,f)\big), \qquad f \in \mathcal{H}(G) \tag{4.8.3}$$

and

$$\widehat{S}'_{\text{disc},t,c}(f'), \qquad f \in \mathcal{H}(G), \tag{4.8.4}$$

in the general c-analogues of (4.1.1) and (4.1.2) into contributions from parameters $\psi \in \Psi(G,t,c)$. The question is embedded in the larger problem of establishing general global intertwining relations for (4.8.3) and stable multiplicity formulas for (4.8.4).

There is no reason to expect that we could establish global intertwining relations in advance. They will no doubt have to be proved by local-global methods, which ultimately rely on the identity of (4.1.1) and (4.1.2). At least this is how we will be dealing with the groups $G \in \widetilde{\mathcal{E}}_{\text{sim}}(N)$ in this volume.

On the other hand, we might often expect to have stable multiplicity formulas already in hand before beginning the comparison. One reason is that they apply only to connected quasisplit groups, while the general object G is much broader. We will not see this distinction for much of the present volume, since the groups $G \in \widetilde{\mathcal{E}}_{\text{sim}}(N)$ of interest in the first eight chapters are in fact quasisplit. Our proof of the stable multiplicity formula for these groups will have to come out of the comparisons of the expansions (4.1.1) and (4.1.2), and will be one of our most difficult tasks. Once we do have a proof, however, it will be easier to deal with inner forms of G.

We shall therefore assume that the linear form (4.8.4) satisfies the stable multiplicity formula of §4.1, or rather the adaptation of (4.1.12) to this general object. We thus assume that for our general triplet G, we can write

$$\widehat{S}'_{\text{disc},t,c}(f') = \sum_{\psi' \in \Psi(G',t,c)} \widehat{S}'_{\text{disc},\psi'}(f'),$$

for any $G' \in \mathcal{E}_{\text{ell}}(G)$, where

$$\widehat{S}'_{\text{disc},\psi'}(f') = |\mathcal{S}_{\psi'}|^{-1}\,\sigma(\overline{S}^0_{\psi'})\,\varepsilon'(\psi')\,f'(\psi'). \tag{4.8.5}$$

The coefficients $|\mathcal{S}_{\psi'}|^{-1}$ and $\sigma(\overline{S}^0_{\psi'})$ in (4.8.5) have already been defined. The sign character $\varepsilon^{G'}_{\psi'}$ in the coefficient

$$\varepsilon'(\psi') = \varepsilon^{G'}_{\psi'}(s_{\psi'})$$

is defined exactly as in (1.5.6). The linear form $f'(\psi')$ is supposed to be a product

$$f'(\psi') = \prod_v f'_v(\psi'_v), \qquad f' = \prod_v f'_v,$$

of stable characters on the groups $\widetilde{G}'(F_v)$, attached to the localizations ψ'_v of ψ'. Part of our assumption here is that we have already defined these local objects.

Given the stable multiplicity formula (4.8.5), the general c-analogue of the endoscopic expansion (4.1.2) will be more accessible than that of the spectral expansion (4.1.1). We shall sketch what extensions of the analysis of §4.4 are needed to deal with it. We do not have to be concerned with the induction hypotheses that complicated the earlier discussion. We do have to address the main complication of the general case here, the fact that the set S_ψ can be larger than the centralizer of the image of ψ. For example, S_ψ contains the full center $Z(\widehat{G}^0)$ of $\widehat{G}^0$, rather than just its subgroup of Γ-invariants. We define the associated two quotients of S_ψ by

$$\overline{S}_\psi = S_\psi / Z(\widehat{G}^0) \tag{4.8.6}$$

and

$$\mathcal{S}_\psi = \pi_0(\overline{S}_\psi) = S_\psi / S_\psi^0 \, Z(\widehat{G}^0).$$

The sets $S_\psi = S_\psi(G)$, $\overline{S}_\psi = \overline{S}_\psi(G)$ and $\mathcal{S}_\psi = \mathcal{S}_\psi(G)$ are then bitorsors under the respective complex groups $S_\psi^* = S_\psi(G^0)$, $\overline{S}_\psi^* = \overline{S}_\psi(G^0)$ and $\mathcal{S}_\psi^* = \mathcal{S}_\psi(G^0)$. In our special cases $G \in \mathcal{E}_{\mathrm{sim}}(N)$, $G = \widetilde{G}(N)$ and $G = \widetilde{G}$ (and so long as $\widehat{G} \neq SO(2,\mathbb{C})$), S_ψ does equal the centralizer, and this notation matches what we have been using.

In the general case, we write $C_\psi = C_\psi(G^0)$ for the actual centralizer of the image of ψ in $\widehat{G}^0$, following Kottwitz [**K3**, §10]. We then define the associated two quotients $\overline{C}_\psi = \overline{C}_\psi(G^0)$ and $\mathcal{C}_\psi = \mathcal{C}_\psi(G^0)$ of C_ψ by

$$\overline{C}_\psi = C_\psi / Z(\widehat{G}^0)^\Gamma$$

and

$$\mathcal{C}_\psi = \pi_0(\overline{C}_\psi) = C_\psi / C_\psi^0 \, Z(\widehat{G}^0)^\Gamma.$$

These are normal subgroups of finite index in the corresponding quotients $\overline{S}_\psi^*$ and $\mathcal{S}_\psi^*$, since

$$Z(\widehat{G}^0) \cap C_\psi = Z(\widehat{G}^0)^\Gamma$$

and

$$C_\psi^0 \, Z(\widehat{G}^0) = S_\psi^0 \, Z(\widehat{G}^0).$$

In particular, the connected components $\overline{C}_\psi^0$ and $\overline{S}_\psi^0$ are equal. We thus obtain a canonical injection

$$\overline{S}_\psi^* / \overline{C}_\psi \cong \mathcal{S}_\psi^* / \mathcal{C}_\psi \hookrightarrow \ker^1\big(F, Z(\widehat{G}^0)\big).$$

We note again that in the special cases above, we have $\overline{C}_\psi = \overline{S}_\psi^*$ and $\mathcal{C}_\psi = \mathcal{S}_\psi^*$, and therefore no need for this supplementary notation.

In the general case, the set $\overline{S}_\psi$ retains its central role. For example, the chain (4.1.13) of subsets of $\Psi(G)$ is defined in terms of $\overline{S}_\psi$, as before. We can of course define a similar chain of subsets of $\Psi(G, t, c)$, as well as an analogous chain of subsets of $\Psi(G', t, c)$ for any endoscopic datum G' with auxiliary datum $(\widetilde{G}', \widetilde{\xi}')$. It is the connected group $\overline{S}^0_\psi$ attached to $\overline{S}_\psi$, or rather its analogue for a parameter ψ', that determines the coefficient $\sigma(\overline{S}^0_{\psi'})$ in (4.8.5). (We were actually anticipating the subsequent definition (4.8.6) of the quotient $\overline{S}_{\psi'}$ when we stated (4.8.5).) Since this coefficient vanishes if the center of $\overline{S}^0_{\psi'}$ is infinite, we can take the sum in the formula for $\widehat{S}'_{\mathrm{disc},t,c}(f')$ above over the subset

$$\Psi_{\mathrm{disc}}(G', t, c) = \{\psi' \in \Psi(G', t, c) : \ |Z(\overline{S}_{\psi'})| < \infty\}$$

of $\Psi(G', t, c)$, or better, the smaller subset

$$\Psi_{s\text{-disc}}(G', t, c) = \{\psi' \in \Psi(G', t, c) : \ |Z(\overline{S}^0_{\psi'})| < \infty\} \tag{4.8.7}$$

of "stable-discrete" parameters. For general perspective, we note that if we were to form the set

$$\Psi_{s\text{-ell}}(G', t, c) = \Psi_{\mathrm{ell}}(G', t, c) \cap \Psi_{s\text{-disc}}(G', t, c)$$

of "stable-elliptic" parameters in $\Psi_{\mathrm{ell}}(G', t, c)$, it would reduce simply to the subset $\Psi_2(G', t, c)$.

We are assuming that c is any class in some given quotient $\widetilde{\mathcal{C}}_{\mathbb{A}}(G, \chi)$. The c-analogue of the endoscopic expansion is the formula

$$\sum_{G' \in \mathcal{E}_{\mathrm{ell}}(G)} \iota(G, G')\, \widehat{S}'_{\mathrm{disc},t,c}(f')$$

for the c-component

$$I_{\mathrm{disc},t,c}(f), \qquad f \in \mathcal{H}(G),$$

of $I_{\mathrm{disc},t}(f)$ given by (3.3.10). We therefore have an expansion

$$\sum_{G'} \sum_{\psi'} \iota(G, G')\, \widehat{S}'_{\mathrm{disc},\psi'}(f') \tag{4.8.8}$$

for $I_{\mathrm{disc},t,c}(f)$, where G' is summed over $\mathcal{E}_{\mathrm{ell}}(G)$, and ψ' is summed over $\Psi_{\mathrm{disc}}(G', t, c)$. The coefficient $\iota(G, G')$ is given in (3.2.4) as

(4.8.9)

$$\big|\pi_0(\kappa_G)\big|^{-1} \big|\ker^1\big(F, Z(\widehat{G}^0)\big)\big|^{-1} \big|\ker^1\big(F, Z(\widehat{G}')\big)\big| \big|\overline{Z}(\widehat{G}')^\Gamma\big|^{-1} \big|\mathrm{Out}_G(G')\big|^{-1},$$

while $\widehat{S}'_{\mathrm{disc},\psi'}(f')$ is defined by (4.8.5). The heart of §4.4 was a transformation of the double sum over (G', ψ') to a double sum over (ψ, s), according to the correspondence (1.4.11). The general process here is a little more complicated. It will be instructive for us to describe the steps in symbolic form, although this should not disguise the fact that we are basically repeating discussion from §4.4.

The general correspondence (1.4.11) can be regarded as an equivariant bijection

$$X(G) \xrightarrow{\sim} Y(G)$$

between left two $\widehat{G}^0$-sets. The domain $X(G)$ is the set of pairs $x = (G', \psi')$, where G' belongs to the set $E(G)$ of endoscopic data for G, taken up to the image $\xi'(\mathcal{G}')$ of $\mathcal{G}'$ in ${}^LG$ and up to translation of the associated semisimple element $s' \in \widehat{G}$ by $Z(\widehat{G}^0)$, while ψ' is an actual L-homomorphism from A_ψ^* to the group ${}^L\widetilde{G}'$, which factors through the L-embedding $\widetilde{\xi}'$. The codomain is the set of pairs $y = (\psi, s)$, where ψ belongs to the set $F(G)$ of actual L-homomorphisms from A_ψ^* to ${}^LG^0$, and s is a semisimple element in $\overline{S}_\psi$. The projections of the two kinds of pairs onto their first components yield a larger diagram

$$\begin{array}{ccc} X(G) & \xrightarrow{\sim} & Y(G) \\ \big\downarrow & & \big\downarrow \\ E(G) & & F(G) \end{array}$$

of $\widehat{G}^0$-equivariant mappings.

We note that the correspondence restricts to a bijection between the $\widehat{G}^0$-invariant subsets

$$X_{\text{disc},t,c}(G) = \{x = (G', \psi') :\ G' \in E_{\text{ell}}(G), \psi' \in F_{s\text{-disc}}(G', t, c)\}$$

and

$$Y_{\text{disc},t,c}(G) = \{y = (\psi, s) :\ \psi \in F_{\text{disc}}(G, t, c),\ s \in \overline{S}_{\psi,\text{ell}})\}$$

of $X(G)$ and $Y(G)$. We are writing $E_{\text{ell}}(G)$, $F_{\text{disc}}(G, t, c)$ and $F_{s\text{-disc}}(G', t, c)$ here for the respective preimages of $\mathcal{E}_{\text{ell}}(G)$, $\Psi_{\text{disc}}(G, t, c)$, and $\Psi_{s\text{-disc}}(G', t, c)$ in sets $E(G)$, $F(G)$ and $F(\widetilde{G}', \widetilde{\xi}')$. This bijection also fits into a diagram, as above, but with $E(G)$ and $F(G)$ replaced by $E_{\text{ell}}(G)$ and $F_{\text{disc}}(G, t, c)$.

The mappings of $\widehat{G}^0$-sets $X(G)$, $X_{\text{disc},t,c}(G)$, etc. all descend to mappings among the corresponding sets $\widehat{G}^0\backslash\backslash X(G)$, $\widehat{G}^0\backslash\backslash X_{\text{disc},t,c}(G)$, etc., of $\widehat{G}^0$-orbits. The set $\widehat{G}^0\backslash\backslash E_{\text{ell}}(G)$ of $\widehat{G}^0$-orbits in $E_{\text{ell}}(G)$ equals $\mathcal{E}_{\text{ell}}(G)$, by definition. The set $\widehat{G}^0\backslash\backslash F_{\text{disc}}(G, t, c)$ of $\widehat{G}^0$-orbits in $F_{\text{disc}}(G, t, c)$ projects onto $\Psi_{\text{disc}}(G, t, c)$, but this mapping is not generally bijective. However, it follows from the definitions that the group $\ker^1\big(F, Z(\widehat{G}^0)\big)$ acts transitively on its fibres, and that the stabilizer of any ψ equals the injective image of $\mathcal{S}_\psi^*/\mathcal{C}_\psi$ in $\ker^1\big(F, Z(\widehat{G}^0)\big)$. The fibre of ψ in $\widehat{G}^0\backslash\backslash F_{\text{disc}}(G, t, c)$ therefore has order

$$\big|\ker^1\big(F, Z(\widehat{G}^0)\big)\big|\,|\mathcal{S}_\psi|^{-1}\,|\mathcal{C}_\psi|, \tag{4.8.10}$$

since $|\mathcal{S}_\psi^*| = |\mathcal{S}_\psi|$. That said, we see that the mappings all fit together into a general diagram

$$
\begin{array}{ccc}
\widehat{G}^0 \backslash\backslash X_{\mathrm{disc},t,c}(G) & \xrightarrow{\ \sim\ } & \widehat{G}^0 \backslash\backslash Y_{\mathrm{disc},t,c}(G) \\
\downarrow & & \downarrow \\
\widehat{G}^0 \backslash\backslash E_{\mathrm{ell}}(G) & & \widehat{G}^0 \backslash\backslash F_{\mathrm{disc}}(G,t,c) \\
\| & & \downarrow \\
\mathcal{E}_{\mathrm{ell}}(G) & & \Psi_{\mathrm{disc}}(G,t,c).
\end{array}
\tag{4.8.11}
$$

The diagram (4.8.11) is the focal point for our transformation of the endoscopic expression (4.8.8). Speaking in symbolic terms, we need to follow the path that leads from the lower left hand corner of the diagram to the lower right hand corner. The first step will be to change the double sum in (4.8.8) to a sum over the set of $\widehat{G}^0$-orbits in $X_{\mathrm{disc},t,c}(G)$. This requires two changes in the inner sum over ψ'.

The variable ψ' in (4.8.8) is supposed to be summed over the set

$$\Psi_{s\text{-disc}}(G',t,c) = \Psi_{s\text{-disc}}(\widetilde{G}',\widetilde{\xi}',t,c).$$

We can instead sum it over the larger set

$$\widehat{\widetilde{G}}' \backslash\backslash F_{s\text{-disc}}(\widetilde{G}',\widetilde{\xi}',t,c) \ \cong\ \widehat{G}' \backslash\backslash F_{s\text{-disc}}(G',t,c),$$

if we multiply the summand by the product

$$\big|\ker^1\big(F, Z(\widehat{G}')\big)\big|^{-1}\, |\mathcal{S}_{\psi'}|\, |\mathcal{C}_{\psi'}|^{-1}. \tag{4.8.12}$$

Indeed, this product equals the inverse of the analogue for $\widetilde{G}'$ of (4.8.10), since the conditions on the extension $\widetilde{G}' \to G'$ imply that the mapping

$$\ker^1(F,\widehat{G}') \longrightarrow \ker^1(F,\widehat{\widetilde{G}}')$$

is an isomorphism. The variable G' in (4.8.8) is supposed to be summed over the set

$$\mathcal{E}_{\mathrm{ell}}(G) = \widehat{G}^0 \backslash\backslash E_{\mathrm{ell}}(G)$$

in the lower left hand corner of the diagram. This is what we want. However, the stabilizer of G' in $\widehat{G}^0$ is the the group $\mathrm{Aut}_G(G')$, rather than its subgroup $\widehat{G}'$. This means that we should really be summing ψ' over the set of $\mathrm{Aut}_G(G')$-orbits in $F_{\mathrm{disc}}(G',t,c)$, rather than $\widehat{G}'$-orbits. We can do this, so long as we multiply the summand by the number

$$|\mathrm{Out}_G(G')| \cdot |\mathrm{Out}_G(G',\psi')|^{-1} \tag{4.8.13}$$

of $\widehat{G}^0$-orbits in the given $\mathrm{Aut}_G(G')$-orbit of ψ'. As earlier, $\mathrm{Out}_G(G',\psi')$ is the stabilizer of ψ', regarded as a $\widehat{G}^0$-orbit, in $\mathrm{Out}_G(G')$.

We have shown how to express (4.8.8) as an iterated sum over the base set $\widehat{G}^0 \backslash\backslash E_{\mathrm{ell}}(G)$ and its pointwise fibres in (4.8.11). That is, we can write

(4.8.8) as a sum over orbits $x = (G', \psi')$ in the upper left hand set in (4.8.11), so long as the summand is rescaled by the product of (4.8.12) and (4.8.13). We write this in turn as a sum over elements $y = (\psi, s)$ in the upper right hand set in (4.8.11), with the expectation of being able to express the summand in terms of the bijective image (ψ, s) of (G', ψ'). In other words, we express (4.8.8) as an iterated sum over the base set $\widehat{G}^0\backslash\backslash F_{\mathrm{disc}}(G,t,c)$ and its pointwise fibres in (4.8.11). The second step is to write this as a double sum over ψ and s in more familiar sets.

As it presently stands, the expression is an iterated sum over orbits of L-homomorphisms $\psi \in F_{\mathrm{disc}}(G,t,c)$ under $\widehat{G}^0$ and orbits of elliptic elements $s \in \overline{S}_{\psi,\mathrm{ell}}$ under the stabilizer of ψ in $\widehat{G}^0$. We can take the first sum over ψ in the quotient $\Psi_{\mathrm{disc}}(G,t,c)$, provided that we multiply the summand by the order (4.8.10) of its fibre. The stabilizer of ψ in $\widehat{G}^0$ is the centralizer C_ψ. The second sum is over the set of orbits in $\overline{S}_{\psi,\mathrm{ell}}$ under C_ψ, or equivalently, the set of orbits under the quotient $\overline{C}_\psi$ of C_ψ. However, we would prefer to take a sum over the set

$$\overline{S}{}^0_\psi\backslash\backslash\overline{S}_{\psi,\mathrm{ell}} = \mathcal{E}(\overline{S}_{\psi,\mathrm{ell}}) = \mathcal{E}_{\psi,\mathrm{ell}}$$

of orbits in $\overline{S}_{\psi,\mathrm{ell}}$ under the subgroup $\overline{C}{}^0_\psi = \overline{S}{}^0_\psi$ of $\overline{C}_\psi$. The $\overline{C}_\psi$-orbit of s is bijective with the quotient of $\overline{C}_\psi$ by the subgroup

$$\overline{C}{}^+_{\psi,s} = \mathrm{Cent}(s, \overline{C}_\psi).$$

The $\overline{C}{}^0_\psi$-orbit of s is bijective with the quotient of $\overline{C}{}^0_\psi$ by the subgroup

$$\overline{C}_{\psi,s} = \mathrm{Cent}(s, \overline{C}{}^0_\psi).$$

We can therefore take the second sum over orbits s in $\mathcal{E}_{\psi,\mathrm{ell}}$, if we multiply the summand by the quotient

$$|\overline{C}{}^+_{\psi,s}/\overline{C}_{\psi,s}|\,|\overline{C}_\psi/\overline{C}{}^0_\psi|^{-1},$$

which is to say the quotient

$$|\overline{C}{}^+_{\psi,s}/\overline{C}_{\psi,s}|\,|\mathcal{C}_\psi|^{-1}. \tag{4.8.14}$$

We have now established that the double sum over G' and ψ' in the endoscopic expression (4.8.8) can be replaced by a double sum over $\psi \in \Psi_{\mathrm{disc}}(G,t,c)$ and $s \in \mathcal{E}_{\psi,\mathrm{ell}}$, provided that the summand is multiplied by the product of the four factors (4.8.10), (4.8.12), (4.8.13) and (4.8.14). The summand itself becomes the product of these four factors with (4.8.9) and the right hand side of (4.8.5). Observe that four of these six factors are given in terms of the pair $x = (G', \psi')$, rather than its image $y = (\psi, s)$. As in §4.4, some of the components of these factors will cancel from the product. Others can be expressed directly in terms of y. For example, the denominator in (4.8.13) equals

$$|\mathrm{Out}_G(G', \psi')| = |C^+_{\psi,s}/C^+_{\psi,s} \cap \widehat{G}'\, Z(\widehat{G}^0)^\Gamma|,$$

where

$$C^+_{\psi,s} = \{g \in C_\psi :\ gsg^{-1} = s\} = C^+_y$$

is the preimage of $\overline{C}^+_{\psi,s}$ in C_ψ. In particular, (4.8.13) specializes to the factor (4.4.5) in §4.4. We conclude that (4.8.8) can be written as the sum over $\psi \in \Psi_{\mathrm{disc}}(G,t,c)$ and $s \in \mathcal{E}_{\psi,\mathrm{ell}}$ of the product of two expressions

$$|C^+_{\psi,s}/C^+_{\psi,s} \cap \widehat{G}' Z(\widehat{G}^0)^\Gamma|^{-1}\, |\mathcal{C}_{\psi'}|^{-1}\, |\overline{Z}(\widehat{G}')^\Gamma|^{-1}\, |\overline{C}^+_{\psi,s}/\overline{C}_{\psi,s}|^{-1} \tag{4.8.15}$$

and

$$|\pi_0(\kappa_G)|^{-1}\, |\mathcal{S}_\psi|^{-1}\, \sigma(\overline{S}^0_{\psi'})\, \varepsilon'(\psi')\, f'(\psi'), \tag{4.8.16}$$

in which $y = (\psi, s)$ is the image of $x = (G', \psi')$ under the bijection in (4.8.11).

The expressions (4.8.15) and (4.8.16) are the two factors (7.9) and (7.10) in [**A9**]. They also reduce to the corresponding two expressions of §4.4, where the group C_ψ is equal to S^*_ψ. (The multiplicity m_ψ in the second expression in §4.4 does not appear here, since ψ now represents an element in a set $\Psi(G)$ rather than an equivalence class in $\widetilde{\Psi}(G)$.) As in the special cases of §4.4, the product of the two expressions simplifies. We shall outline the steps.

The product of the first two factors in (4.8.15) has a reduction

$$|C^+_{\psi,s}/C^+_{\psi,s} \cap \widehat{G}'\, Z(\widehat{G}^0)^\Gamma|^{-1}\, |\mathcal{C}_{\psi'}|^{-1} = |\overline{C}^+_{\psi,s}/\overline{C}^0_{\psi,s}\, \overline{Z}(\widehat{G}')^\Gamma|^{-1},$$

as in the discussion in §4.4. (See also [**A9**, p. 48–49].) The entire expression (4.8.15) can then be written as

$$\begin{aligned}
&|\overline{C}_{\psi,s}/\overline{C}^0_{\psi,s}\, \overline{Z}(\widehat{G}')^\Gamma|^{-1}\, |\overline{Z}(\widehat{G}')^\Gamma|^{-1}\\
=&|\pi_0(\overline{C}_{\psi,s})|^{-1}\, |\overline{C}^0_{\psi,s} \cap \overline{Z}(\widehat{G}')^\Gamma|^{-1}\\
=&|\pi_0(\overline{S}_{\psi,s})|^{-1}\, |\overline{S}^0_{\psi,s} \cap \overline{Z}(\widehat{G}')^\Gamma|^{-1},
\end{aligned}$$

since $\overline{C}^0_{\psi,s} = \overline{S}^0_{\psi,s}$. In the expression (4.8.16), the coefficient

$$\sigma(\overline{S}^0_{\psi'}) = \sigma\big((S_{\psi'}/Z(\widehat{G})^\Gamma)^0\big) = \sigma\big(\overline{S}^0_{\psi,s}/\overline{S}^0_{\psi,s} \cap \overline{Z}(\widehat{G}')^\Gamma\big)$$

satisfies an identity

$$\sigma(\overline{S}^0_{\psi'}) = \sigma(\overline{S}^0_{\psi,s})\, |\overline{S}^0_{\psi,s} \cap \overline{Z}(\widehat{G}')^\Gamma|,$$

by (4.1.9). Lemma 4.4.1, whose proof in §4.6 can be carried over to the general case, allows us to write the coefficient $\varepsilon'(\psi')$ in (4.8.16) as

$$\varepsilon'(\psi') = \varepsilon^{\widetilde{G}'}_{\psi'}(s_{\psi'}) = \varepsilon^G_\psi(s_\psi\, x_s).$$

Finally, we can again write

$$f'(\psi') = f'_G(\psi, s), \tag{4.8.17}$$

to denote the dependence of the linear form in (4.8.16) on ψ and s.

All of the components in (4.8.15) and (4.8.16) have now been expressed in terms of ψ and s. With no further need to refer to the original pair

$x = (G', \psi')$, we can allow x to denote an element in the group $\mathcal{S}_\psi$ of connected components of $\overline{S}_\psi$. We can then write the endoscopic expression (4.8.8) as a triple sum

$$(4.8.18)\quad |\pi_0(\kappa_G)|^{-1}\,|\mathcal{S}_\psi|^{-1}\sum_\psi\sum_x\sum_s |\pi_0(\overline{S}_{\psi,s})|^{-1}\,\sigma(\overline{S}^0_{\psi,s})\,\varepsilon^G_\psi(s_\psi x)\,f'_G(\psi, s)$$

over $\psi \in \Psi_{\mathrm{disc}}(G, t, c)$, $x \in \mathcal{S}_\psi$ and $s \in \mathcal{E}_{\psi,\mathrm{ell}}(x)$.

The transformation (4.8.18) of the endoscopic expression (4.8.8) is the general analogue of the formula of Lemma 4.4.2 (with allowance for our slightly different use of the symbol ψ). We would again expect the linear form

$$f'_G(\psi, x) = f'_G(\psi, s), \qquad x \in \mathcal{S}_\psi,\ s \in \overline{S}_{\psi,\mathrm{ss}}(x),$$

to depend only on the component x of s. If this is so, we can compress the sum over s in (4.8.18) to the coefficient $e(x) = e_\psi(x)$ of Proposition 4.1.1. After a change of variables in the sum over x, we will then be able to write (4.8.18) as a double sum

$$(4.8.19)\qquad |\pi_0(\kappa_G)|^{-1}\,|\mathcal{S}_\psi|^{-1}\sum_\psi\sum_x e_\psi(x)\,\varepsilon^G_\psi(x)\,f'_G(\psi, s_\psi x)$$

over $\psi \in \Psi_{\mathrm{disc}}(G, t, c)$ and $x \in \mathcal{S}_\psi$. This is the general analogue of the expression (4.4.11) of Corollary 4.4.3.

The c-discrete part $I_{\mathrm{disc},t,c}(f)$ of the trace formula is thus equal to the elementary endoscopic expression (4.8.19). The other half of the problem would be to extend the analysis of §4.3 to the spectral expansion of $I_{\mathrm{disc},t,c}(f)$. The formal aspect of this process can be carried out. For example, the commutative diagram (4.2.3) makes sense for any parameter ψ in the general set $\Psi(G)$. However, one is immediately forced to deal with the linear form (4.8.3). One would need to decompose (4.8.3) explicitly into suitable linear forms

$$f_G(\psi, u), \qquad \psi \in \Psi_{\mathrm{disc}}(G, t, c),\ u \in \mathfrak{N}_\psi,$$

defined explicitly in terms of local normalized intertwining operators. This would be the general analogue of Corollary 4.2.4. We would again expect the linear form

$$f_G(\psi, x) = f_G(\psi, u), \qquad x \in \mathcal{S}_\psi,\ u \in \mathfrak{N}_\psi(x),$$

to depend only on the image x of u in $\mathcal{S}_\psi$. If this is so, the methods in §4.3 would lead to an elementary spectral expression for $I_{\mathrm{disc},t,c}(f)$ as a double sum

$$(4.8.20)\qquad |\pi_0(\kappa_G)|^{-1}\,|\mathcal{S}_\psi|^{-1}\sum_\psi\sum_x i_\psi(x)\,\varepsilon^G_\psi(x)\,f_G(\psi, x)$$

over $\psi \in \Psi_{\mathrm{disc}}(G, t, c)$ and $x \in \mathcal{S}_\psi$.

We will say no more about the expected expression (4.8.20), except to point out that it is the general analogue of the expression (4.3.9) of Corollary 4.3.3. In particular, the coefficient $i(x) = i_\psi(x)$ in the expression equals the coefficient $e(x) = e_\psi(x)$ in (4.8.19), by Proposition 4.1.1. Moreover, the linear form $f_G(\psi, x)$ in (4.8.20) ought to be equal to the linear form $f'_G(\psi, s_\psi x)$ in (4.8.19), by some general version of the global intertwining relation. This would establish a term by term identification of the two expressions (4.8.19) and (4.8.20).

We have noted that the chain of subsets (4.1.13) of $\Psi(G)$ also makes sense in this general setting. Of most significance is the subset $\Psi_2(G)$ of $\Psi(G)$, which ought to govern the discrete spectrum. To be more specific, we could say that the comparison of (4.8.19) and (4.8.20) has two purposes. One is to prove that parameters in the complement of $\Psi_2(G)$ do not contribute to the discrete spectrum. The other is to compute the contribution of parameters ψ that do lie in $\Psi_2(G)$. The expected contribution of ψ will be a multiplicity formula for representations π in a global packet Π_ψ, in terms of associated characters $\langle x, \pi \rangle$ on the finite group $\mathcal{S}_\psi$. This would be the natural generalization of the multiplicity formula (4.7.9) of Theorem 1.5.2, with allowance made for the larger group $\mathcal{S}_\psi$, the fact that $\mathcal{S}_\psi$ could be nonabelian, and the possibility that G not be quasisplit (or equal to G^0). It could also be regarded as the degenerate case of the proposed global intertwining relation. Given the stable multiplicity formula (4.8.5) for $\psi \in \Psi(G)$, one would try to deduce the actual spectral multiplicity formula for representations π by a comparison of (4.8.19) with (4.8.20), as in our special case discussed at the end of the last section.

We add one final comment – on the title of this chapter. It seems apt, despite being a possible case of cultural appropriation! The contents of the chapter do represent a model of sorts. The model is standard, in the sense that it governs the general endoscopic comparison of trace formulas. However, we are still a very long way from being able to apply it to arbitrary groups. We can hope that the results we obtain for the groups $G \in \widetilde{\mathcal{E}}_{\mathrm{sim}}(N)$, combined with similar results for other classical groups, will lead to an endoscopic classification of a much broader class of groups, which share only the property of being endoscopically related to general linear groups. What about more general groups? Langlands has outlined a remarkable, though still speculative, strategy for applying the trace formula to cases of functoriality that are not tied to endoscopy [**L13**], [**L14**], [**FLN**]. His proposal is probably best suited to the stable trace formula. If it works, it might ultimately yield a general stable multiplicity formula as a byproduct. One could then imagine applying this formula in the standard model, as above, to a general endoscopic classification of representations.

CHAPTER 5

A Study of Critical Cases

5.1. The case of square integrable ψ

In the last chapter, we studied the contribution to the trace formula of many global parameters ψ. We did so by applying induction arguments to what we called the standard model. The culmination of these efforts was Proposition 4.5.1, which established the stable multiplicity formula in a number of cases. It also yielded the multiplicity formula of Theorem 1.5.2 for these cases, with its interpretation as a vanishing condition.

We recall that the stable multiplicity formula is the assertion of Theorem 4.1.2. It is at the heart of all of our global theorems. We have now established it for global parameters $\psi \in \widetilde{\Psi}(N)$ that do not lie in $\widetilde{\Psi}_{\mathrm{ell}}(N)$, or in any of the subsets $\widetilde{\Psi}_{\mathrm{ell}}(G)$ of $\widetilde{\Psi}(N)$ attached to elements $G \in \widetilde{\mathcal{E}}_{\mathrm{sim}}(N)$. In this chapter we shall begin the study of the cases that remain.

These cases are of course the most important. They are also the most difficult. The questions become increasingly subtle as we consider successive subsets of parameters in the chain

$$\widetilde{\Psi}_{\mathrm{sim}}(G) \subset \widetilde{\Psi}_2(G) \subset \widetilde{\Psi}_{\mathrm{ell}}(G), \qquad G \in \widetilde{\mathcal{E}}_{\mathrm{sim}}(N), \tag{5.1.1}$$

attached to a given G. The answers will draw upon new techniques, which we will introduce in this and future chapters. These will be founded on local-global methods, where we use local and global techniques to reinforce each other.

We begin by setting up a general family of global parameters, on which we will later impose local constraints. Throughout the chapter, we will let

$$\widetilde{\mathcal{F}} = \coprod_{N=1}^{\infty} \widetilde{\mathcal{F}}(N) \tag{5.1.2}$$

denote a fixed family of global parameters

$$\psi = \ell_1\psi_1 \boxplus \cdots \boxplus \ell_r\psi_r$$

in the general set

$$\widetilde{\Psi} = \coprod_{N=1}^{\infty} \widetilde{\Psi}(N).$$

We assume that $\widetilde{\mathcal{F}}$ is closed under the projections

$$\psi \longrightarrow \psi_i, \qquad 1 \leqslant i \leqslant r,$$

and under direct sums

$$(\psi^1, \psi^2) \longrightarrow \psi^1 \boxplus \psi^2, \qquad \psi^1, \psi^2 \in \widetilde{\mathcal{F}}.$$

In other words, $\widetilde{\mathcal{F}}$ is the graded semigroup generated by its simple components

$$\psi = \mu \boxtimes \nu, \qquad \psi \in \widetilde{\mathcal{F}}_{\mathrm{sim}},$$

with the grading

$$\deg \psi = N, \qquad \psi \in \widetilde{\Psi}(N),$$

inherited from $\widetilde{\Psi}$. We have obviously written

$$\widetilde{\mathcal{F}}_{\mathrm{sim}} = \widetilde{\mathcal{F}} \cap \widetilde{\Psi}_{\mathrm{sim}}$$

here. We shall also write $\widetilde{\mathcal{F}}_{\mathrm{ell}}$, $\widetilde{\mathcal{F}}_2(N)$, $\widetilde{\mathcal{F}}_{\mathrm{sim}}(G)$, etc., for the intersection of $\widetilde{\mathcal{F}}$ with the corresponding subset of $\widetilde{\Psi}$.

Throughout this chapter, the field F will remain global, and N will again be fixed. We shall assume inductively that all of the local and global theorems hold for any $\psi \in \widetilde{\mathcal{F}}$ with $\deg(\psi) < N$. This is the blanket induction hypothesis we have carried since §4.3. We are introducing it formally within the family $\widetilde{\mathcal{F}}$, since it is in the limited context of such families that we will eventually establish the general local theorems. Its interpretation here will be for the most part obvious.

There is one point in the induction hypothesis for $\widetilde{\mathcal{F}}$ that does call for an explanation. It concerns the assertion of Theorem 1.5.3(b). Consider a pair

$$\psi_i = \mu_i \boxtimes \nu_i, \qquad i = 1, 2,$$

of distinct self-dual, simple parameters in $\widetilde{\mathcal{F}}$. Motivated by the remark following the proof of Lemma 4.3.1 in §4.6, we shall call (ψ_1, ψ_2) an *ε-pair*, for want of a better term, if the following condition holds: the generic pair (μ_1, μ_2) is orthogonal, in that it satisfies the condition imposed on the pair (ϕ_1, ϕ_2) in Theorem 1.5.3(b), and the tensor product $\nu_1 \otimes \nu_2$ is the sum of an *odd* number of irreducible representations of $SL(2, \mathbb{C})$ of *even* dimension. The corresponding sum

$$\psi = \psi_1 \boxplus \psi_2, \tag{5.1.3}$$

which we will call an *ε-parameter*, is a self-dual, elliptic parameter for a general linear group. If w is the associated twisted Weyl element for the relevant maximal Levi subset, it follows from the analysis of (4.6.9) and (4.6.10) in the proof of Lemma 4.3.1 that the corresponding global normalizing factor satisfies

$$r_\psi(w) = \varepsilon(\tfrac{1}{2}, \mu_1 \times \mu_2). \tag{5.1.4}$$

Theorem 1.5.3(b) asserts that this number equals 1. Our induction hypothesis for $\widetilde{\mathcal{F}}$ will be that the assertion is valid for any ε-pair (ψ_1, ψ_2) with

$$\deg(\psi_1) + \deg(\psi_2) < N.$$

It then follows from the remark in §4.6 mentioned above that Lemma 4.3.1 is valid for any $\psi \in \widetilde{\mathcal{F}}(N)$ that is not an ε-parameter.

To make further progress, we shall also take on the following temporary hypothesis on the elements in $\widetilde{\mathcal{F}}_{\text{ell}}(N)$.

ASSUMPTION 5.1.1. *Suppose that $G \in \widetilde{\mathcal{E}}_{\text{ell}}(N)$, and that ψ belongs $\widetilde{\mathcal{F}}_2(G)$. Then there is a unique stable linear form*

$$f \longrightarrow f^G(\psi), \qquad f \in \widetilde{\mathcal{H}}(G),$$

on $\widetilde{\mathcal{H}}(G)$ with the general property

$$\widetilde{f}^G(\psi) = \widetilde{f}_N(\psi), \qquad \widetilde{f} \in \widetilde{\mathcal{H}}(N),$$

together with a secondary property that

$$f^G(\psi) = f^S(\psi_S) f^O(\psi_O), \qquad f \in \widetilde{\mathcal{H}}(G),$$

in case

$$G = G_S \times G_O,$$
$$\psi = \psi_S \times \psi_O,$$

and

$$f^G = f^S \times f^O$$

are composite.

The reader will recognize in this hypothesis a global analogue of the local assertion of Theorem 2.2.1(a). It would obviously follow from the local assertion, but this has yet to be established. We note that if ψ lies in the complement of $\widetilde{\mathcal{F}}_2(G)$ in $\widetilde{\mathcal{F}}(G)$, the global assertion can be reduced to the corresponding assertion for a Levi subgroup of $GL(N)$, which then follows from our induction hypothesis. In other words, the assertion of Assumption 5.1.1 is valid more generally for any $\psi \in \widetilde{\mathcal{F}}(G)$. (If $G = G_S \times G_O$, we have naturally to define $\widetilde{\mathcal{F}}(G)$ as the product of $\widetilde{\mathcal{F}}(G_S)$ and $\widetilde{\mathcal{F}}(G_O)$, rather than as a subset of $\widetilde{\mathcal{F}}(N)$.) This is the way Theorem 2.2.1(a) was stated.

We need to pause here for some further discussion of the condition $\psi \in \widetilde{\mathcal{F}}_2(G)$ in Assumption 5.1.1. For it raises a second point concerning our induction hypothesis for $\widetilde{\mathcal{F}}$. If ψ is not simple generic, which is to say that it does not lie in the set $\widetilde{\Phi}_{\text{sim}}(N)$ of self-dual cuspidal automorphic representations of $GL(N)$, the datum $G \in \widetilde{\mathcal{E}}_{\text{ell}}(N)$ such that ψ belongs to $\widetilde{\mathcal{F}}_2(G)$ is given by the induction hypothesis and the original constructions of §1.4. However, we will obviously have to deal also with parameters that are simple generic.

According to the global theorems we have stated, there will be three equivalent ways to characterize the subset $\widetilde{\Phi}_{\text{sim}}(G)$ of $\widetilde{\Phi}_{\text{sim}}(N)$ attached to a given $G \in \widetilde{\mathcal{E}}_{\text{sim}}(N)$. Theorem 1.4.1, which we have used for the formal definition, characterizes $\widetilde{\Phi}_{\text{sim}}(G)$ in terms of the discrete spectrum of G. Theorem

1.5.3(a), together with the fact that $\eta_\psi = \eta_G$ for any $\psi \in \widetilde{\Phi}(G)$, describes $\widetilde{\Phi}_{\text{sim}}(G)$ in terms of the poles of L-functions. The third characterization is provided by the stable multiplicity formula of Theorem 4.1.2. It tells us that $\widetilde{\Phi}_{\text{sim}}(G)$ is the subset of elements $\psi \in \widetilde{\Phi}_{\text{sim}}(N)$ such that the linear form $S^G_{\text{disc},\psi}$ on $\widetilde{\mathcal{H}}(G)$ is nonzero. We have of course not proved any of these theorems, regarded as conditions for the generic elements in $\widetilde{\mathcal{F}}_{\text{sim}}(N)$. But we could ultimately still take the assertion of any one of them as a definition of the subset $\widetilde{\mathcal{F}}_{\text{sim}}(G)$ of $\widetilde{\mathcal{F}}_{\text{sim}}(N)$. The third characterization is the least elementary. However, with its roots in harmonic analysis, it would be the most natural one to work with as we try to extend the global part of our induction hypothesis here to N. We shall therefore take it as the basis for a temporary definition, which we will later show is equivalent to the one in §1.4.

Suppose that $\widetilde{\mathcal{F}}$ is *generic*, in the sense that it is contained in the subset of generic global parameters $\widetilde{\Phi}$. If G belongs to $\widetilde{\mathcal{E}}_{\text{sim}}(N)$, we could simply define $\widetilde{\mathcal{F}}_{\text{sim}}(G)$ by the third condition above, namely as the subset of elements $\psi \in \widetilde{\mathcal{F}}_{\text{sim}}(N)$ with $S^G_{\text{disc},\psi} \neq 0$. For we know from (3.3.14) that if ψ belongs to $\widetilde{\mathcal{F}}_{\text{sim}}(N)$, there is a $G \in \widetilde{\mathcal{E}}_{\text{sim}}(N)$ with this property. Indeed, the left hand side of (3.3.14) is nonzero, while it follows from Proposition 3.4.1 and Theorem 1.3.2 that the contribution to the right hand side of (3.3.14) of any datum in the complement of $\widetilde{\mathcal{E}}_{\text{sim}}(N)$ in $\widetilde{\mathcal{E}}_{\text{ell}}(N)$ vanishes. We expect $G \in \widetilde{\mathcal{E}}_{\text{sim}}(N)$ to be uniquely determined by ψ. We know this to be so if $\eta_\psi \neq 1$ or N is odd. But if $\eta_\psi = 1$ and N is even, we would not be able to rule out the possibility at present that ψ could lie in two different sets $\widetilde{\mathcal{F}}_{\text{sim}}(G)$. A more serious problem with this definition is the possibility of a conflict with Assumption 5.1.1 in some of our later constructions. We shall therefore adopt a slightly more flexible (though ultimately equivalent) convention.

For the generic family $\widetilde{\mathcal{F}}$, we simply assume that we have attached global subsets

$$\widetilde{\mathcal{F}}_{\text{sim}}(G) \subset \{\psi \in \widetilde{\mathcal{F}}_{\text{sim}}(N) : \ S^G_{\text{disc},\psi} \neq 0\} \tag{5.1.5}$$

of $\widetilde{\mathcal{F}}_{\text{sim}}(N)$ to the simple endoscopic data $G \in \widetilde{\mathcal{E}}_{\text{sim}}(N)$, so that

$$\widetilde{\mathcal{F}}_{\text{sim}}(N) = \bigcup_{G \in \widetilde{\mathcal{E}}_{\text{sim}}(N)} \widetilde{\mathcal{F}}_{\text{sim}}(G). \tag{5.1.6}$$

Assumption 5.1.1, as it applies to simple pairs

$$(G, \psi), \qquad G \in \widetilde{\mathcal{E}}_{\text{sim}}(N),\ \psi \in \widetilde{\mathcal{F}}_{\text{sim}}(G),$$

represents an axiom that is imposed afterwards. Once it is granted, we can then make our temporary definition precise by taking $\widetilde{\mathcal{F}}_{\text{sim}}(G)$ to be the set of *all* parameters $\psi \in \widetilde{\mathcal{F}}_{\text{sim}}(N)$ such that $S^G_{\text{disc},\psi} \neq 0$, and such that the linear form $\widetilde{f}_N(\psi)$ transfers to G. Assumption 5.1.1 for simple pairs (G, ψ) then becomes the assertion (5.1.6). In working with this temporary definition, we must keep in mind the hypothetical possibility that the inclusion (5.1.5) might be proper, or the possibility that of the union (5.1.6) might not be disjoint. But in any case, we have now agreed on an ad hoc convention for the simple parameters in $\widetilde{\mathcal{F}}(N)$, thereby introducing a new definition of the sets $\widetilde{\mathcal{F}}_2(G)$ to which Assumption 5.1.1 is supposed to apply. We shall resolve it in terms of the original definition (for a particular family $\widetilde{\mathcal{F}}$) at the end of the chapter.

If $\widetilde{\mathcal{F}}$ is *nongeneric*, in the sense that it contains elements in the complement of $\widetilde{\Phi}$ in $\widetilde{\Psi}$, we will assume that the relevant global theorems hold for the generic part μ of *any* simple parameter $\psi \in \widetilde{\mathcal{F}}_{\mathrm{sim}}$. That is, we assume that the three ways of defining $\widetilde{\Phi}_{\mathrm{sim}}(G)$ are equivalent for such parameters. The subsets $\widetilde{\mathcal{F}}_2(G)$ and $\widetilde{\mathcal{F}}(G)$ of $\widetilde{\mathcal{F}}$ are then all well defined.

We have thus refined the global induction hypothesis, as it applies to parameters in a given family $\widetilde{\mathcal{F}}$. We observe that the two adjustments concern the two parts of Theorem 1.5.3. Part (a) of the theorem represents an induction hypothesis for generic families that is assumed to have been resolved in the nongeneric case. Part (b) represents an induction hypothesis in the nongeneric case that is irrelevant for generic families. We will ultimately study the two kinds of families $\widetilde{\mathcal{F}}$ separately, the generic case being the topic of Chapter 6, and the nongeneric case the topic of Chapter 7.

In this chapter, however, we will generally be able to treat the two cases together. Having assumed inductively that for a given family $\widetilde{\mathcal{F}}$ the relevant theorems hold for elements $\psi_* \in \widetilde{\mathcal{F}}(N_*)$ with $N_* < N$, we must try to show that they hold also for parameters $\psi \in \widetilde{\mathcal{F}}(N)$. This is a long term proposition. It will not be settled in general until the later chapters, after we have constructed families $\widetilde{\mathcal{F}}$ with specified local conditions. In the meantime, we shall see what can be deduced from the induction hypotheses, when used in combination with Assumption 5.1.1.

In this section, we shall look at the basic case that ψ lies in $\widetilde{\mathcal{F}}_{\mathrm{ell}}(N)$. With our convention above for the simple generic elements, $\widetilde{\mathcal{F}}_{\mathrm{ell}}(N)$ is the union over $G \in \widetilde{\mathcal{E}}_{\mathrm{ell}}(N)$ of the sets $\widetilde{\mathcal{F}}_2(G)$. However, this temporary convention still leaves open the question of how to assign a meaning to the centralizers $\mathcal{S}_\psi^G$ attached to simple pairs (G,ψ), with ψ a generic element in $\widetilde{\mathcal{F}}_{\mathrm{sim}}(G)$. In situations where (G,ψ) is understood to have been fixed, we shall implicitly agree that

$$\mathcal{S}_\psi^* = \mathcal{S}_\psi^{G^*} = \begin{cases} 1, & \text{if } G^* = G, \\ \varnothing, & \text{otherwise,} \end{cases} \tag{5.1.7}$$

for any $G^* \in \widetilde{\mathcal{E}}_{\mathrm{sim}}(N)$. This is what we expect, since G should be the unique datum with $\psi \in \widetilde{\mathcal{F}}_{\mathrm{sim}}(G)$. In any case, the difference

$${}^0S^*_{\mathrm{disc},\psi}(f^*) = {}^0S^{G^*}_{\mathrm{disc},\psi}(f^*), \qquad f^* \in \widetilde{\mathcal{H}}(G^*),$$

between $S^*_{\mathrm{disc},\psi}(f^*)$ and its expected value will then be defined as in (4.4.9) (with G^* in place of G'). We will of course also be interested in the case that ψ is not simple generic. In this case, $\mathcal{S}^{G^*}_\psi$ is given by induction, and the linear form ${}^0S^*_{\mathrm{disc},\psi}(f^*)$ is then defined by Assumption 5.1.1. We fix $\psi \in \widetilde{\mathcal{F}}_{\mathrm{ell}}(N)$, and take G to be a fixed datum in $\widetilde{\mathcal{E}}_{\mathrm{ell}}(N)$ with $\psi \in \widetilde{\mathcal{F}}_2(G)$. We remind ourselves that G is uniquely determined by ψ, except possibly in the case that ψ is simple generic.

The centralizer set

$$\widetilde{S}_\psi(N) = S_\psi\big(\widetilde{G}(N)\big)$$

for $\widetilde{G}(N)$ is a connected abelian bitorsor. It follows that the global intertwining relation for $\widetilde{G}(N)$ and ψ reduces simply to the assertion of Assumption 5.1.1. In particular, the stated conditions of Corollaries 4.3.3 and 4.4.3 (with $\widetilde{G}(N)$ in place of G) both hold. In the case of Corollary 4.3.3, there is also an implicit condition, namely that the proof of Lemma 4.3.1 is valid under the induction hypotheses now in force. If we assume that ψ is not an ε-parameter (5.1.3), this condition holds, and both corollaries are valid for $\widetilde{G}(N)$ and ψ. We then see from the identity of Proposition 4.1.1, which is completely elementary in this case, that the corresponding expressions (4.3.9) and (4.4.11) are equal. It follows that for $\widetilde{f} \in \widetilde{\mathcal{H}}(N)$, the difference between $I_{\mathrm{disc},\psi}(\widetilde{f})$ and the linear form

$${}^0\widetilde{r}^N_{\mathrm{disc},\psi}(\widetilde{f}) = {}^0r^{\widetilde{G}(N)}_{\mathrm{disc},\psi}(\widetilde{f})$$

equals the difference between $I_{\mathrm{disc},\psi}(\widetilde{f})$ and the linear form

$${}^0\widetilde{s}^N_{\mathrm{disc},\psi}(\widetilde{f}) = {}^0s^{\widetilde{G}(N)}_{\mathrm{disc},\psi}(\widetilde{f}).$$

Since $\widetilde{G}(N)^0 = GL(N)$, we know that ${}^0\widetilde{r}^N_{\mathrm{disc},\psi}(\widetilde{f}) = 0$. It follows from the definition (4.4.10) (again with $\widetilde{G}(N)$ in place of G) that

$$\sum_{G^* \in \widetilde{\mathcal{E}}_{\mathrm{sim}}(N)} \widetilde{\iota}(N, G^*)\, {}^0\widehat{S}^*_{\mathrm{disc},\psi}(\widetilde{f}^*) = 0, \tag{5.1.8}$$

where $\widetilde{f}^*$ is the image of $\widetilde{f}$ in $\widehat{S}(G)$.

If G belongs to the complement of $\widetilde{\mathcal{E}}_{\mathrm{sim}}(N)$ in $\widetilde{\mathcal{E}}_{\mathrm{ell}}(N)$, we know from our induction hypothesis that ${}^0S^G_{\mathrm{disc},\psi}(f)$, the difference between $S^G_{\mathrm{disc},\psi}(f)$ and its expected value, vanishes. One of our long term goals is to show that the same is true if G belongs to $\widetilde{\mathcal{E}}_{\mathrm{sim}}(N)$, or indeed if G is replaced by any element $G^* \in \widetilde{\mathcal{E}}_{\mathrm{sim}}(N)$. In other words, we would like to establish the stable multiplicity formula of Theorem 4.1.2 for the given ψ, and any $G^* \in \widetilde{\mathcal{E}}_{\mathrm{sim}}(N)$. The next lemma gives some preliminary information.

LEMMA 5.1.2. *Assume that for the given pair*

$$(G, \psi), \qquad G \in \widetilde{\mathcal{E}}_{\mathrm{ell}}(N),\ \psi \in \widetilde{\mathcal{F}}_2(G),$$

ψ is not an ε-parameter (5.1.3). *Then the stable multiplicity formula* (4.1.11) *holds for G and ψ if and only if*

$$S^*_{\mathrm{disc},\psi}(f^*) = {}^0S^*_{\mathrm{disc},\psi}(f^*) = 0, \qquad f^* \in \widetilde{\mathcal{H}}(G^*),$$

for every $G^* \in \widetilde{\mathcal{E}}_{\mathrm{sim}}(N)$ with $G^* \neq G$.

PROOF. One direction is clear. If ${}^0S^*_{\mathrm{disc},\psi} = 0$ for every $G^* \neq G$ in $\widetilde{\mathcal{E}}_{\mathrm{sim}}(N)$, (5.1.8) tells us that ${}^0\widehat{S}^G_{\mathrm{disc},\psi}(\widetilde{f}^G)$ vanishes for any $\widetilde{f} \in \widetilde{\mathcal{H}}(N)$. The required assertion in this direction already being known if G is composite, we can assume that $G \in \widetilde{\mathcal{E}}_{\mathrm{sim}}(N)$. The correspondence $\widetilde{f} \to \widetilde{f}^G$ from $\widetilde{\mathcal{H}}(N)$ to $\widetilde{\mathcal{S}}(G)$ is then surjective, by Corollary 2.1.2. It follows that ${}^0S^G_{\mathrm{disc},\psi}$ vanishes, as an $\widetilde{\mathrm{Out}}_N(G)$-symmetric stable linear form. In other words, the stable multiplicity formula for

$$S^G_{\mathrm{disc},\psi}(f), \qquad f \in \widetilde{\mathcal{H}}(G),$$

is valid.

Conversely, suppose that ${}^0S^G_{\mathrm{disc},\psi} = 0$. Then (5.1.8) reduces to

$$\sum_{G^* \neq G} \widetilde{\iota}(N, G^*)\, {}^0\widehat{S}^*_{\mathrm{disc},\psi}(\widetilde{f}^*) = 0.$$

But if $G^* \in \widetilde{\mathcal{E}}_{\mathrm{sim}}(N)$ is distinct from G, (4.4.12) tells us that

$${}^0S^*_{\mathrm{disc},\psi}(f^*) = S^*_{\mathrm{disc},\psi}(f^*) = \mathrm{tr}\big(R^*_{\mathrm{disc},\psi}(f^*)\big), \qquad f^* \in \widetilde{\mathcal{H}}(G^*),$$

since ψ does not lie in $\widetilde{\Psi}(G^*)$ unless ψ is generic and simple, in which case the convention (5.1.7) leads to the same outcome. Applying Proposition 3.5.1 to the resulting formula

$$\sum_{G^* \neq G} \widetilde{\iota}(N, G^*)\, \mathrm{tr}\big(R^*_{\mathrm{disc},\psi}(f^*)\big) = 0,$$

in which $\{\widetilde{f}^*\}$ is replaced by any compatible family of functions $\{f^*\}$, we deduce that

$$\mathrm{tr}\big(R^*_{\mathrm{disc},\psi}(f^*)\big) = S^*_{\mathrm{disc},\psi}(f^*) = 0, \qquad G^* \neq G,$$

as required. □

COROLLARY 5.1.3. *Suppose that G and ψ are as in the lemma, and that one of the further conditions holds:*

(i) *N is odd,*

(ii) *$\eta_\psi \neq 1$,*

or

(iii) *$G \notin \widetilde{\mathcal{E}}_{\mathrm{sim}}(N)$.*

Then

$$S^G_{\mathrm{disc},\psi}(f) = m_\psi |\mathcal{S}_\psi|^{-1}\, \varepsilon^G(\psi)\, f^G(\psi), \qquad f \in \widetilde{\mathcal{H}}(G), \tag{5.1.9}$$

while $S^{G^*}_{\mathrm{disc},\psi} = 0$ for every $G^* \in \widetilde{\mathcal{E}}_{\mathrm{sim}}(N)$ distinct from G. In other words, the stable multiplicity formula holds for ψ and any $G^* \in \widetilde{\mathcal{E}}_{\mathrm{sim}}(N)$.

PROOF. If (iii) holds, the stable multiplicity formula for $S^G_{\mathrm{disc},\psi}(f)$ is a consequence of our induction hypotheses, as we noted above. It then follows from the lemma that $S^{G^*}_{\mathrm{disc},\psi} = 0$ for all $G^* \neq G$, which is to say, every $G^* \in \mathcal{E}_{\mathrm{sim}}(N)$.

We can therefore assume that $G \in \widetilde{\mathcal{E}}_{\mathrm{sim}}(N)$. Recall that $S^{G^*}_{\mathrm{disc},\psi}$ vanishes unless the characters η_ψ and η_{G^*} on Γ_F attached to G^* and ψ are equal. But $\eta_\psi = \eta_G$, since $\psi \in \widetilde{\Psi}_2(G)$, and we know that η_G determines G uniquely as a simple twisted endoscopic datum, if it is nontrivial or if N is odd. In these cases then, $S^{G^*}_{\mathrm{disc},\psi} = 0$ for every $G^* \in \widetilde{\mathcal{E}}_{\mathrm{sim}}(N)$ distinct from G. The required stable multiplicity formula (5.1.9) for G then follows from the lemma. □

The remaining case is for N even, $\eta_\psi = 1$, and $G \in \widetilde{\mathcal{E}}_{\mathrm{sim}}(N)$. It is not so easily resolved. There is now a second group $G^\vee \in \widetilde{\mathcal{E}}_{\mathrm{sim}}(N)$ with $\eta_{G^\vee} = \eta_\psi = \eta_G$, and even though ψ does not belong to $\widetilde{\mathcal{F}}_2(G^\vee)$ (except possibly when ψ is generic and simple), we cannot say a priori that ${}^0S^{G^\vee}_{\mathrm{disc},\psi}$ vanishes. In this case, G and $G^\vee$ are split groups, whose dual groups are $Sp(N,\mathbb{C})$ and $SO(N,\mathbb{C})$. We write L for the group, isomorphic to $GL(\frac{1}{2}N)$, that represents a Siegel maximal Levi subgroup of both G and $G^\vee$. For convenience, we shall also write

$$\widetilde{o}(G^*) = |\widetilde{\mathrm{O}}\mathrm{ut}_N(G^*)|, \qquad G^* \in \widetilde{\mathcal{E}}_{\mathrm{sim}}(N),$$

an integer equal to either 1 or 2. There are two transfer mappings

$$f \longrightarrow f^L = f_L, \qquad f \in \widetilde{\mathcal{H}}(G)$$

and

$$f^\vee \longrightarrow f^{\vee,L} = f^\vee_L, \qquad f^\vee \in \widetilde{\mathcal{H}}(G^\vee),$$

from $\widetilde{\mathcal{H}}(G)$ and $\widetilde{\mathcal{H}}(G^\vee)$ respectively to the space $\mathcal{S}(L) = \mathcal{I}(L)$. It is not hard to see from the definitions that they have a common image in $\mathcal{S}(L)$, which we denote by $\widetilde{\mathcal{S}}^0(L)$.

LEMMA 5.1.4. *Suppose that G and ψ are as in Lemma 5.1.2, and also that N is even, $\xi_\psi = 1$, and $G \in \widetilde{\mathcal{E}}_{\mathrm{sim}}(N)$. Then there is a linear form*

$$h^L \longrightarrow h^L(\Lambda), \qquad h^L \in \widetilde{\mathcal{S}}^0(L),$$

on $\widetilde{\mathcal{S}}^0(L)$ such that

$$S^G_{\mathrm{disc},\psi}(f) = m_\psi\, |\mathcal{S}_\psi|^{-1}\, \varepsilon^G(\psi)\, f^G(\psi) - \widetilde{o}(G)\, f^L(\Lambda), \qquad f \in \widetilde{\mathcal{H}}(G),$$

and

$$S^{G^\vee}_{\mathrm{disc},\psi}(f^\vee) = \widetilde{o}(G^\vee)\, f^{\vee,L}(\Lambda), \qquad f^\vee \in \widetilde{\mathcal{H}}(G^\vee),$$

while

$$S^*_{\mathrm{disc},\psi}(f^*) = {}^0S^*_{\mathrm{disc},\psi}(f^*) = 0, \qquad f^* \in \widetilde{\mathcal{H}}(G^*),$$

for every $G^* \in \widetilde{\mathcal{E}}_{\mathrm{sim}}(N)$ distinct from G and $G^\vee$.

PROOF. If $G^* \in \widetilde{\mathcal{E}}_{\mathrm{sim}}(N)$ is distinct from G and $G^\vee$, the character η_{G^*} is distinct from η_ψ, and

$$S^*_{\mathrm{disc},\psi} = {}^0S^*_{\mathrm{disc},\psi} = 0.$$

The corresponding summand in the identity (5.1.8) thus vanishes, and the left hand side of (5.1.8) reduces to a sum of two terms. It follows from (3.2.6) that

$$\widetilde{\iota}(N,G)\,\widetilde{o}(G) = \widetilde{\iota}(N,G^\vee)\,\widetilde{o}(G^\vee)$$

since

$$|\overline{Z}(\widehat{G})^\Gamma| = |\overline{Z}(\widehat{G}^\vee)^\Gamma| = 1.$$

The identity then takes the form

$$\widetilde{o}(G)^{-1}\big({}^0S^G_{\mathrm{disc},\psi}(f)\big) + \widetilde{o}(G^\vee)^{-1}\big({}^0S^{G^\vee}_{\mathrm{disc},\psi}(f^\vee)\big) = 0, \tag{5.1.10}$$

for any compatible pair of functions

$$\{f, f^\vee : \ f \in \widetilde{\mathcal{H}}(G),\ f^\vee \in \widetilde{\mathcal{H}}(G^\vee)\}.$$

As members of a compatible family of functions, f and $f^\vee$ are not independent. Their stable orbital integrals have the same values on Levi subgroups that are common to G and $G^\vee$. But any such Levi subgroup is conjugate to a subgroup of L. Since this is the only obstruction to the independence of f and $f^\vee$, the two summands on the left hand side of (5.1.10) can differ only by the pullback of a linear form on $\widetilde{\mathcal{S}}^0(L)$. In other words, we can write

$${}^0S^{G^\vee}_{\mathrm{disc},\psi}(f^\vee) = S^{G^\vee}_{\mathrm{disc},\psi}(f^\vee) = \widetilde{o}(G^\vee)\,f^{\vee,L}(\Lambda), \qquad f^\vee \in \widetilde{\mathcal{H}}(G^\vee),$$

where Λ represents a linear form on $\widetilde{\mathcal{S}}^0(L)$.

The required formula for $S^G_{\mathrm{disc},\psi}(f)$ then follows from (5.1.10), and the specialization

$${}^0S^G_{\mathrm{disc},\psi}(f) = S^G_{\mathrm{disc},\psi}(f) - m_\psi\,|\mathcal{S}_\psi|^{-1}\varepsilon^G(\psi)\,f^G(\psi)$$

of the definition (4.4.9). □

COROLLARY 5.1.5. *With the conditions of the lemma, the linear form*

$$f^\vee \longrightarrow \widetilde{o}(G^\vee)\,f^{\vee,L}(\Lambda), \qquad f^\vee \in \widetilde{\mathcal{H}}(G^\vee),$$

is the restriction to $\widetilde{\mathcal{H}}(G^\vee)$ *of a unitary character on* $G^\vee(\mathbb{A})$.

PROOF. Since ψ does not belong to $\widetilde{\Psi}(G^\vee)$ (except possibly when ψ is generic and simple), (4.4.12) tells us that

$${}^0S^{G^\vee}_{\mathrm{disc},\psi}(f^\vee) = S^{G^\vee}_{\mathrm{disc},\psi}(f^\vee) = \mathrm{tr}\big(R^{G^\vee}_{\mathrm{disc},\psi}(f^\vee)\big).$$

Since the given linear form equals $S^{G^\vee}_{\mathrm{disc},\psi}(f^\vee)$, the corollary follows. □

The last two lemmas and their corollaries came with the condition that $\psi \in \widetilde{\mathcal{E}}_{\mathrm{ell}}(N)$ not be an ε-parameter (5.1.3). We shall now see what happens in this supplementary case.

LEMMA 5.1.6. *Suppose that for the given pair*

$$(G, \psi), \qquad G \in \widetilde{\mathcal{E}}_{\mathrm{ell}}(N),\ \psi \in \widetilde{\mathcal{F}}_2(G),$$

ψ is an ε-parameter (5.1.3). *Then*

$$\mathrm{tr}\big(R^*_{\mathrm{disc},\psi}(f^*)\big) = 0 = {}^0S^*_{\mathrm{disc},\psi}(f^*), \qquad f^* \in \widetilde{\mathcal{H}}(G^*),$$

for every $G^* \in \widetilde{\mathcal{E}}_{\mathrm{sim}}(N)$. Moreover, the generic components μ_1 and μ_2 of ψ_1 and ψ_2 satisfy

$$\varepsilon\big(\tfrac{1}{2}, \mu_1 \times \mu_2\big) = 1,$$

in accordance with Theorem 1.5.3*(b).*

PROOF. We are assuming that ψ is a sum of an ε-pair (ψ_1, ψ_2) of simple parameters. By assumption, the generic components μ_1 and μ_2 are either both orthogonal or both symplectic, while the unipotent components ν_1 and ν_2 of ψ_1 and ψ_2 are of opposite type. Therefore ψ_1 and ψ_2 are themselves of opposite type. This implies that the datum $G = G_1 \times G_2$ is composite. It follows that

$$(5.1.11)\quad \widetilde{f}^G(\psi) = \mathrm{tr}\big(\pi_\psi(\widetilde{f}^G)\big) = \mathrm{tr}\big(\pi_{\psi_1}(\widetilde{f}^1)\big)\,\mathrm{tr}\big(\pi_{\psi_2}(\widetilde{f}^2)\big), \qquad \widetilde{f} \in \widetilde{\mathcal{H}}(N),$$

where $\pi_\psi = \pi_{\psi_1} \otimes \pi_{\psi_2}$ is the automorphic representation in the packet $\widetilde{\Pi}_\psi$ for G obtained from Assumption 5.1.1 and our induction hypothesis.

We would like to argue as in the proof of Lemma 5.1.2. However, we do not know that the identity (5.1.8) is valid. The problem is that we do not have a proof of Lemma 4.3.1 (with $\widetilde{G}(N)$ in place of G) for the ε-parameter ψ. This alters the contribution of Corollary 4.3.3 to (5.1.8). We must describe the resulting modification of the identity.

With the limited information presently at our disposal, we can make use of Corollary 4.3.3 (with $\widetilde{G}(N)$ in place of G) only if the summand in (4.3.9) is multiplied by the quotient

$$r^N_\psi(w_u)\,\varepsilon^{N,1}_\psi(u)\,\varepsilon^N_\psi(x_u)^{-1}\,s^{N,0}_\psi(u)^{-1}$$

of the two sides of the analogue for $\widetilde{G}(N)$ of the unproven identity of Lemma 4.3.1. There is of course only one summand, since the centralizer $\widetilde{S}^N_\psi$ is connected. Given that $\widetilde{S}^N_\psi$ is also an abelian torsor, we deduce that

$$\varepsilon^{N,1}_\psi(u) = \varepsilon^N_\psi(x_u) = s^{N,0}_\psi(u) = 1.$$

The remaining factor in the quotient satisfies

$$r_\psi(w_u) = r^N_\psi(w) = \varepsilon\big(\tfrac{1}{2}, \mu_1 \times \mu_2\big),$$

according to (5.1.4). By Assumption 5.1.1, the linear forms in (4.3.9) and (4.4.11) again satisfy the global intertwining relation

$$f'_G(\psi, s_\psi x) = f'_G(\psi, x) = f_G(\psi, x),$$

but with $\widetilde{G}(N)$ and $\widetilde{f}$ again in place of G and f. This equals the trace (5.1.11). We can now compare the modified form of (4.3.9) with the corresponding expression (4.4.11). The difference now becomes an explicit, possibly nonzero, multiple of (5.1.11). In fact, by substituting the formulas for the coefficients above, we find that the left hand side of (5.1.8) equals

$$C'_\psi\big(\varepsilon(\tfrac{1}{2},\, \mu_1 \times \mu_2) - 1\big)\,\mathrm{tr}\big(\pi_\psi(\widetilde{f}^G)\big), \tag{5.1.12}$$

where C'_ψ is the product of the positive constants C_ψ and $e_\psi(x)$. This is the modification of (5.1.8), an identity that could also have been obtained directly from first principles rather than the general formulas of Corollaries 4.3.3 and 4.4.3.

We have shown that the left hand side of (5.1.8) equals (5.1.12). In the summand of $G^* \in \widetilde{\mathcal{E}}_{\mathrm{sim}}(N)$ in (5.1.8), we can use (4.4.12) to write

$$ {}^0S^*_{\mathrm{disc},\psi}(f^*) = \mathrm{tr}\big(R^*_{\mathrm{disc},\psi}(f^*)\big), \qquad f^* \in \widetilde{\mathcal{H}}(G^*), \tag{5.1.13}$$

since ψ does not belong to $\widetilde{\Psi}(G^*)$. In the identity itself, we can replace the family $\{\widetilde{f}^*\}$ by an arbitrary compatible family of functions

$$\big\{f^* \in \widetilde{\mathcal{H}}(G^*) :\ G^* \in \widetilde{\mathcal{E}}_{\mathrm{ell}}(N)\big\},$$

writing $f = f^*$ in case $G^* = G$. We conclude that

$$\sum_{G^* \in \widetilde{\mathcal{E}}_{\mathrm{sim}}(N)} \widetilde{\iota}(N, G^*)\,\mathrm{tr}\big(R^*_{\mathrm{disc},\psi}(f^*)\big) + C''_\psi\,\mathrm{tr}\big(\pi_\psi(f)\big) = 0,$$

for the nonnegative coefficient

$$C''_\psi = C'_\psi\big(1 - \varepsilon(\tfrac{1}{2},\, \mu_1 \times \mu_2)\big).$$

The left hand side of this expression can be written as a nonnegative linear combination of irreducible characaters. Proposition 3.5.1 then tells us that each of the coefficients vanishes. In particular, the right hand side of (5.1.13) equals 0. This gives the first assertion of the lemma. From the vanishing of the remaining coefficient C''_ψ, and the fact that C'_ψ is strictly positive, we see also that

$$1 - \varepsilon\big(\tfrac{1}{2},\, \mu_1 \times \mu_2\big) = 0.$$

This is the second assertion of the lemma. □

Lemma 5.1.6 is an important complement to the previous lemmas and their corollaries. It completes our discussion of square integrable parameters $\psi \in \widetilde{\mathcal{F}}_2(G)$ attached to composite endoscopic data $G \in \widetilde{\mathcal{E}}_{\mathrm{ell}}(N)$. We now know that these parameters do not contribute to the discrete spectrum of any simple endoscopic datum $G \in \widetilde{\mathcal{E}}_{\mathrm{sim}}(N)$. Lemma 5.1.6 also gives a reduction of our induction hypothesis, as it applies to Theorem 1.5.3(b). To be precise, we have shown that the number (5.1.4) attached to an ε-pair (ψ_1, ψ_2) is equal to 1 whenever

$$\deg(\psi_1) + \deg(\psi_2) = N.$$

With this step, we can retire the terms "ε-pair" and "ε-parameter" from our lexicon!

5.2. The case of elliptic ψ

In the last section, we looked at square-integrable parameters

$$\{\psi \in \widetilde{\Psi}_2(G) : \ G \in \widetilde{\mathcal{E}}_{\mathrm{ell}}(N)\}. \tag{5.2.1}$$

We resolved what remained to be done when G lies in the complement of $\widetilde{\mathcal{E}}_{\mathrm{sim}}(N)$ in $\widetilde{\mathcal{E}}_{\mathrm{ell}}(N)$. We shall now add some observations for the other critical case, that of elliptic parameters

$$\{\psi \in \widetilde{\Psi}_{\mathrm{ell}}(G) : \ G \in \widetilde{\mathcal{E}}_{\mathrm{sim}}(N)\}, \tag{5.2.2}$$

or more precisely, parameters in (5.2.2) that lie in the complement of (5.2.1). Recall that the sets (5.2.1) and (5.2.2) can both be identified with subsets of $\widetilde{\Psi}(N)$. Their union is precisely the set of parameters that were not treated by the general induction arguments of Chapter 4. Their intersection, the set of square integrable parameters ψ for simple data G, will be the most difficult family to treat. Among other things, we will eventually have to show that the linear form Λ of Lemma 5.1.4 vanishes.

The elliptic parameters $\psi \in \widetilde{\Psi}_{\mathrm{ell}}(G)$ that are not square-integrable have their own set of difficulties. These stem from the global intertwining relation for G, since we are not yet in a position to prove this identity. We will therefore not be able to establish the predicted formula for $S^G_{\mathrm{disc},\psi}(f)$, which reduces to 0 in this case, or to establish the expected property that the representation $R^G_{\mathrm{disc},\psi}(f)$ vanishes. Another consequence is that we will not yet be able to resolve the case $r = 1$ of (4.5.11) (with $\psi \in \widetilde{\Psi}_{\mathrm{ell}}(G^\vee)$) that was left open from §4.5.

In this section, we shall put together what can be obtained from the standard model. We will introduce formulas to which the unresolved intertwining relation contributes an obstruction term. These formulas will eventually be strengthened to yield what we want. They will also be applied slightly differently in the next section to give more information about the square integrable parameters of §5.1.

For the rest of the chapter, we will be following the global notation of the last section (and the previous two chapters). In particular, F is a global field, while $G \in \widetilde{\mathcal{E}}_{\mathrm{ell}}(N)$ and $\psi \in \widetilde{\Psi}(G)$ are of course to be understood as global objects over F. In this section, we will be assuming that G is simple. It will then be convenient to denote the complement of an embedded set in (5.1.1) in its successor by the appropriate superscript. Thus $\widetilde{\Psi}_2^{\mathrm{sim}}(G)$ is the complement of $\widetilde{\Psi}_{\mathrm{sim}}(G)$ in $\widetilde{\Psi}_2(G)$, while $\widetilde{\Psi}^2_{\mathrm{ell}}(G)$ is the complement of $\widetilde{\Psi}_2(G)$ in $\widetilde{\Psi}_{\mathrm{ell}}(G)$. We therefore have a decomposition

$$\widetilde{\Psi}_{\mathrm{ell}}(G) = \widetilde{\Psi}_{\mathrm{sim}}(G) \coprod \widetilde{\Psi}_2^{\mathrm{sim}}(G) \coprod \widetilde{\Psi}^2_{\mathrm{ell}}(G). \tag{5.2.3}$$

It is the set $\widetilde{\Psi}^2_{\mathrm{ell}}(G)$ that we will be looking at in this section.

We will ultimately be focused on a family $\widetilde{\mathcal{F}}$ of parameters (5.1.2), on which we will later impose local constraints. Our present concern will therefore be with the subset

$$\widetilde{\mathcal{F}}^2_{\text{ell}}(G) = \widetilde{\mathcal{F}} \cap \widetilde{\Psi}^2_{\text{ell}}(G), \qquad G \in \widetilde{\mathcal{E}}_{\text{sim}}(N),$$

of $\widetilde{\mathcal{F}}$. There is actually no compelling reason to work with $\widetilde{\mathcal{F}}$ in much of this section, since Assumption 5.1.1, the one axiom we have imposed beyond our blanket induction hypothesis, is irrelevant to parameters in $\widetilde{\mathcal{F}}^2_{\text{ell}}(G)$. However, we may as well state the results for $\widetilde{\mathcal{F}}^2_{\text{ell}}(G)$, since it is in this context that they will be applied.

Suppose then that N and $G \in \widetilde{\mathcal{E}}_{\text{sim}}(N)$ are fixed, and that ψ is a given parameter in the set $\widetilde{\mathcal{F}}^2_{\text{ell}}(G)$. Then ψ is characterized by the conditions

$$\begin{cases} \psi = 2\psi_1 \boxplus \cdots \boxplus 2\psi_q \boxplus \psi_{q+1} \boxplus \cdots \boxplus \psi_r, \\ S_\psi = \big(O(2,\mathbb{C})^q \times O(1,\mathbb{C})^{r-q}\big)^+_\psi, \qquad q \geqslant 1, \end{cases} \tag{5.2.4}$$

along with the requirement that the Weyl group W_ψ contain an element w in the regular set $W_{\psi,\text{reg}}$. Recall that $(\,\cdot\,)^+_\psi$ denotes the kernel of a sign character, which was defined in terms of the degrees N_i of the simple components ψ_i of ψ. The condition on the existence of w is that there be an element in S_ψ whose projection onto any factor $O(2,\mathbb{C})$ belong to the nonidentity connected component. This holds either if $q \neq r$, or if there is an even number of indices i with N_i odd. In general, we may as well write

$$\mathcal{S}_{\psi,\text{ell}} = \big\{x \in \mathcal{S}_\psi : \ \mathcal{E}_{\psi,\text{ell}}(x) \neq \varnothing\big\}$$

for the subset of components in the quotient $\mathcal{S}_\psi$ of S_ψ that contain elliptic elements. Then $\mathcal{S}_{\psi,\text{ell}}$ is also the set of components x which contain a regular Weyl element $w_x \in W_{\psi,\text{reg}}(x)$. For our parameter (5.2.4), this element is unique.

We write M for a Levi subgroup of G such that ψ belongs to $\widetilde{\Psi}_M(G)$. We can then identify the multiplicity free parameter

$$\psi_M = \psi_1 \times \cdots \times \psi_r$$

with an element in $\widetilde{\Psi}_2(M,\psi)$. We can also write

$$M \cong GL(N_1) \times \cdots \times GL(N_q) \times G_-$$

as in the local notation (2.3.4). Then if

$$\psi_- = \psi_{q+1} \boxplus \cdots \boxplus \psi_r$$

and

$$N_- = N_{q+1} + \cdots + N_r,$$

the group $G_- = G_{\psi_-}$ represents the unique element in $\widetilde{\mathcal{E}}_{\text{sim}}(N_-)$ such that ψ_- lies in $\widetilde{\mathcal{F}}_2(G_-)$. Our condition $q \geqslant 1$ is of course because ψ does not lie in $\widetilde{\mathcal{F}}_2(G)$. It implies that M is proper in G, and that we can apply our induction hypotheses to M and ψ_M.

The first step is to apply the results of Chapter 4 to the component $\widetilde{G}(N)$, as we did in the last section. The centralizer set $\widetilde{S}_{\psi}(N)$ for $\widetilde{G}(N)$ is connected, for any parameter ψ at all, so the quotient $\widetilde{\mathcal{S}}_{\psi}(N)$ contains one element x. The global intertwining relation for $\widetilde{G}(N)$ and ψ is still valid in this case. It is the formula (4.5.1), with $\widetilde{G}(N)$ in place of G, which we established inductively for any parameter in the complement of $\widetilde{\Psi}_{\mathrm{ell}}(N)$, near the beginning of the proof of Proposition 4.5.1. (It was of course the exceptional set $\widetilde{\Psi}_{\mathrm{ell}}(N)$ that we treated in the last section, where we relied on Assumption 5.1.1.) In particular, the conditions of Corollaries 4.3.3 and 4.4.3 apply to $\big(\widetilde{G}(N), \psi\big)$, and the corresponding terms in the two expansions (4.3.9) and (4.4.11) are equal. The expansions are themselves then equal, and the two corollaries yield the familiar identity

$$ {}^0\widetilde{r}^{N}_{\mathrm{disc},\psi}(\widetilde{f}) = {}^0\widetilde{s}^{N}_{\mathrm{disc},\psi}(\widetilde{f}), \qquad \widetilde{f} \in \widetilde{\mathcal{H}}(N). $$

Since

$$ {}^0\widetilde{r}^{N}_{\mathrm{disc},\psi}(\widetilde{f}) = \mathrm{tr}\big(R^{N}_{\mathrm{disc},\psi}(\widetilde{f})\big) = 0 $$

in this case, it follows from the definition (4.4.10) that

$$ \sum_{G^* \in \widetilde{\mathcal{E}}_{\mathrm{sim}}(N)} \widetilde{\iota}(N, G^*)\, {}^0\widehat{S}^{*}_{\mathrm{disc},\psi}(\widetilde{f}^*) = 0. \tag{5.2.5} $$

This is the same as the identity (5.1.8), but for the parameter $\psi \in \widetilde{\mathcal{F}}^{2}_{\mathrm{ell}}(G)$ rather than the parameter in $\widetilde{\mathcal{F}}_2(G)$ of §5.1. We shall use it in much the same way.

The next lemma is stated in terms of a compatible family

$$ \{f^* \in \widetilde{\mathcal{H}}(G^*) : \; G^* \in \widetilde{\mathcal{E}}_{\mathrm{ell}}(N)\}, \tag{5.2.6} $$

or more precisely, the subset of functions in a compatible family parametrized by the subset $\widetilde{\mathcal{E}}_{\mathrm{sim}}(N)$ of $\widetilde{\mathcal{E}}_{\mathrm{ell}}(N)$. If $G^* = G$, we write $f = f^*$. In case N is even and $\eta_G = 1$, so that there is a second group $G^\vee \in \widetilde{\mathcal{E}}_{\mathrm{sim}}(N)$ with $\eta_{G^\vee} = 1$, we will also write $f^\vee = f^*$, if $G^* = G^\vee$.

LEMMA 5.2.1. *Suppose that for the given pair*

$$ (G, \psi), \qquad G \in \widetilde{\mathcal{E}}_{\mathrm{sim}}(N), \; \psi \in \widetilde{\mathcal{F}}^{2}_{\mathrm{ell}}(G), $$

the index r in (5.2.4) is greater than 1. Then the sum

$$ \sum_{G^* \in \widetilde{\mathcal{E}}_{\mathrm{sim}}(N)} \widetilde{\iota}(N, G^*)\, \mathrm{tr}\big(R^{*}_{\mathrm{disc},\psi}(f^*)\big) \tag{5.2.7} $$

equals

$$ c \sum_{x \in \mathcal{S}_{\psi,\mathrm{ell}}} \varepsilon^{G}_{\psi}(x)\, \big(f'_G(\psi, s_\psi x) - f_G(\psi, x)\big), \tag{5.2.8} $$

for a positive constant $c = c(G, \psi)$, and any compatible family of functions (5.2.6).

PROOF. The general strategy of proof is pretty clear by now. We need to establish a formula for the sum (5.2.7). We have a formula for the parallel sum on the left hand side of (5.2.5). For each index of summation $G^* \in \widetilde{\mathcal{E}}_{\text{sim}}(N)$, we have then to obtain an expression for the difference

$$\text{(5.2.9)} \qquad \operatorname{tr}\big(R^*_{\text{disc},\psi}(f^*)\big) - {}^0S^*_{\text{disc},\psi}(f^*), \qquad f^* \in \widetilde{\mathcal{H}}(G),$$

of the corresponding two summands. We shall apply the standard model, specifically Corollaries 4.3.3 and 4.4.3, to the pair (G^*, ψ). We observe that the two corollaries would have been easy to establish directly in the case at hand, since the centralizer $\overline{S}^*_\psi = \overline{S}^{G^*}_\psi$ for (5.2.9) is considerably simpler than the general object $\overline{S}_\psi$ treated in §4.3 and §4.4.

The most important case is that of $G^* = G$. It is responsible for the linear forms $f'_G(\psi, s_\psi x)$ and $f_G(\psi, x)$ in (5.2.8). We recall that these objects are defined by (4.2.4) and (4.2.5). They satisfy the conditions of Corollaries 4.4.3 and 4.3.3, with (G, ψ) being our given pair. In other words, they are well defined functions of the component x in $\mathcal{S}_\psi$. For the first function $f'_G(\psi, s_\psi x)$, this fact was established under general conditions prior to the proof of Proposition 4.5.1. For the function $f_G(\psi, x)$, it follows from the special nature of ψ, specifically, the fact that there is only one element w_x in the Weyl set $W_\psi(x)$.

We can therefore apply Corollaries 4.3.3 and 4.4.3 to our pair (G, ψ). Together, they tell us that for any $f \in \widetilde{\mathcal{H}}(G)$, the difference

$$\begin{aligned} &\big(I^G_{\text{disc},\psi}(f) - {}^0s^G_{\text{disc},\psi}(f)\big) - \big(I^G_{\text{disc},\psi}(f) - {}^0r^G_{\text{disc},\psi}(f)\big) \\ &= {}^0r^G_{\text{disc},\psi}(f) - {}^0s^G_{\text{disc},\psi}(f) \end{aligned}$$

equals

$$C_\psi \sum_{x \in \mathcal{S}_\psi} i_\psi(x)\, \varepsilon^G_\psi(x)\, \big(f'_G(\psi, s_\psi x) - f_G(\psi, x)\big).$$

On the one side of the identity, we can write

$${}^0r^G_{\text{disc},\psi}(f) - {}^0s^G_{\text{disc},\psi}(f) = \operatorname{tr}\big(R^G_{\text{disc},\psi}(f)\big) - {}^0S^G_{\text{disc},\psi}(f).$$

This follows from the definitions, since the expected value of the trace of $R^G_{\text{disc},\psi}(f)$ is 0. It also follows from the definitions that the coefficient $i_\psi(x)$ on the other side vanishes unless x belongs to the subset $\mathcal{S}_{\psi,\text{ell}}$ of $\mathcal{S}_\psi$. In this case, we have

$$\begin{aligned} i_\psi(x) &= |W^0_\psi|^{-1} \sum_{w \in W_{\psi,\text{reg}}(x)} s^0_\psi(w)\, |\det(w-1)|^{-1} \\ &= 1 \cdot s^0_\psi(w_x)\, |\det(w_x - 1)|^{-1} \\ &= 1 \cdot 1 \cdot \big(\tfrac{1}{2}\big)^q. \end{aligned}$$

The other side of the identity therefore equals

$$C_\psi \big(\tfrac{1}{2}\big)^q \sum_{x \in \mathcal{S}_{\psi,\text{ell}}} \varepsilon^G_\psi(x)\, \big(f'_G(\psi, s_\psi x) - f_G(\psi, x)\big).$$

This gives a formula for (5.2.9) in case $G^* = G$.

Suppose that G^* is an element in $\widetilde{\mathcal{E}}_{\mathrm{sim}}(N)$ that is distinct from G. We claim that the difference (5.2.9) vanishes. If ψ does not belong to $\widetilde{\Psi}(G^*)$, this is just the identity (4.4.12). In fact, we already know that both terms in (5.2.9) vanish unless $\eta_{G^*} = \eta_\psi$. Suppose then that ψ does belong to $\widetilde{\Psi}(G^*)$. Then

$$\eta_{G^*} = \eta_\psi = \eta_G.$$

Since G and G^* are distinct elements in $\widetilde{\mathcal{E}}_{\mathrm{sim}}(N)$, this implies that $\eta_{G^*} = 1$, $G^* = G^\vee$, $r = q$, and

$$S_\psi^\vee = S_\psi(G^\vee) = \big(Sp(2,\mathbb{C})\big)^q.$$

By assumption, $r = q \geqslant 2$. We are therefore dealing with a case that was treated by induction in §4.5. Indeed, the pair $(G^\vee, \psi)$ satisfies the condition

$$\dim(\overline{T}_\psi^\vee) = \dim(\overline{T}_{\psi,x}^\vee) = q \geqslant 2, \qquad x \in \mathcal{S}_\psi^\vee,$$

since the subtorus $\overline{T}_{\psi,x}^\vee$ of the maximal torus $\overline{T}_\psi^\vee$ of $\overline{S}_\psi^\vee$ defined in §4.5 actually equals $\overline{T}_\psi^\vee$ in this case. As in the discussion of the conditions (i) and (ii) in the proof of Proposition 4.5.1, we see that the global intertwining relation (4.5.1) holds for $(G^\vee, \psi)$. The difference (5.2.9) therefore vanishes in the remaining case $G^* = G^\vee$, by Corollaries 4.3.3 and 4.4.3.

The last step is to apply the identity (5.2.5). We can replace the functions $\{\widetilde{f}^*\}$ in (5.2.5) by any compatible family (5.2.6). Substituting the formulas we have obtained in the cases $G^* = G$ and $G^* \neq G$, we conclude that (5.2.7) equals (5.2.8), with

$$c = \widetilde{\iota}(N,G)\, C_\psi \left(\tfrac{1}{2}\right)^q. \qquad \square$$

REMARK. The only property of the constant c in (5.2.8) we will use is its positivity. We could of course write it more explicitly. It follows from (3.2.6) and (4.3.5) that

$$\begin{aligned} c &= \widetilde{\iota}(N,G)\, C_\psi \left(\tfrac{1}{2}\right)^q \\ &= \tfrac{1}{2}\, |\widetilde{\mathrm{Out}}_N(G)|^{-1}\, m_\psi\, |\mathcal{S}_\psi|^{-1} \left(\tfrac{1}{2}\right)^q \\ &= \tfrac{1}{2}\, |\widetilde{\mathrm{Out}}_N(G,\psi)|^{-1}\, |\mathcal{S}_\psi|^{-1} \left(\tfrac{1}{2}\right)^q, \end{aligned}$$

and one can check that this equals $\left(\frac{1}{2}\right)^{q+r+\varepsilon}$, where $\varepsilon = 0$ if N is odd and 1 if n is even.

We have to treat the case of $r = q = 1$ in (5.2.4) separately. The underlying object of course remains a parameter ψ in the complement $\widetilde{\mathcal{F}}^2_{\mathrm{ell}}(G)$ of $\widetilde{\mathcal{F}}_2(G)$, for some $G \in \widetilde{\mathcal{E}}_{\mathrm{sim}}(N)$. However, we now suppose that it falls into the special case

$$\begin{cases} \psi = 2\psi_1 \\ S_\psi = O(2,\mathbb{C}) \end{cases} \tag{5.2.10}$$

of (5.2.4). This implies that N is even and that $\eta_G = 1$. We thus have a pair of split groups G and $G^\vee$ in $\widetilde{\mathcal{E}}_{\rm sim}(N)$, with

$$S_\psi^\vee = S_\psi(G^\vee) = Sp(2,\mathbb{C}).$$

In this case, the group

$$M \cong GL(N_1)$$

can be identified with a Levi subgroup of both G and $G^\vee$. If G^* equals either G or $G^\vee$, the subset

$$\mathcal{S}^*_{\psi,\rm ell} = \{x \in \mathcal{S}^*_\psi : \ \mathcal{E}_{\psi,\rm ell}(x) \neq \varnothing\}$$

of the group $\mathcal{S}^*_\psi = \mathcal{S}_\psi(G^*)$ consists of one element, which we denote simply by x_1. In each case, the Weyl set $W_{\psi,\rm reg}(x_1)$ consists of one element w_1.

LEMMA 5.2.2. *Suppose that for the given pair*

$$(G,\psi), \qquad G \in \widetilde{\mathcal{E}}_{\rm sim}(N),\ \psi \in \widetilde{\mathcal{F}}^2_{\rm ell}(G),$$

the index r in (5.2.4) equals 1. Then the sum of

$$\sum_{G^*\in\widetilde{\mathcal{E}}_{\rm sim}(N)} \widetilde{\iota}(N,G^*)\,\mathrm{tr}\big(R^*_{\mathrm{disc},\psi}(f^*)\big) \tag{5.2.11}$$

and

$$\tfrac{1}{8}\big((f^\vee)^M(\psi_1) - f^\vee_{G^\vee}(\psi,x_1)\big) \tag{5.2.12}$$

equals

$$\tfrac{1}{8}\big(f'_G(\psi,s_\psi x_1) - f_G(\psi,x_1)\big), \tag{5.2.13}$$

for any compatible family of functions (5.2.6).

PROOF. The proof is essentially the same as that of the last lemma. For each $G^* \in \widetilde{\mathcal{E}}_{\rm sim}(N)$, we have to obtain an expression for the difference (5.2.9) that we can substitute into the identity (5.2.5). In particular, we have to apply Corollaries 4.3.3 and 4.4.3 to the data $G^* = G$ and $G^* = G^\vee$. We require separate discussion here simply because we know less about the contribution from $G^* = G^\vee$. That is, the inductive arguments of §4.5 are not strong enough to establish the expected global intertwining relation for $G^\vee$. This will account for the extra term (5.2.12).

We can say that the conditions of the two corollaries are applicable to both G and $G^\vee$. The justification follows from the remarks at the beginning of the proof of the last lemma, except in the application of Corollary 4.3.3 to $G^\vee$. In this case, we simply adopt the convention used to treat (4.5.11) in the proof of Proposition 4.5.1, which relies on the fact that the element $w_x \in W_{\psi,\rm reg}(x)$ introduced earlier is unique. Once again, we shall apply both corollaries together. Assume that G^* is one of the groups G or $G^\vee$. For any $f^* \in \widetilde{\mathcal{H}}(G^*)$, the difference

$${}^0r^*_{\mathrm{disc},\psi}(f^*) - {}^0s^*_{\mathrm{disc},\psi}(f^*) = \mathrm{tr}\big(R^*_{\mathrm{disc},\psi}(f)\big) - {}^0S^*_{\mathrm{disc},\psi}(f^*)$$

then equals

$$C_\psi^* \sum_{x \in \mathcal{S}_\psi^*} i_\psi(x)\, \varepsilon_\psi^{G^*}(x)\, \big((f^*)'_{G^*}(\psi, s_\psi x) - f^*_{G^*}(\psi, x)\big),$$

where C_ψ^* is the positive constant (4.3.5) attached to G^*. It follows easily from Corollary 4.6.2 and the original definition (1.5.6) that the sign character $\varepsilon_\psi^{G^*}$ equals 1. Since $i_\psi(x)$ vanishes unless x belongs to $\mathcal{S}^*_{\psi,\mathrm{ell}}$, a subset of $\mathcal{S}_\psi^*$ that consists of the one element x_1, the last expression therefore reduces to

$$C_\psi^*\, i_\psi(x_1)\, \big((f^*)'_{G^*}(\psi, s_\psi x_1) - f^*_{G^*}(\psi, x_1)\big). \tag{5.2.14}$$

If $G^* = G$, the identity component of the group $\overline{S}_\psi^* = \overline{S}_\psi$ is abelian. We have

$$i_\psi(x_1) = \left(\tfrac{1}{2}\right)^q = \tfrac{1}{2},$$

as in the proof of the last lemma. Since

$$\widetilde{\imath}(N, G) \cdot C_\psi \cdot \tfrac{1}{2} = \tfrac{1}{8},$$

the product of $\widetilde{\imath}(N, G)$ with (5.2.14) in this case is equal to the expression (5.2.13), with $f = f^*$. If $G^* = G^\vee$, the group $\overline{S}_\psi^* = \overline{S}_\psi^\vee$ is connected, and simple of rank 1. We obtain

$$\begin{aligned} i_\psi(x_1) &= |(W_\psi^\vee)^0|^{-1} \sum_{w \in W^\vee_{\psi,\mathrm{reg}}(x_1)} s_\psi^0(w)\, |\det(w-1)|^{-1} \\ &= \tfrac{1}{2}\, s_\psi^0(w_x)\, |\det(w_x - 1)|^{-1} = -\tfrac{1}{4}, \end{aligned}$$

as in the proof of the case (4.5.11) in Proposition 4.5.1. We also have the reduction

$$(f^*)'_{G^*}(\psi, s_\psi x_1) = (f^*)'_{G^*}(\psi, 1) = (f^*)^M(\psi_1)$$

in (5.2.14). Since

$$\widetilde{\imath}(N, G^\vee) \cdot C_\psi^\vee \cdot \tfrac{1}{4} = \tfrac{1}{8},$$

the product of $\widetilde{\imath}(N, G^\vee)$ with (5.2.14) in this case is equal to (-1) times the expression (5.2.12), with $f^\vee = f^*$. The minus sign here is critical. It is what forces us to place (5.2.12) and (5.2.13) on opposite sides of the proposed identity.

We continue the argument at this stage as in the proof of the last lemma. If G^* belongs to $\widetilde{\mathcal{E}}_{\mathrm{sim}}(N)$, but is not equal to G and $G^\vee$, ψ does not belong to $\widetilde{\Psi}(G^*)$. The difference (5.2.9) then vanishes by (4.4.12) (or simply from the fact that $\eta_{G^*} \neq \eta_\psi$). It remains only to apply (5.2.5), with a general compatible family (5.2.6) in place of the functions $\{\widetilde{f}^*\}$. Substituting the formulas we have obtained in the cases $G^* = G$, $G^* = G^\vee$, and G^* distinct from G and $G^\vee$, we conclude that the sum of (5.2.11) and (5.2.12) equals (5.2.13), as required. □

We have so far worked exclusively with parameters of degree equal to the fixed integer N of our running induction hypotheses. This will not be sufficient. We will soon have to consider some elliptic parameters of degree greater than N. For such objects, we will of course not be able to rely on the results of this section. We will need to see what can be salvaged.

Suppose that

$$\psi_+ = \ell_1\psi_1 \boxplus \cdots \boxplus \ell_r\psi_r$$

is a general parameter in our family $\widetilde{\mathcal{F}}(N_+)$, with degree

$$N_+ = \ell_1 N_1 + \cdots + \ell_r N_r$$

greater than N. We may as well assume that ψ_+ belongs to the subset $\widetilde{\mathcal{F}}_{\mathrm{disc}}(N_+)$ of $\widetilde{\mathcal{F}}(N_+)$. In other words, the simple components ψ_i of ψ_+ are self-dual. We define

$$\psi_{+,-} = \boxplus_{\ell_i \text{ odd}} \psi_i,$$

an auxiliary parameter in $\widetilde{\mathcal{F}}_{\mathrm{ell}}(N_{+,-})$, where

$$N_{+,-} = \sum_{\ell_i \text{ odd}} N_i.$$

Suppose that ψ_+ belongs to $\widetilde{\Psi}(G_+)$, for a given simple datum $G_+ \in \widetilde{\mathcal{E}}_{\mathrm{sim}}(N_+)$. Following earlier notation, we write M_+ for a Levi subgroup of G_+ such that ψ_+ belongs to $\widetilde{\Psi}_{M_+}(G_+)$. Then

$$M_+ \cong GL(N_1)^{\ell_1'} \times \cdots \times GL(N_r)^{\ell_r'} \times G_{+,-},$$

where $\ell_i' = [\ell_i/2]$ is the greatest integer in $\ell_i/2$, and $G_{+,-}$ is the unique element in $\widetilde{\mathcal{E}}_{\mathrm{sim}}(N_{+,-})$ such that $\psi_{+,-}$ lies in $\widetilde{\Psi}_2(G_{+,-})$. We can identify the corresponding parameter

$$\psi_{+,M_+} = \psi_1^{\ell_1'} \times \cdots \times \psi_r^{\ell_1'} \times \psi_{+,-}$$

with an element in $\widetilde{\Psi}_2(M_+, \psi_+)$. We will naturally have an interest in the stable linear form on $\widetilde{\mathcal{H}}(G_+)$ attached to ψ_+. It can be written as the pullback

$$f^{G_+}(\psi_+) = f^{M_+}(\psi_{M_+}), \qquad f \in \widetilde{\mathcal{H}}(G_+),$$

of a stable linear form on $M_+(\mathbb{A})$. This linear form will thus be well defined by our Assumption 5.1.1 on $\widetilde{\mathcal{F}}(N)$, so long as $N_{+,-} \leqslant N$.

There is one part of the standard model that can be established for many ψ_+. It is a variant of (5.2.5), in which the sum is taken over $\widetilde{\mathcal{E}}_{\mathrm{ell}}(N_+)$ instead of its subset $\widetilde{\mathcal{E}}_{\mathrm{sim}}(N_+)$. For any $G^* \in \widetilde{\mathcal{E}}_{\mathrm{ell}}(N_+)$, the stable linear form $S^*_{\mathrm{disc},\psi_+}$ on $\widetilde{\mathcal{H}}(G^*)$ is defined by (3.3.9), (3.3.13) and Corollary 3.4.2. Its expected value is given by the analogue of Corollary 4.1.3, as a linear combination of linear forms

$$f^*(\psi^*), \qquad f^* \in \widetilde{\mathcal{S}}(G^*),\ \psi^* \in \Psi(G^*, \psi_+).$$

These objects are again well defined by Assumption 5.1.1 whenever $N_{+,-} \leqslant N$. We can therefore define the stable linear form

$${}^0S^*_{\mathrm{disc},\psi_+}(f^*), \qquad f^* \in \widetilde{\mathcal{H}}(G^*),$$

as the difference between $S^*_{\mathrm{disc},\psi_+}(f^*)$ and its expected value, so long as $N_{+,-} \leqslant N$.

LEMMA 5.2.3. *Suppose that $\psi_+ \in \widetilde{\mathcal{F}}_{\mathrm{disc}}(N_+)$ is as above, and that*

$$N_1 + \cdots + N_r \leqslant N. \tag{5.2.15}$$

Then

$$\sum_{G^* \in \widetilde{\mathcal{E}}_{\mathrm{ell}}(N_+)} \widetilde{\iota}(N_+, G^*)\, {}^0\widehat{S}^*_{\mathrm{disc},\psi_+}(\widetilde{f}^*) = 0, \qquad \widetilde{f} = \widetilde{\mathcal{H}}(N_+). \tag{5.2.16}$$

PROOF. We are taking the sum in (5.2.16) over the full set $\widetilde{\mathcal{E}}_{\mathrm{ell}}(N_+)$ for the obvious reason that there is only limited scope for the inductive application of Theorem 4.1.2. That is, we cannot say that ${}^0\widehat{S}_{\mathrm{disc},\psi_+}(\widetilde{f}^*)$ vanishes for general elements G^* in the complement of $\widetilde{\mathcal{E}}_{\mathrm{sim}}(N_+)$. We do know that this linear form is well defined, since the condition (5.2.15) implies that $N_{+,-} \leqslant N$. We also know that Lemma 4.3.1 is valid for the pair $\big(\widetilde{G}(N_+), \psi_+\big)$, which is to say that the application of Theorem 1.5.3 to the proof of Lemma 4.3.1 in §4.6 remains valid. The assertion (a) of Theorem 1.5.3 is not actually relevant to $\widetilde{G}(N_+)$, since in this case it is only Rankin-Selberg L-functions that contribute to the factor $s^0_\psi(w)$ in (4.6.13). The condition on the other assertion (b) still holds, by virtue of its extension in Lemma 5.1.6 and the condition (5.2.15). Lemma 4.4.1 is also valid for $\big(\widetilde{G}(N_+), \psi_+\big)$, since there were no implicit conditions in its proof from §4.6.

With these remarks, we can see that the discussion preceding the special case (5.2.5) for $\widetilde{G}(N)$ applies here as well. In particular, the conditions of Corollaries 4.3.3 and 4.4.3 remain in force, and the analogues of these corollaries hold for the general case $\big(\widetilde{G}(N_+), \psi_+\big)$ at hand. Moreover, the analogues for $\big(\widetilde{G}(N_+), \psi_+\big)$ for the expressions (4.3.9) and (4.4.11) are equal. It follows that

$${}^0\widetilde{r}^{N_+}_{\mathrm{disc},\psi_+}(\widetilde{f}) = {}^0\widetilde{s}^{N_+}_{\mathrm{disc},\psi_+}(\widetilde{f}), \qquad \widetilde{f} \in \widetilde{\mathcal{H}}(N_+),$$

where ${}^0\widetilde{s}^{N_+}_{\mathrm{disc},\psi_+}(\widetilde{f})$ equals the left hand side of (5.2.16). Since

$${}^0\widetilde{r}^{N_+}_{\mathrm{disc},\psi_+}(\widetilde{f}) = 0,$$

we obtain the required identity (5.2.16). □

5.3. A supplementary parameter ψ_+

In this section, we shall return to the "square integrable" parameters $\psi \in \widetilde{\mathcal{F}}_{\mathrm{ell}}(N)$ of §5.1. Recall that ψ takes the general form

$$\psi = \psi_1 \boxplus \cdots \boxplus \psi_r, \qquad \psi_i = \widetilde{\mathcal{F}}_{\mathrm{sim}}(N_i), \tag{5.3.1}$$

and lies in the subset $\widetilde{\mathcal{F}}_2(G)$ of $\widetilde{\mathcal{F}}_{\mathrm{ell}}(N)$ attached to some $G \in \widetilde{\mathcal{E}}_{\mathrm{ell}}(N)$. (Recall also that G is unique, except possibly if ψ is generic and simple. In the latter case, we simply fix a G with $\psi \in \widetilde{\mathcal{F}}_{\mathrm{sim}}(G)$ and treat it as if it were unique, knowing that this is what we will eventually establish.) Our goal is to establish the stable multiplicity formula for $S^G_{\mathrm{disc},\psi}(f)$. In §5.1, we found that the proof was rather direct under any of the conditions N odd, $\eta_G \neq 1$ or $G \neq \widetilde{\mathcal{E}}_{\mathrm{sim}}(N)$. However, the remaining case of N even and $G \in \widetilde{\mathcal{E}}_{\mathrm{sim}}(N)$ split is considerably harder. We will need to establish it for the relevant parameters in $\widetilde{\mathcal{F}}_{\mathrm{ell}}(N)$ in order to complete the induction argument of this chapter.

In the special case that ψ is simple generic, we will also need to deal with the first part (a) of Theorem 1.5.3. This is the condition on the poles of L-functions, which we recall is intimately tied up with the definition of $\widetilde{\mathcal{F}}_{\mathrm{sim}}(G)$ and the corresponding special case of the stable multiplicity formula. It represents a second critical global assertion that will have to be established in the context of the family $\widetilde{\mathcal{F}}_{\mathrm{ell}}(N)$. The other global assertions are less pressing. They will be resolved later, in terms of the stable multiplicity formula and the local results we establish. We recall that the second part (b) of Theorem 1.5.3 has already been resolved for $\widetilde{\mathcal{F}}(N)$. It was established as one of the assertions of Lemma 5.1.6.

To deal with these questions, we can assume that $G \in \widetilde{\mathcal{E}}_{\mathrm{sim}}(N)$ is simple. Given the parameter $\psi \in \widetilde{\mathcal{F}}_2(G)$, we set

$$N_+ = N_1 + N,$$

where the constituents ψ_i of ψ have been ordered so that $N_1 \leqslant N_i$ for each i. We then introduce the supplementary parmeter

$$\psi_+ = 2\psi_1 \boxplus \psi_2 \boxplus \cdots \boxplus \psi_r \tag{5.3.2}$$

in $\widetilde{\mathcal{F}}(N_+)$. There is a unique element G_+ in $\widetilde{\mathcal{E}}_{\mathrm{sim}}(N_+)$ such that $\psi_+ \in \widetilde{\Psi}(G_+)$, and such that $\widehat{G}_+$ contains the product

$$\widehat{G}_1 \times \widehat{G} = \widehat{G}_{\psi_1} \times \widehat{G},$$

where $G_1 = G_{\psi_1}$ is the element in $\widetilde{\mathcal{E}}_{\mathrm{sim}}(N_1)$ such that ψ_1 is contained in $\widetilde{\mathcal{F}}_{\mathrm{sim}}(G_1)$. Then ψ_+ lies in the subset $\widetilde{\mathcal{F}}_{\mathrm{ell}}(G_+)$ of $\widetilde{\mathcal{F}}(G_+)$.

We will need to work with two maximal Levi subgroups of G_+. The first applies to the case of N even and $\eta_\psi = 1$, in which we have a maximal Levi subgroup $L \cong GL(\frac{1}{2}N)$ of G. We then have the corresponding Levi subgroup

$$L_+ = G_1 \times L \cong GL\big(\tfrac{1}{2}N\big) \times G_1$$

of G_+. It is adapted to the decomposition

$$\psi_+ = \psi_1 \boxplus \psi = \psi \boxplus \psi_1$$

of ψ, and comes with the linear form

$$f^{L_+} \longrightarrow f^{L_+}(\psi_1 \times \Lambda), \qquad f \in \widetilde{\mathcal{H}}(G_+),$$

where Λ is the linear form attached to L in Lemma 5.1.4. The second is a special case of the general definition prior to Lemma 5.2.3. This is the Levi subgroup

$$M_+ \cong GL(N_1) \times G_{+,-}, \qquad G_{+,-} \in \widetilde{\mathcal{E}}_{\text{sim}}(N_{+,-}),$$

of G_+ such that ψ_+ belongs to $\widetilde{\Psi}_{M_+}(G_+)$. It is adapted to the decomposition

$$\psi_+ = 2\psi_1 \boxplus \psi_{+,-}, \qquad \psi_{+,-} = \psi_2 \boxplus \cdots \boxplus \psi_r$$

of ψ, and comes with the linear form

$$f \longrightarrow f^{M_+}(\psi), \qquad f \in \widetilde{\mathcal{H}}(G_+),$$

in which ψ is identified with the product $\psi_1 \times \psi_{+,-}$.

One of our ultimate goals is to show that Λ vanishes. An essential step will be to establish analogues of Lemmas 5.2.1 and 5.2.2 for the parameter ψ_+. These will be needed for the induction argument that is to drive the proof of the local theorems over the next two chapters. They will also be used directly for the general proof of the global theorems in Chapter 8. As with their predecessors, we shall state the lemmas in terms of a compatible family of functions

$$\{f^* \in \widetilde{\mathcal{H}}(G^*) : G^* \in \widetilde{\mathcal{E}}_{\text{ell}}(N_+)\}, \tag{5.3.3}$$

with the same notation $f = f^*$ and $f^\vee = f^*$ in the cases $G^* = G_+$ and $G^* = G_+^\vee$. We shall again deal separately with the cases $r > 1$ and $r = 1$.

Suppose first that $r > 1$. This is the case that ψ lies in the complement $\widetilde{\mathcal{F}}_2^{\text{sim}}(G)$ of $\widetilde{\mathcal{F}}_{\text{sim}}(G)$ in $\widetilde{\mathcal{F}}_2(G)$. The problem is to understand the linear form Λ of Lemma 5.1.4, and to show ultimately that it vanishes. We can therefore assume that N is even and $\eta_\psi = 1$, as in the earlier lemma, and in particular that G is split. We then have the maximal Levi subgroups L and L_+ of G and G_+ respectively. To deal with this case, we shall also have to strengthen our induction hypothesis. We shall assume that in dealing with the square integrable parameter here, we have already been able to treat the elliptic parameters of §5.2.

LEMMA 5.3.1. *Suppose that for the given pair*

$$(G, \psi), \qquad G \in \widetilde{\mathcal{E}}_{\text{sim}}(N),\ \psi \in \widetilde{\mathcal{F}}_2(G),$$

the index r in (5.3.1) is greater than 1, while N is even and $\eta_\psi = 1$. Assume also that the global theorems are valid for parameters in the complement of $\widetilde{\mathcal{F}}_{\text{ell}}(N)$ in $\widetilde{\mathcal{F}}(N)$. Then the sum

$$\sum_{G^* \in \widetilde{\mathcal{E}}_{\text{sim}}(N_+)} \widetilde{\iota}(N_+, G^*) \operatorname{tr}\big(R^*_{\text{disc},\psi_+}(f^*)\big) + \big(b_+ f^{L_+}(\psi_1 \times \Lambda)\big) \tag{5.3.4}$$

equals

$$c_+ \sum_{x \in \mathcal{S}_{\psi_+,\text{ell}}} \varepsilon^{G_+}_{\psi_+}(x) \big(f'_{G_+}(\psi_+, s_{\psi_+}x) - f_{G_+}(\psi_+, x)\big), \tag{5.3.5}$$

for positive constants $b_+ = b(G_+, \psi_+)$ and $c_+ = c(G_+, \psi_+)$, and any compatible family of functions (5.3.3).

PROOF. This is the first of two lemmas in which we have to work with parameters of rank greater than N. The proofs require extra discussion, since they are beyond the scope of our running induction hypotheses. The supplementary arguments are usually not difficult, but they do contain a number of points to be checked. We shall include more detail here than in the next lemma.

We may as well start with a general compatible family (5.3.3). We then fix a function $\widetilde{f} \in \widetilde{\mathcal{H}}(N_+)$ that for any $G^* \in \widetilde{\mathcal{E}}_{\text{ell}}(N_+)$ has the same image in $\widetilde{\mathcal{S}}(G^*)$ as the corresponding function f^* in (5.3.3). We can then work interchangeably with the function $\widetilde{f}$ or the family (5.3.3). This will simplify the notation at times, particularly when G^* equals G.

In principle, the lemma should simply be the formula for

$$\sum_{G^* \in \widetilde{\mathcal{E}}_{\text{sim}}(N_+)} \widetilde{\iota}(N_+, G^*)\,\text{tr}\big(R^*_{\text{disc},\psi_+}(f^*)\big) \tag{5.3.6}$$

that would be the direct analogue for ψ_+ of Lemma 5.2.1. In particular, the coefficient in (5.3.5) is to be the analogue

$$c_+ = \tfrac{1}{2}\,\widetilde{\iota}(N_+, G_+)\,C_{\psi_+}$$

of the coefficient c in the corresponding expression (5.2.8) of Lemma 5.2.1. However, the induction assumptions hold only proper subparameters of ψ. We will see that most proper subparameters of ψ_+ can either be ruled out by general considerations, or expressed in terms of proper subparameters of ψ. For the ones that cannot, we will be able to use Λ to describe the defect in the associated terms. In other words, we will keep track of what changes the existence of Λ makes in the arguments used to derive the earlier formula.

The discussion will again center around a sum

$$\sum_{G^* \in \widetilde{\mathcal{E}}_{\text{ell}}(N_+)} \widetilde{\iota}(N_+, G^*)\,{}^0\widehat{S}^*_{\text{disc},\psi_+}(\widetilde{f}^*), \qquad \widetilde{f} \in \widetilde{\mathcal{H}}(N_+), \tag{5.3.7}$$

over endoscopic data. Recall that this is the left hand side of the formula (5.2.16), established for more general ψ_+ in Lemma 5.2.3. In the case at hand, the sum

$$N_1 + N_2 + \cdots + N_r$$

in (5.2.15) equals N. The formula therefore holds, and (5.3.7) vanishes. The problem is to analyze the summands in (5.3.7). We shall apply what we can of the standard model for G^* to the stable linear form

$${}^0S^*_{\text{disc},\psi_+}(f^*) = {}^0S^{G^*}_{\text{disc},\psi_+}(f^*), \qquad f^* \in \widetilde{\mathcal{H}}(G^*),$$

where f^* maps to the function $\widetilde{f}^* \in \widetilde{\mathcal{S}}(G^*)$ in (5.3.7). The problem now is complicated by the fact that G^* ranges over $\widetilde{\mathcal{E}}_{\text{ell}}(N_+)$, rather than the subset $\widetilde{\mathcal{E}}_{\text{sim}}(N_+)$ of simple data whose analogue for N indexed the sum in the original formula (5.2.5). We recall that ${}^0S^*_{\text{disc},\psi_+}(f^*)$ is defined as the difference between the original stable form $S^*_{\mathrm{disc},\psi_+}(f^*)$, and its expected formula

$$\sum_{\psi^* \in \mathcal{F}(G^*,\psi_+)} |\mathcal{S}_{\psi^*}|^{-1} \sigma(\overline{S}^0_{\psi^*}) \, \varepsilon^*(\psi^*) \, f^*(\psi^*) \tag{5.3.8}$$

given by the twisted analogue of Corollary 4.1.3.

Consider a general element

$$G^* = G^1 \times G^2, \qquad G^k \in \widetilde{\mathcal{E}}_{\mathrm{sim}}(N^k), \ k = 1, 2,$$

in $\widetilde{\mathcal{E}}_{\mathrm{ell}}(N_+)$. Thus, $N_+ = N^1 + N^2$, and G^1 and G^2 represent the two data denoted G_S and G_O in Assumption 5.1.1. We can write

$$S^*_{\mathrm{disc},\psi_+}(f^*) = \sum_{\psi^*} S^*_{\mathrm{disc},\psi^*}(f^*) \tag{5.3.9(a)}$$

and

$${}^0S^*_{\mathrm{disc},\psi_+}(f^*) = \sum_{\psi^*} {}^0S^*_{\mathrm{disc},\psi^*}(f^*), \tag{5.3.9(b)}$$

where the sum in each case is over parameters

$$\psi^* = \psi^1 \times \psi^2, \qquad \psi^k \in \widetilde{\mathcal{F}}(N^k), \ k = 1, 2,$$

with

$$\psi_+ = \psi^1 \boxplus \psi^2.$$

The decomposition (5.3.9)(a) follows from (3.3.9), (3.3.13) and Corollary 3.4.2. Since the expected value (5.3.8) of the left hand side of (5.3.9)(a) has a similar decomposition, with the summand of any ψ^* in the complement of $\widetilde{\mathcal{F}}(G^*, \psi_+)$ taken to be 0, the difference does satisfy (5.3.9)(b). We observe also that the summand of ψ^* in (5.3.9)(a) has an obvious product decomposition

$$S^*_{\mathrm{disc},\psi^*}(f^*) = S^1_{\mathrm{disc},\psi^1}(f^1) \, S^2_{\mathrm{disc},\psi^2}(f^2), \tag{5.3.10}$$

if $f^* = f^1 \times f^2$.

We can assume that the two nonnegative integers N^1 and N^2 in the partition of N_+ attached to G^* satisfy $N^1 \leqslant N^2$. The most important cases will then be the trivial partition

(i) $N^1 = 0$, $N^2 = N_+$,

in which G^* lies in the subset $\widetilde{\mathcal{E}}_{\mathrm{sim}}(N_+)$ of simple data, and the partition

(ii) $N^1 = N_1$, $N^2 = N$

attached to ψ_1 and ψ. Before we discuss these, however, we shall first take care of the others.

If $0 < N^1 < N_1$, there is no index of summation ψ^* in (5.3.9), since $N_1 \leqslant N_i$ for each i. It follows that

$$ {}^0S^*_{\mathrm{disc},\psi_+}(f^*) = 0, \tag{5.3.11}$$

in this case. In the remaining case that $N^1 > N_1$, both N^1 and N^2 are less than N. We can then apply our induction hypothesis to the factors on the right hand side of (5.3.10). This tells us that each factor equals its expected value, and so therefore does the product. It follows that the summands in (5.3.9)(b) all vanish, and that (5.3.11) holds in this case as well. Thus, the summand of G^* in (5.3.7) vanishes unless the corresponding partition $N_+ = N^1 + N^2$ satisfies (i) or (ii).

Consider the case (ii). This of course does not characterize G^* uniquely. Moreover, there are several indices $\psi^* = \psi^1 \times \psi^2$ in (5.3.9)(b) attached to the given G^*. Among these, it is the pair

$$ (G^*, \psi^*) = (G_1 \times G^\vee, \psi_1 \times \psi), \qquad G_1 = G_{\psi_1}, \tag{5.3.12}$$

that is the most significant. We shall deal first with the other pairs.

Assume that (G^*, ψ^*) is any pair associated with the partition (ii) that is not equal to (5.3.12). If the parameter $\psi^* = \psi^1 \times \psi^2$ is not equal to $\psi_1 \times \psi$, the second component ψ^2 contains the factor ψ_1 with multiplicity two. This means that ψ^2 lies in the complement of $\widetilde{\mathcal{F}}_{\mathrm{ell}}(N)$ in $\widetilde{\mathcal{F}}(N)$. The given condition of the lemma then tell us that the global theorems are valid for (G^2, ψ^2), and in particular, that the factor $S^2_{\mathrm{disc},\psi^2}(f^2)$ in (5.3.10) equals its expected value. Our induction hypothesis tells us that the other factor $S^1_{\mathrm{disc},\psi^1}(f^1)$ in (5.3.10) also equals its expected value, since $N^1 = N_1$ is less than N. It follows that the summand of ψ^* in (5.3.9(b)) vanishes. Assume next that $\psi^* = \psi_1 \times \psi$, but that the datum $G^* = G^1 \times G^2$ is not equal to $G_1 \times G^\vee$. If $G^1 \neq G_1$ the parameter $\psi^1 = \psi_1$ does not lie $\widetilde{\mathcal{F}}(G^1)$. It then follows by induction that

$$ {}^0S^1_{\mathrm{disc},\psi^1}(f^1) = S^1_{\mathrm{disc},\psi^1}(f^1) = 0, $$

and therefore that the summand of ψ^* again vanishes. On the other hand, if G^1 equals G_1, the second component G^2 cannot equal G. This is because $\widehat{G}_1$ and $\widehat{G}$ are both orthogonal or both symplectic (by definition, since the same is true of the parameters ψ_1 and ψ), whereas the factors $\widehat{G}^1$ and $\widehat{G}^2$ of the dual twisted endoscopic group $\widehat{G}^*$ must be of opposite type. Since we have ruled out $G^2 = G^\vee$, the only other possibility is a datum that satisfies

$$ \eta_{G^2} \neq \eta_{\psi^2} = 1. $$

In this case,

$$ {}^0S^{G^2}_{\mathrm{disc},\psi^2}(f^2) = S^{G^2}_{\mathrm{disc},\psi^2}(f^2) = 0, $$

by (3.4.7), and so the summand of ψ^* in (5.3.9(b)) vanishes once again.

Suppose then that (G^*, ψ^*) equals (5.3.12). The decomposition (5.3.10) takes the form

$$S^*_{\mathrm{disc},\psi^*}(f^*) = S^{G_1}_{\mathrm{disc},\psi_1}(f_1)\, S^{G^\vee}_{\mathrm{disc},\psi}(f^\vee), \qquad G_1 = G_{\psi_1},$$

where $f^* = f_1 \times f^\vee$. The first factor becomes

$$S^{G_1}_{\mathrm{disc},\psi_1}(f_1^{G_1}) = f_1^{G_1}(\psi_1),$$

with an inductive application of Theorem 4.1.2 to G_1. According to Lemma 5.1.4, the second factor satisfies

$$S^{G^\vee}_{\mathrm{disc},\psi}(f^\vee) = \widetilde{o}(G^\vee)\, f^{\vee,L}(\Lambda).$$

It follows that

$${}^0S^*_{\mathrm{disc},\psi^*}(f^*) = S^*_{\mathrm{disc},\psi^*}(f^*) = \widetilde{o}(G^\vee)\, f^{L_+}(\psi_1 \times \Lambda),$$

since the expected value of $S^{G^*}_{\mathrm{disc},\psi^*}(f^*)$ is 0. This provides the contribution to the sum (5.3.7) of $G^* = G_1 \times G^\vee$. In fact, from what we have shown, it represents the only contribution from composite endoscopic data G^*. To be precise, the sum of those terms in (5.3.7) corresponding to data G^* in the complement of $\widetilde{\mathcal{E}}_{\mathrm{sim}}(N_+)$ in $\widetilde{\mathcal{E}}_{\mathrm{ell}}(N_+)$ equals

$$\widetilde{\iota}(N_+, G_1 \times G^\vee)\, \widetilde{o}(G^\vee)\, f^{L_+}(\psi_1 \times \Lambda). \tag{5.3.13}$$

It remains to consider the partition (i) attached to simple data $G^* \in \widetilde{\mathcal{E}}_{\mathrm{sim}}(N_+)$. This represents the point at which we began the proof of Lemma 5.2.1, the model we are following from §5.2. For each such G^*, we have to obtain an expression for the difference

$$\mathrm{tr}\big(R^*_{\mathrm{disc},\psi_+}(f^*)\big) - {}^0S^*_{\mathrm{disc},\psi_+}(f^*), \qquad f^* \in \widetilde{\mathcal{H}}(G^*), \tag{5.3.14}$$

of the corresponding two summands in (5.3.6) and (5.3.7).

The most important case is the datum $G^* = G_+$. It has the property that η_{G^*} equals the sign character

$$\eta_{G^+} = \eta_{\eta_1}\eta_\psi = \eta_{\psi_1}$$

of η_+. It is in fact the only element in $\widetilde{\mathcal{E}}_{\mathrm{sim}}(N_+)$ with this property unless N_1 is even and $\eta_{\psi_1} = 1$, in which case G_+ is split, and there is a second split datum $G_+^\vee \in \widetilde{\mathcal{E}}_{\mathrm{sim}}(N_+)$. The remaining elements $G^* \in \widetilde{\mathcal{E}}_{\mathrm{sim}}(N_+)$ can be ignored, since

$${}^0S^*_{\mathrm{disc},\psi_+}(f^*) = S^*_{\mathrm{disc},\psi_+}(f^*) = \mathrm{tr}\big(R^*_{\mathrm{disc},\psi_+}(f^*)\big) = 0,$$

by (3.4.7). In particular, for the elements $G^* \in \widetilde{\mathcal{E}}_{\mathrm{sim}}(N)$ distinct from G_+ and $G_+^\vee$, the difference (5.3.14) vanishes.

Suppose first that $G^* = G_+^\vee$ (so in particular, N_1 is even and $\eta_{\psi_1} = 1$). We claim that (5.3.14) vanishes in this case as well. Since ψ_+ does not lie in $\widetilde{\Psi}(G_+^\vee)$, we would expect both terms in (5.3.14) to vanish, but we are of course not in a position to prove this. We also cannot appeal to (4.4.12), since it applies only to parameters in $\widetilde{\Psi}(N)$.

To study (5.3.14), we need to apply the expansions (4.1.1) and (4.1.2) (with $G_+^\vee$ in place of G). The proper spectral terms in (4.1.1) correspond to proper Levi subgroups $M_+^\vee$ of $G_+^\vee$ of which ψ_+ contributes to the discrete spectrum, in the sense that the representation $R^{M_+^\vee}_{\mathrm{disc},\psi_+}$ is nonzero. Upon reflection, we see that there is no $M_+^\vee$ with this property. Indeed, any $M_+^\vee$ is either attached to a partition of N_+ that is not compatible with ψ_+, in which case we apply Corollary 3.4.3, or $M_+^\vee$ is a product of groups for which we can combine our induction hypothesis with the fact that the set $\widetilde{\Psi}_2(M_+^\vee, \psi_+)$ is empty. It follows from (4.1.1) that

$$I^{G_+^\vee}_{\mathrm{disc},\psi_+}(f^\vee) - \mathrm{tr}\big(R^{G_+^\vee}_{\mathrm{disc},\psi_+}(f^\vee)\big) = 0.$$

The proper endoscopic terms in (4.1.2) are indexed by composite endoscopic data $(G_+^\vee)'$ in $\widetilde{\mathcal{E}}_{\mathrm{ell}}(G_+^\vee)$. After further reflection, we see that there is no composite $(G_+^\vee)'$ for which the associated linear form $S'_{\mathrm{disc},\psi_+}$ is nonzero. Indeed, either the two factors of $(G_+^\vee)'$ give a partition of N_+ that is incompatible with ψ_+, in which case we apply Proposition 3.4.1, or they are data for which we can combine our induction hypothesis with the fact that the set $\widetilde{\Psi}\big((G_+^\vee)', \psi\big)$ is empty. It follows from (4.1.2) that

$$I^{G_+^\vee}_{\mathrm{disc},\psi_+}(f^\vee) - S^{G_+^\vee}_{\mathrm{disc},\psi_+}(f^\vee) = 0.$$

The vanishing of (5.3.14) then follows from the fact that the expected value of $S^{G_+^\vee}_{\mathrm{disc},\psi_+}$ is 0.

Suppose finally that $G^* = G_+$. In this case we have again to apply (4.1.1) and (4.1.2) (with G_+ in place of G). The expansions will now contain proper terms that are nonzero, so we will want to work with the refinements represented by Corollaries 4.3.3 and 4.4.3. This will lead to a formula for the value

$$\mathrm{tr}\big(R^{G_+}_{\mathrm{disc},\psi_+}(f)\big) - {}^0S^{G_+}_{\mathrm{disc},\psi_+}(f), \qquad f \in \widetilde{\mathcal{H}}(G_+), \tag{5.3.15}$$

of (5.3.14) at $G^* = G_+$. The connected group $S^0_{\psi_+}$ is just $SO(2,\mathbb{C})$, so we are still dealing with a rather elementary form of the standard model. However, we will have to be careful in applying our induction hypotheses to derive analogues of the corollaries for ψ_+, since the condition

$$N_+ = \deg \psi_+ > N$$

is more serious in this case.

The proper spectral terms for G_+ in (4.1.1) are attached to proper Levi subgroups. We claim that these terms vanish for Levi subgroups that are not conjugate to the group M_+ described prior to the statement of the lemma. We recall that the orthogonal or symplectic factor $G_{+,-}$ of M_+ is the unique datum in $\widetilde{\mathcal{E}}_{\mathrm{sim}}(N_{+,-})$, such that $\widetilde{\Psi}_2(G_{+,-})$ contains the subparameter

$$\psi_{+,-} = \psi_2 \boxplus \cdots \boxplus \psi_r.$$

The claim follows easily from the original classification Theorem 1.3.2 for $GL(N)$, given the structure of Levi subgroups of G_+, and the nature of ψ_+. This leaves the terms attached to conjugates of M_+. Corollary 4.3.3 is designed express the sum of such terms. With a little thought, and a brief review of our later justification of Lemma 5.2.1, we see that its proof in §4.3 extends to G_+, since it requires only that the degree $N_{+,-}$ of $\psi_{+,-}$ be less than N. The expansion (4.1.1) therefore behaves as expected. We see that the difference

$$I^{G_+}_{\mathrm{disc},\psi_+}(f) - \mathrm{tr}\big(R^{G_+}_{\mathrm{disc},\psi_+}(f)\big), \qquad f \in \widetilde{\mathcal{H}}(G_+), \tag{5.3.16}$$

equals the corresponding expansion (4.3.9) of Corollary 4.3.3. Arguing then as in the proof of Lemma 5.2.1, we conclude that if we take the product of (5.3.16) with (-1) times the corresponding coefficient $\widetilde{\iota}(N_+, G_+)$ in (5.3.6), we obtain the contribution of the linear forms $f_{G_+}(\psi_+, x)$ to the given expression (5.3.5).

The proper endoscopic terms for G_+ in (4.1.2) are attached to composite endoscopic data

$$G'_+ = G'_1 \times G'_2, \qquad G'_k \in \widetilde{\mathcal{E}}_{\mathrm{sim}}(N'_k),\ k = 1, 2,$$

in $\widetilde{\mathcal{E}}_{\mathrm{ell}}(G_+)$. Corollary 4.4.3 is designed to express the sum of such terms. However, its extension to G_+ would require the stable multiplicity formula for the linear forms

$$S'_{\mathrm{disc},\psi'_+}(f') = S'_{\mathrm{disc},\psi'_1}(f'_1)\, S'_{\mathrm{disc},\psi'_2}(f'_2), \tag{5.3.17}$$

attached to parameters

$$\psi'_+ = \psi'_1 \times \psi'_2, \qquad \psi'_k \in \widetilde{\mathcal{F}}(N'_k),\ k = 1, 2,$$

such that

$$\psi_+ = \psi'_1 \boxplus \psi'_2,$$

and functions $f' = f'_1 \times f'_2$ in $\widetilde{\mathcal{H}}(G'_+)$. It is possible to treat most composite pairs (G'_+, ψ'_+) with our induction hypotheses, augmented by the given assumption on parameters in the complement of $\widetilde{\mathcal{F}}_{\mathrm{ell}}(N)$ in $\widetilde{\mathcal{F}}(N)$. In fact, after a moment's consideration, we see that the stable multiplicity formula holds in all cases, with the possible exception of the pair

$$(G'_+, \psi'_+) = (G_1 \times G,\ \psi_1 \times \psi).$$

Moreover, Lemma 5.1.4 can be applied to this exceptional pair. It tells us in this case that the linear form (5.3.17) equals the difference between its expected value and the linear form

$$\widetilde{o}(G)\, f^{L_+}(\psi_1 \times \Lambda). \tag{5.3.18}$$

Recalling our justification of Lemma 5.2.1 as needed, we see that apart from this defect, the proof of Corollary 4.4.3 in §4.4 extends to G_+. It follows that the sum of the difference

$$I^{G_+}_{\mathrm{disc},\psi_+}(f) - {}^0S^{G_+}_{\mathrm{disc},\psi_+}(f), \qquad f \in \widetilde{\mathcal{H}}(G_+), \tag{5.3.19}$$

with (5.3.18) equals the corresponding expansion (4.4.11). Arguing then as in the proof of Lemma 5.2.1, we conclude that if we add the expression

$$\widetilde{\iota}(N_+, G_+)\,\widetilde{o}(G)\, f^{L_+}(\psi_1 \times \Lambda) \tag{5.3.20}$$

to the product of (5.3.19) with $\widetilde{\iota}(N_+, G_+)$, we obtain the rest of the given expression (5.3.5), namely the contribution from the linear forms $f'_{G_+}(\psi_+, s_{\psi_+}x)$.

We have now finished our discussion of the difference (5.3.15). It is of course equal to the difference between (5.3.19) and (5.3.16). Our conclusion is that the sum of (5.3.20) with the product of $\widetilde{\iota}(N_+, G_+)$ and (5.3.15) equals the entire expression (5.3.5).

This in turn completes our discussion of the full set simple endoscopic data $G^* \in \widetilde{\mathcal{E}}_{\mathrm{sim}}(N_+)$. For we have already seen that the difference (5.3.14) vanishes if $G^* \neq G_+$. Therefore the sum of (5.3.20) with the sum over G^* of the product of $\widetilde{\iota}(N_+, G^*)$ and (5.3.14) equals (5.3.5). It thus follows that the sum of (5.3.20) with the spectral sum (5.3.6) equals the sum of (5.3.5) with the contribution of the simple data G^* to the endoscopic sum (5.3.7). We observed earlier that the contribution of the remaining composite data G^* to (5.3.7) equals (5.3.13). The sum of the two terms (5.3.13) and (5.3.20) equals

$$b_+\, f^{L_+}(\psi_1 \times \Lambda),$$

for the positive coefficient

$$b_+ = \widetilde{\iota}(N_+, G_1 \times G^\vee)\,\widetilde{o}(G^\vee) + \widetilde{\iota}(N_+, G_+)\,\widetilde{o}(G).$$

This represents the supplementary summand in the original given expression (5.3.4).

We have now shown that (5.3.4) equals the sum of (5.3.5) with the endoscopic sum (5.3.7). We have also observed that (5.3.7) vanishes, by Lemma 5.2.3. We have therefore completed the proof of Lemma 5.3.1, at last. □

Suppose now that $r = 1$. Then $\psi = \psi_1$ belongs to the subset $\widetilde{\mathcal{F}}_{\mathrm{sim}}(G)$ of the $\widetilde{\mathcal{F}}_2(G)$. In this case, $\psi_+ = 2\psi$, so that $N_+ = 2N$ is even and $\eta_{\psi_+} = 1$. In particular, we have the two split groups G_+ and $G_+^\vee$ in $\widetilde{\mathcal{E}}_{\mathrm{sim}}(N_+)$. The maximal Levi subgroup $M_+ \cong GL(N)$ of G_+ becomes also a maximal Levi subgroup of $G_+^\vee$.

This case carries the added burden of the assertion of Theorem 1.5.3(a) on the poles of L-functions. We shall set $\delta_\psi = 1$ if ψ is not generic, or in case ψ is generic, the Langlands-Shahidi L-function $L(s, \psi, \rho_+^\vee)$ attached to a maximal parabolic subgroup $P_+^\vee = M_+ N_+^\vee$ of $G_+^\vee$ has a pole at $s = 1$. This is what Theorem 1.5.3(a) predicts. We set $\delta_\psi = -1$ if ψ is generic, and it is the L-function $L(s, \psi, \rho_+)$ attached to a maximal parabolic subgroup $P_+ = M_+ N_+$ of G_+ that has a pole at $s = 1$. In other words, $\delta_\psi = -1$ if Theorem 1.5.3(a) does not hold for ψ.

We recall that G_+ is defined in terms of the group $G = G_\psi$ assigned to ψ. We recall further that G is defined by induction unless ψ is (simple) generic, in which case it was taken to be some preassigned element in $\widetilde{\mathcal{E}}_{\mathrm{sim}}(N)$ such that ψ lies in the subset $\widetilde{\mathcal{F}}_{\mathrm{sim}}(G)$ of $\widetilde{\mathcal{F}}_{\mathrm{sim}}(N)$ specified by the temporary definition of §5.1. In the latter instance, there remains some possible ambiguity in the choice of G. It will be resolved by the stable multiplicity formula, or equally well, the formula $\delta_\psi = 1$. We will of course have to establish both formulas.

If N is even and $\eta_\psi = 1$, we have the two split groups G and $G^\vee$ in $\widetilde{\mathcal{E}}_{\mathrm{sim}}(N)$, with a common maximal Levi subgroup $L \cong GL(\frac{1}{2}N)$. In this case, we must also be concerned about the possible existence of the linear form Λ. Recall that

$$S^\vee_{\mathrm{disc},\psi}(f^\vee) = \widetilde{o}(G^\vee) f^{\vee,L}(\Lambda), \qquad f^\vee \in \widetilde{\mathcal{H}}(G^\vee),$$

and that the right hand side represents a unitary character on $G^\vee(\mathbb{A})$. It will be convenient also to write

$$S^G_{\mathrm{disc},\psi}(f) = \widetilde{o}(G)\, f^G(\Gamma), \qquad f \in \widetilde{\mathcal{H}}(G),$$

for a stable linear form Γ on $\widetilde{\mathcal{H}}(G)$. The formula of Lemma 5.1.4 reduces to

$$S^G_{\mathrm{disc},\psi}(f) = \widetilde{o}(G)\, f^G(\psi) - \widetilde{o}(G)\, f^L(\Lambda),$$

since the coefficients in the formula satisfy

$$m_\psi\, |\mathcal{S}_\psi|^{-1}\, \varepsilon^G(\psi) = \widetilde{o}(G)\cdot 1\cdot 1 = \widetilde{o}(G),$$

in the case at hand. We then obtain

$$f^G(\psi) = f^G(\Gamma) + f^L(\Lambda), \qquad f \in \widetilde{\mathcal{H}}(G).$$

As in the case of $r > 1$ above, we also have the maximal Levi subgroup

$$L_+ = G \times L \cong G \times GL\big(\tfrac{1}{2}N\big) \cong GL\big(\tfrac{1}{2}N\big) \times G$$

of G_+. It comes with the stable linear form

$$f \longrightarrow f^{L_+}(\Gamma \times \Lambda), \qquad f \in \widetilde{\mathcal{H}}(G_+).$$

LEMMA 5.3.2. *Suppose that for the given pair*

$$(G,\psi), \qquad G \in \widetilde{\mathcal{E}}_{\mathrm{sim}}(N),\ \psi \in \widetilde{\mathcal{F}}_2(G),$$

the index r in (5.3.1) *equals* 1, *while N is even and $\eta_\psi = 1$. Then the sum of*

$$\sum_{G^* \in \widetilde{\mathcal{E}}_{\mathrm{sim}}(N_+)} \widetilde{\iota}(N_+, G^*)\, \mathrm{tr}\big(R^*_{\mathrm{disc},\psi_+}(f^*)\big) + \big(\tfrac{1}{2}\, f^{L_+}(\Gamma \times \Lambda)\big) \tag{5.3.21}$$

and

$$\tfrac{1}{8}\,\big((f^\vee)^{M_+}(\psi) - \delta_\psi\, f_{G_+^\vee}^\vee(\psi_+, x_1)\big) \tag{5.3.22}$$

equals

$$\tfrac{1}{8}\,\big(f'_{G_+}(\psi_+, s_{\psi_+}x_1) - \delta_\psi\, f_{G_+}(\psi_+, x_1)\big), \tag{5.3.23}$$

for any compatible family of functions (5.3.3), *and the element $x_1 \in \mathcal{S}^*_{\psi_+,\mathrm{ell}}$ as in Lemma* 5.2.2.

PROOF. The lemma will again be a modification of its counterpart from §5.2. We have thus to see what can be salvaged of the proof of Lemma 5.2.2. This time there will be modifications introduced by the sign δ_ψ, as well as from the possible existence of Λ. As in the proof of the last lemma, we shall work interchangeably with a general compatible family (5.3.3) and a fixed function $\widetilde{f} \in \widetilde{\mathcal{H}}(N_+)$, which for any $G^* \in \widetilde{\mathcal{E}}_{\text{ell}}(N_+)$ have the same image in $\widetilde{\mathcal{S}}(G^*)$.

The sign δ_ψ is not hard to keep track of. The formula of Corollary 4.3.3 is based on the inductive assumption that the sign obeys Theorem 1.5.3(a). Its direct analogues for G_+ and $G_+^\vee$ would hold only if δ_ψ equals 1. If we look to see what changes δ_ψ makes if the proof of Lemma 5.2.2 is applied to ψ_+, we observe that we need only insert a coefficient δ_ψ in the negative summands in the expressions (5.2.12) and (5.2.13) (or rather their analogues for $G_+^\vee$ and G_+). The expressions so obtained are just (5.3.22) and (5.3.23).

Our task then is to describe what change the possible presence of Λ makes in the analogue

$$\sum_{G^* \in \widetilde{\mathcal{E}}_{\text{sim}}(N_+)} \widetilde{\imath}(N_+, G^*)\, \text{tr}\big(R^*_{\text{disc},\psi_+}(f^*)\big) \tag{5.3.24}$$

of (5.2.11). It amounts to a re-examination of the relevant parts of the proof of Lemma 5.3.1, specifically the finer analysis of the summands in

$$\sum_{G^* \in \widetilde{\mathcal{E}}_{\text{ell}}(N_+)} \widetilde{\imath}(N_+, G^*)\, {}^0\widehat{S}^*_{\text{disc},\psi_+}(\widetilde{f}^*).$$

The sum is the left hand side of (5.2.16). It vanishes, since the condition of Lemma 5.2.3 remains valid, so we can write (5.3.24) as the sum of

$$\sum_{G^* \in \widetilde{\mathcal{E}}_{\text{sim}}(N_+)} \widetilde{\imath}(N_+, G^*)\, \big(\text{tr}(R^*_{\text{disc},\psi_+}(f^*)) - {}^0\widehat{S}^*_{\text{disc},\psi_+}(f^*)\big) \tag{5.3.25}$$

with

$$-\sum_{G^*} \widetilde{\imath}(N_+, G^*)\, {}^0\widehat{S}^*_{\text{disc},\psi_+}(f^*), \qquad G^* \in \widetilde{\mathcal{E}}_{\text{ell}}(N_+) - \widetilde{\mathcal{E}}_{\text{sim}}(N_+). \tag{5.3.26}$$

This was of course also the foundation of the proof of Lemma 5.3.1. Having outlined the argument for this last lemma in considerable detail, we can afford to be briefer here.

The effect of Λ will be seen in the endoscopic contributions to (5.3.26) or (5.3.25) of proper products $G_1 \times G_2$ in any of the sets $\widetilde{\mathcal{E}}_{\text{ell}}(N_+)$, $\widetilde{\mathcal{E}}_{\text{ell}}(G_+)$ or $\widetilde{\mathcal{E}}_{\text{ell}}(G_+^\vee)$. (Products from the sets $\widetilde{\mathcal{E}}_{\text{ell}}(N_+)$ and $\widetilde{\mathcal{E}}_{\text{ell}}(G_+)$ were denoted $G^1 \times G^2$ and $G_1' \times G_2'$) in the proof of Lemma 5.3.1, while products from the set $\widetilde{\mathcal{E}}_{\text{ell}}(G_+^\vee)$ were seen to be innocuous.) The presence of Λ will be a consideration only if both G_1 and G_2 are taken from the set $\{G, G^\vee\}$. We have then to describe the change Λ makes in the expected formula for

$$S^{G_1 \times G_2}_{\text{disc},\psi_+}(f_1 \times f_2) = S^1_{\text{disc},\psi}(f_1)\, S^2_{\text{disc},\psi}(f_2), \qquad f_k \in \widetilde{\mathcal{H}}(G_k), \tag{5.3.27}$$

where $G_1 \times G_2$ is the group $G^*_+ = G \times G^\vee$ in $\widetilde{\mathcal{E}}_{\text{ell}}(N_+)$, the group $G'_+ = G \times G$ in $\mathcal{E}_{\text{ell}}(G_+)$, and the group $(G^\vee_+)' = G^\vee \times G^\vee$ in $\mathcal{E}_{\text{ell}}(G^\vee_+)$.

Now the formula for (5.3.24) that would be given by the analogue of Lemma 5.2.2 for ψ_+ (modified by the signs δ_ψ as above) is just the expression (5.3.23) minus the expression (5.3.22). However, this formula is predicated on the compound linear form $S(f_1 \times f_2)$ in each of the three cases of (5.3.27) being equal to its expected value. Suppose that the discrepancy ${}^0S(f_1 \times f_2)$ is in fact nonzero. In the first case, it contributes a nonzero summand to (5.3.26), which must then be added to (5.3.24) as a correction term for the formula to continue to hold. In the second and third cases, it alters the proper endoscopic component of the corresponding term ${}^0\widehat{S}^*_{\text{disc},\psi}(f^*)$ in (5.3.25). The change in the associated summand of (5.3.25) must then be subtracted from (5.3.24) as a correction term, for the formula to remain valid. These remarks are of course with the understanding that the function $f_1 \times f_2$ in each case has the same image in $\widetilde{\mathcal{S}}(G_1 \times G_2)$ as the chosen compatible family.

In the first case, (5.3.27) equals the product of $\widetilde{o}(G)\widetilde{o}(G^\vee)$ with the linear form

$$f_1^G(\Gamma)\, f_2^L(\Lambda) = f^{G\times L}(\Gamma \times \Lambda) = f^{L_+}(\Gamma \times \Lambda).$$

The expected value of this linear form is 0. We have then to *add* a term equal to the product of the coefficient

$$\widetilde{\iota}(N_+, G^*_+)\, \widetilde{o}(G)\, \widetilde{o}(G^\vee) = \tfrac{1}{4}$$

with

$$f^{L_+}(\Gamma \times \Lambda)$$

to (5.3.24).

In the second case, (5.3.27) equals the product of $\widetilde{o}(G)^2$ with the linear form

$$f_1^G(\Gamma)\, f_2^G(\Gamma) = f^{G\times G}(\Gamma \times \Gamma).$$

The expected value of this linear form equals

$$\begin{aligned} & f^{G\times G}(\psi \times \psi) \\ &= f_1^G(\psi)\, f_2^G(\psi) = \big(f_1^G(\Gamma) + f_1^L(\Lambda)\big)\big(f_2^G(\Gamma) + f_2^L(\Lambda)\big) \\ &= f^{G\times G}(\Gamma \times \Gamma) + f^{L\times G}(\Lambda \times \Gamma) + f^{G\times L}(\Gamma \times \Lambda) + f^{L\times L}(\Lambda \times \Lambda). \end{aligned}$$

We note that

$$f^{L\times G}(\Lambda \times \Gamma) = f^{G\times L}(\Gamma \times \Lambda) = f^{L_+}(\Gamma \times \Lambda),$$

since $f^{G\times G}$ acquires a symmetry from the automorphism in $\text{Aut}(G_+, G'_+)$ that interchanges the two factors of $G'_+ = G \times G$. We must therefore *subtract* the product of the coefficient

$$\widetilde{\iota}(N_+, G_+)\, \iota(G_+, G \times G)\, \widetilde{o}(G)^2 = \tfrac{1}{8}$$

with the difference

$$f^{G\times G}(\Gamma \times \Gamma) - f^{G\times G}(\psi \times \psi) = -\big(2f^{L_+}(\Lambda \times \Gamma) + f^{L\times L}(\Lambda \times \Lambda)\big)$$

from (5.3.24).

In the third case, (5.3.27) equals the product of $\widetilde{\sigma}(G^\vee)^2$ with the linear form

$$f_1^L(\Lambda)\, f_2^L(\Lambda) = f^{L\times L}(\Lambda \times \Lambda).$$

The expected value of this linear form is 0. We must therefore subtract the product of the coefficient

$$\widetilde{\iota}(N_+, G_+^\vee)\, \iota(G_+^\vee,\, G^\vee \times G^\vee)\, \widetilde{\sigma}(G^\vee)^2 = \tfrac{1}{8}$$

with

$$f^{L\times L}(\Lambda \times \Lambda)$$

from (5.3.24). We note that it is the coefficient $\frac{1}{8}$ rather than $\frac{1}{4}$ that appears in the latter two cases, since the associated two endoscopic data $G \times G$ and $G^\vee \times G^\vee$ each come with an extra automorphism that interchanges the two factors.

Combining the three cases, we see that the sum total of what must be added to (5.3.24) is the linear form

$$\begin{aligned}&\tfrac{1}{4}\, f^{L_+}(\Gamma \times \Lambda) + \tfrac{1}{8}\big(2f^{L_+}(\Gamma \times \Lambda) + f^{L\times L}(\Lambda \times \Lambda)\big) - \tfrac{1}{8}\, f^{L\times L}(\Lambda \times \Lambda)\\ =&\tfrac{1}{2}\, f^{L_+}(\Gamma \times \Lambda).\end{aligned}$$

Its sum with (5.3.24) is the given expression (5.3.21). The formula of Lemma 5.2.2, adjusted for the lack of information we have about ψ_+, thus tells us that (5.3.21) equals the difference between (5.3.23) and (5.3.22). The proof is complete. □

From the last proof, we can extract formulas for the stable distributions attached to (G_+, ψ_+) and $(G_+^\vee, \psi_+)$. We record them here for use later, when we come to the general global classification in Chapter 8.

COROLLARY 5.3.3. *Under the conditions of the lemma, we have*

$$S^{G_+}_{\mathrm{disc},\psi_+}(f) = \mathrm{tr}\big(R^{G_+}_{\mathrm{disc},\psi_+}(f)\big) + \tfrac{1}{4}\,\widetilde{\sigma}(G_+)\big(\delta_\psi\, f_{G_+}(\psi_+, x_1) - f^{G\times G}(\Gamma \times \Gamma)\big)$$

and

$$S^{G_+^\vee}_{\mathrm{disc},\psi_+}(f^\vee) = \mathrm{tr}\big(R^{G_+^\vee}_{\mathrm{disc},\psi_+}(f^\vee)\big) - \tfrac{1}{4}\,\widetilde{\sigma}(G_+^\vee)\big(\delta_\psi f^\vee_{G_+^\vee}(\psi_+, x_1) + f^{\vee, L\times L}(\Lambda \times \Lambda)\big).$$

PROOF. These formulas are implicit in the proof of the lemma. If G^* is equal to either G_+ or $G_+^\vee$, it would not be hard to compute the difference

$$S^*_{\mathrm{disc},\psi_+}(f^*) - \mathrm{tr}\big(R^*_{\mathrm{disc},\psi_+}(f^*)\big)$$

directly from the general expansions (4.1.1) and (4.1.2). There will be one spectral term, corresponding to the Levi subgroup M_+ of G_+ or $G_+^\vee$, and one endoscopic term, corresponding to the proper endoscopic datum G'_+ of G_+ or $(G_+^\vee)'$ of $G_+^\vee$. The point is that we have already done the calculations, and can in principle read these terms off from the lemma.

The spectral terms are in the second summands of (5.3.23) and (5.3.22) respectively, the contributions of the terms with coefficient $(-\delta_\psi)$. These summands must be taken over to the left hand side of the relevant part of the formula given by the lemma, and then multiplied by the inverse of the coefficient $\tilde{\iota}(N_+, G^*)$ that occurs in (5.3.21). Since

$$\tfrac{1}{8}\,\tilde{\iota}(N_+, G^*)^{-1} = \tfrac{1}{4}\,\tilde{\sigma}(G^*),$$

we see that the spectral term in the resulting formula for

$$S^*_{\mathrm{disc},\psi_+}(f^*), \qquad G^* \in \{G_+, G_+^\vee\}, \tag{5.3.28}$$

will be

$$\tfrac{1}{4}\,\tilde{\sigma}(G_+)\,\delta_\psi\, f_{G_+}(\psi_+, x_1)$$

if $G^* = G_+$, and

$$-\tfrac{1}{4}\,\tilde{\sigma}(G_+^\vee)\,\delta_\psi\, f^\vee_{G_+^\vee}(\psi_+, x_1)$$

if $G^* = G_+^\vee$.

The endoscopic terms do not come from the first summands in (5.3.23) and (5.3.22). For these summands represent only expected values. Moreover in the case of (5.3.22), the summand corresponds to the simple endoscopic datum $G_+^\vee$ for $G_+^\vee$ rather than $(G'_+)^\vee$. However, the two endoscopic terms are easily extracted from the proof of the lemma. They are essentially the formulas for (5.3.27), in the cases that $G_1 \times G_2$ equals G'_+ and $(G_+^\vee)'$. We of course have again to place each of these ingredients on the left hand side of the relevant part of the formula of the lemma. Accounting for the coefficients, and inserting the extra sign (-1) in the case of (5.3.22), we then see that the endoscopic term in the formula for (5.3.28) will be

$$-\tfrac{1}{4}\,\tilde{\sigma}(G_+)\, f^{G\times G}(\Gamma \times \Gamma)$$

if $G^* = G_+$,

$$-\tfrac{1}{4}\,\tilde{\sigma}(G_+^\vee)\, f^{L\times L}(\Lambda \times \Lambda)$$

if $G^\vee = G_+^\vee$. The stated formulas follow. □

REMARK. If we change the conditions N even and $\eta_\psi = 1$ in the statement of the lemma, the formula simplifies. For without these conditions, Λ does not exist, and the last summand in (5.3.21) can be taken to be 0. In other words, (5.3.21) reduces to the sum (5.3.24) over G^*. In the particular case that N is odd, the term (5.3.23) can also be taken to be 0, since the set $\mathcal{S}_{\psi_+,\mathrm{ell}}$ is then empty. It follows from a simplified form of the proof of the lemma that the sum of (5.3.24) with (5.3.22) vanishes if N is odd, and equals (5.3.23) if N is even and $\eta_\psi \neq 1$.

5.4. Generic parameters with local constraints

Our aim is to apply the discussion of the last three sections to the proof of the local theorems. We shall do so in due course, by constructing particular families $\widetilde{\mathcal{F}}$ that among other things, satisfy the conditions of §5.1. We begin here by imposing some ad hoc local conditions on the members of a general family $\widetilde{\mathcal{F}}$, with a view to strengthening the results obtained so far.

Suppose that

$$\widetilde{\mathcal{F}} = \coprod_N \widetilde{\mathcal{F}}(N), \qquad \widetilde{\mathcal{F}}(N) \subset \widetilde{\Psi}(N),$$

is a family of parameters in the general set $\widetilde{\Psi}$. We assume as in §5.1 that $\widetilde{\mathcal{F}}$ is the graded semigroup generated by its subset $\widetilde{\mathcal{F}}_{\mathrm{sim}}$ of simple elements. However, we shall also take on a temporary hypothesis of a different sort. It consists of three local constraints on elements in $\widetilde{\mathcal{F}}$, which we state as follows in terms of notation for $\widetilde{\mathcal{F}}$ introduced in §5.1 and a positive integer N.

ASSUMPTION 5.4.1. *There is a finite, nonempty set* $V = V(\widetilde{\mathcal{F}})$ *of archimedean valuations of* F *for which the following three conditions hold.*
(5.4.1)(a) *Suppose that* $\psi \in \widetilde{\mathcal{F}}_{\mathrm{sim}}(G)$, *for some* $G \in \widetilde{\mathcal{E}}_{\mathrm{sim}}(N)$, *and that* $v \in V$. *Then* $\psi_v \in \widetilde{\Psi}_2(G_v)$.
(5.4.1)(b) *Suppose that* $\psi \in \widetilde{\mathcal{F}}_2^{\mathrm{sim}}(G)$, *for some* $G \in \widetilde{\mathcal{E}}_{\mathrm{ell}}(N)$. *Then there is a valuation* $v \in V$ *such that* ψ_v *does not lie in* $\widetilde{\Psi}^+(G_v^*)$ *for any* $G_v^* \in \widetilde{\mathcal{E}}_{\mathrm{sim}}(N)$ *over* v *with* $\widehat{G}_v^* \neq \widehat{G}$.
(5.4.1)(c) *Suppose that* $\psi \in \widetilde{\mathcal{F}}_{\mathrm{disc}}^2(G)$, *for some* $G \in \widetilde{\mathcal{E}}_{\mathrm{ell}}(N)$. *Then there is a valuation* $v \in V$ *such that the kernel of the composition of mappings*

$$\mathcal{S}_\psi \longrightarrow \mathcal{S}_{\psi_v} \longrightarrow R_{\psi_v}, \qquad \mathcal{S}_\psi = \mathcal{S}_\psi(G),$$

contains no element whose image in the global R*-group* $R_\psi = R_\psi(G)$ *is regular.*

The intent of the first condition (5.4.1)(a) is that Theorem 1.4.2 should hold for simple generic parameters $\psi \in \widetilde{\mathcal{F}}_{\mathrm{sim}}(N)$ and valuations $v \in V$. The condition in fact will allow us to bypass the possible ambiguity of the provisional definition of $\widetilde{\mathcal{F}}_{\mathrm{sim}}(G)$, adopted for families $\widetilde{\mathcal{F}}$ in §5.1. The remaining two conditions contain technical assertions that are less clear. They will make better sense once we see how to apply them later in this section, and after we have constructed families for which they hold in §6.3.

We fix a family $\widetilde{\mathcal{F}}$ that satisfies Assumption 5.4.1. We may as well assume at this point that the elements in $\widetilde{\mathcal{F}}$ are all generic, since it is to the local classification of tempered representations in Chapter 6 that they will be applied. However, we will continue to denote them by ψ (rather than ϕ). This preserves the notation of the last three sections, and will also ease the transition to the family that will be applied to nontempered representations in Chapter 7.

We do not assume a priori that the elements in $\widetilde{\mathcal{F}}$ satisfy Assumption 5.1.1. We instead fix the positive integer N, and as in §5.1, assume inductively that all the local and global theorems hold for any $\psi \in \widetilde{\mathcal{F}}$ with $\deg(\psi) < N$. We shall combine this inductive property with Assumption 5.4.1 to establish the original conditions in Assumption 5.1.1.

LEMMA 5.4.2. *Suppose that $G \in \mathcal{E}_{\mathrm{ell}}(N)$, and that ψ belongs to $\widetilde{\mathcal{F}}_2(G)$. Then the conditions of Assumption* 5.1.1 *hold for the pair* (G, ψ).

PROOF. The assertion of course includes the case that ψ is simple. Expanding on the remark concerning (5.4.1)(a) above, we recall that $\widetilde{\mathcal{F}}_{\mathrm{sim}}(G)$ was defined provisionally in (5.1.5) and (5.1.6) as a set of parameters $\psi \in \widetilde{\mathcal{F}}_{\mathrm{sim}}(N)$ for which the corresponding stable linear form $S^G_{\mathrm{disc},\psi}$ is nonzero. The problem is that for a given ψ, we do not know yet that this characterizes the datum $G \in \widetilde{\mathcal{E}}_{\mathrm{sim}}(N)$ uniquely. We do know from our remarks in §3.4 that if $G^* \in \widetilde{\mathcal{E}}_{\mathrm{sim}}(N)$ is another such datum, and if η_{G^*} is distinct from the quadratic character $\eta_\psi = \eta_G$, then $S^*_{\mathrm{disc},\psi}$ vanishes. Moreover, (5.4.1)(a) asserts that for any $v \in V$, ψ_v lies in $\widetilde{\Phi}_{\mathrm{ell},v}(N)$, and that G_v is the unique datum in $\widetilde{\mathcal{E}}_v(N)$ such that $\widetilde{\Phi}(G_v)$ contains ψ_v. This condition will fail if G is replaced by G^*, unless G^*_v equals G_v, which in turn implies that $\widehat{G}^* = \widehat{G}$. Since the properties $\widehat{G}^* = \widehat{G}$ and $\eta_{G^*} = \eta_G$ imply that $G^* = G$, the condition (5.4.1)(a) does indeed characterize the set $\widetilde{\mathcal{F}}_{\mathrm{sim}}(G)$.

If ψ is not simple, we recall that the condition $\psi \in \widetilde{\mathcal{F}}_2(G)$ (or the more general condition $\psi \in \widetilde{\mathcal{F}}(G)$) is defined inductively. In this case, our induction hypotheses imply that ψ_v belongs to $\widetilde{\Phi}(G_v)$, for *any* valuation v of F.

Given ψ and G, we write the set of valuations of F as a disjoint union

$$V \coprod U \coprod V_{\mathrm{un}},$$

where V is given by Assumption 5.4.1 and U is a finite set, and where (G, ψ) is assumed to be unramified at every v in the remaining set V_{un}. We will apply the twisted endoscopic decomposition (3.3.14) of Corollary 3.3.2 to a function

$$\widetilde{f} = \widetilde{f}_V \cdot \widetilde{f}_U \cdot \widetilde{f}_{\mathrm{un}}, \qquad \widetilde{f} \in \widetilde{\mathcal{H}}(N),$$

that is compatible with this decomposition. In so doing, we will first fix the functions

$$\widetilde{f}_V = \prod_{v \in V} \widetilde{f}_v, \qquad \widetilde{f}_v \in \widetilde{\mathcal{H}}_v(N),$$

and

$$\widetilde{f}_{\mathrm{un}} = \prod_{v \in V_{\mathrm{un}}} \widetilde{f}_v,$$

and then allow $\widetilde{f}_U \in \widetilde{\mathcal{H}}_U(N)$ to vary. We are writing $\widetilde{\mathcal{H}}_v(N)$ here for the Hecke module on $\widetilde{G}(N, F_v)$, and $\widetilde{\mathcal{H}}_U(N)$ for the $\mathbb{C}$-tensor product over $v \in U$ of $\widetilde{\mathcal{H}}_v(N)$.

We are assuming that ψ belongs to $\widetilde{\mathcal{F}}_2(G)$. We shall use (5.4.1)(b) to define $\widetilde{f}_V$. This condition, together with the stronger condition (5.4.1)(a) in case ψ lies in the subset $\widetilde{\mathcal{F}}_{\mathrm{sim}}(G)$ of $\widetilde{\mathcal{F}}_2(G)$, tells us that there is a valuation $w \in V$ such that ψ_w does not lie in $\widetilde{\Phi}(G^*_w)$, for any datum G^*_w in $\widetilde{\mathcal{E}}_{\mathrm{sim}}(N)$ over w with $\widehat{G}^*_w \neq \widehat{G}$. We then choose a function $\widetilde{f}_w \in \widetilde{\mathcal{H}}_w(N)$ such that

$\widetilde{f}_{w,N}(\psi_w)$ is nonzero, but so that the image $f_w^* = \widetilde{f}_w^{G_w^*}$ of $\widetilde{f}_w$ in $\widetilde{\mathcal{S}}(G_w^*)$ vanishes for any such G_w^*. The existence of $\widetilde{f}_w$ follows easily from Proposition 2.1.1, and the trace Paley-Wiener theorems [**CD**] and [**DM**] at w. At the other places $v \in V$, we choose $\widetilde{f}_v$ subject only to the requirement that $\widetilde{f}_{v,N}(\psi_v)$ not vanish. The function $\widetilde{f}_V$ then has the property that its value

$$\widetilde{f}_{V,N}(\psi_V) = \prod_v \widetilde{f}_{v,N}(\psi_v)$$

is nonzero, but its transfer

$$\widetilde{f}_V^* = \prod_{v \in V} \widetilde{f}_v^{G_v^*}$$

vanishes for any $G^* \in \widetilde{\mathcal{E}}_{\mathrm{sim}}(N)$ with $\widehat{G}^* \neq \widehat{G}$.

The other function to be fixed is $\widetilde{f}_{\mathrm{un}}$. We take it simply to be any decomposable function

$$\widetilde{f}_{\mathrm{un}} = \prod_{v \in V_{\mathrm{un}}} \widetilde{f}_v$$

in the unramified Hecke algebra $\widetilde{\mathcal{H}}(N, \widetilde{K}_{\mathrm{un}}(N))$ such that the value

$$\widetilde{f}_{\mathrm{un},N}(\psi_{\mathrm{un}}) = \prod_{v \in V_{\mathrm{un}}} \widetilde{f}_{v,N}(\psi_v),$$

is nonzero. Of course, $\widetilde{K}_{\mathrm{un}}(N)$ stands here for the product over $v \in V_{\mathrm{un}}$ of the standard maximal compact subgroups $\widetilde{K}_v(N)$ of $GL(N, F_v)$.

We have thus to apply the identity

$$I_{\mathrm{disc},\psi}^N(\widetilde{f}) = \sum_{G^* \in \widetilde{\mathcal{E}}_{\mathrm{ell}}(N)} \widetilde{\iota}(N, G^*)\, \widehat{S}_{\mathrm{disc},\psi}^*(\widetilde{f}^*) \tag{5.4.2}$$

to the variable function

$$\widetilde{f} = \widetilde{f}_V \cdot \widetilde{f}_U \cdot \widetilde{f}_{\mathrm{un}}, \qquad \widetilde{f}_U \in \widetilde{\mathcal{H}}_U(N). \tag{5.4.3}$$

We have not asked that the given datum G be simple. However, if $G^* \in \widetilde{\mathcal{E}}_{\mathrm{ell}}(N)$ is a datum that is not simple, and distinct from G, we can say that

$$S_{\mathrm{disc},\psi}^{G^*} = S_{\mathrm{disc},\psi}^* = 0.$$

This follows from the application of our induction hypothesis to the composite factors of G^*, since as an element in $\widetilde{\Psi}_2(G)$, ψ cannot also belong to $\widetilde{\Psi}(G^*)$. Suppose next that $G^* \in \widetilde{\mathcal{E}}_{\mathrm{sim}}(N)$, but is again distinct from G. If $\widehat{G}^* \neq \widehat{G}$, we obtain

$$\widehat{S}_{\mathrm{disc},\psi}^*(\widetilde{f}^*) = \widehat{S}_{\mathrm{disc},\psi}^*(\widetilde{f}_V^* \cdot \widetilde{f}_U^* \cdot \widetilde{f}_{\mathrm{un}}^*) = 0,$$

by our choice of $\widetilde{f}_V$. If $\widehat{G}^* = \widehat{G}$, we have

$$\eta_{G^*} \neq \eta_G = \eta_\psi,$$

given that ψ lies in $\widetilde{\Psi}(G)$. This implies that $S^*_{\mathrm{disc},\psi} = 0$, as we agreed in §3.4. The summands in (5.4.2) with $G^* \neq G$ therefore all vanish, and we obtain

$$I^N_{\mathrm{disc},\psi}(\widetilde{f}) = \widetilde{\imath}(N,G)\,\widehat{S}^G_{\mathrm{disc},\psi}(\widetilde{f}^G). \tag{5.4.4}$$

We shall compare the last formula with the spectral expansion of $I^N_{\mathrm{disc},\psi}(\widetilde{f})$. This expansion is the analogue of (4.1.1) for $\widetilde{G}(N)$, described first at the end of §3.3. It is a sum of terms parametrized by Levi subgroups

$$\widetilde{M}^0 = GL(N_1) \times \cdots \times GL(N_r)$$

of $GL(N)$. From the classification of automorphic representations of $GL(N)$ (Theorem 1.3.3), we know that there is at most one $\widetilde{M}$ (taken up to conjugacy) for which the corresponding term is nonzero. It is then a consequence of the definitions (or if one prefers, the twisted global intertwining relation discussed at the end of §4.2) that this term is a nonzero multiple of the linear form

$$\begin{aligned}\widetilde{f}_N(\psi) &= \widetilde{f}_{V,N}(\psi_V)\,\widetilde{f}_{U,N}(\psi_U)\,\widetilde{f}_{\mathrm{un},N}(\psi_{\mathrm{un}})\\ &= c(\psi_V, \psi_{V_{\mathrm{un}}})\,\widetilde{f}_{U,N}(\psi_U),\end{aligned}$$

where $c(\psi_V, \psi_{V_{\mathrm{un}}})$ is a nonzero constant. Combined with (5.4.4), this tells us that $\widetilde{f}_{U,N}(\psi_U)$ is a nonzero multiple of the linear form

$$\widehat{S}^G_{\mathrm{disc},\psi}(\widetilde{f}^G) = \widehat{S}^G_{\mathrm{disc},\psi}(\widetilde{f}^G_V \cdot f^G_U \cdot \widetilde{f}^G_{\mathrm{un}}),$$

and hence depends only on $\widetilde{f}^G_U$. Therefore, $\widetilde{f}_{U,N}(\psi_U)$ is the pullback of a linear form on $\widetilde{\mathcal{S}}(G_U)$.

We are now free to take $\widetilde{f}_V$ to be a variable function. It follows from the conditions (5.4.1)(a) and (5.4.1)(b), and the properties of the specific (archimedean) function $\widetilde{f}_V$ above, that the corresponding linear form $\widetilde{f}_{V,N}(\psi_V)$ is also the pullback of a linear form on $\widetilde{\mathcal{S}}(G_V)$. Enlarging the finite set U, if necessary, we can assume that $\widetilde{f}_{\mathrm{un}}$ is the characteristic function of $\widetilde{K}_{\mathrm{un}}(N)$. Having treated the function $\widetilde{f}_U$, we conclude that $\widetilde{f}_N(\psi)$ is indeed the pullback of a linear form on $\widetilde{\mathcal{S}}(G)$. In other words, we can write

$$\widetilde{f}_N(\psi) = \widetilde{f}^G(\psi), \qquad \widetilde{f} \in \widetilde{\mathcal{H}}(N), \tag{5.4.5}$$

where the right hand side represents the restriction to the image of $\widetilde{\mathcal{H}}(N)$ of some linear form on $\widetilde{\mathcal{S}}(G)$.

If G is simple, the linear form on $\widetilde{\mathcal{S}}(G)$ is uniquely determined by (5.4.5), since Corollary 2.1.2 tells us that the twisted transfer mapping is surjective. In particular, it can be identified with a unique *stable* linear form on $\widetilde{\mathcal{H}}(G)$. If $G = G_S \times G_O$ is composite, and $\psi = \psi_S \times \psi_O$ is the corresponding decomposition of ψ, we define the linear form on $\widetilde{\mathcal{S}}(G)$ as the tensor product of the linear forms on $\widetilde{\mathcal{S}}(G_S)$ and $\widetilde{\mathcal{S}}(G_O)$ given by ψ_S and ψ_O. We then have to check that this definition is compatible with (5.4.5). In other words, we need to show that

$$\widetilde{f}^G(\psi) = \widetilde{f}^G(\psi_S \times \psi_O), \qquad \widetilde{f} \in \widetilde{\mathcal{H}}(N), \tag{5.4.6}$$

where the left hand side is defined by (5.4.5), and the right hand side is defined by the linear forms attached to ψ_S and ψ_O.

To establish (5.4.6), we return briefly to the discussion surrounding the function (5.4.3) above. For this particular function $\widetilde{f}$, we know that $I^N_{\mathrm{disc},\psi}(\widetilde{f})$ is on the one hand equal to

$$\widetilde{\iota}(N,G)\, \widehat{S}^G_{\mathrm{disc},\psi}(\widetilde{f}^G),$$

and on the other, is a constant multiple of the value at $\widetilde{f}$ of the left hand side of (5.4.5). This constant is easy to compute, as an elementary special case of the general remarks in §4.7 for example. The reader can check that its product with $\widetilde{\iota}(N,G)^{-1}$ equals the product of $m(\psi)$ with $|\mathcal{S}|^{-1}$. In other words, the stable multiplicity formula

$$\widehat{S}^G_{\mathrm{disc},\psi}(\widetilde{f}^G) = m(\psi)\, |\mathcal{S}_\psi|^{-1}\, \widetilde{f}^G(\psi)$$

holds for the given function $\widetilde{f}$. We are using the fact that ψ is generic here, so there is no ε-factor in the computation, and no sign $\varepsilon^G(\psi)$.

Suppose again that $G = G_S \times G_O$ and $\psi = \psi_S \times \psi_O$. Our general induction hypothesis then tells us that

$$\begin{aligned}\widehat{S}^G_{\mathrm{disc},\psi}(f^G) &= \widehat{S}^S_{\mathrm{disc},\psi_S}(f^S)\, \widehat{S}^0_{\mathrm{disc},\psi_O}(f^O)\\ &= m(\psi_S)\, |\mathcal{S}_{\psi_S}|^{-1}\, f^S(\psi_S)\cdot m(\psi_O)\, |\mathcal{S}_{\psi_O}|^{-1}\, f^O(\psi_O),\end{aligned}$$

for any decomposable function $f^G = f^S \times f^O$ in $\widehat{\mathcal{S}}(G)$. We thus have two formulas for $\widehat{S}^G_{\mathrm{disc},\psi}$. Replacing f^G by the image $\widetilde{f}^G$ of the function (5.4.3), and noting that

$$m(\psi)\, |\mathcal{S}_\psi|^{-1} = m(\psi_S)\, |\mathcal{S}_{\psi_S}|^{-1}\, m(\psi_O)\, |\mathcal{S}_{\psi_O}|^{-1},$$

we see that (5.4.6) holds if $\widetilde{f}$ is the function (5.4.3). Now the local analogue of (5.4.6) holds if $\widetilde{f}$ is replaced by an arbitrary function $\widetilde{f}_V \in \widetilde{\mathcal{H}}(G_V)$ (this is Theorem 2.2.1(a) for the archimedean valuations $v \in V$, which is part of what we are taking for granted from [**Me**] and [**S8**]), or by the unit $\widetilde{f}_{\mathrm{un}}$ in the Hecke algebra (by definition). Since the components $\widetilde{f}^G_V(\psi_V)$ and $\widetilde{f}^G_{\mathrm{un}}(\psi_{\mathrm{un}})$ are nonzero for the functions $\widetilde{f}_V$ and $\widetilde{f}_{\mathrm{un}}$ in (5.4.3), the local analogue of (5.4.6) holds also if $\widetilde{f}$ is replaced by an arbitrary function $\widetilde{f}_U \in \widetilde{\mathcal{H}}(G_U)$. With this fact in hand, we replace the fixed component $\widetilde{f}_V$ in (5.4.3) by a general function in $\widetilde{\mathcal{H}}(G_V)$. The required identity (5.4.6) then follows for a general decomposable function $\widetilde{f} \in \widetilde{\mathcal{H}}(N)$, since it holds for its local factors $\widetilde{f}_V$, $\widetilde{f}_U$ and $\widetilde{f}_{\mathrm{un}}$. This completes the last step in the proof of the lemma. □

REMARK. The assertion of the lemma holds more generally if ψ is any element in $\widetilde{\mathcal{F}}(G)$. For as we remarked after stating Assumption 5.1.1, the assertion for parameters ψ in the complement of $\widetilde{\mathcal{F}}_2(G)$ follows from the corresponding assertion for a Levi subgroup of $GL(N)$.

We shall now establish some global identities for $\widetilde{\mathcal{F}}(N)$. Keep in mind that $\widetilde{\mathcal{F}}(N)$ is composed of generic global parameters with rather serious local constraints. For this reason, the identities we obtain will not be general enough to have much interest in their own right. Their role is intended rather to be local. They will drive the local classification we are going to establish in the next chapter. We will return to global questions later in Chapter 8, where having armed ourselves with the required local results, we will be able to establish the global theorems in general.

We shall prove four lemmas, which represent partial resolutions of the four main lemmas from §5.2 and §5.3, and apply to parameters in the family $\widetilde{\mathcal{F}}(N)$. We continue to denote these generic parameters by ψ rather than ϕ, in order to match the notation from the two earlier sections.

LEMMA 5.4.3. *Suppose that*

$$(G,\psi), \qquad G \in \widetilde{\mathcal{E}}_{\mathrm{sim}}(N),\ \psi \in \widetilde{\mathcal{F}}^2_{\mathrm{ell}}(G),$$

is as in Lemma 5.2.1, *but with* $\widetilde{\mathcal{F}}$ *being our family of generic parameters that satisfy Assumption* 5.4.1. *Then*

$$R^*_{\mathrm{disc},\psi}(f^*) = 0 = {}^0S^*_{\mathrm{disc},\psi}(f^*), \qquad f^* \in \widetilde{\mathcal{H}}(G^*),$$

for every $G^* \in \widetilde{\mathcal{E}}_{\mathrm{sim}}(N)$, *while the expression* (5.2.8) *vanishes for any* $f \in \widetilde{\mathcal{H}}(G)$.

PROOF. Given the pair (G,ψ), we choose a place $v \in V(\widetilde{\mathcal{F}})$ that satisfies (5.4.1)(c). We then consider the expression (5.2.8) in Lemma 5.2.1, for a decomposable function

$$f = f_v\, f^v, \qquad f_v \in \widetilde{\mathcal{H}}(G_v),\ f^v \in \widetilde{\mathcal{H}}\big(G(\mathbb{A}^v)\big),$$

in $\widetilde{\mathcal{H}}(G)$, and a corresponding decomposition

$$\psi = \psi_v\, \psi^v$$

of ψ. The main terms in (5.2.8) become

$$f_G(\psi,x) = f_{v,G}(\psi_v,x_v)\, f^v_G(\psi^v,x^v)$$

and

$$f'_G(\psi,x) = f'_{v,G}(\psi_v,x_v)\,(f^v)'_G\,(\psi^v,x^v),$$

by (4.2.4) and (4.2.5). Since ψ is generic, we also have reductions $s_\psi = 1$ and $\varepsilon_\psi(x) = 1$. The expression (5.2.8) therefore equals

$$c \sum_{x\in\mathcal{S}_{\psi,\mathrm{ell}}} \big((f^v)'_G\,(\psi^v,x^v)\, f'_{v,G}(\psi_v,x_v) - f^v_G(\psi^v,x^v)\, f_{v,G}(\psi_v,x_v)\big). \tag{5.4.7}$$

We will derive the required properties from Corollary 3.5.3. This entails identifying (5.4.7) with the supplementary expression (3.5.12) of the corollary. We have therefore to expand the linear forms in f_v from (5.4.7) in terms of the basis $T(G_v)$ described at the beginning of the proof of Proposition 3.5.1.

The element x_v in (5.4.7) is the image of x in $\mathcal{S}_{\psi_v}$. We shall write $r_v(x)$ for its image in the local R-group R_{ψ_v}. We shall also identify ψ_v with a fixed local preimage in the set $\widetilde{\Psi}_2(M_v, \psi_v)$, where M_v is a fixed Levi subgroup over G_v such that this set is not empty. (We are thus following the local notation (2.3.4) rather than letting M_v denote the localization of the global Levi subgroup M attached to ψ, a group that could properly contain M_v.) The linear form $f_{v,G}(\psi_v, x_v)$ in (5.4.7) then equals

$$\sum_{\pi_v \in \widetilde{\Pi}_{\psi_v}(M_v)} \langle \widetilde{x}_v, \widetilde{\pi}_v \rangle \operatorname{tr}\big(R_{P_v}\big(r_v(x), \widetilde{\pi}_v, \psi_v\big)\, \mathcal{I}_{P_v}(\pi_v, f_v)\big), \tag{5.4.8}$$

in the notation (2.4.5).

The other term in (5.4.7) that depends on f_v is the linear form $f'_{v,G}(\psi_v, x_v)$. We are not assuming the local intertwining relation of Theorem 2.4.1, which asserts that it equals $f_{v,G}(\psi_v, x_v)$. However, our assumption that ψ_v is archimedean and generic, and hence that Theorem 2.2.1 holds, leads to a weaker relation between the two linear forms. For if the assertion (b) of this theorem is combined with the theory of the R-groups $R(\pi_v)$, attached to representations $\pi_v \in \widetilde{\Pi}_{\psi_v}(M_v)$ and reviewed in §3.5, one obtains an endoscopic characterization of the irreducible constituents of the induced representation $\mathcal{I}_{P_v}(\pi_v)$. They correspond to characters on the group $\mathcal{S}_{\psi_v}(G)$ whose restriction to the subgroup $\mathcal{S}_{\psi_v}(M_v)$ equals the character attached to π_v. It is then not hard to see that $f'_{v,G}(\psi_v, x_v)$ has an expansion

$$\sum_{\pi_v \in \widetilde{\Pi}_{\psi_v}(M_v)} \langle \widetilde{x}_v, \widetilde{\pi}_v \rangle \operatorname{tr}\big(R'_{P_v}\big(r_v(x), \widetilde{\pi}_v, \psi_v\big)\, \mathcal{I}_{P_v}(\pi_v, f_v)\big), \tag{5.4.9}$$

where

$$R'_{P_v}(r_v, \widetilde{\pi}_v, \psi_v) = \varepsilon_{\pi_v}(r_v)\, R_{P_v}(r_v, \widetilde{\pi}_v, \psi_v), \qquad \pi_v \in \Pi_{\psi_v}(M_v),\ r_v \in R_{\psi_v},$$

for a (sign) character

$$r_v \longrightarrow \varepsilon_{\pi_v}(r_v) = \varepsilon_{\pi_v}(r_v)^{-1}$$

on the 2-group R_{ψ_v}. This is included in Shelstad's endoscopic classification of representations for real groups, as we will recall (6.1.5) in the next section. The local intertwining relation would tell us that the character ε_{π_v} is trivial for each π_v.

As it applies to the archimedean parameter ψ_v, Theorem 2.2.1(b) can thus be summarized as the existence of a natural isomorphism from the endoscopic R-group R_{ψ_v} onto any of the associated representation theoretic

R-groups $R(\pi_v)$. A triplet

$$\tau_v = \big(M_v, \pi_v, r_v(x)\big), \qquad \pi_v \in \Pi_{\psi_v}(M_v),$$

or rather its orbit under the local Weyl group $W_0^{G_v}$, can therefore be identified with an element in the basis $T(G_v)$. It follows from the definition (3.5.3) that we can write (5.4.8) and (5.4.9) respectively as

$$f_{v,G}(\psi_v, x_v) = \sum_{\pi_v \in \widetilde{\Pi}_{\psi_v}(M_v)} a(\pi_v, x_v)\, f_{v,G}\big(M_v, \pi_v, r_v(x)\big)$$

and

$$f'_{v,G}(\psi_v, x_v) = \sum_{\pi_v \in \widetilde{\Pi}_{\psi_v}(M_v)} a'(\pi_v, x_v)\, f_{v,G}\big(M_v, \pi_v, r_v(x)\big),$$

for coefficients $a(\pi_v, x_v)$ and $a'(\pi_v, x_v) = \varepsilon_{\pi_v}\big(r_v(x)\big)a(\pi_v, x_v)$. The expression (5.4.7) can therefore be written as

$$\sum_{\tau_v \in T(G_v)} d(\tau_v, f^v)\, f_{v,G}(\tau_v), \tag{5.4.10}$$

where $d(\tau_v, f^v)$ equals the sum

$$c \sum_x \sum_{\pi_v} \big(a'(\pi_v, x_v)\, (f^v)'_G(\psi^v, x^v) - a(\pi_v, x_v)\, f^v_G(\psi^v, x^v)\big)$$

over elements $x \in \mathcal{S}_{\psi,\mathrm{ell}}$ and $\pi_v \in \widetilde{\Pi}_{\psi_v}(M_v)$ such that the triplet $\big(M_v, \pi_v, r_v(x)\big)$ belongs to the $W_0^{G_v}$-orbit represented by τ_v.

Our expression (5.4.10) for (5.2.8) matches the supplementary expression (3.5.12) of Corollary 3.5.3 (with the group G_1 in (3.5.12) being the group G here). Observe also that any $x \in \mathcal{S}_{\psi,\mathrm{ell}}$ maps to a regular element in the global R-group R_ψ. It follows from the condition (5.4.1(c)) on v that the image $r_v(x)$ of x in the local R-group R_{ψ_v} in nontrivial. In other words, the coefficient $d(\tau_v, f^v)$ vanishes if τ_v is represented by a triplet of the form $(M_v, \pi_v, 1)$. We can therefore apply Corollary 3.5.3 to the formula given by Lemma 5.2.1. We find that the coefficents $d(\tau_v, f^v)$ all vanish, as do the multiplicities of any irreducible constituents of representations $R^*_{\mathrm{disc},\psi}$ in (5.2.7). It follows that the expression (5.2.8) vanishes , as required, and that the representations $R^*_{\mathrm{disc},\psi}$ are all zero. Finally, as we observed in the proof of Lemma 5.2.1, the difference

$$R^*_{\mathrm{disc},\psi}(f^*) - {}^0S_{\mathrm{disc},\psi}(f^*), \qquad G^* \in \widetilde{\mathcal{E}}_{\mathrm{sim}}(N),\ f^* \in \widetilde{\mathcal{H}}(G^*),$$

vanishes unless $G = G^*$, in which case it equals (5.2.8). Having just shown that (5.2.8) itself vanishes, we can say that the difference vanishes also in the case $G^* = G$. We conclude that

$$R^*_{\mathrm{disc},\psi}(f^*) = 0 = {}^0S^*_{\mathrm{disc},\psi}(f^*), \qquad G^* \in \widetilde{\mathcal{E}}_{\mathrm{sim}}(N),\ f^* \in \widetilde{\mathcal{H}}(G^*),$$

as required. This completes the proof of the lemma. □

LEMMA 5.4.4. *Suppose that*

$$(G,\psi), \qquad G \in \widetilde{\mathcal{E}}_{\mathrm{sim}}(N),\ \psi \in \widetilde{\mathcal{F}}^2_{\mathrm{ell}}(G),$$

is as in Lemma 5.2.2, *with* $\widetilde{\mathcal{F}}$ *again being our family of generic parameters that satisfy Assumption* 5.4.1. *Then*

$$R^*_{\mathrm{disc},\psi}(f^*) = 0 = {}^0S^*_{\mathrm{disc}}(f^*), \qquad f^* \in \widetilde{\mathcal{H}}(G^*),$$

for every $G^* \in \widetilde{\mathcal{E}}_{\mathrm{sim}}(N)$, *while the expressions* (5.2.13) *and* (5.2.12) *vanish for any functions* $f \in \widetilde{\mathcal{H}}(G)$ *and* $f^\vee \in \widetilde{\mathcal{H}}(G^\vee)$ *respectively.*

PROOF. This is the case of a parameter (5.2.4) with $r = 1$. Apart from the analysis of the new term (5.2.12), which is attached to $G^\vee$ and M, the proof is similar to that of the last lemma. For the given pair (G,ψ), we choose $v \in V(\widetilde{\mathcal{F}})$ so that the condition (5.4.1)(c) holds. We then write the expression (5.2.13) in a form (5.4.10) that matches the supplementary expression (3.5.12) of Corollary 3.5.3.

To deal with (5.2.12), we can argue as in the treatment of the parameter (4.5.11) from the proof of Proposition 4.5.1. In the simple case here, the restricted parameter $\psi_M \in \widetilde{\Psi}_2(M,\psi)$ for the Levi subgroup $M \cong GL(N)$ is just the simple component ψ_1 of ψ. We can therefore identify ψ_M with the automorphic representation π_{ψ_1} of $GL(N)$. It follows from the definitions of its two terms that (5.2.12) equals

$$\frac{1}{8}\,\mathrm{tr}\big((1 - R_{P^\vee}(w_1,\widetilde{\pi}_{\psi_1},\psi_1))\mathcal{I}_{P^\vee}(\pi_{\psi_1},f^\vee)\big). \tag{5.4.11}$$

This is essentially the analogue for $G^\vee$ of the general expression (4.5.14) obtained in the proof of Proposition 4.5.1.

As in the earlier discussion of (4.5.14), we observe that (5.4.11) is a nonnegative integral combination of irreducible characters. Its sum with the first term (5.2.11) in the formula given by Lemma 5.2.2 is therefore of the general form (3.5.1). Since the other expression (5.2.13) is of the form (3.5.12), we can apply Corollary 3.5.3 to the formula. We conclude that the expressions (5.2.13) and (5.2.12) both vanish, as required, and that the representations $R^*_{\mathrm{disc},\psi}$ are all zero. We then recall that as in the proof of Lemma 5.2.2, the difference

$$R^*_{\mathrm{disc},\psi}(f^*) - {}^0S_{\mathrm{disc},\psi}(f^*), \qquad G^* \in \widetilde{\mathcal{E}}_{\mathrm{sim}}(N), \qquad f^* \in \widetilde{\mathcal{H}}(G^*),$$

equals a positive multiple of either (5.2.13) or (5.2.12) if G^* is equal to the associated group G or $G^\vee$, and vanishes otherwise. The difference therefore vanishes in all cases, and

$$R^*_{\mathrm{disc},\psi}(f^*) = 0 = {}^0S^*_{\mathrm{disc},\psi}(f^*), \qquad G^* \in \widetilde{\mathcal{E}}_{\mathrm{sim}}(N),\ f^* \in \widetilde{\mathcal{H}}(G),$$

as required. □

LEMMA 5.4.5. *Suppose that*

$$(G,\psi), \qquad G \in \widetilde{\mathcal{E}}_{\mathrm{sim}}(N),\ \psi \in \widetilde{\mathcal{F}}_2(G),$$

is as in Lemma 5.3.1, *for our family* $\widetilde{\mathcal{F}}$ *of generic parameters with local constraints. Then the linear form* Λ *in* (5.3.4) *vanishes.*

PROOF. Recall that Λ was defined in Lemma 5.1.4 as a linear form on $\widetilde{S}^0(L)$. It represents the defect in the stable multiplicity formula for each of the pairs (G, ψ) and $(G^\vee, \psi)$. The lemma thus asserts that the stable multiplicity formula is valid for either of these pairs.

We observe also that the premise of Lemma 5.3.1 holds here. It amounts to the first vanishing assertion of Lemma 5.4.3 for parameters in the complement of $\widetilde{\mathcal{F}}_{\mathrm{ell}}(N)$ in $\widetilde{\mathcal{F}}(N)$. (It was this global theorem that was required in the proof of Lemma 5.3.1.) Therefore Lemma 5.3.1 is valid in the case at hand.

The supplementary term

$$b_+ f^{L_+}(\psi_1 \times \Lambda), \qquad L_+ = G_1 \times L,\ f \in \widetilde{\mathcal{H}}(G_+), \tag{5.4.12}$$

in (5.3.4) is clearly the main point. If this linear form were a nonnegative linear combination of characters in f, the entire expression (5.3.4) would be of the general form (3.5.1), and we could argue as in the proof of Lemma 5.4.3. According to Corollary 5.1.5, the pullback of Λ from $L(\mathbb{A})$ to $G^\vee(\mathbb{A})$ is a unitary character. However, it is not clear that the pullback of Λ to $G(\mathbb{A})$ has the same property. To sidestep this question, we introduce the composite endoscopic datum $G_1^\vee = G_1 \times G^\vee$ in $\widetilde{\mathcal{E}}_{\mathrm{ell}}(N_+)$. The group L_+, regarded itself as an endoscopic datum in $\widetilde{\mathcal{E}}(N_+)$, has a global Levi embedding into both G_+ and $G_1^\vee$. If $f_1 \in \widetilde{\mathcal{H}}(G_1^\vee)$ is the associated function in the compatible family (5.3.3), it follows from the definitions that $f_1^{L_+}$ equals f^{L_+}. The supplementary term in (5.3.4) therefore equals

$$b_+ f_1^{L_+}(\psi_1 \times \Lambda), \qquad f_1 \in \widetilde{\mathcal{H}}(G_1^\vee),\ G_1^\vee = G_1 \times G^\vee. \tag{5.4.13}$$

Since ψ_1 is assumed to be a proper subparameter of ψ, its degree N_1 is less than N. It follows from our induction hypothesis that the stable linear form on $\widetilde{\mathcal{H}}(G_1)$ attached to ψ_1 is a nonzero unitary character. Corollary 5.1.5 then tells us that $f_1^{L_+}(\psi_1 \times \Lambda)$ represents a positive multiple of a unitary character (possibly 0) in f_1. It follows that if we replace the supplementary term (5.4.12) by (5.4.13), the expression (5.3.4) will indeed be of the general form (3.5.1).

With this modification, we choose $v \in V(\widetilde{\mathcal{F}})$ so that the condition (5.4.1)(c) holds for the pair (G_+, ψ_+). Following the proof of Lemma 5.4.3, we then write the expression (5.3.5) in Lemma 5.3.1 as a sum (3.5.12) for which the required condition of Corollary 3.5.3 holds. We can therefore apply Corollary 3.5.3 to the formula provided by Lemma 5.3.1. It then follows that the coefficients in the irreducible decomposition of the modified form of (5.3.4) vanish, as does the entire expression (5.3.5). In particular, the summand (5.4.13) vanishes. Since the stable linear form on $\widetilde{\mathcal{H}}(G_1)$ attached to ψ_1 is nonzero, this can only happen if Λ vanishes, as claimed. □

LEMMA 5.4.6. *Suppose that*

$$(G, \psi), \qquad G \in \widetilde{\mathcal{E}}_{\mathrm{sim}}(N), \ \tilde{\psi} \in \widetilde{\mathcal{F}}_{\mathrm{sim}}(G),$$

is as in Lemma 5.3.2, *for our family* $\widetilde{\mathcal{F}}$ *of generic parameters with local constraints. Then the linear form* Λ *in* (5.3.21) *vanishes, while the sign* δ_ψ *in* (5.3.22) *and* (5.3.23) *equals* 1.

PROOF. We will have to draw on the techniques of all of the last three lemmas, since all of the corresponding unknown quantities occur here. We will also have to include something more to take care of the sign δ_ψ. However, the basic idea is the same. We shall apply Corollary 3.5.3 to the formula provided by Lemma 5.3.2.

We choose $v \in V(\widetilde{\mathcal{F}})$ so that (5.4.1)(c) holds for the pair (G_+, ψ_+). As in the proof of Lemma 5.4.3, we then write the expression (5.3.23) on the right hand side of the formula of Lemma 5.3.2 in the general form (3.5.12). The fact that there is now an unknown sign δ_ψ in the expression is of no consequence. For the condition on the coefficients in (3.5.12) required for Corollary 3.5.3 remains a consequence of our given condition (4.5.1)(c).

The left hand side of the formula is the sum of two expressions (5.3.21) and (5.3.22). The first of these contains the supplementary summand

$$\tag{5.4.14} \tfrac{1}{2} f^{L_+}(\Gamma \times \Lambda), \qquad L_+ = G \times L, \ f \in \widetilde{\mathcal{H}}(G_+).$$

Following the proof of the last lemma, we introduce the composite endoscopic datum $G_1^\vee = G \times G^\vee$ in $\widetilde{\mathcal{E}}_{\mathrm{ell}}(N_+)$, with a corresponding function $f_1 \in \widetilde{\mathcal{H}}(G_1^\vee)$ in the compatible family (5.3.3). Since L_+ again has a global Levi embedding into both G_+ and $G_1^\vee$, the supplementary summand (5.4.14) equals

$$\tag{5.4.15} \tfrac{1}{2} f_1^{L_+}(\Gamma \times \Lambda), \qquad f_1 \in \widetilde{\mathcal{H}}(G_1^\vee), \ G_1^\vee = G \times G^\vee.$$

It then follows from Corollary 5.1.5 and the definition of Γ that $f_1^{L_+}(\Gamma \times \Lambda)$ represents a positive multiple of a unitary character (possibly 0) in f_1. The expression (5.3.21), written with (5.4.15) in place of (5.4.14), is consequently of the general form (3.5.1). As for the second expression (5.3.22), we note that the parameter ψ_{+,M_+} for $M_+ \cong GL(N)$ can be identified with the automorphic representation π_ψ of $GL(N)$. Appealing to the definitions, as in the proof of Lemma 5.4.4, we can write the expression in the form

$$\tag{5.4.16} \tfrac{1}{8}\,\mathrm{tr}\big((1 - \delta_\psi\, R_{P_+^\vee}(w^\vee, \tilde{\pi}_\psi, \psi))\, \mathcal{I}_{P_+^\vee}(\pi_\psi, f^\vee)\big), \qquad f^\vee \in \widetilde{\mathcal{H}}(G_+^\vee).$$

Since δ_ψ equals either $(+1)$ or (-1), (5.4.14) is a nonnegative linear combination of characters on the group $G_+^\vee(\mathbb{A})$. It follows that the sum of (5.3.21) and (5.3.22) can be written as an expression of the general form (3.5.1).

We can thus apply Corollary 3.5.3 to the formula given by Lemma 5.3.2. We then see that, among other things, the expressions (5.4.14) and (5.4.16) each vanish.

From the vanishing of (5.4.14), we deduce that either Γ or Λ equals zero. Suppose that Γ is the factor that vanishes. The linear form

$$f^G(\psi) = f^G(\Gamma) + f^L(\Lambda) = f^L(\Lambda), \qquad f \in \widetilde{\mathcal{H}}(G),$$

on $\widetilde{\mathcal{H}}(G)$ is then induced from the Levi subgroup $L(\mathbb{A})$. This contradicts the local condition (5.4.1(a)) for the simple parameter ψ. It is therefore Λ that vanishes, as claimed.

The operator

$$\delta_\psi \, R_{P_+^\vee}(w^\vee, \widetilde{\pi}_\psi, \psi)$$

in (5.4.16) is unitary, and commutes with $\mathcal{I}_{P_+^\vee}(\pi_\psi, f^\vee)$. The vanishing of (5.4.16) then implies that this operator equals 1. In particular, our normalized intertwining operator here is a scalar, as we expect from the fact that the centralizer group $S_{\psi_+}^\vee$ is isomorphic to $SL(2,\mathbb{C})$. To show that it actually equals 1, we appeal to the relation between normalized intertwining operators and Whittaker models. Since the parameter ψ is generic, case (iii) of Corollary 2.5.2 tells us that

$$R_{P_+^\vee}(w^\vee, \pi_{\psi,v}, \psi_v) = 1$$

for any v. Therefore

$$R_{P_+^\vee}(w^\vee, \pi_\psi, \psi) = \prod_v R_{P_+^\vee}(w^\vee, \pi_{\psi,v}, \psi_v) = 1.$$

We conclude that the sign δ_ψ equals 1, and therefore that the second assertion of the lemma also holds. □

In the last lemma, N is even and η_ψ equals 1. If one of these conditions is not satisfied, the formula of Lemma 5.3.2 becomes simpler, as we noted at the end of the last section. In particular, Λ is not defined, and does not contribute a summand to (5.3.21). If N is odd, the summands in (5.3.23) are also not defined, and the term (5.3.23) does not exist. However, the formula of Lemma 5.3.2 is otherwise valid, and the sign δ_ψ is invariably present in the term (5.3.22). The simplified analogue Lemma 5.4.6 asserts that $\delta_\psi = 1$. It is proved as above.

We can now clarify the temporary definition of the sets $\widetilde{\mathcal{F}}_{\mathrm{sim}}(G)$ adopted in §5.1. We shall show that the three possible ways to characterize $\widetilde{\mathcal{F}}_{\mathrm{sim}}(G)$ are all equivalent.

COROLLARY 5.4.7. *Suppose that ψ belongs to the subset $\widetilde{\mathcal{F}}_{\mathrm{sim}}(N)$ of our family $\widetilde{\mathcal{F}}$ of generic parameters with local constraints, and that G belongs to $\widetilde{\mathcal{E}}_{\mathrm{sim}}(N)$. Then the following conditions on (G,ψ) are equivalent.*

(i) *The linear form $S^G_{\mathrm{disc},\psi}$ on $\widetilde{\mathcal{H}}(G)$ does not vanish.*

(ii) *Theorem 1.4.1 holds for ψ, with $G_\psi = G$.*

(iii) *The global quadratic characters η_G and η_ψ are equal, and the global L-function condition $\delta_\psi = 1$ holds.*

PROOF. The first assertion $\Lambda = 0$ of Lemma 5.4.6 tells us that condition (i) characterizes G uniquely in terms of ψ. If G^* is any element in $\widetilde{\mathcal{E}}_{\mathrm{sim}}(N)$, we know that

$$S^*_{\mathrm{disc},\psi}(f^*) = \mathrm{tr}\big(R^*_{\mathrm{disc},\psi^*}(f^*)\big), \qquad f^* \in \widetilde{\mathcal{H}}(G^*),$$

since the proper terms in the expansions (4.1.1) and (4.1.2) all vanish by our induction hypotheses. It follows that the condition of Theorem 1.4.1 characterizes the postulated datum G_ψ uniquely, and that $G_\psi = G$. The conditions (i) and (ii) are therefore equivalent.

The third assertion $\delta_\psi = 1$ of Corollary 5.4.7 (together with (3.4.7), and the remark above in case N is odd or $\eta_\psi \neq 1$) tells us that (i) implies (iii). But we know that for the given ψ, there is at most one G that satisfies (iii). The conditions (i) and (iii) are therefore also equivalent. □

The condition (i) was the basis for our temporary definition in §5.1. It has been the pivotal condition for our arguments in this chapter. The condition (ii) was used for the original definition from Chapter 1. Knowing now that the three conditions are equivalent for the family $\widetilde{\mathcal{F}}$ of this section, we can define $\widetilde{\mathcal{F}}_{\mathrm{sim}}(G)$ to be the set of $\psi \in \widetilde{\mathcal{F}}_{\mathrm{sim}}(N)$ such that any one of the conditions holds for (G, ψ). It is clear from the discussion that $\widetilde{\mathcal{F}}_{\mathrm{sim}}(N)$ is a disjoint union over $G \in \widetilde{\mathcal{E}}_{\mathrm{sim}}(N)$ of the sets $\widetilde{\mathcal{F}}_{\mathrm{sim}}(G)$.

We shall add a comment to what we have established in this section. The preceding Lemmas 5.4.3, 5.4.4, 5.4.5 and 5.4.6 apply to the respective formulas of Lemmas 5.2.1, 5.2.2, 5.3.1 and 5.3.2. We recall that the four formulas became increasingly complex as each one took on another unknown quantity. The four unknowns were the global intertwining relation for G (or G_+), the global intertwining relation for $G^\vee$ (or $G^\vee_+$), the linear form Λ, and the sign δ_ψ. By the time we reached the last Lemma 5.3.2, all four of these unknowns were present as terms in the associated formula. Their resolution, under the local constraints of this section, rests exclusively on Corollary 3.5.3 of Proposition 3.5.1.

The basic principle behind Proposition 3.5.1 can be described as follows: if a linear combination of reducible characters vanishes, *and* if the coefficients are all positive, then the characters are themselves equal to zero. This is an obvious simplification, but it serves to illustrate the main idea for the proofs of the last four lemmas. The critical point is that the unknown quantities, or rather the ones that do not go into the supplementary expression (3.5.12) of Corollary 3.5.3, must all occur with the appropriate signs. They do – most fortunately for us! This phenomenon seems quite remarkable. We have seen it before, in the treatment of the parameters (4.5.11) and (4.5.12) of Proposition 4.5.1 and in Lemma 5.1.6, and we will see it again, most vividly perhaps in our completion of the global classification in §8.2. I do not have an explanation for it.

CHAPTER 6

The Local Classification

6.1. Local parameters

The next two chapters will be devoted to the proof of the local theorems. Our methods will be global. Given the appropriate local data, we shall construct a family of global parameters to which we can apply the results of Chapter 5.

Since we are aiming for local results, we take F to be a local field in Chapters 6 and 7. We will then use the symbol $\dot{F}$ to denote the auxiliary global field over which the global arguments take place. This follows a general convention of representing global objects by placing a dot over symbols that stand for the given local objects.

To obtain a general idea of the strategy, consider a local parameter

$$\psi = \ell_1\psi_1 \oplus \cdots \oplus \ell_r\psi_r, \qquad \psi_i \in \widetilde{\Psi}_{\mathrm{sim}}(G_i),\ G_i \in \widetilde{\mathcal{E}}_{\mathrm{sim}}(N_i),$$

where the sets $\widetilde{\Psi}_{\mathrm{sim}}(G_i)$ and $\widetilde{\mathcal{E}}_{\mathrm{sim}}(N_i)$ are understood to be over the local field F. We are proposing to construct suitable endoscopic data $\dot{G}_i \in \widetilde{\mathcal{E}}_{\mathrm{sim}}(N_i)$ and parameters $\dot{\psi}_i \in \widetilde{\Psi}_{\mathrm{sim}}(\dot{G}_i)$ over a global field $\dot{F}$ such that

$$(F, G_i, \psi_i) = (\dot{F}_u, \dot{G}_{i,u}, \dot{\psi}_{i,u}), \qquad 1 \leqslant i \leqslant r,$$

at some place u of $\dot{F}$. This will lead to a family,

$$\dot{\widetilde{\mathcal{F}}} = \widetilde{\mathcal{F}}(\dot{\psi}_i, \ldots, \dot{\psi}_r) = \{\dot{\ell}_1\dot{\psi}_1 \oplus \cdots \oplus \dot{\ell}_r\dot{\psi}_r :\ \dot{\ell}_i \geqslant 0\},$$

of global parameters to which we can try to apply the methods of Chapter 5. In this chapter we will treat the case that $\psi = \phi$ is generic. We will show that the resulting global family $\dot{\widetilde{\mathcal{F}}}$ has local constraints (5.4.1), where $V = V(\dot{\widetilde{\mathcal{F}}})$ is a set of archimedean places of $\dot{F}$. In the next chapter, we will deal with general parameters. There we will establish local constraints that are roughly parallel to (5.4.1), but with $V = V(\dot{\widetilde{\mathcal{F}}})$ being a set of p-adic places.

We shall prepare for this analysis with some remarks on local Langlands parameters. We shall state them in the setting of a given group $G \in \widetilde{\mathcal{E}}_{\mathrm{sim}}(N)$, but we will apply them presently to the localizations $\dot{\phi}_v$ of global parameters $\dot{\phi} \in \dot{\widetilde{\mathcal{F}}}$. They fall into three categories. We will have separate observations for the completions $\dot{F}_v$ at $\dot{F}$ at $v = u$, at nonarchimedean places $v \neq u$, or at archimedean places $v \neq u$.

The first remarks are aimed at general Langlands parameters over $F = \dot{F}_u$. They pertain thus to the objects for which we are trying to establish the local theorems. Since the main questions are nonarchimedean, we assume for the time being that F is p-adic.

Our purpose is to introduce a temporary modification of one of our definitions. It will be a local p-adic analogue of the temporary global definition we adopted after Assumption 5.1.1, and later resolved in terms of the original definition in Corollary 5.4.7. At issue is the definition of the local parameter sets $\widetilde{\Phi}(G)$.

We begin by stating one more theorem, which can be regarded as a local analogue of Theorem 1.4.1. This last of our collection of stated theorems amounts to a special case of Theorem 2.2.1, just as the initial Theorem 1.4.1 was a special case of Theorem 1.5.2. We will resolve it, along with the generic case of all the other local theorems, in §6.8.

THEOREM 6.1.1. *Suppose that $\phi \in \widetilde{\Phi}_{\mathrm{sim}}(N)$ is a simple generic local parameter. Then there is a unique $G_\phi \in \widetilde{\mathcal{E}}_{\mathrm{ell}}(N)$ such that*

$$\widetilde{f}_N(\phi) = \widetilde{f}^{G_\phi}(\phi), \qquad \widetilde{f} \in \widetilde{\mathcal{H}}(N),$$

for a linear form $\widetilde{f}^{G_\phi}(\phi)$ on $\widetilde{\mathcal{S}}(G_\phi)$. Moreover, G_ϕ is simple.

We consider now the definition of the set $\widetilde{\Phi}(G)$ attached to a given simple datum $G \in \widetilde{\mathcal{E}}_{\mathrm{sim}}(N)$. As in the global case, there will be three equivalent ways to characterize the subset $\widetilde{\Phi}_{\mathrm{sim}}(G)$ of simple local parameters in $\widetilde{\Phi}_{\mathrm{sim}}(N)$. Our original definition gives $\widetilde{\Phi}_{\mathrm{sim}}(G)$ as the subset of irreducible N-dimensional representations of L_F that factor through the image of ${}^LG$ in $GL(N,\mathbb{C})$. The second characterization is in terms of poles of local L-functions, and will be postponed until §6.8. The third is provided by Theorem 6.1.1. It characterizes $\widetilde{\Phi}_{\mathrm{sim}}(G)$ as the subset of local parameters $\phi \in \widetilde{\Phi}_{\mathrm{sim}}(N)$ for which the datum G_ϕ of the theorem equals G. The first characterization is the simplest. However, the local Langlands parameters for G are quite removed from the local harmonic analysis that governs the classification of Theorem 2.2.1. The third is more technical, but is formulated purely in terms of local harmonic analysis. For this reason, we shall temporarily replace the original definition by one that is based on Theorem 6.1.1. We shall make it parallel to the temporary global definition from the beginning of §5.1.

Let $\widetilde{\mathcal{H}}_{\mathrm{cusp}}(N)$ be the subspace of functions $\widetilde{f} \in \widetilde{\mathcal{H}}(N)$ such that for each $G \in \widetilde{\mathcal{E}}_{\mathrm{ell}}(N)$, $\widetilde{f}^G$ lies in the subspace $\widetilde{\mathcal{S}}_{\mathrm{cusp}}(G)$ of $\widetilde{\mathcal{S}}(G)$. It follows from Proposition 2.1.1 that the mapping

$$\widetilde{f} \longrightarrow \bigoplus_{G \in \widetilde{\mathcal{E}}_{\mathrm{ell}}(N)} \widetilde{f}^G, \qquad \widetilde{f} \in \widetilde{\mathcal{H}}_{\mathrm{cusp}}(N),$$

descends to an isomorphism from the image $\widetilde{\mathcal{I}}_{\mathrm{cusp}}(N)$ of $\widetilde{\mathcal{H}}_{\mathrm{cusp}}(N)$ in $\widetilde{\mathcal{I}}(N)$ onto the direct sum over $G \in \widetilde{\mathcal{E}}_{\mathrm{ell}}(N)$ of the spaces $\widetilde{\mathcal{S}}_{\mathrm{cusp}}(G)$. If ϕ is any element in $\widetilde{\Phi}(N)$, we can then write

$$\widetilde{f}_N(\phi) = \sum_{G \in \widetilde{\mathcal{E}}_{\mathrm{ell}}(N)} \widetilde{f}^G(\phi), \qquad \widetilde{f} \in \widetilde{\mathcal{H}}_{\mathrm{cusp}}(N), \tag{6.1.1}$$

for uniquely determined linear forms $\widetilde{f}^G(\phi)$ on each of the spaces $\widetilde{\mathcal{S}}_{\mathrm{cusp}}(G)$.

The formula (6.1.1) is a rough local analogue of the global decomposition (3.3.14). It tells us that we can attach local subsets,

$$\widetilde{\Phi}_{\mathrm{sim}}(G) \subset \{\phi \in \widetilde{\Phi}_{\mathrm{sim}}(N) : \ f^G(\phi) \neq 0 \text{ for some } f \in \widetilde{\mathcal{H}}_{\mathrm{cusp}}(G)\}, \tag{6.1.2}$$

of $\widetilde{\Phi}_{\mathrm{sim}}(N)$ to the simple endoscopic data $G \in \widetilde{\mathcal{E}}_{\mathrm{sim}}(N)$ so that

$$\widetilde{\Phi}_{\mathrm{sim}}(N) = \bigcup_{G \in \widetilde{\mathcal{E}}_{\mathrm{sim}}(N)} \widetilde{\Phi}_{\mathrm{sim}}(G). \tag{6.1.3}$$

To be definite, and in minor contrast to the global convention of §5.1, we define $\widetilde{\Phi}_{\mathrm{sim}}(G)$ by taking the inclusion in (6.1.2) to be equality. That is, we define $\widetilde{\Phi}_{\mathrm{sim}}(G)$ to be the set of *all* parameters $\phi \in \widetilde{\Phi}_{\mathrm{sim}}(N)$ such that the linear form $f^G(\phi)$ on $\widetilde{\mathcal{H}}_{\mathrm{cusp}}(G)$ does not vanish, as on the right hand side of (6.1.2). This temporary definition will be in force throughout Chapter 6. During this period, the local theorems will be interpreted accordingly. For example, Theorem 6.1.1 will now include the assertion that the union (6.1.3) is actually disjoint. If we assume inductively that the local theorems (including Theorem 6.1.1) are all valid for parameters in $\widetilde{\Psi}(N_1)$ with $N_1 < N$, we can define the full set $\widetilde{\Phi}(G)$ in terms of the fundamental sets $\widetilde{\Phi}_{\mathrm{sim}}(G_1)$ with $G_1 \in \widetilde{\mathcal{E}}_{\mathrm{sim}}(N_1)$, exactly as we did for the global case in §1.4. This in turn allows us to describe the centralizers S_ϕ by the global prescription of §1.4 and the convention (5.1.7), and then to introduce the subsets

$$\widetilde{\Phi}_{\mathrm{sim}}(G) \subset \widetilde{\Phi}_2(G) \subset \widetilde{\Phi}_{\mathrm{ell}}(G) \subset \widetilde{\Phi}(G),$$

according to the later global definitions of §4.1.

We will thus temporarily forget about general Langlands parameters for the p-adic group G (though we will still be able to treat elements in $\widetilde{\Phi}(N)$ as representations of L_F). We will come back to them in §6.8, after proving the local theorems in their interpretation above. We will then be able to resolve the temporary definition of $\widetilde{\Phi}(G)$ in terms of the original one. This will yield the general local theorems for generic ϕ in their original form.

The next remarks are very elementary. They concern special elements in $\widetilde{\Phi}(G)$, which will still be identified with local Langlands parameters, and which will apply to localizations $\dot{\phi}_v$ at p-adic places v of $\dot{F}$ distinct from u. The parameters we have in mind are *spherical*, in the sense that they are attached to the p-adic spherical functions of [**Ma**]. In particular, they are unramified if the given group $G \in \widetilde{\mathcal{E}}_{\mathrm{sim}}(N)$ over F is unramified.

The group G is quasisplit over F. It is unramified, by definition, if it splits over an unramified extension of F. This is the case unless N is even, $\widehat{G} = SO(N, \mathbb{C})$, and the quadratic character η_G is ramified. The group itself is split over F unless N is even, $\widehat{G} = SO(N, \mathbb{C})$ and $\eta_G \neq 1$. Let K be our fixed maximal compact subgroup of $G(F)$. Then K is special in general, and hyperspecial if G is unramified. An irreducible representation of $G(F)$ is said to be *K-spherical* (or simply spherical) if its restriction to K contains the trivial representation. A local Langlands parameter ϕ will be called *spherical* if its L-packet $\widetilde{\Pi}_\phi$ contains a K-spherical representation. We recall the description of these familiar objects.

Let M_0 be the standard minimal Levi subgroup of G. Then M_0 is a maximally split, maximal torus over F, whose dual $\widehat{M}_0$ is the group of diagonal matrices in $\widehat{G}$. Spherical parameters factor through the image of ${}^L M_0$ in ${}^L G$, and in particular descend to the quotient W_F of L_F. Let

$$\phi_0 : \, W_F \longrightarrow {}^L M_0 = \widehat{M}_0 \rtimes W_F$$

be the canonical splitting of ${}^L M_0$, which is to say the identity mapping of W_F to the second factor of ${}^L M_0$. If λ is a point in the complex vector space

$$\mathfrak{a}^*_{M_0,\mathbb{C}} = X^*(M_0)_F \otimes \mathbb{C},$$

we write $|w|^\lambda$ for the point in the complex torus

$$A_{\widehat{M}_0} = (\widehat{M}_0^\Gamma)^0 = X^*(M_0)_F \otimes \mathbb{C}^*$$

attached to λ and the absolute value $|w|$ of an element $w \in W_F$. The mapping

$$\phi_\lambda(w) = \phi_0(w)\,|w|^\lambda, \qquad w \in W_F,$$

is then a Langlands parameter for M_0, whose image in $\Phi(G)$ we continue to denote by ϕ_λ. The elements in the family

$$\{\phi_\lambda : \, \lambda \in \mathfrak{a}^*_{M_0,\mathbb{C}}\}$$

are the spherical parameters for G.

Under the Langlands correspondence for the torus M_0, the spherical parameter $\phi_\lambda \in \Phi(M_0)$ maps to the spherical character

$$\pi_\lambda(t) = e^{\lambda(H_{M_0}(t))}, \qquad t \in M_0(F),$$

on $M_0(F)$. If λ is purely imaginary, π_λ is unitary and ϕ_λ has bounded image in $\widehat{M}_0$. Treating ϕ_λ as an element in $\Phi(G)$, we write Π_{ϕ_λ} in this case for the set of irreducible constituents of the induced representation $\mathcal{I}_{P_0}(\pi_\lambda)$, where $P_0 \in \mathcal{P}(M_0)$ is the standard Borel subgroup of G. This is the L-packet of ϕ_λ. For any λ, it contains a unique K-spherical representation.

Our final remarks concern archimedean parameters. These will be applied to localizations $\dot{\phi}_v$ at the archimedean places v of $\dot{F}$ distinct from u. The complex case presents little difficulty, and we will in any case be choosing the global field $\dot{F}$ in §6.2 to be totally real. We therefore assume for the rest of this section that $F = \mathbb{R}$, with $G \in \widetilde{\mathcal{E}}_{\mathrm{sim}}(N)$ continuing to be a simple endoscopic datum over F. We shall review a few points in Shelstad's general classification of the representations of $G(F)$ in terms of local Langlands parameters $\phi \in \widetilde{\Phi}(G)$.

The archimedean valuations $v \neq u$ of $\dot{F}$ will in fact play a central role in the global arguments of this chapter. To prepare ourselves, it will be helpful to review the archimedean parameters $\phi \in \widetilde{\Phi}_2(G)$ attached to discrete series representations of $G(F)$. This amounts to an explicit specialization of some of the observations in §1.2.

The local Langlands group for F is just the real Weil group $W_F = W_{\mathbb{R}}$. It is generated by the group $\mathbb{C}^*$ and an element σ_F, with relations $\sigma_F^2 = -1$ and

$$\sigma_F \, z \, \sigma_F^{-1} = \overline{z}, \qquad z \in \mathbb{C}^*.$$

A parameter

$$\phi : \; W_F \to {}^L G,$$

when composed with our mapping of ${}^L G$ into $GL(N, \mathbb{C})$, becomes a self-dual, N-dimensional representation of W_F. We are interested in the case that ϕ lies in $\widetilde{\Phi}_2(G)$, which is to say that the group

$$S_\phi = S_\phi(G) = \mathrm{Cent}\big(\phi(W_F), \widehat{G}\big)$$

is finite. As we noted in the general discussion of §1.2, this is equivalent to the condition that the irreducible constituents of the representation of W_F attached to ϕ are all distinct and self-dual.

Recall [**T2**] that the irreducible representations of W_F are all of one or two dimensions. There are two self-dual one-dimensional representations, the trivial representation and the sign character of $\mathbb{R}^* \cong (W_F)^{\mathrm{ab}}$. The irreducible self-dual representations of W_F of dimension two are parametrized by positive half integers $\mu \in \frac{1}{2}\mathbb{N}$, and are given by

$$z \longrightarrow \begin{pmatrix} (z\overline{z}^{-1})^{\mu} & 0 \\ 0 & (z\overline{z}^{-1})^{-\mu} \end{pmatrix}, \qquad z \in \mathbb{C}^*,$$

and

$$\sigma_F \longrightarrow \begin{pmatrix} 0 & 1 \\ (-1)^{2\mu} & 0 \end{pmatrix}.$$

The image of any such representation lies in $O(2, \mathbb{C})$ or $Sp(2, \mathbb{C})$, according to whether μ is an integer or not. To construct all the parameters in $\Phi_2(G)$, we simply select families of distinct irreducible representations of W_F that are of the same type (either symplectic or orthogonal) as $\widehat{G}$.

To be more precise, let $p = [N/2]$ be the greatest integer in $N/2$. If $\widehat{G} = Sp(N, \mathbb{C})$, any $\phi \in \widetilde{\Phi}_2(G)$ is a direct sum of p distinct two-dimensional representations of symplectic type. It can therefore be identified with a set

$$\mu_\phi = (\mu_1, \ldots, \mu_p), \qquad \mu_i \in \tfrac{1}{2}\mathbb{N} - \mathbb{N},$$

of p distinct, positive, *proper* half integers. If $\widehat{G} = SO(N, \mathbb{C})$, with N odd, a parameter $\phi \in \widetilde{\Phi}_2(G)$ is a direct sum of p distinct two-dimensional representations of orthogonal type and a uniquely determined one-dimensional representation. It can therefore be identified with a set

$$\mu_\phi = (\mu_1, \ldots, \mu_p), \qquad \mu_i \in \mathbb{N},$$

of p distinct positive integers. The one-dimensional representation is trivial if p is even, and is the sign character if p is odd. If $\widehat{G} = SO(N, \mathbb{C})$, with N even, we of course want the set $\widetilde{\Phi}_2(G)$ to be nonempty. We therefore assume that G is split if p is even, and nonsplit if p is odd. There remain two possibilities in this case, which we characterize by setting p' equal either to p or $p-1$. A parameter $\phi \in \widetilde{\Phi}_2(G)$ can then be identified with a set

$$\mu_\phi = (\mu_1, \dots, \mu_{p'}), \qquad \mu_i \in \mathbb{N},$$

of p' distinct positive integers. It corresponds to a direct sum of p' distinct two-dimensional representations of orthogonal type, augmented by the sum of the trivial and sign characters in case $p' = p - 1$.

The notation μ_ϕ was used before. In the proof of Lemma 2.2.2, it was used for a vector in $\left(\frac{1}{2}\mathbb{Z}\right)^N$ that represented the infinitesimal character of ϕ as an element in $\widetilde{\Phi}_{\mathrm{ell}}(N)$. In this section, μ_ϕ is a vector in $\left(\frac{1}{2}\mathbb{N}\right)^{p'}$ (where $p' = p$ in the first two cases above) attached to $\phi \in \widetilde{\Phi}_2(G)$. Having long ago agreed to identify $\widetilde{\Phi}_2(G)$ with a subset of $\widetilde{\Phi}_{\mathrm{ell}}(N)$, we must now be prepared to identify μ_ϕ with the vector

$$\mu_\phi \oplus (-\mu_\phi) \in \left(\tfrac{1}{2}\mathbb{Z}\right)^{2p'} \subset \left(\tfrac{1}{2}\mathbb{Z}\right)^N$$

in $\left(\frac{1}{2}\mathbb{Z}\right)^N$. In any case, μ_ϕ in this section represents the infinitesimal character of ϕ as an element in $\widetilde{\Phi}_2(G)$. We shall say that ϕ is in *general position* if its infinitesimal character is highly regular, in the sense that the positive half-integers

$$\{\mu_i,\ |\mu_i - \mu_j| :\ 1 \leqslant i \neq j \leqslant p'\} \tag{6.1.4}$$

are all large.

We should also note that if $\widehat{G} = SO(N, \mathbb{C})$, for N even *or* odd, $G(F)$ has a central subgroup of order 2. In this case, we can speak of the *central character* of ϕ on this subgroup, with the understanding that it refers to the representations in the corresponding L-packet $\widetilde{\Pi}_\phi$. One shows that it is trivial if and only if the integer

$$c_\phi = \sum_i \mu_i + \frac{p'(p'+1)}{2}$$

is *even*.

It would not have been difficult to describe more general Langlands parameters in explicit terms. (See [**L11**, §3] for the case of a general group over $\mathbb{R}$.) We have treated the square integrable (archimedean) parameters $\phi \in \widetilde{\Phi}_2(G)$ because they are the objects we will use in the next section to construct automorphic representations. We recall in any case that Theorem 2.2.1 is valid for general archimedean parameters $\phi \in \widetilde{\Phi}(G)$ (assuming the two special cases for the twisted groups $\widetilde{G}(N)$ and $\widetilde{G}$ of the work in progress by Mezo and Shelstad, which would also give Theorem 2.2.4).

The results for generic archimedean parameters (including their projected analogues for $\widetilde{G}(N)$ and $\widetilde{G}$) do not include the local intertwining relation of Theorem 2.4.1 (or of Theorem 2.4.4). They do, however, implicitly contain a weaker form of the identity. It is a consequence of Shelstad's proof of the generic case

$$f'_G(\phi, x) = f'(\phi') = \sum_{\pi \in \widetilde{\Pi}_\phi} \langle x, \pi \rangle f_G(\pi), \qquad \phi \in \widetilde{\Phi}_{\mathrm{bdd}}(G),\ x \in \mathcal{S}_\phi,$$

of the formula (2.2.6) of Theorem 2.2.1(b), specifically the fact that the pairing $\langle x, \pi \rangle$ in this formula is compatible with the short exact sequence (2.4.9). In particular, if $\phi_M \in \Phi_2(M, \phi)$ is a square integrable parameter that maps to ϕ, Π_ϕ is a disjoint union over $\pi_M \in \Pi_{\phi_M}$ of sets $\Pi_{\pi_M}(G)$ on which the dual group $\widehat{R}_\phi$ of R_ϕ acts simply transitively. As the irreducible constituents of induced representations $\mathcal{I}_P(\pi_M)$, these sets are in turn compatible with the self-intertwining operators $R_P(w, \widetilde{\pi}_M, \phi_M)$, in the sense that follows from the work of Harish-Chandra [**Ha4**, Theorem 38.1] and Knapp and Stein [**KnS**, Theorem 13.4]. (See [**A10**, (2.3)], for example, and the general review in Chapter 3 prior to Lemma 3.5.2.) It then follows from [**A10**, (2.3)], the formula above for $f'_G(\phi, x)$, and the definition (2.4.5) of $f_G(\phi, x)$ (which includes the definition of the factor $\langle \widetilde{u}, \widetilde{\pi} \rangle$ on the right hand side of (2.4.5)) that

$$f_G(\phi, x) = \sum_{\pi \in \widetilde{\Pi}_\phi} \varepsilon_{\pi_M}(x) \langle x, \pi \rangle f_G(\pi), \qquad x \in \mathcal{S}_\phi, \tag{6.1.5}$$

where π_M is the projection of π onto $\widetilde{\Pi}_{\phi_M}$, and ε_{π_M} is the pullback to $\mathcal{S}_\phi$ of a character on R_ϕ. The as yet unknown characters $\varepsilon_{\pi_M}(x)$ of course are what weaken (6.1.5). They free the identity from its dependence on the finer normalization of intertwining operators in §2.3–§2.4, and the particular pairing $\langle x, \pi \rangle$.

Shelstad's proof of the formula for $f'_G(\phi, x)$ is contained in [**S3**] and [**S6**]. It follows from her proof of [**S3**, Corollary 5.3.16 and Theorem 5.4.27], the discussion of R-groups in [**S3**, §5.3], and her conversion of the spectral transfer factors of [**S5**] to a pairing $\langle x, \pi \rangle$ represented by the specialization of [**S6**, Theorem 7.5] to connected quasisplit groups. It is clear that the resulting identity (6.1.5) is indeed a weaker version of the local intertwining identity. It reduces Theorem 2.4.1, in the case of archimedean F and generic $\phi = \psi$, to a question of base points for the simply transitive action of $\widehat{R}_\psi$ on each of the sets $\widetilde{\Pi}_{\pi_M}(G)$. We recall, however, that Shelstad has also shown that if π is the unique (B, χ)-generic representation in $\widetilde{\Pi}_\phi$, then $\langle \cdot, \pi \rangle = 1$ [**S6**, Theorem 11.5]. It then follows from Theorem 2.5.1(b) that $\varepsilon_{\pi_M}(\cdot) = 1$ in this case. The formula (6.1.5) will be our starting point for the general proof of Theorem 2.4.1.

Recall that the mapping

$$\widetilde{\Pi}_\phi \hookrightarrow \widehat{\mathcal{S}}_\phi$$

for G is generally a proper injection. In using the archimedean parameters with our global arguments, we will need the following lemma.

LEMMA 6.1.2. *Suppose that $G \in \widetilde{\mathcal{E}}_{\mathrm{sim}}(N)$ is a simple endoscopic datum over our archimedean field F, and that $\phi \in \widetilde{\Phi}(G)$ is a corresponding generic parameter. Then the image of $\widetilde{\Pi}_\phi$ in $\widehat{\mathcal{S}}_\phi$ generates $\widehat{\mathcal{S}}_\phi$.*

PROOF. We will prove the lemma with G being any connected quasisplit group over F. We will also take the opportunity to recall the initial foundations of real endoscopy, on which the results of this chapter ultimately depend.

We first choose a parameter $\phi_M \in \Phi_2(M, \phi)$, for some Levi subgroup M of G. As we noted above, Π_ϕ is a disjoint union over $\pi_M \in \Pi_{\phi_M}$ of sets $\Pi_{\pi_M}(G)$ on which the group $\widehat{R}_\phi$ acts simply transitively. The set of characters $\langle \cdot, \pi \rangle$ on $\mathcal{S}_\phi$ attached to representations $\pi \in \Pi_{\pi_M}(G)$ is thus a coset $\xi \widehat{R}_\phi$, where ξ is the character $\langle \cdot, \pi_M \rangle$ on $\mathcal{S}_{\phi_M}$. From the short exact sequence (2.4.9) for G, it is clear that we need only show that the injective image of Π_{ϕ_M} generates $\widehat{\mathcal{S}}_{\phi_M}$. We may therefore assume that $M = G$, or in other words, that ϕ lies in $\Phi_2(G)$.

Since $\Phi_2(G)$ is assumed not to be empty, G has a discrete series, and therefore a maximal torus T over F that is anisotropic modulo $Z(G)$. As in the examples for G described explicitly above, ϕ factors through the image of ${}^LT$ under an admissible L-embedding into ${}^LG$. Moreover, as a general consequence of the fact that ϕ belongs to $\Phi_2(G)$, the centralizer of its image in $\widehat{G}$ is contained in $\widehat{T}$. This implies that $S_\phi = \widehat{T}^\Gamma$ and $\mathcal{S}_\phi = \widehat{T}^\Gamma / Z(\widehat{G})^\Gamma$. It is then not hard to see from the definitions that there is a canonical isomorphism from the dual $\widehat{\mathcal{S}}_\phi$ of $\mathcal{S}_\phi$ onto the finite 2-group

$$\Sigma^\vee(G,T)/\Sigma^\vee(G,T) \cap \{\mu - \sigma\mu : \ \mu \in X_*(T), \ \sigma \in \Gamma_F\}, \tag{6.1.6}$$

where $\Sigma^\vee(G,T)$ is the set of co-roots for (G,T). (This assertion relies on the fact that $T/Z(G)$ is anisotropic, and hence that the norm of any element in $\Sigma^\vee(G,T)$ vanishes.)

Following Langlands and Shelstad, we write

$$\mathcal{D}(T) = W(G,T)/W\big(G(\mathbb{R}), T(\mathbb{R})\big),$$

where $W\big(G(\mathbb{R}), T(\mathbb{R})\big)$ is the subgroup of the full Weyl group induced by elements in $G(\mathbb{R})$. This set is bijective with Π_ϕ. More precisely, one sees from Harish-Chandra's classification of discrete series, and the resulting natural grouping of these objects into packets according to their infinitesimal characters, that if a base point $\pi_0 \in \Pi_\phi$ is fixed, there is a *canonical* bijection

$$\pi \longrightarrow \mathrm{inv}(\pi_0, \pi), \qquad \pi \in \Pi_\phi,$$

from Π_ϕ to $\mathcal{D}(T)$. We also write

$$\mathcal{E}(T) = \mathrm{im}\big(H^1(F, T_{\mathrm{sc}}) \longrightarrow H^1(F,T)\big),$$

where $T_{\rm sc}$ is the preimage of T in the simply connected cover of the derived group of G. This group is canonically isomorphic with $\widehat{\mathcal{S}}_\phi$. For by the remarks in [**K5**, §1.1] based on Tate-Nakayama duality, one can write

$$\begin{aligned}\mathcal{E}(T) &\cong \mathrm{im}\big(\pi_0(\widehat{T}^\Gamma_{\rm sc})^\wedge \longrightarrow \pi_0(\widehat{T}^\Gamma)^\wedge\big)\\ &= \big(\widehat{T}^\Gamma/Z(\widehat{G})^\Gamma\big)^\wedge = \widehat{\mathcal{S}}_\phi.\end{aligned}$$

But the mapping

$$g \longrightarrow g\sigma(g)^{-1}, \qquad \sigma \in \Gamma_F,$$

in which $g \in G$ normalizes T, descends to a canonical bijection

$$\mathcal{D}(T) \xrightarrow{\sim} \ker\big(H^1(F,T) \longrightarrow H^1(F,G)\big)$$

from $\mathcal{D}(T)$ onto a set that is in fact contained in $\mathcal{E}(T)$. (See [**L8**, p. 702].) If $\pi_0 \in \Pi_\phi$ is fixed, the mapping of $\mathcal{D}(T)$ into $\mathcal{E}(T)$ thus gives an embedding of Π_ϕ into $\widehat{\mathcal{S}}_\phi$.

We take π_0 to be the unique (B,χ)-generic representation in Π_ϕ. It then follows from [**S6**, Theorem 11.5] that the embedding of Π_ϕ into $\widehat{\mathcal{S}}_\phi$ gives the pairing

$$\langle x, \pi\rangle, \qquad x \in \mathcal{S}_\phi,\ \pi \in \Pi_\phi,$$

of Theorem 2.2.1(b). (See also [**S5**, §13].) We are trying to show that the image of the set Π_ϕ in $\widehat{\mathcal{S}}_\phi$ generates the group $\widehat{\mathcal{S}}_\phi$. It follows from Shelstad's initial observation [**S2**, Lemma 2.1] on the set $\mathcal{D}(T)$ that the image in (6.1.6) of the coroot $\alpha^\vee$ of any noncompact root of (G,T) is contained in the image of Π_ϕ. We have only to show that such coroots generate the group (6.1.6).

Having recalled the background, it will now be easy for us to prove the lemma. We choose a base for the roots $\Sigma(G,T)$ of (G,T) by taking the chamber in the Lie algebra of T that contains a Harish-Chandra parameter of π_0. It then follows from [**V1**, Theorem 6.2(f)] that the simple roots are all noncompact. The simple coroots $\alpha^\vee \in \Sigma^\vee(G,T)$ therefore map into the image of Π_ϕ in (6.1.6). Any coroot $\beta^\vee \in \Sigma^\vee(G,T)$ is of course an integral combination of simple coroots. The group (6.1.6) is therefore generated by the images of simple coroots, and hence also by the image of Π_ϕ. We conclude that $\widehat{\mathcal{S}}_\phi$ is generated by the image of Π_ϕ, as claimed. □

6.2. Construction of global representations $\dot{\pi}$

We now begin the construction of the global objects needed to establish the local theorems. The methods of this section will be representation theoretic. They will also be more elementary than in earlier chapters, to the extent that they do not depend on the running induction hypotheses we have been carrying since §4.3. We shall construct global representations from local ones as an exercise in the invariant trace formula.

We fix a simple endoscopic datum $G \in \widetilde{\mathcal{E}}_{\rm sim}(N)$ over our local field F. The main goal of this section will be to inflate any square-integrable representation of $G(F)$ to an automorphic representation. The square-integrability condition rules out the case that $F = \mathbb{C}$, so we assume that F is either real or p-adic.

We have first to identify F with a localization of some global field $\dot{F}$. The construction of suitable global fields from local fields is well known, but we may as well formulate what we need as a lemma, for the convenience of the reader.

LEMMA 6.2.1. *Assume that the local field F is either real or p-adic, and that r_0 is a positive integer. Then there is a totally real number field $\dot{F}$ with the following properties.*

(i) *$\dot{F}_u = F$ for some valuation u of $\dot{F}$.*

(ii) *There are at least r_0 Archimedean valuations on $\dot{F}$.*

PROOF. This is a simple exercise in Galois theory. The main point is to construct a totally real field $\dot{F}$ so that (i) holds. There is no problem if $F = \mathbb{R}$, since we can then set $\dot{F} = \mathbb{Q}$. We can therefore take F to be a finite extension of the p-adic field $\mathbb{Q}_p$. Then $F = \mathbb{Q}_p(\alpha)$, where α is a root of an irreducible monic polynomial q over $\mathbb{Q}_p$ of degree n.

We identify the space of monic polynomials of degree n over a field E with E^n. Since F is open in the union of the finite set of extensions of $\mathbb{Q}_p$ of degree n, we can replace q by any polynomial in some open neighborhood U_p of q in $\mathbb{Q}_p^n$. Consider also the set of monic polynomials of degree n over $\mathbb{R}$. The subspace of such polynomials with distinct real roots is an open subset U_∞ of $\mathbb{R}^n$. Since the diagonal image of $\mathbb{Q}^n$ in $\mathbb{R}^n \times \mathbb{Q}_p^n$ is dense, the intersection

$$\mathbb{Q}^n \cap (U_\infty \times U_p)$$

is nonempty. We can therefore assume that the coefficients of q lie in $\mathbb{Q}$, and that the roots of q in $\mathbb{C}$ are distinct and real. We then see that $\dot{F} = \mathbb{Q}(\alpha)$ is a totally real field, and that $F = \dot{F}_u$ for some place u of $\dot{F}$ over p.

To construct a totally real field with many Archimedean places, we replace $\dot{F}$ by a field $\dot{F}(\sqrt{q_1}, \dots, \sqrt{q_k})$, where $q_1, \dots, q_k$ are distinct (positive) prime numbers. These primes can be arbitrary if $F = \mathbb{R}$, but must be chosen so that $\mathbb{Q}_p(\sqrt{q_i}) = \mathbb{Q}_p$ if F is p-adic as above. There are infinitely many such primes, as one sees readily from Dirichlet's theorem on primes in arithmetic progression and the law of quadratic reciprocity (or its second supplement in case $p = 2$). The field $\dot{F}(\sqrt{q_1}, \dots, \sqrt{q_k})$ then satisfies the required conditions. □

We can now construct the automorphic representation. We are assuming that F is real or p-adic, and that G is the fixed datum in $\widetilde{\mathcal{E}}_{\mathrm{sim}}(N)$. We take $\dot{F}$ to be a totally real global field with $\dot{F}_u = F$, as in the lemma, for which the set S_∞ of archimedean places is large. We then set

$$S_\infty(u) = S_\infty \cup \{u\}$$

and

$$S_{\infty}^{u} = S_{\infty} - \{u\}.$$

To compactify the notation, it will be convenient to denote other objects associated with $\dot{F}$ by a dot. For example, we have the adeles $\dot{\mathbb{A}}$ over $\dot{F}$, the endoscopic sets $\dot{\widetilde{\mathcal{E}}}(N) = \mathcal{E}(\dot{\widetilde{G}}(N))$ and $\dot{\widetilde{\mathcal{E}}}_v(N) = \mathcal{E}(\dot{\widetilde{G}}_v(N))$ over $\dot{F}$ and $\dot{F}_v$, for any valuation v on $\dot{F}$, as well as the Hecke modules $\dot{\widetilde{\mathcal{H}}}(N) = \mathcal{H}(\dot{\widetilde{G}}(N))$ and $\dot{\widetilde{\mathcal{H}}}_v(N) = \mathcal{H}(\dot{\widetilde{G}}_v(N))$ on $\dot{\widetilde{G}}(N, \dot{\mathbb{A}})$ and $\dot{\widetilde{G}}(N, \dot{F}_v)$. We can write $\widetilde{\mathcal{E}}(N) = \dot{\widetilde{\mathcal{E}}}_u(N)$ and $\widetilde{\mathcal{H}}(N) = \dot{\widetilde{\mathcal{H}}}_u(N)$ as usual for corresponding objects attached to $F = \dot{F}_u$.

LEMMA 6.2.2. *Suppose that $\pi \in \Pi_2(G)$ is a square integrable representation of $G(F)$. We can then find a simple endoscopic datum $\dot{G} \in \dot{\widetilde{\mathcal{E}}}_{\rm sim}(N)$, together with an automorphic representation $\dot{\pi}$ of $\dot{G}(\dot{\mathbb{A}})$ that occurs discretely in $L^2(\dot{G}(\dot{F})\backslash\dot{G}(\dot{\mathbb{A}}))$, with the following properties:*

(i) $(\dot{G}_u, \dot{\pi}_u) = (G, \pi)$.

(ii) *For any valuation $v \notin S_{\infty}(u)$, $\dot{\pi}_v$ is spherical.*

(iii) *For any $v \in S_{\infty}^{u}$, $\dot{\pi}_v$ is a square integrable representation of $\dot{G}(\dot{F}_v)$ whose Langlands parameter $\phi_v \in \Phi_2(\dot{G}_v)$ is in general position.*

PROOF. We have first to define the endoscopic datum $\dot{G}$ over $\dot{F}$. The complex dual group $\widehat{\dot{G}}$ must be equal to $\widehat{G}$. This determines G uniquely if $\widehat{G} = Sp(N, \mathbb{C})$, and up to a quadratic idèle class character $\dot{\eta} = \eta_{\dot{G}}$ if $\widehat{G} = SO(N, \mathbb{C})$. There are two kinds of local constraints on $\dot{\eta}$. The component $\dot{\eta}_u$ of $\dot{\eta}$ at u must equal the quadratic character η_G on F^* determined by the given local endoscopic datum. The other constraints are at the archimedean places $v \in S_{\infty}^{u}$, and are imposed by the requirement that $\dot{G}(\dot{F}_v)$ have a discrete series. The question is relevant here only if N is even and $\widehat{G} = SO(N, \mathbb{C})$. As we noted in §6.1, the requirement is that

$$\dot{\eta}_v = \varepsilon_v^{p},$$

where $p = \frac{1}{2}N$, and ε_v is the sign character on $F_v^* = \mathbb{R}^*$. We take $\dot{\eta} = \eta_{\dot{G}}$ to be any quadratic idèle class character with the given finite set of local constraints. (We take $\dot{\eta}$ to be equal to 1 if all of the constrained local characters $\dot{\eta}_v$ equal 1.) This determines an endoscopic datum $\dot{G}$ over $\dot{F}$ such that $\dot{G}_u = G$, and such that at each $v \in S_{\infty}$, the real group $\dot{G}_v(\dot{F}_v)$ has a discrete series.

To construct the automorphic representation $\dot{\pi}$, we turn to a natural generalization of an argument applied by Langlands to the group $GL(2)$ [**L9**, p. 229]. We shall apply the trace formula to a suitable function $\dot{f}$ on $\dot{G}(\mathbb{A})$. (See also [**CC**] for a similar application of the twisted trace formula for $GL(N)$.)

Let

$$\dot{K}^{\infty,u} = \prod_{v \notin S_\infty(u)} \dot{K}_v$$

be a standard maximal compact subgroup of $\dot{G}(\dot{\mathbb{A}}^{\infty,u})$, the group of points $\dot{G}$ with values in the finite adèles of $\dot{F}$ that are 0 at u. We take $\dot{f}^{\infty,u}$ to be the characteristic function of $\dot{K}^{\infty,u}$ on $\dot{G}(\dot{\mathbb{A}}^{\infty,u})$. At u, we take $\dot{f}_u \in \mathcal{H}(G)$ to be a pseudocoefficient f_π of π. In other words, $\dot{f}_u$ has the property that for any irreducible, tempered representation π^* of $G(F)$, the relation

$$\dot{f}_{u,G}(\pi^*) = \operatorname{tr}\big(\pi^*(f_\pi)\big) = \begin{cases} 1, & \text{if } \pi^* \cong \pi, \\ 0, & \text{otherwise,} \end{cases} \tag{6.2.1}$$

holds. At the remaining set of places S^u_∞, we will take a function

$$\dot{f}^u_\infty(x^u_\infty) = \prod_{v \in S^u_\infty} \dot{f}_v(x_v)$$

on the group

$$\dot{G}(\dot{F}^u_\infty) = \prod_{v \in S^u_\infty} \dot{G}(\dot{F}_v)$$

obtained by letting each $\dot{f}_v$ be a stable sum of pseudocoefficients corresponding to a Langlands parameter $\phi_v \in \Phi_2(\dot{G}_v)$ in general position. In other words,

$$\dot{f}_v = \sum_{\pi_v \in \Pi_{\phi_v}} f_{\pi_v},$$

where f_{π_v} is a pseudocoefficient of the square integrable representation π_v of $\dot{G}(F_v)$. The existence of pseudocoefficients follows from [**BDK**] in the p-adic case, and [**CD**] in the archimedean case.

Let $\dot{Z}^{\infty,u}$ be the intersection of $\dot{K}^{\infty,u}$ with (the diagonal image in $\dot{G}(\dot{\mathbb{A}}^{\infty,u})$ of) the center of $\dot{G}(\dot{F})$. This actually equals the center of $\dot{G}(\dot{F})$, a group of order 1 or 2, except in the abelian case $\widehat{G} = SO(2,\mathbb{C})$. In case $\widehat{G}$ does equal to $SO(2,\mathbb{C})$, the existence of discrete series implies that $\dot{G}(\dot{F}_v)$ is compact if v either belongs to S^u_∞ or equals u, and therefore that $\dot{Z}^{\infty,u}$ is a finite group. We require that the function $\dot{f}^u_\infty \dot{f}_u$ on $\dot{G}(\dot{F}^u_\infty) \times G(F)$ be constant on (the diagonal image of) $\dot{Z}^{\infty,u}$. We are also requiring that ϕ_v be in general position for each $v \in S^u_\infty$. Finally, it will be convenient to fix a place $v \in S^u_\infty$, and then require that regularity of the infinitesimal character of ϕ_v dominate that of the other parameters $\phi_{v'}$. We can meet these conditions by first choosing the parameters $\phi_{v'}$ and the corresponding function

$$\dot{f}^{u,v}_\infty = \prod_{v' \in S^{u,v}_\infty} \dot{f}_{v'}$$

so that the product $\dot{f}^{u,v}_\infty \dot{f}_u$ is constant on $\dot{Z}^{\infty,u}$. At the place v, we then define ϕ_v in terms of its infinitesimal character by setting

$$\mu_{\phi_v} = n\mu_v$$

where μ_v is fixed, and n is a large integer such that μ_{ϕ_v} represents an element in $\Phi_2(\dot{G}_v)$ whose central character on $\dot{Z}^{\infty,u}$ is trivial. The resulting function

$$\dot{f}^u_\infty \, \dot{f}_u = \dot{f}^{u,v}_\infty \, \dot{f}_v \, \dot{f}_u$$

then has the required properties.

We shall apply the trace formula to the function

$$\dot{f}(x) = \dot{f}^u_\infty(x^u_\infty) \, \dot{f}_u(x_u) \, \dot{f}^{\infty,u}(x^{\infty,u}), \qquad x \in \dot{G}(\dot{\mathbb{A}}),$$

in $\mathcal{H}(\dot{G})$. Since $\dot{f}$ is cuspidal at more than one place, we can use the simple version [**A5**, Theorem 7.1] of the invariant trace formula for $\dot{G}$. Since $\dot{f}$ is both cuspidal and stable at some archimedean place, the invariant orbital integrals on the geometric side vanish on conjugacy classes that are not semisimple. (See [**A6**, Theorem 5.1].) The geometric side of the trace formula therefore reduces to a finite sum

$$\sum_\gamma \operatorname{vol}\big(\dot{G}_\gamma(\dot{F})\backslash \dot{G}_\gamma(\dot{\mathbb{A}})\big) \, \dot{f}_{\dot{G}}(\gamma) \tag{6.2.2}$$

of invariant orbital integrals

$$\dot{f}_{\dot{G}}(\gamma) = \int_{\dot{G}_\gamma(\dot{\mathbb{A}})\backslash \dot{G}(\dot{\mathbb{A}})} \dot{f}(x^{-1}\gamma x) \, dx,$$

taken over the semisimple conjugacy classes γ in $\dot{G}(\dot{F})$ that are $\mathbb{R}$-elliptic at each place in S_∞.

Consider the place $v \in S^u_\infty$ chosen above. The infinitesimal character of ϕ_v, or more precisely, of any of the representations in the packet of ϕ_v that define $\dot{f}_v$ as a sum of pseudocoefficients, depends linearly on n. It follows from fundamental results of Harish-Chandra [**Ha1**, Lemma 23], [**Ha2**, Lemmas 17.4 and 17.5] that

$$\dot{f}_G(\gamma) = c(\gamma) \, n^{q(\dot{G}_\gamma)}, \tag{6.2.3}$$

where $c(\gamma)$ is independent of n, and

$$q(\dot{G}_\gamma) = \tfrac{1}{2}\big(\dim(\dot{G}_\gamma) - \operatorname{rank}(\dot{G}_\gamma)\big).$$

The constant $c(\gamma)$ vanishes for all but a finite set of γ, but is nonzero at $\gamma = 1$. The exponent $q(\dot{G}_\gamma)$ is maximal when the centralizer $\dot{G}_\gamma$ of γ in $\dot{G}$ equals $\dot{G}$, which is to say that γ is central. In this case,

$$\dot{f}_{\dot{G}}(\gamma) = \dot{f}_{\dot{G}}(1),$$

by our condition on $\dot{f}^u_\infty$. The dominant term in (6.2.2) therefore corresponds to $\gamma = 1$. It follows that (6.2.2) is nonzero whenever n is sufficiently large.

The geometric expansion (6.2.2) equals the spectral side of the trace formula. The parabolic terms in [**A5**, Theorem 7.1(a)], namely those corresponding to proper Levi subgroups M in the general expansion (3.1.1), vanish on functions that are pseudocoefficients of discrete series. (See [**A6**, p. 268].) The spectral side therefore reduces simply to the trace

$$\mathrm{tr}\big(R^{\dot G}_{\mathrm{disc}}(\dot f)\big) = \sum_{\dot\pi} m_{\mathrm{disc}}(\dot\pi)\, \dot f_G(\dot\pi)$$

of $\dot f$ (operating by right convolution) on the automorphic discrete spectrum of $\dot G$. It follows that for n sufficiently large, there is an irreducible representation $\dot\pi$ of $\dot G(\dot{\mathbb{A}})$ with positive multiplicity $m_{\mathrm{disc}}(\dot\pi)$, and with

$$\dot f_{\dot G}(\dot\pi) = \mathrm{tr}\big(\dot\pi(\dot f)\big) \neq 0.$$

It remains to establish that the automorphic representation

$$\dot\pi = \bigotimes_v \dot\pi_v = \dot\pi^u_\infty \otimes \dot\pi_u \otimes \dot\pi^{\infty,u}$$

has the properties (i)–(iii). At first glance, these properties appear to be consequences of the definition of $\dot f$, and the fact that

$$\mathrm{tr}\big(\dot f^u_\infty(\dot\pi^u_\infty)\big)\,\mathrm{tr}\big(\dot f_u(\dot\pi_u)\big)\,\mathrm{tr}\big(\dot f^{\infty,u}(\dot\pi^{\infty,u})\big) = \mathrm{tr}\big(\dot\pi(\dot f)\big) \neq 0. \tag{6.2.4}$$

For example, (ii) follows immediately from the definition of $\dot f^{\infty,u}$ as the characteristic function of $\dot K^{\infty,u}$. However, something more is needed for (i) and (iii). The problem is that the definition of a pseudocoefficient (6.2.1) (in the case of (i)) does not rule out the possibility that the localization $\dot\pi_u$ could be a nontempered representation whose character on the elliptic set matches that of π.

We can treat (iii) by an argument based on the infinitesimal character. As a constituent of the automorphic discrete spectrum of $\dot G$, $\dot\pi$ is unitary, and so therefore are its localizations $\dot\pi_v$. If v belongs to S^u_∞, $\dot\pi_v$ has the same infinitesimal character as ϕ_v, since the trace of $\dot\pi_v(\dot f_v)$ is nonzero when $\dot f_v$ is the stable sum of pseudocoefficients for ϕ_v above. It follows from basic estimates for unitary representations, and the fact that ϕ_v is in general position, that $\dot\pi_v$ must be tempered. Appealing again to the definition of $\dot f_v$, we conclude that $\dot\pi_v$ itself belongs to the L-packet Π_{ϕ_v}, for the Langlands parameter $\phi_v \in \Phi_2(\dot G_v)$ in general position we have chosen. This is the condition (iii).

Our proof of (iii) does not apply to (i). For the infinitesimal character of $\dot\pi_u$ need not be in general position if $F = \mathbb{R}$, and is also too weak to help us if F is p-adic. We shall turn instead to a separate global argument, which would also give a different proof of (iii). While the argument is somewhat more sophisticated, it makes use only of the results of §3.4 and §3.5. It therefore does not rely on any induction hypotheses (in the form of yet unproved theorems), even though it seems also to have inherited some of the complexity of the arguments of §3.5!

Corollary 3.4.3 tells us that $\dot\pi$ occurs in the subspace

$$L^2_{\mathrm{disc},\dot\psi}\big(\dot G(\dot F)\backslash \dot G(\dot{\mathbb{A}})\big) \subset L^2_{\mathrm{disc}}\big(\dot G(\dot F)\backslash \dot G(\dot{\mathbb{A}})\big)$$

of the discrete spectrum attached to a (uniquely determined) global parameter $\dot{\psi} \in \tilde{\dot{\Psi}}(N)$. Given $\dot{\psi}$, we then use the formula (3.3.14) to write

$$I^N_{\mathrm{disc},\dot{\psi}}(\tilde{\dot{f}}) = \sum_{G^* \in \tilde{\dot{\mathcal{E}}}_{\mathrm{ell}}(N)} \widetilde{\iota}(N, G^*)\, \widehat{S}^*_{\mathrm{disc},\dot{\psi}}(\tilde{\dot{f}}^*), \qquad \tilde{\dot{f}} \in \tilde{\dot{\mathcal{H}}}(N).$$

The quadratic character $\eta_{\dot{\psi}}$ attached to $\dot{\psi}$ satisfies

$$\eta_{\dot{\psi}} = \eta_{\dot{\pi}} = \eta_{\dot{G}},$$

according to the definitions. It follows that the stable linear form $S^*_{\mathrm{disc},\dot{\psi}}$ attached to any $G^* \in \tilde{\dot{\mathcal{E}}}_{\mathrm{ell}}(N)$ vanishes unless η_{G^*} equals $\eta_{\dot{G}}$. This yields the usual simplification in the right hand side of the last formula. To reduce it further, we have to specialize the function $\tilde{\dot{f}}$ in the global Hecke module $\tilde{\dot{\mathcal{H}}}(N)$.

We will take new functions

$$\tilde{\dot{f}} = \prod_v \tilde{\dot{f}}_v = \tilde{\dot{f}}^u_\infty \cdot \tilde{\dot{f}}_u \cdot \tilde{\dot{f}}^{\infty,u} \tag{6.2.5}$$

and

$$\dot{f} = \prod_v \dot{f}_v = \dot{f}^u_\infty \cdot \dot{f}_u \cdot \dot{f}^{\infty,u} \tag{6.2.6}$$

in $\tilde{\dot{\mathcal{H}}}(N)$ and $\widetilde{\mathcal{H}}(\dot{G})$ respectively so that for any v, $\tilde{\dot{f}}_v$ and $\dot{f}_v$ have the same images in $\widetilde{\mathcal{S}}(\dot{G}_v)$. We impose no further conditions on the components $\tilde{\dot{f}}_u$ and $\dot{f}_u$ at u. However, we will constrain the other components of $\dot{f}$ as above. We take the component $\dot{f}^{\infty,u}$ of $\dot{f}$ away from $S_\infty(u)$ to be the characteristic function of $\dot{K}^{\infty,u}$ in $\widetilde{\mathcal{H}}\big(\dot{G}(\dot{\mathbb{A}}^{\infty,u})\big)$. If v belongs to S^u_∞, we take $\dot{f}_v$ to be the image in the symmetric Hecke algebra $\widetilde{\mathcal{H}}(\dot{G}_v)$ of the stable sum of pseudo-coefficients for ϕ_v. In addition, we choose $\tilde{\dot{f}}_v$ so that its image $\tilde{\dot{f}}^*_v$ in $\widetilde{\mathcal{S}}(G^*_v)$ vanishes for any $G^*_v \in \tilde{\dot{\mathcal{E}}}_{\mathrm{ell},v}(N)$ distinct from $\dot{G}_v$. The existence of $\tilde{\dot{f}}_v$ in this case is a consequence of Proposition 2.1.1, as we noted at a similar juncture in the proof of Lemma 5.4.2.

The condition on a function $\tilde{\dot{f}}_v$ at $v \in S^u_\infty$ implies that the global transfer $\tilde{\dot{f}}^*$ of $\tilde{\dot{f}}$ vanishes for any $G^* \in \tilde{\dot{\mathcal{E}}}_{\mathrm{ell}}(N)$ with $\widehat{G}^* \neq \widehat{\dot{G}}$. Combined with the property of η_{G^*} above, it tells us that

$$\widehat{S}^*_{\mathrm{disc},\dot{\psi}}(\tilde{\dot{f}}^*) = 0,$$

for any $G^* \in \dot{\widetilde{\mathcal{E}}}_{\mathrm{ell}}(N)$ distinct from $\dot{G}$. The formula given by (3.3.14) therefore reduces to

$$\begin{aligned} I^N_{\mathrm{disc},\dot{\psi}}(\dot{\widetilde{f}}) &= \widetilde{\iota}(N,\dot{G})\, \widehat{S}^{\dot{G}}_{\mathrm{disc},\dot{\psi}}(\dot{\widetilde{f}}^{\dot{G}}) \\ &= \widetilde{\iota}(N,\dot{G})\, S^{\dot{G}}_{\mathrm{disc},\dot{\psi}}(\dot{f}). \end{aligned}$$

We shall combine this with the expansions (4.1.1) and (4.1.2) for $\dot{G}$.

Suppose $G' \in \mathcal{E}_{\mathrm{ell}}(\dot{G})$ is a global elliptic endoscopic datum for $\dot{G}$ that is distinct from $\dot{G}$. Since the dual group $\widehat{G}'$ is then composite, the localization G'_v of G' at any v is distinct from $\dot{G}_v$. If v belongs to S^u_∞, the transfer $\dot{f}'_v$ of $\dot{f}_v$ to $\widetilde{\mathcal{S}}(G'_v)$ therefore vanishes, since $\dot{f}_v$ is the image in $\widetilde{\mathcal{H}}(\dot{G}_v)$ of the stable sum of pseudocoefficients for the square integrable parameter $\phi_v \in \Phi_2(\dot{G}_v)$. It follows that

$$\dot{f}' = \dot{f}^{G'} = 0.$$

The proper terms in the endoscopic expansion (4.1.2) of $I_{\mathrm{disc},\dot{\psi}}(\dot{f})$ therefore vanish. Similarly, the proper terms in the spectral expansion (4.1.1) of $I_{\mathrm{disc},\dot{\psi}}(\dot{f})$ also vanish, as we noted earlier in the proof. It follows that

$$\mathrm{tr}\big(R_{\mathrm{disc},\dot{\psi}}(\dot{f})\big) = I_{\mathrm{disc},\dot{\psi}}(\dot{f}) = S^{\dot{G}}_{\mathrm{disc},\dot{\psi}}(\dot{f}).$$

Combining this with the reduction above, we conclude that

$$\mathrm{tr}\big(R_{\mathrm{disc},\dot{\psi}}(\dot{f})\big) = \widetilde{\iota}(N,\dot{G})^{-1}\, I^N_{\mathrm{disc},\dot{\psi}}(\dot{\widetilde{f}}). \tag{6.2.7}$$

We can regard the two sides of (6.2.7) as linear forms in the corresponding pair of variable functions $\dot{f}_u$ and $\dot{\widetilde{f}}_u$. From (6.2.4) and the fact that $\dot{\pi}$ is a constituent of $R_{\mathrm{disc},\dot{\psi}}$, we know that if $\dot{f}_u$ is a pseudocoefficient of π, the left hand side of (6.2.7) is nonzero. Indeed, it is a nontrivial sum of irreducible characters on the group $\dot{G}(\dot{\mathbb{A}})$, taken at the positive definite function $\dot{f}$. The same is therefore true of the right hand side. It then follows from the definition at any $v \in S^u_\infty$ of $\dot{f}_v$ in terms of ϕ_v, together with the relation of the corresponding function $\dot{\widetilde{f}}_v$ with $\dot{\psi}_v$, that the infinitesimal characters of $\dot{\psi}_v$ and ϕ_v correspond. At this point we are taking for granted a property of archimedean infinitesimal characters, namely that their twisted transfer is compatible with the Kottwitz-Shelstad transfer of functions. The property can be regarded as a part of the work [**Me**], [**S8**] in progress. It could also be treated as a more elementary but nonetheless interesting exercise in twisted transfer factors, following its untwisted analogue [**A13**, (2.6)], for example.

The infinitesimal character of $\dot{\psi}_v$ is by definition that of the Langlands parameter $\phi_{\dot{\psi}_v}$. It cannot be in general position if $\dot{\psi}_v$ is not generic. On the other hand, we have agreed that it equals the infinitesimal character of the original parameter ϕ_v, regarded as a generic element in $\dot{\widetilde{\Psi}}_v(N)$. Since ϕ_v was chosen to be in general position, $\dot{\psi}_v$ must therefore be generic. The nongeneric component of any global parameter is of course unaffected by localization. It follows that the global parameter $\dot{\phi} = \dot{\psi}$ is generic. Its localization $\dot{\phi}_u = \dot{\psi}_u$ is consequently also generic.

We have shown that as a self-dual, local parameter for $GL(N)$, $\dot{\phi}_u = \dot{\psi}_u$ is generic. We will need to know further that it is tempered. We have therefore to show that as an N-dimensional representation of the local Langlands group $L_F = L_{\dot{F}_v}$, $\dot{\phi}_u$ is unitary.

Consider the expansion (4.1.1) for the term

$$I^N_{\mathrm{disc},\dot{\phi}}(\dot{\tilde{f}}) = I^N_{\mathrm{disc},\dot{\psi}}(\dot{\tilde{f}})$$

on the right hand side of (6.2.7). Appealing to the general discussion of §4.3, or simply to the local definition (2.2.1) and the properties in §2.5 of local intertwining operators for $GL(N)$, we see that it is a scalar multiple of $\dot{\tilde{f}}_N(\dot{\phi})$. As a function of the variable component $\dot{\tilde{f}}_u \in \widetilde{\mathcal{H}}(N)$ of $\dot{\tilde{f}}$, the right hand side of (6.2.7) therefore takes the form

$$c^u\, \dot{\tilde{f}}_{u,N}(\dot{\phi}_u), \qquad c^u \in \mathbb{C}. \tag{6.2.8}$$

Since the left hand side of (6.2.7) does not vanish if the corresponding function $\dot{f}_u \in \widetilde{\mathcal{H}}(G)$ is a pseudocoefficient of π, the coefficient c^u is nonzero. In addition, there is a (twisted) cuspidal function $\dot{\tilde{f}}_u \in \widetilde{\mathcal{H}}_{\mathrm{cusp}}(N)$ such that $\dot{\tilde{f}}_{u,N}(\dot{\phi}_u)$ is nonzero. (Recall that $\widetilde{\mathcal{H}}_{\mathrm{cusp}}(N)$ is the subspace of functions $\tilde{f} \in \widetilde{\mathcal{H}}(N)$ such that $\tilde{f}^*$ belongs to $\tilde{\mathcal{S}}_{\mathrm{cusp}}(G^*)$ for every $G^* \in \tilde{\mathcal{E}}_{\mathrm{ell}}(N)$, or equivalently, such that $\tilde{f}_{\widetilde{M}} = 0$ for any proper Levi subset $\widetilde{M}$ of $\widetilde{G}(N)$. That we can take $\dot{\tilde{f}}_u$ above to be in $\widetilde{\mathcal{H}}_{\mathrm{cusp}}(N)$ follows from Proposition 2.1.1.) We conclude that $\dot{\phi}_u$ lies in the subset $\widetilde{\Phi}_{\mathrm{ell}}(N)$ of $\widetilde{\Phi}(N)$. Indeed, the linear form on $\widetilde{\mathcal{H}}(N)$ attached to a local parameter in the complement of $\widetilde{\Phi}_{\mathrm{ell}}(N)$ vanishes on $\widetilde{\mathcal{H}}_{\mathrm{cusp}}(N)$. We have been making free use here of the local Langlands correspondence for $GL(N)$. In particular, the original interpretation of $\widetilde{\Phi}_{\mathrm{ell}}(N)$ in §1.2 tells us that as an N-dimensional representation of L_F, $\dot{\phi}_u$ decomposes into irreducible constituents that are self-dual and multiplicity free. It follows that $\dot{\phi}_u$ is unitary. In other words, the linear form $\dot{\tilde{f}}_{u,N}(\dot{\phi}_u)$ in $\dot{\tilde{f}}_u \in \widetilde{\mathcal{H}}(N)$ is tempered.

We have shown that the right hand side (6.2.8) of (6.2.7) is tempered in the variable component $\tilde{f}$ of $\dot{\tilde{f}}_u$. We want to conclude that the same is then true of the left hand side of (6.2.7). At this point, it would be convenient to be able to say that the local twisted transfer mapping

$$\tilde{f} = \dot{\tilde{f}}_u \longrightarrow \tilde{f}^G$$

from $\widetilde{\mathcal{H}}(N)$ onto $\tilde{\mathcal{S}}(G)$ is represented by a correspondence that extends to the underlying Schwartz spaces. A proof of this property would no doubt be accessible. It is perhaps already known, but I do not have a reference. We need only the special case represented by the twisted tempered character in (6.2.8), and this is not hard to obtain directly. Let me give a very brief sketch of the argument.

The twisted tempered character in (6.2.8) has *exponents*. They are attached to the chamber $\mathfrak{a}^+_{\widetilde{P}}$ of any given cuspidal parabolic subset $\widetilde{P}$ of $\widetilde{G}(N)$, and are linear forms λ on the complex vector space $\mathfrak{a}_{\widetilde{P},\mathbb{C}}$. If we evaluate the character at functions in the subspace $\widetilde{\mathcal{C}}_{\mathrm{reg}}(N)$ of Schwartz functions on $\widetilde{G}(N,F)$ with strongly regular support, we find that the $\widetilde{P}$-exponents satisfy the inequality

$$\mathrm{Re}\big(\lambda(H)\big) \leqslant 0, \qquad H \in \mathfrak{a}^+_{\widetilde{P}}.$$

On the other hand, one can describe the image of $\widetilde{\mathcal{C}}_{\mathrm{reg}}(N)$ under the transfer mapping $\widetilde{f} \to \widetilde{f}^G$. Given Corollary 2.1.2 and the explicit nature of the twisted transfer factors for $\big(\widetilde{G}(N), G\big)$, we see that the image of $\widetilde{\mathcal{C}}_{\mathrm{reg}}(N)$ is the space of functions of δ whose restriction to any maximal torus $T(F) \subset G(F)$ lies in the subspace of $\widetilde{\mathcal{C}}_{N\text{-reg}}(T)$ of Schwartz functions of strongly $\widetilde{G}(N)$-regular support. The variable δ here represents an $\widetilde{\mathrm{Out}}_N(G)$-orbit of strongly $\widetilde{G}(N)$-regular, stable conjugacy classes in $G(F)$. It is then not hard to show that the P-exponents implicit in the left hand side of (6.2.7) satisfy the analogue of the inequality above. This in turn implies that the left hand side of (6.2.7) does indeed represent a tempered linear form in the variable component $f = \dot{f}_u$ of $\dot{f}$.

Since the global representation $\dot{\pi}$ is a constituent of the representation $R_{\mathrm{disc},\dot{\phi}}$ on the left hand side of (6.2.7), its local u-component $\dot{\pi}_u$ must then be tempered. It remains only to apply (6.2.4) one more time. If $\dot{f}_u$ is specialized to a pseudocoefficient of the original representation π of $G(F)$, we see that

$$\dot{f}_{u,G}(\dot{\pi}_u) = \mathrm{tr}\big(\dot{\pi}_u(\dot{f}_u)\big) \neq 0,$$

for the irreducible representation $\dot{\pi}_u$ of $G(F)$ we now know is tempered. It then follows from the definition of a pseudocoefficient that $\dot{\pi}_u = \pi$, as required. We have therefore established the remaining condition (i) of Lemma 6.2.2. Our proof of the lemma is at last complete. □

COROLLARY 6.2.3. *Suppose that (F, G, π) and $(\dot{F}, \dot{G}, \dot{\pi})$ are as in the statement of the lemma, and that $\dot{\phi} \in \widetilde{\Phi}(N)$ is the generic global parameter obtained from $\dot{\pi}$ in the proof of the lemma. Then for any valuation $v \neq u$ of $\dot{F}$, the ($\widetilde{\mathrm{Out}}_N(\dot{G}_v)$-orbit of the) localization $\dot{\pi}_v$ lies in the local packet $\dot{\Pi}_{\dot{\phi}_v}$. In particular, the archimedean parameters ϕ_v fixed at the beginning of the construction represent the localizations $\dot{\phi}_v$ of $\dot{\phi}$.*

PROOF. We choose a pair $(\dot{\widetilde{f}}, \dot{f})$ of functions (6.2.5) and (6.2.6) as in the proof of lemma, but with two minor differences. We take $\dot{f}_u$ to be the fixed pseudocoefficient f_π rather than a variable function in $\widetilde{\mathcal{H}}(\dot{G}_u)$, and we take

$\dot{f}^{\infty,u}$ to be a variable function in the $\dot{K}^{\infty,u}$-spherical Hecke algebra, rather than the unit. With the two functions being otherwise exactly as before, the earlier conclusions remain valid. In particular, the identity (6.2.7) holds. Moreover, the right hand side of (6.2.7) is a nonzero multiple of $\tilde{\dot{f}}_N(\dot{\phi})$. It follows that

$$\tilde{\dot{f}}_N(\dot{\phi}) = c(\dot{\phi})\,\mathrm{tr}\big(R_{\mathrm{disc},\dot{\phi}}(\dot{f})\big), \tag{6.2.9}$$

for a nonzero constant $c(\dot{\phi})$.

The right hand side of (6.2.9) is nonzero as a linear form in $\dot{f}^{\infty,u}$, since $\dot{f}_{\dot{G}}(\dot{\pi})$ is nonzero if $\dot{f}^{\infty,u}$ is a unit. Since $\dot{f}^{\infty,u}$ is allowed to vary, we see from the left hand side that for any $v \notin S_\infty(u)$, $\dot{\pi}_v$ does belong to $\widetilde{\Pi}_{\dot{\phi}_v}$. If v belongs to S^u_∞, we combine (6.2.9) with the twisted archimedean character relations for $GL(N)$ and the fact that $\tilde{\dot{f}}_v$ transfers to $\dot{f}^{\dot{G}}_v$. It follows that $\dot{\pi}_v$ belongs to $\widetilde{\Pi}_{\dot{\phi}_v}$ in this case as well. Therefore ϕ_v does equal $\dot{\phi}_v$. □

Recall that we have been carrying an unproven lemma from §2.3. It is Lemma 2.3.2, which we used to bring global methods to bear on the normalization of local intertwining operators. We can now see that Lemma 2.3.2 is a special case of the lemma we have just established (with the factor G_- of §2.3 in place of the group G here). Its two assertions (i) and (ii) are just the corresponding assertions of Lemma 6.2.2. Lemma 2.3.2 has therefore now been proved.

Returning to Lemma 6.2.2 itself, we note from the early part of the construction that the global pair $(\dot{G},\dot{F})$ depends only on the given local pair (G,F). In particular, it is independent of the local representation π. Bearing this in mind, we formulate an important corollary that is based on a separate hypothesis.

COROLLARY 6.2.4. *Suppose that F, G, $\dot{F}$ and $\dot{G}$ are as in Lemma 6.2.2, and that the local theorems (interpreted as in §6.1) are valid for generic parameters $\phi_1 \in \widetilde{\Phi}(N_1)$ over F with $N_1 \leqslant N$. Then for any simple local parameter $\phi \in \widetilde{\Phi}_{\mathrm{sim}}(G)$, there is a simple global parameter $\dot{\phi} \in \widetilde{\Phi}_{\mathrm{sim}}(\dot{G})$ with the following properties.*

(i) $(\dot{F}_u, \dot{G}_u, \dot{\phi}_u) = (F, G, \phi)$.

(ii) *For any valuation $v \notin S_\infty(u)$, the localization $\dot{\phi}_v$ is spherical.*

(iii) *For any $v \in S^u_\infty$, $\dot{\phi}_v$ is a local parameter in general position in $\widetilde{\Phi}_2(\dot{G}_v)$.*

PROOF. Given ϕ, we obtain a representation $\pi \in \widetilde{\Pi}_2(G)$ with

$$\mathrm{tr}\big(\pi(\dot{f}_u)\big) = \tilde{\dot{f}}^G_u(\phi) = \tilde{\dot{f}}_{u,N}(\phi),$$

for any functions $\dot{f}_u \in \widetilde{\mathcal{H}}(G)$ and $\tilde{\dot{f}}_u \in \widetilde{\mathcal{H}}(N)$ with the same image in $\widetilde{\mathcal{S}}(G)$. This follows from our assumption that the local theorems hold for G. We can then construct the automorphic representation $\dot{\pi}$ of $\dot{G}(\mathbb{A})$ of the lemma.

In proving Lemma 6.2.2, we obtained a global parameter $\dot{\phi} = \dot{\psi}$ in $\tilde{\dot{\Psi}}(N)$ from Corollary 3.4.3, which we then showed was generic. We will see in a moment that $\dot{\phi}$ is simple. Since $\dot{\pi}$ is a constituent of the representation $R^{\dot{G}}_{\mathrm{disc},\dot{\phi}}$ attached to $\dot{\phi}$, this will tell us that $\dot{G}$ satisfies the condition of the global datum $G_{\dot{\phi}}$ of Theorem 1.4.1.

The condition (i) applies to the valuation $v = u$. In the proof of the theorem, we showed that

$$\tilde{\dot{f}}_{u,N}(\dot{\phi}_u) \neq 0,$$

if $\tilde{\dot{f}}_u$ and $\dot{f}_u$ have the same image in $\widetilde{\mathcal{S}}(G)$, and $\dot{f}_u$ is a pseudocoefficient of π. The local theorems we are assuming tell us that the generic parameter $\dot{\phi}_u$ is uniquely determined by this condition. Since the condition also holds for ϕ, we conclude that

$$\dot{\phi}_u = \phi.$$

This completes the proof of (i). It also tells us that the local parameter $\dot{\phi}_u$ is simple. Therefore the global parameter $\dot{\phi}$ is also simple, as an element in $\tilde{\dot{\Phi}}(N)$.

We have shown that $\dot{G}$ satisfies the condition of the global datum $G_{\dot{\phi}}$ imposed on the simple datum $\dot{\phi}$ by Theorem 1.4.1. If $G^* \in \tilde{\dot{\mathcal{E}}}_{\mathrm{ell}}(N)$ is another global datum that satisfies this condition, one can show that G^*_u equals $\dot{G}_u$. One applies (3.3.14) to any function $\tilde{\dot{f}}^* \in \tilde{\dot{\mathcal{H}}}(N)$ with $\tilde{\dot{f}}_{u,N}(\dot{\phi}_u) = 0$, taking account of Propositions 2.1.1 and 3.4.1. If we grant this property, we can then combine the resulting identity

$$\widehat{G}^* = \widehat{G}^*_u = \widehat{G} = \widehat{\dot{G}}$$

of dual groups with the identity

$$\eta_{G^*} = \eta_{\dot{\phi}} = \eta_{\dot{G}}$$

of quadratic characters to deduce that $G^* = \dot{G}$. This is the uniqueness assertion of Theorem 1.4.1. It implies that $\dot{\phi} \in \widetilde{\Phi}_{\mathrm{sim}}(\dot{G})$, as claimed.

The conditions (ii) and (iii) apply to valuations v of $\dot{F}$ that are distinct from u. They follow from Corollary 6.2.3. □

REMARKS. 1. We have not yet made any induction assumption on the global theorems. Our claim that $\dot{\phi}$ belongs to $\widetilde{\Phi}_{\mathrm{sim}}(\dot{G})$ is based on the original definition provided by conditions of Theorem 1.4.1. We did not fully justify the uniqueness assertion of this theorem in the proof above. There is really no call to do so. For we will impose a global induction hypothesis in §6.4, in the context of a generic family of global parameters $\tilde{\dot{\mathcal{F}}}$ of the kind considered in §5.4, which will include whatever is needed. We will construct the family $\tilde{\dot{\mathcal{F}}}$ in the next section. In the process, we will apply Corollary 6.2.4 as stated, knowing that $\dot{\phi}$ does determine $\dot{G}$ by the given conditions

$\hat{\dot{G}} = \widehat{G}$ and $\eta_{\dot{G}} = \eta_{\dot{\phi}}$, and that this will give the uniqueness assertion of Theorem 1.4.1 once the global induction hypothesis is imposed.

2. In the proof of Lemma 6.2.2, we chose a variable archimedean parameter ϕ_v in order to work with the simple formula (6.2.3). A reader familar with the identities of Harish-Chandra on which (6.2.3) is based will observe that our condition of general position is sufficient. Suppose that

$$\phi_v \in \widetilde{\Phi}_2(\dot{G}_v), \qquad v \in S_\infty^u,$$

are archimedean parameters in general position, and that the product of their corresponding central characters (on $\dot{Z}^{\infty,u}$ if $\widehat{G} = SO(2,\mathbb{C})$, or simply $Z(\dot{F})$ otherwise) with that of ϕ is trivial. Then we can choose the global parameter $\dot{\phi}$ in Corollary 6.2.4 so that $\dot{\phi}_v = \phi_v$ for each v in S_∞^u.

The point here is that the formula (6.2.3) is just a special case of a more general formula, in which the variable n is replaced by a tensor product over v of vectors composed of the half integers (6.1.4) attached to the infinitesimal characters of the parameters ϕ_v. Our condition on the general position of each ϕ_v means that these half integers are all larger than some preassigned constant. This forces the terms with γ central to dominate the others, and the sum (6.2.2) to be nonzero.

3. There are other variants of Lemma 6.2.2, which could be proved in the same way. For example, we could construct the automorphic representation $\dot{\pi}$ so that $\dot{\pi}_v$ equals a given square-integrable representation for every v in some finite set V of p-adic places, and so that $\dot{\pi}_v$ is as in (ii) for each v not in $S_\infty(u) \cup V$. Similarly, we could arrange for the global parameter $\dot{\phi}$ of Corollary 6.2.4 to be equal to a prescribed element in $\widetilde{\Phi}_2(\dot{G}_v)$ at each $v \in V$, and to be as in (ii) at each v outside $S_\infty(u) \cup V$.

6.3. Construction of global parameters $\dot{\phi}$

Our goal for Chapter 6 will be to establish the local theorems for Langlands parameters in the general set

$$\widetilde{\Phi} = \coprod_N \widetilde{\Phi}(N),$$

taken over the fixed local field F. To this end, we now take on the natural induction hypothesis that will carry us through the chapter. We fix N, and assume inductively that the local theorems (interpreted as in §6.1) all hold for generic parameters $\phi \in \widetilde{\Phi}$ with $\deg(\phi) < N$.

In this section, F is assumed to be real or p-adic, as in §6.2. We fix an elliptic endoscopic datum $G \in \widetilde{\mathcal{E}}_{\mathrm{ell}}(N)$ over F, which in general need not be in the subset $\widetilde{\mathcal{E}}_{\mathrm{sim}}(N)$ of simple data. We then fix a local parameter

$$\phi = \ell_1\phi_1 \oplus \cdots \oplus \ell_r\phi_r \tag{6.3.1}$$

in $\widetilde{\Phi}(G)$, with simple components

$$\phi_i \in \widetilde{\Phi}_{\mathrm{sim}}(G_i), \qquad G_i \in \widetilde{\mathcal{E}}_{\mathrm{sim}}(N_i).$$

(Our requirement here that the simple components be self-dual means that ϕ lies in the subset

$$\widetilde{\Phi}_{\text{disc}}(G) = \{\phi \in \widetilde{\Phi}(G) : \ |Z(\overline{S}_\phi)| < \infty\}$$

of $\widetilde{\Phi}(G)$. This is the local, generic analogue of corresponding global set in the chain (4.1.13).) We assume that ϕ is not simple, or equivalently, that each N_i is less than N. The induction hypothesis above then tells us that the data $G_i \in \widetilde{\mathcal{E}}_{\text{sim}}(N_i)$ are determined by the uniqueness assertion of Theorem 6.1.1. For clarity we also assume that G is in fact simple if ϕ does not belong to $\widetilde{\Phi}_{\text{ell}}(N)$, or in other words, if $\ell_i > 1$ for some i. This insures that $\widetilde{\Phi}(G)$ injects as a subset of $\widetilde{\Phi}(N)$, which is implicit in the assertion that the compound parameter ϕ in (6.3.1) lies in $\widetilde{\Phi}(G)$.

According to the induction hypothesis, we can apply Corollary 6.2.4 to the pairs (G_i, ϕ_i). If $\dot{F}$ is the totally real field of Lemma 6.2.1, we will obtain global pairs

$$(\dot{G}_i, \dot{\phi}_i), \qquad \dot{G}_i \in \widetilde{\dot{\mathcal{E}}}_{\text{sim}}(N_i), \ \dot{\phi}_i \in \widetilde{\Phi}_{\text{sim}}(\dot{G}_i),$$

that satisfy the conditions (i)–(iii) of the corollary. These will determine a global endoscopic datum $\dot{G} \in \widetilde{\dot{\mathcal{E}}}_{\text{ell}}(N)$, which will be simple if some $\ell_i > 1$, such that the resulting global parameter

$$\dot{\phi} = \ell_1\dot{\phi}_1 \boxplus \cdots \boxplus \ell_r\dot{\phi}_r \tag{6.3.2}$$

belongs to $\widetilde{\Phi}(\dot{G})$. Our aim is to establish some finer properties of $\dot{\phi}$, for suitable choices of its simple components $\dot{\phi}_i$.

The case that each $\ell_i = 1$ in (6.3.1) is at the heart of things. However, we will also have to be concerned with higher multiplicities in dealing with the local intertwining relation. We fix a Levi subgroup M of G such that the set $\widetilde{\Phi}_2(M, \phi)$ is nonempty. For a given choice of the global datum $\dot{G}$, we will also choose a global Levi subgroup $\dot{M}$ such that $\widetilde{\Phi}_2(\dot{M}, \dot{\phi})$ is nonempty. The general notation, first introduced for the global parameter ψ_+ of Lemma 5.2.3, is relevant here. For example, if

$$N_- = \sum_{\ell_i \text{ odd}} N_i$$

and

$$\phi_- = \bigoplus_{\ell_i \text{ odd}} \phi_i,$$

we have

$$M \cong GL(N_1)^{\ell_1'} \times \cdots \times GL(N_r)^{\ell_r'} \times G_-,$$

where $\ell_i' = [\ell_i/2]$, and G_- is the unique element in $\widetilde{\mathcal{E}}_{\text{ell}}(N_-)$ such that ϕ_- lies in in $\widetilde{\Phi}_2(G_-)$. We can then identify the local parameter

$$\phi_M = \phi_1^{\ell_1'} \times \cdots \times \phi_r^{\ell_r'} \times \phi_- \tag{6.3.3}$$

with an element in $\widetilde{\Phi}_2(M,\phi)$. Similarly, we can identify the corresponding global parameter

$$\dot{\phi}_M = \dot{\phi}_1^{\ell_1'} \times \cdots \times \dot{\phi}_r^{\ell_r'} \times \dot{\phi}_- \tag{6.3.4}$$

with an element in $\widetilde{\Phi}_2(\dot{M},\dot{\phi})$.

The properties we will use are summarized in the multiple conditions of the statement of the following proposition.

PROPOSITION 6.3.1. *Given the local objects G, ϕ, M and ϕ_M over F as in (6.3.1) and (6.3.3), we can choose corresponding global objects $\dot{G}$, $\dot{\phi}$, $\dot{M}$ and $\dot{\phi}_M$ over $\dot{F}$ as in (6.3.2) and (6.3.4) such that the following conditions are satisfied.*

(i) *There is a valuation u of $\dot{F}$ such that*

$$(\dot{F}_u, \dot{G}_u, \dot{\phi}_u, \dot{M}_u, \dot{\phi}_{M,u}) = (F, G, \phi, M, \phi_M),$$

and such that the canonical maps

$$\mathcal{S}_{\dot{\phi}_{\dot{M}}} \longrightarrow \mathcal{S}_{\phi_M}$$

and

$$\mathcal{S}_{\dot{\phi}} \longrightarrow \mathcal{S}_{\phi}$$

are isomorphisms.

(ii) *For any valuation v outside the set $S_\infty(u)$, the local Langlands parameter*

$$\dot{\phi}_v = \ell_1\dot{\phi}_{1,v} \oplus \cdots \oplus \ell_r\dot{\phi}_{r,v}$$

is a direct sum of quasicharacters of $\dot{F}_v^*$, while the corresponding decomposition of the subparameter

$$\dot{\phi}_{1,v} \oplus \cdots \oplus \dot{\phi}_{r,v}$$

contains at most one ramified quasicharacter.

(iii)(a) *Set $V = S_\infty^u$. Then for any $v \in V$, the parameters $\dot{\phi}_{i,v}$ lie in $\widetilde{\Phi}_2(\dot{G}_{i,v})$, and are in relative general position. Moreover, the canonical mapping*

$$\Pi_{\dot{\phi}_{M,V}} \longrightarrow \widehat{\mathcal{S}}_{\dot{\phi}_M} \cong \widehat{\mathcal{S}}_{\phi_M},$$

obtained from the combined places $v \in V$, is surjective.

(iii)(b) *Suppose that each ℓ_i equals 1. Then there is a $v \in V$ with the property that if $\dot{\phi}_v$ lies in $\widetilde{\Phi}(G_v^*)$ for some $G_v^* \in \dot{\widetilde{\mathcal{E}}}_{\mathrm{sim},v}(N)$, the dual group $\widehat{G}_v^*$ equals $\widehat{G}$.*

(iii)(c) *Suppose that some ℓ_i is greater than 1. Then there is a $v \in V$ such that the kernel of the composition of mappings*

$$\mathcal{S}_{\dot{\phi}} \longrightarrow \mathcal{S}_{\dot{\phi}_v} \longrightarrow R_{\dot{\phi}_v}$$

contains no element whose image in the global R-group $R_{\dot{\phi}} = R_{\dot{\phi}}(\dot{G})$ belongs to $R_{\dot{\phi},\mathrm{reg}}$.

REMARKS. 1. The statement of the proposition does contain a rather large number of conditions. I have tried to organize them in a way that makes sense. For example, they represent refinements of the conditions (i)–(iii) of Corollary 6.2.4. In addition, the three parts (a)–(c) of (iii) are parallel to the three conditions (5.4.1)(a)–(5.4.1)(c) of Assumption 5.4.1.

2. The first assertion of (iii)(a) means that the integers

$$\coprod_{i=1}^{r}(\mu_{\dot\phi_{i,v}}) = \{\dot\mu_{i,v,k} :\ 1 \leqslant k \leqslant p_i = [N_i/2],\ 1 \leqslant i \leqslant r\} \tag{6.3.5}$$

are all large, and that their differences are large in absolute value. The second assertion of (iii)(a) is that for any character ξ on the abelian group $\mathcal{S}_{\dot\phi_M}$, there is a representation

$$\dot\pi_V = \bigotimes_{v\in V}\dot\pi_v, \qquad \dot\pi_v \in \widetilde\Pi_{\dot\phi_{M,v}},$$

in the product $\widetilde\Pi_{\dot\phi_{M,V}}$ of local archimedean packets such that

$$\xi(s) = \prod_{v\in V}\langle s, \dot\pi_v\rangle, \qquad s \in \mathcal{S}_{\dot\phi_M}.$$

3. The conditions (b) and (c) of (iii) are perhaps more palatable here than in the original forms (5.4.1)(b) and (5.4.1)(c), since we have now seen their application to the lemmas of §5.4.

PROOF. We choose the totally real field $\dot F$ according to Lemma 6.2.1. The only requirements are that $\dot F_u = F$ for a fixed place u, and that $\dot F$ have sufficiently many real places. We will then apply Corollary 6.2.4 inductively to each of the local pairs (G_i, ϕ_i), as agreed above. This will yield global endoscopic data $\dot G_i \in \widetilde{\dot{\mathcal{E}}}_{\mathrm{sim}}(N_i)$ and generic global parameters $\dot\phi_i \in \widetilde\Phi_{\mathrm{sim}}(\dot G_i)$. Once chosen, the global pairs $(\dot G_i, \dot\phi_i)$ determine a global endoscopic datum $\dot G \in \widetilde{\mathcal{E}}_{\mathrm{ell}}(N)$ with $\widehat{\dot G} = \widehat G$, together with a generic global parameter $\dot\phi \in \widetilde\Phi(\dot G)$ as in (6.3.2). They also determine a global Levi subgroup $\dot M$ of $\dot G$ with $\widehat{\dot M} = \widehat M$, together with a generic global parameter $\dot\phi_M \in \widetilde\Phi(\dot M, \dot\phi)$ as in (6.3.4).

The conditions $\dot G_u = G$, $\dot\phi_u = \phi$, $\dot M_u = M$ and $\dot\phi_{M,u} = \phi_M$ in (i) follow directly from the corresponding conditions of Corollary 6.2.4 for the pairs $(\dot G_i, \dot\phi_i)$. The isomorphisms of centralizers in (i) are also consequences of the construction. They follow easily from the mappings

$$\begin{array}{ccc} {}^L G_i & \longrightarrow & {}^L \dot G_i \\ \big\downarrow & & \big\downarrow \\ \Gamma & \hookrightarrow & \dot\Gamma \end{array}$$

and the definitions of the local and global centralizers.

Before we consider the remaining conditions (ii) and (iii)(a)–(iii)(c), we recall that the construction of each $\dot{\phi}_i$, and hence of $\dot{\phi}$, entails a number of choices. These include a choice of quadratic idèle class character $\dot{\eta}_i = \eta_{\dot{G}_i}$ for each i with $\widehat{G}_i = SO(N_i, \mathbb{C})$. If N_i is even, $\dot{\eta}_i$ determines the outer twisting of $\dot{G}_i$. In this case, the local values $\dot{\eta}_{i,v}$ of $\dot{\eta}_i$ are predetermined at each v in the set $S_\infty(u)$. If N_i is odd, $\dot{\eta}_i$ determines the L-embedding of ${}^L G_i$ into $GL(N_i, \mathbb{C})$. Its local value is fixed at u, but is otherwise arbitrary.

We impose two further local conditions on the family $\{\dot{\eta}_i\}$. The first is a constraint on the characters $\dot{\eta}_i$ with N_i odd, at the supplementary archimedean places. We fix a set of *distinct* places

$$V^o = \{v_i : \; N_i \text{ odd}\}$$

in the set $V = S_\infty^u$. We then require that for any $v \in V$,

$$\dot{\eta}_{i,v} = \begin{cases} (\varepsilon_v)^{p_i+1}, & \text{if } v = v_i, \\ (\varepsilon_v)^{p_i}, & \text{if } v \neq v_i, \end{cases} \tag{6.3.6}$$

where ε_v is the nontrivial quadratic character on $\dot{F}_v^* = \mathbb{R}^*$, and $p_i = \frac{1}{2}(N_i - 1)$. This of course requires that there be at least as many places in V as there are indices i with N_i odd, one of our reasons for insisting that $\dot{F}$ have sufficiently many real places. The second constraint applies to all characters $\dot{\eta}_i$, at the places $v \notin S_\infty(u)$. We require that for any such v, there be at most one i such that $\dot{\eta}_{i,v}$ ramifies, and for good measure, that the ramification be tame. This is certainly possible. It is well known that one can choose a quadratic extension of any number field with arbitrarily prescribed localizations at finitely many places. If we choose the characters $\dot{\eta}_i$ successively with increasing i, we can insist that $\dot{\eta}_{i,v}$ be unramified at any place $v \notin S_\infty(u)$ which divides 2, or at which $\dot{\eta}_{j,v}$ ramifies for some $j < i$.

We can now establish (ii). Suppose that v lies in the complement of $S_\infty(u)$. Corollary 6.2.4 tells us that for any i, the parameter $\dot{\phi}_{i,v}$ in the subset $\widetilde{\Phi}(\dot{G}_{i,v})$ of $\widetilde{\dot{\Phi}}_v(N_i)$ is spherical. According to the remarks in §6.1, this means that it is a direct sum of $(N_i - 1)$ unramified quasicharacters of $\dot{F}_v^*$, together with another quasicharacter that is unramified if and only if $\dot{\eta}_{i,v}$ is unramified. In particular, the sum $\dot{\phi}_v$ is a direct sum of quasicharacters, as claimed. Our condition above on $\dot{\eta}_{i,v}$ implies further that there is at most one i such that $\dot{\phi}_{i,v}$ contains a ramified quasicharacter. This gives the second claim of (ii).

It remains to establish the property (iii), with its three parts (a)–(c), for the valuations v in $V = S_\infty^u$. We first recall that for any $v \in V$, we have a decomposition

$$\dot{\phi}_{i,v} = \bigoplus_{k=0}^{p_i} \dot{\phi}_{i,v,k}$$

for any index i, where

$$\{\dot{\phi}_{i,v,k}: \ 1 \leqslant k \leqslant p_i = [N_i/2]\}$$

are irreducible, self-dual, two-dimensional representations of $W_{\dot{F}_v}$, and $\dot{\phi}_{i,v,0}$ is a one-dimensional quadratic character that occurs only when N_i is odd. In case N_i is odd, the two-dimensional representations $\dot{\phi}_{i,v,k}$ are orthogonal, and hence have a nontrivial determinant. It follows that

$$\begin{aligned}\dot{\phi}_{i,v,0} &= \det(\dot{\phi}_{i,v}) \cdot \prod_{k=1}^{p_i} \det(\dot{\phi}_{i,v,k})^{-1} \\ &= \dot{\eta}_{i,v} \cdot (\varepsilon_v)^{p_i} = \begin{cases} \varepsilon_v, & \text{if } v = v_i, \\ 1, & \text{otherwise,} \end{cases}\end{aligned}$$

where $v_i \in V$ is the valuation fixed above. For any i, the two-dimensional representations

$$\{\dot{\phi}_{i,v,k}: \ 1 \leqslant k \leqslant p_i, \ 1 \leqslant i \leqslant r\}$$

are all mutually inequivalent, as we recall from their origins in the proof of Lemma 6.2.2. Apart from a harmless condition on the central character, the only constraint was a regularity condition on the infinitesimal characters of these parameters. For a given v, we choose the parameters successively with increasing i. We can then insist that the infinitesimal character of $\dot{\phi}_{i,v}$ be highly regular, in a sense that dominates the condition already imposed on the parameters $\dot{\phi}_{j,v}$ with $j < i$. The first assertion of (iii)(a) is then valid.

The second assertion of (iii)(a) requires more discussion. We are certainly free to replace $\dot{M}$ by the subgroup $\dot{G}_-$, since the general linear factors of $\dot{M}$ are connected. In fact, we may as well assume simply that $M = G_- = G$, to save notation. In other words, we shall assume that the group $S_\phi = S_{\dot{\phi}}$ is finite. We have then to show that the global objects can be chosen so that the mapping

$$\Pi_{\dot{\phi}_V} \longrightarrow \widehat{\mathcal{S}}_{\dot{\phi}}, \qquad \mathcal{S}_{\dot{\phi}} = S_{\dot{\phi}}/Z(\hat{\dot{G}})^\Gamma, \tag{6.3.7}$$

is surjective. Given Lemma 6.1.2, the main step will be to verify that the dual mapping

$$\mathcal{S}_{\dot{\phi}} = (S_{\dot{\phi}}/Z(\hat{\dot{G}})^\Gamma) \longrightarrow \mathcal{S}_{\dot{\phi}_V} = \prod_{v \in V} \mathcal{S}_{\dot{\phi}_v} \tag{6.3.8}$$

is injective. The center

$$Z(\hat{\dot{G}})^\Gamma = Z(\hat{\dot{G}})^{\Gamma_v}, \qquad v \in V,$$

plays no role in this, as we will be able to verify that the mapping

$$S_{\dot{\phi}} \longrightarrow \pi_0(S_{\dot{\phi}_V}) = \prod_{v \in V} (\pi_0(S_{\dot{\phi}_v}))$$

is injective. To do so, we have to show that if $s \in S_{\dot{\phi}}$ lies in the identity component $S^0_{\dot{\phi}_v}$ for each $v \in V$, then $s = 1$.

We are regarding $\dot{\phi}_v$ as an N-dimensional representation of $W_{\dot{F}_v} = W_{\mathbb{R}}$. Since its two-dimensional constituents are distinct, they each contribute a discrete factor $O(1,\mathbb{C})$ to the centralizer $S_{\dot{\phi}_v}$. The one-dimensional constituents are parametrized by the subset I^o of indices i with N_i odd. They of course need not be distinct. In fact, for any $v \in V$, they are all trivial unless $v = v_i$ for some $i \in I^o$, in which case the representation corresponding to i is the sign character ε_v of $\dot{F}_v^* = \mathbb{R}^*$. It is these repeated factors that define the identity component of $\mathcal{S}_{\dot{\phi}_v}$. In particular, in the case $v = v_i$, we see that

$$S^0_{\dot{\phi}_v} \cong SO(|I^o| - 1, \mathbb{C}).$$

The intersection of $S_{\dot{\phi}}$ with $S^0_{\dot{\phi}_v}$ is given by the factors of $S_{\dot{\phi}}$ attached to the subset I^1 of indices $i \in I^o$ with $N_i = 1$, since the factors corresponding to other indices $i \in I^o$ map to nontrivial components of $S_{\dot{\phi}_v}$ attached to two-dimensional representations of $W_{\dot{F}_v}$. If $v = v_{i_1}$, for some $i_1 \in I^1$, the intersection of $S_{\dot{\phi}}$ with $S^0_{\dot{\phi}_v}$ takes the form

$$\Big(\prod_{\{i\in I^1:\ i\neq i_1\}} O(1,\mathbb{C})\Big)^+_{\dot{\phi}_v} \cong O(1,\mathbb{C})^k, \qquad k = |I^1| - 2,$$

where the subscript on the left follows the notation defined prior to (1.4.8), and the exponent on the right is understood to be 0 if $|I^1| \leqslant 2$. The intersection of these groups, as i_1 ranges over I^1, is trivial. In particular, any element in the intersection of $S_{\dot{\phi}}$ with all of the groups $S^0_{\dot{\phi}_v}$ is trivial.

We have shown that the mapping (6.3.8) is injective. The dual mapping of abelian character groups

$$\widehat{\mathcal{S}}_{\dot{\phi}_V} \longrightarrow \widehat{\mathcal{S}}_{\dot{\phi}}$$

is therefore surjective. Recall that we have an injection

$$\Pi_{\dot{\phi}_V} = \prod_{v\in V} (\Pi_{\dot{\phi}_v}) \hookrightarrow \widehat{\mathcal{S}}_{\dot{\phi}_V} = \prod_{v\in V} (\widehat{\mathcal{S}}_{\dot{\phi}_v}).$$

It follows from Lemma 6.1.2, applied to each field $\dot{F}_v$, that the image of the set $\Pi_{\dot{\phi}_V}$ generates the finite group $\widehat{\mathcal{S}}_{\dot{\phi}_V}$. Its image under the mapping (6.3.7) therefore generates $\widehat{\mathcal{S}}_{\dot{\phi}}$. To insure that the image of (6.3.7) actually equals $\widehat{\mathcal{S}}_{\dot{\phi}}$, we simply enlarge $\dot{F}$. That is, we carry out the entire argument anew, with a totally real extension $\dot{F}'$ in place of $\dot{F}$, and with $\dot{\phi}'_{v'} = \dot{\phi}_v$ for each place v' of $\dot{F}'$ over v. (In allowing different completions $\dot{\phi}'_{v'}$ to be equal, we are actually applying the variant of Corollary 6.2.4 described in Remark 2 following its proof.) If $\dot{F}'$ is large enough, the global parameter $\dot{\phi}'$ obtained as above gives rise to a composition

$$\Pi_{\dot{\phi}'_{V'}} \longrightarrow \widehat{\mathcal{S}}_{\phi_V} \longrightarrow \widehat{\mathcal{S}}_{\dot{\phi}} \cong \widehat{\mathcal{S}}_{\dot{\phi}'}$$

of surjective mappings. Denoting $\dot{F}'$ and $\dot{\phi}'$ by the original symbols $\dot{F}$ and $\dot{\phi}$, we obtain global data for which the mapping (6.3.7) is indeed surjective. This completes the proof of the second assertion of (iii)(a).

Consider next the condition (iii)(b). Then each ℓ_i equals 1, and the group S_ϕ is finite. This is the case in which the datum $G \in \widetilde{\mathcal{E}}_{\mathrm{ell}}(N)$ is not required to be simple. If ϕ has any one-dimensional constituents, we take $v = v_1 = v_{i_1}$, for any fixed index $i_1 \in I^1$ with $N_{i_1} = 1$. Otherwise, we take v to be any fixed valuation in V. Suppose that $\dot{\phi}_v$ lies in $\widetilde{\Phi}(G_v^*)$, for some simple datum $G_v^* \in \mathcal{E}_{\mathrm{sim}}(N)$ over $\dot{F}_v$. If $\widehat{G}_v^*$ equals $SO(N,\mathbb{C})$, the two-dimensional constituents of $\dot{\phi}_v$ are all orthogonal, since they are mutually inequivalent. By our local induction hypothesis, the localization assertion of Theorem 1.4.2 is valid for each $\dot{G}_i$. (See also Remark 1 after the proof of Corollary 6.2.4.) This implies that the group $\widehat{G}_i = \widehat{\dot{G}}_i$ equals $SO(N_i,\mathbb{C})$, for any i with $N_i > 1$. The same being trivially true if $N_i = 1$, it follows that

$$\widehat{G} = SO(N,\mathbb{C}) = \widehat{G}_v^*,$$

as required. The other possibility is that $\widehat{G}_v^* = Sp(N,\mathbb{C})$. Then all of the two-dimensional constituents of $\dot{\phi}_v$ are symplectic, which implies that $\widehat{G}_i$ equals $Sp(N_i,\mathbb{C})$ for any i with $N_i > 1$. If there were any remaining one-dimensional constituents of $\dot{\phi}_v$, one of them, namely

$$\dot{\phi}_{i_1,v,0} = \dot{\eta}_{i_1,v},$$

would be equal to ε_v, while all of the others would be trivial. Under such circumstances, the image of $\dot{\phi}_v$ cannot be contained in $Sp(N,\mathbb{C})$. Therefore, there are no one-dimensional representations, and $\widehat{G}_i$ equals $Sp(N_i,\mathbb{C})$ for all i. It follows that

$$\widehat{\dot{G}} = Sp(N,\mathbb{C}) = \widehat{G}^*,$$

and in particular, that $\dot{G} \in \widetilde{\dot{\mathcal{E}}}_{\mathrm{sim}}(N)$ is simple. We have established (iii)(b).

Consider finally the condition (iii)(c). Then some ℓ_i is greater than 1, and the group S_ϕ is infinite. In this case, G is required to be simple. We choose v as above, namely to be v_{i_1} if there is an index $i = i_1$ with $N_i = 1$, and to be any fixed valuation in V otherwise. Let x be an element in $\mathcal{S}_{\dot{\phi}}$ whose image in the global R-group $R_{\dot{\phi}}$ lies in $R_{\dot{\phi},\mathrm{reg}}$. We have to show that its image in the local R-group $R_{\dot{\phi}_v}$ is nontrivial. The fact that $R_{\dot{\phi},\mathrm{reg}}$ is nonempty implies that $S_{\dot{\phi}}$ is of the form

$$O(2,\mathbb{C})^q \times O(1,\mathbb{C})^{r'}.$$

In particular, $\ell_i \leqslant 2$ for each i. If there is an i with $N_i = 1$ and $\ell_i = 2$, we choose the index i_1 above so that $\ell_{i_1} = 2$. In this case, i_1 contributes a factor $O(2,\mathbb{C})$ to the local group $S_{\dot{\phi}_v}$. The image of x in $R_{\dot{\phi}}$ then projects onto the nonidentity component of this factor, and is nontrivial as required. If i_1 cannot be chosen in this way, there is an i with $N_i > 1$ and $\ell_i = 2$. In this case, i contributes a product of several factors $O(2,\mathbb{C})$ to $S_{\dot{\phi}_v}$. The image of x in $R_{\dot{\phi}_v}$ then projects onto the product of nonidentity components, and again is nontrivial as required. This establishes the last of the conditions (iii)(c), completing the proof of the lemma. □

The construction of Proposition 6.3.1 will allow us to apply global methods to the proof of the local theorems. The essential case is that of p-adic F. However, we will still have work to do in the case $F = \mathbb{R}$. This will be based on a variant of the construction, which entails a more restrictive choice of the global field $\dot{F}$.

LEMMA 6.3.2. *Suppose that the given local objects F, G, ϕ, M and ϕ_M are such that $F = \mathbb{R}$, and such that the two-dimensional constituents*

$$\{\phi_i : \ N_i = 2\} \tag{6.3.9}$$

of ϕ are in relative general position. Then we can choose global objects $\dot{F}$, $\dot{G}$, $\dot{\phi}$, $\dot{M}$ and $\dot{\phi}_M$ that satisfy the conditions of Proposition 6.3.1, and so that

$$\dot{\phi}_{i,v} = \phi_i, \qquad 1 \leqslant i \leqslant r, \ v \in S^u_\infty. \tag{6.3.10}$$

PROOF. The first step is to choose the global field $\dot{F}$. Since $F = \mathbb{R}$, we can start with the rational field $\mathbb{Q}$. We will then take $\dot{F}$ to be the totally real Galois extension $\mathbb{Q}(\sqrt{q_1}, \dots, \sqrt{q_k})$ of $\mathbb{Q}$, for distinct (positive) prime numbers $q_1, \dots, q_k$, as in the proof of Lemma 6.2.1. We need to take k to be large, in order that the set S_∞ of archimedean valuations of $\dot{F}$ be large. Once k is fixed, we choose the primes q_α to be large relative to the degree 2^k of $\dot{F}$ over $\mathbb{Q}$, and so that

$$q_\alpha \equiv 1 (\text{mod } 4), \qquad 1 \leqslant \alpha \leqslant k.$$

Then the discriminant of $\dot{F}$ equals $q_1 \cdots q_k$, so that $\dot{F}/\mathbb{Q}$ is unramified at any prime p that is not large relative to 2^k. Taking u to be any fixed archimedean valuation of $\dot{F}$, we obtain

$$\dot{F}_u = F = \mathbb{R},$$

as required.

The next step is to choose global endoscopic data $\dot{G}_i \in \tilde{\dot{\mathcal{E}}}_{\text{ell}}(N_i)$ with $\dot{G}_{i,u} = G_i$. This amounts to the choice of quadratic idèle class characters $\dot{\eta}_i$ for $\dot{F}$, which are 1 by definition unless $\widehat{G}_i = SO(N_i, \mathbb{C})$. The degrees N_i here are always equal to 1 or 2, since $F = \mathbb{R}$. The archimedean local conditions

$$\dot{\eta}_{i,v} = \eta_i, \qquad v \in S^u_\infty, \ 1 \leqslant i \leqslant r, \tag{6.3.11}$$

are forced on us by (6.3.10). As in the proof of Proposition 6.3.1, we chose $\dot{\eta}_i$ successively with increasing i so that $\eta_{i,v}$ is unramified at any place $v \notin S_\infty$ such that $\dot{\eta}_{j,v}$ ramifies for some $j < i$. In addition to this condition, we can also insist that $\dot{\eta}_{i,v}$ be unramified for any $v \notin S_\infty$ whose norm is not large relative to 2^k. We choose each $\dot{\eta}_i$ subject to these conditions, and thereby obtain the required global endoscopic data $\dot{G}_i \in \tilde{\dot{\mathcal{E}}}(N_i)$.

Consider an index i such that $\widehat{G}_i = SO(2,\mathbb{C})$. The archimedean localizations (6.3.11) of $\dot{\eta}_i$ then each equal the nontrivial quadratic character $\varepsilon_{\mathbb{R}}$ of $\mathbb{R}^*$. It follows that the quadratic extension $\dot{E}_i$ of $\dot{F}$ attached to $\dot{\eta}_i$ by class field theory is totally imaginary. The group $\dot{G}_i$ is of course the corresponding anisotropic form of $SO(2)$. Let $\dot{Z}_i^\infty = Z^\infty(\dot{G}_i)$ be the (central) subgroup of

$$\dot{G}_{i,\infty} = \prod_{v\in S_\infty} \dot{G}_i(\dot{F}_v),$$

defined as in the proof of Lemma 6.2.2. It is the diagonal image in $\dot{G}_{i,\infty}$ of the intersection of $\dot{G}_i(\dot{F})$ with the maximal compact subgroup $\dot{K}^\infty$ of $\dot{G}_i(\dot{\mathbb{A}}^\infty)$. We shall show that $\dot{Z}_i^\infty$ equals $\mathbb{Z}/2\mathbb{Z}$.

We shall first observe that $\dot{Z}_i^\infty$ is isomorphic to the cyclic subgroup $\dot{C}_i$ of roots of unity in $\dot{E}_i^*$. Obviously

$$\dot{G}_i(\dot{F}) = \{x \in \dot{E}_i^* : \sigma(x) = x^{-1}\},$$

where σ generates the Galois group $\dot{E}_i/\dot{F}$. Therefore $\dot{Z}_i^\infty$ is the group of units u of $\dot{E}_i$ such that $\sigma(u) = u^{-1}$. The Dirichlet unit theorem asserts that the group of all units is the direct product of $\dot{C}_i$ with a free abelian group of rank r [**Cassels**, §18]. Since the group $\dot{Z}_i^\infty$ is finite, it can contain no unit of infinite order, and is therefore contained in $\dot{C}_i$. Since $\dot{E}_i$ is totally complex, σ acts by complex conjugation on $\dot{Z}_i^\infty$. Therefore $\dot{Z}_i^\infty$ does equal $\dot{C}_i$

Let $m \geqslant 2$ be the positive integer such that $\dot{C}_i$ is generated by a primitive mth root of unity

$$\zeta_m = e^{\frac{2\pi i}{m}}.$$

Then $\mathbb{Q}[\zeta_m]$ is the maximal cyclotomic subfield of $\dot{E}_i$. If m is greater than 2, there is a prime number $p \leqslant m$ that ramifies in $\mathbb{Q}[\zeta_m]$, and hence also in $\dot{E}_i$. On the other hand, we know from the choice of $\dot{F}$ and $\dot{\eta}_i$ that $\dot{E}_i/\mathbb{Q}$ can ramify at p only if p is very large. Indeed, it follows from the construction that any such p must be large relative to the degree 2^{k+1} of $\dot{E}_i$ over $\mathbb{Q}$. But the degree $\phi(m)$ of $\mathbb{Q}[\zeta_m]/\mathbb{Q}$ both divides 2^{k+1}, and approaches infinity as m becomes large. This contradicts the existence of p. Therefore $m = 2$, and $\dot{E}_i^*$ contains only the square root (-1) of unity. We have therefore established that

$$\dot{Z}_i^\infty = \dot{C}_i = \mathbb{Z}/2\mathbb{Z}.$$

We note in passing that our ramification condition on the primes q_α is more stringent than necessary. The argument we have just given would apply to any totally real extension $\dot{F}/\mathbb{Q}$ that ramifies only at primes that are large relative its degree.

We now use Corollary 6.2.4 to construct global parameters $\dot{\phi}_i \in \dot{\widetilde{\Phi}}_{\mathrm{sim}}(\dot{G}_i)$. Since $|S_\infty| = 2^k$ is even, the product of $|S_\infty|$-copies of the central character of ϕ_i on the group $\dot{Z}_i^\infty = \mathbb{Z}/2\mathbb{Z}$ equals 1. According to Remark 2 following Corollary 6.2.4, we can then choose the simple parameters $\dot{\phi}_i$ so that

(6.3.10) holds. This construction is not quite the same as that of the proof of Proposition 6.3.1. For example, the earlier property (6.3.6) is incompatible with the requirement (6.3.10) here. However, the conditions (i), (ii) and (iii) of the proposition remain valid for the resulting compound parameters $\dot{\phi}$ and $\dot{\phi}_M$ here. Indeed, the assertions (iii) that relied on (6.3.6) in the earlier construction are now trivial consequences of (6.3.10). The lemma follows. □

REMARK. In following parts of the proof of Proposition 6.3.1, we have made the argument above a little more elaborate than necessary. It would have been sufficient to take

$$\dot{E}_i = \dot{F} \cdot \dot{Q}_i,$$

for a suitable imaginary quadratic extension $\dot{Q}_i$ of $\mathbb{Q}$. However, the argument we have given could conceivably be of interest in its own right.

The primary objects in both Proposition 6.3.1 and Lemma 6.3.2 are the simple global parameters

$$\dot{\phi}_i, \qquad 1 \leqslant i \leqslant r.$$

They are in turn given by Corollary 6.2.4, which relies on the application of the simple trace formula in Lemma 6.2.2 that is the source of the underlying global construction. Once chosen, they determine the required global pair $(\dot{G}, \dot{\phi})$ from the given local pair (G, ϕ).

We took the secondary objects G and ϕ as the given data in the proposition for the obvious reason that they are the objects to which the local theorems apply. Let us put them aside for a moment. Suppose instead that we have been given only the simple constituents ϕ_i of ϕ, and that we have constructed corresponding simple global parameters $\dot{\phi}_i$ by Corollary 6.2.4. We can then form the associated family

$$\dot{\widetilde{\mathcal{F}}} = \widetilde{\mathcal{F}}(\dot{\phi}_1, \dots, \dot{\phi}_r) = \{\dot{\ell}_1\dot{\phi}_1 \boxplus \cdots \boxplus \dot{\ell}_r\dot{\phi}_r : \dot{\ell}_i \geqslant 0\} \tag{6.3.12}$$

of compound global parameters. The supplementary constraints we imposed on the parameters during the proofs of Proposition 6.3.1 and Lemma 6.3.2 are independent of G and $\dot{G}$. More precisely, the primary objects can be chosen independently of any pair

$$(\dot{G}, \dot{\phi}), \qquad \dot{\phi} \in \dot{\mathcal{F}}(\dot{G}),\ \dot{G} \in \dot{\widetilde{\mathcal{E}}}_{\mathrm{ell}}(\dot{N}),\ \dot{N} \geqslant 0,$$

(with $\dot{G}$ being simple if $\dot{\phi} \notin \widetilde{\Phi}_{\mathrm{ell}}(\dot{N})$). The conditions (iii)(a)–(iii)(c) of Proposition 6.3.1 then imply that $\dot{\widetilde{\mathcal{F}}}$ satisfies Assumption 5.4.1, with $V = V(\dot{\widetilde{\mathcal{F}}})$ being the set of archimedean places S_∞^u. We will therefore be able to apply the global results of §5.4 to $\dot{\widetilde{\mathcal{F}}}$.

6.4. The local intertwining relation for ϕ

We turn now to our main task, the proof of the local theorems. We are treating local generic (Langlands) parameters ϕ in this chapter. Our general goal is to apply the inductive procedure of Chapter 5 to the global parameters constructed in the last section. In this section, we shall establish the local intertwining identity of Theorem 2.4.1.

We are taking F to be a local field. If F is the complex field $\mathbb{C}$, the case (ii) of Corollary 2.5.2 tells us that the relevant intertwining operator equals 1. Theorem 2.4.1 then follows from the definitions, and the fact that any centralizer S_ϕ is connected in this case. We can therefore assume that F is real or p-adic as in the last two sections. We also have the simplification of Lemma 2.4.2, which reduces the problem to the case that $\phi = \psi$ belongs to the subset $\widetilde{\Phi}_2(M)$ of $\widetilde{\Phi}(M)$. We will therefore use the notation introduced prior to Proposition 2.4.3, in which ϕ represents a parameter in $\widetilde{\Phi}(G)$, and (M, ϕ_M) is a fixed pair with $\phi_M \in \widetilde{\Phi}_2(M, \phi)$.

We fix a datum $G \in \widetilde{\mathcal{E}}_{\mathrm{sim}}(N)$ over the local field F and a local Langlands parameter

$$\phi = \ell_1\phi_1 \oplus \cdots \oplus \ell_r\phi_r$$

in $\widetilde{\Phi}(G)$, as in (6.3.1). In this section we will assume that some ℓ_i is greater than 1, or in other words, that ϕ does not lie in $\widetilde{\Phi}_2(G)$. Given G and ϕ, we choose the global objects $\dot{F}$, $\dot{G}$, $\dot{\phi}$, $\dot{M}$ and $\dot{\phi}_M$ that satisfy the conditions of Proposition 6.3.1. Having fixed the simple global constituents $\dot{\phi}_1, \dots, \dot{\phi}_r$ of $\dot{\phi}$, we then form the family

$$\dot{\widetilde{\mathcal{F}}} = \widetilde{\mathcal{F}}(\dot{\phi}_1, \dots, \dot{\phi}_r)$$

of global parameters (6.3.12). As we have noted, the conditions of Assumption 5.4.1 follow for $\dot{\widetilde{\mathcal{F}}}$ and $V = S^u_\infty$ from the conditions (iii) of Proposition 6.3.1.

The construction of the global objects relies on the induction hypothesis from §6.3 that the local theorems are valid for all parameters of degree less than N. To this, we now add the formal global induction hypothesis promised in *Remark* 1 following Corollary 6.2.4. We assume that the global theorems hold for global parameters of degree less than N in the set $\dot{\widetilde{\mathcal{F}}}$. The induction arguments of Chapter 5 therefore apply to the parameters in $\dot{\widetilde{\mathcal{F}}}$. In particular, they will allow us to extend our global hypothesis to the compound parameter $\dot{\phi} \in \dot{\widetilde{\Phi}}(N)$ of this section. The first step in the global induction argument, the case of simple parameters, will be resolved in §6.7.

We studied the global intertwining relation in §4.5. We established it directly, by induction arguments that can clearly be carried out in the context of global parameters in the set $\dot{\widetilde{\mathcal{F}}}$, in the cases outlined in Corollary 4.5.2. As we remarked after this corollary, the reduction also carries over without change to the local intertwining relation of Theorem 2.4.1. In other words, the local intertwining relation holds for our generic parameter ϕ, unless it belongs to $\widetilde{\Phi}_{\mathrm{ell}}(G^*)$ for some $G^* \in \widetilde{\mathcal{E}}_{\mathrm{sim}}(N)$, or unless it belongs to the local analogue for ϕ of one of the two exceptional classes (4.5.11) or (4.5.12). In each of these cases, it suffices to consider elements s and u that map to a point x in the subset $\mathcal{S}_{\phi,\mathrm{ell}}$ of $\mathcal{S}_{\phi}$. (We understand $\mathcal{S}_{\phi,\mathrm{ell}}$ to be the local generic analogue of the global set $\mathcal{S}_{\psi,\mathrm{ell}}$ introduced near the beginning of §5.2.) We observe that Proposition 4.5.1 also yields the other global theorems for $\dot{\phi}$, unless ϕ falls into the first of the three cases above.

We thus have three exceptional cases to deal with. We shall study them together. In each case, we shall apply the results of Chapter 5 (and the assertion of Proposition 4.5.1) to the global parameter $\dot{\phi}$. Keep in mind that since $\dot{\phi} = \dot{\psi}$ is generic, the point $s_{\dot{\psi}}$ and the sign character $\varepsilon^G_{\dot{\psi}}$ are both trivial.

The case of $\phi \in \widetilde{\Phi}_{\mathrm{ell}}(G)$ is characterized by the local generic form of (5.2.4). The conditions are

$$(6.4.1)\qquad \begin{cases} \phi = 2\phi_1 \oplus \cdots \oplus 2\phi_q \oplus \phi_{q+1} \oplus \cdots \oplus \phi_r, \\ S_\phi \cong \big(O(2,\mathbb{C})^q \times O(1,\mathbb{C})^{r-q}\big)^+_\phi, \qquad q \geqslant 1, \end{cases}$$

with the requirement that the Weyl group W_ϕ contain an element w in the regular set $W_{\phi,\mathrm{reg}}$. The other two cases are represented by (4.5.11) and (4.5.12) (which is to say the local generic forms of (4.5.11) and (4.5.12) with ϕ and $\oplus$ in place of ψ and $\boxplus$). In all three cases, the associated global pair $(\dot{G},\dot{\phi})$ satisfies an identity

$$(6.4.2)\qquad \sum_{x\in\mathcal{S}_{\phi,\mathrm{ell}}} \big(\dot{f}'_{\dot{G}}(\dot{\phi},\dot{x}) - \dot{f}_{\dot{G}}(\dot{\phi},\dot{x})\big) = 0, \qquad \dot{f} \in \widetilde{\mathcal{H}}(\dot{G}),$$

where $\dot{x}$ is the isomorphic image of x in $\mathcal{S}_{\dot{\phi}}$. This follows from Lemmas 5.4.3 and 5.4.4 in the case (6.4.1), and Corollary 4.5.2 (together with the assertion of Lemma 5.4.4) in the cases (4.5.11) and (4.5.12). Recall that if $q = r = 1$ in (6.4.1), there is a second group $G^\vee \in \widetilde{\mathcal{E}}_{\mathrm{sim}}(N)$ such that $\phi \in \widetilde{\Phi}(G^\vee)$, and such that $S^\vee_\phi = Sp(2,\mathbb{C})$. The corresponding global pair $(\dot{G}^\vee,\dot{\phi})$ actually belongs to the second case (4.5.11) (with (G,ψ) equal to $(\dot{G},\dot{\phi}^\vee)$). It represents that part of the specialization to $r = 1$ of (4.5.11) that was excluded by the hypotheses of Proposition 4.5.1. Since this was later covered by Lemma 5.4.4, in the assertion that (5.2.13) vanishes, we can include it now in our discussion of (4.5.11). In the cases (4.5.11) and (4.5.12), we set $q = 1$.

We assume that

$$\dot{f} = \prod_v \dot{f}_v$$

is a decomposable function. The summand in (6.4.2) can then be decomposed into a difference of products

$$\prod_v \dot{f}'_{v,\dot{G}}(\dot{\phi}_v, \dot{x}_v) - \prod_v \dot{f}_{v,\dot{G}}(\dot{\phi}_v, \dot{x}_v)$$

over all valuations v. In order to remove the places v outside $S_\infty(u)$ from this equation, we have first to establish a very special case of the local intertwining relation.

LEMMA 6.4.1. *Suppose that F is nonarchimedean. Then the local intertwining relation of Theorem* 2.4.1 *holds for ϕ in case $r \leqslant 3$ and*

$$N_i = 1, \qquad 1 \leqslant i \leqslant r.$$

PROOF. The constraints we have imposed here are obviously pretty stringent. Applying them to the cases (6.4.1), (4.5.11) and (4.5.12), we see that the group M is particularly simple. In fact, M is abelian except in the case (4.5.12), with $r = 3$, where we have

$$G_- = M_{\text{der}} = Sp(2).$$

Notice that this last case is the setting of Corollary 2.5.2(v). One could perhaps use the earlier corollary to give a direct local proof of the lemma, but there would still be a number of supplementary points that I have not tried to verify. We shall instead give a less direct proof based on the application of Lemma 2.5.5.

We must show that the objects G, M, ϕ and ϕ_M over F satisfy the conditions (i)–(iii) of Lemma 2.5.5. The condition (ii) follows from [**KS**, §5.3]. (We assume here that the stable distribution on $G_-(F)$ attached to ϕ_M in [**LL**] is the same as the linear form given by Theorem 2.2.1(a), an induction hypothesis we will resolve in Lemma 6.6.2.) Before we consider the condition (i) on char F, we shall establish the final condition (iii) of Lemma 2.5.5.

The global identity (6.4.2) in the restricted case here is based on the idèle class characters

$$\dot{\eta}_i = \dot{\pi}_i, \qquad 1 \leqslant i \leqslant r,$$

of order 1 or 2, which are obtained in Proposition 6.3.1 from the given characters $\eta_i = \pi_i$ on F^*. Suppose that v lies in the set $S_\infty = S_\infty^u$ of archimedean places. Since a completion $\dot{\eta}_v$ at v must be equal to 1 or $\varepsilon_{\mathbb{R}}$, the representations in $\widetilde{\Pi}_{\dot{\phi}_v}$ are all induced from a minimal parabolic subgroup. It then follows from the remark on [**S6**, Lemma 11.5] following (6.1.5) that the characters $\varepsilon_{\dot{\pi}_M,v}(\,\cdot\,)$ in (6.1.5) are all trivial. Therefore

$$\dot{f}'_{v,\dot{G}}(\dot{\phi}_v, \dot{x}_v) = \dot{f}_{v,G}(\dot{\phi}_v, \dot{x}_v), \qquad x \in \mathcal{S}_\phi.$$

The identity (6.4.2) becomes

$$\sum_{x \in \mathcal{S}_{\phi,\text{ell}}} \Big(\prod_{v \notin S_\infty} \dot{f}'_{v,\dot{G}}(\dot{\phi}_v, \dot{x}_v) - \prod_{v \notin S_\infty} \dot{f}_{v,\dot{G}}(\dot{\phi}_v, \dot{x}_v) \Big) \dot{f}_\infty(\dot{\phi}_\infty, \dot{x}_\infty),$$

where

$$\dot{f}_\infty(\dot{\phi}_\infty, \dot{x}_\infty) = \prod_{v\in S_\infty} \dot{f}_{v,\dot{G}}(\dot{\phi}_v, \dot{x}_v).$$

It is a direct consequence of the condition (iii)(a) of Proposition 6.3.1 that the linear mapping from functions $\dot{f}_\infty$ in $\widetilde{\mathcal{H}}(\dot{G}_\infty)$ to functions

$$x \longrightarrow \dot{f}_\infty(\dot{\phi}_\infty, \dot{x}_\infty), \qquad x \in \mathcal{S}_\phi,$$

on $\mathcal{S}_\phi$ is surjective. This implies that

$$\prod_{v\notin S_\infty} \dot{f}'_{v,\dot{G}}(\dot{\phi}_v, \dot{x}_v) = \prod_{v\notin S_\infty} \dot{f}_{v,\dot{G}}(\dot{\phi}_v, \dot{x}_v), \tag{6.4.3}$$

for any $x \in \mathcal{S}_{\dot{\phi},\mathrm{ell}}$.

It follows from the fundamental lemma and Theorem 2.5.1(b) that almost all of the factors on each side of (6.4.3) equal 1. From the definitions and the local correspondence for $\dot{M}_v$ (which we are assuming by induction), it follows that for any $v \notin S_\infty(u)$ and $x \in \mathcal{S}_{\phi,\mathrm{ell}}$, we can choose $\dot{f}_v$ so that the corresponding factor on the right hand side of (6.4.3) is nonzero. This allows us to reduce (6.4.3) to an identity in $f = \dot{f}_u$ and $\phi = \dot{\phi}_u$. We obtain

$$f_G(\phi, x) = e(x)\, f'_G(\phi, x), \qquad x \in \mathcal{S}_{\phi,\mathrm{ell}},\ f \in \widetilde{\mathcal{H}}(G), \tag{6.4.4}$$

for a complex number

$$e(x) = e_u(\dot{x}_u) = \prod_{v\notin S_\infty(u)} \big(\dot{f}'_{v,\dot{G}}(\dot{\phi}_v, \dot{x}_v)\, \dot{f}_{v,\dot{G}}(\dot{\phi}_v, \dot{x}_v)^{-1}\big)$$

that is independent of f.

The identity (6.4.4) would seem to be a little weaker than the condition (iii) of Lemma 2.5.5. However, with the elliptic elements $x \in \mathcal{S}_{\phi,\mathrm{ell}}$ acounted for, the condition follows easily for a general pair (s, u) by induction from the usual arguments of descent. If we impose the condition (i) that $\mathrm{char}(F) \neq 2$, ϕ will then satisfy the three requirements (i), (ii) and (iii) of Lemma 2.5.5. The lemma then tells us that the coefficient $e(x)$ equals 1, and therefore that the local intertwining relation is valid if $\mathrm{char}(F) \neq 2$.

It remains to deal with the case that $\mathrm{char}(F) = 2$. From (6.4.3), (6.4.4), and what we have just established, we see that

$$\prod_{v\in S_2} e_v(\dot{x}_v) = 1,$$

where S_2 is the set of valuations of $\dot{F}$ with $\mathrm{char}(\dot{F}_v) = 2$. The totally real field $\dot{F}$ constructed in the proof of Lemma 6.2.1 has the property that $\dot{F}_v = F$ for each $v \in S_2$. A slightly more sophisticated construction, which we leave to the reader, would give another totally real field $\dot{F}^*$ with the property $\dot{F}^*_{v^*} = F$ for every $v^* \in S^*_2$, but such that

$$|S^*_2| = |S_2| + 1 = 2^k + 1.$$

(The point is that any totally real field has a totally real extension of any given degree, in which a given prime splits completely.) In applying Proposition 6.3.1 to both $\dot{F}$ and $\dot{F}^*$, we can choose the quadratic characters $\dot{\eta}_i$ and $\dot{\eta}_i^*$ so that $\dot{\eta}_{i,v} = \dot{\eta}_{i,v^*}^*$ for any valuations $v \in S_2$ and $v^* \in S_2^*$ not equal to the distinguished valuations $u \in S_2$ and $u^* \in S_2^*$. It follows that

$$e_v(\dot{x}_v) = e_{v^*}(\dot{x}_{v^*}) = e^u(x), \qquad x \in \mathcal{S}_{\phi,\mathrm{ell}},$$

for any such v and v^*, where $e^u(x) \neq 0$ is a fixed complex number. We conclude that

$$e^u(x) = \Big(\prod_{v^* \in S_2^*} e_{v^*}(\dot{x}_{v^*}^*)\Big)\Big(\prod_{v \in S_2} e_v(\dot{x}_v)\Big)^{-1} = 1,$$

since

$$\varepsilon_{u^*}(\dot{x}_{u^*}^*) = e_u(\dot{x}_u)$$

by construction, and therefore that

$$e(x) = e(x)\Big(\prod_{v \neq u} e^u(x)\Big) = \prod_{v \in S_2} e_v(\dot{x}_v) = 1.$$

The local intertwining relation for $\mathrm{char}(F) = 2$ then follows again from (6.4.4). □

We now return to the general case. We are trying to extract the local intertwining relation from the global identity (6.4.2), which we recall applies to the global parameter $\dot{\phi}$ attached to ϕ by Proposition 6.3.1. We shall exploit the different factors of the terms of (6.4.2) attached to a decomposable function $\dot{f} = \prod \dot{f}_v$.

The first step is to remove the factors at places $v \notin S_\infty(u)$. We claim that they satisfy the local intertwining relation

$$\dot{f}'_{v,\dot{G}}(\dot{\phi}_v, \dot{x}_v) = \dot{f}_{v,\dot{G}}(\dot{\phi}_v, \dot{x}_v), \qquad v \notin S_\infty(u). \tag{6.4.5}$$

To see this, we recall that $\dot{\phi}_v$ is a direct sum of quasicharacters of $\dot{F}_v^*$ by Proposition 6.3.1(ii). If one of these constituents is not self-dual, $\dot{\phi}_v$ does not factor through an element in $\widetilde{\Phi}_2(\dot{M}_v)$. It is then easy to see that $\dot{x}_v$ does not belong to $\mathcal{S}_{\dot{\phi}_v,\mathrm{ell}}$, or equivalently, that the Weyl element w_x does not belong to $W_{\dot{\phi}_v,\mathrm{reg}}$. The identity (6.4.5) then follows inductively by reductions similar to those in §4.5 prior to Proposition 4.5.1. If all of the constituents of $\dot{\phi}_v$ are self-dual, it follows from the second assertion of Proposition 6.3.1(ii) that there are at most three distinct constituents. The local pair $(\dot{G}_v, \dot{\phi}_v)$ then satisfies the condition of Lemma 6.4.1, and so (6.4.5) again follows. Since we can choose the functions $\dot{f}_v$ in (6.4.5) so that each side is nonzero, we can remove these factors from the global identity (6.4.2). The identity becomes

$$\sum_{x \in \mathcal{S}_{\phi,\mathrm{ell}}} \Big(\prod_{v \in S_\infty(u)} \dot{f}'_{v,\dot{G}}(\dot{\phi}_v, \dot{x}_v) - \prod_{v \in S_\infty(u)} \dot{f}_{v,\dot{G}}(\dot{\phi}_v, \dot{x}_v)\Big) = 0. \tag{6.4.6}$$

For what remains to be proved of the local intertwining relation postulated in Theorem 2.4.1, the main step turns out to be the case of certain archimedean parameters ϕ. There ought to be a local proof in this case, perhaps following from the methods of [**KnZ2**] and [**S4**]–[**S6**]. However, this still seems illusive, to me at least, so we shall continue to apply the global methods above. We do know that there is nothing further to prove in case $F = \mathbb{C}$. We shall therefore assume for the present that $F = \mathbb{R}$. The degrees of the irreducible constituents ϕ_i of ϕ are then equal to 1 and 2.

We are writing ϕ_M as usual for a fixed element in $\widetilde{\Phi}_2(M, \phi)$, and we identify $\mathcal{S}_{\phi_M}$ with a subgroup of $\mathcal{S}_\phi$. The action of $\mathcal{S}_{\phi_M}$ on $\mathcal{S}_\phi$ by translation stabilizes the subset $\mathcal{S}_{\phi,\mathrm{ell}}$, and gives a simply transitive action

$$x \longrightarrow x\, x_M, \qquad x \in \mathcal{S}_{\phi,\mathrm{ell}},\ x_M \in \mathcal{S}_{\phi_M},$$

of $\mathcal{S}_{\phi_M}$ on $\mathcal{S}_{\phi,\mathrm{ell}}$. In the case $F = \mathbb{R}$ under present consideration, we already have much information about the associated L-packets. According to the results of Shelstad reviewed in §6.1, we already have the archimedean packets $\widetilde{\Pi}_\phi$ and $\widetilde{\Pi}_{\phi_M}$ of irreducible representations, and a surjective mapping $\pi \to \pi_M$ from $\widetilde{\Pi}_\phi$ to $\widetilde{\Pi}_{\phi_M}$. We also have a canonical pairing $\langle \cdot, \cdot \rangle$ on $\mathcal{S}_\phi \times \widetilde{\Pi}_\phi$ such that

$$f'_G(\phi, x) = \sum_{\pi \in \widetilde{\Pi}_\phi} \langle x, \pi \rangle f_G(\pi), \qquad x \in \mathcal{S}_\phi \tag{6.4.7}$$

and

$$f_G(\phi, x) = \sum_{\pi \in \widetilde{\Pi}_\phi} \varepsilon(\pi_M) \langle x, \pi \rangle f_G(\pi), \qquad x \in \mathcal{S}_{\phi,\mathrm{ell}}, \tag{6.4.8}$$

for any function $f \in \widetilde{\mathcal{H}}(G)$ in the associated archimedean Hecke algebra. In the second equation, x has been restricted to the subset $\mathcal{S}_{\phi,\mathrm{ell}}$ of $\mathcal{S}_\phi$, and

$$\varepsilon(\pi_M) = \varepsilon_{\pi_M}(x) = \pm 1$$

is the corresponding restriction of the character ε_{π_M} in (6.1.5), which then depends only on the image π_M of π in $\widetilde{\Pi}_{\phi_M}$. The local intertwining relation asserts that $\varepsilon(\pi_M)$ equals 1.

LEMMA 6.4.2. *Assume that $F = \mathbb{R}$, and that ϕ is in general position. Then there is an element $\varepsilon_1 \in \mathcal{S}_{\phi_M}$ such that*

$$f_G(\phi, x) = f'_G(\phi, x\,\varepsilon_1), \qquad f \in \widetilde{\mathcal{H}}(G),\ x \in \mathcal{S}_{\phi,\mathrm{ell}}.$$

PROOF. The lemma asserts simply that the unknown sign takes the special form

$$\varepsilon(\pi_M) = \langle \varepsilon_1, \pi_M \rangle, \qquad \pi_M \in \widetilde{\Pi}_{\phi_M}.$$

In other words, $\varepsilon(\pi_M)$ can be identified with the restriction to the image of $\widetilde{\Pi}_{\phi_M}$ in $\widehat{\mathcal{S}}_{\phi_M}$ of a character on $\widehat{\mathcal{S}}_{\phi_M}$. Since it treats only a pretty severe specialization of both F and ϕ, the lemma would appear to be a pretty minor step. In point of fact, it represents essential preparation for the general argument. Its proof amounts to an exercise in the Fourier analysis on a finite abelian group.

We choose global data $(\dot{F}, \dot{G}, \dot{\phi}, \dot{M}, \dot{\phi}_M)$ as in Lemma 6.3.2. Then $\dot{F}$ is a totally real field, whose set S_∞ of archimedean places is large. The valuation u belongs to S_∞, and we have

$$(\dot{F}_v, \dot{G}_v, \dot{\phi}_v, \dot{M}_v, \dot{\phi}_{M,v}) = (F, G, \phi, M, \phi_M),$$

for each $v \in S_\infty$. The identity (6.4.6) becomes

$$\sum_{x \in \mathcal{S}_{\phi,\mathrm{ell}}} \Big(\prod_{v \in S_\infty} \dot{f}'_{v,G}(\phi, x) - \prod_{v \in S_\infty} \dot{f}_{v,G}(\phi, x) \Big) = 0, \tag{6.4.9}$$

for a test function

$$\dot{f}_\infty = \prod_{v \in S_\infty} \dot{f}_v, \qquad \dot{f}_v \in \widetilde{\mathcal{H}}(G),$$

which is now a finite product of functions from the same space $\widetilde{\mathcal{H}}(G)$.

If $f \in \widetilde{\mathcal{H}}(G)$ is fixed, we can regard the function (6.4.7) of $x \in \mathcal{S}_\phi$ as an inverse Fourier transform of the function $f_G(\pi)$ of $\pi \in \widetilde{\Pi}_\phi$. In particular, we can recover $f_G(\pi)$ from $f'_G(\phi, x)$ as a Fourier transform on $\mathcal{S}_\phi$. It will be convenient to take a partial Fourier transform, relative to the variable represented by the $\mathcal{S}_{\phi_M}$-torsor $\mathcal{S}_{\phi,\mathrm{ell}}$ in $\mathcal{S}_\phi$. Bearing in mind the general lack of a canonical splitting for the sequence

$$1 \longrightarrow \mathcal{S}_{\phi_M} \longrightarrow \mathcal{S}_\phi \longrightarrow R_\phi \longrightarrow 1,$$

we define

$$f'_G(\phi, \xi) = |\mathcal{S}_{\phi_M}|^{-1} \sum_{x \in \mathcal{S}_{\phi,\mathrm{ell}}} f'_G(\phi, x)\, \xi(x)^{-1},$$

for any character $\xi \in \widehat{\mathcal{S}}_\phi$. The function

$$a(\xi) = f'_G(\phi, \xi), \qquad \xi \in \widehat{\mathcal{S}}_\phi,$$

obviously satisfies the equivariance relation

$$a(\xi \cdot \xi_R) = \xi_R(r_{\mathrm{ell}})^{-1}\, a(\xi), \qquad \xi_R \in \widehat{R}_\phi, \tag{6.4.10}$$

under translation by the subgroup $\widehat{R}_\phi$ of $\widehat{\mathcal{S}}_\phi$, where $r_{\mathrm{ell}} = r(x)$ is the image in R_ϕ of (any) $x \in \mathcal{S}_{\phi,\mathrm{ell}}$. We can therefore write

$$f'_G(\phi, x) = a^\vee(x), \qquad x \in \mathcal{S}_{\phi,\mathrm{ell}},$$

for the inverse transform

$$a^\vee(x) = \sum_{\xi \in \widehat{\mathcal{S}}_{\phi_M}} a(\xi)\, \xi(x), \qquad x \in \mathcal{S}_{\phi,\mathrm{ell}}. \tag{6.4.11}$$

The last summand is obviously invariant under translation of $\xi \in \widehat{\mathcal{S}}_\phi$ by $\widehat{R}_\phi$, and therefore does represent a function of ξ in the quotient $\widehat{\mathcal{S}}_{\phi_M} = \widehat{\mathcal{S}}_\phi / \widehat{R}_\phi$.

We define the partial Fourier transform $f_G(\phi,\xi)$ of $f_G(\phi,x)$ in the same way. It is then easy to see that

$$f_G(\phi,\xi) = \varepsilon(\xi)\, f'_G(\phi,\xi), \qquad \xi \in \widehat{\mathcal{S}}_\phi,$$

for the function

$$\varepsilon(\xi) = \begin{cases} \varepsilon(\pi_M), & \text{if } \xi = \langle\,\cdot\,,\pi_M\,\rangle, \text{ for some } \pi_M \in \widetilde{\Pi}_{\phi_M}, \\ 0, & \text{otherwise,} \end{cases}$$

on $\widehat{\mathcal{S}}_{\phi_M} = \widehat{\mathcal{S}}_\phi/\widehat{R}_\phi$. (As above, we are making no distinction between a function on $\widehat{\mathcal{S}}_{\phi_M}$ and an $\widehat{R}_\phi$-invariant function on $\widehat{\mathcal{S}}_\phi$.) The original function is therefore the inverse transform

$$f_G(\phi,x) = (\varepsilon a)^\vee(x)$$

of the product

$$(\varepsilon a)(\xi) = \varepsilon(\xi)\, a(\xi) = \varepsilon(\xi)\, f'_G(\phi,\xi).$$

These definitions are of course examples of formal operations in finite abelian Fourier analysis. In general, any function a on $\widehat{\mathcal{S}}_\phi$ that satisfies (6.4.10) has an inverse transform (6.4.11) on $\mathcal{S}_{\phi,\mathrm{ell}}$. It satisfies the inversion formula

$$a(\xi) = |\mathcal{S}_{\phi_M}|^{-1} \sum_{x\in\mathcal{S}_{\phi,\mathrm{ell}}} a^\vee(x)\,\xi(x)^{-1},$$

which specializes to

$$\sum_{x\in\mathcal{S}_{\phi,\mathrm{ell}}} a^\vee(x) = |\mathcal{S}_{\phi_M}|\, a(1)$$

at $\xi = 1$. Any two functions have a convolution

$$(a_1 * a_2)(\xi) = \sum_{\xi_1\in\widehat{\mathcal{S}}_{\phi_M}} a_1(\xi_1)\, a_2(\xi_1^{-1}\xi),$$

which satisfies

$$(a_1 * a_2)^\vee = a_1^\vee\, a_2^\vee.$$

We shall use these operations to transform (6.4.9) to the assertion of the lemma.

We order the valuations in S_∞ as $v_1,\dots,v_n$, and set

$$a_i(\xi) = a_{v_i}(\xi) = \dot{f}'_{v_i,G}(\phi,\xi), \qquad \xi\in\widehat{\mathcal{S}}_\phi,$$

for $1 \leqslant i \leqslant n$. The left hand side of (6.4.9) can then be written

$$\begin{aligned}
&\sum_{x \in \mathcal{S}_{\phi,\mathrm{ell}}} \Big(\prod_{i=1}^{n} a_i^{\vee}(x) - \prod_{i=1}^{n} (\varepsilon a_i)^{\vee}(x) \Big) \\
=&\sum_{x} \big((a_1 * \cdots * a_n)^{\vee}(x) - \big((\varepsilon a_1) * \cdots * (\varepsilon a_n)\big)^{\vee}(x) \big) \\
=&|\mathcal{S}_{\phi_M}| \big((a_1 * \cdots * a_n)(1) - \big((\varepsilon a_1) * \cdots * (\varepsilon a_n)\big)(1) \big) \\
=&|\mathcal{S}_{\phi_M}| \sum_{\xi_1,\dots,\xi_n} \big(a_1(\xi_1) \cdots a_n(\xi_n) - \varepsilon(\xi_1) a_1(\xi_1) \cdots \varepsilon(\xi_n) a(\xi_n) \big) \\
=&|\mathcal{S}_{\phi_M}| \sum_{\xi_1,\dots,\xi_n} \big(1 - \varepsilon(\xi_1) \cdots \varepsilon(\xi_n) \big) \, a_1(\xi_1) \cdots a_n(\xi_n),
\end{aligned}$$

where the last two sums are taken over the set

$$\big\{ (\xi_1, \dots, \xi_n) \in (\widehat{\mathcal{S}}_{\phi_M})^n : \ \xi_1 \cdots \xi_n = 1 \big\}.$$

Suppose that $(\xi_1, \dots, \xi_n)$ is a multiple index such that for each i, ξ_i belongs to the image of $\widetilde{\Pi}_{\phi_M}$ in $\widehat{\mathcal{S}}_{\phi_M}$. We can then choose the functions $\dot{f}_{v_i} \in \widetilde{\mathcal{H}}(G)$ so that for any $\xi \in \widehat{\mathcal{S}}_{\phi}$, $a_i(\xi)$ is nonzero if ξ lies in the coset of ξ_i, and is zero otherwise. This follows from the results of Shelstad, by which R_ϕ is identified with the representation theoretic R-group $R(\pi_M)$ of any $\pi_M \in \widetilde{\Pi}_{\phi_M}$. The functions $\dot{f}_{v_i}$ serve to isolate the summand of $(\xi_1, \dots, \xi_n)$ as the only nonvanishing term in our expression for the left hand side of (6.4.9). Since the right hand side of (6.4.9) vanishes, we conclude that

$$\varepsilon(\xi_1) \cdots \varepsilon(\xi_n) = 1, \qquad \text{if } \xi_1 \cdots \xi_n = 1. \tag{6.4.12}$$

The character $\xi = 1$ in $\widehat{\mathcal{S}}_{\phi_M}$ belongs to the image of $\widetilde{\Pi}_{\phi_M}$. In fact, it follows from [**S6**, Theorem 11.5] that ξ corresponds to the (B_M, χ_M)-generic representation $\pi_M \in \widetilde{\Pi}_{\phi_M}$. Applying the identity (2.5.4) of Theorem 2.5.1(b) for the (B, χ)-generic constituent of $\mathcal{I}_P(\pi_M)$ to the definition of the sign $\varepsilon(\pi_M)$, we see that $\varepsilon(1) = 1$. We are assuming that the integer $n = |S_\infty|$ in (6.4.12) is fixed and large. Having now established that $\varepsilon(1) = 1$, we are free to replace n by any integer $m \leqslant n$, since we can always set $\xi_i = 1$ for $i > m$. We can therefore assume also that the mapping ε preserves any relation

$$\xi_1 \cdots \xi_m = 1, \qquad m \geqslant 1,$$

among elements ξ in the image of $\widetilde{\Pi}_{\phi_M}$ in $\widehat{\mathcal{S}}_{\phi_M}$. Since this image generates $\widehat{\mathcal{S}}_{\phi_M}$, by condition (iii)(a) of Proposition 6.3.1, ε extends uniquely to a character on the group $\widehat{\mathcal{S}}_{\phi_M}$. Therefore

$$\varepsilon(\xi) = \xi(\varepsilon_1), \qquad \xi \in \widehat{\mathcal{S}}_{\phi_M},$$

for a unique element $\varepsilon_1 \in \mathcal{S}_{\phi_M}$. It follows from (6.4.7) and (6.4.8) that

$$\begin{aligned} f_G(\phi, x) &= \sum_{\pi \in \widetilde{\Pi}_\phi} \varepsilon(\pi_M)\langle x, \pi\rangle f_G(\pi) \\ &= \sum_{\pi \in \widetilde{\Pi}_\phi} \langle \varepsilon_1, \pi_M\rangle\langle x, \pi\rangle f_G(\pi) \\ &= \sum_{\pi \in \widetilde{\Pi}_\phi} \langle x\,\varepsilon_1, \pi\rangle f_G(\pi) = f'_G(\phi, x\,\varepsilon_1), \end{aligned}$$

as required. □

LEMMA 6.4.3. *The formula of the last lemma holds with* $\varepsilon_1 = 1$.

PROOF. We are taking ϕ to be a parameter (6.4.1), (4.5.11) or (4.5.12) in $\widetilde{\Phi}(G)$ over $F = \mathbb{R}$. Our assumption is that ϕ is in general position, in the sense that its simple constituents ϕ_i with $N_i = 2$ are in relative general position. In the proof of the last lemma, we applied the global construction of Lemma 6.3.2. In particular, we worked with identical archimedean parameters

$$\dot{\phi}_{i,v} = \phi_i, \qquad 1 \leqslant i \leqslant r,\ v \in S_\infty, \tag{6.4.13}$$

attached to the global field $\dot{F}$. In the proof of this lemma, we shall use the p-adic form of the construction in Proposition 6.3.1. This will give us supplementary information about ϕ. The logic of the construction will actually be slightly convoluted, since we will choose the objects of the proposition in a different order.

The irreducible degrees N_i of ϕ equal 1 or 2. We introduce new degrees by setting

$$N_j^\# = \begin{cases} N_j, & \text{if } j \leqslant q, \\ N_{q+1} + \cdots + N_r, & \text{if } j = q+1, \end{cases}$$

where j ranges from 1 to $(q+1)$ and where we recall that $q = 1$ in the exceptional cases (4.5.11) and (4.5.12). In principle, we will be applying Proposition 6.3.1 to a local parameter

$$\phi^\# = \ell_1\phi_1^\# \oplus \cdots \oplus \ell_q\phi_q^\# \oplus \phi_{q+1}^\#,$$

with simple components

$$\phi_j^\# \in \widetilde{\Phi}_{\mathrm{sim}}(G_j^\#), \qquad G_j^\# \in \widetilde{\mathcal{E}}_{\mathrm{sim}}(N_j^\#),$$

over a p-adic field $F^\#$, and multiplicities $\ell_1, \ldots, \ell_q$ that match those of ϕ. However, we will need to choose endoscopic data $\dot{G}_j^\# \in \dot{\widetilde{\mathcal{E}}}_{\mathrm{sim}}(N_j^\#)$ over a global field $\dot{F}^\#$ before we introduce $\phi^\#$.

We let $\dot{F}^\#$ be the global field $\dot{F}$ chosen in the proof of Lemma 6.3.2. For each $j \leqslant q$, we then take $\dot{G}_j^\#$ to be the global endoscopic datum $\dot{G}_j$ over $\dot{F}$

that was constructed also in the proof of the last lemma. The remaining datum $\dot{G}^{\sharp}_{q+1}$ is determined by the requirements that

$$\hat{\dot{G}}^{\sharp}_{q+1} \supset \hat{\dot{G}}_{q+1} \times \cdots \times \hat{\dot{G}}_r = \widehat{G}_{q+1} \times \cdots \times \widehat{G}_r$$

and

$$\dot{\eta}^{\sharp}_{q+1} = \eta_{\dot{G}^{\sharp}_{q+1}} = \prod_{i=q+1}^{r} \dot{\eta}_i.$$

(By convention, $\dot{\eta}_i = 1$ in case $\hat{\dot{G}}_i = Sp(N_i, \mathbb{C})$.) Once these objects are chosen, we fix a nonarchimedean place $u^{\sharp}$ of $\dot{F}$ such that for each j, the local endoscopic datum

$$G^{\sharp}_j = \dot{G}^{\sharp}_{j,u^{\sharp}}$$

is elliptic, and hence simple, over the local field $F^{\sharp} = \dot{F}_{u^{\sharp}}$. (This last condition is needed only in case $\hat{\dot{G}}^{\sharp}_j = SO(2, \mathbb{C})$.) We then choose local parameters $\phi^{\sharp}_j \in \widetilde{\Phi}_{\mathrm{sim}}(N^{\sharp}_j)$ over the p-adic field $F^{\sharp}$ whose central character on the finite group $\dot{Z}^{\infty}_j = Z^{\infty}(\dot{G}_j)$ is trivial. The existence of $\phi^{\sharp}_j$ follows easily from our induction hypothesis that the local theorems hold for parameters in $\widetilde{\Phi}(N^{\sharp}_j)$ over $F^{\sharp}$.

Once we have the local parameters $\phi^{\sharp}_j$ and the global data $\dot{G}^{\sharp}_j$, we choose global parameters $\dot{\phi}^{\sharp}_j \in \widetilde{\Phi}_{\mathrm{sim}}(\dot{G}^{\sharp}_j)$ over $\dot{F}$ according to Corollary 6.2.4. In fact, by Remark 2 following Corollary 6.2.4, we can choose them so that

$$\dot{\phi}^{\sharp}_{j,v} = \begin{cases} \phi_j, & \text{if } j \leqslant q, \\ \phi_{q+1} \oplus \cdots \oplus \phi_r, & \text{if } j = q+1, \end{cases}$$

for any $v \in S_{\infty}$. This is because the product over v of the associated central characters is trivial on $\dot{Z}^{\infty}_j$, as in the proof of Lemma 6.3.2, while the central character of $\phi^{\sharp}_j$ is also trivial. The global pairs $(\dot{G}^{\sharp}_j, \dot{\phi}^{\sharp}_j)$ then determine a global endoscopic datum $\dot{G}^{\sharp} \in \dot{\widetilde{\mathcal{E}}}_{\mathrm{sim}}(N)$, as at the beginning of §6.3, together with a global parameter

$$\dot{\phi}^{\sharp} = \ell_1 \dot{\phi}^{\sharp}_1 \boxplus \cdots \boxplus \ell_r \dot{\phi}^{\sharp}_q \boxplus \dot{\phi}^{\sharp}_{q+1} \tag{6.4.14}$$

in $\widetilde{\Phi}(\dot{G}^{\sharp})$ such that

$$\dot{\phi}^{\sharp}_v = \phi, \qquad v \in S_{\infty}.$$

The global objects $\dot{G}^{\sharp}$ and $\dot{\phi}^{\sharp}$ are not those of the proof of Proposition 6.3.1, since the archimedean components of $\dot{\phi}^{\sharp}$ are all equal. They nonetheless satisfy the conditions of the proposition (with $\dot{M}^{\sharp}$ and $\dot{\phi}^{\sharp}_M$ obtained from $\dot{G}^{\sharp}$ and $\dot{\phi}^{\sharp}$ as at the beginning of §6.3). The conditions are in fact straightforward consequences of the restricted nature of the multiplicities ℓ_i in (6.4.14). For example, in the cases (6.4.1) or (4.5.11) that the multiplicities are all 2, the datum $\dot{G}^{\sharp}_{q+1}$ is just the factor $\dot{G}^{\sharp}_{-}$ of $\dot{M}^{\sharp}$ that is classical, namely an orthogonal or symplectic group. Since the associated parameter $\dot{\phi}^\sharp_{q+1} = \dot{\phi}^\sharp_-$ is then simple, the group $\mathcal{S}_{\dot{\phi}^\sharp_M} = \mathcal{S}_{\phi^\sharp_M}$ in condition (iii)(a) actually equals 1, making the condition trivial. The other conditions are either equally clear or as in the proof of the proposition. It therefore follows that the relation (6.4.6) holds for $\dot{\phi}^\sharp$. We shall apply it to the case that the group $\mathcal{S}_{\phi^\sharp_M}$ is trivial.

With the assumption that $\mathcal{S}_{\phi^\sharp_M} = \{1\}$, the sum over $x \in \mathcal{S}_{\phi^\sharp,\mathrm{ell}}$ in (6.4.6) reduces to a single element x_1. We identify x_1 with the base point in the larger set $\mathcal{S}_{\phi,\mathrm{ell}}$. Since $\dot{M}^\sharp_v = M$ for any $v \in S_\infty$, the identity reduces in this case to

$$\Big(\prod_v \dot{f}_{v,G}(\phi, x_1\varepsilon_1)\Big) f^\sharp_G(\phi^\sharp, x_1) = \Big(\prod_v \dot{f}_{v,G}(\phi, x_1)\Big) (f^\sharp)'_G(\phi^\sharp, x_1), \tag{6.4.15}$$

for any function

$$\Big(\prod_v \dot{f}_v\Big) f^\sharp, \qquad f^\sharp \in \widetilde{\mathcal{H}}(G^\sharp),\ \dot{f}_v \in \widetilde{\mathcal{H}}(\dot{G}_v),\ v \in S_\infty.$$

Suppose that the point ε_1 is not equal to 1. Then the linear forms $f_G(\phi, x_1)$ and $f_G(\phi, x_1\varepsilon_1)$ in $f \in \widetilde{\mathcal{H}}(G)$ are linearly independent. This follows from the definition (2.4.5) of $f_G(\phi, x)$, the disjointness of the irreducible constituents of induced tempered representations, and the fact that the image of $\widetilde{\Pi}_{\phi_M}$ generates $\widehat{\mathcal{S}}_{\phi_M}$. Let $f = f_{v_1}$ be a variable function in $\widetilde{\mathcal{H}}(G) = \widetilde{\mathcal{H}}(G_{v_1})$ at some fixed place $v_1 \in S_\infty$. At the other places $v \in S_\infty$, and also at the place $u^\sharp$, we fix functions so that the contributions to the left hand side of (6.4.15) do not vanish. It follows that

$$f_G(\phi, x_1) = c_1 f_G(\phi, x_1\varepsilon_1), \qquad f \in \widetilde{\mathcal{H}}(G),$$

for a constant c_1 that is independent of f. This is a contradiction.

We have established the required condition that $\varepsilon_1 = 1$, for parameters ϕ such that the group $\mathcal{S}_{\dot{\phi}^\sharp_M}$ equals $\{1\}$. This includes the basic elliptic case (6.4.1), and the first supplementary case (4.5.11). The remaining supplementary case (4.5.12), in which $q = 1$ and $\ell_1 = 3$, is slightly different. In this case, the classical factor $\dot{G}^\sharp_-$ of $\dot{M}^\sharp$ is larger than the group $\dot{G}^\sharp_{q+1} = \dot{G}^\sharp_2$. The associated global parameter equals

$$\dot{\phi}^\sharp_- = \dot{\phi}^\sharp_q \oplus \dot{\phi}^\sharp_{q+1} = \dot{\phi}^\sharp_1 \oplus \dot{\phi}^\sharp_2,$$

and is not simple. If it satisfies the further two conditions that $N_1 = N^\sharp_1 = 1$ and $N^\sharp_2$ is odd, the rank

$$N = 3N_1 + N_2 + \cdots + N_r = 2 + N^\sharp_1 + N^\sharp_2$$

is even, and the character $\xi^+_{\dot{\psi}^\sharp_M}$ on the group $\big(O(1,\mathbb{C}) \times O(1,\mathbb{C})\big)$ is nontrivial. We then see from (1.4.9) that $S_{\dot{\phi}^\sharp_M}$ is isomorphic to $\mathbb{Z}/2\mathbb{Z}$, and also that there is a nontrivial central element in the dual group $\widehat{G}$. The quotient $\mathcal{S}_{\dot{\phi}^\sharp_M}$ of

$S_{\dot\phi^\sharp_M}$ is therefore trivial, and we still have $\varepsilon_1 = 1$. If these two conditions are not met, however, $\mathcal{S}_{\dot\phi^\sharp_M}$ has order 2, and the argument fails.

It therefore remains to establish the lemma in the case (4.5.12), under the further restriction that either $N_1^\sharp = 2$ or $N_2^\sharp$ is even. Since there are only two self-dual, one-dimensional characters on $F^* = \mathbb{R}^*$, this restriction implies that if $N_1 = N_1^\sharp = 1$, then $N_i = 2$ for every $i > 1$.

There are several ways to proceed. The simplest perhaps is to introduce a new parameter

$$\phi^\flat = 2\phi_1 \oplus \phi_1^\flat \oplus \phi_2 \oplus \cdots \oplus \phi_r, \qquad \phi_1^\flat \in \widetilde{\Phi}_{\mathrm{sim}}(N_1),$$

in $\widetilde{\Phi}(G)$ over $F = \mathbb{R}$. That is, we replace one of the three copies of ϕ_1 in the original parameter ϕ with a simple parameter $\phi_1^\flat$ that is different from all the others. If $N_1 = 1$, for example, the last restriction allows us to set $\phi_1^\flat$ equal to $\phi_1\varepsilon_{\mathbb{R}}$. If we apply Proposition 6.3.1 to $\phi^\flat$, we obtain a parameter $\dot\phi^\flat$ over the global field $\dot F$. In fact, with a little thought, one can modify the argument in such a way as to choose $\dot\phi^\flat$ so that

$$\dot\phi_v^\flat = \begin{cases} \phi, & \text{if } v = v_0, \\ \phi^\flat, & \text{if } v \neq v_0, \end{cases}$$

for any $v \in S_\infty$, where v_0 is a fixed valuation in S_∞, and so that the conditions of Proposition 6.3.1 remain valid for $\dot\phi^\flat$. Then for any $v \neq v_0$, $\dot\phi_v^\flat$ is an archimedean parameter for which we have already established the lemma. We leave the reader to check that the resulting application of (6.4.6) yields the required property $\varepsilon_1 = 1$ for the remaining parameter $\phi = \dot\phi_{v_0}^\flat$. □

The last two lemmas represent the main step. They yield the local intertwining relation of Theorem 2.4.1 in the special case that $F = \mathbb{R}$ and ϕ is in general position. With this information, it will be possible to deduce the general case.

PROPOSITION 6.4.4. *Theorem* 2.4.1 *is valid for generic parameters* $\phi \in \widetilde{\Phi}(G)$ *over the general local field* F.

PROOF. As we agreed at the beginning of the section, the theorem holds unless ϕ belongs to one of the three exceptional cases (6.4.1), (4.5.11) and (4.5.12), and the given elements s and u map to a point x in the set $\mathcal{S}_{\phi,\mathrm{ell}}$. We must still allow F to be either real or p-adic, since the parameters ϕ over $\mathbb{R}$ treated above were required to be in general position. We therefore take $\phi \in \widetilde{\Phi}(G)$ to be an arbitrary generic parameter over the given field F, of the form (6.4.1), (4.5.11) or (4.5.12).

We shall apply Proposition 6.3.1 directly to ϕ. From the objects G, ϕ, M and ϕ_M over F attached to ϕ, we obtain global objects $\dot G$, $\dot\phi$, $\dot M$ and $\dot\phi_M$ over $\dot F$ that satisfy the conditions of the proposition. We then obtain the identity (6.4.2) from the results of §5.4 and Corollary 4.5.2, and its reduction

(6.4.6) implied by Lemma 6.4.1. It follows that the sum

$$\sum_{x\in\mathcal{S}_{\phi,\mathrm{ell}}}\Big(\Big(\prod_{v\in S^u_\infty}\dot f'_{v,G}(\dot\phi_v,\dot x_v)\Big)f'_G(\phi,x)-\Big(\prod_{v\in S^u_\infty}\dot f_{v,G}(\dot\phi_v,\dot x_v)\Big)f_G(\phi,x)\Big)$$

vanishes, for any decomposable function

$$\Big(\prod_{v\in S^u_\infty}\dot f_v\Big)f=\dot f^u_\infty f,\qquad \dot f_v\in\widetilde{\mathcal{H}}(\dot G_v),\ f\in\widetilde{\mathcal{H}}(G).$$

For any $v\in S^u_\infty$, the parameter $\dot\phi_v\in\widetilde\Phi(\dot G_v)$ is in general position. It follows from Lemmas 6.4.2 and 6.4.3 that

$$\dot f'_{v,G}(\dot\phi_v,\dot x_v)=\dot f_{v,G}(\dot\phi_v,\dot x_v),\qquad x\in\mathcal{S}_{\phi,\mathrm{ell}}.$$

The last identity then reduces to a linear relation

$$\sum_{x\in\mathcal{S}_{\phi,\mathrm{ell}}}\dot f^u_{\infty,\dot G}(\phi,x)\big(f'_G(\phi,x)-f_G(\phi,x)\big)=0$$

in f, with coefficients

$$\dot f^u_{\infty,\dot G}(\phi,x)=\prod_{v\in S^u_\infty}\dot f_{v,\dot G}(\dot\phi_v,\dot x_v).$$

As linear forms in $\dot f^u_\infty$, the coefficients are linearly independent. This follows from the definition (2.4.5), the disjointness of constituents of induced tempered representations, and the condition (iii)(a) of Proposition 6.3.1. We therefore reach the conclusion that

$$f'_G(\phi,x)=f_G(\phi,x),\qquad x\in\mathcal{S}_{\phi,\mathrm{ell}},\ f\in\widetilde{\mathcal{H}}(G).$$

This is the identity to which we had reduced Theorem 2.4.1. We have now established it simultaneously for each of the three critical cases (6.4.1), (4.5.11) and (4.5.12). Therefore Theorem 2.4.1 holds for each of the three cases, and hence for any generic parameter. □

COROLLARY 6.4.5. *The canonical self-intertwining operator attached any $\phi\in\widetilde\Phi(G)$, $w\in W^0_\phi$ and $\pi_M\in\widetilde\Pi_{\phi_M}$ satisfies*

$$R_P(w,\widetilde\pi_M,\phi)=1. \tag{6.4.16}$$

PROOF. As we noted near the end of §2.5, the extension $\widetilde\pi_M$ of π_M is canonical in this case, and so therefore is the intertwining operator in (6.4.16). To establish (6.4.16), it suffices to consider the two special cases represented by the local analogues of (4.5.11) and (4.5.12). We shall sketch a proof for these cases that combines what we have just established with another global observation.

From the local objects G, ϕ, M and ϕ_M over F, we obtain global objects $\dot G$, $\dot\phi$, $\dot M$ and $\dot\phi_M$ over $\dot F$. It is not hard to see from the global identity (4.5.14) (which holds for both (4.5.11) and (4.5.12)) and the condition (iii)(a) of Proposition 6.3.1 that

$$R_P(w,\widetilde\pi_M,\phi)=e(w)\,1,$$

for a complex number $e(w)$. The right hand side of the local intertwining relation (2.4.7) then satisfies

$$f_G(\phi, u) = f_G(\phi, wu) = e(w)\, f_G(\phi, u),$$

since u and wu have the same image in $\mathcal{S}_\phi$. It follows that $e(w) = 1$. □

6.5. Orthogonality relations for ϕ

We are working towards the local classification of tempered representations. We will continue to use global methods, and in particular, the constructions of global data in §6.2 and §6.3. However, the global methods will have to be supplemented with some new local information. In this section, we shall stabilize the orthogonality relations given by the local trace formula, and as well as their twisted analogues. In particular, we shall use the local intertwining relation we have now established to refine the general results of [**A11**, §1–3].

We suppose henceforth that our local field F is nonarchimedean. We are assuming inductively that the local theorems are valid for generic parameters of degree less than our fixed positive integer N. The theorems now have the interpretation of §6.1. However, we recall that this entails essentially no change in their application.

Let us be a little more explicit about the last point. Suppose that ϕ lies in the complement of $\widetilde{\Phi}_{\mathrm{sim}}(N)$ and $\widetilde{\Phi}(N)$. We can then use our induction hypothesis to define a substitute of the local Langlands group L_F, exactly as in the global case treated in §1.4. It is a complex reductive group

$$\mathcal{L}_\phi \longrightarrow \Gamma_F$$

over Γ_F, which comes with a $GL(N, \mathbb{C})$-orbit of L-embeddings into ${}^L GL(N)$. If G belongs to $\widetilde{\mathcal{E}}_{\mathrm{sim}}(N)$, ϕ lies in the subset $\widetilde{\Phi}(G)$ of $\widetilde{\Phi}(N)$ if and only if one of these embeddings factors through the L-image of ${}^L G$ in ${}^L GL(N)$. The local centralizer sets S_ϕ, $\overline{S}_\phi$ and $\mathcal{S}_\phi$ are then given as usual in terms of the image of $\mathcal{L}_\phi$ in ${}^L G$. We can also form the local set $\Phi(G, \phi)$ of $\widehat{G}$-orbits of L-homomorphisms from $\mathcal{L}_\phi$ to ${}^L G$ that map to ϕ. This set has order

$$m_\phi = m(\phi) = |\Phi(G, \phi)|$$

equal to 1 or 2, like its global analogue from the statement of Theorem 1.5.2.

Our concern in this section will be the subset $\widetilde{\Phi}_{\mathrm{ell}}(G)$ of elliptic parameters in $\widetilde{\Phi}(G)$. These are the elements $\phi \in \widetilde{\Phi}(G)$ such that the subset

$$\overline{S}_{\phi,\mathrm{reg,ell}} = \{s \in \overline{S}_{\phi,\mathrm{ss}} : \ |\overline{S}_{\phi,s}| < \infty\}$$

of $\overline{S}_\phi$ is nonempty. We have taken the liberty here of writing

$$\overline{S}_{\phi,\mathrm{reg,ell}} = \overline{S}_{\phi,\mathrm{reg}} \cap \overline{S}_{\phi,\mathrm{ell}},$$

where

$$\overline{S}_{\phi,\mathrm{reg}} = \{s \in \overline{S}_{\phi,\mathrm{ss}} : \ \overline{S}^0_{\phi,s} \ \text{ is a torus}\}$$

conforms to the usual definition of regular semisimple. It is then clear that the two characterizations of $\overline{S}_{\phi,\mathrm{reg,ell}}$ are equivalent. It is also easy to see that if $\overline{S}_{\phi,\mathrm{reg,ell}}$ is nonempty, it actually equals $\overline{S}_{\phi,\mathrm{ell}}$.

Observe that ϕ is elliptic if and only if there is an elliptic pair

$$(G', \phi'), \qquad G' \in \mathcal{E}_{\mathrm{ell}}(G),\ \phi' \in \widetilde{\Phi}_2(G'),$$

such that the image of ϕ' in $\widetilde{\Phi}(G)$ equals ϕ. In fact, it follows from the definitions that our basic correspondence

$$(G', \phi') \longrightarrow (\phi, s)$$

defines a bijection between local, generic, elliptic variants

$$X_{\mathrm{ell}}(G) = \{(G', \phi') : \ G' \in E_{\mathrm{ell}}(G),\ \phi' \in F_2(G')\}$$

and

$$Y_{\mathrm{ell}}(G) = \{(\phi, s) : \phi \in F_{\mathrm{ell}}(G),\ \ s \in \overline{S}_{\phi,\mathrm{ell}}\}$$

of the general sets introduced in §4.8. This descends to a bijection between their quotients

$$\mathfrak{X}_{\mathrm{ell}}(G) = \{(G', \phi') : \ G' \in \mathcal{E}_{\mathrm{ell}}(G),\ \phi' \in \widetilde{\Phi}_2(G')\}$$

and

$$\mathcal{Y}_{\mathrm{ell}}(G) = \{(\phi, x) : \ \phi \in \widetilde{\Phi}_{\mathrm{ell}}(G),\ x \in \mathcal{S}_{\phi,\mathrm{ell}}\}.$$

Our primary goal is to describe the linear form

$$f \longrightarrow f'(\phi'), \qquad f \in \widetilde{\mathcal{H}}(G),\ (G', \phi') \in \mathfrak{X}_{\mathrm{ell}}(G), \tag{6.5.1}$$

in terms of (ϕ, s). This is the content of Theorem 2.2.1(b), the central assertion of the local theorems. The local intertwining relation provides a solution of the problem when ϕ lies in the complement of $\widetilde{\Phi}_2(G)$ in $\widetilde{\Phi}_{\mathrm{ell}}(G)$, as we saw in Proposition 2.4.3. Part of the remaining problem, in which ϕ lies in $\widetilde{\Phi}_2(G)$, will be to complete the inductive definition of $f'(\phi')$ by establishing Theorem 2.2.1(a). This includes Theorem 6.1.1, which will be needed to refine our temporary definition of the subset $\widetilde{\Phi}_{\mathrm{sim}}(G)$ of $\widetilde{\Phi}_2(G)$ and complete the inductive definition of the sets $\widetilde{\Phi}_2(G)$ and $\widetilde{\Phi}_{\mathrm{ell}}(G)$. We will deal with these matters in §6.6 and §6.7.

For the orthogonality relations of this section, it is best to work with the set $T_{\mathrm{ell}}(G)$ that was part of the proof of Proposition 3.5.1. Recall that $T_{\mathrm{temp}}(G)$ is the set of $G(F)$-orbits of triplets $\tau = (M, \sigma, r)$, where M is a Levi subgroup of G, σ is a unitary, square-integrable representation of $M(F)$, and r is an element in the R-group $R(\sigma)$ of σ. The subset $T_{\mathrm{ell}}(G)$ consists of (orbits of) triplets τ such that r belongs to the subset $R_{\mathrm{reg}}(\sigma)$ of elements in $R(\sigma)$ for which the determinant

$$d(\tau) = \det(r - 1)_{\mathfrak{a}_M}$$

is nonzero. Extending the convention we are using in this volume (in contrast to that of [**A10**]), we write $\widetilde{T}_{\mathrm{temp}}(G)$ and $\widetilde{T}_{\mathrm{ell}}(G)$ for the sets of $\widetilde{\mathrm{Out}}_N(G)$-orbits of elements in $T_{\mathrm{temp}}(G)$ and $T_{\mathrm{ell}}(G)$ respectively.

We recall that there is a linear form $f_G(\tau)$ attached to any

$$\tau = (M, \sigma, r).$$

For general groups G, $f_G(\tau)$ is determined only up to a scalar multiple, which depends on various choices related to the normalization of intertwining operators. Having made such choices for $G \in \widetilde{\mathcal{E}}_{\mathrm{sim}}(N)$ in Chapter 2, we may now be more precise, even though this is not really necessary for orthogonality relations. The case that $M = G$ is not at issue, since $f_G(\tau)$ is then just equal to $f_G(\sigma)$. If M is proper in G, we apply our induction hypothesis to the tempered representation $\sigma \in \widetilde{\Pi}_2(M)$. We obtain a unique bounded, generic parameter $\phi_M \in \widetilde{\Phi}_2(M)$, whose packet contains the representation $\pi_M = \sigma$, and whose image in $\widetilde{\Phi}(G)$ we denote by ϕ. We can then set

$$f_G(\tau) = \mathrm{tr}\big(R_P(r, \widetilde{\pi}_M, \phi)\mathcal{I}_P(\pi_M, f)\big), \qquad f \in \widetilde{\mathcal{H}}(G), \tag{6.5.2}$$

in the notation (2.4.2) (but with π there written as π_M here).

We note that the R-group $R(\pi_M)$ of π_M defined in [**A10**] is a priori different from the R-group R_ϕ of ϕ defined in §2.4. We have of course agreed to identify the stabilizer W_ϕ of ϕ_M with a subgroup of $W(M)$, the Weyl group that contains the stabilizer $W(\pi_M)$ of π_M. It is a consequence of the disjointness of tempered L-packets for M that W_ϕ contains $W(\pi_M)$. On the other hand, it follows from the form (2.3.4) of M, the corresponding form of the Weyl group $W(M)$, and our induction assumption that Theorem 2.2.4(b) holds for the factor G_- of M, that any element in W_ϕ stabilizes π_M. Therefore W_ϕ equals $W(\pi_M)$, in the generic case at hand. Moreover, we now know that elements in the subgroup W_ϕ^0 of W_ϕ give scalar intertwining operators for the induced representation $\mathcal{I}_P(\pi_M)$. It follows that W_ϕ^0 is contained in the subgroup $W^0(\pi_M)$ of $W(\pi_M)$. We thus have a surjective mapping

$$R_\phi \cong W_\phi/W_\phi^0 \;\mapsto\; W(\pi_M)/W^0(\pi_M) \cong R(\pi_M), \qquad \pi_M \in \widetilde{\Pi}_{\phi_M}, \tag{6.5.3}$$

which makes sense for any parameter $\phi \in \widetilde{\Phi}(G)$. At the beginning of the next section, we shall observe that this mapping is an isomorphism, and hence that $R(\pi_M)$ equals R_ϕ.

The essential part of the study of (6.5.1) turns out to be the special case in which f lies in the subspace $\widetilde{\mathcal{H}}_{\mathrm{cusp}}(G)$ of cuspidal functions in $\widetilde{\mathcal{H}}(G)$. It follows from [**Ka**, Theorem A] that the mapping that sends $f_G \in \mathcal{I}_{\mathrm{cusp}}(G)$ to the function

$$\tau \longrightarrow f_G(\tau), \qquad \tau \in T_{\mathrm{ell}}(G),$$

on $T_{\mathrm{ell}}(G)$ is a linear isomorphism from $\mathcal{I}_{\mathrm{cusp}}(G)$ onto the space of complex valued functions of finite support on $T_{\mathrm{ell}}(G)$. (This result is also a special case of the orthogonality relations in [**A10**, §6].) A similar assertion applies to the space $\widetilde{\mathcal{I}}_{\mathrm{cusp}}(G)$ of $\widetilde{\mathrm{Out}}(G)$-invariant functions in $\mathcal{I}_{\mathrm{cusp}}(G)$, if we replace $T_{\mathrm{ell}}(G)$ by $\widetilde{T}_{\mathrm{ell}}(G)$. Suppose that (G', ϕ') and (ϕ, s) are corresponding pairs in the elliptic sets $X_{\mathrm{ell}}(G)$ and $Y_{\mathrm{ell}}(G)$. Observe that G' equals G if and only if s equals 1, in which case ϕ maps to the subset $\widetilde{\Phi}_2(G)$ of $\widetilde{\Phi}_{\mathrm{ell}}(G)$. When this is so, we suppose that ϕ is such that the linear form $f^G(\phi)$ is defined, according to the prescription of Theorem 2.2.1(a), since this is the case not covered by our induction hypothesis. In all cases, we then have an expansion

$$f'(\phi') = \sum_{\tau \in \widetilde{T}_{\mathrm{ell}}(G)} c_\phi(x,\tau)\, f_G(\tau), \qquad f \in \widetilde{\mathcal{H}}_{\mathrm{cusp}}(G), \tag{6.5.4}$$

for uniquely determined complex numbers $c_\phi(x,\tau)$ that depend only on the image x of s in $\mathcal{S}_{\phi,\mathrm{ell}}$.

The expansion (6.5.4) was a central object of study in [**A11**], but with the different notation

$$\Delta(\phi',\tau) = c_\phi(x,\tau).$$

The article [**A11**] was aimed at general groups. The family of functionals

$$f^G \longrightarrow f^G(\phi), \qquad \phi \in \Phi_2(G),$$

could be defined there only as an abstract basis of the space of stable linear forms on $\mathcal{I}_{\mathrm{cusp}}(G)$, rather than by local Langlands parameters (or their makeshift analogues from §6.1 we have temporarily taken on for this chapter). In particular, there were no groups $\overline{S}_\phi$ in [**A11**] associated to the linear forms ϕ, and hence no pairs (ϕ, s). This accounts for the earlier notation, and the cruder results of that paper.

The local intertwining relation gives a description of the coefficients in (6.5.4) if ϕ lies in the complement $\widetilde{\Phi}^2_{\mathrm{ell}}(G)$ of $\widetilde{\Phi}_2(G)$ in $\widetilde{\Phi}_{\mathrm{ell}}(G)$. In this case, ϕ and S_ϕ are given by (6.4.1). We can then take

$$\phi_M = \phi_1 \times \cdots \times \phi_q \times \cdots \times \phi_r$$

to be the parameter in $\widetilde{\Phi}_2(M,\phi)$ that was part of the discussion of the last section. Since F is nonarchimedean, our induction hypothesis gives a perfect pairing $\langle \cdot,\cdot \rangle$ on $\mathcal{S}_{\phi_M} \times \widetilde{\Pi}_{\phi_M}$. We form the subset

$$\widetilde{T}_{\phi,\mathrm{ell}}(G) = \{\tau = (M,\pi_M,r) : \ \pi_M \in \widetilde{\Pi}_{\phi_M},\ r \in R_{\mathrm{reg}}(\pi_M)\}$$

of $\widetilde{T}_{\mathrm{ell}}(G)$ attached to ϕ, a set that could a priori be empty. The coefficients in (6.5.4) can then be expressed in terms of the pairing

$$\langle x,\tau \rangle = \langle x, \widetilde{\pi}_M \rangle, \qquad \tau = (M,\pi_M,r_{\mathrm{reg}}),$$

on $\mathcal{S}_{\phi,\mathrm{ell}} \times \widetilde{T}_{\phi,\mathrm{ell}}$, where as in §2.4, $\langle \cdot, \widetilde{\pi}_M \rangle$ is an extension of the character $\langle \cdot, \pi_M \rangle$ on $\mathcal{S}_{\phi_M}$ to the $\mathcal{S}_{\phi_M}$-torsor $\mathcal{S}_{\phi,\mathrm{ell}}$. To see this, we consider the local intertwining relation (2.4.7), as f varies over the space $\widetilde{\mathcal{H}}_{\mathrm{cusp}}(G)$. After expanding the two sides of (2.4.7), according to the definitions (2.4.6), (6.5.4), (2.4.5) and (6.5.2), we need only compare the coefficients of $f_G(\tau)$. We see that

$$c_\phi(x,\tau) = \begin{cases} \langle x,\tau \rangle, & \text{if } \tau \in \widetilde{T}_{\phi,\mathrm{ell}}(G), \\ 0, & \text{otherwise,} \end{cases} \tag{6.5.5}$$

for any $x \in \mathcal{S}_{\phi,\mathrm{ell}}$ and $\tau \in \widetilde{T}_{\mathrm{ell}}(G)$, and for the given parameter $\phi \in \widetilde{\Phi}^2_{\mathrm{ell}}(G)$. We note that the two sides of (6.5.5) have a parallel dependence on the extension $\widetilde{\pi}_M$ of the component π_M of τ, a choice we have not built into our notation for τ.

We turn now to orthogonality relations. The starting point is the canonical Hermitian inner product

$$I(f,g) = \widehat{I}(f_G, g_G) = \int_{\Gamma_{\mathrm{ell}}(G)} f_G(\gamma)\, \overline{g_G(\gamma)}\, d\gamma$$

on $\widetilde{\mathcal{I}}_{\mathrm{cusp}}(G)$. Recall that $\Gamma_{\mathrm{ell}}(G)$ denotes the set of elliptic conjugacy classes in $G(F)$, and $d\gamma$ is a canonical measure that is supported on a set of strongly regularly classes in $\Gamma_{\mathrm{ell}}(G)$. According to the local trace formula, this inner product has a spectral expansion

$$I(f,g) = \sum_{\tau \in \widetilde{T}_{\mathrm{ell}}(G)} m(\tau)\, |d(\tau)|^{-1}\, |R(\tau)|^{-1} f_G(\tau)\, \overline{g_G(\tau)}, \tag{6.5.6}$$

where $m(\tau)$ denotes the number of elements in $T_{\mathrm{ell}}(G)$ that map to τ, as an equivalence class $\{(M,\sigma,r)\}$ in $\widetilde{T}_{\mathrm{ell}}(G)$, and $R(\tau)$ is the R-group $R(\sigma)$. The formula (6.5.6) is a specialization of [**A10**, Corollary 3.2], in which we have written

$$\overline{g_G(\tau)} = (\overline{g})_G(\tau^\vee).$$

The R-groups are all abelian here, and the 2-cocycles introduced in [**A10**, p. 86] all split. The supplementary coefficient $m(\tau)$ occurs in (6.5.6) because we have summed over $\widetilde{T}_{\mathrm{ell}}(G)$ instead of $T_{\mathrm{ell}}(G)$.

A similar identity holds for the twisted component $\widetilde{G}(N)$. There is a canonical twisted Hermitian inner product

$$I(\widetilde{f}, \widetilde{g}) = \widehat{I}(\widetilde{f}_N, \widetilde{g}_N) = \int_{\widetilde{\Gamma}_{\mathrm{ell}}(N)} \widetilde{f}_N(\gamma)\, \overline{\widetilde{g}_N(\gamma)}\, d\gamma$$

on the space $\widetilde{\mathcal{I}}_{\mathrm{cusp}}(N) = \mathcal{I}_{\mathrm{cusp}}(\widetilde{G}(N))$, where $\widetilde{f}_N = \widetilde{f}_{\widetilde{G}(N)}$ is the usual image of a given function $\widetilde{f} \in \widetilde{\mathcal{H}}_{\mathrm{cusp}}(N)$ in the space $\widetilde{\mathcal{I}}_{\mathrm{cusp}}(N)$, while $\widetilde{\Gamma}_{\mathrm{ell}}(N) = \Gamma_{\mathrm{ell}}(\widetilde{G}(N))$ is the set of elliptic $\widetilde{G}^0(N,F)$-orbits in $\widetilde{G}(N,F)$, and $d\gamma$ is a canonical measure on $\widetilde{\Gamma}_{\mathrm{ell}}(N)$ that is supported on the strongly regular classes in $\widetilde{\Gamma}_{\mathrm{ell}}(N)$. This inner product also has a spectral expansion. It is a sum of terms over the analogue $\widetilde{T}_{\mathrm{ell}}(N) = T_{\mathrm{ell}}(\widetilde{G}(N))$ for $\widetilde{G}(N)$ of the set $\widetilde{T}_{\mathrm{ell}}(G)$. By the special properties of $\widetilde{G}(N)^0 = GL(N)$, $\widetilde{T}_{\mathrm{ell}}(N)$ is our set $\widetilde{\Phi}_{\mathrm{ell}}(N) = \Phi_{\mathrm{ell}}(\widetilde{G}(N))$ of multiplicity free elements

$$\phi = \phi_1 \oplus \cdots \oplus \phi_r$$

in $\widetilde{\Phi}(N)$. The twisted spectral expansion is then

$$I(\widetilde{f}, \widetilde{g}) = \sum_{\phi \in \widetilde{\Phi}_{\mathrm{ell}}(N)} |d(\phi)|^{-1}\, \widetilde{f}_N(\phi)\, \overline{\widetilde{g}_N(\phi)}, \tag{6.5.7}$$

where $|d(\phi)| = 2^r$. Twisted orthogonality relations for $GL(N)$, which include the cases treated in [**W4**], are a part of work in progress by Waldspurger on the twisted local trace formula for $\widetilde{G}(N)$. Like its global analogue (3.1.1) (with $\widetilde{G}(N)$ in place of G), the discrete part of the twisted local trace formula contains a sum over elements $w \in W(M)_{\mathrm{reg}}$. In the case at hand, $M \subset \widetilde{G}(N)^0$ is the Levi subgroup such that ϕ belongs to the set $\widetilde{\Phi}_M(N) = \Phi_M(\widetilde{G}(N))$, and the sum over $W(M)_{\mathrm{reg}}$ can be restricted to the elements w_{reg} that act on the space $\mathfrak{a}_M \cong \mathbb{R}^r$, and for which

$$|\det(w_{\mathrm{reg}} - 1)_{\mathfrak{a}_M}| = |d(\phi)| = 2^r.$$

For any $G \in \widetilde{\mathcal{E}}_{\mathrm{sim}}(N)$, there is also a stable Hermitian inner product

$$S(f,g) = \widehat{S}(f^G, g^G) = \int_{\Delta_{\mathrm{reg}}(G)} f^G(\delta)\, \overline{g^G(\delta)}\, d\delta$$

on the space $\widetilde{\mathcal{S}}_{\mathrm{cusp}}(G)$. We recall that $\Delta_{\mathrm{ell}}(G)$ denotes the set of elliptic, stable conjugacy classes in $G(F)$, and $d\delta$ is a canonical measure supported on the strongly regular classes in $\Delta_{\mathrm{ell}}(G)$. The definition obviously extends to products of groups, such as the compound elements in $\widetilde{\mathcal{E}}_{\mathrm{ell}}(N)$ and $\mathcal{E}_{\mathrm{ell}}(G)$. It can be used to stabilize the inner products $I(f,g)$ an $I(\widetilde{f},\widetilde{g})$. One obtains two expansions

$$I(f,g) = \sum_{G' \in \mathcal{E}_{\mathrm{ell}}(G)} \iota(G,G')\, \widehat{S}(f',g'), \qquad f,g \in \widetilde{\mathcal{H}}_{\mathrm{cusp}}(G) \tag{6.5.8}$$

and

$$I(\widetilde{f},\widetilde{g}) = \sum_{G \in \widetilde{\mathcal{E}}_{\mathrm{ell}}(N)} \widetilde{\iota}(N,G)\, \widehat{S}(\widetilde{f}^G, \widetilde{g}^G), \qquad \widetilde{f},\widetilde{g} \in \widetilde{\mathcal{H}}_{\mathrm{cusp}}(N), \tag{6.5.9}$$

for local coefficients

$$\iota(G,G') = |\overline{Z}(G')^\Gamma|^{-1}\, |\mathrm{Out}_G(G')|^{-1}$$

and

$$\widetilde{\iota}(N,G) = \iota(\widetilde{G}(N),G) = \tfrac{1}{2}|\overline{Z}(\widehat{G})^\Gamma|^{-1}\, |\widetilde{\mathrm{O}}\mathrm{ut}_N(G)|^{-1}$$

analogous to their global counterparts described at the end of §3.2. The linear mappings

$$\widetilde{\mathcal{I}}_{\mathrm{cusp}}(G) \longrightarrow \bigoplus_{G' \in \mathcal{E}_{\mathrm{ell}}(G)} \widetilde{\mathcal{S}}_{\mathrm{cusp}}(G') \tag{6.5.10}$$

and

$$\widetilde{\mathcal{I}}_{\mathrm{cusp}}(N) \longrightarrow \bigoplus_{G \in \widetilde{\mathcal{E}}_{\mathrm{ell}}(N)} \widetilde{\mathcal{S}}_{\mathrm{cusp}}(G), \tag{6.5.11}$$

given by $f_G \to \bigoplus_{G'} f'$ and $\widetilde{f}_N \to \bigoplus_{G} \widetilde{f}^G$ are thus isometric isomorphisms, relative to the natural inner products on the right hand spaces. For the ordinary expansion (6.5.8), we refer the reader to [**A11**, §2], which is the restriction to cuspidal functions of the general stabilization of [**A16**, §10]. The treatment of the twisted expansion (6.5.9) will be similar.

We need to derive an explicit spectral expansion for $S(f,g)$. We shall use the spectral expansion (6.5.7) of the left hand side of (6.5.9) to deduce information about the right hand side. However, in order to accommodate the limited understanding we have at this point, we need to adopt some temporary terminology.

We shall say an element $\phi \in \widetilde{\Phi}_{\mathrm{ell}}(N)$ is a *cuspidal lift* if there is a $G \in \widetilde{\mathcal{E}}_{\mathrm{ell}}(N)$ such that

$$\text{(i)}\qquad \widetilde{f}_N(\phi) = \widetilde{f}^G(\phi), \qquad \widetilde{f} \in \widetilde{\mathcal{H}}_{\mathrm{cusp}}(N),$$

where $\widetilde{f}^G(\phi)$ is the linear form on $\widetilde{\mathcal{S}}_{\mathrm{cusp}}(G)$ defined by (6.1.1), and

$$\text{(ii)}\qquad \phi \in \widetilde{\Phi}_2(G).$$

It is a consequence of (6.1.1) that G is uniquely determined by (i). If ϕ satisfies both (i) and (ii), and lies in the subset $\widetilde{\Phi}_{\mathrm{sim}}(N)$ of $\widetilde{\Phi}_{\mathrm{ell}}(N)$, then G lies in the subset $\widetilde{\mathcal{E}}_{\mathrm{sim}}(N)$ of $\widetilde{\mathcal{E}}_{\mathrm{ell}}(N)$. This is a formal consequence of our temporary definition of the subset $\widetilde{\Phi}_{\mathrm{sim}}(G)$ of $\widetilde{\Phi}_2(G)$. Conversely, if $\phi \in \widetilde{\Phi}_{\mathrm{sim}}(N)$ satisfies (i), and the corresponding datum G lies in $\widetilde{\mathcal{E}}_{\mathrm{sim}}(N)$, (ii) itself follows from the definition. Observe however that if ϕ lies in the complement of $\widetilde{\Phi}_{\mathrm{sim}}(N)$ in $\widetilde{\Phi}_{\mathrm{ell}}(N)$, the condition (ii) is based on the inductive definition of §6.1, and is therefore independent of (i).

Theorem 2.2.1(a) asserts that every $\phi \in \widetilde{\Phi}_{\mathrm{ell}}(N)$ is a cuspidal lift, but we have yet to establish this. If ϕ is not simple, there is a unique G with $\phi \in \widetilde{\Phi}_2(G)$ by our induction assumption, but it remains to establish the first condition (i). This amounts to the secondary assertion (2.2.4) of Theorem 2.2.1(a). In any case, we can certainly assume inductively that for each $N' < N$, every element $\phi' \in \widetilde{\Phi}_{\mathrm{ell}}(N')$ is a cuspidal lift.

Suppose that we are given a subset

$$\widetilde{\Phi}^c_{\mathrm{ell}}(N) \subset \Phi_{\mathrm{ell}}(N)$$

that consists entirely of cuspidal lifts. Then $\Phi^c_{\mathrm{ell}}(N)$ is a disjoint union over $G \in \widetilde{\mathcal{E}}_{\mathrm{ell}}(N)$ of the subsets

$$\widetilde{\Phi}^c_2(G) = \widetilde{\Phi}^c_{\mathrm{ell}}(N) \cap \widetilde{\Phi}_2(G).$$

As in the untwisted case, one can show that the mapping

$$\widetilde{f}_N \longrightarrow \widetilde{f}_N(\phi), \qquad \phi \in \widetilde{\Phi}_{\mathrm{ell}}(N),\ \widetilde{f}_N \in \widetilde{\mathcal{I}}_{\mathrm{cusp}}(N),$$

is a linear isomorphism from $\widetilde{\mathcal{I}}_{\mathrm{cusp}}(N)$ onto the space of complex valued functions of finite support on the set $\widetilde{\Phi}_{\mathrm{ell}}(N) = \widetilde{T}_{\mathrm{ell}}(N)$. Let $\widetilde{\mathcal{I}}^c_{\mathrm{cusp}}(N)$ denote the subspace of elements in $\widetilde{\mathcal{I}}_{\mathrm{cusp}}(N)$ that as functions on $\widetilde{\Phi}_{\mathrm{ell}}(N)$ are supported on the subset $\widetilde{\Phi}^c_{\mathrm{ell}}(N)$. Then for any $G \in \widetilde{\mathcal{E}}_{\mathrm{ell}}(N)$, the mapping (6.5.11) sends

$\widetilde{\mathcal{I}}^c_{\mathrm{cusp}}(N)$ onto a subspace of $\widetilde{\mathcal{S}}_{\mathrm{cusp}}(G)$, which we shall denote by $\widetilde{\mathcal{S}}^c_{\mathrm{cusp}}(G)$. We note that $\widetilde{\mathcal{S}}^c_{\mathrm{cusp}}(G)$ can be identified with the space of functions of finite support on $\widetilde{\Phi}^c_2(G)$.

PROPOSITION 6.5.1. *Suppose that $G \in \widetilde{\mathcal{E}}_{\mathrm{ell}}(N)$, and that f and g are functions in $\widetilde{\mathcal{H}}_{\mathrm{cusp}}(G)$ such that f^G and g^G lie in $\widetilde{\mathcal{S}}^c_{\mathrm{cusp}}(G)$. Then*

$$S(f,g) = \sum_{\phi \in \widetilde{\Phi}^c_2(G)} m(\phi)\, |\mathcal{S}_\phi|^{-1} f^G(\phi)\, \overline{g^G(\phi)}. \tag{6.5.12}$$

Moreover, the formula (6.5.8) can be written as

$$I(f,g) = \sum_{\phi \in \widetilde{\Phi}_{\mathrm{ell}}(G)} m(\phi)\, |\mathcal{S}_\phi|^{-1} \sum_{x \in \mathcal{S}_{\phi,\mathrm{ell}}} f'(\phi')\, \overline{g'(\phi')}, \tag{6.5.13}$$

where in the last summand, (G', ϕ') corresponds to (ϕ, x).

PROOF. Let $\widetilde{f}$ and $\widetilde{g}$ be functions in $\widetilde{\mathcal{H}}_{\mathrm{cusp}}(N)$ such that $\widetilde{f}_N$ and $\widetilde{g}_N$ are the preimages of f^G and g^G under the mapping (6.5.11). Then the only nonvanishing term on the right hand side of (6.5.9) is that of the given group G. Using (6.5.7) to rewrite the left hand side of (6.5.9), we obtain

$$\sum_{\phi \in \widetilde{\Phi}^c_{\mathrm{ell}}(N)} |d(\phi)|^{-1} \widetilde{f}_N(\phi)\, \overline{\widetilde{g}_N(\phi)} = \widetilde{\iota}(N,G)\, S(f,g).$$

By assumption, each ϕ in the sum on the left is the pullback of a linear form on $\widetilde{\mathcal{S}}_{\mathrm{cusp}}(G)$, which we are also denoting by ϕ. That is,

$$\widetilde{f}_N(\phi)\, \overline{\widetilde{g}_N(\phi)} = f^G(\phi)\, \overline{g^G(\phi)}.$$

From the explicit formula for $\widetilde{\iota}(N,G)$ above, it is easy to check that

$$|d(\phi)|^{-1}\, \widetilde{\iota}(N,G)^{-1} = m(\phi)\, |\mathcal{S}_\phi|^{-1},$$

for any $\phi \in \widetilde{\Phi}^c_2(G)$. The identity (6.5.12) follows. Observe that we can enlarge the index of summation in this formula from $\widetilde{\Phi}^c_2(G)$ to $\widetilde{\Phi}_2(G)$, since elements in the subspace $\widetilde{\mathcal{S}}^c_{\mathrm{cusp}}(G)$ of $\widetilde{\mathcal{S}}_{\mathrm{cusp}}(G)$ can be regarded as functions on $\widetilde{\Phi}_2(G)$ that vanish on the complement of $\widetilde{\Phi}^c_2(G)$. In fact, we are free to replace $\widetilde{\Phi}_2(G)$ by the subset $\Phi_2(G)$ of $\widehat{G}$-orbits of local parameters (rather than $\widetilde{\mathrm{O}}\mathrm{ut}(G)$-orbits), so long as we remove the coefficient $m(\phi)$.

To establish the second assertion, we apply the formula we have just obtained to any $G' \in \mathcal{E}_{\mathrm{ell}}(G)$. If $G' \neq G$, we can assume inductively that every element in $\widetilde{\Phi}_2(G')$ is a cuspidal lift. In other words, f' and g' are images of functions in $\widetilde{\mathcal{H}}_{\mathrm{cusp}}(G')$ that satisfy the condition of the proposition, with $\widetilde{\Phi}^c_2(G') = \widetilde{\Phi}_2(G')$. We can therefore write (6.5.8) as

$$I(f,g) = \sum_{G' \in \mathcal{E}_{\mathrm{ell}}(G)} \iota(G,G') \sum_{\phi' \in \widetilde{\Phi}_2(G')} m(\phi')\, |\mathcal{S}_{\phi'}|^{-1} f'(\phi')\, \overline{g'(\phi')}.$$

The process of converting this to the formula (6.5.13) is an elementary local analogue of the discussion of the global multiplicity formula in §4.7, itself a very simple case of the standard model of §4.3–4.4. Indeed, the last double sum is over pairs (G', ϕ') in the local set $\mathfrak{X}_{\mathrm{ell}}(G)$ defined at the beginning of the section. We have only to rewrite it in terms of the bijective image (ϕ, x) of (G', ϕ') in the second local set $\mathcal{Y}_{\mathrm{ell}}(G)$. From the explicit formula for $\iota(G, G')$, it is easy to see that

$$m(\phi)\,|\mathcal{S}_\phi|^{-1} = \iota(G, G')\,m(\phi')\,|\mathcal{S}_{\phi'}|^{-1}.$$

The identity (4.5.13) follows. □

We shall actually use (6.5.13) in its interpretation as a set of orthogonality relations for the matrix coefficients $\{c_\phi(x, \tau)\}$ in (6.5.4). We of course have to require that the linear form (6.5.4) attached to a given pair (ϕ, x) in $\mathcal{Y}_{\mathrm{ell}}(G)$ be defined, which is to say that for the corresponding pair (G', ϕ') in $\mathfrak{X}_{\mathrm{ell}}(G)$, the element $\phi' \in \widetilde{\Phi}_2(G')$ is a cuspidal lift. By induction, this is only at issue if $x = 1$, which in turn implies that ϕ lies in the subset $\widetilde{\Phi}_2(G)$ of $\widetilde{\Phi}_{\mathrm{ell}}(G)$. To describe the orthogonality relations, we write

$$a(\phi, x) = m(\phi)^{-1}\,|\mathcal{S}_\phi|$$

and

$$b(\tau) = m(\tau)^{-1}\,|d(\tau)|\,|R(\tau)|, \qquad \tau \in \widetilde{T}_{\mathrm{ell}}(G).$$

COROLLARY 6.5.2. *Suppose that*

$$y_i = (\phi_i, x_i), \qquad i = 1, 2,$$

are two pairs in $\mathcal{Y}_{\mathrm{ell}}(G)$ *such that the associated linear forms* (6.5.4) *are defined. Then*

$$\sum_{\tau \in \widetilde{T}_{\mathrm{ell}}(G)} b(\tau)\,c_{\phi_1}(x_1, \tau)\,\overline{c_{\phi_2}(x_2, \tau)} = \begin{cases} a(\phi_1, x_1), & \text{if } y_1 = y_2, \\ 0, & \text{otherwise}. \end{cases} \tag{6.5.14}$$

PROOF. We identify $\widetilde{\mathcal{I}}_{\mathrm{cusp}}(G)$ with the space of finitely supported functions of $\{\tau\}$, and we identify the right hand side of (6.5.10) with the space of finitely supported functions of an index set $\{y\}$ that includes all pairs $(\phi, x) \in \mathcal{Y}_{\mathrm{ell}}(G)$ for which (6.5.4) is defined. Then the isomorphism (6.5.10) is given by a matrix $C = \{c(y, \tau)\}$ such that $c(y, \tau) = c_\phi(x, \tau)$ if $y = (\phi, x)$. We can assume that the basis of $\widetilde{S}(G)$ defined by the appropriate subset of $\{y\}$ is orthogonal. We can then write the two expressions (6.5.6) and (6.5.13) for $I(f, g)$ as an orthogonality relation

$$\sum_y a(y)^{-1}\,c(y, \tau_1)\,\overline{c(y, \tau_2)} = b(\tau_1)^{-1}\,\delta_{\tau_1, \tau_2}$$

for the matrix C, where $A = \{a(y)\}$ and $B = \{b(\tau)\}$ are the invertible diagonal matrices that represent the two inner products. The matrix form of this identity is

$$C^* A^{-1} C = B^{-1},$$

which becomes

$$CBC^* = A$$

upon inversion. The required formula follows. □

As a preliminary application of the orthogonality relations of the corollary, we shall show that the expansion (6.5.4) simplifies in the case not covered by (6.5.5), namely that $\phi \in \widetilde{\Phi}_2(G)$.

LEMMA 6.5.3. (a) *Suppose that* (ϕ, x) *is a pair in* $\mathcal{Y}_{\mathrm{ell}}(G)$ *such that* ϕ *lies in the subset* $\widetilde{\Phi}_2(G)$ *of* $\widetilde{\Phi}_{\mathrm{ell}}(G)$, *and such that the associated linear form* (6.5.4) *is defined. Then*

$$f'(\phi') = \sum_{\pi \in \widetilde{\Pi}_2(G)} c_\phi(x, \pi)\, f_G(\pi), \qquad f \in \widetilde{\mathcal{H}}_{\mathrm{cusp}}(G). \tag{6.5.15}$$

(b) *Suppose that*

$$y_i = (\phi_i, x_i), \qquad i = 1, 2,$$

are two pairs in $\mathcal{Y}_{\mathrm{ell}}(G)$ *that both satisfy the conditions of* (a). *Then*

(6.5.16)

$$\sum_{\pi \in \widetilde{\Pi}_2(G)} m(\pi)^{-1}\, c_{\phi_1}(x_1, \pi)\, \overline{c_{\phi_2}(x_2, \pi)} = \begin{cases} m(\phi_1)^{-1} |\mathcal{S}_{\phi_1}|, & \text{if } y_1 = y_2, \\ 0, & \text{otherwise.} \end{cases}$$

PROOF. The first part (a) asserts that the definition (6.5.4) simplifies in the case that ϕ belongs to $\widetilde{\Phi}_2(G)$. We have to show $c_\phi(x, \tau_1) = 0$ for any fixed element τ_1 in the complement of $\widetilde{\Pi}_2(G)$ in $\widetilde{T}_{\mathrm{ell}}(G)$.

We know that τ_1 lies in a packet $\widetilde{T}_{\phi_1,\mathrm{ell}}$, for a unique parameter ϕ_1 in the complement of $\widetilde{\Phi}_2(G)$ in $\widetilde{\Phi}_{\mathrm{ell}}(G)$. We shall apply the formula (6.5.5) for the coefficients

$$c_{\phi_1}(x_1, \tau), \qquad x_1 \in \mathcal{S}_{\phi_1,\mathrm{ell}},\ \tau \in \widetilde{T}_{\mathrm{ell}}(G),$$

or rather the inversion

$$|\mathcal{S}_{\phi_1,\mathrm{ell}}|^{-1} \sum_{x_1 \in \mathcal{S}_{\phi_1,\mathrm{ell}}} \langle x_1, \tau_1 \rangle\, \overline{c_{\phi_1}(x_1, \tau)} = \begin{cases} 1, & \text{if } \tau = \tau_1, \\ 0, & \text{otherwise,} \end{cases}$$

of this formula that is a consequence of the definition of the pairing $\langle \cdot, \cdot \rangle$. It allows us to write

$$\begin{aligned} & b(\tau_1)\, c_\phi(x, \tau_1) \\ & = \sum_{\tau \in \widetilde{T}_{\mathrm{ell}}(G)} b(\tau)\, c_\phi(x, \tau) \cdot |\mathcal{S}_{\phi_1,\mathrm{ell}}|^{-1} \sum_{x_1 \in \mathcal{S}_{\phi_1,\mathrm{ell}}} \langle x_1, \tau_1 \rangle\, \overline{c_{\phi_1}(x_1, \tau)} \\ & = |\mathcal{S}_{\phi_1,\mathrm{ell}}|^{-1} \sum_{x_1} \langle x_1, \tau_1 \rangle \sum_{\tau} b(\tau)\, c_\phi(x, \tau)\, \overline{c_{\phi_1}(x_1, \tau)}. \end{aligned}$$

Since $\phi_1 \neq \phi$, Corollary 6.5.2 tells us that the last sum over τ vanishes. Since $b(\tau_1) \neq 0$, we then see that $c_{\phi,x}(\tau_1)$ vanishes, as required. The definition (6.5.4) thus reduces to the required formula (6.5.15) of (a).

Having established (a), we substitute (6.5.15) into the formula (6.5.14) of Corollary 6.5.2. The required formula (6.5.16) of (b) follows. □

6.6. Local packets for composite ϕ

We are now ready to start proving the theorems that give the local classification. They consist of the five local theorems we stated in Chapters 1 and 2, specialized to generic local parameters $\psi = \phi$. We have already established the local intertwining relation Theorem 2.4.1 for pairs (G, ϕ). What remains is the classification of general L-packets $\widetilde{\Pi}_\phi$ described by Theorems 1.5.1 and 2.2.1. We will also have to deal with the supplementary Theorems 2.2.4 and 2.4.4 for pairs $(\widetilde{G}, \phi)$. These will be treated afterwards in §6.8, as will the questions in §6.1 related to Theorem 6.1.1.

We note again that in the archimedean case, the local classification has been established. With the exception of our formulation of Theorem 2.4.1, the theorems for archimedean parameters ϕ are included in the general results of Shelstad for real groups (and work in progress by Mezo for twisted real groups). We recall that our proof of Theorem 2.4.1 for generic parameters, completed by global means in §6.4, applies to all local fields F.

We thus continue to assume that the local field F is nonarchimedean, as in the last section. We also maintain our induction hypothesis that the local theorems hold for generic parameters of degree less than the fixed integer N. Suppose that G is a fixed datum in $\widetilde{\mathcal{E}}_{\mathrm{sim}}(N)$. We are trying to classify the irreducible tempered representations of $G(F)$. In particular, we want to attach a packet $\widetilde{\Pi}_\phi$ of $(\widetilde{\mathrm{Out}}_N(G)$-orbits of) irreducible representations to any $\phi \in \widetilde{\Phi}_{\mathrm{bdd}}(G)$, which satisfies the endoscopic character identity (2.2.6) of Theorem 2.2.1.

Suppose for the moment that ϕ lies in the complement of $\widetilde{\Phi}_2(G)$ in $\widetilde{\Phi}_{\mathrm{bdd}}(G)$. Then ϕ is the image of a parameter $\phi_M \in \widetilde{\Phi}_2(M, \phi)$, for a proper Levi subgroup M of G. The assertion of Theorem 2.2.1(a) actually applies to data G in the larger set $\widetilde{\mathcal{E}}(N)$. However, as we have noted before, the usual argument of descent on any such G reduces the statement to a corresponding assertion for M, which then follows from our induction hypothesis. For Theorem 2.2.1(b), we can appeal to Proposition 2.4.3, since we have established the local intertwining relation of Theorem 2.4.1 for generic parameters. We obtain a packet $\widetilde{\Pi}_\phi$ of (orbits of) of irreducible tempered representations of $G(F)$, and a pairing on $\mathcal{S}_\phi \times \widetilde{\Pi}_\phi$ that satisfies the endoscopic character identity (2.2.6). Therefore, Theorem 2.2.1 holds for ϕ. So does Theorem 1.5.1(a), since apart from the assertion for unramified pairs (G, π) that we mentioned also in §6.1, it represents only a less precise form of Theorem 2.2.1. As for this remaining assertion, Corollary 2.5.2(iv) tells us that the relevant intertwining operators are trivial. The local intertwining relation then tells us that the associated unramified character $\langle \cdot, \pi \rangle$ is trivial.

However, since we are dealing with tempered representations, we have more to establish. We must show that the packet $\widetilde{\Pi}_\phi$ also satisfies the supplementary conditions of Theorem 1.5.1(b). These are tied to the question on the relation between the endoscopic and representation theoretic R-groups R_ϕ and $R(\pi_M)$, raised in our discussion of (6.5.3) from the last section.

The question is now easy to resolve. We can identify (6.5.10) as a mapping from the space of finitely supported functions of $\tau \in \widetilde{T}_{\mathrm{ell}}(G)$ to the space of finitely supported functions of $(\phi, x) \in \mathcal{Y}_{\mathrm{ell}}(G)$. It follows from (6.5.4) that the mapping factors through the projection (6.5.3), or rather the restriction of (6.5.3) to the subset $R_{\phi,\mathrm{reg}}$ of R_ϕ. But we also know that the mapping (6.5.10) is an isomorphism, and so in particular, is surjective. It follows that the projection (6.5.3) takes $R_{\phi,\mathrm{reg}}$ bijectively onto $R_{\mathrm{reg}}(\pi_M)$, for any $\pi_M \in \widetilde{\Pi}_{\phi_M}$. This statement amounts to the assertion that the set $T_{\phi,\mathrm{ell}}$ is not empty if ϕ belongs to $\widetilde{\Phi}_{\mathrm{ell}}(G)$. It might appear to be insignificant, since $R_{\phi,\mathrm{reg}}$ consists of only one element if ϕ lies in $\widetilde{\Phi}_{\mathrm{ell}}(G)$, and is in fact empty otherwise. Nonetheless, its application to Levi subgroups $L \supset M$ gives us what we want. Indeed, any element in R_ϕ belongs to the subset $R_{\phi_L,\mathrm{reg}}$ attached to some L and some parameter $\phi_L \in \widetilde{\Phi}_{\mathrm{ell}}(L)$ that maps to ϕ. It follows that the projection (6.5.3) is an isomorphism, and hence that $R(\pi_M)$ equals R_ϕ for any $\pi_M \in \widetilde{\Pi}_{\phi_M}$.

In the proof of Proposition 3.5.1, we described the general classification of $\Pi_{\mathrm{temp}}(G)$ by harmonic analysis, based largely on the results of Harish-Chandra. This characterizes $\Pi_{\mathrm{temp}}(G)$ as the image of a bijection

$$\{(L, \sigma, \xi)\} \xrightarrow{\sim} \{\pi_\xi\}, \qquad L \in \mathcal{L},\ \sigma \in \Pi_{2,\mathrm{temp}}(L),\ \xi \in \Pi\big(R(\sigma)\big),$$

where the left hand side represents a W_0^G-orbit of triplets, and the right hand side is given by the irreducible constituent $(\xi \otimes \pi_\xi)$ of the character

$$(r, f) \longrightarrow \mathrm{tr}\big(R_P(r, \widetilde{\pi}_M, \phi)\,\mathcal{I}_P(\pi_M, f)\big)$$

of $R(\sigma) \times \mathcal{H}(G)$. Our goal here is to establish the finer endoscopic classification of Theorem 1.5.1. In the proof of Proposition 2.4.3, we constructed the packet $\widetilde{\Pi}_\phi$ over $\widetilde{\Pi}_{\mathrm{temp}}(G)$, for the given parameter ϕ in the complement of $\widetilde{\Phi}_2(G)$. It is a priori a multiset, equipped with a mapping to $\widehat{\mathcal{S}}_\phi$ given by the irreducible constituents $(\xi \otimes \pi)$ of the character

$$(x, f) \longrightarrow f_G(\phi, x) = \sum_{\pi_M \in \widetilde{\Pi}_{\phi_M}} \langle x, \widetilde{\pi}_M\rangle\, \mathrm{tr}\big(R_P(w_x, \widetilde{\pi}_M, \phi)\,\mathcal{I}_P(\pi_M, f)\big)$$

of $\mathcal{S}_\phi \times \widetilde{\mathcal{H}}(G)$. We have just seen that we can identify $R(\pi_M)$ with the R-group $R_\phi = \mathcal{S}_\phi/\mathcal{S}_\phi^1$ of ϕ. It follows that the two kinds of classification are compatible. In particular, the packet structure of the second imposes an endoscopic, character theoretic interpretation on the first.

The general classification, for its part, yields finer properties of the packets. This is the implication we need to exploit here. It tells us that the elements in $\widetilde{\Pi}_\phi$ are multiplicity free. Combined with the properties of $\widetilde{\Pi}_{\phi_M}$

that are part of our induction hypothesis, it also confirms that the mapping from $\widetilde{\Pi}_\phi$ to $\widehat{\mathcal{S}}_\phi$ is a bijection. Finally, it tells us that the complement of $\widetilde{\Pi}_2(G)$ in $\widetilde{\Pi}_{\mathrm{temp}}(G)$ is a disjoint union over local parameters ϕ in the complement of $\widetilde{\Phi}_2(G)$ in $\widetilde{\Phi}_{\mathrm{bdd}}(G)$ of the packets $\widetilde{\Pi}_\phi$. These are the conditions of Theorem 1.5.1(b), for the p-adic case at hand.

We have established

PROPOSITION 6.6.1. *Theorems* 1.5.1 *and* 2.2.1 *hold for generic parameters* ϕ *in the complement of* $\widetilde{\Phi}_2(G)$ *in* $\widetilde{\Phi}_{\mathrm{bdd}}(G)$. □

Having dealt with the complementary set, we assume from this point on that ϕ lies in $\widetilde{\Phi}_2(G)$. Then

$$\phi = \phi_1 \oplus \cdots \oplus \phi_r, \qquad \phi_i \in \widetilde{\Phi}_{\mathrm{sim}}(G_i),\ G_i \in \widetilde{\mathcal{E}}_{\mathrm{sim}}(N_i), \tag{6.6.1}$$

for distinct simple parameters ϕ_i. In this section, we shall treat the case that $r > 1$, or equivalently, that ϕ lies in the complement $\widetilde{\Phi}_2^{\mathrm{sim}}(G)$ of $\widetilde{\Phi}_{\mathrm{sim}}(G)$. From Proposition 6.3.1 we obtain corresponding global objects $(\dot{F}, \dot{G}, \dot{\phi})$ that satisfy the given list of conditions. In particular,

$$\dot{\phi} = \dot{\phi}_1 \boxplus \cdots \boxplus \dot{\phi}_r$$

is a global parameter which lies in $\widetilde{\Phi}_2(\dot{G})$. Moreover, the set

$$\dot{\widetilde{\mathcal{F}}} = \widetilde{\mathcal{F}}(\dot{\phi}_1, \dots, \dot{\phi}_r) = \{\dot{\ell}_1\dot{\phi}_1 \boxplus \cdots \boxplus \dot{\ell}_r\dot{\phi}_r :\ \dot{\ell}_i \geqslant 0\} \tag{6.6.2}$$

is a family of global parameters that satisfies the conditions of Assumption 5.4.1. We can therefore apply the relevant lemmas of Chapter 5 to the pair $(\dot{G}, \dot{\phi})$.

In order to extract local information for (G, ϕ) from the global properties of $(\dot{G}, \dot{\phi})$, we need to be able to remove valuations $v \notin S_\infty(u)$. The role of the following lemma will be similar to that of Lemma 6.4.1.

LEMMA 6.6.2. *Theorem* 2.2.1 *holds for the nonarchimedean field* F *and the parameter* ϕ *in case* $r \leqslant 3$ *and*

$$N_i = 1, \qquad 1 \leqslant i \leqslant r.$$

PROOF. We are assuming that ϕ belongs to $\widetilde{\Phi}_2^{\mathrm{sim}}(G)$. It is therefore a direct sum N inequivalent characters of order 1 or 2, where

$$1 < N = r \leqslant 3.$$

Therefore $\widehat{G}$ is orthogonal, and G is simple. The secondary condition of Theorem 2.2.1(a) consequently does not apply here. The primary condition of Theorem 2.2.1(a) follows from the global property

$$\dot{\widetilde{f}}_N(\dot{\phi}) = \dot{\widetilde{f}}^{\dot{G}}(\dot{\phi}), \qquad \dot{\widetilde{f}} \in \dot{\widetilde{\mathcal{H}}}(N),$$

established in Lemma 5.4.2. For we can fix the component $\dot{\widetilde{f}}^u$ of $\dot{\widetilde{f}}$ away from u so that $\dot{\widetilde{f}}_N(\dot{\phi})$ is a nonzero multiple of its component $\widetilde{f}_N(\phi) = \dot{\widetilde{f}}_{u,N}(\dot{\phi}_u)$

at u. In other words, the linear form $\widetilde{f}_N(\phi)$ in $\widetilde{f}$ is the pullback of a stable linear form $f^G(\phi)$ in $f \in \widetilde{\mathcal{H}}(\dot{G})$.

For the main assertion (b) of Theorem 2.2.1, we shall be content to treat the case $N = 3$. Then $\widehat{G} = SO(3,\mathbb{C})$ and $G = Sp(2)$. We may as well also assume that the quadratic idèle class character

$$\eta_{\dot{\phi}} = \dot{\phi}_1\dot{\phi}_2\dot{\phi}_3$$

equals 1, since we could otherwise replace each $\dot{\phi}_i$ by its product with $\eta_{\dot{\phi}} = \eta_{\dot{\phi}}^{-1}$. Then $\dot{\phi}$ corresponds to a homorphism of $W_{\dot{F}}$ to $\widehat{G}$. We shall reduce the assertion to results of [**LL**], using the properties of Whittaker models implicit in Lemma 2.5.5.

We do not know a priori that the stable distribution attached to ϕ in [**LL**] is the same as the linear form $f^G(\phi)$. We therefore set

$$f_*^G(\phi) = \sum_{\pi\in\Pi_\phi} f_G(\pi), \qquad f \in \mathcal{H}(G),$$

where Π_ϕ is the packet attached to the local parameter ϕ in [**LL**]. Since ϕ is a local factor $\dot{\phi}_u$ of the global parameter $\dot{\phi}$, the global results of [**LL**] implicitly exhibit $f_*^G(\phi)$ as a local factor of the $\dot{\phi}$-component $S^{\dot{G}}_{\mathrm{disc},\dot{\phi}}(\dot{f})$ of the stable trace formula for $\dot{G}$. In particular, if

$$\dot{f} = \dot{f}_u\,\dot{f}^u = f\,\dot{f}^u,$$

we can fix $\dot{f}^u$ so that $S^{\dot{G}}_{\mathrm{disc},\dot{\phi}}(\dot{f})$ is a nonzero multiple of $f_*^G(\phi)$. On the other hand, it follows easily from the fact that $\widehat{G} = SO(3,\mathbb{C})$ that the formula (3.3.14) reduces to an identity

$$I^N_{\mathrm{disc},\dot{\phi}}(\dot{\widetilde{f}}) = \widetilde{\iota}(N,\dot{G})\,\widehat{S}^{\dot{G}}_{\mathrm{disc},\dot{\phi}}(\dot{\widetilde{f}}^{\dot{G}}), \qquad \dot{\widetilde{f}} \in \dot{\widetilde{\mathcal{H}}}(N),$$

of which the left hand side is a nonzero multiple of $\dot{\widetilde{f}}_N(\dot{\phi})$. We choose the variable component $\widetilde{f} = \dot{\widetilde{f}}_u$ of $\dot{\widetilde{f}}$ at u so that it has the same image in $\widetilde{\mathcal{S}}(G)$ as f. We can then fix the complementary component $\dot{\widetilde{f}}^u$ so that it has the same image in $\widetilde{\mathcal{S}}(\dot{G}^u)$ as $\dot{f}^u$, and so that $\dot{\widetilde{f}}_N(\dot{\phi})$ is a nonzero multiple of $f^G(\phi)$. It follows that

$$f_*^G(\phi) = e_*(\phi)\,f^G(\phi), \qquad f \in \mathcal{H}(G),$$

for a nonzero constant $e_*(\phi)$.

The last equation is the analogue of the identity (6.4.4) from the proof of Lemma 6.4.1. If $\mathrm{char}(F) \neq 2$, we can combine it with a corresponding analogue of Lemma 2.5.5, which is very simple in this case and will appear in [**A27**]. It follows that $e_*(\phi) = 1$. If $\mathrm{char}(F) = 2$, we can vary the global field $\dot{F}$ as in the proof of Lemma 6.4.1. It follows that $e_*(\phi) = 1$ in this case as well. We see therefore that

$$f_*^G(\phi) = f^G(\phi), \qquad f \in \mathcal{H}(G),$$

in all cases. The possible ambiguity between [**LL**] and Theorem 2.2.1(a) thus resolved, we conclude that the packet Π_ϕ of [**LL**] satisfies the required identity of Theorem 2.2.1(b). □

We shall now apply the general results from Chapter 5. Recall that Lemma 5.4.2 affirms the validity of Assumption 5.1.1 for the parameter $\dot{\phi} \in \widetilde{\mathcal{F}}_2(\dot{G})$. As in the special case established in the last lemma, we need to convert this global property to the local assertion of Theorem 2.2.1(a). To do so, we temporarily allow G to be an element in the larger set $\widetilde{\mathcal{E}}_{\mathrm{ell}}(N)$. In other words,

$$(G, \phi) = (\dot{G}_u, \dot{\phi}_u)$$

is the local component of a pair

$$(\dot{G}, \dot{\phi}), \qquad \dot{G} \in \dot{\widetilde{\mathcal{E}}}_{\mathrm{ell}}(N),\ \dot{\phi} \in \dot{\widetilde{\mathcal{F}}}_2^{\mathrm{sim}}(\dot{G}).$$

This is compatible with our understanding that the basic objects in the construction of §6.3 are really the simple parameters $\{\phi_1, \dots, \phi_r\}$ and $\{\dot{\phi}_1, \dots, \dot{\phi}_r\}$ rather than the pairs (G, ϕ) and $(\dot{G}, \dot{\phi})$.

LEMMA 6.6.3. *The linear form on $\widetilde{\mathcal{H}}(N)$ attached to the parameter $\phi \in \widetilde{\Phi}_2^{\mathrm{sim}}(G)$ is the pullback of a stable linear form*

$$f \longrightarrow f^G(\phi), \qquad f \in \widetilde{\mathcal{H}}(G),$$

on $\widetilde{\mathcal{H}}(G)$. If $G = G^1 \times G^2$ is not simple, and

$$\phi = \phi^1 \times \phi^2, \qquad \phi^i \in \widetilde{\Phi}_2(G^i),\ i = 1, 2,$$

then

$$f^G(\phi) = f_1^{G^1}(\phi^1)\, f_2^{G^2}(\phi^2), \qquad f = f_1 f_2.$$

PROOF. This is the assertion of Theorem 2.2.1(a), for the generic local parameter $\phi \in \widetilde{\Phi}_2^{\mathrm{sim}}(G)$. It is also the local form of Lemma 5.4.2, which we shall apply to the global parameter $\dot{\phi}$. Recall that $\phi = \dot{\phi}_u$, for a valuation u of $\dot{F}$. If v is an archimedean valuation, the analogue of the lemma for $\dot{F}_v$ will be assumed as part of the theory of twisted endoscopy in [**Me**]. Suppose that $v \neq u$ is not archimedean. If the corresponding centralizer $\overline{S}_{\dot{\phi}_v}$ is infinite, the analogue of the lemma for $\dot{F}_v$ follows by induction. In view of condition (ii) of Proposition 6.3.1, this leaves only the case that the integer N is less than 4, and $\dot{\phi}_v$ is a sum of N-inequivalent characters of order at most 2. In these cases, $\widehat{G}$ must be orthogonal, and the analogue of the lemma follows from Lemma 6.4.1. The assertions of the lemma for $F = \dot{F}_u$ thus follow from the global assertion of Lemma 5.4.2, and their analogues for $v \neq u$. □

The product formula from the lemma allows us to apply our induction hypotheses to any datum $G \in \widetilde{\mathcal{E}}_{\mathrm{ell}}(N)$ that is composite. We can therefore assume henceforth that $G \in \widetilde{\mathcal{E}}_{\mathrm{sim}}(N)$ and $\dot{G} \in \dot{\widetilde{\mathcal{E}}}_{\mathrm{sim}}(N)$, as before. Observe that the lemma includes the assertion ϕ is cuspidal lift, which will allow us to make use of the orthogonality relations of the last section.

We are now ready to exploit the essential global properties of Chapter 5. Corollary 5.1.3, together with the application to Lemma 5.1.4 of its supplement Lemma 5.4.5 in case N is even and $\eta_{\dot{\phi}} = 1$, tells us that the stable multiplicity formula of Theorem 4.1.2 is valid for $\dot{G}$ and $\dot{\phi}$. We thus have

$$S^{\dot{G}}_{\mathrm{disc},\dot{\phi}}(\dot{f}) = |\Phi(\dot{G},\dot{\phi})|\,|\mathcal{S}_{\dot{\phi}}|^{-1}\,\dot{f}^{\dot{G}}(\dot{\phi}), \qquad \dot{f} \in \widetilde{\mathcal{H}}(\dot{G}), \tag{6.6.3}$$

since $\varepsilon^{\dot{G}}(\dot{\phi}) = 1$ in the generic case at hand. We shall apply this to the $\dot{\phi}$-component

$$I^{\dot{G}}_{\mathrm{disc},\dot{\phi}}(\dot{f}) = \sum_{\dot{G}' \in \mathcal{E}_{\mathrm{ell}}(\dot{G})} \iota(\dot{G},\dot{G}')\,\widehat{S}'_{\mathrm{disc},\dot{\phi}}(\dot{f}') \tag{6.6.4}$$

of the stabilized trace formula given by (3.3.15). More precisely, we apply the corollary of Theorem 4.1.2 to the groups $\dot{G}'$ that index the sum in (6.6.4).

It thus follows from Corollary 4.1.3 that the right hand side of (6.6.4) equals

$$\sum_{\dot{G}' \in \mathcal{E}_{\mathrm{ell}}(\dot{G})} \iota(\dot{G},\dot{G}') \sum_{\dot{\phi}' \in \Phi(\dot{G}',\dot{\phi})} |\mathcal{S}_{\dot{\phi}'}|^{-1}\,\dot{f}'(\dot{\phi}').$$

This is the expression (4.7.7) (with $\psi = \dot{\phi}$) from the beginning of the elementary proof of Lemma 4.7.1, which we recall was the conditional justification in §4.7 of the spectral multiplicity formula. The conditions from §4.7 included the local theorems, which of course are what we are now trying to establish (for generic parameters). However, the simple arguments from the proof of Lemma 4.7.1 are still available to us here. We shall revisit them very briefly, in order to obtain a modified global formula that will help us prove the local results.

Arguing as in the proof of Lemma 4.7.1, we convert the last expression to the $\dot{\phi}$-analogue

$$\sum_{(\dot{G}',\dot{\phi}')} |\mathcal{S}_{\dot{\phi}'}|^{-1}\,|\overline{Z}(\widehat{\dot{G}'})^{\Gamma}|^{-1}\,|\mathrm{Out}_{\dot{G}}(\dot{G}',\dot{\phi}')|^{-1}\,\dot{f}'(\dot{\phi}')$$

of (4.7.8). The double sum is over the set of $\widehat{\dot{G}}$-orbits of pairs $(\dot{G}',\dot{\phi}')$ in the family $X(\dot{G})$ of §4.8 such that $\dot{\phi}'$ maps to $\dot{\phi}$. It can be replaced by a double sum over the set of $\widehat{\dot{G}}$-orbits of pairs $(\dot{\phi}_G,\dot{x}_G)$ in the associated family $Y(\dot{G})$ such that $\dot{\phi}_G$ maps to $\dot{\phi}$. The expression becomes

$$\sum_{(\dot{\phi}_G,\dot{x}_G)} |\mathcal{S}_{\dot{\phi}_G}|^{-1}\,\dot{f}'(\dot{\phi}'),$$

since we can write

$$|\mathcal{S}_{\dot{\phi}'}|^{-1}\,|\overline{Z}(\widehat{\dot{G}}')^{\Gamma}|^{-1}\,|\mathrm{Out}_G(\dot{G}',\dot{\phi}')|^{-1} = |\mathcal{S}_{\dot{\phi}_G}|^{-1},$$

as again in the proof of Lemma 4.7.1. This summand depends only on the image $\dot{x}$ of $\dot{x}_G$ in $\mathcal{S}_{\dot{\phi}}$. The right hand side of (6.6.4) therefore equals

$$|\Phi(\dot{G},\dot{\phi})|\,|\mathcal{S}_{\dot{\phi}}|^{-1}\sum_{\dot{x}\in\mathcal{S}_{\dot{\phi}}}\dot{f}'(\dot{\phi}'),$$

where $(\dot{G}',\dot{\phi}')$ maps to the pair $(\dot{\phi},\dot{x})$, and where we recall that $\Phi(\dot{G},\dot{\phi})$ is the preimage of $\dot{\phi}$ in $\Phi(\dot{G})$, a set of order 1 or 2.

Consider next the formula (4.1.1) (with $\dot{\psi}=\dot{\phi}$) for the left hand side of (6.6.4). We claim that $\dot{\phi}$ cannot contribute to the discrete spectrum of any proper Levi subgroup $\dot{M}$ of $\dot{G}$. For as we have argued in the past, our induction hypothesis on the factor $\dot{G}_-\in\tilde{\dot{\mathcal{E}}}(N_-)$ of $\dot{M}$ prevents any parameter $\dot{\psi}\in\tilde{\dot{\Psi}}(N)$ that does contribute to the discrete spectrum of $\dot{M}$ from lying in the subset $\tilde{\dot{\Phi}}_{\mathrm{ell}}(N)$ that contains $\dot{\phi}$. Such assertions are of course implicit in the definition of the set $\tilde{\dot{\Phi}}(N)$, which rests ultimately on the general classification of Theorems 1.3.2 and 1.3.3. There is consequently no contribution from $\dot{M}$ to (4.1.1). The left hand side of (6.6.4) then equals

$$I^{\dot{G}}_{\mathrm{disc},\dot{\phi}}(\dot{f}) = \mathrm{tr}\big(R^{\dot{G}}_{\mathrm{disc},\dot{\phi}}(f)\big) = \sum_{\dot{\pi}_G} n_{\dot{\phi}}(\dot{\pi}_G)\,\dot{f}_{\dot{G}}(\dot{\pi}_G),$$

where $\dot{\pi}_G$ ranges over the set $\Pi(\dot{G})$ of irreducible unitary representations of $\dot{G}(\dot{\mathbb{A}})$, and $n_{\dot{\phi}}(\dot{\pi}_G)$ are nonnegative integers. We are assuming that $\dot{f}$ lies in $\widetilde{\mathcal{H}}(\dot{G})$. This means that $\dot{f}_{\dot{G}}(\dot{\pi}_G)$ depends only on the image $\dot{\pi}$ of $\dot{\pi}_G$ in the set $\widetilde{\Pi}(\dot{G})$ of orbits in $\Pi(\dot{G})$ under the restricted direct product

$$\widetilde{\mathrm{O}}\mathrm{ut}_N(\dot{G}_{\dot{\mathbb{A}}}) = \prod_v \widetilde{\mathrm{O}}\mathrm{ut}_N(\dot{G}_v).$$

We can therefore write the left hand side of (6.6.4) as

$$|\Phi(\dot{G},\dot{\phi})|\sum_{\dot{\pi}\in\widetilde{\Pi}(\dot{G})} n_{\dot{\phi}}(\dot{\pi})\,\dot{f}_{\dot{G}}(\dot{\pi}),$$

for a modified coefficient

$$n_{\dot{\phi}}(\dot{\pi}) = |\Phi(\dot{G},\dot{\phi})|^{-1}\sum_{\dot{\pi}_G\in\Pi(\dot{G},\dot{\pi})} n_{\dot{\phi}}(\dot{\pi}_G) \tag{6.6.5}$$

in which $\Pi(\dot{G},\dot{\pi})$ is the preimage of $\dot{\pi}$ in $\Pi(\dot{G})$.

We have converted (6.6.4) to an identity

$$\sum_{\dot{\pi}\in\widetilde{\Pi}(\dot{G})} n_{\dot{\phi}}(\dot{\pi})\,\dot{f}_{\dot{G}}(\dot{\pi}) = |\mathcal{S}_{\dot{\phi}}|^{-1}\sum_{\dot{x}\in\mathcal{S}_{\dot{\phi}}}\dot{f}'(\dot{\phi}'),\qquad \dot{f}\in\widetilde{\mathcal{H}}(\dot{G}), \tag{6.6.6}$$

where $(\dot{G}', \dot{\phi}')$ maps to $(\dot{\phi}, \dot{x})$. We shall apply this identity when $\dot{f}$ equals a product

$$\dot{f} = \dot{f}_\infty \cdot \dot{f}_u \cdot \dot{f}^{\infty,u},$$

relative to the decomposition

$$\dot{G}(\dot{\mathbb{A}}) = \dot{G}_\infty \times \dot{G}_u \times \dot{G}^{\infty,u} = \dot{G}(\dot{F}_\infty) \times \dot{G}(\dot{F}_u) \times \dot{G}(\dot{\mathbb{A}}^{\infty,u})$$

of $\dot{G}(\dot{\mathbb{A}})$.

Suppose that v is a valuation of $\dot{F}$ in the complement of $S_\infty(u)$. The packet $\widetilde{\Pi}_{\dot{\phi}_v}$ is then defined, and satisfies the conditions of Theorem 2.2.1. For if $S_{\dot{\phi}_v}$ is infinite, $\dot{\phi}_v$ lies in the complement of $\widetilde{\Phi}_2(\dot{G}_v)$, and therefore represents the case we have already established. If $S_{\dot{\phi}_v}$ is finite, Proposition 6.3.1 tells us that $N \leqslant 3$, in which case the local classification follows from Lemma 6.6.2. We can therefore write

$$(\dot{f}^{\infty,u})'((\dot{\phi}^{\infty,u})') = \sum_{\dot{\pi}^{\infty,u}} \langle \dot{x}, \dot{\pi}^{\infty,u}\rangle (\dot{f}^{\infty,u})_{\dot{G}}(\dot{\pi}^{\infty,u}), \tag{6.6.7}$$

where $\dot{\pi}^{\infty,u}$ ranges over elements in the packet

$$\Big\{\dot{\pi}^{\infty,u} = \bigotimes_{v \notin S_\infty(u)} \dot{\pi}_v : \dot{\pi}_v \in \widetilde{\Pi}_{\dot{\phi}_v}, \langle \cdot, \pi_v \rangle = 1 \text{ for almost all } v\Big\}$$

of

$$\dot{\phi}^{\infty,u} = \prod_{v \notin S_\infty(u)} \dot{\phi}_v,$$

and

$$\langle \dot{x}, \dot{\pi}^{\infty,u}\rangle = \prod_{v \notin S_\infty(u)} \langle \dot{x}_v, \dot{\pi}_v\rangle$$

is the corresponding product of local pairings. If v is archimedean, we have

$$(\dot{f}_v')(\dot{\phi}_v') = \sum_{\dot{\pi}_v \in \widetilde{\Pi}_{\dot{\phi}_v}} \langle \dot{x}_v, \dot{\pi}_v\rangle \dot{f}_{v,\dot{G}}(\dot{\pi}_v),$$

by the results of Shelstad. We can therefore write

$$(\dot{f}_\infty')(\dot{\phi}_\infty') = \sum_{\dot{\pi}_\infty} \langle \dot{x}, \dot{\pi}_\infty\rangle (\dot{f}_\infty)_{\dot{G}}(\dot{\pi}_\infty), \tag{6.6.8}$$

where $\dot{\pi}_\infty$ ranges over representations in the packet

$$\Big\{\dot{\pi}_\infty = \bigotimes_{v \in S_\infty} \dot{\pi}_v : \dot{\pi}_v \in \widetilde{\Pi}_{\dot{\phi}_v}\Big\}$$

of

$$\dot{\phi}_\infty = \prod_{v \in S_\infty} \dot{\phi}_v,$$

and

$$\langle \dot{x}, \dot{\pi}_\infty\rangle = \prod_{v \in S_\infty} \langle \dot{x}_v, \dot{\pi}_v\rangle$$

is the associated product of local pairings.

Consider the remaining valuation $v = u$ of $\dot{F}$. Then $\dot{F}_u = F$, $\dot{G}_u = G$ and $\dot{\phi}_u = \phi$. At this point we assume that the function $\dot{f}_u$ lies in the subspace $\widetilde{\mathcal{H}}_{\mathrm{cusp}}(G)$ of $\widetilde{\mathcal{H}}(G)$. We can then write

$$\dot{f}'_u(\dot{\phi}'_u) = \sum_{\pi \in \widetilde{\Pi}_2(G)} c_\phi(x,\pi)\,(\dot{f}_{u,\dot{G}})(\pi), \tag{6.6.9}$$

in the notation of Lemma 6.5.3. We shall use the global identity (6.6.6) to extract information about the coefficients $c_\phi(x,\pi)$. More precisely, we shall exploit the identity obtained by substituting the formulas (6.6.7), (6.6.8) and (6.6.9) for the three factors of the summand

$$\dot{f}'(\dot{\phi}') = \dot{f}'_\infty(\dot{\phi}'_\infty)\,\dot{f}'_u(\dot{\phi}'_u)\,(\dot{f}^{\infty,u})'\,((\dot{\phi}^{\infty,u})')$$

in (6.6.6).

The right hand side of (6.6.6) becomes

$$|\mathcal{S}_\phi|^{-1} \sum_{x \in \mathcal{S}_\phi} \sum_{\dot{\pi}} \langle \dot{x}, \dot{\pi}_\infty \rangle\, c_{\phi,x}(\pi)\, \langle \dot{x}, \dot{\pi}^{\infty,u} \rangle\, \dot{f}_{\dot{G}}(\dot{\pi}),$$

where the inner sum is over products

$$\dot{\pi} = \dot{\pi}_\infty \otimes \pi \otimes \dot{\pi}^{\infty,u}$$

of representations $\dot{\pi}_\infty$ in the packet of $\dot{\phi}_\infty$, π in $\widetilde{\Pi}_2(G)$ and $\dot{\pi}^{\infty,u}$ in the packet of $\dot{\phi}^{\infty,u}$, and $\dot{x}$ is the isomorphic image of x in $\mathcal{S}_{\dot{\phi}}$. Suppose that $\xi \in \widehat{\mathcal{S}}_\phi$ is a character on the 2-group $\mathcal{S}_{\dot{\phi}} \cong \mathcal{S}_\phi$. It follows from the condition (iii)(a) of Proposition 6.3.1 that there is a representation $\dot{\pi}_{\infty,\xi}$ in the packet of $\dot{\phi}_\infty$ such that

$$\langle \dot{x}, \dot{\pi}_{\infty,\xi} \rangle = \xi(x)^{-1}, \qquad x \in \mathcal{S}_\phi.$$

For the places outside $S_\infty(u)$, we take the representation

$$\dot{\pi}^{\infty,u}(1) = \bigotimes_{v \notin S_\infty(u)} \dot{\pi}_v(1)$$

in the packet of $\dot{\phi}^{\infty,u}$ such that for each v, the character $\langle\,\cdot\,, \dot{\pi}_v(1)\rangle$ on $\mathcal{S}_{\dot{\phi}_v}$ is 1. With an appropriate choice of the functions $\dot{f}_\infty$ and $\dot{f}^{\infty,u}$, the formula (6.6.6) then reduces to an identity

$$\sum_\pi n_\phi(\xi,\pi)\, f_G(\pi) = \sum_\pi |\mathcal{S}_\phi|^{-1} \sum_{x \in \mathcal{S}_\phi} c_\phi(x,\pi)\, \xi(x)^{-1}\, f_G(\pi) \tag{6.6.10}$$

for any $f \in \widetilde{\mathcal{H}}_{\mathrm{cusp}}(G)$, where π is summed over $\widetilde{\Pi}_2(G)$ on each side, and

$$n_\phi(\xi,\pi) = n_{\dot{\phi}}\big(\dot{\pi}_{\infty,\xi} \otimes \pi \otimes \dot{\pi}^{\infty,u}(1)\big) \tag{6.6.11}$$

Since the characters of representations $\pi \in \widetilde{\Pi}_2(G)$ are linearly independent on $\widetilde{\mathcal{H}}_{\mathrm{cusp}}(G)$, we conclude that

$$n_\phi(\xi,\pi) = |\mathcal{S}_\phi|^{-1} \sum_{x \in \mathcal{S}_\phi} c_\phi(x,\pi)\, \xi(x)^{-1}, \tag{6.6.12}$$

for any $\pi \in \widetilde{\Pi}_2(G)$. In particular, the number (6.6.11) does depend only on the local data (ϕ, ξ, π) at u, as the notation suggests.

LEMMA 6.6.4. (a) *For any $\pi \in \widetilde{\Pi}_2(G)$, $n_\phi(\xi, \pi)$ is a nonnegative integer.*

(b) *If $m(\phi)$ and $m(\pi)$ are the orders of the preimages of ϕ and π in $\Phi_2(G)$ and $\Pi_2(G)$ respectively, the product*

$$\widetilde{n}_\phi(\xi, \pi) = m(\pi)^{-1} m(\phi)\, n_\phi(\xi, \pi)$$

is also a nonnegative integer.

PROOF. By definition, $n_\phi(\xi, \pi)$ is the number $n_{\dot{\phi}}(\dot{\pi})$ given by (6.6.5), with

$$\dot{\pi} = \dot{\pi}_{\infty,\xi} \otimes \pi \otimes \dot{\pi}^{\infty,u}(1).$$

The summands $n_{\dot{\phi}}(\dot{\pi}_G)$ on the right hand side of (6.6.5) are nonnegative integers. To prove (a), we need to show that the sum itself is divisible by the positive integer $|\Phi(\dot{G}, \dot{\phi})|$. We can therefore assume that $\widehat{G} = SO(N, \mathbb{C})$, with N even, since the group $\widetilde{\mathrm{Out}}_N(\dot{G})$ under which $\Phi(\dot{G}, \dot{\phi})$ is an orbit would otherwise be trivial. Then $\widetilde{\mathrm{Out}}_N(\dot{G})$ is a group of order 2, the nontrivial element being represented by a point in $O(N, \mathbb{C})$ with determinant equal to (-1). We can also assume that the rank N_i of each component $\dot{\phi}_i$ of $\dot{\phi}$ is even, since the stabilizer of $\dot{\phi}$ would otherwise be the group $\widetilde{\mathrm{Out}}_N(\dot{G})$ itself. With this condition, we consider the archimedean component $\dot{\pi}_v \in \widetilde{\Pi}_{\dot{\phi}_v}$ of $\dot{\pi}$ at any $v \in S_\infty$. The irreducible components of $\dot{\phi}_v$ are distinct and of degree 2, by the condition (iii)(a) of Proposition 6.3.1. It follows easily that as an $\widetilde{\mathrm{Out}}_N(\dot{G})$-orbit of irreducible representations of $\dot{G}(\dot{F}_v)$, $\dot{\pi}_v$ contains two elements $\dot{\pi}_{v,G}$. This implies that $\widetilde{\mathrm{Out}}_N(\dot{G})$ acts freely on the indices of summation $\dot{\pi}_G$ in (6.6.5). Since the original multiplicity $n_{\dot{\psi}}(\dot{\pi}_G)$ is invariant under the action of any $\dot{F}$-automorphism of $\dot{G}$, the sum in (6.6.5) is an even integer, and is hence divisible by the order 2 of $\Phi(\dot{G}, \dot{\phi})$. This establishes (a).

The proof of (b) is simpler. Since

$$m(\phi) = m(\dot{\phi}) = |\Phi(\dot{G}, \dot{\phi})|,$$

we need only show that the sum in (6.6.5) is in all cases divisible by the number of elements π_G in the $\widetilde{\mathrm{Out}}_N(G)$-orbit of π. This number equals either 1 or 2. If it equals 2, $\widetilde{\mathrm{Out}}_N(\dot{G})$ acts freely on the indices of summation $\dot{\pi}_G$ in (6.6.5). It follows that the product

$$\widetilde{n}_\phi(\xi, \pi) = m(\pi)^{-1} \sum_{\dot{\pi}_G} n_{\dot{\phi}}(\dot{\pi}_G),$$

is also a nonnegative integer, as required. □

We can now establish the essential properties of composite parameters for our group $G \in \widetilde{\mathcal{E}}_{\mathrm{sim}}(N)$ over the local p-adic field F.

PROPOSITION 6.6.5. (a) *For every* $\phi \in \widetilde{\Phi}_2^{\mathrm{sim}}(G)$, *there is a finite subset* $\widetilde{\Pi}_\phi$ *of* $\widetilde{\Pi}_2(G)$, *together with a bijection*

$$\pi \longrightarrow \langle \cdot, \pi \rangle, \qquad \pi \in \widetilde{\Pi}_\phi,$$

from $\widetilde{\Pi}_\phi$ *onto* $\widehat{\mathcal{S}}_\phi$, *such that*

$$f'(\phi') = \sum_{\pi \in \widetilde{\Pi}_\phi} \langle x, \pi \rangle f_G(\pi), \qquad f \in \widetilde{\mathcal{H}}_{\mathrm{cusp}}(G),\ x \in \mathcal{S}_\phi, \tag{6.6.13}$$

where (G', ϕ') *is the pair (in* $X_{\mathrm{ell}}(G)$*) that maps to* (ϕ, x).

(b) *As* ϕ *ranges over* $\widetilde{\Phi}_2^{\mathrm{sim}}(G)$, *the subsets* $\widetilde{\Pi}_\phi$ *of* $\widetilde{\Pi}_2(G)$ *are disjoint.*

PROOF. The key ingredient will be the orthogonality relations of Lemma 6.5.3(b). We shall combine them with the formula (6.6.12) for the nonnegative integer $n_\phi(\xi, \pi)$.

Suppose that $\xi \in \widehat{\mathcal{S}}_\phi$ and $\xi_1 \in \widehat{\mathcal{S}}_{\phi_1}$ are characters of the finite groups attached to two parameters ϕ and ϕ_1 in $\widetilde{\Phi}_2^{\mathrm{sim}}(G)$. Applying (6.6.12) to each of the integers $n_\phi(\xi, \pi)$ and $n_{\phi_1}(\xi_1, \pi)$, we write

$$\begin{aligned} &\sum_{\pi \in \widetilde{\Pi}_2(G)} \widetilde{n}_\phi(\xi, \pi)\, n_{\phi_1}(\xi_1, \pi) \\ &= \sum_{\pi} m(\pi)^{-1} m(\phi)\, n_\phi(\xi, \pi)\, \overline{n_{\phi_1}(\xi_1, \pi)} \\ &= m(\phi)\, |\mathcal{S}_\phi|^{-1}\, |\mathcal{S}_{\phi_1}|^{-1} \sum_{x, x_1} \xi(x)^{-1} \xi_1(x_1) \Big(\sum_{\pi} m(\pi)^{-1}\, c_\phi(x, \pi)\, \overline{c_{\phi_1}(x_1, \pi)} \Big), \end{aligned}$$

where x and x_1 are summed over $\mathcal{S}_\phi$ and $\mathcal{S}_{\phi_1}$ respectively. The orthogonality relation (6.5.16) then tells us that the last sum over π vanishes unless (ϕ, x) equals (ϕ_1, x_1), in which case it equals $m(\phi)^{-1} |\mathcal{S}_\phi|$. The original sum over π therefore vanishes unless $\phi_1 = \phi$, in which case it equals

$$|\mathcal{S}_\phi|^{-1} \sum_{x \in \mathcal{S}_\phi} \xi(x)^{-1}\, \xi_1(x).$$

This in turn vanishes unless $\xi_1 = \xi$, in which case it equals 1. We have established an identity

$$\sum_{\pi \in \widetilde{\Pi}_2(G)} \widetilde{n}_\phi(\xi, \pi)\, n_{\phi_1}(\xi_1, \pi) = \begin{cases} 1, & \text{if } (\phi_1, \xi_1) = (\phi, \xi), \\ 0, & \text{otherwise.} \end{cases} \tag{6.6.14}$$

Suppose first that $(\phi_1, \xi_1) = (\phi, \xi)$. Then the nonnegative integers $n_\phi(\xi, \pi)$ and $\widetilde{n}_\phi(\xi, \pi)$ are either both zero or both nonzero. Since the sum over π of their product equals 1, we see that there is a unique element $\pi(\xi)$ in $\widetilde{\Pi}_2(G)$ such that

$$\widetilde{n}_\phi(\xi, \pi) = n_\phi(\xi, \pi) = \begin{cases} 1, & \text{if } \pi = \pi(\xi), \\ 0, & \text{otherwise,} \end{cases} \tag{6.6.15}$$

for any $\pi \in \widetilde{\Pi}_2(G)$. Then taking the case that $(\phi_1, \xi_1) \neq (\phi, \xi)$, we see from (6.6.14) that the mapping

$$(\phi, \xi) \longrightarrow \pi(\xi)$$

is injective. In other words, if we define

$$\widetilde{\Pi}_\phi = \{\pi(\xi) : \xi \in \widehat{\mathcal{S}}_\phi\}, \qquad \phi \in \widetilde{\Phi}_2^{\text{sim}}(G),$$

the subsets $\{\widetilde{\Pi}_\phi\}$ of $\widetilde{\Pi}_2(G)$ are mutually disjoint. This is the assertion (b) of the proposition.

To establish the remaining identity (6.6.13), we have only to invert (6.6.12). It follows from (6.6.15) that $c_\phi(x, \pi)$ equals $\xi(x)$ if $\pi = \pi(\xi)$ for some $\xi \in \widehat{\mathcal{S}}_\phi$, and that $c_\phi(x, \pi) = 0$ otherwise. The required formula (6.6.13) then follows from (6.5.15), if we set

$$\langle x, \pi(\xi) \rangle = \xi(x) = c_\phi\big(x, \pi(\xi)\big), \qquad x \in \mathcal{S}_\phi, \ \xi \in \widehat{\mathcal{S}}_\phi.$$

This completes the proof of the proposition. □

COROLLARY 6.6.6. *If π belongs to $\widetilde{\Pi}_\phi$, then*

$$m(\pi) = m(\phi).$$

PROOF. The corollary follows from the equality (6.6.15) of $\widetilde{n}_\phi(\xi, \pi)$ with $n_\phi(\xi, \pi)$. □

6.7. Local packets for simple ϕ

We continue with the discussion of §6.6. In this section we shall complete the local classification of tempered representations. We are taking $G \in \widetilde{\Phi}_{\text{sim}}(N)$ to be a simple datum over the local p-adic field F, for the fixed positive integer N. It remains to treat the simple generic parameters $\phi \in \widetilde{\Phi}_{\text{sim}}(G)$. However, to exploit the global construction of §6.2, we will have to start with a representation π instead of a parameter ϕ.

We shall write $\widetilde{\Pi}_{\text{sim}}(G)$ for the set of elements in $\widetilde{\Pi}_2(G)$ that do not lie in the union over $\phi \in \widetilde{\Phi}_2^{\text{sim}}(G)$ of the disjoint sets $\widetilde{\Pi}_\phi$ we constructed in the last section. Let us also write $\widetilde{\mathcal{I}}_{\text{sim}}(G)$ for the subspace of functions $f_G \in \widetilde{\mathcal{I}}_{\text{cusp}}(G)$ such that $f_G(\tau) = 0$ for every τ in the complement of $\widetilde{\Pi}_{\text{sim}}(G)$ in $\widetilde{T}_{\text{ell}}(G)$, and $\widetilde{\mathcal{S}}_{\text{sim}}(G)$ for the subspace of functions $f^G \in \widetilde{\mathcal{S}}_{\text{cusp}}(G)$ such that $f^G(\phi) = 0$ for every ϕ in $\Phi_2^{\text{sim}}(G)$.

LEMMA 6.7.1. *The isomorphism (6.5.10) maps $\widetilde{\mathcal{I}}_{\text{sim}}(G)$ isomorphically onto $\widetilde{\mathcal{S}}_{\text{sim}}(G)$.*

PROOF. In the statement, $\widetilde{\mathcal{S}}_{\text{sim}}(G)$ is to be understood as a subspace of collections of functions $\{f' = f^{G'}\}$ in

$$\bigoplus_{G' \in \mathcal{E}_{\text{ell}}(G)} \widetilde{\mathcal{S}}_{\text{cusp}}(G')$$

such that $f' = 0$ for $G' \neq G$. We observe from (6.5.6) that the orthogonal complement of $\widetilde{\mathcal{I}}_{\mathrm{sim}}(G)$ in $\widetilde{\mathcal{I}}_{\mathrm{cusp}}(G)$ may be identified with the space of functions on $\widetilde{T}_{\mathrm{ell}}(G)$ that are supported on the complement of $\widetilde{\Pi}_{\mathrm{sim}}(G)$. By (6.5.13), the orthogonal complement of $\widetilde{\mathcal{S}}_{\mathrm{sim}}(G)$ can be identified with the space of collections $\{f'\}$ such that f^G is supported on $\widetilde{\Phi}_2^{\mathrm{sim}}(G)$. It follows easily from the formulas (6.5.5) and (6.6.13) that the mapping (6.5.10) sends the first orthogonal complement onto the second. Since the mapping is an isometry, it does indeed send $\widetilde{\mathcal{I}}_{\mathrm{sim}}(G)$ isomorphically onto $\widetilde{\mathcal{S}}_{\mathrm{sim}}(G)$. □

REMARK. Suppose that $f_\pi \in \widetilde{\mathcal{H}}_{\mathrm{cusp}}(G)$ is a pseudocoefficient of a representation $\pi \in \widetilde{\Pi}_{\mathrm{sim}}(G)$. This means that

$$f_{\pi,G}(\tau) = \begin{cases} 1, & \text{if } \tau = \pi, \\ 0, & \text{otherwise,} \end{cases}$$

for any $\tau \in \widetilde{T}_{\mathrm{ell}}(G)$. Then $f_{\pi,G}$ lies in $\widetilde{\mathcal{I}}_{\mathrm{sim}}(G)$, according to the definition above. It follows that

$$f'_\pi = f_\pi^{G'} = 0,$$

for any datum $G' \in \widetilde{\mathcal{E}}_{\mathrm{ell}}(G)$ distinct from G.

We will now turn to the global arguments. We cannot use the discussion of §6.3, since it was based on the inductive application of Corollary 6.2.4. Corollary 6.2.4 is not available because it requires that the local theorems be valid for parameters $\phi \in \widetilde{\Phi}(N)$. We have therefore to go back to the original Lemma 6.2.2, and the arguments from its proof.

We fix a representation $\pi \in \widetilde{\Pi}_{\mathrm{sim}}(G)$. Having chosen this object, we apply Lemmas 6.2.1 and 6.2.2. We obtain global objects $(\dot{F}, \dot{G}, \dot{\pi})$ that satisfy the conditions of Lemma 6.2.2. In particular, (F, G, π) equals $(\dot{F}_u, \dot{G}_u, \dot{\pi}_u)$, for a place u of $\dot{F}$. If v belongs to S_∞, $\dot{\pi}_v$ belongs to the L-packet of a Langlands parameter ϕ_v in general position. If v lies in the complement of $S_\infty(u)$, $\dot{\pi}_v$ is a spherical representation, corresponding to a spherical Langlands parameter ϕ_v. The key property of $\dot{\pi}$ is of course that it occurs in the automorphic discrete spectrum of $\dot{G}$. This fact was established in the first half of the proof of Lemma 6.2.2. It then led naturally to the main point of the second half of the proof, the application of Corollary 3.4.3 to $\dot{\pi}$. This in turn allowed us to attach a global parameter $\dot{\phi} \in \widetilde{\dot{\Psi}}(N)$ to $\dot{\pi}$, which we then showed was in the subset of $\widetilde{\dot{\Phi}}(N)$ of generic parameters.

In Corollary 6.2.3, we saw that for any valuation $v \neq u$, the Langlands parameter of the localization $\dot{\pi}_v$ of $\dot{\pi}$ is the localization $\dot{\phi}_v$ of $\dot{\phi}$. Our task here will be to establish properties of the localization

$$\phi = \dot{\phi}_u$$

of $\dot{\phi}$ at u. These will be consequences of the fact that π lies in $\widetilde{\Pi}_{\mathrm{sim}}(G)$, together with the results we have established in sections following Lemma 6.2.2.

We did see in proof of Lemma 6.2.2 that ϕ lies in $\widetilde{\Phi}_{\mathrm{ell}}(N)$. (See the discussion following (6.2.8).) Moreover, it is a consequence of (6.2.9) (in the proof of Corollary 6.2.3) that there is a constant $c(\phi) \neq 0$ such that

$$\widetilde{f}_{\pi,N}(\phi) = c(\phi),$$

for any function $\widetilde{f}_\pi = \dot{\widetilde{f}}_{u,\pi}$ in $\mathcal{H}_{\mathrm{cusp}}(N)$ with $\widetilde{f}_\pi^G = f_\pi^G$. It then follows that the linear form $\widetilde{f} \to \widetilde{f}_N(\phi)$ on $\mathcal{H}_{\mathrm{cusp}}(N)$ vanishes on the kernel of the transfer mapping from $\widetilde{\mathcal{H}}_{\mathrm{cusp}}(N)$ to $\widetilde{\mathcal{S}}_{\mathrm{cusp}}(G)$. This in turn implies that for any $\widetilde{f} \in \widetilde{\mathcal{H}}_{\mathrm{cusp}}(N)$ at all, the component $\widetilde{f}^*(\phi)$ of $G^* \in \widetilde{\mathcal{E}}_{\mathrm{ell}}(N)$ in the decomposition (6.1.1) vanishes if G^* is distinct from G. Therefore, ϕ satisfies the condition (i) of the definition of a cuspidal lift in §6.5.

We do not know a priori that ϕ is simple. Suppose however that it lies in the complement $\widetilde{\Phi}_{\mathrm{ell}}^{\mathrm{sim}}(N)$ of $\widetilde{\Phi}_{\mathrm{sim}}(N)$. Then Lemma 6.6.3 implies that the conditions (i) and (ii) in the definition of a cuspidal lift are equivalent. More precisely, if we apply this lemma to the unique datum $G^* \in \widetilde{\mathcal{E}}_{\mathrm{ell}}(N)$ with $\phi \in \widetilde{\Phi}_2^{\mathrm{sim}}(G^*)$ (the existence of which is guaranteed by our induction hypotheses), we see that $G^* = G$. In other words, ϕ lies in the subset $\widetilde{\Phi}_2^{\mathrm{sim}}(G)$ of $\widetilde{\Phi}_{\mathrm{ell}}^{\mathrm{sim}}(N)$. It then follows from the formula (6.6.13) of Proposition 6.6.5 that

$$f_\pi^G(\phi) = \sum_{\pi^* \in \widetilde{\Pi}_\phi} f_{\pi,G}(\pi^*).$$

Since f_π is a pseudocoefficient of π, a representation in a set $\widetilde{\Pi}_{\mathrm{sim}}(G)$ that is disjoint from $\widetilde{\Pi}_\phi$, the sum on the right vanishes. On the other hand, since ϕ satisfies the condition (i), we have

$$f_\pi^G(\phi) = \widetilde{f}_{\pi,N}(\phi) = c(\phi) \neq 0.$$

This is a contradiction.

We have shown that ϕ lies in the subset $\widetilde{\Phi}_{\mathrm{sim}}(N)$ of $\widetilde{\Phi}_{\mathrm{ell}}(N)$. In this case, the first condition (i) for ϕ to be a cuspidal lift implies the second condition (ii). In other words, ϕ lies in the subset $\widetilde{\Phi}_{\mathrm{sim}}(G)$ of $\widetilde{\Phi}_{\mathrm{sim}}(N)$ by virtue of the fact that the definition in §6.1 is implied by the condition we have already established. The local parameter $\phi \in \widetilde{\Phi}_{\mathrm{sim}}(N)$ is therefore a cuspidal lift.

Consider the global parameter $\dot{\phi}$. It represents a generic automorphic representation of $GL(N)$ whose component at u lies in the discrete series of $GL(N,F)$, since the corresponding component $\phi = \dot{\phi}_u$ of $\dot{\phi}$ belongs to $\widetilde{\Phi}_{\mathrm{sim}}(N)$. The automorphic representation is therefore cuspidal, and $\dot{\phi}$ consequently lies in the set $\dot{\widetilde{\Phi}}_{\mathrm{sim}}(N)$ of simple global parameters. This argument is familiar from the proof of Corollary 6.2.4, where we were working with the benefit of the stronger hypothesis. As in the earlier case, $\dot{\phi}$ also satisfies the condition

$$c(\dot{\phi}) = \xi_{\dot{\phi}}\big(c(\dot{\pi})\big), \qquad \dot{\pi} \in \mathcal{A}_2(\dot{G}),$$

of Theorem 1.4.1. It therefore meets the formal requirement for belonging to the subset $\widetilde{\Phi}_{\mathrm{sim}}(\dot{G})$ of $\dot{\widetilde{\Phi}}_{\mathrm{sim}}(N)$, according to the original definition of §1.4. In the case here, however, we will have to work for a moment with the provisional global definition of §5.1.

Having shown that $\dot{\phi}$ lies in $\dot{\widetilde{\Phi}}_{\mathrm{sim}}(N)$, we are now free to form the associated family

$$\dot{\widetilde{\mathcal{F}}} = \widetilde{\mathcal{F}}(\dot{\phi}) = \{\ell\dot{\phi}\} \tag{6.7.1}$$

of global parameters. To make use of the results of Chapter 5, we need to know that the conditions of Assumption 5.4.1 hold, with V again being the set S_∞ of archimedean valuations. We cannot appeal to Proposition 6.3.1, as we did for the earlier compound parameter (6.6.1). However, the properties we need are already in the earlier Lemma 6.2.2. We know that $\dot{\phi}_v$ equals the original archimedean parameter ϕ_v. Conditions (5.4.1(a)) and (5.4.1(c)) then follow from the constraints we placed on ϕ_v in the construction of $\dot{\pi}$ in the first half of the proof of Lemma 6.2.2. Since the condition (5.4.1(b)) is not relevant to this case, Assumption 5.4.1 is valid for $\dot{\widetilde{\mathcal{F}}}$.

We can therefore apply the results of Chapter 5 to the family $\dot{\widetilde{\mathcal{F}}}$. We note that the local and global induction hypotheses of Chapter 5 are vacuous here, since the basic simple parameters ϕ and $\dot{\phi}$ have degree N. In particular, we have to rely on the provisional definition of §5.1 for the set $\widetilde{\mathcal{F}}_{\mathrm{sim}}(\dot{G})$. This poses no difficulty. For we established earlier that

$$S_{\mathrm{disc},\dot{\phi}}(\dot{f}) = \mathrm{tr}\big(R_{\mathrm{disc},\dot{\phi}}(\dot{f})\big),$$

for a function $\dot{f} \in \widetilde{\mathcal{H}}(\dot{G})$, chosen prior to (6.2.7) in the proof of Lemma 6.2.2, with the property that the right hand side is nonzero. Therefore, the left hand side is not identically 0. Therefore $\dot{\phi}$ belongs to $\widetilde{\mathcal{F}}_2(\dot{G})$, according to any of the three equivalent conditions of Corollary 5.4.7.

Among the results of Chapter 5 we can use is Lemma 5.4.2, which affirms that Assumption 5.1.1 is valid. Then, following the proof of Lemma 6.6.3, we can establish the local assertion that

$$\widetilde{f}_N(\phi) = \widetilde{f}^G(\phi), \qquad \widetilde{f} \in \widetilde{\mathcal{H}}(N),$$

for a stable linear form $f^G(\phi)$ on $\widetilde{\mathcal{H}}(G)$. We note that we have already proved the weaker assertion for cuspidal functions $\widetilde{f} \in \widetilde{\mathcal{H}}_{\mathrm{cusp}}(N)$, namely that ϕ is a cuspidal lift. We had to do so independently, in order to show that the parameters ϕ and $\dot{\phi}$ are simple. The main result from Chapter 5 is the stable multiplicity formula

$$S^{\dot{G}}_{\mathrm{disc},\dot{\phi}}(\dot{f}) = |\Phi(\dot{G},\dot{\phi})|\, \dot{f}^{\dot{G}}(\dot{\phi}), \qquad \dot{f} \in \widetilde{\mathcal{H}}(\dot{G}), \tag{6.7.2}$$

for $\dot{\phi}$. It follows from Corollary 5.1.3, and the application to Lemma 5.1.4 of its supplement Lemma 5.4.6, in case N is even and $\eta_{\dot{\phi}} = 1$.

We can now follow the arguments of §6.6, but with one essential difference. In §6.6, we started with a local parameter ϕ. Through a correspondence

$$\phi \longrightarrow \dot{\phi} \longrightarrow \dot{\pi} \longrightarrow \pi$$

based on our global construction, we then obtained a local packet $\widetilde{\Pi}_\phi$ of representations π in $\widetilde{\Pi}_2(G)$. In this section, the correspondence has had to go in the opposite direction. Starting with a representation π in the subset $\widetilde{\Pi}_{\text{sim}}(G)$ of $\widetilde{\Pi}_2(G)$, the global construction has led us through a correspondence

$$\pi \longrightarrow \dot{\pi} \longrightarrow \dot{\phi} \longrightarrow \phi$$

to a local parameter ϕ.

Since π has been fixed from the beginning, we shall have to write π^* for the variable representation in $\widetilde{\Pi}_2(G)$ that accompanies our attempt to reverse the correspondence. We note that the group $\mathcal{S}_\phi$ is now trivial. The sums over $x \in \mathcal{S}_\phi$ that were a part of the discussion of the last section will therefore not occur here. In particular, the analogue of the formula (6.6.6) is just

$$\sum_{\dot{\pi}^* \in \widetilde{\Pi}(\dot{G})} n_{\dot{\phi}}(\dot{\pi}^*)\, \dot{f}_{\dot{G}}(\dot{\pi}^*) = \dot{f}(\dot{\phi}), \qquad \dot{f} \in \widetilde{\mathcal{H}}(\dot{G}), \tag{6.7.3}$$

where

$$n_{\dot{\phi}}(\dot{\pi}^*) = |\Phi(\dot{G}, \dot{\phi})|^{-1} \sum_{\dot{\pi}^*_G \in \Pi(\dot{G}, \dot{\pi}^*)} n_{\dot{\phi}}(\dot{\pi}^*_G)$$

as in (6.6.5). It follows directly from the stable multiplicity formula above.

We know that ϕ is a cuspidal lift. We can therefore consider the expansion (6.5.15) of

$$f^G(\phi), \qquad f \in \widetilde{\mathcal{H}}_{\text{cusp}}(G),$$

in terms of elliptic values of discrete series characters

$$f_G(\pi^*), \qquad f \in \widetilde{\mathcal{H}}_{\text{cusp}}(G),\ \pi^* \in \widetilde{\Pi}_2(G).$$

We would like to describe the coefficients

$$c_\phi(\pi^*) = c_\phi(1, \pi^*), \qquad \pi^* \in \widetilde{\Pi}_2(G).$$

Retracing the steps from §6.6, we arrive at a reduction

$$n_\phi(\pi^*) = c_\phi(\pi^*) \tag{6.7.4}$$

of the formula (6.6.12), for nonnegative numbers

$$n_\phi(\pi^*) = n_\phi(1, \pi^*), \qquad \pi^* \in \widetilde{\Pi}_2(G),$$

as in (6.6.11). Thus

$$n_\phi(\pi^*) = n_{\dot{\phi}}(\dot{\pi}^*)$$

is defined by the analogue of (6.6.5) above, in terms of the global representation

$$\dot{\pi}^* = \dot{\pi}_\infty(1) \otimes \pi^* \otimes \dot{\pi}^{\infty,u}(1)$$

in $\widetilde{\Pi}(\dot{G}(\dot{\mathbb{A}}))$, and the global multiplicities $n_{\dot{\phi}}(\dot{\pi}^*_G)$ in the automorphic discrete spectrum of $\dot{G}$. The arguments 1 are of no significance in $c_\phi(1, \pi^*)$ and $n_\phi(1, \pi^*)$. They are of course just simply references to the trivial character on the trivial group $\mathcal{S}_{\dot{\phi}}$. We observe that the global representation

$$\dot{\pi} = \dot{\pi}_\infty \otimes \pi \otimes \dot{\pi}^{\infty,u}$$

attached to the original representation π is of the required form. Since it occurs in the discrete spectrum of $\dot{G}$, the number $n_\phi(\pi)$ attached to π is *strictly* positive. Finally, the analogue of Lemma 6.6.4 is valid for any π^*. It tells us that both $n_\phi(\pi^*)$ and the product

$$\widetilde{n}_\phi(\pi^*) = m(\pi^*)^{-1}\, m(\phi)\, n_\phi(\pi^*)$$

are actually nonnegative *integers.*

The next step is to apply the orthogonality relations (6.5.16), as at the beginning of the proof of Proposition 6.6.5. The properties of the coefficients $c_\phi(\pi^*)$ in Lemma 6.5.3 were stated with the understanding that ϕ is a cuspidal lift. Suppose that ϕ^* is an arbitrary parameter in $\widetilde{\Phi}_{\mathrm{sim}}(G)$ that is also a cuspidal lift. The corresponding expansion (6.5.15) supplies ϕ^* with its own set of coefficients $c_{\phi^*}(\pi^*)$. It then follows from (6.5.16) that the sum

$$\sum_{\pi^* \in \widetilde{\Pi}_2(G)} m(\pi^*)^{-1}\, c_\phi(\pi^*)\, \overline{c_{\phi^*}(\pi^*)} \tag{6.7.5}$$

vanishes unless ϕ^* equals ϕ, in which case it equals $m(\phi)^{-1}$. Setting $\phi^* = \phi$, and making the substitution (6.7.4), we obtain

$$\sum_{\pi^*} \widetilde{n}_\phi(\pi^*)\, n_\phi(\pi^*) = 1.$$

It then follows from the fact that $\widetilde{n}_\phi(\pi^*)$ and $n_\phi(\pi^*)$ are nonnegative integers that

$$\widetilde{n}_\phi(\pi^*) = n_\phi(\pi^*) = \begin{cases} 1, & \text{if } \pi^* = \pi, \\ 0, & \text{otherwise,} \end{cases} \tag{6.7.6}$$

for any representation $\pi^* \in \widetilde{\Pi}_2(G)$.

It thus follows from (6.7.4) and (6.7.6) that

$$c_\phi(\pi^*) = \begin{cases} 1, & \pi^* = \pi, \\ 0, & \text{otherwise,} \end{cases} \tag{6.7.7}$$

for any $\pi^* \in \widetilde{\Pi}_2(G)$. The vanishing assertion in this formula leads in turn to a collapse in the sum (6.7.5). The formula for (6.7.5) reduces simply to the relation

$$m(\pi)^{-1}\, c_\phi(\pi)\, c_{\phi^*}(\pi) = 0,$$

in case $\phi^* \neq \phi$. It follows that

$$c_{\phi^*}(\pi) = \begin{cases} 1, & \text{if } \phi^* = \phi, \\ 0, & \text{otherwise,} \end{cases} \tag{6.7.8}$$

for any cuspidal lift $\phi^* \in \widetilde{\Phi}_{\mathrm{sim}}(G)$. In particular, the parameter ϕ is uniquely determined by the representation π.

We see now that the indirect global process by which we associated the parameter ϕ to π gives a canonical local construction. For it follows from (6.7.8) that the correspondence $\pi \to \phi$ is a well defined mapping from $\widetilde{\Pi}_{\mathrm{sim}}(G)$ to $\widetilde{\Phi}_{\mathrm{sim}}(G)$. We have next to show that this mapping is a bijection. We shall formulate the assertion in terms of the inverse mapping $\phi \to \pi_\phi$, as a complement to the correspondence of Proposition 6.6.5.

PROPOSITION 6.7.2. (a) *Every parameter ϕ in the subset $\widetilde{\Phi}_{\mathrm{sim}}(G)$ of $\widetilde{\Phi}_{\mathrm{sim}}(N)$ is a cuspidal lift.*

(b) *There is a canonical bijection*

$$\phi \longrightarrow \pi_\phi, \qquad \phi \in \widetilde{\Phi}_{\mathrm{sim}}(G),$$

from $\widetilde{\Phi}_{\mathrm{sim}}(G)$ onto $\widetilde{\Pi}_{\mathrm{sim}}(G)$ such that

$$f^G(\phi) = f_G(\pi_\phi), \qquad f \in \widetilde{\mathcal{H}}_{\mathrm{cusp}}(G),\ \phi \in \widetilde{\Phi}_{\mathrm{sim}}(G). \tag{6.7.9}$$

PROOF. The relation (6.7.9) is simply the general expansion (6.5.15) of Lemma 6.5.3, together with the identity (6.7.7) for the coefficients. To establish the proposition, however, we must prove the prior assertion that the original mapping $\pi \to \phi$ is a bijection from $\widetilde{\Pi}_{\mathrm{sim}}(G)$ onto $\widetilde{\Phi}_{\mathrm{sim}}(G)$. The condition (6.7.8) implies that the mapping is injective. Let us write $\widetilde{\Phi}^c_{\mathrm{sim}}(G)$ for its image. Our task is then to show that $\widetilde{\Phi}^c_{\mathrm{sim}}(G)$ equals $\widetilde{\Phi}_{\mathrm{sim}}(G)$. The problem is that we do not know that every element $\phi \in \widetilde{\Phi}_{\mathrm{sim}}(G)$ satisfies the first condition (i) of a cuspidal lift. It is essentially the question of showing that the union (6.1.2) is disjoint.

Consider the subspaces $\widetilde{\mathcal{I}}_{\mathrm{sim}}(G)$ and $\widetilde{\mathcal{S}}_{\mathrm{sim}}(G)$ of $\widetilde{\mathcal{I}}_{\mathrm{cusp}}(G)$ and $\widetilde{\mathcal{S}}_{\mathrm{cusp}}(G)$ introduced at the beginning of the section. Lemma 6.7.1(a) asserts that the linear mapping

$$f_G \longrightarrow f^G, \qquad f \in \widetilde{\mathcal{I}}_{\mathrm{cusp}}(G),$$

takes $\widetilde{\mathcal{I}}_{\mathrm{sim}}(G)$ isomorphically onto $\widetilde{\mathcal{S}}_{\mathrm{sim}}(G)$. It is clear from the original definition that the invariant pseudocoefficients

$$f_{\pi,G}, \qquad \pi \in \widetilde{\Pi}_{\mathrm{sim}}(G),$$

are a basis of the complex vector space $\widetilde{\mathcal{I}}_{\mathrm{sim}}(G)$. For any $\pi \in \widetilde{\Pi}_{\mathrm{sim}}(G)$, set

$$f^\phi = f^G_\pi,$$

where $\phi \in \widetilde{\Phi}^c_{\text{sim}}(G)$ is the image of π. Then f^ϕ can be regarded as a stable pseudocoefficient of ϕ, in the sense that

$$f^\phi(\phi^*) = \begin{cases} 1, & \text{if } \phi^* = \phi, \\ 0, & \text{otherwise,} \end{cases}$$

for any $\phi^* \in \widetilde{\Phi}^c_{\text{sim}}(G)$. The stable pseudocoefficients

$$f^\phi, \qquad \phi \in \widetilde{\Phi}^c_{\text{sim}}(G),$$

are then a basis of the complex vector space $\widetilde{\mathcal{S}}_{\text{sim}}(G)$.

We can augment this basis with objects from the last section. The parameters ϕ in $\widetilde{\Phi}^{\text{sim}}_2(G)$ are all cuspidal lifts. They also provide stable pseudocoefficients f^ϕ, which can be expressed in terms of the corresponding packets by

$$f^\phi = |\mathcal{S}_\phi|^{-1}\Big(\sum_{\pi \in \widetilde{\Pi}_\phi} f^G_\pi\Big),$$

and which span the orthogonal complement of $\widetilde{\mathcal{S}}_{\text{sim}}(G)$ in $\widetilde{\mathcal{S}}_{\text{cusp}}(G)$. The expanded family of pseudocoefficients

$$f^\phi, \qquad \phi \in \widetilde{\Phi}^c_2(G),$$

parametrized by the set

$$\widetilde{\Phi}^c_2(G) = \widetilde{\Phi}^{\text{sim}}_2(G) \coprod \widetilde{\Phi}^c_{\text{sim}}(G),$$

then forms a basis of $\widetilde{\mathcal{S}}_{\text{cusp}}(G)$.

Suppose that ϕ^* is *any* element in $\widetilde{\Phi}_{\text{sim}}(G)$. By definition, ϕ^* is a parameter in $\widetilde{\Phi}_{\text{sim}}(N)$ such that the linear form $\widetilde{f}^G(\phi^*)$ on $\widetilde{\mathcal{S}}_{\text{cusp}}(G)$ defined by (6.1.1) is nonzero. Suppose that ϕ is any parameter in $\widetilde{\Phi}^c_2(G)$ that is distinct from ϕ^*. By Proposition 2.1.1, we can choose a function $\widetilde{f}^\phi \in \widetilde{\mathcal{H}}_{\text{cusp}}(N)$ that is a preimage of ϕ, in the sense that

$$(\widetilde{f}^\phi)^* = \begin{cases} f^\phi, & \text{if } G^* = G, \\ 0, & \text{otherwise,} \end{cases}$$

for any $G^* \in \widetilde{\mathcal{E}}_{\text{ell}}(N)$. Since ϕ is a cuspidal lift, we can regard $\widetilde{f}^\phi$ as a twisted pseudocoefficient of ϕ. In particular, $(\widetilde{f}^\phi_N)(\phi^*)$ vanishes, since ϕ and ϕ^* represent distinct elements in $\widetilde{\Phi}_{\text{ell}}(N)$. It follows from (6.1.1) that

$$(f^\phi)(\phi^*) = (\widetilde{f}^\phi)^G(\phi^*) = \widetilde{f}^\phi_N(\phi^*) = 0.$$

On the other hand, there must be a ϕ such that $f^\phi(\phi^*)$ is nonzero, since $\widetilde{\mathcal{S}}_{\text{cusp}}(G)$ is spanned by the functions f^ϕ. There must therefore be a parameter $\phi \in \widetilde{\Phi}^c_2(G)$ that equals ϕ^*. Since $\phi^* \in \widetilde{\Phi}_{\text{sim}}(G)$, ϕ must lie in $\widetilde{\Phi}^c_{\text{sim}}(G)$.

We have shown that $\widetilde{\Phi}_{\text{sim}}(G)$ equals $\widetilde{\Phi}^c_{\text{sim}}(G)$. This gives the assertion (a) of the proposition, since the elements in $\widetilde{\Phi}^c_{\text{sim}}(G)$ are cuspidal lifts. It also tells us that the original mapping $\pi \to \phi$ from $\widetilde{\Pi}_{\text{sim}}(G)$ to $\widetilde{\Phi}_{\text{sim}}(G)$ is surjective, and hence a bijection. The inverse mapping $\phi \to \pi_\phi$ therefore exists, as asserted in (b). The proof of the proposition is complete. □

COROLLARY 6.7.3. *If π equals π_ϕ, then $m(\pi) = m(\phi)$.*

PROOF. The corollary follows from the equality of $\widetilde{n}_\phi(\pi)$ with $n_\phi(\pi)$ in (6.7.6). □

We define the L-packet of any simple parameter $\phi \in \widetilde{\Phi}_{\mathrm{sim}}(G)$ to be the singleton

$$\widetilde{\Pi}_\phi = \{\pi_\phi\}.$$

Propositions 6.6.5 and 6.7.2 together thus attach an L-packet $\widetilde{\Pi}_\phi$ to any square integrable Langlands parameter $\phi \in \widetilde{\Phi}_2(G)$. However, there is still something more to be said about the corresponding character identities. We shall state it as a joint corollary of the two propositions.

COROLLARY 6.7.4. *Suppose that ϕ is any parameter in $\widetilde{\Phi}_2(G)$. Then the character identity*

$$f'(\phi') = \sum_{\pi \in \widetilde{\Pi}_\phi} \langle x, \pi \rangle f_G(\pi), \qquad x \in \mathcal{S}_\phi, \tag{6.7.10}$$

established in Propositions 6.6.5 and 6.7.2 for a cuspidal function f, remains valid if f is an arbitrary function in $\widetilde{\mathcal{H}}(G)$.

PROOF. At first glance, the assertion might seem to be immediate. In the case that ϕ is not simple, for example, the global formula (6.6.6) from which we derived the cuspidal version (6.6.13) of the required character identity is valid for any function $\dot{f} \in \widetilde{\mathcal{H}}(\dot{G})$. Likewise, the local transfer formula

$$\widetilde{f}_N(\phi) = \widetilde{f}^G(\phi) \tag{6.7.11}$$

of Lemma 6.6.3 holds for any function $f \in \widetilde{\mathcal{H}}(G)$. However, we need to be careful. For we do not know a priori that the right hand side of (6.7.11) equals the value at $x = 1$ of the right hand side of (6.6.13), if the function f belongs to the complement of $\widetilde{\mathcal{H}}_{\mathrm{cusp}}(G)$. Nor can we expect an immediate rescue from (6.6.6). For we cannot a priori rule out the possible existence of global representations $\dot{\pi}$ on the left hand side for which $\dot{\pi}_u$ is not elliptic.

The simplest way to establish the corollary is to appeal to the two general theorems of [**A11**], and their twisted analogue for $GL(N)$ (which we discussed briefly in the proof of Proposition 2.1.1). The proof of these theorems is not elementary, and has not yet been written down in the twisted case. However, we have already used them in the proof of Proposition 2.1.1. We may as well use the theorems again here, since it is exactly for this sort of problem that they were intended. In particular, they serve as a substitute for the fundamental lemma for the full spherical Hecke algebra [**Hal**].

The coefficients on the right hand side of (6.7.10), regarded as a sum over the full set $\widetilde{\Pi}_{\mathrm{temp}}(G)$ whose summands are supported on the subset $\widetilde{\Pi}_\phi$, are already determined from the case that f is cuspidal. This is because the characters of discrete series are determined by their values on the strongly elliptic set. Suppose then that f is a general function in $\widetilde{\mathcal{H}}(G)$. Suppose also that we define $f^G(\phi)$ alternatively as the value at $x = 1$ of the right hand side of (6.7.10), rather than the left hand side of (6.7.11). Theorem 6.1 of [**A11**] then asserts that the resulting linear form in f is stable. Theorem 6.2 of [**A11**] asserts in addition that (6.7.10), with the left hand side being the analogue for G' of the stable linear form just defined, is valid for f and for any element x in $\mathcal{S}_\phi$. Lastly, the twisted analogue for $GL(N)$ of Theorem 6.2 of [**A11**] asserts that (6.7.11) holds for f. In other words, the alternative definition for $f^G(\phi)$ just given matches the one we have always used. We conclude that the character identity (6.7.10) does hold for the general function f. □

We have established the general character identity of Theorem 2.2.1(b), for parameters $\phi \in \widetilde{\Phi}_2(G)$. Since Lemma 6.6.3 takes care of its first assertion (a), Theorem 2.2.1 itself is valid for any such parameter. We note that the character identity (6.7.10) also allows us to complete some of the global induction hypothesis for the family $\dot{\tilde{\mathcal{F}}}$. Applied to the multiplicity formulas (6.6.6) and (6.7.3) we have deduced, it gives the contribution of $\dot{\phi}$ to the discrete spectrum postulated in Theorem 1.5.2.

Observe that Corollary 6.6.6 of Proposition 6.6.5 can be regarded as a proof of the assertion (b) of Theorem 2.2.4. Together with Corollary 6.7.3 of Proposition 6.7.2, it resolves an important induction hypothesis from Proposition 6.6.1. We are speaking of the hypothesis that for any parameter ϕ in the complement of $\widetilde{\Phi}_2(G)$ in $\widetilde{\Phi}_{\mathrm{bdd}}(G)$, the group $W(\pi_M)$ equals W_ϕ, in the notation of (6.5.3). This is what allowed us to conclude that the mapping (6.5.3) was surjective, which we recall was a starting point for Proposition 6.6.1. Proposition 6.6.1 itself tells us that Theorem 2.2.1 holds for any such ϕ.

Theorem 2.2.1 is thus valid for any generic, bounded parameter $\phi \in \widetilde{\Phi}_{\mathrm{bdd}}(G)$. Recall that this includes the assertion (a) of Theorem 1.5.1. We shall state what remains of Theorem 1.5.1(b) as a joint corollary of the three Propositions 6.6.1, 6.6.5 and 6.7.2.

COROLLARY 6.7.5. *The set $\widetilde{\Pi}_{\mathrm{temp}}(G)$ of irreducible, tempered ($\widetilde{\mathrm{Out}}_N(G)$-orbits of) representations of $G(F)$ is a disjoint union over $\phi \in \widetilde{\Phi}_{\mathrm{bdd}}(G)$ of the packets $\widetilde{\Pi}_\phi$.*

PROOF. Proposition 6.6.1 includes the assertion that the complement of $\widetilde{\Pi}_2(G)$ in $\widetilde{\Pi}_{\mathrm{temp}}(G)$ is a disjoint union of packets $\widetilde{\Pi}_\phi$, in which ϕ ranges over the complement of $\widetilde{\Phi}_2(G)$ in $\widetilde{\Phi}_{\mathrm{bdd}}(G)$. Propositions 6.6.5(b) and 6.7.2(b), combined with the definition of the set $\widetilde{\Pi}_{\mathrm{sim}}(G)$, tell us that $\widetilde{\Pi}_2(G)$ is the disjoint union over $\phi \in \widetilde{\Phi}_2(G)$ of the remaining packets $\widetilde{\Pi}_\phi$. The corollary follows. □

We have now proved both of the local Theorems 1.5.1 and 2.2.1 for parameters $\phi \in \widetilde{\Phi}_{\mathrm{bdd}}(G)$. This gives the local Langlands correspondence (up to the action of the group $\widetilde{\mathrm{Out}}_N(G)$) for any quasisplit orthogonal or symplective group $G \in \widetilde{\mathcal{E}}_{\mathrm{sim}}(N)$ over the p-adic field F. Actually, we should really say that the local classification will be complete once we have resolved the induction assumptions on which the arguments of the chapter have been based. We shall do so now.

6.8. Resolution

We have established Theorems 1.5.1, 2.2.1 and 2.4.1 for generic local parameters ϕ. These are the theorems that characterize the tempered representations of our classical group $G \in \widetilde{\mathcal{E}}_{\mathrm{sim}}(N)$ over the p-adic field F. We are not quite done, however. As we have just noted, we have still to resolve the induction hypotheses on which the proof of the three theorems has been based.

There are actually three kinds of hypotheses. First and foremost, we have the supplementary local Theorems 2.2.4 and 2.4.4 to prove. These theorems are of interest in their own right, since they give new information about characters and local harmonic analysis. But they were also forced on us for the statement of Theorem 2.4.1. We must prove them for the relevant parameters in $\widetilde{\Phi}(N)$ in order to complete the local hypothesis of this chapter. Secondly, we must also complete the global induction hypothesis of the chapter. That is, we must convince ourselves that the global theorems hold for parameters in $\dot{\widetilde{\mathcal{F}}}(N)$. Lastly, we need to resolve the temporary definition of the local set $\widetilde{\Phi}_{\mathrm{sim}}(G)$ in §6.1 in terms of the natural definition provided by local Langlands parameters. This will take care of Theorem 6.1.1, the last of the local assertions.

Consider then the supplementary local Theorems 2.2.4 and 2.4.4. They concern the case of a datum $G \in \widetilde{\mathcal{E}}_{\mathrm{sim}}(N)$, where N is even and $\widehat{G} = SO(N, \mathbb{C})$. As we recall, the standard outer automorphism of G gives a bitorsor $\widetilde{G}$ under G. The theorems apply in this case to a generic parameter $\phi = \psi$ in the subset of $\Phi(\widetilde{G})$ of $\widetilde{\Phi}(G)$. The proofs are similar to those of Theorems 2.2.1 and 2.4.1. In particular, they rely on the global methods that originate with the construction of §6.3. However, the arguments now are considerably easier. For one thing, the parameter ϕ is never simple, so we have no need of the general discussion of §6.7. For another, we will be able to use *ab initio* a number of properties that were established in the earlier arguments.

Suppose for a moment that

$$\psi = \ell_1\psi_1 \oplus \cdots \oplus \ell_r\psi_r$$

is a general local parameter in $\Psi(\widetilde{G})$. Recall that the centralizer $\widetilde{S}_\psi$ in $\widehat{\widetilde{G}}$ of the image of ψ is an S_ψ-torsor. It has a decomposition

$$\widetilde{S}_\psi = \Big(\prod_{i\in I_\psi^+(G)} O(\ell_i,\mathbb{C})\Big)_\psi^- \times \Big(\prod_{i\in I_\psi^-(G)} Sp(\ell_i,\mathbb{C})\Big) \times \Big(\prod_{j\in J_\psi} GL(\ell_j,\mathbb{C})\Big) \tag{6.8.1}$$

that is parallel to (1.4.8). We have written $(\,\cdot\,)_\psi^-$ here for the preimage of (-1) under the sign character ξ_ψ^+ (noting that this minus sign is not related to the superscript in the earlier set $I_\psi^-(G)$). Similar remarks hold if the local pair (G,ψ) is replaced by a pair $(\dot{G},\dot{\psi})$ over a global field $\dot{F}$. In the global case, the definitions of §3.3 apply to the $\dot{G}$-bitorsor $\dot{\widetilde{G}}$. In particular, we can form the $\dot{\psi}$-component

$$\widetilde{I}_{\mathrm{disc},\dot{\psi}}(\dot{\widetilde{f}}) = I^{\dot{\widetilde{G}}}_{\mathrm{disc},\dot{\psi}}(\dot{\widetilde{f}}), \qquad \dot{\widetilde{f}} \in \mathcal{H}(\dot{\widetilde{G}}),$$

of the discrete part of the twisted trace formula for $\dot{\widetilde{G}}$, following (3.3.12). It satisfies the analogue

$$\widetilde{I}_{\mathrm{disc},\dot{\psi}}(\dot{\widetilde{f}}) = \sum_{\dot{G}'\in\widetilde{\mathcal{E}}_{\mathrm{ell}}(\dot{\widetilde{G}})} \iota(\dot{\widetilde{G}},\dot{\widetilde{G}}')\,\widehat{\widetilde{S}}'_{\mathrm{disc},\dot{\psi}}(\dot{\widetilde{f}}') \tag{6.8.2}$$

of (3.3.15). We can also form the $\dot{\psi}$-component

$$\dot{\widetilde{R}}_{\mathrm{disc},\dot{\psi}}(\dot{\widetilde{f}}) = R^{\dot{\widetilde{G}}}_{\mathrm{disc},\dot{\psi}}(\dot{\widetilde{f}}), \qquad \dot{\widetilde{f}} \in \mathcal{H}(\dot{\widetilde{G}}),$$

of the canonical extension of the representation $R^{\dot{G}}_{\mathrm{disc},\dot{\psi}}$ of $\dot{G}(\dot{\mathbb{A}})$.

Returning to the focus of this chapter, we consider a generic local parameter

$$\phi = \ell_1\phi_1 \oplus \cdots \oplus \ell_r\phi_r, \qquad \phi_i \in \widetilde{\Phi}_{\mathrm{sim}}(G_i),\ G_i \in \widetilde{\mathcal{E}}_{\mathrm{sim}}(N_i),$$

in $\Phi_{\mathrm{disc}}(\widetilde{G})$. We use Proposition 6.3.1 to construct simple global parameters $\dot{\phi}_i \in \widetilde{\Phi}_{\mathrm{sim}}(\dot{G}_i)$, and the associated family

$$\dot{\widetilde{\mathcal{F}}} = \widetilde{\mathcal{F}}(\dot{\phi}_1,\dots,\dot{\phi}_r)$$

of compound global parameters. If $(\dot{G},\dot{\phi})$ is the global pair attached to (G,ϕ), then $\dot{\phi}$ lies in $\Phi(\dot{\widetilde{G}})$. The corresponding centralizers $\widetilde{S}_\phi$ and $\widetilde{S}_{\dot{\phi}}$ are isomorphic. They are given by a product (6.8.1), but without the general linear factors, since the simple constituents ϕ_i of ϕ are assumed here to be self dual. The other conditions of Proposition 6.3.1, which were used to validate the results of Chapter 5, will not be needed here.

Having described the underlying global machinery, we can be brief in our treatment of the two local theorems for $\widetilde{G}$. We will be content simply to sketch a proof of the main cases, those of square integrable parameters

$\phi \in \Phi_2(\widetilde{G})$ for Theorem 2.2.4, and elliptic parameters $\phi \in \Phi_{\mathrm{ell}}(\widetilde{G})$ for Theorem 2.4.4. The various arguments of descent needed to deal then with the general cases are similar to those of the corresponding Theorems 2.2.1 and 2.4.1, and will be left to the reader.

We consider Theorem 2.2.4 first, for a parameter ϕ in $\Phi_2(\widetilde{G})$. In the case $F = \mathbb{R}$, we treat $\widetilde{G}$ as we have $\widetilde{G}(N)$, as an object that is part of the work in progress by Shelstad and Mezo. We shall therefore assume that F is a p-adic field, as in §6.6. We then have

$$(6.8.3) \qquad \begin{cases} \phi = \phi_1 \oplus \cdots \oplus \phi_r, \\ \widetilde{S}_\phi = \big(O(1,\mathbb{C})^r\big)^-_\phi, \end{cases}$$

for $\phi_i \in \widetilde{\Phi}_{\mathrm{sim}}(G_i)$ and $G_i \in \widetilde{\mathcal{E}}_{\mathrm{sim}}(N_i)$, with the requirement that some N_i be odd. The proof of Theorem 2.2.4 will be a straightforward application of the discussion of §6.6.

We first apply (6.8.2) and Corollary 4.1.3 to a function $\dot{\widetilde{f}} \in \mathcal{H}(\dot{\widetilde{G}})$. As in the remarks preceding (6.6.6), we obtain

$$\begin{aligned} \widetilde{I}_{\mathrm{disc},\dot{\phi}}(\dot{\widetilde{f}}) &= \sum_{\dot{\widetilde{G}}'} \iota(\dot{\widetilde{G}},\dot{\widetilde{G}}')\, \widehat{\widetilde{S}}{}'_{\mathrm{disc},\dot{\phi}}(\dot{\widetilde{f}}') \\ &= |\widetilde{S}_{\dot{\phi}}|^{-1} |\Phi(\dot{G},\dot{\phi})| \sum_{\widetilde{x}\in\widetilde{S}_\phi} \dot{\widetilde{f}}'(\dot{\phi}'), \end{aligned}$$

where $(\dot{\widetilde{G}}',\dot{\phi}')$ maps to the pair $(\dot{\phi},\dot{\widetilde{x}})$. The condition that $\dot{\phi}$ lies in $\Phi_2(\dot{\widetilde{G}})$ is equivalent to the property

$$|\Phi(\dot{G},\dot{\phi})| = 1.$$

The twisted analogue of (6.6.6) then becomes

$$(6.8.4) \qquad \sum_{\dot{\pi}\in\Pi(\dot{\widetilde{G}})} n_{\dot{\phi}}(\dot{\pi})\, \dot{\widetilde{f}}_{\dot{\widetilde{G}}}(\dot{\widetilde{\pi}}) = |\mathcal{S}_{\dot{\phi}}|^{-1} \sum_{\widetilde{x}\in\widetilde{\mathcal{S}}_\phi} \dot{\widetilde{f}}'(\dot{\phi}'),$$

where $n_{\dot{\phi}}(\dot{\pi})$ is the multiplicity of $\dot{\pi}$ in $R^{\dot{G}}_{\mathrm{disc},\dot{\phi}}$, and $\dot{\widetilde{\pi}}$ is the canonical extension (4.2.7) of $\dot{\pi}$ to $\dot{\widetilde{G}}(\dot{\mathbb{A}})$ defined whenever $n_{\dot{\phi}}(\dot{\pi})$ is nonzero. Recall that in the proof of the local results in §6.6 and §6.7, we evaluated $n_{\dot{\phi}}(\dot{\pi})$. We showed that it satisfies the formula of Theorem 1.5.2, with $m_{\dot{\phi}}$ being equal to 1, at least if the component $\dot{\pi}^{\infty,u}$ of $\dot{\pi}$ outside $S_\infty(u)$ equals $\dot{\pi}^{\infty,u}(1)$. With this condition on $\dot{\widetilde{\pi}}^{\infty,u}$, $n_{\dot{\phi}}(\dot{\pi})$ vanishes unless the representation $\dot{\pi}_\infty$ lies in the product $\Pi_{\dot{\phi}_\infty}$ of the archimedean packets $\Pi_{\dot{\phi}_v}$, the representation $\pi = \dot{\pi}_u$ lies in the packet $\Pi_\phi = \Pi_{\dot{\phi}_u}$ at $F = \dot{F}_u$, and the character

$$\langle \dot{x}, \dot{\pi}_\infty \rangle \langle x, \pi \rangle, \qquad x \in \mathcal{S}_\phi,$$

is trivial, in which case $n_{\dot{\phi}}(\dot{\pi}) = 1$. We have only to investigate the form taken by (6.8.4) for a suitable product

$$\dot{\tilde{f}} = \dot{\tilde{f}}_{\infty} \cdot f \cdot \dot{\tilde{f}}^{\infty,u}, \qquad f = \dot{\tilde{f}}_u.$$

Let ξ be a fixed character on $\mathcal{S}_{\phi}$. By the condition (iii)(a) of Proposition 6.3.1, we can choose a representation $\dot{\pi}_{\infty,\xi}$ in $\Pi_{\dot{\phi}_{\infty}}$ that maps to the character ξ^{-1} in $\mathcal{S}_{\phi}$. The archimedean form of Theorem 2.2.4, which we are taking for granted, implies that any extension $\dot{\tilde{\pi}}_{\infty,\xi}$ of $\dot{\pi}_{\infty}$ to $\dot{\tilde{G}}_{\infty}$ determines an extension $\tilde{\xi}$ of ξ to $\tilde{\mathcal{S}}_{\phi}$. We choose the first factor $\dot{\tilde{f}}_{\infty} \in \mathcal{H}(\dot{\tilde{G}}_{\infty})$ so that

$$\dot{\tilde{f}}_{\infty,\tilde{G}}(\dot{\tilde{\pi}}_{\infty}) = \begin{cases} 1, & \text{if } \dot{\pi}_{\infty} = \dot{\pi}_{\infty,\xi}, \\ 0, & \text{otherwise,} \end{cases}$$

for any representation $\dot{\pi}_{\infty}$ in the packet $\Pi_{\dot{\phi}_{\infty}}$. We choose the third factor $\dot{\tilde{f}}^{\infty,u} \in \mathcal{H}(\dot{\tilde{G}}^{\infty,u})$ so that

$$\dot{\tilde{f}}^{\infty,u}_{\tilde{G}}(\dot{\tilde{\pi}}^{\infty,u}) = \begin{cases} 1, & \text{if } \dot{\pi}^{\infty,u} = \dot{\pi}^{\infty,u}(1), \\ 0, & \text{otherwise,} \end{cases}$$

for any representation $\dot{\pi}^{\infty,u}$ in the packet $\Pi_{\dot{\phi}^{\infty,u}}$. We recall that $\dot{\pi}^{\infty,u}(1)$ is the product over $v \notin S_{\infty}(u)$ of those representations $\dot{\pi}_v \in \Pi_{\dot{\phi}_v}$ with $\langle \cdot, \dot{\pi}_v \rangle = 1$. It therefore has a canonical extension $\dot{\tilde{\pi}}^{\infty,u}(1)$ to $\dot{\tilde{G}}(\dot{\mathbb{A}}^{\infty,u})$. At the remaining place u, we take $f = \dot{f}_u$ to be any function in $\mathcal{H}(\tilde{G}) = \mathcal{H}(\dot{\tilde{G}}_u)$. This represents a minor departure from this stage of the discussion in §6.6, where f was chosen to be cuspidal.

Consider first the right hand side

$$|\mathcal{S}_{\phi}|^{-1} \sum_{\tilde{x}} \dot{\tilde{f}}'(\dot{\phi}') = |\mathcal{S}_{\phi}|^{-1} \sum_{\tilde{x}} \dot{\tilde{f}}'_{\infty}(\dot{\phi}'_{\infty})\, \tilde{f}'(\phi')\, (\dot{\tilde{f}}^{\infty,u})'\left((\dot{\phi}^{\infty,u})'\right)$$

of (6.8.4). Ruling out the trivial case that $N = 2$, we shall assume that $N \geqslant 4$, since N is even. It follows from the condition (ii) of Proposition 6.3.1 that $\dot{\phi}_v$ does not lie in $\Phi_2(\dot{\tilde{G}}_v)$ for any $v \notin S_{\infty}(u)$. It then follows from Theorem 2.4.4 (which we will discuss in a moment), and the appropriate descent argument that

$$(\dot{\tilde{f}}^{\infty,u})'\left((\dot{\phi}^{\infty,u})'\right) = \sum_{\dot{\pi}^{\infty,u} \in \Pi_{\dot{\phi}^{\infty,u}}} \langle \tilde{x}, \dot{\tilde{\pi}}^{\infty,u} \rangle\, \dot{\tilde{f}}^{\infty,u}_{\tilde{G}}(\dot{\tilde{\pi}}^{\infty,u}) = 1.$$

We see also from the definitions that

$$\dot{\tilde{f}}'_{\infty}(\dot{\phi}'_{\infty}) = \sum_{\dot{\pi}_{\infty} \in \Pi_{\dot{\phi}_{\infty}}} \langle \tilde{x}, \dot{\tilde{\pi}}_{\infty} \rangle\, \dot{\tilde{f}}_{\infty,\tilde{G}}(\dot{\tilde{\pi}}_{\infty}) = \tilde{\xi}(\tilde{x})^{-1}.$$

The right hand side of (6.8.4) becomes

$$|\mathcal{S}_\phi|^{-1} \sum_{\widetilde{x}} \widetilde{\xi}(\widetilde{x})^{-1}\, \widetilde{f}'(\phi'),$$

where $(\widetilde{G}', \phi')$ is the preimage of $(\phi, \widetilde{x})$. To describe the other side, we write $\pi(\xi)$ as in §6.6 for the representation of $G(F)$ in the packet Π_ϕ with $\langle\,\cdot\,, \pi\rangle = \xi$, and we take $\widetilde{\pi}(\xi) = \pi(\widetilde{\xi})$ to be the extension of $\pi(\xi)$ determined by the canonical extension (4.2.7) of

$$\dot{\pi} = \dot{\pi}_{\infty,\xi} \otimes \pi(\xi) \otimes \dot{\pi}^{\infty,u}(1),$$

the canonical extension of $\dot{\pi}^{\infty,u}(1)$, and the extension we have chosen for $\dot{\pi}_{\infty,\xi}$. The left hand side of (6.8.4) becomes

$$\begin{aligned}&\sum_{\dot{\pi}\in\Pi(\dot{G})} n_{\dot{\phi}}(\dot{\pi})\, \dot{\widetilde{f}}_{\dot{\widetilde{G}}}(\dot{\widetilde{\pi}})\\ &= \sum_{\dot{\pi}} n_{\dot{\phi}}(\dot{\pi})\, \dot{\widetilde{f}}_{\infty,\dot{\widetilde{G}}}(\dot{\widetilde{\pi}}_\infty)\, \widetilde{f}_{\widetilde{G}}(\widetilde{\pi})\, \dot{\widetilde{f}}^{\infty,u}_{\dot{\widetilde{G}}}(\dot{\widetilde{\pi}}^{\infty,u}) = \widetilde{f}_{\widetilde{G}}(\pi(\widetilde{\xi})).\end{aligned}$$

The formula (6.8.4) therefore reduces to

$$\widetilde{f}_{\widetilde{G}}(\pi(\widetilde{\xi})) = |\mathcal{S}_\phi|^{-1} \sum_{\widetilde{x}\in\widetilde{\mathcal{S}}_\phi} \widetilde{\xi}(\widetilde{x})^{-1}\, \widetilde{f}'(\phi'), \tag{6.8.5}$$

for the given character $\xi \in \widehat{\mathcal{S}}_\phi$.

If $\widetilde{x}^*$ is any fixed point in $\widetilde{\mathcal{S}}_\phi$, the product of $\widetilde{\xi}(\widetilde{x}^*)$ with either side of (6.8.5) is independent of the extension $\widetilde{\xi}$ of ξ. We sum each of the two products over $\xi \in \widehat{\mathcal{S}}_\phi$. Since

$$|\mathcal{S}_\phi|^{-1} \sum_{\xi\in\widehat{\mathcal{S}}_\phi} \widetilde{\xi}(\widetilde{x}^*)\, \widetilde{\xi}(\widetilde{x})^{-1} = \begin{cases}1, & \text{if } \widetilde{x} = \widetilde{x}^*,\\ 0, & \text{otherwise,}\end{cases}$$

the right hand simplifies. We conclude that

$$\widetilde{f}'(\phi') = \sum_{\xi\in\widehat{\mathcal{S}}_\phi} \langle \widetilde{x}, \widetilde{\xi}\rangle\, \widetilde{f}_{\widetilde{G}}(\pi(\widetilde{\xi})).$$

The extension $\widetilde{\xi}$ of ξ to $\widetilde{\mathcal{S}}_\phi$ is arbitrary. It determines the extension $\widetilde{\pi}(\xi) = \pi(\widetilde{\xi})$ of $\pi(\xi)$ to $\widetilde{G}(F)$ on the right hand side. The formula becomes

$$\widetilde{f}'(\phi') = \sum_{\pi\in\Pi_\phi} \langle \widetilde{x}, \widetilde{\pi}\rangle\, \widetilde{f}_{\widetilde{G}}(\widetilde{\pi}), \qquad \widetilde{f} \in \mathcal{H}(\widetilde{G}).$$

This is the assertion (a) of Theorem 2.2.4. The assertion (b) of the theorem was Corollary 6.6.6, as we noted at the end of the last section. We have therefore established Theorem 2.2.4, the first supplementary local theorem for parameters $\phi \in \Phi_2(\widetilde{G})$.

Consider next the assertion of Theorem 2.4.4, for a parameter ϕ in the complement of $\Phi_2(\widetilde{G})$ in $\Phi_{\mathrm{ell}}(\widetilde{G})$. The conditions on ϕ are

$$(6.8.6)\qquad \begin{cases} \phi = 2\phi_1 \oplus \cdots \oplus 2\phi_q \oplus \phi_{q+1} \oplus \cdots \oplus \phi_r, \\ \widetilde{S}_\phi = \big(O(2,\mathbb{C})^q \times O(1,\mathbb{C})^{r-q}\big)^-_\psi, \qquad q \geqslant 1, \end{cases}$$

with the requirement that there be a regular element $w \in \widetilde{W}_{\phi,\mathrm{reg}}$ in the Weyl set $\widetilde{W}_\phi$ for $(\widetilde{G},\phi)$. This is the twisted analogue for $\widetilde{G}$ of the main case (6.4.1) treated in §6.4. There is no need to be concerned with the other two exceptional cases (4.5.11) and (4.5.12). Their local analogues for $\widetilde{G}$ follow from Corollary 6.4.5, together with a natural descent argument. We are also taking for granted the analogue for $\widetilde{G}$ of Lemma 2.4.2, which represents a reduction of M by descent. In other words, we shall assume as in §6.4 that M is minimal with respect to ϕ, in the sense that there is a parameter $\phi_M \in \widetilde{\Phi}_2(M,\phi)$.

Theorem 2.4.4 can be proved in the same way as Theorem 2.4.1. Given the local objects G, ϕ, M and ϕ_M, we again choose global objects $\dot{G}$, $\dot{\phi}$, $\dot{M}$ and $\dot{\phi}_M$ according to Proposition 6.3.1. Then $\dot{\phi}$ lies in the subset $\Phi_{\mathrm{ell}}(\dot{\widetilde{G}})$ of $\widetilde{\Phi}_{\mathrm{ell}}(\dot{G})$ over the global field $\dot{F}$. The first step is to establish the $\dot{\widetilde{G}}$-analogue

$$(6.8.7)\qquad \sum_{\dot{\tilde{x}} \in \widetilde{\mathcal{S}}_{\dot\phi,\mathrm{ell}}} \big(\dot{f}'_{\dot{\widetilde{G}}}(\dot\phi,\dot{\tilde{x}}) - \dot{f}_{\dot{\widetilde{G}}}(\dot\phi,\dot{\tilde{x}})\big) = 0, \qquad \dot{\tilde{f}} \in \mathcal{H}(\dot{\widetilde{G}}),$$

of the global identity (6.4.2). Recall that (6.4.2) was one of the assertions of the refinement Lemma 5.4.3 of Lemma 5.2.1. The original lemma was given by a very simple case of the standard model, which applies equally well to $\dot{\widetilde{G}}$. Another consequence of Lemma 5.4.3 is that $\dot{\phi}$ does not contribute to the discrete spectrum of any datum $G^* \in \dot{\widetilde{\mathcal{E}}}_{\mathrm{sim}}(N)$. This implies that the analogue for $\dot{\widetilde{G}}$ of the sum (5.2.7) in Lemma 5.2.1 vanishes. The formula (6.8.7) then follows from the analogue for $\dot{\widetilde{G}}$ of the lemma itself.

The next step is to remove the contributions to (6.8.7) of the valuations v in the complement of $S_\infty(u)$. This reduces to the analogue for $\widetilde{G}$ of Lemma 6.4.1, which amounts to the assertion of Theorem 2.4.4 in case

$$N_i = 1, \qquad 1 \leqslant i \leqslant r \leqslant 3.$$

Since the Levi subgroup M of G can have no symplectic factors, we see from the first observation in the proof of Lemma 6.4.1 that M is actually abelian. It is then easy to see that G is a quasisplit form of either $SO(4)$ or $SO(2)$. We leave to the reader the exercise of establishing Theorem 2.4.4 in either of these cases. Once we have the $\widetilde{G}$-analogue of the lemma, we can use the descent argument following its proof to treat the contributions to (6.8.7) of valuations $v \notin S_\infty(u)$. This reduces (6.8.7) to the $\dot{\widetilde{G}}$ analogue

$$\sum_{\widetilde{x}\in\widetilde{\mathcal{S}}_{\phi,\mathrm{ell}}}\Big(\prod_{v\in S_\infty(u)}\dot{\widetilde{f}}'_{v,\dot{\widetilde{G}}}(\dot\phi_v,\dot{\widetilde{x}}_v)-\prod_{v\in S_\infty(u)}\dot{\widetilde{f}}_{v,\dot{\widetilde{G}}}(\dot\phi_v,\dot{\widetilde{x}}_v)\Big)=0 \tag{6.8.8}$$

of (6.4.6), for functions $\dot{\widetilde{f}}_v \in \mathcal{H}(\dot{\widetilde{G}}_v)$.

The assertion of Theorem 2.4.4 for ϕ is the local identity

$$\widetilde{f}'_{\widetilde{G}}(\phi,\widetilde{x}) = f_{\widetilde{G}}(\phi,\widetilde{x}), \qquad \widetilde{x}\in\widetilde{\mathcal{S}}_{\phi,\mathrm{ell}},\ \widetilde{f}\in\mathcal{H}(\widetilde{G}). \tag{6.8.9}$$

As in the proof of Theorem 2.4.1, the main point is to treat the special case that F equals $\mathbb{R}$, and ϕ is in general position. To do so, we need to persuade ourselves that Lemmas 6.4.2 and 6.4.3 remain valid if G is replaced by $\widetilde{G}$. For Lemma 6.4.2, we use the simply transitive action

$$\widetilde{x} \longrightarrow \widetilde{x}\,x_M, \qquad \widetilde{x}\in\widetilde{\mathcal{S}}_{\phi,\mathrm{ell}},\ x_M\in\mathcal{S}_{\phi_M},$$

of the group $\mathcal{S}_{\phi_M}$ on $\widetilde{\mathcal{S}}_{\phi,\mathrm{ell}}$ in place of its action on $\mathcal{S}_{\phi,\mathrm{ell}}$. The proof for G then carries over directly to $\widetilde{G}$. For Lemma 6.4.3, we again introduce supplementary degrees $N_j^\sharp$. The original degrees N_i of ϕ of course equal 1 or 2, since F is archimedean. If $N_i = 2$ for each $i > q$, we set

$$N_j^\sharp = \begin{cases} N_j, & \text{if } i\leqslant q,\\ N_{q+1}+\cdots+N_r, & \text{if } i = q+1,\end{cases}$$

as in the original proof. If $N_i = 1$ for some index $i > q$, which we can take to be $q+1$, we set

$$N_j^\sharp = \begin{cases} N_j, & \text{if } i\leqslant q,\\ 1, & \text{if } i=q+1,\\ N_{q+1}+\cdots+N_r, & \text{if } i = q+2.\end{cases}$$

We leave the reader to verify that with this modification, the proof of Lemma 6.4.3 for G carries over directly to $\widetilde{G}$.

Once we have established (6.8.9) for $F=\mathbb{R}$ and ϕ in general position, we can apply the arguments Proposition 6.4.4. These arguments carry over directly from G to $\widetilde{G}$. They yield the formula (6.8.9) for any F and ϕ. This completes our discussion of Theorem 2.4.4, the remaining supplementary local theorem.

The second question to resolve is the global induction hypothesis for $\dot{\widetilde{\mathcal{F}}}$. The global assertions are Theorems 1.5.2, 1.5.3, 4.1.2 and 4.2.2 (as well as Theorems 1.4.1 and 1.4.2, taken as implicit foundations for Theorem 1.5.2). We must check that they are valid for parameters $\dot\phi$ in the set $\dot{\widetilde{\mathcal{F}}}(N)$. As we noted at the end of §6.3, and in §6.6 and §6.7, the family $\dot{\widetilde{\mathcal{F}}}$ satisfies Assumption 5.4.1. It consequently also satisfies Assumption 5.1.1, by Lemma 5.4.2. We can therefore apply any of the results of Chapter 5 to parameters $\dot\phi\in\dot{\widetilde{\mathcal{F}}}$.

The central global assertion is the stable multiplicity formula of Theorem 4.1.2. The general global induction argument used in Proposition 4.5.1 can be restricted to the family $\dot{\tilde{\mathcal{F}}}$. It yields the stable multiplicity formula for parameters $\dot{\phi} \in \dot{\tilde{\mathcal{F}}}(N)$ that do not fall into the critical cases studied in Chapter 5. For these remaining cases, the formula was established in Lemmas 5.4.3–5.4.6. Theorem 4.1.2 therefore holds for any parameter $\dot{\phi} \in \dot{\tilde{\mathcal{F}}}(N)$.

Theorem 1.5.2 is of course the actual multiplicity formula. It describes representations in the discrete spectrum of a group $\dot{G} \in \dot{\tilde{\mathcal{E}}}_{\mathrm{sim}}(N)$. For the packet attached to a parameter $\dot{\phi} \in \widetilde{\mathcal{F}}_2(\dot{G})$, the formula follows from the local classification and the stable multiplicity formula, as we noted after Corollary 6.7.4, or as we can see from a direct appeal to the obvious variant of Lemma 4.7.1. If $\dot{\phi}$ lies in the complement of $\widetilde{\mathcal{F}}_2(\dot{G})$ in $\dot{\tilde{\mathcal{F}}}(N)$, Theorem 1.5.2 is a vanishing assertion. It follows from (4.4.12), (4.5.5) and the stable multiplicity formula.

Theorem 1.5.3 has two parts. The assertion (a) follows from Lemma 5.4.6. The assertion (b) is not part of the induction argument for generic parameters, as we can see from the remarks at the beginning of §5.1. It need not concern us here. For Theorem 4.2.2, the essential assertion (b) applies to parameters $\dot{\phi} \in \mathcal{F}_2(\dot{\tilde{G}})$. It is a natural biproduct of the discussion of the local Theorem 2.2.4 above. Finally, Theorems 1.4.1 and 1.4.2 follow from our resolution of the definition of the global set $\widetilde{\mathcal{F}}_{\mathrm{sim}}(\dot{G})$ in Corollary 5.4.7.

The last question concerns the definition of the local sets $\widetilde{\Phi}_{\mathrm{sim}}(G)$. We need to resolve the temporary definition in §6.1 in terms of the natural definition in terms of local Langlands parameters. To do so, we introduce a third condition in terms of local L-functions.

We assume for the rest of the section that the local field F is p-adic, although the discussion would also be valid in the archimedean case. Suppose that $\phi \in \widetilde{\Phi}_{\mathrm{sim}}(N)$ is a simple datum. The question is whether ϕ lies in the subset $\widetilde{\Phi}_{\mathrm{sim}}(G)$ of $\widetilde{\Phi}_{\mathrm{sim}}(N)$. Following the global construction from Lemma 5.3.2, we introduce the local parameter $\phi_+ = \phi \oplus \phi$ in the set $\widetilde{\Phi}(N_+)$. Then $N_+ = 2N$ is even and $\eta_{\psi_+} = 1$, and there are two split groups G_+ and $G_+^\vee$ in $\widetilde{\mathcal{E}}_{\mathrm{sim}}(N_+)$ over F. They share a maximal Levi subgroup $M_+ \cong GL(N)$. As in the global case, G_+ is distinguished from $G_+^\vee$ by the condition that $\widehat{G}_+$ contain the product $\widehat{G} \times \widehat{G}$.

To emphasize the analogy with the global resolution given by Corollary 5.4.7, let us write $\delta_\phi = 1$ if the local Langlands-Shahidi L-function $L(s, \phi, \rho_+^\vee)$ attached to a maximal parabolic subgroup $P_+^\vee = M_+ N_+^\vee$ of $G_+^\vee$ has a pole at $s = 0$. This is what we expect if ϕ maps L_F into ${}^L G$. For if $\widehat{G}$ is orthogonal, $\rho_+^\vee$ is the symmetric square representation of $\widehat{M}_+$. The restriction of $\rho_+^\vee$ to the image of ${}^L G$ in $\widehat{M}_+$ then contains the trivial representation. Similarly, if $\widehat{G}$ is symplectic, $\rho_+^\vee$ is the skew-symmetric square representation of $\widehat{M}_+$, whose restriction to the image of ${}^L G$ again contains the trivial representation. We can also write $\delta_\phi = -1$ in case it is the L-function $L(s, \phi, \rho_+)$ attached to a maximal parabolic subgroup $P_+ = M_+ N_+$ that has a pole at $s = 0$. This of course is what we have to rule out.

COROLLARY 6.8.1. *Suppose that $\phi \in \widetilde{\Phi}_{\mathrm{sim}}(N)$ and that G belongs to $\widetilde{\mathcal{E}}_{\mathrm{sim}}(N)$. Then the following conditions on the pair (G, ϕ) over our local p-adic field F are equivalent.*

(i) *The linear form $\widetilde{f}^G(\phi)$ on $\widetilde{\mathcal{S}}_{\mathrm{cusp}}(G)$ in (6.1.1) does not vanish.*

(ii) *As a local Langlands parameter for $GL(N)$, ϕ factors through the image of ${}^L G$ in $GL(N, \mathbb{C})$.*

(iii) *The local quadratic characters η_G and η_ψ are equal, and the local L-function condition $\delta_\phi = 1$ holds.*

PROOF. The conditions (i) and (ii) represent our two possible definitions of $\widetilde{\Phi}_{\mathrm{sim}}(G)$. They will be linked by the last condition (iii).

We recall that Henniart [**He2**] has established the identities

$$L(s, \phi, S^2) = L(s, S^2 \circ \phi)$$

and

$$L(s, \phi, \Lambda^2) = L(s, \Lambda^2 \circ \phi)$$

of local L-functions. The left hand sides are Langlands-Shahidi L-functions attached to the square integrable representation π_ϕ of $GL(N, F)$, while the right hand sides are the L-functions attached to representations of L_F. With these relations, the corollary becomes a consequence of the isomorphism between the representation theoretic and endoscopic R-groups of the parameter $\phi_+ \in \widetilde{\Phi}(G_+)$. For it follows from the remarks preceding the statement of the corollary that (ii) is equivalent to (iii), with L-functions interpreted as on the right. It follows from the local intertwining relation that (i) is also equivalent to (iii). More precisely, (i) is valid if and only if the induced representation $\mathcal{I}_{P_+}(\pi_\phi)$ is reducible, since the left hand side of the analogue of (2.4.7) for (G_+, ϕ_+) then represents the expected nonzero linear form $\phi \times \phi$ on $\widetilde{\mathcal{S}}_{\mathrm{cusp}}(G \times G)$. By the definition of the representation theoretic R-group $R(\pi_\phi)$, $\mathcal{I}_{P_+}(\pi_\phi)$ is reducible if and only if the Plancherel density $\mu_{P_+}(\pi_{\phi,\lambda})$ is nonzero at $\lambda = 0$, or equivalently, the normalizing factor

$$r_{P_+|\overline{P}_+}(\pi_{\phi,\lambda}) = r_{P_+|\overline{P}_+}(\phi_\lambda)$$

has no pole at $\lambda = 0$. The equivalence of (i) with (iii) then follows from the definition (2.3.3) of the normalizing factors in terms of L-functions. □

Corollary 6.8.1 is clearly parallel to its global counterpart Corollary 5.4.7, from Chapter 5. Together, they resolve the temporary definitions of the local and global sets $\widetilde{\Phi}_{\mathrm{sim}}(G)$ from §5.1 and §6.1 in terms of the original definitions. In each case, it is the condition (i) that gives the temporary definition in terms of harmonic analysis, and condition (ii) that gives the original definition. Notice, however, that while the global conditions (i) and (ii) of Corollary 5.4.7 are closely related, the local conditions (ii) and (iii) of Corollary 6.8.1 are the ones that are most closely related to each other. There is also another difference. Corollary 5.4.7 was proved, for the general family $\widetilde{\mathcal{F}}$ with local constraints in §5.4, as part of a running induction argument on N. Corollary 6.8.1 was proved, with the help of the family $\dot{\widetilde{\mathcal{F}}}$ constructed in this chapter, only after the induction hypothesis on N had been resolved.

With the proof of Corollary 6.8.1, we have reconciled the temporary definition of $\widetilde{\Phi}_{\mathrm{sim}}(G)$ in terms of harmonic analysis with the natural definition in terms of local Langlands parameters. In particular, we have established Theorem 6.1.1. This completes our proof of the local Langlands correspondence for any group $G \in \widetilde{\mathcal{E}}_{\mathrm{sim}}(N)$ (up to the action of the group $\widetilde{\mathrm{Out}}_N(G)$ of order 2, in case G is of type D_n).

CHAPTER 7

Local Nontempered Representations

7.1. Local parameters and duality

We come now to the local theorems for nongeneric parameters ψ. Our general strategy will be similar to that used in the last section to treat generic parameters ϕ. Given local objects (F, G, ψ), we shall construct corresponding global objects $(\dot{F}, \dot{G}, \dot{\psi})$, and a family of global parameters $\dot{\tilde{\mathcal{F}}}$ to which we can apply the results of Chapter 5.

This section is parallel to §6.1. Its purpose is to examine local parameters that occur as completions $\dot{\psi}_v$ of what will be our special global parameters $\dot{\psi}$. These completions will again fall into different categories. First of all, we will have the place $v = u$ at which $\dot{\psi}_v$ equals the given local parameter ψ. To obtain information about $\dot{\psi}_u$, we will again need to control the behaviour of completions $\dot{\phi}_v$ at a set of places $v \in V$ distinct from u. In contrast to the last chapter, however, V will be a finite set of p-adic places rather than the set S^u_∞ of complementary archimedean places.

We are assuming in this chapter that F is a local field. Suppose that $\psi \in \widetilde{\Psi}(G)$ is a general local parameter for a simple endoscopic datum $G \in \widetilde{\mathcal{E}}_{\text{sim}}(N)$ over F. The main local result to establish is Theorem 2.2.1. The first assertion (a) of this theorem is now known. It was reduced in Lemma 2.2.2 to the generic case, which we resolved in Lemma 6.6.3. In particular, the $\widetilde{\text{Out}}_N(G)$-symmetric linear form

$$f'(\psi'), \qquad f \in \widetilde{\mathcal{H}}(G), \tag{7.1.1}$$

of the second assertion (b) of the theorem is well defined. The main point is to show that it has an expansion (2.2.6).

We shall introduce a formal expansion of (7.1.1). Recall that the pair (G', ψ') is the preimage of (ψ, s), for a semisimple element s in S_ψ. We first observe that $f'(\psi')$ depends only on the image x of s in the quotient $\mathcal{S}_\psi$ of S_ψ. This would of course be a consequence of the local intertwining relation for ψ, which we have yet to prove. However, the claim is easy to establish directly. It follows from the local form of the discussion of the left hand side of (4.5.1), given prior to the statement of Proposition 4.5.1.

Knowing that (7.1.1) depends only on x, we can expand it formally as a linear combination of irreducible characters on $\mathcal{S}_\psi$. Building a translation of $\mathcal{S}_\psi$ by s_ψ into the definition, we obtain a general set of $\widetilde{\text{Out}}_N(G)$-symmetric linear forms

$$f \longrightarrow f_G(\sigma), \qquad \sigma \in \widetilde{\Sigma}_\psi,\ f \in \widetilde{\mathcal{H}}(G),$$

which is in bijection with the set of irreducible characters

$$x \longrightarrow \langle x, \sigma\rangle, \qquad x \in \mathcal{S}_\psi,$$

on $\mathcal{S}_\psi$, such that

$$f'(\psi') = \sum_{\sigma \in \widetilde{\Sigma}_\psi} \langle s_\psi x, \sigma\rangle f_G(\sigma), \tag{7.1.2}$$

for any $x \in \mathcal{S}_\psi$ and $f \in \widetilde{\mathcal{H}}(G)$. The assertion (b) of Theorem 2.2.1 is that each $\sigma \in \widetilde{\Sigma}_\psi$ is a nonnegative, integral linear combination of ($\widetilde{\mathrm{O}}\mathrm{ut}_N(G)$-orbits of) irreducible unitary characters on $G(F)$.

We shall take the liberty of calling $\widetilde{\Sigma}_\psi$ the packet of ψ, in anticipation of what we expect to prove. For $\widetilde{\Sigma}_\psi$ and $\widetilde{\Pi}_\psi$ will then represent the same object, differing only in their interpretations as packets over the respective sets $\widehat{\mathcal{S}}_\psi$ and $\widetilde{\Pi}_{\mathrm{unit}}(G)$.

We will use the expansion (7.1.2) to study the localization $\dot{\psi}_u$ of $\dot{\psi}$ we want to understand. It will also be necessary to study localizations that are archimedean as well as p-adic. This is because the archimedean packets of [**ABV**], which exist for general groups, are not defined by twisted transfer to $GL(N)$. In fact, it will be valuations in the union

$$U = S_\infty(u) = S_\infty \cup \{u\}$$

of S_∞ and $\{u\}$ that parametrize completions $\dot{\phi}_v$ for which we have to establish Theorem 2.2.1. The set U will in some sense be the analogue of the singleton $\{u\}$ from Chapter 6.

The remainder of this section will be devoted to parameters that are to be localizations $\dot{\psi}_v$ at places v in complement of U. Such valuations are of course nonarchimedean. We will break them into a disjoint union of a finite set V, and the complement $S^{V,U}$ of V and U. The localizations $\dot{\phi}_v$ at $v \in S^{V,U}$ will play the role of the spherical parameters described in §6.1. As we have already said, the remaining set V will be the analogue of the set S^u_∞ in Chapter 6. We will take it to be any large finite set of valuations v of $\dot{F}$ for which the degrees $\dot{q}_v$ of the corresponding residue fields are all large.

We assume for the rest of the section that the local field F is nonarchimedean. Suppose for the moment that G is a general connected reductive group over F. An irreducible representation π of $G(F)$ determines a connected component in the Bernstein center of G. This is given by a pair

$$(M_\pi, r_\pi),$$

where M_π is a Levi subgroup of G, and r_π is a supercuspidal representation of the group

$$M_\pi(F)^1 = \{x \in M_\pi(F):\ H_{M_\pi}(x) = 0\}$$

such that π is a subquotient of an induced representation

$$\mathcal{I}_{P_\pi}(r_{\pi,\lambda}), \qquad P_\pi \in \mathcal{P}(M_\pi),$$

where

$$r_{\pi,\lambda}, \qquad \lambda \in \mathfrak{a}^*_{M_\pi,\mathbb{C}},$$

is the $\mathfrak{a}^*_{M_\pi,\mathbb{C}}$-orbit of supercuspidal representations of $M_\pi(F)$ attached to r_π. We write

$$\beta(\pi) = (-1)^{\dim(A_{M_0}/A_{M_\pi})},$$

where M_0 is a minimal Levi subgroup contained in M_π. Since (M_π, r_π) is determined up to conjugacy, the sign $\beta(\pi)$ is an invariant of π. It plays an important role in the duality operator of Aubert-Schneider-Stuhler.

The duality operator $D = D_G$ is an involution on the Grothendieck group $\mathcal{K}(G)$ of the category of $G(F)$-modules of finite length. It is defined [**Au**, 1.5] by

$$D_G = \sum_{P \supset P_0} (-1)^{\dim(A_{P_0}/A_P)}\, i_P^G \circ r_P^G, \tag{7.1.3}$$

where $P = MN_P$ ranges over standard parabolic subgroups of G, and i_P^G and r_P^G are the functors of induction and restriction between $\mathcal{K}(G)$ and $\mathcal{K}(M)$ [**BZ**, §2.3]. A key property of D_G is that it preserves irreducibility up to sign. More precisely, suppose that π is an irreducible representation of $G(F)$, with image $[\pi]$ in $\mathcal{K}(G)$. Then

$$D[\pi] = \beta(\pi)\,[\widehat{\pi}], \tag{7.1.4}$$

for an irreducible representation $\widehat{\pi}$ of $G(F)$. (See [**Au**, Corollaire 3.9 and Erratum] and [**ScS**, Proposition IV.5.1].)

Duality also has the property of being compatible with endoscopy. Suppose that G' is an endoscopic datum for G, equipped with auxiliary datum $(\widetilde{G}', \widetilde{\xi}')$ and corresponding transfer factors, as in §2.1. The transpose of the associated transfer mapping $f \to f'$ of functions is a transfer $S' \to S'_G$ of distributions. Here S' is any stable, $(\widetilde{\chi}')^{-1}$-equivariant distribution on $\widetilde{G}'(F)$, and S'_G is the invariant, χ^{-1}-equivariant distribution on $G(F)$ defined by

$$S'_G(f) = \widehat{S}'(f'), \qquad f \in \mathcal{H}(G,\zeta),$$

as in (2.1.2), and in the local form of the general global notation of §3.2. Now the representation theoretic character provides an isomorphism from the complex vector space

$$\mathcal{K}(G)_{\mathbb{C}} = \mathcal{K}(G) \otimes_{\mathbb{Z}} \mathbb{C}$$

onto the space of invariant distributions on $G(F)$ that are *admissible*, in the sense that they are finite linear combinations of irreducible characters. The duality mapping D_G can therefore be regarded as an involution on the space of admissible invariant distributions on $G(F)$. The compatibility property is a formula

$$(D'S')_G = \alpha(G,G')\,(D_G S'_G), \tag{7.1.5}$$

in which S' is assumed to be admissible, D' is the duality involution for $\widetilde{G}'$, and

$$\alpha(G, G') = (-1)^{\dim(A_{M_0}/A_{M_0'})}.$$

It comes with the implicit assertion that D' maps the subspace of stable distributions to itself. (See [**Hi**], [**A25**].)

More generally, suppose that G is a twisted triplet (G^0, θ, ω), as in §2.1. One can define the complex vector space $\mathcal{K}(G) \otimes \mathbb{C}$ as the span of objects (2.1.1). The duality involution (7.1.3) makes sense in this generality, provided that the indices of summation P are treated as standard parabolic *subsets* of G [**A4**, §1]. It acts on the Grothendieck group $\mathcal{K}(G)$ of the subcategory of $G^+(F)$-modules of finite length whose irreducible constituents are of the form (2.1.1). The compatibility formula (7.1.5) then extends to this setting, with the appropriate definition of the sign $\alpha(G, G')$. (For a proof of this and other properties discussed below, we refer the reader to the paper [**A25**] in preparation.) In the case of $\widetilde{G}(N) = (GL(N), \widetilde{\theta}(N), 1)$, we denote the sign by

$$\widetilde{\alpha}(N, G) = \alpha(\widetilde{G}(N), G), \qquad G \in \widetilde{\mathcal{E}}(N).$$

We note that we still have local parameters $\psi \in \Psi(G)$, if G represents a general triplet (G^0, θ, ω). For any such ψ, we write

$$\widehat{\psi}(w, u_1, u_2) = \psi(w, u_2, u_1), \qquad w \in W_F, \ u_1, u_2 \in SU(2),$$

for the dual parameter that interchanges the two $SU(2)$-factors in the group

$$L_F \times SU(2) = W_F \times SU(2) \times SU(2).$$

We also form the sign

$$\beta(\psi) = (-1)^{\dim(A_{M_0}/A_{M_\psi})},$$

where M_0 is a minimal Levi subset of G, and M_ψ is a minimal Levi subset for which the L-group ${}^L M_\psi^0$ contains the image of ψ.

Assume now that G belongs to $\widetilde{\mathcal{E}}_{\mathrm{sim}}(N)$. At this point, we are thinking of particular parameters that are to be localizations $\dot{\psi}_v$ at places v outside of U. They will be of the form $\psi = \widehat{\phi}$, for generic parameters $\phi \in \widetilde{\Phi}(G)$.

Any pair of local parameters $\phi \in \widetilde{\Phi}(G)$ and $\psi = \widehat{\phi}$ corresponds to a pair of irreducible, self-dual representations π_ϕ and π_ψ of $GL(N, F)$. It follows from the conjecture of Zelevinsky [**Z**, 9.17], proved by Aubert [**Au**, Théorème 2.3] and Schneider and Stuhler [**ScS**], that $\widehat{\pi}_\phi$ equals π_ψ. However, our real interest is in the twisted characters

$$\widetilde{f}_N(\phi) = \mathrm{tr}\big(\widetilde{\pi}_\phi(\widetilde{f})\big)$$

and

$$\widetilde{f}_N(\psi) = \mathrm{tr}\big(\widetilde{\pi}_\psi(\widetilde{f})\big)$$

in $\widetilde{f} \in \widetilde{\mathcal{H}}(N)$. These are admissible invariant distributions on $\widetilde{G}(N, F)$, which we can denote by $\widetilde{\phi}_N$ and $\widetilde{\psi}_N$. They behave in a simple way under the duality operator $\widetilde{D}$ for $\widetilde{G}(N)$. For it can be seen from twisted extensions of the arguments of [**Au**], applied to the very special case at hand, that

$$\widetilde{D}\widetilde{\phi}_N = \beta(\widetilde{\phi})\,\widetilde{\psi}_N.$$

(See the general construction of [**MW4**, §3].) The parameters ϕ and $\psi = \widehat{\phi}$ also provide admissible, stable distributions on $G(F)$. We may as well denote them again by ϕ and ψ, since they are the unique $\widetilde{\mathrm{Out}}_N(G)$-preimages of $\widetilde{\phi}_N$ and $\widetilde{\psi}_N$ under the twisted transfer mapping of distributions

$$S \longrightarrow \widetilde{S}_N = S_{\widetilde{G}(N)}.$$

To describe their behaviour under the duality operator $D = D_G$, we apply the analogue for $\widetilde{G}(N)$ of (7.1.5) and the formula for $\widetilde{D}\widetilde{\phi}_N$ above. We obtain

$$(D\phi)_{\widetilde{N}} = \widetilde{\alpha}(N,G)\,\widetilde{D}\widetilde{\phi}_N = \widetilde{\alpha}(N,G)\,\beta(\widetilde{\phi})\,\widetilde{\psi}_N.$$

The mapping $S \to \widetilde{S}$ is an injection if S is required to be $\widetilde{\mathrm{Out}}_N(G)$-invariant. Since

$$\widetilde{\alpha}(N,G)\,\beta(\widetilde{\phi}) = \beta(\phi)$$

in this case, we conclude that

$$D\phi = \beta(\phi)\,\psi. \tag{7.1.6}$$

We shall apply these remarks to the case $\psi = \widehat{\phi}$ of the general expansion (7.1.2). Since the groups $\mathcal{S}_\phi$ and $\mathcal{S}_\psi$ are the same, we can define a canonical bijection $\pi \to \sigma_\pi$ from $\widetilde{\Pi}_\phi$ and $\widetilde{\Pi}_\psi$, with matching characters

$$\langle x, \sigma_\pi\rangle = \langle x, \pi\rangle, \qquad x \in \mathcal{S}_\phi,\ \pi \in \widetilde{\Pi}_\phi.$$

LEMMA 7.1.1. *Suppose that $\psi = \widehat{\phi}$, for $\phi \in \widetilde{\Phi}(G)$ and $G \in \widetilde{\mathcal{E}}_{\mathrm{sim}}(N)$. Then*

$$\langle s_\psi, \pi\rangle\,\sigma_\pi = \beta(\phi)\,\beta(\pi)\,\widehat{\pi}, \tag{7.1.7}$$

for any $\pi \in \widetilde{\Pi}_\phi$.

PROOF. The formula (2.2.6) is valid for the generic parameter ϕ. It can be written

$$\phi'_G = \sum_{\pi\in\widetilde{\Pi}_\phi} \langle x, \pi\rangle\,\pi,$$

where (G',ϕ') corresponds to (ϕ, x), and π is regarded as an $\widetilde{\mathrm{Out}}_N(G)$-invariant distribution on the right hand side. We shall apply the operator $D = D_G$ to each side of this identity.

It follows from (7.1.5) and the analogue for G' of (7.1.6) that

$$D_G\phi'_G = \alpha(G,G')\,(D'\phi')_G = \alpha(G,G')\,\beta(\phi')\,(\psi')_G.$$

Since

$$\alpha(G,G')\,\beta(\phi') = \beta(\phi),$$

we obtain

$$D_G\phi'_G = \beta(\phi)(\psi')_G = \beta(\phi)\,\psi'_G.$$

The original identity becomes

$$\begin{aligned}\psi'_G &= \sum_{\pi\in\widetilde{\Pi}_\phi} \langle x,\pi\rangle\,\beta(\phi)\,D\pi \\ &= \sum_{\pi}\langle x,\pi\rangle\,\big(\beta(\phi)\,\beta(\pi)\,\widehat{\pi}\big).\end{aligned}$$

On the other hand, the expansion (7.1.2) can be written

$$\psi'_G = \sum_{\sigma\in\widetilde{\Sigma}_\psi} \langle s_\psi x,\sigma\rangle\,\sigma,$$

where σ is again understood as an $\widetilde{\mathrm{Out}}_N(G)$-invariant distribution on the right hand side. This identity then becomes

$$\begin{aligned}\psi'_G &= \sum_{\sigma\in\widetilde{\Sigma}_\psi} \langle x,\sigma\rangle\big(\langle s_\psi,\sigma\rangle\,\sigma\big) \\ &= \sum_{\pi\in\widetilde{\Pi}_\phi} \langle x,\pi\rangle\big(\langle s_\psi,\pi\rangle\,\sigma_\pi\big),\end{aligned}$$

since s_ψ can be regarded as an element in $\mathcal{S}_\phi$. It gives us a second expansion of ψ'_G, as a function of x into characters $\langle x,\pi\rangle$. Identifying the coefficients of $\langle x,\pi\rangle$, we obtain the required formula (7.1.7). □

Theorem 2.2.1 asserts that the distribution σ_π attached to any $\pi\in\widetilde{\Pi}_\phi$ is a character (up to the action of $\widetilde{\mathrm{Out}}_N(G)$). Combined with Lemma 7.1.1, it implies the stronger assertion that σ_π is actually equal to the irreducible character $\widehat{\pi}$. It also implies the identity of signs

$$\beta(\phi)\,\beta(\pi) = \langle s_\psi,\pi\rangle, \tag{7.1.8}$$

for every representation $\pi\in\Pi_\phi$. Conversely, suppose that the identity (7.1.8) is valid for all π. Then σ_π equals $\widehat{\pi}$, so the assertion of Theorem 2.2.1 holds for π in this stronger form.

Given the generic parameter $\phi\in\widetilde{\Phi}(G)$, we write $\widetilde{\Pi}^G_\phi$ for the subset of representations $\pi\in\widetilde{\Pi}_\phi$ such that the identity (7.1.8) holds. We then define the corresponding subset

$$\widetilde{\Pi}^G_\psi = \{\widehat{\pi} :\ \pi\in\widetilde{\Pi}^G_\phi\} \tag{7.1.9}$$

of $\widetilde{\Pi}_\psi$. The subset $\widetilde{\Sigma}^G_\psi = \widetilde{\Pi}^G_\psi$ of the packet $\widetilde{\Pi}_\psi = \widetilde{\Sigma}_\psi$ thus consists of ($\mathrm{Out}_N(G)$-orbits of) irreducible representations $\widehat{\pi} = \sigma_\pi$. The assertion of Theorem 2.2.1 for the dual parameter $\psi = \widehat{\phi}$ amounts to the equality

$$\widetilde{\Pi}^G_\psi = \widetilde{\Pi}_\psi \tag{7.1.10}$$

of the two sets.

Lemma 7.1.1 also has implications for the function $f'_G(\psi,\cdot)$ on the left hand side of the nongeneric local intertwining relation (2.4.7) of Theorem

2.4.1. To describe it, we first transfer the duality operator to the function spaces $\widetilde{\mathcal{I}}(G)$ and $\widetilde{\mathcal{S}}(G)$.

Recall that $\widetilde{\mathcal{I}}(G)$ is a space of functions on $\widetilde{\Pi}_{\text{temp}}(G)$. Any such function extends by analytic continuation to a function on the corresponding set of standard representations, and hence by linearity, to a function on $\widetilde{\Pi}(G)$ and on the complex Grothendieck group $\mathcal{K}(G)_{\mathbb{C}}$. We can therefore regard $\widetilde{\mathcal{I}}(G)$ as a space of linear functions on the vector space of $\widetilde{\mathrm{Out}}_N(G)$-symmetric, admissible invariant distributions on $G(F)$. Recall also that $\widetilde{\mathcal{S}}(G)$ is a space of functions on $\widetilde{\Phi}_{\text{bdd}}(G)$. These objects can also be extended by analytic continuation and linearity. The generic form of Theorem 2.2.1, established in Chapter 6, can be used to characterize the space of stable distributions within the larger space of $\widetilde{\mathrm{Out}}_N(G)$-symmetric, admissible invariant distributions. It allows us to regard $\widetilde{\mathcal{S}}(G)$ as a space of linear functions on the space of $\widetilde{\mathrm{Out}}_N(G)$-symmetric, admissible stable distributions of $G(F)$.

These descriptions of $\widetilde{\mathcal{I}}(G)$ and $\widetilde{\mathcal{S}}(G)$ are compatible with the convention above of treating a parameter $\psi \in \widetilde{\Psi}(G)$ as an admissible stable distribution (as in (7.1.6)), and a representation $\pi \in \widetilde{\Pi}(G)$ as an admissible invariant distribution (as in the proof of Lemma 7.1.1). With this understanding, we define duality operators $D = D_G$ on both $\widetilde{\mathcal{I}}(G)$ and $\widetilde{\mathcal{S}}(G)$ by the adjoint actions

$$(Da)(\pi) = a(D\pi), \qquad a \in \widetilde{\mathcal{I}}(G),\ \pi \in \widetilde{\Pi}_{\text{temp}}(G)$$

and

$$(Db)(\phi) = b(D\phi), \qquad b \in \widetilde{\mathcal{S}}(G),\ \phi \in \widetilde{\Phi}_{\text{bdd}}(G).$$

This gives continuous involutions on both $\widetilde{\mathcal{I}}(G)$ and $\widetilde{\mathcal{S}}(G)$.

Suppose that G, ϕ and $\psi = \widehat{\phi}$ are as in Lemma 7.1.1. It is easy to deduce from the lemma that

$$f'_G(\psi, s_\psi x) = \beta(\phi)\,(Df_G)'\,(\phi, s_\psi x), \qquad f \in \widetilde{\mathcal{H}}(G), \tag{7.1.11}$$

for any element x in the group $\mathcal{S}_\psi = \mathcal{S}_\phi$. To do so, we first observe that

$$(Df_G)'(\phi, x) = \sum_{\pi \in \widetilde{\Pi}_\phi} \langle x, \pi\rangle (Df_G)(\pi), \qquad x \in \mathcal{S}_\phi,$$

according to the definition (2.4.6) and the generic form of (2.2.6) we have established. We can therefore write

$$\begin{aligned}(Df_G)'(\phi, x) &= \sum_{\pi} \langle x, \pi\rangle f_G(D\pi) \\ &= \beta(\phi) \sum_{\pi \in \widetilde{\Pi}_\phi} \langle s_\psi x, \pi\rangle f_G(\sigma_\pi),\end{aligned}$$

since the lemma tells us that

$$D\pi = \beta(\pi)\,\widehat{\pi} = \beta(\phi)\,\langle s_\psi, \pi\rangle\,\sigma_\pi.$$

It then follows from (2.4.6) and (7.1.2) that

$$(Df_G)'(\phi, x) = \beta(\phi) \sum_{\sigma \in \widetilde{\Sigma}_\psi} \langle s_\psi x, \sigma \rangle f_G(\sigma) = \beta(\phi)\, f'_G(\psi, x).$$

Replacing x by $s_\psi x$, we obtain (7.1.11). Now the generic form of Theorem 2.4.1 we have established implies that

$$\beta(\phi)\,(Df_G)'(\phi, s_\psi x) = \beta(\phi)\,(Df_G)\,(\phi, s_\psi u),$$

where u is any element in $\mathfrak{N}_\phi$ whose image x_u in $\mathcal{S}_\psi$ equals x. Given the reduction of Lemma 2.4.2, the assertion of Theorem 2.4.1 for the dual parameter $\psi = \widehat{\phi}$ is consequently equivalent to the relation

$$\beta(\phi)\,(Df_G)\,(\phi, s_\psi u) = f_G(\psi, u), \qquad u \in \mathfrak{N}_\psi,\ f \in \widetilde{\mathcal{H}}(G). \tag{7.1.12}$$

This is a statement about the behaviour under duality of intertwining operators, equipped with the normalization (2.4.4) of Chapter 2. To be more precise, we fix a pair

$$(M, \psi_M), \qquad \psi_M \in \widetilde{\Psi}_2(M, \psi),$$

as in the preamble to Proposition 2.4.3. Then ψ_M equals $\widehat{\phi}_M$, for a fixed parameter $\phi_M \in \widetilde{\Phi}_2(M, \phi)$. We assume that M is proper in G, and that the analogue of Theorem 2.2.1 holds inductively for (M, ψ_M). According to the definition (2.4.5), $f_G(\psi, u)$ is the sum, over σ_M in the set $\widetilde{\Pi}_{\psi_M} = \widetilde{\Sigma}_{\psi_M}$, of the linear forms

$$f_G(\psi, u, \sigma_M) = \langle \widetilde{u}, \widetilde{\sigma}_M \rangle \operatorname{tr}\big(R_P(w_u, \widetilde{\sigma}_M, \psi)\, \mathcal{I}_P(\sigma_M, f)\big). \tag{7.1.13}$$

With this notation, it is not hard to see that Theorem 2.4.1 implies the identity of linear forms

$$\beta(\phi)\,(Df_G)\,(\phi, s_\psi u, \pi_M) = f_G(\psi, u, \widehat{\pi}_M), \qquad u \in \mathfrak{N}_\phi,\ f \in \widetilde{\mathcal{H}}(G), \tag{7.1.14}$$

for every representation $\pi_M \in \widetilde{\Pi}_{\phi_M}$. This is a straightforward consequence of the fact that either side of (7.1.12) equals the sum over $\pi_M \in \widetilde{\Pi}_{\phi_M}$ of the linear forms on the corresponding side of (7.1.14). Conversely, it is clear that if (7.1.14) is valid for all π_M, then Theorem 2.4.1 holds for ψ.

It is clear from the definition (7.1.13) that the putative identity (7.1.14) includes a condition on the representation theoretic R-groups of ψ and ϕ. D. Ban [**Ban**] has shown that for the groups $G \in \widetilde{\mathcal{E}}_{\mathrm{sim}}(N)$ in question, the two R-groups are indeed the same. We recall that the endoscopic R-groups of ψ and ϕ are equal by definition. These are in turn isomorphic to any of the representation theoretic R-groups of ϕ, by the results of Chapter 6. Theorem 8.1 of [**Ban**], which relies on the results of [**G**], implies that the same is true of the representation theoretic R-groups of ψ. More precisely, the normalized intertwining operators for representations $\pi_M \in \widetilde{\Pi}_{\phi_M}$ and $\widehat{\pi}_M \in \widetilde{\Pi}_{\psi_M}$ correspond, up to a scalar multiple, under the action of D. Stated in terms of the notation (7.1.13), this amounts to a weaker form

$$f_G(\psi, u, \widehat{\pi}_M) = \varepsilon_{\pi_M}(u)(Df_G)(\phi, u, \pi_M), \qquad u \in \mathfrak{N}_\phi,\ f \in \widetilde{\mathcal{H}}(G), \tag{7.1.15}$$

of (7.1.14), for a sign $\varepsilon_{\pi_M}(u)$.

The formula (7.1.15) is reminiscent of (6.1.5), the weaker form of the local intertwining relation for archimedean F and generic ϕ that follows from Shelstad's work. In fact, the two formulas have almost identical roles. They apply to elements v in the set V of special valuations of $\dot{F}$ (archimedean in Chapter 6, p-adic here) by which the local intertwining relation (for generic parameters in Chapter 6, and general parameters here) is deduced by global means. In each case, ε_{π_M} represents a character on the subgroup W_ψ of $W(M)$. Its inclusion in the formula has the effect of making the left hand side insensitive to the specific normalizations introduced in Chapter 2. The formulas would both hold (with different choices of characters ε_{π_M}) for any normalizations $R(w, \pi_M)$ of the intertwining operators that are multiplicative in w. They follow directly from the property that the operators $R(w, \pi_M)$ are scalars for $w \in W^0_\psi$, and form a basis of the space of all intertwining operators of the induced representation $\mathcal{I}_P(\pi_M)$ as w ranges over R_ψ. It is this property that follows from [**Ban**]. We note that Theorem 8.1 of [**Ban**] requires that $\widehat{\pi}_M$ be unitary. This condition is included in our induction assumption that Theorem 2.2.1 is valid for ψ_M. (The results in [**Ban**] and [**G**] were actually stated for G split, but with our knowledge of the groups $R(\pi_M)$ from §6.6, they extend easily to the case that $G \in \widetilde{\mathcal{E}}_{\mathrm{sim}}(N)$ is quasisplit.)

Given the generic parameter $\phi \in \widetilde{\Phi}(G)$, we write $\widetilde{\Pi}^G_{\phi_M}$ for the subset of representations $\pi_M \in \widetilde{\Pi}_{\phi_M}$ such that the identity (7.1.14) holds. We then define the corresponding subset

$$\widetilde{\Pi}^G_{\psi_M} = \{\widehat{\pi}_M : \ \pi_M \in \widetilde{\Pi}^G_{\phi_M}\} \tag{7.1.16}$$

of $\widetilde{\Pi}_{\psi_M}$. The assertion of Theorem 2.4.1 for the dual parameter $\psi = \widehat{\phi}$ amounts to the equality

$$\widetilde{\Pi}^G_{\psi_M} = \widetilde{\Pi}_{\psi_M} \tag{7.1.17}$$

of the two sets.

Our discussion of duality has been designed for application to local completions $\dot{\psi}_v$ at places $v \notin U$. The role of such completions will be similar to that of their analogues from Chapter 6. To emphasize this point, we can specialize the local parameters ϕ and ψ further.

A finite dimensional representation of W_F is said to be *tamely ramified* if its kernel contains the wildly ramified inertia group. In other words, the representation is trivial on the pro-p part P_F of the inertia subgroup I_F of W_F, p being the residual characteristic of F. We shall say that the representation is *quadratic* if its irreducible constituents have dimension at most two. We thus observe a further symmetry with the discussion of §6.1. For it is clear that the tamely ramified, quadratic representations of W_F that lie in $\widetilde{\Phi}_2(G)$ are in some sense analogues of the archimedean parameters described prior to Lemma 6.1.2.

We shall say that the local parameters ϕ and $\psi = \widehat{\phi}$ are tamely ramified and quadratic if their restrictions to W_F have these properties.

LEMMA 7.1.2. *Suppose that $\psi = \widehat{\phi}$ and $\phi \in \widetilde{\Phi}(G)$ are tamely ramified, quadratic parameters for our simple endoscopic datum $G \in \widetilde{\mathcal{E}}_{\text{sim}}(N)$ over the p-adic field F.*

(a) *The image of $\widetilde{\Pi}_\psi^G$ in the dual group $\widehat{\mathcal{S}}_\psi$ generates $\widehat{\mathcal{S}}_\psi$, and contains the trivial character 1.*

(b) *The image of $\widetilde{\Pi}_{\psi_M}^G$ in the dual group $\widehat{\mathcal{S}}_{\psi_M}$ generates $\widehat{\mathcal{S}}_{\psi_M}$, and contains the trivial character 1.*

Lemma 7.1.2 is intended as an analogue of Lemma 6.1.2. It provides the limited local information that will be used to deduce the general local theorems by global means. The two assertions of the lemma are obviously equivalent to their analogues for ϕ. It is really the first assertion (i) (taken for ϕ) that is parallel to Lemma 6.1.2. The second assertion (ii) is actually a stronger analogue of what was available in §6.1. It represents a refinement (for the p-adic field F here) of the formula (6.1.5), in which the characters $\varepsilon_{\pi_M}(x)$ are all trivial. As we shall see in §7.3, this refinement will greatly simplify our proof of the local intertwining relation. We shall give a proof of Lemma 7.1.2 in [**A25**] that is based on the representations of Hecke algebras, as well as a proof of a natural variant of the lemma for the bitorsor $\widetilde{G}$ attached to an even orthogonal group $G \in \widetilde{\mathcal{E}}_{\text{sim}}(N)$.

We should note that analogues of Lemma 7.1.2(i) have been established in complete generality by Moeglin [**M1**] [**M2**] [**M3**]. Using a remarkable family of partial duality involutions, she has established results that among other things imply that $\widetilde{\Pi}_\phi^G$ equals $\widetilde{\Pi}_\phi$, for *any* generic parameter $\phi \in \widetilde{\Phi}(G)$. Her results are partially global, in that they depend on the generic form of Theorem 2.2.1, which was established by global means in Chapter 6.

7.2. Construction of global parameters $\dot{\psi}$

We will now introduce the families of global parameters needed to establish the general nontempered local theorems. The construction is essentially that of §6.2–§6.3. However, we shall have to impose local constraints at a finite set of nonarchimedean places. These will be used to apply the duality properties described in the last section. In particular, they will be used in the next section to modify the discussion of §5.4.

Suppose that $G \in \widetilde{\mathcal{E}}_{\text{sim}}(N)$ is a fixed simple twisted endoscopic datum for $\widetilde{G}(N)$ over the given local field F. Consider a nongeneric parameter

$$\psi = \ell_1\psi_1 \oplus \cdots \oplus \ell_r\psi_r, \tag{7.2.1}$$

in $\widetilde{\Psi}(G)$, with simple components

$$\psi_i \in \widetilde{\Psi}_{\text{sim}}(G_i), \qquad G_i \in \widetilde{\mathcal{E}}_{\text{sim}}(N_i).$$

Given the local pairs (G_i, ψ_i) over F, we will want to choose global pairs

$$(\dot{G}_i, \dot{\psi}_i), \qquad \dot{G}_i \in \dot{\widetilde{\mathcal{E}}}_{\text{sim}}(N_i),\ \dot{\psi}_i \in \widetilde{\Psi}_{\text{sim}}(\dot{G}_i),$$

over a suitable global field $\dot{F}$ such that

$$(\dot{F}_u, \dot{G}_{i,u}, \dot{\psi}_{i,u}) = (F, G_i, \psi_i), \qquad 1 \leqslant i \leqslant r,$$

for some valuation u of $\dot{F}$. These will determine a global endoscopic datum $\dot{G} \in \dot{\widetilde{\mathcal{E}}}_{\text{sim}}(N)$ such that the global parameter

$$\dot{\psi} = \ell_1 \dot{\psi}_1 \boxplus \cdots \boxplus \ell_r \dot{\psi}_r \tag{7.2.2}$$

belongs to $\widetilde{\Psi}(\dot{G})$. The process will be similar to that of Proposition 6.3.1, except that supplementary constraints on the global data will have to be slightly different. There is also the minor difference that G was taken from the larger set $\widetilde{\mathcal{E}}_{\text{ell}}(N)$ in Proposition 6.3.1, rather than the set of simple data $\widetilde{\mathcal{E}}_{\text{sim}}(N)$ we have specified here.

As in the generic case in §6.3, we fix a Levi subgroup M of G such that the set $\widetilde{\Psi}_2(M, \psi)$ is nonempty. Given $\dot{G}$, we will also choose a global Levi subgroup $\dot{M}$ of $\dot{G}$ such that $\widetilde{\Psi}_2(\dot{M}, \dot{\psi})$ is nonempty. The analogues

$$\psi_M = \psi_1^{\ell_1'} \times \cdots \times \psi_r^{\ell_r'} \times \psi_- \tag{7.2.3}$$

and

$$\dot{\psi}_M = \dot{\psi}_1^{\ell_1'} \times \cdots \times \dot{\psi}_r^{\ell_r'} \times \dot{\psi}_- \tag{7.2.4}$$

of (6.3.3) and (6.3.4) can then be identified with fixed elements in the respective sets $\widetilde{\Psi}_2(M, \psi)$ and $\widetilde{\Psi}_2(\dot{M}, \dot{\psi})$. As before, we will have a special interest in the case that each ℓ_i equals 1. This of course is the case that $M = G$ and $\psi_M = \psi = \psi_-$, with $\dot{M} = \dot{G}$ and $\dot{\psi}_M = \dot{\psi} = \dot{\psi}_-$.

Recall that there are decompositions

$$\psi_i = \mu_i \otimes \nu_i, \qquad N_i = m_i n_i,\ 1 \leqslant i \leqslant r,$$

where $\mu_i \in \widetilde{\Phi}_{\text{sim}}(H_i)$ is a simple generic parameter for a datum $H_i \in \widetilde{\mathcal{E}}_{\text{sim}}(m_i)$ over F, and ν_i is an irreducible representation of $SU(2)$ of dimension n_i. Set

$$I^{m,n} = \{i :\ m_i = m,\ n_i = n\},$$

for any positive integers m and n. We then have another decomposition

$$\psi = \bigoplus_{n=1}^{\infty} (\mu^n \otimes \nu^n), \tag{7.2.5}$$

where

$$\mu^n = \bigoplus_{m=1}^{\infty} \bigoplus_{i \in I^{m,n}} (\ell_i \mu_i),$$

and ν^n is the irreducible representation of $SU(2)$ of dimension n. The global constituents of $\dot{\psi}$ will be of the form

$$\dot{\psi}_i = \dot{\mu}_i \boxtimes \dot{\nu}_i, \qquad 1 \leqslant i \leqslant r,$$

where $\dot{\mu}_i \in \widetilde{\Phi}_{\text{sim}}(\dot{H}_i)$ is a simple generic parameter for some $\dot{H}_i \in \tilde{\dot{\mathcal{E}}}_{\text{sim}}(m_i)$ over $\dot{F}$, and $\dot{\nu}_i$ is the extension of ν_i to the group $SL(2,\mathbb{C})$. As we noted back in §1.4, these objects determine the simple data $\dot{G}_i$ with $\dot{\psi}_i \in \widetilde{\Psi}_{\text{sim}}(\dot{G}_i)$. For the construction, we will apply a variant of Proposition 6.3.1 directly to the local generic parameter

$$\mu = \bigoplus_{n=1}^{\infty} (\mu^n) = \bigoplus_{i=1}^{r} (\ell_i \mu_i).$$

There will be four possibilities for the global field $\dot{F}$. The simplest case is when the local field F equals $\mathbb{C}$. In this case we take $\dot{F}$ to be any imaginary quadratic extension of $\mathbb{Q}$. If $F = \mathbb{R}$, and if the two-dimensional generic constituents

$$\{\mu_i : \; m_i = 2, \; 1 \leqslant i \leqslant r\}$$

of ψ are in relative general position, we simply take $\dot{F} = \mathbb{Q}$. If $F = \mathbb{R}$, but these two-dimensional generic constituents are not in relative general position, we take $\dot{F}$ to be a totally real field with several archimedean places. Finally, if F is p-adic, we take $\dot{F}$ to be any totally real field of which F equals some completion. In all cases, we fix a valuation u of $\dot{F}$ such that $\dot{F}_u = F$. As agreed in the last section, we also fix a large finite set V of nonarchimedean places, which does not contain u, and for which $\dot{q}_v$ is large for any $v \in V$. This set will assume the role played in Proposition 6.3.1 by the complementary set

$$S_\infty^u = S_\infty - \{u\}$$

of archimedean places. According to our remarks in §7.1, it will be the set

$$U = S_\infty(u) = S_\infty \cup \{u\}$$

that often takes the place of the individual valuation u in Chapter 6, while it will be the union

$$S_\infty(u,V) = S(U,V) = U \cup V$$

that assumes the earlier role played by $S_\infty(u)$.

PROPOSITION 7.2.1. *Given the local objects G, ψ, M and ψ_M over F as in (7.2.1) and (7.2.3), we can choose corresponding global objects $\dot{G}$, $\dot{\psi}$, $\dot{M}$ and $\dot{\psi}_M$ over $\dot{F}$ as in (7.2.2) and (7.2.4) such that the following conditions are satisfied.*

(i) *There is a valuation u of $\dot{F}$ such that*

$$(\dot{F}_u, \dot{G}_u, \dot{\psi}_u, \dot{M}_u, \dot{\psi}_{M,u}) = (F, G, \psi, M, \psi_M),$$

and such that the canonical maps

$$\mathcal{S}_{\dot{\psi}_M} \longrightarrow \mathcal{S}_{\psi_M}$$

and

$$\mathcal{S}_{\dot{\psi}} \longrightarrow \mathcal{S}_{\psi}$$

are isomorphisms.

(ii) *For any valuation v outside the set $S_\infty(u,V)$, the local Langlands parameter*

$$\dot\mu_v = \ell_1\dot\mu_{1,v} \oplus \cdots \oplus \ell_r\dot\mu_{r,v},$$

is a direct sum of tamely ramified quasicharacters of $\dot F_v^*$, while the corresponding decomposition of the subparameter

$$\dot\mu_{1,v} \oplus \cdots \oplus \dot\mu_{r,v}$$

contains at most one ramified quasicharacter.

(iii)(a) *For any $v \in V$, the parameters $\dot\mu_{i,v}$ lie in $\widetilde\Phi_2(\dot H_{i,v})$ and are each a direct sum of tamely ramified representations of $W_{\dot F_v}$ of dimension one and two, so in particular, $\dot\psi_v$ equals the dual $\widehat\phi_v$ of a generic parameter $\phi_v \in \widetilde\Phi(\dot G_v)$. Moreover, the canonical mappings*

$$\widetilde\Pi^M_{\dot\psi_{M,V}} = \prod_{v\in V} (\widetilde\Pi^M_{\dot\psi_{M,v}}) \longrightarrow \widehat{\mathcal S}_{\dot\psi_M} \cong \widehat{\mathcal S}_{\psi_M}$$

and

$$\widetilde\Pi^G_{\dot\psi_{M,V}} = \prod_{v\in V} (\widetilde\Pi^G_{\dot\psi_{M,v}}) \longrightarrow \widehat{\mathcal S}_{\dot\psi_M} \cong \widehat{\mathcal S}_{\psi_M}$$

are surjective.

(iii)(b) *Suppose that each ℓ_i equals 1. Then there is a $v \in V$ with the property that if $\dot\psi_v$ lies in $\widetilde\Psi(G_v^*)$ for some $G_v^* \in \dot{\widetilde{\mathcal E}}_{\mathrm{sim},v}(N)$, the dual group $\widehat G_v^*$ equals $\widehat G$.*

(iii)(c) *Suppose that some ℓ_i is greater than 1. Then there is a $v \in V$ such that the kernel of the composition of mappings*

$$\mathcal S_{\dot\psi} \longrightarrow \mathcal S_{\dot\psi_v} \longrightarrow R_{\dot\psi_v}$$

contains no element whose image in the global R-group $R_{\dot\psi} = R_{\dot\psi}(\dot G)$ belongs to $R_{\dot\psi,\mathrm{reg}}$.

PROOF. The statement is clearly very close to that of Proposition 6.3.1. Notice that the condition (iii)(a) here applies to the two sets of p-adic representations from Lemma 7.1.2 (but with G replaced by M in the first set). The condition (iii)(a) of Proposition 6.3.1 applies to the L-packets over $\mathbb R$ from Lemma 6.1.2.

We have included the possibility here that $F = \mathbb C$. In this case, however, each constituent is one-dimensional. Since it is also self-contragredient, μ_i must then be the trivial character on $\mathbb C^*$. We simply take $\dot\mu_i$ to be the trivial idèle class character for $\dot F$. The various conditions then follow without difficulty.

We can therefore assume that $F \neq \mathbb C$. Then $\dot F$ is a totally real field, chosen according to the description prior to the statement of the proposition. There will be some superficial differences from the setting of Proposition 6.3.1. For example, we are allowing a case $\dot F = \mathbb Q$ in which there is only one real valuation. Moreover, the constituents μ_i of μ need not be distinct, since they could be matched with different representations ν_i of $SU(2)$. Finally, we will be applying the construction of Proposition 6.3.1 as it applies to the larger set of valuations

$$S_\infty^u(V) = S_\infty^u \cup V$$

in place of S_∞^u. However, the general structure of the proof will be similar.

The first step is to construct the primary global pairs

$$(\dot{H}_i, \dot{\mu}_i), \qquad \dot{H}_i \in \dot{\widetilde{\mathcal{E}}}_{\text{sim}}(m_i),\ \dot{\mu}_i \in \widetilde{\Phi}_{\text{sim}}(\dot{H}_i),$$

over $\dot{F}$ from the given local pairs

$$(H_i, \mu_i), \qquad H_i \in \widetilde{\mathcal{E}}_{\text{sim}}(m_i),\ \mu_i \in \widetilde{\Phi}_{\text{sim}}(H_i),$$

over F. We apply Corollary 6.2.4, supplemented by Remark 3 following its proof. For every i, the global pair then has the properties that

$$(\dot{F}_u, \dot{H}_{i,u}, \dot{\mu}_{i,u}) = (F, H_i, \mu_i),$$

that $\dot{\mu}_{i,v}$ is spherical for every $v \notin S_\infty(u, V)$, and that $\dot{\mu}_{i,v}$ is an archimedean parameter in general position for every $v \in S_\infty^u$. The new condition applies to the finite set V of nonarchimedean places. According to the supplementary remark, we can arrange that for each $v \in V$, the local Langlands parameter $\dot{\mu}_{i,v}$ belongs to $\widetilde{\Phi}_2(\dot{H}_{i,v})$. In fact, we can assume that $\dot{\mu}_{i,v}$ is a direct sum of distinct, irreducible, self-dual, tamely ramified representations of $W_{\dot{F}_v}$ of dimension two with unramified determinant, together with an unramified character of $W_{\dot{F}_v}$ of order 1 or 2, in case m_i is odd. Our hypothesis that the residual characteristics of valuations $v \in V$ are large insures that there are enough such two-dimensional representations of $W_{\dot{F}_v}$.

As we noted above, the simple generic pairs $(\dot{H}_i, \dot{\mu}_i)$ can then be combined with the representations $\dot{\nu}_i$ of $SL(2, \mathbb{C})$ to give nongeneric pairs

$$(\dot{G}_i, \dot{\psi}_i), \qquad \dot{G}_i \in \dot{\widetilde{\mathcal{E}}}_{\text{sim}}(N_i),\ \dot{\psi}_i \in \dot{\mathcal{E}}_{\text{sim}}(\dot{G}_i).$$

These objects in turn determine a global endoscopic datum $\dot{G} \in \dot{\widetilde{\mathcal{E}}}_{\text{sim}}(N)$ with $\widehat{\dot{G}} = \widehat{G}$, together with a global parameter $\dot{\psi} \in \widetilde{\Psi}(\dot{G})$ as in (7.2.2). They also determine a global Levi subgroup $\dot{M}$ of $\dot{G}$ with $\widehat{\dot{M}} = \widehat{M}$, together with a global parameter $\dot{\psi}_M \in \widetilde{\Psi}(\dot{M}, \dot{\psi})$ as in (7.2.4). The condition (i) follows.

To obtain the remaining conditions of the proposition, we need to impose more local constraints on the basic global pairs $(\dot{H}_i, \dot{\mu}_i)$. We can by and large follow the prescription from the proof of Proposition 6.3.1, with V now being the chosen set of nonarchimedean valuations instead of S_∞^u. In particular, we will choose the global pairs successively with increasing i.

Consider the quadratic idèle class character

$$\dot{\theta}_i = \eta_{\dot{H}_i} = \eta_{\dot{\mu}_i} = \det(\dot{\mu}_i)$$

attached to any $\dot{H}_i$ with $\widehat{\dot{H}}_i = SO(m_i, \mathbb{C})$. Its local values $\dot{\theta}_{i,v}$ at places $v \in V$ are predetermined if m_i is even. In this case, the conditions on $\dot{\mu}_{i,v}$

imply that

$$\dot{\theta}_{i,v} = (\varepsilon_v)^{q_i}, \qquad q_i = \tfrac{1}{2}\, m_i,$$

where ε_v is the nontrivial unramified quadratic character on $\dot{F}_v^*$. If m_i is odd, we apply the criterion of Proposition 6.3.1 to V. Namely, we fix a set of *distinct* places

$$V^o = \{v_i : \ m_i \text{ odd}\},$$

and then require that

$$\dot{\theta}_{i,v} = \begin{cases} (\varepsilon_v)^{q_i+1}, & \text{if } v = v_i, \\ (\varepsilon_v)^{q_i}, & \text{if } v \neq v_i, \end{cases} \qquad q_i = \tfrac{1}{2}\,(m_i - 1), \tag{7.2.6}$$

for any $v \in V$ and i with m_i odd. Recall that the local values $\dot{\theta}_{i,v}$ are also determined at places $v \in S_\infty(u)$ if m_i is even, and at $v = u$ in general. To be definite, we arbitrarily fix $\dot{\theta}_{i,v}$ in the remaining case of $v \in S_\infty^u$ and m_i odd. (We cannot use the criterion of Proposition 6.3.1, which we have just applied to our p-adic set V here, since we are not assuming that S_∞^u is large. There is no need to do so in any case, since V is now taking the earlier role of S_∞^u.)

Having fixed the local values $\dot{\theta}_{i,v}$ at any $v \in S_\infty(u,V)$ and i with $\widehat{H}_i = SO(m_i, \mathbb{C})$, we choose the global quadratic characters as in the proof of Proposition 6.3.1. We require that for any $v \notin S_\infty(u,V)$ there be at most one i such that $\dot{\theta}_{i,v}$ ramifies, and that the ramification be tame. Recall that this is done by insisting that $\dot{\theta}_{i,v}$ be unramified at any place $v \notin S_\infty(u,V)$, which divides 2, or at which $\dot{\theta}_{j,v}$ ramifies for some $j < i$. Once we have chosen the quadratic character $\dot{\theta}_i$, we obtain the global endoscopic datum $\dot{H}_i$ from the dual group $\widehat{\dot{H}}_i = \widehat{H}_i$. We then construct the global parameters $\dot{\mu}_i \in \widetilde{\Phi}_{\mathrm{sim}}(\dot{H}_i)$ according to Lemma 6.2.2 and Corollary 6.2.4 (modified by the accompanying Remark 3). At the places v in either S_∞^u or V, we first specify the localizations $\dot{\phi}_{i,v}$ in terms of their predecessors $\dot{\phi}_{j,v}$, with $j < i$. This is how we arrange that as i varies, the constituents of $\dot{\phi}_{i,v}$ are mutually disjoint, and in fact, in relative general position if $v \in S_\infty^v$.

The required property (ii) follows just as in the proof of Proposition 6.3.1. The same is also largely true of the remaining assertions in (iii)(a)–(iii)(c). They are for the most part trivial unless the subset I^1 of indices $i \in I^o$ with $m_i = 1$ is not empty. As in the earlier proof, it is really only for this possibility that the property (7.2.6) was needed.

The first assertion of (iii)(a) is an immediate consequence of the construction. In the other assertion of (iii)(a), the first mapping is analogous to that of Proposition 6.3.1(iii)(a). As before, it suffices to treat the case that $M = G$. Working with the valuations v_i as i varies over I^1, we again show that the mapping

$$\widehat{\mathcal{S}}_{\dot{\psi}_V} \longrightarrow \widehat{\mathcal{S}}_{\dot{\psi}}$$

is surjective. We then establish this part of the assertion, using Lemma 7.1.2(a) in place of Lemma 6.1.2, from the fact that the set V is large. The second mapping in (iii)(a) has no counterpart in Proposition 6.3.1. However, the required surjectivity follows in the same way from Lemma 7.1.2(b).

For the conditions (b) and (c) of (iii), we can again take v to be any valuation in V if I^1 is empty. Otherwise, we must take $v = v_1 = v_{i_1}$, for any fixed index i_1 in I^1. It might appear that (iii)(b) is complicated here by the existence of simple components $\psi_i = \mu_i \otimes \nu_i$ with $i \in I^1$ and $n_i \neq n_{i_1}$. Indeed, the condition would fail on this account if we were allowing G to be a general element in $\widetilde{\mathcal{E}}_{\mathrm{ell}}(N)$, as in Proposition 6.3.1. However, we have deliberately taken $G \in \widetilde{\mathcal{E}}_{\mathrm{sim}}(N)$ to be simple here. The condition (iii)(b) then follows easily. The proof of the last condition (iii)(c) is identical to that of Proposition 6.3.1. □

The discussion of this section has been parallel to that of §6.3. Having no need of a nontempered analogue of Lemma 6.3.2, we are now at the point we reached at the end of §6.3. Given simple local parameters $\psi_i = \mu_i \otimes \nu_i$ over F, we have constructed global parameters $\dot{\psi}_i = \dot{\mu}_i \otimes \dot{\nu}_i$ over $\dot{F}$. We can then form the associated family

$$\dot{\tilde{\mathcal{F}}} = \tilde{\mathcal{F}}(\dot{\psi}_1, \dots, \dot{\psi}_r) = \{\dot{\ell}_1 \dot{\psi}_1 \boxplus \cdots \boxplus \dot{\ell}_r \dot{\psi}_r : \dot{\ell}_i \geqslant 0\} \tag{7.2.7}$$

of compound global parameters. This represents another family of the sort treated in Chapter 5. In contrast with the earlier generic case, Assumption 5.1.1 will require no further effort to justify here. It will be a consequence of Lemma 2.2.2 and what we have established in the generic case. We recall that this assumption was the basic premise of Chapter 5.

We now formally adopt the induction hypotheses for $\dot{\tilde{\mathcal{F}}}$ that were in force in Chapter 5. We fix the positive integer N, and assume that the local theorems all hold for general parameters $\psi \in \widetilde{\Psi}$ with $\deg(\psi) < N$. To this, we add the induction assumption that the global theorems all hold for parameters $\dot{\psi}$ in the given family $\dot{\tilde{\mathcal{F}}}$ with $\deg(\dot{\psi}) < N$.

The global hypothesis is understood to include the conditions on Theorem 1.5.3, described near the beginning of §5.1. The assertion of Theorem 1.5.3(b) thus represents an induction assumption for the nontempered family $\dot{\tilde{\mathcal{F}}}$, which is resolved in Lemma 5.1.6. The assertion of Theorem 1.5.3(a) represents a condition on the simple generic constituents $\dot{\mu}_i$ of parameters $\dot{\psi} \in \dot{\tilde{\mathcal{F}}}$, which has also been treated. It is part of the resolution from Corollary 5.4.7, which among other things, is the foundation for the definition of the subsets $\tilde{\mathcal{F}}(\dot{G})$ of $\dot{\tilde{\mathcal{F}}}$. We recall that the global theorems for simple generic parameters with local constraints were established in Lemma 5.4.6. The local constraints for the simple generic parameters $\dot{\mu}_i$ here are admittedly at the nonarchimedean places V, rather than the archimedean places S^u_∞ of §5.4. But with the local theorems in place now for all of the constituents of

$\dot{\mu}_i$, the proof of Lemma 5.4.6 carries over in a straightforward way. Alternatively, we observe that the general proof of the global theorems, which we will complete in §8.2, can be carried out for generic parameters $\dot{\mu}$ with no reference to the local results for nongeneric parameters of this chapter.

We would like to apply the results of Chapter 5 to the nongeneric family (7.2.7) with local constraints. As we have just observed, Assumption 5.1.1 is easy to verify in this case. It follows from the reduction in Lemma 2.2.2 of the local assertion of Theorem 2.2.1(a) and our induction assumption on the generic constituents $\dot{\mu}_i$ of parameters in (7.2.7). (The argument here is similar to that of the early part of the proof of Lemma 8.1.1, at the beginning of the next chapter.) This places the family (7.2.7) in the general framework of §5.1. In particular, the general lemmas of §§5.1–5.3 are valid (as applicable) for its parameters. However, their stronger refinements in §5.4, which we used to establish the local theorems of Chapter 6, hold only for the generic family of §5.4. We are going to have to extend them to the nongeneric family (7.2.7).

To deal with Theorem 2.4.1, we will need nongeneric refinements of Lemmas 5.2.1 and 5.2.2. These apply to parameters $\dot{\psi}$ in the subset $\widetilde{\mathcal{F}}^2_{\mathrm{ell}}(\dot{G})$ of $\dot{\widetilde{\mathcal{F}}}$ attached to a simple datum $\dot{G} \in \dot{\widetilde{\mathcal{E}}}_{\mathrm{sim}}(N)$. Any such $\dot{\psi}$ takes the form

$$(7.2.8) \qquad \begin{cases} \dot{\psi} = 2\dot{\psi}_1 \boxplus \cdots \boxplus 2\dot{\psi}_q \boxplus \dot{\psi}_{q+1} \boxplus \cdots \boxplus \dot{\psi}_r, \\ S_{\dot{\psi}} = \big(O(2,\mathbb{C})^q \times O(1,\mathbb{C})^{r-q}\big)^+_{\dot{\psi}}, \qquad q \geqslant 1, \end{cases}$$

as in (5.2.4). In contrast to the earlier generic refinements of §5.4, we will not need a separate argument for the case that the index r equals 1.

To deal with Theorem 2.2.1, we will need nongeneric refinements of Lemmas 5.3.1 and 5.3.2. These apply to parameters

$$(7.2.9) \qquad \dot{\psi} = \dot{\psi}_1 \boxplus \cdots \boxplus \dot{\psi}_r, \qquad \dot{\psi}_i \in \dot{\widetilde{\mathcal{F}}}_{\mathrm{sim}}(N_i),$$

in the subset $\widetilde{\mathcal{F}}_2(\dot{G})$ of $\dot{\widetilde{\mathcal{F}}}$. We will again not require a separate argument for the case that r equals 1. However, we will otherwise follow the general strategy of Chapter 5. In particular, we shall introduce the simple datum $\dot{G}_+ \in \dot{\widetilde{\mathcal{E}}}_{\mathrm{sim}}(N_+)$, for $N_+ = N_1 + N$, and the parameter

$$\dot{\psi}_+ = \dot{\psi}_1 \boxplus \dot{\psi},$$

as in §5.3.

We shall establish the necessary refinements in the next section. The argument will be similar to that of §5.4, except that the exceptional place v will be taken from the set of p-adic valuations V. In order to reduce the problem to Corollary 3.5.3, which was the underlying tool in §5.4, we shall apply the duality theory of the last section to the parameters $\dot{\psi}_v$ and $\dot{\psi}_{+,v}$.

7.3. The local intertwining relation for ψ

The nongeneric form of Theorem 2.4.1 will be easier than the generic case treated in §6.4. This is because we will have some special cases to work with. They consist of a limited number of local intertwining relations that can be obtained by duality from the generic form of the theorem already established. We will be able to combine them with a simplified form of the global argument from §6.4.

First, however, we will have to deal with the question raised at the end of the last section. We shall establish a pair of lemmas that extend the global results of §5.4.

The first lemma gives the nongeneric refinements of Lemmas 5.2.1 and 5.2.2, for parameters (7.2.8) in the subset $\widetilde{\mathcal{F}}^2_{\text{ell}}(\dot{G})$ of $\dot{\widetilde{\mathcal{F}}}$. It will be used later in the section to derive the general local intertwining relation.

LEMMA 7.3.1. *Suppose that*

$$(\dot{G}, \dot{\psi}), \qquad \dot{G} \in \dot{\widetilde{\mathcal{E}}}_{\text{sim}}(N), \ \dot{\psi} \in \widetilde{\mathcal{F}}^2_{\text{ell}}(\dot{G}), \tag{7.3.1}$$

is an analogue for the nongeneric family $\dot{\widetilde{\mathcal{F}}}$ *of the pair given in either Lemmas* 5.2.1 *or* 5.2.2. *Then the expressions*

$$c \sum_{x \in \mathcal{S}_{\psi,\text{ell}}} \varepsilon^{\dot{G}}_{\dot{\psi}}(\dot{x}) \left(\dot{f}'_{\dot{G}}(\dot{\psi}, s_{\dot{\psi}}\dot{x}) - \dot{f}_{\dot{G}}(\dot{\psi}, \dot{x})\right), \qquad c > 0 \tag{7.3.2}$$

and

$$\tfrac{1}{8}\left((\dot{f}^\vee)^{\dot{M}}(\dot{\psi}_1) - \dot{f}^\vee_{\dot{G}^\vee}(\dot{\psi}, \dot{x}_1)\right), \qquad \textit{if } r = 1, \tag{7.3.3}$$

from these lemmas vanish, for any functions $\dot{f} \in \widetilde{\mathcal{H}}(\dot{G})$ *and* $\dot{f}^\vee \in \widetilde{\mathcal{H}}(\dot{G}^\vee)$.

PROOF. This lemma represents an analogue for $\dot{\widetilde{\mathcal{F}}}$ of Lemmas 5.4.3 and 5.4.4, which we recall were the refinements of Lemmas 5.2.1 and 5.2.2 for generic parameters. In particular, (7.3.2) corresponds to one of the two expressions (5.2.8) or (5.2.13) (according to whether $r > 1$ and $r = 1$), while (7.3.3) corresponds to (5.2.12). As in the two refinements for generic parameters, we actually have to establish several vanishing assertions. In stating this lemma, we have included only those assertions that will be used directly in the coming proof of the nontempered local intertwining relation.

Following the argument from the two lemmas of §5.4, we shall reduce the proof to the assertion of Corollary 3.5.3. The main step is to write the expression (7.3.2) in the form (5.4.10) obtained in the proof of Lemma 5.4.3. We choose $v \in V$ according to the condition (iii)(c) of Proposition 7.2.1. In particular, $\dot{\psi}_v$ is the dual $\widehat{\phi}_v$ of a generic parameter $\phi_v \in \widetilde{\Phi}(\dot{G}_v)$, as prescribed by the condition (iii)(a) of the proposition. We take

$$\dot{f} = \dot{f}_v \, \dot{f}^v, \qquad \dot{f}_v \in \widetilde{\mathcal{H}}(\dot{G}_v), \ \dot{f}^v \in \widetilde{\mathcal{H}}(\dot{G}^v),$$

to be decomposible. Then (7.3.2) equals

$$c \sum_{x \in \mathcal{S}_{\psi,\mathrm{ell}}} \varepsilon_{\dot\psi}^{\dot G}(\dot x)\big(f'_{v,\dot G}(\dot\psi_v, s_{\dot\psi}\dot x_v)(\dot f^v)'_{\dot G}(\dot\psi^v, s_{\dot\psi}\dot x^v) - \dot f_{v,\dot G}(\dot\psi_v, \dot x_v)\, \dot f^v_{\dot G}(\dot\psi^v, \dot x^v)\big),$$

in the notation (5.4.7). There are two linear forms in $\dot f_v$ in each summand of this expression. To write them in terms of the generic parameter ϕ_v, we shall apply the duality operator D_v for the group $\dot G_v$.

The first linear form is $\dot f'_{v,\dot G}(\dot\psi_v, s_{\dot\psi}\dot x_v)$. It equals

$$\beta(\phi_v)(D_v \dot f_{v,\dot G})'(\phi_v, s_{\dot\psi}\dot x_v) = \beta(\phi_v)(D_v \dot f_{v,\dot G})(\phi_v, s_{\dot\psi}\dot x_v),$$

by (7.1.11) and the local intertwining relation for the generic parameter ϕ_v. This in turn can be written as

$$\beta(\phi_v) \sum_{\pi_v \in \widetilde\Pi_{\phi_v}(M_v)} (D_v\, \dot f_{v,\dot G})\,(\phi_v, s_{\dot\psi}\dot x_v, \pi_v)$$

in the notation (7.1.13), or what is the same thing, the product of $\beta(\phi_v)$ with the analogue for $D_v\dot f_{v,\dot G}$, ϕ_v and $s_{\dot\psi}\dot x_v$ of the expression (5.4.8). (We are writing $\widetilde\Pi_{\phi_v}(M_v)$ for the M_v-packet here, as in (5.4.8), rather than $\widetilde\Pi_{\phi_v,M_v}$ as in (7.1.13). The Levi subgroup M_v is chosen in terms of ϕ_v, and is not generally equal to the localization $\dot M_v$ of Proposition 7.2.1.) As in the earlier discussion of (5.4.8), we can then express $\dot f'_{v,\dot G}(\dot\psi_v, s_{\dot\psi}\dot x_v)$ as a general sum

$$\sum_{\pi_v \in \widetilde\Pi_{\phi_v}(M_v)} \alpha'(\pi_v, \dot x_v)\,(D_v\dot f_{v,\dot G})\big(M_v, \pi_v, r_v(x)\big),$$

following the definition (3.5.3). It is understood that $\alpha'(\pi_v, \dot x_v)$ is a complex coefficient, that

$$r_v(x) = r_v(s_\psi x)$$

is the image of x in the local R-group R_{ϕ_v}, and that $\big(M_v, \pi_v, r_v(x)\big)$ represents an element in the set $T(\dot G_v)$, as in the proof of Lemma 5.4.3.

The other linear form is $\dot f_{v,\dot G}(\dot\psi_v, \dot x_v)$. It can be written as

$$\sum_{\pi_v \in \widetilde\Pi_{\phi_v}(M_v)} \dot f_{v,\dot G}(\dot\psi_v, \dot x_v, \widehat\pi_v)$$

in the notation (7.1.13), or equivalently, the analogue for $\dot f_{v,\dot G}$, $\dot\psi_v$ and $\dot x_v$ of (5.4.8). Our expectation (7.1.14) is that these expressions are equal to their analogues above. However, we do not need this finer property here. We can instead use the weaker formula (7.1.15) obtained from [**Ban**]. It asserts that

$$\dot f_{v,\dot G}(\dot\psi_v, \dot x_v, \widehat\pi_v) = \varepsilon_{\pi_v}(x)\,(D_v\dot f_{v,\dot G})\,(\phi_v, \dot x_v, \pi_v),$$

for any $\pi_v \in \widetilde{\Pi}_{\phi_v}(M_v)$ with associated sign $\varepsilon_{\pi_v}(x)$. We may therefore express $\dot{f}_{v,\dot{G}}(\dot{\psi}_v, \dot{x}_v)$ as a general sum

$$\sum_{\pi_v \in \widetilde{\Pi}_{\phi_v}(M_v)} \alpha(\pi_v, \dot{x}_v)\,(D_v\, \dot{f}_{v,\dot{G}})\,\big(M_v, \pi_v, r_v(x)\big),$$

according again to the definition (3.5.3), and for another complex coefficient $\alpha(\pi_v, \dot{x}_v)$.

We can now write the original expression (7.3.2) in the desired form. For we see from the formulas above that (7.3.2) is equal to

$$\sum_{\tau_v \in T(\dot{G}_v)} \delta(\tau_v, \dot{f}^v)(D_v \dot{f}_{v,\dot{G}})(\tau_v), \tag{7.3.4}$$

where $\delta(\tau_v, \dot{f}^v)$ equals the sum

$$c \sum_{x} \sum_{\pi_v} \varepsilon^{\dot{G}}_{\dot{\psi}}(\dot{x})\,\big(\alpha'(\pi_v, \dot{x}_v)(\dot{f}^v)'_{\dot{G}}(\dot{\psi}^v, s_{\dot{\psi}}\dot{x}^v) - \alpha(\pi_v, \dot{x}_v)\, \dot{f}^v_{\dot{G}}(\dot{\psi}^v, \dot{x}^v)\big),$$

over elements $x \in \mathcal{S}_{\psi,\mathrm{ell}}$ and $\pi_v \in \widetilde{\Pi}_{\phi_v}(M_v)$ such that the triplet $\big(M_v, \pi_v, r_v(x)\big)$ belongs to the $W_0^{\dot{G}_v}$-orbit represented by τ_v.

We now apply Lemmas 5.2.1 and 5.2.2. Together, they assert that if $\{f^*\}$ is any global compatible family of functions (5.2.6) on the groups $G^* \in \dot{\widetilde{\mathcal{E}}}_{\mathrm{sim}}(N)$, the sum

$$\sum_{G^* \in \dot{\widetilde{\mathcal{E}}}_{\mathrm{sim}}(N)} \tilde{\iota}(N, G^*)\, \mathrm{tr}\big(R^*_{\mathrm{disc},\dot{\psi}}(f^*)\big), \tag{7.3.5}$$

added to (7.3.3) in case $r = 1$, equals (7.3.2). We are following the earlier convention of writing $\dot{f}$ or $\dot{f}^\vee$ in place of f^* in case G^* equals $\dot{G}$ or $\dot{G}^\vee$. The expression (7.3.5) is of course a nonnegative linear combination of irreducible characters on the groups $G^*(\dot{\mathbb{A}})$. In the case $r = 1$, the expression (7.3.3) equals the analogue

$$\tfrac{1}{8}\,\mathrm{tr}\big(\big(1 - R_{P^\vee}(w_1, \widetilde{\pi}_{\dot{\psi}_1}, \dot{\psi}_1)\big)\, \mathcal{I}_{P^\vee}(\dot{\pi}_{\dot{\psi}_1}, \dot{f}^\vee)\big) \tag{7.3.6}$$

for $\dot{\widetilde{\mathcal{F}}}$ of the expression (5.4.11) obtained in the proof of Lemma 5.4.4. It too is a nonnegative linear combination of irreducible characters, in this case on the group $\dot{G}^\vee(\dot{\mathbb{A}})$. We can therefore write the sum of (7.3.5) with (7.3.3) as

$$\sum_{G^*} \sum_{\pi^*} c_{G^*}(\pi^*)\, f^*_{G^*}(\pi^*), \qquad G^* \in \dot{\widetilde{\mathcal{E}}}_{\mathrm{sim}}(N),\ \pi^* \in \widetilde{\Pi}(G^*),$$

for nonnegative coefficients $c_{G^*}(\pi^*)$.

For any $G^* \in \dot{\widetilde{\mathcal{E}}}_{\mathrm{ell}}(N)$, let D^*_v be the duality operator on G^*_v. At this point, we will actually be concerned with the "absolute value"

$$\widehat{D}^*_v(\pi^*_v) = \beta(\pi^*_v)\, D^*_v(\pi^*_v) = \widehat{\pi}^*_v, \qquad \pi^*_v \in \Pi(G^*_v),$$

of D_v^*, where $\widehat{\pi}_v^*$ is the dual (7.1.4) of the representation π_v^*. This extends to an involution

$$\widehat{D}_v^*(\pi^*) = \widehat{D}_v^*(\pi_v^*) \otimes \pi^{*,v} = \widehat{\pi}_v^* \otimes \pi^{*,v}, \qquad \pi^* = \pi_v^* \otimes \pi^{*,v},$$

on $\widetilde{\Pi}(G^*)$, and an adjoint involution

$$(\widehat{D}_v^* f_{G^*}^*)(\pi^*) = f_{G^*}^*(\widehat{D}_v^* \pi^*) = f_{G^*}^*(\widehat{\pi}_v^* \otimes \pi^{*,v})$$

on $\widetilde{\mathcal{I}}(G^*)$. Our sum of (7.3.5) and (7.3.3) can then be written as

$$\sum_{G^*}\sum_{\pi^*} \widehat{c}_{G^*}(\pi^*)\,(\widehat{D}_v^* f_{G^*}^*)(\pi^*), \qquad G^* \in \dot{\widetilde{\mathcal{E}}}_{\mathrm{sim}}(N),\ \pi^* \in \widetilde{\Pi}(G^*), \tag{7.3.7}$$

for nonnegative coefficients

$$\widehat{c}_{G^*}(\pi^*) = c_{G^*}(\widehat{D}_v^* \pi^*).$$

The earlier expression (7.3.4) takes the form

$$\sum_{\tau_v \in T(\dot{G}_v)} \widehat{\delta}(\tau_v, \dot{f}^v)(\widehat{D}_v \dot{f}_{v,\dot{G}})(\tau_v), \tag{7.3.8}$$

where

$$\widehat{\delta}(\tau_v, \dot{f}^v) = \beta(\tau_v)\,\delta(\tau_v, \dot{f}^v)$$

and

$$\beta(\tau_v) = \beta(\pi_v), \qquad \tau_v = \big(M_v, \pi_v, r_v(x)\big).$$

The assertions of Lemmas 5.2.1 and 5.2.2 may thus be characterized as an identity between the expressions (7.3.7) and (7.3.8), for any compatible family of functions $\{f^*\}$. But it is easy to see from the definitions in §2.1 and §3.4 that the family of operators $\{\widehat{D}_v^*\}$ defines a correspondence of compatible families. To be precise, suppose that $\{f^*\}$ and $\{\widehat{f}^*\}$ are two families of functions on the spaces $\widetilde{\mathcal{H}}(G^*)$ such that

$$\widehat{D}_v^* f_{G^*}^* = \widehat{f}_{G^*}^*, \qquad G^* \in \dot{\widetilde{\mathcal{E}}}_{\mathrm{sim}}(N).$$

Then if $\{f^*\}$ is a compatible family, the same is true of $\{\widehat{f}^*\}$. The identity between (7.3.7) and (7.3.8) therefore remains valid, with the functions $\widehat{D}_v^* f_{G^*}^*$ replaced by $\widehat{f}_{G^*}^*$, for any compatible family $\{\widehat{f}^*\}$.

We can now apply Corollary 3.5.3. It tells us that the coefficients $\widehat{c}_{G^*}(\pi^*)$ and $\widehat{\delta}(\dot{\tau}_v, \dot{f}^v)$ all vanish. Since (7.3.2) equals (7.3.8), it vanishes, as asserted. Corollary 3.5.3 likewise tells us that (7.3.6) vanishes, as do any of the summands in (7.3.5). Since (7.3.3) equals (7.3.6), it also vanishes. This is the second assertion of the lemma. □

The second lemma gives the nongeneric refinements of Lemmas 5.3.1 and 5.3.2, for parameters (7.2.9) in the subset $\widetilde{\mathcal{F}}_2(\dot{G})$ of $\dot{\widetilde{\mathcal{F}}}$. It will be used in the next section to establish what is left of Theorem 2.2.1.

LEMMA 7.3.2. *Suppose that*

$$(\dot{G}, \dot{\psi}), \qquad \dot{G} \in \dot{\widetilde{\mathcal{E}}}_{\mathrm{sim}}(N),\ \dot{\psi} \in \widetilde{\mathcal{F}}_2(\dot{G}),$$

is an analogue for the nongeneric family $\dot{\widetilde{\mathcal{F}}}$ *of the pair given in either Lemmas 5.3.1 or 5.3.2. Then*

$$S^{\dot{G}}_{\mathrm{disc},\dot{\psi}}(\dot{f}) = |\Psi(\dot{G}, \dot{\psi})|\, |\mathcal{S}_{\dot{\psi}}|^{-1}\, \varepsilon^{\dot{G}}(\dot{\psi})\, \dot{f}^{\dot{G}}(\dot{\psi}), \tag{7.3.9}$$

for any function $\dot{f} \in \widetilde{\mathcal{H}}(\dot{G})$.

PROOF. This lemma is an analogue for $\dot{\widetilde{\mathcal{F}}}$ of Lemmas 5.4.5 and 5.4.6. It is stated with the implicit assumption of the condition of Lemma 5.3.1, namely that the global theorems are valid for parameters in the complement of $\dot{\mathcal{F}}_{\mathrm{ell}}(N)$ in $\dot{\widetilde{\mathcal{F}}}(N)$.

The required identity (7.3.9) is the stable multiplicity formula for $\dot{\psi}$. We can assume that N is even and $\eta_{\dot{\psi}} = 1$, since the formula is otherwise valid by Corollary 5.1.3. It then follows from Lemma 5.1.4 that the left hand side minus the right hand side is a multiple of the linear form

$$\dot{f} \longrightarrow \dot{f}^{L}(\dot{\Lambda}), \qquad \dot{f} \in \widetilde{\mathcal{H}}(\dot{G}),$$

on $\widetilde{\mathcal{H}}(\dot{G})$ obtained from the maximal Levi subgroup $L(\dot{\mathbb{A}})$ of $\dot{G}(\dot{\mathbb{A}})$. (We are writing $\dot{L}$ and $\dot{\Lambda}$ in place of L and Λ here, to be consistent with the ambient notation.) We must show that $\dot{\Lambda}$ vanishes.

As we agreed at the end of the last section, we introduce the simple datum $\dot{G}_+ \in \dot{\widetilde{\mathcal{E}}}_{\mathrm{sim}}(N_+)$, for $N_+ = N_1 + N$, and the parameter

$$\dot{\psi}_+ = \dot{\psi}_1 \boxplus \dot{\psi}$$

in $\dot{\widetilde{\mathcal{F}}}_{\mathrm{ell}}(\dot{G}_+)$. We have then to modify the proofs of Lemmas 5.4.5 and 5.4.6 in the manner of the last lemma. This entails working with compatible families of functions (5.3.3) on the groups $G^* \in \dot{\widetilde{\mathcal{E}}}_{\mathrm{sim}}(N_+)$, and the correspondence on such objects under the local duality operators $\widehat{D}^*_{+,v}$. We shall be brief.

We first introduce the composite endoscopic datum $\dot{G}_1^\vee = \dot{G}_1 \times \dot{G}^\vee$ in $\dot{\widetilde{\mathcal{E}}}_{\mathrm{sim}}(N_+)$, with corresponding function $\dot{f}_1 \in \widetilde{\mathcal{H}}(\dot{G}_1^\vee)$ in the compatible family (5.3.3). We are including the case $r = 1$ here, in which $\dot{G}_1 = \dot{G}$. It follows from the definitions that the supplementary summands in the expressions (5.3.4) and (5.3.21) of Lemmas 5.3.1 and 5.3.2 can be written

$$b_+ \dot{f}_1^{\dot{L}_+}(\dot{\psi}_1 \times \dot{\Lambda})$$

and

$$\tfrac{1}{2}\, \dot{f}_1^{\dot{L}_+}(\dot{\Gamma} \times \dot{\Lambda}),$$

Respectively. We then apply Corollary 3.5.3 as in Lemmas 5.4.5 and 5.4.6, in combination with the operators $\widehat{D}^*_v$. This leads to the conclusion that the linear forms $\dot{\psi}_1 \times \dot{\Lambda}$ and $\dot{\Gamma} \times \dot{\Lambda}$, attached to the maximal Levi subgroup $\dot{L}_+ = \dot{G}_1 \times \dot{L}$ of $\dot{G}_1^\vee$ in the respective cases $r > 1$ and $r = 1$, vanish.

If $r > 1$, the linear form $\dot{\psi}_1$ attached to $\dot{G}_1$ is nonzero, according to our induction hypothesis. Therefore $\dot{\Lambda}$ vanishes in this case, as required. Suppose then that $r = 1$. If $\dot{\Gamma}$ vanishes, the linear form

$$\dot{f}(\dot{\psi}) = \dot{f}^{\dot{G}}(\dot{\Gamma}) + \dot{f}^{\dot{L}}(\dot{\Lambda}) = \dot{f}^{\dot{L}}(\dot{\Lambda}), \qquad \dot{f} \in \widetilde{\mathcal{H}}(\dot{G}),$$

is induced from the Levi subgroup $\dot{L}(\mathbb{A})$. This contradicts the first assertion of Proposition 7.2.1(iii)(a), which is the analogue of the local generic condition (5.4.1)(a), as one sees easily from the properties of $\widehat{D}_v^*$. Therefore, it is again the linear form $\dot{\Lambda}$ that vanishes, as required. □

We shall now establish Theorem 2.4.1 for nongeneric parameters $\psi \in \widetilde{\Psi}(G)$. To recall the general strategy, we go back to the discussion for generic parameters at the beginning of §6.4. As we noted there, the reduction of the global intertwining relation in §4.5 can be made in the context of the family $\dot{\widetilde{\mathcal{F}}}$, defined now by (7.2.7). The reduction also carries over to the local intertwining relation of Theorem 2.4.1. In other words, the local intertwining relation holds for a parameter $\psi \in \widetilde{\Psi}(N)$, unless ψ belongs to $\widetilde{\Psi}_{\mathrm{ell}}(G^*)$ for some $G^* \in \mathcal{E}_{\mathrm{sim}}(N)$, or belongs to the local analogue for ψ of one of the two exceptional cases (4.5.11) or (4.5.12). As before, we prove the three exceptional cases together. Again, it is sufficient to establish the assertion of Theorem 2.4.1 for elements s and u that map to a point x in the set $\mathcal{S}_{\psi,\mathrm{ell}}$.

In the first case, we fix $G \in \widetilde{\mathcal{E}}_{\mathrm{sim}}(N)$ with $\psi \in \widetilde{\Psi}_{\mathrm{ell}}(G)$. The conditions are

$$\begin{cases} \psi = 2\psi_1 \oplus \cdots \oplus 2\psi_q \oplus \psi_{q+1} \oplus \cdots \oplus \psi_r, \\ S_\psi = \big(O(2,\mathbb{C})^q \times O(1,\mathbb{C})^{r-q}\big)_\psi^+, \qquad q \geqslant 1, \end{cases} \tag{7.3.10}$$

with the requirement that the Weyl group W_ψ contain an element w in $W_{\psi,\mathrm{reg}}$. Proposition 7.2.1 then gives us the global family $\dot{\widetilde{\mathcal{F}}}$, together with a global pair $(\dot{G}, \dot{\psi})$ such that $\dot{\psi} \in \widetilde{\mathcal{F}}^2_{\mathrm{ell}}(\dot{G})$ is of the form (7.2.8). The global pair satisfies the condition of Lemma 7.3.1. This lemma then gives us an identity

$$\sum_{x \in \mathcal{S}_{\psi,\mathrm{ell}}} \varepsilon_{\dot{\psi}}^{\dot{G}}(\dot{x}) \big(\dot{f}'_{\dot{G}}(\dot{\psi}, s_{\dot{\psi}}\dot{x}) - \dot{f}_{\dot{G}}(\dot{\psi}, \dot{x})\big) = 0, \qquad \dot{\widetilde{f}} \in \widetilde{\mathcal{H}}(\dot{G}). \tag{7.3.11}$$

The second and third cases apply to local parameters $\psi \in \widetilde{\Psi}(G)$, for $G \in \widetilde{\mathcal{E}}_{\mathrm{sim}}(N)$, that are of the local forms of (4.5.11) and (4.5.12). In each of these two cases, Proposition 7.2.1 gives a global family $\dot{\widetilde{\mathcal{F}}}$, together with a global pair $(\dot{G}, \dot{\psi})$ such that $\dot{\psi} \in \widetilde{\mathcal{F}}(\dot{G})$ has the corresponding global form (4.5.11) or (4.5.12). In both cases, Corollary 4.5.2 again tells us that (7.3.11) is valid as stated. Recall that if $q = r = 1$ in the original case (7.3.10), there is a second group $G^\vee \in \widetilde{\mathcal{E}}_{\mathrm{sim}}(N)$ such that $\psi \in \widetilde{\Psi}(G^\vee)$. The corresponding global pair $(\dot{G}^\vee, \dot{\psi})$ then belongs to the second exceptional case (4.5.11), and the global formula (7.3.11) in this case amounts to the vanishing of (7.3.3) established in Lemma 7.3.1. We can therefore include it in our discussion of (4.5.11), as we did for the generic parameters treated in §6.4.

We have thus to extract the remaining local intertwining relation

$$f'_G(\psi, s_\psi x) = f_G(\psi, x), \qquad x \in \mathcal{S}_{\psi,\mathrm{ell}},\ f \in \widetilde{\mathcal{H}}(G),$$

from the global identity (7.3.11), when ψ represents one of the three exceptional cases above. As in §6.4, we take

$$\dot{f} = \prod_v \dot{f}_v$$

to be a decomposable function. The summand of x in (7.3.11) then decomposes into a difference of products

$$\varepsilon_{\dot{\psi}}^{\dot{G}}(\dot{x})\Big(\prod_v \dot{f}'_{v,\dot{G}}(\dot{\psi}_v, s_{\dot{\psi}}\dot{x}_v) - \prod_v \dot{f}_{v,\dot{G}}(\dot{\psi}_v, \dot{x}_v)\Big) \tag{7.3.12}$$

over all valuations v. The first step is to remove the contributions to these products from valuations in the complement of the finite set $S(U,V)$.

We shall establish a local lemma that can be applied to the places $v \in V$, as well as those in the complement of $S(U,V)$. To simplify the notation, we shall formulate it in terms of the basic local pair (G,ψ) over F, specialized by the requirement that $\psi = \widehat{\phi}$ for some $\phi \in \widetilde{\Phi}(G)$. This of course implies that F is p-adic. For any such ϕ, we write $\widetilde{\Pi}^G_{\phi,M}$ for the subset of representations $\pi \in \widetilde{\Pi}_\phi$ that lie in $\widetilde{\Pi}_\phi(\pi_M)$, the set of irreducible constituents of the induced representation $\mathcal{I}_P(\pi_M)$, for some π_M in the set $\widetilde{\Pi}^G_{\phi_M}$ defined prior to (7.1.16).

LEMMA 7.3.3. *Assume that F is p-adic, and that $\psi = \widehat{\phi}$ for some $\phi \in \widetilde{\Phi}(G)$. Then if $f \in \widetilde{\mathcal{H}}(G)$ is such that the function $f_G(\widehat{\pi})$ of $\pi \in \widetilde{\Pi}_\phi$ is supported on the subset $\widetilde{\Pi}^G_{\phi,M}$, we have*

$$f'_G(\psi, s_\psi x) = f_G(\psi, x), \qquad x \in \mathcal{S}_\psi.$$

Moreover, if ψ is tamely ramified and quadratic, and f satisfies the stronger condition that

$$f_G(\widehat{\pi}) = \begin{cases} 1, & \text{if } \langle\,\cdot\,,\pi\rangle = 1,\\ 0, & \text{otherwise,}\end{cases} \tag{7.3.13}$$

for any $\pi \in \widetilde{\Pi}_\phi$, then

$$f'_G(\psi, s_\psi x) = f_G(\psi, x) = 1, \qquad x \in \mathcal{S}_\psi.$$

PROOF. Using (7.1.11) and the generic form of Theorem 2.4.1, we write

$$\begin{aligned} f'_G(\psi, s_\psi x) &= \beta(\phi)\,(Df_G)'\,(\phi, s_\psi x)\\ &= \beta(\phi)\,(Df_G)\,(\phi, s_\psi x), \end{aligned}$$

as in §7.1. Our condition on f implies that the function

$$(Df_G)(\pi) = f_G(D\pi) = \beta(\pi)\, f_G(\widehat{\pi}), \qquad \pi \in \widetilde{\Pi}_\phi,$$

is supported on the subset $\widetilde{\Pi}^G_{\phi,M}$ of $\widetilde{\Pi}_\phi$. It then follows from the formula (7.1.13), together with the definition of $\widetilde{\Pi}^G_{\phi,M}$, that the function

$$(Df_G)(\phi, s_\psi u, \pi_M), \qquad \pi_M \in \widetilde{\Pi}_{\phi_M},$$

is supported on the subset $\widetilde{\Pi}^G_{\phi_M}$ of $\widetilde{\Pi}_{\phi_M}$. (We recall that u is any element in the group $\mathfrak{N}_\psi = \mathfrak{N}_\phi$ whose image in $\mathcal{S}_\psi = \mathcal{S}_\phi$ equals x.) We can therefore write

$$\begin{aligned}
&\beta(\phi)\,(Df_G)(\phi, s_\psi u) \\
&= \beta(\phi) \sum_{\pi_M \in \widetilde{\Pi}_{\phi_M}} (Df_G)(\phi, s_\psi u, \pi_M) \\
&= \beta(\phi) \sum_{\pi_M \in \widetilde{\Pi}^G_{\phi_M}} (Df_G)(\phi, s_\psi u, \pi_M) \\
&= \sum_{\pi_M \in \widetilde{\Pi}^G_{\phi_M}} f_G(\psi, u, \widehat{\pi}_M) \\
&= \sum_{\pi_M \in \widetilde{\Pi}_{\phi_M}} f_G(\psi, u, \widehat{\pi}_M) \\
&= f_G(\psi, u),
\end{aligned}$$

since the identity (7.1.14) holds by definition for representations $\pi_M \in \widetilde{\Pi}^G_{\phi_M}$. We have established that

$$f'_G(\psi, s_\psi x) = f_G(\psi, u) = f_G(\psi, x),$$

as required.

Suppose now that ψ is tamely ramified and quadratic. It then follows from Lemma 7.1.2(b) that (7.3.13) does indeed imply the first condition on f. In particular, the functions $f'_G(\psi, s_\psi x)$ and $f_G(\psi, x)$ are equal. To see that they are equal to 1, we have only to observe that

$$f'_G(\psi, s_\psi x) = \sum_{\pi \in \widetilde{\Pi}_\phi} \langle x, \widehat{\pi}\rangle\, f_G(\widehat{\pi}) = 1,$$

by (7.1.2), Lemma 7.1.1 and the fact that (7.1.8) holds by our induction hypothesis on M, which implies in turn that it is also valid for the constituents of representations induced from M. This completes the proof of the lemma. □

REMARK. Suppose that $\psi = \widehat{\phi}$ for an arbitrary parameter $\phi \in \widetilde{\Phi}(G)$, as in the lemma. Then for any complex valued function a on $\widehat{\mathcal{S}}_\phi$, there is a function $f \in \widetilde{\mathcal{H}}(G)$ such that

$$a(\langle \cdot, \pi \rangle) = f_G(\widehat{\pi}), \qquad \pi \in \widetilde{\Pi}_\phi.$$

This follows by duality from the assertion (b) of Theorem 1.5.1 (which we have now proved). In particular, we can always choose f so that the condition (7.3.13) holds.

We will need another lemma for the unramified places v, which will again be stated in terms of (G, ψ). Its proof does not rely on the induction hypothesis that $M \neq G$.

LEMMA 7.3.4. *Suppose that (G, ψ) is unramified over F, and that f is the characteristic function of a hyperspecial maximal compact subgroup of $G(F)$. Then f satisfies the condition (7.3.13) of Lemma 7.3.3.*

PROOF. The condition that ψ be unramified means that the restriction of ψ to L_F is an unramified unitary representation of W_F (and in particular, is trivial on the $SU(2)$ factor of L_F). This implies that ψ is the dual of a tamely ramified, quadratic parameter in $\widetilde{\Phi}_{\mathrm{bdd}}(G)$.

We shall first establish that

$$f'_G(\psi, x) = 1, \qquad x \in \mathcal{S}_\psi. \tag{7.3.14}$$

We recall that $f'_G(\psi, x)$ equals $f'(\psi')$, for any pair (G', ψ') that maps to (ψ, x). The fundamental lemma for (G, G') asserts that f' is the image in $\widetilde{\mathcal{S}}(G')$ of the characteristic function of a hyperspecial maximal compact subgroup of $G'(F)$. Since we can assume inductively that $f'(\psi') = 1$ whenever $x \neq 1$, it will be enough to show that

$$f'_G(\psi, 1) = f^G(\psi) = 1.$$

Let $\widetilde{f} = \widetilde{f}^0 \rtimes \widetilde{\theta}$, where $\widetilde{f}^0$ is the characteristic function of the standard maximal compact subgroup of $GL(N, F)$. The fundamental lemma for $(\widetilde{G}(N), G)$ asserts that $\widetilde{f}^G$ equals the image f^G of f in $\widetilde{\mathcal{H}}(G)$. It follows that

$$f^G(\psi) = \widetilde{f}^G(\psi) = \widetilde{f}_N(\psi).$$

But it is easy to see from the definition (2.2.1), and the properties of Whittaker functionals for $GL(N, F)$ discussed in §2.5, that $\widetilde{f}_N(\psi)$ equals 1. The identity (7.3.14) follows.

We need to be a little careful here, as I was reminded by the referee. The proof of the two fundamental lemmas by Ngo and Waldspurger applies only to the case that the residual characteristic of F is large. However, it is easily extended to the general case by using the generic forms of the local theorems we have now proved (results which of course would have been unavailable without the original two fundamental lemmas). We shall sketch a separate argument, even though it would also be possible to appeal directly to the unramified assertion of Theorem 1.5.1(a) we established at the beginning of §6.6.

If we combine the identity (2.2.6) of Theorem 2.2.1 with the last assertion of Theorem 1.5.1(a), keeping in mind the remarks in §2.5 and §6.1 on spherical functions, we see that

$$f^G(\phi) = \begin{cases} 1, & \text{if } \phi \text{ is unramified,} \\ 0, & \text{otherwise,} \end{cases}$$

for any $\phi \in \widetilde{\Phi}_{\mathrm{bdd}}(G)$. This characterizes the image of f in $\widetilde{\mathcal{S}}(G)$ as a function on $\widetilde{\Phi}_{\mathrm{bdd}}(G)$. We also see more generally that

$$f'(\phi') = \begin{cases} 1, & \text{if } \phi \text{ is unramified,} \\ 0, & \text{otherwise,} \end{cases}$$

if (G', ϕ') is a preimage of a given pair (ϕ, x). Since the pair (G', ϕ') is unramified if and only if the same is true of ϕ, we obtain the formula

$$f'(\phi') = \begin{cases} 1, & \text{if } \phi' \text{ is unramified,} \\ 0, & \text{otherwise,} \end{cases}$$

for $G' \in \mathcal{E}(G)$ unramified and any $\phi' \in \widetilde{\Phi}_{\mathrm{bdd}}(G')$. Therefore, f' is indeed the image in $\widetilde{\mathcal{S}}(G')$ of the characteristic function of a hyperspecial subgroup of $G'(F)$, as asserted by the fundamental lemma for (G, G'). To treat the twisted fundamental lemma, we use the formula

$$\widetilde{f}_N(\phi) = \begin{cases} 1, & \text{if } \phi \text{ is unramified,} \\ 0, & \text{otherwise,} \end{cases}$$

which holds for any $\phi \in \widetilde{\Phi}_{\mathrm{bdd}}(N)$. This is not hard to establish from the unramified case of (i) from Corollary 2.5.2. It tells us that $\widetilde{f}^G$ equals the image f^G of f in $\widetilde{\mathcal{S}}(G)$, as asserted by the fundamental lemma for $\big(\widetilde{G}(N), G\big)$. The two fundamental lemmas thus hold in general, and the identity (7.3.14) is therefore valid with no restriction on the residual characteristic of F.

To finish the proof of the lemma, we take ϕ to be the tamely ramified, quadratic parameter in $\widetilde{\Phi}_{\mathrm{bdd}}(G)$ such that $\psi = \widehat{\phi}$. We must show that f satisfies (7.3.13). To this end, we write

$$f'(\psi, x) = f'(\psi') = \sum_{\pi \in \widetilde{\Pi}_\phi} \langle s_\psi x, \pi \rangle f_G(\sigma_\pi),$$

by (7.1.2). It then follows from (7.3.14) that

$$f_G(\sigma_\pi) = \begin{cases} 1, & \text{if } \langle \cdot, \pi \rangle = 1, \\ 0, & \text{otherwise,} \end{cases}$$

for any $\pi \in \widetilde{\Pi}_\phi$. The required condition (7.3.13) is thus a consequence of Lemma 7.1.2(a), which tells us that $\sigma_\pi = \widehat{\pi}$ if $\langle \cdot, \pi \rangle = 1$. □

We return to our proof of the local intertwining relation, in which $\psi \in \widetilde{\Psi}(G)$ falls into one of the three exceptional cases (7.3.10), (4.5.11) or (4.5.12). The global identity (7.3.11) asserts that the sum over $x \in \mathcal{S}_{\psi,\mathrm{ell}}$ of (7.3.12) vanishes. We will use this identity to show that the two factors for $v = u$ in the two products in (7.3.12) are equal. The process will be considerably easier than its counterpart from §6.4, thanks to Lemma 7.3.3.

If v is a valuation of $\dot{F}$ in the complement of U, Proposition 7.2.1 asserts that $\dot{\psi}_v = \phi_v$, for a tamely ramified, quadratic, generic parameter $\phi_v \in \widetilde{\Phi}(\dot{G}_v)$. Suppose first that v lies in the complement of both U and V. In this case, we choose the function $\dot{f}_v \in \widetilde{\mathcal{H}}(\dot{G}_v)$ so that it satisfies the analogue for ϕ_v of the condition (7.3.13). In the special case that $\dot{\psi}_v$ is unramified, we can also assume that $\dot{f}_v$ is the characteristic function of a hyperspecial maximal compact subgroup of $\dot{G}(\dot{F}_v)$, by Lemma 7.3.4. It then follows from Lemma 7.3.3 that the corresponding two factors in (7.3.12) are each equal to 1. We may therefore remove them from the products.

We have established that the two products in (7.3.12) can each be taken over the finite set of valuations $v \in S(U, V)$. If v belongs to V, we choose $\dot{f}_v$ so that it satisfies the first condition of Lemma 7.3.3, namely that the function $\dot{f}_{v,\dot{G}}(\widetilde{\pi}_v)$ of $\pi_v \in \widetilde{\Pi}_{\phi_v}$ is supported on the subset $\widetilde{\Pi}^G_{\phi_v,M}$. It then follows from the lemma that

$$\dot{f}'_{v,\dot{G}}(\dot{\psi}_v, s_{\dot{\psi}}\dot{x}_v) = \dot{f}_{v,\dot{G}}(\dot{\psi}_v, \dot{x}_v), \qquad x \in \mathcal{S}_{\psi,\mathrm{ell}}.$$

The identity (7.3.11) reduces to

$$\sum_{x \in \mathcal{S}_{\dot{\psi},\mathrm{ell}}} \varepsilon^{\dot{G}}_{\dot{\psi}}(\dot{x}) \left(\dot{f}'_{U,\dot{G}}(\dot{\psi}_U, s_{\dot{\psi}}\dot{x}) - \dot{f}_{U,\dot{G}}(\dot{\psi}_U, \dot{x})\right) \dot{f}'_V(\dot{\psi}_V, s_{\dot{\psi}}\dot{x}),$$

where

$$\dot{f}'_V(\dot{\psi}_V, s_{\dot{\psi}}\dot{x}) = \prod_{v \in V} \dot{f}'_{v,\dot{G}}(\dot{\psi}_v, s_{\dot{\psi}}\dot{x}_v),$$

while $\dot{f}'_{U,\dot{G}}(\dot{\psi}_U, s_{\dot{\psi}}\dot{x})$ and $\dot{f}_{U,\dot{G}}(\dot{\psi}_U, \dot{x})$ are defined by corresponding products over $v \in U$. It is a consequence of Proposition 7.2.1(iii)(a) that as $\dot{f}_V$ varies under the given constraints, the functions

$$x \longrightarrow \dot{f}'_V(\dot{\psi}_V, s_{\dot{\psi}}\dot{x}), \qquad x \in \mathcal{S}_{\psi,\mathrm{ell}},$$

span the space of all functions on $\mathcal{S}_{\psi}$. It follows that

$$\dot{f}'_{U,\dot{G}}(\dot{\psi}_U, s_{\dot{\psi}}\dot{x}) = \dot{f}_{U,\dot{G}}(\dot{\psi}_U, \dot{x}), \qquad x \in \mathcal{S}_{\psi,\mathrm{ell}}, \tag{7.3.15}$$

for any function $\dot{f}_U \in \widetilde{\mathcal{H}}(\dot{G}_U)$.

It remains to separate the contribution of u to each side of (7.3.15) from that of its complement S^u_∞ in U. To do so, we need only recall how we chose the global field $\dot{F}$ prior to Proposition 7.2.1.

If $F = \mathbb{C}$, $\dot{F}$ is any imaginary quadratic field. Then U consists of the one valuation u, and (7.3.15) is simply the required identity (2.4.7) for the local parameter $\psi = \dot{\psi}_u$. If $F = \mathbb{R}$, and the simple generic constituents

$$\{\mu_i : \ m_i = 2, \ 1 \leqslant i \leqslant r\} \tag{7.3.16}$$

of ψ are in relative general position, $\dot{F}$ equals $\mathbb{Q}$. Again U consists of the one valuation u, and (7.3.15) is the required identity for $\psi = \dot{\psi}_u$. If $F = \mathbb{R}$, but the constituents (7.3.16) are not in relative general position, or if F is p-adic, we can take $\dot{F}$ to be a totally real field with several archimedean places. In these two cases, the corresponding global parameter $\dot{\psi}$ was chosen in the proof of Proposition 7.2.1 so that for any $v \in S^v_\infty$, the analogues for $\dot{\psi}_v$ of the generic constituents (7.3.16) are in general position. It follows from the case treated above that

$$\dot{f}'_v(\dot{\psi}_v, s_{\dot{\psi}}\dot{x}_v) = \dot{f}_v(\dot{\psi}_v, \dot{x}_v), \qquad \dot{f}_v \in \widetilde{\mathcal{H}}(\dot{G}_v),$$

for any $v \in S^u_\infty$ and $x \in \mathcal{S}_{\psi,\mathrm{ell}}$. These linear forms are all nonzero, as one can see by relating the left hand side to $GL(N, \dot{F}_v)$, according to the definitions (2.4.6), (2.2.3) and (2.2.1). We may therefore remove their contributions to (7.3.15). This leaves only the contribution of the complementary valuation $v = u$. The formula (7.3.15) thus reduces to the required identity for $\psi = \dot{\psi}_u$ in these remaining two cases.

We have now established the local intertwining relation for parameters $\psi \in \widetilde{\Psi}(G)$ in any of the three outstanding cases (7.3.10), (4.5.11) or (4.5.12). This completes the proof of Theorem 2.4.1 for any parameter $\psi \in \widetilde{\Psi}(G)$ over F.

We also obtain the following corollary, which is proved in the same way as its generic analogue Corollary 6.4.5.

COROLLARY 7.3.5. *The canonical self-intertwining operator attached to any* $\psi \in \widetilde{\Psi}(G)$, $w \in W^0_\psi$ *and* $\pi_M \in \widetilde{\Pi}_{\psi_M}$ *satisfies*

$$R_P(w, \widetilde{\pi}_M, \psi) = 1. \qquad \square$$

7.4. Local packets for composite and simple ψ

We are at last ready to complete our proof of the local theorems. They consist of five local theorems stated in Chapters 1 and 2. Having established everything for generic local parameters $\psi = \phi$ in Chapter 6 (including the generic supplement Theorem 6.1.1), we are working now with nongeneric local parameters ψ. We dealt with the local intertwining relation of Theorem 2.4.1 in the last section. We shall sketch a proof of its supplement Theorem 2.4.4 at the end of this section. Since the nongeneric form of Theorem 1.5.1 amounts to a less precise form of Theorem 2.2.1, we have then only to establish this last theorem, and its supplement Theorem 2.2.4.

The general assertion (a) of Theorem 2.2.1 is now known. It was reduced in Lemma 2.2.2 to the generic case, which we resolved in Lemma 6.6.3. The assertion (b) of this theorem is the crux of the matter. For any pair

$$(G, \psi), \qquad G \in \widetilde{\mathcal{E}}_{\mathrm{sim}}(N), \ \psi \in \widetilde{\Psi}(G),$$

over the local field F, it postulates the existence of a packet $\widetilde{\Pi}_\psi$, together with a pairing on $\mathcal{S}_\psi \times \widetilde{\Pi}_\psi$ that satisfies (2.2.6). Before we begin its proof, however, we shall derive a consequence of Theorem 2.2.1. We shall establish a proposition that provides some elementary constituents of the packet $\widetilde{\Pi}_\psi$. A special case of the proposition will then be used in the general proof of Theorem 2.2.1.

We should first recall that if $\phi \in \widetilde{\Phi}(G)$ is a general Langlands parameter for G, the L-packet $\widetilde{\Pi}_\phi$ is defined. It consists of Langlands quotients π_ρ of elements ρ in a corresponding packet $\widetilde{P}_\phi$ of standard representations. We introduced the set $P(G)$ of standard representations of an arbitrary group $G(F)$ in §3.5. The packet $\widetilde{P}_\phi$ here is composed as usual of $\widetilde{\mathrm{Out}}_N(G)$-orbits in $P(G)$. It is defined by analytic continuation from the case that ϕ is bounded, as is the pairing

$$\langle x, \rho\rangle = \langle x, \pi_\rho\rangle, \qquad x \in \mathcal{S}_\phi,\ \rho \in \widetilde{P}_\phi.$$

However, the endoscopic identity (2.2.6) will be valid for general ϕ only if $\widetilde{\Pi}_\phi$ is replaced by the packet $\widetilde{P}_\phi$ of standard representations. Its failure for the packet $\widetilde{\Pi}_\phi$ of irreducible representations is of course one of the reasons we have had to introduce other parameters ψ. This notation, incidentally, appears to be at odds with that of the original (nontempered) packet (1.5.1), in case the parameter $\phi = \psi$ in (1.5.1) is generic. But since ϕ then belongs to the subset $\widetilde{\Phi}_{\mathrm{unit}}(G)$ of $\widetilde{\Phi}(G)$, the representations in $\widetilde{P}_\phi$ should be irreducible (according to Conjecture 8.3.1 in our next chapter). This would imply that the packet $\widetilde{\Pi}_\phi$ of (1.5.1) is the same as the packet $\widetilde{P}_\phi$ (and the packet $\widetilde{\Pi}_\phi$) defined here.

Suppose for a moment that Theorem 2.2.1 is valid for the given pair (G, ψ), and that x is an element in $\mathcal{S}_\psi$. We can then choose a semisimple element $s \in S_\psi$ with image x in $\mathcal{S}_\psi$, such that if (G', ψ') is the preimage of the pair (ψ, s), the endoscopic datum $G' \in \mathcal{E}(G)$ is elliptic. The endoscopic identity (2.2.6) of Theorem 2.2.1(b) is

$$f'(\psi') = \sum_{\pi \in \widetilde{\Pi}_\psi} \langle s_\psi x, \pi\rangle f_G(\pi), \qquad f \in \widetilde{\mathcal{H}}(G).$$

Recall that $\widetilde{\Pi}_\psi$ is a finite $\widetilde{\Pi}_{\mathrm{unit}}(G)$-packet, which is to say that it fibres over the set $\widetilde{\Pi}_{\mathrm{unit}}(G)$. Let us write $\widetilde{\Pi}_\psi(\pi)$ for the fibre in $\widetilde{\Pi}_\psi$ of any element $\pi \in \widetilde{\Pi}_{\mathrm{unit}}(G)$. The endoscopic identity can then be written

$$f'(\psi') = \sum_{\pi \in \widetilde{\Pi}_{\mathrm{unit}}(G)} \Big(\sum_{\tilde{\pi} \in \widetilde{\Pi}_\psi(\pi)} \langle s_\psi x, \tilde{\pi}\rangle \Big) f_G(\pi), \qquad f \in \widetilde{\mathcal{H}}(G). \tag{7.4.1}$$

We shall combine it with the definition in Theorem 2.2.1(a) to give a formal construction of the packet $\widetilde{\Pi}_\psi$.

Given (G', ψ'), we write

$$G' = G_1' \times G_2', \qquad G_i' \in \widetilde{\mathcal{E}}_{\mathrm{sim}}(N_i'),$$

and

$$\psi' = \psi_1' \times \psi_2', \qquad \psi_i' \in \widetilde{\Psi}(G_i'),\ i = 1, 2,$$

where $N = N_1' + N_2'$. We can then write

$$f'(\psi') = \sum_{\phi' \in \widetilde{\Phi}(G')} \widetilde{n}(\psi', \phi')\, f'(\phi'),$$

for generic parameters

$$\phi' = \phi_1' \times \phi_2', \qquad \phi_i' \in \widetilde{\Phi}(G_i'),$$

in $\widetilde{\Phi}(G')$, and integers

$$\widetilde{n}(\psi', \phi') = \widetilde{n}(\psi_1', \phi_1')\, \widetilde{n}(\psi_2', \phi_2'),$$

by applying (2.2.12) separately to G_1' and G_2'. For any such ϕ', (G', ϕ') maps to a pair

$$\big(\phi(\phi'), s(\phi')\big), \qquad \phi(\phi') \in \widetilde{\Phi}(G),\ s(\phi') \in S_{\phi(\phi')}.$$

The analogue of (2.2.6) for standard representations is then

$$f'(\phi') = \sum_{\rho \in \widetilde{P}_{\phi(\phi')}} \langle s(\phi'), \rho \rangle\, f_G(\rho).$$

Lastly, we have the decomposition

$$f_G(\rho) = \sum_{\pi \in \widetilde{\Pi}(G)} m(\rho, \pi)\, f_G(\pi)$$

of a standard representation $\rho \in \widetilde{P}_{\phi(\phi')}$ into irreducible representations π, for nonnegative integers $m(\rho, \pi)$. Combining the three decompositions, we obtain a second endoscopic identity

$$f'(\psi') = \sum_{\pi \in \widetilde{\Pi}(G)} \lambda(\psi', \pi)\, f_G(\pi), \qquad f \in \widetilde{\mathcal{H}}(G), \tag{7.4.2}$$

where

$$\lambda(\psi', \pi) = \sum_{\phi' \in \widetilde{\Phi}(G')} \sum_{\rho \in \widetilde{P}_{\phi(\phi')}} \widetilde{n}(\psi', \phi')\, \langle s(\phi'), \rho \rangle\, m(\rho, \pi), \qquad \pi \in \widetilde{\Pi}(G). \tag{7.4.3}$$

It then follows from (7.4.1) that

$$\sum_{\widetilde{\pi} \in \widetilde{\Pi}_\psi(\pi)} \langle s_\psi x, \widetilde{\pi} \rangle = \lambda(\psi', \pi), \qquad \pi \in \widetilde{\Pi}(G). \tag{7.4.4}$$

We recall that there is a generic parameter $\phi_\psi \in \widetilde{\Phi}(G)$ attached to ψ. It is defined by

$$\phi_\psi(w) = \psi\left(w, \begin{pmatrix} |w|^{\frac{1}{2}} & 0 \\ 0 & |w|^{-\frac{1}{2}} \end{pmatrix}\right), \qquad w \in L_F,$$

where $|w|$ is the extension to L_F of the absolute value on W_F. The embedding of the centralizer S_ψ into S_{ϕ_ψ} factors to a surjective homomorphism from $\mathcal{S}_\psi$ to $\mathcal{S}_{\phi_\psi}$. This is dual to an injection of the character group $\widehat{\mathcal{S}}_{\phi_\psi}$ into $\widehat{\mathcal{S}}_\psi$. The next proposition asserts that the representations in $\widetilde{\Pi}_{\phi_\psi}$ take on an endoscopic interpretation as elements in the larger packet $\widetilde{\Pi}_\psi$.

PROPOSITION 7.4.1. *Assume that Theorem 2.2.1 is valid for the pair (G,ψ). Then there is an injection from $\widetilde{\Pi}_{\phi_\psi}$ into the set of multiplicity free elements in the packet $\widetilde{\Pi}_\psi$ such that the diagram*

$$\begin{array}{ccc} \widetilde{\Pi}_{\phi_\psi} & \hookrightarrow & \widetilde{\Pi}_\psi \\ \downarrow & & \downarrow \\ \widehat{\mathcal{S}}_{\phi_\psi} & \hookrightarrow & \widehat{\mathcal{S}}_\psi \end{array}$$

is commutative. In particular, the elements in the nontempered packet $\widetilde{\Pi}_{\phi_\psi}$ are unitary.

PROOF. As we observed in §2.2 and §3.5, an irreducible representation π of $GL(N,F)$ comes with a linear form Λ_π in the closure $\overline{(\mathfrak{a}_B^*)^+}$ of the associated maximal dual chamber. It is a measure of the failure of π to be tempered. Given parameters $\phi \in \Phi(N)$ and $\psi \in \Psi(N)$ for $GL(N)$, we have been writing $\Lambda_\phi = \Lambda_{\pi_\phi}$ and $\Lambda_\psi = \Lambda_{\phi_\psi} = \Lambda_{\pi_\psi}$. In particular, we have linear forms Λ_ϕ and Λ_ψ for parameters ϕ and ψ in the subsets $\widetilde{\Phi}(G)$ and $\widetilde{\Psi}(G)$ of $\Phi(N)$ and $\Psi(N)$. The maximal closed chamber $\overline{(\mathfrak{a}_{P_0}^*)^+}$ for G embeds canonically in the chamber $\overline{(\mathfrak{a}_B^*)^+}$ for $GL(N)$. With this understanding, it is clear that the linear form Λ_π attached to a representation $\pi \in \widetilde{\Pi}_\phi$ equals Λ_ϕ, for the parameter $\phi \in \widetilde{\Phi}(G)$ such that $\widetilde{\Pi}_\phi$ contains π.

The proposition will be a consequence of the identity (7.4.4) applied to a fixed element $\pi \in \widetilde{\Pi}(G)$, together with elementary properties of the factors in the summands of (7.4.3). It might be a little simpler to argue in terms of the norm on $\overline{(\mathfrak{a}_B^*)^+}$, as we did in the proof of Lemma 3.5.2, rather than the underlying partial order. The factor $\widetilde{n}(\psi',\phi')$ in (2.4.3) represents the general coefficient in the expansion of a stable linear form ψ' on $\widetilde{\mathcal{H}}(G')$ in terms of "standard" stable linear forms ϕ'. If it is nonzero, we have

$$\|\Lambda_{\phi'}\| \leqslant \|\Lambda_{\psi'}\| = \|\Lambda_\psi\|,$$

with equality if and only if $\phi' = \phi_{\psi'}$. We also have

$$\Lambda_{\phi(\phi')} = \Lambda_{\phi'}$$

for any parameter $\phi' \in \widetilde{\Phi}(G')$, and

$$\Lambda_\phi = \Lambda_\rho = \Lambda_\pi, \qquad \rho \in \widetilde{P}_\phi,\ \pi \in \widetilde{\Pi}_\phi,$$

for any parameter $\phi \in \widetilde{\Phi}(G)$. The factor $m(\rho, \pi)$ in (7.4.3) is the general coefficient in the expansion of a standard representation ρ of $G(F)$ in terms of irreducible representations π. If it is nonzero, we have

$$\|\Lambda_\pi\| \leqslant \|\Lambda_\rho\|,$$

with equality if and only if π is the Langlands quotient π_ρ of ρ. Taken together, these conditions place a similar restriction on the coefficient $\lambda(\psi', \pi)$ defined by (7.4.3). Our conclusion is that if $\lambda(\psi', \pi)$ is nonzero, then

$$\|\Lambda_\pi\| \leqslant \|\Lambda_{\psi'}\|,$$

with equality if and only if π lies in $\widetilde{\Pi}_\phi$, for parameters $\phi = \phi(\phi')$ and $\phi' = \phi_{\psi'}$. We note that if ϕ' does equal $\phi_{\psi'}$, then $\phi(\phi') = \phi_\psi$ and $s(\phi') = s$. In this case, the original pair (ψ, s) is transformed to (ϕ_ψ, s) under the composition

$$(\psi, s) \longrightarrow (G', \psi') \longrightarrow (G', \phi_{\psi'}) \longrightarrow \big(\phi(\phi_{\psi'}), s(\phi_{\psi'})\big) = (\phi_\psi, s).$$

Suppose now that π belongs to $\widetilde{\Pi}_{\phi_\psi}$. The double sum in (7.4.3) then reduces to a sum over the single pair $(\phi', \rho) = (\phi_{\psi'}, \rho_\pi)$, where ρ_π is the representation in $\widetilde{P}_{\phi_\psi}$ corresponding to π. As the leading coefficients in their respective expansions, $\widetilde{n}(\psi', \phi_{\psi'})$ and $m(\rho_\pi, \pi)$ both equal 1. We see also that

$$\langle s(\phi'), \rho\rangle = \langle s, \rho_\pi\rangle = \langle x, \pi\rangle.$$

The formula (7.4.3) thus reduces to

$$\lambda(\psi', \pi) = \langle x, \pi\rangle.$$

It then follows from (7.4.4) that

$$\sum_{\widetilde{\pi} \in \widetilde{\Pi}_\psi(\pi)} \langle s_\psi x, \widetilde{\pi}\rangle = \langle x, \pi\rangle.$$

The point $x \in \mathcal{S}_\psi$ is arbitrary, and has been identified with its image in $\mathcal{S}_{\phi_\psi}$ on the right hand side of the formula. Since the image of s_ψ in $\mathcal{S}_{\phi_\psi}$ is trivial, the formula can be written

$$\sum_{\widetilde{\pi} \in \widetilde{\Pi}_\psi(\pi)} \langle x, \widetilde{\pi}\rangle = \langle x, \pi\rangle, \qquad x \in \mathcal{S}_\psi, \tag{7.4.5}$$

after translation of x by $s_\psi = s_\psi^{-1}$.

If we set $x = 1$ in (7.4.5), we see that the index of summation is trivial. In other words, the fibre $\widetilde{\Pi}_\psi(\pi)$ of π in $\widetilde{\Pi}_\psi$ consists of one element, which we simply identify with π. This is the first assertion of the proposition. The other assertions follow from the resulting simplification of (7.4.5), and the fact that the representations in $\widetilde{\Pi}_\psi$ are unitary. □

We can now go back to our main topic, the proof of Theorem 2.2.1. Our task is to establish assertion (b) of the theorem for the given pair (G, ψ). If ψ lies in the complement of $\widetilde{\Psi}_2(G)$, the assertion follows from the local intertwining relation we have now established, or rather its embodiment in Proposition 2.4.3. We have therefore only to consider the case that ψ belongs to $\widetilde{\Psi}_2(G)$.

We take

$$\psi = \psi_1 \oplus \cdots \oplus \psi_r, \qquad \psi_i \in \widetilde{\Psi}_{\mathrm{sim}}(G_i),\ G_i \in \widetilde{\mathcal{E}}_{\mathrm{sim}}(N_i), \tag{7.4.6}$$

to be a fixed parameter in $\widetilde{\Psi}_2(G)$. Applying Proposition 7.2.1, we obtain an endoscopic datum $\dot{G} \in \widetilde{\mathcal{E}}_{\mathrm{sim}}(N)$ over the global field $\dot{F}$ and a global parameter $\dot{\psi} \in \widetilde{\mathcal{F}}_2(\dot{G})$ of the corresponding form (7.2.9). The pair $(\dot{G}, \dot{\psi})$ is as in Lemma 7.3.2. It follows from this lemma that the stable multiplicity formula (7.3.9) holds for the pair.

By the global induction hypothesis on $\dot{\widetilde{\mathcal{F}}}$, (7.3.9) applies more generally to the groups $\dot{G}'$ that occur in

$$I_{\mathrm{disc},\dot{\psi}}(\dot{f}) = \sum_{\dot{G}' \in \widetilde{\mathcal{E}}_{\mathrm{ell}}(\dot{G})} \iota(\dot{G}, \dot{G}')\, \widehat{S}'_{\mathrm{disc},\dot{\psi}}(\dot{f}'), \tag{7.4.7}$$

the endoscopic decomposition (4.1.2) of the $\dot{\psi}$-component of the discrete part of the trace formula. The specialization of (7.3.9) to any linear form on the right hand side was expressed in Corollary 4.1.3. Following the conditional proof of the spectral multiplicity formula in Lemma 4.7.1, as we did in the discussion of (6.6.4), we obtain a further decomposition

$$I^{\dot{G}}_{\mathrm{disc},\dot{\psi}}(\dot{f}) = |\Psi(\dot{G}, \dot{\psi})|\, |\mathcal{S}_{\dot{\psi}}|^{-1} \sum_{\dot{x} \in \mathcal{S}_{\dot{\psi}}} \varepsilon'(\dot{\psi}')\, \dot{f}'(\dot{\psi}'),$$

where $(\dot{G}', \dot{\psi}')$ maps to the pair $(\dot{\psi}, \dot{x})$.

Continuing as in the relevant passage from §6.6, we turn next to the spectral decomposition (4.1.1) of $I_{\mathrm{disc},\dot{\psi}}(\dot{f})$. Since the parameter $\dot{\psi}$ belongs to $\widetilde{\Psi}_2(\dot{G})$, it cannot factor through any proper Levi subgroup $\dot{M}$ of $\dot{G}$. Applying our global induction hypothesis to the factor $\dot{G}_-$ of $\dot{M}$ as in the discussion of §6.6, we see that the summand of $\dot{M}$ in (4.1.1) vanishes. The left hand side of (7.4.7) therefore equals

$$I^{\dot{G}}_{\mathrm{disc},\dot{\psi}}(\dot{f}) = \mathrm{tr}\big(R^{\dot{G}}_{\mathrm{disc}}(\dot{f})\big) = \sum_{\dot{\pi}_G} n_{\dot{\psi}}(\dot{\pi}_G)\, \dot{f}_{\dot{G}}(\dot{\pi}_G),$$

where $\dot{\pi}_G$ ranges over the set $\Pi(\dot{G})$ of irreducible representations of $\dot{G}(\dot{\mathbb{A}})$, and $n_{\dot{\psi}}(\dot{\pi}_G)$ are nonnegative integers. Since $\dot{f}$ lies in $\widetilde{\mathcal{H}}(\dot{G})$, $\dot{f}_{\dot{G}}(\pi_{\dot{G}})$ depends only on the image $\dot{\pi}$ of $\dot{\pi}_G$ in the set $\widetilde{\Pi}(\dot{G})$ of orbits in $\Pi(\dot{G})$ under the restricted direct product

$$\widetilde{\mathrm{O}}\mathrm{ut}_N(\dot{G}_{\mathbb{A}}) = \prod_v \widetilde{\mathrm{O}}\mathrm{ut}_N(\dot{G}_v).$$

We can therefore write

$$I^{\dot{G}}_{\mathrm{disc},\dot{\psi}}(\dot{f}) = |\Psi(\dot{G},\dot{\psi})| \sum_{\dot{\pi}\in\widetilde{\Pi}(\dot{G})} n_{\dot{\psi}}(\dot{\pi})\, \dot{f}_{\dot{G}}(\dot{\pi}),$$

for a modified coefficient

$$n_{\dot{\psi}}(\dot{\pi}) = |\Psi(\dot{G},\dot{\psi})|^{-1} \sum_{\dot{\pi}_G\in\Pi(\dot{G},\dot{\pi})} n_{\dot{\psi}}(\dot{\pi}_G),$$

in which $\Pi(\dot{G},\dot{\pi})$ is the preimage of $\dot{\pi}$ in $\Pi(\dot{G})$, as in (6.6.5).

We have converted (7.4.7) to an identity

$$\sum_{\dot{\pi}\in\widetilde{\Pi}(\dot{G})} n_{\dot{\psi}}(\dot{\pi})\, \dot{f}_{\dot{G}}(\dot{\pi}) = |\mathcal{S}_{\dot{\psi}}|^{-1} \sum_{\dot{x}\in\mathcal{S}_{\dot{\psi}}} \varepsilon'(\dot{\psi}')\, \dot{f}'(\dot{\psi}'), \qquad \dot{f}\in\widetilde{\mathcal{H}}(\dot{G}), \tag{7.4.8}$$

where $(\dot{G}',\dot{\psi}')$ maps to $(\dot{\psi},\dot{x})$. This formula is the analogue of the identity (6.6.6) in §6.6. We shall apply it to a product

$$\dot{f} = \dot{f}_V \cdot \dot{f}_U \cdot \dot{f}^{V,U},$$

relative to the decomposition

$$\dot{G}(\dot{\mathbb{A}}) = \dot{G}_V \times \dot{G}_U \times \dot{G}^{V,U} = \dot{G}(\dot{F}_V) \times \dot{G}(\dot{F}_U) \times \dot{G}(\dot{\mathbb{A}}^{V,U}).$$

Recall that the set $U = S_\infty(u)$ is composed of places at which we have little information, the analogue of the singleton $\{u\}$ in §6.6. Recall also that the set V of p-adic places here plays the role of the set S^u_∞ of archimedean places in §6.6. It is what we will use to extract information about the places in U.

We have been trying to follow the discussion of §6.6 as closely as possible, in order to clarify the structure of the general argument. At this point we come to a minor bifurcation. Our concern is the linear form

$$f'(\psi') = \dot{f}'_u(\dot{\psi}'_u), \qquad f = \dot{f}_u,\ x\in\mathcal{S}_\psi,$$

in which (G',ψ') maps to (ψ,x). We want to express it in terms of irreducible characters. In §6.6, we proceeded directly. We decomposed the linear form (with ϕ in place of ψ) into a linear combination of irreducible characters π in f, with formal coefficients $c_\phi(x,\pi)$ in x that had then to be determined. It is convenient here to approach the question from the opposite direction. We shall decompose $f'(\psi')$ into a linear combination of irreducible characters in x, with formal coefficients $f_G(\sigma)$ in f that have yet to be understood. In other words, we shall follow the general definition (7.1.2). In fact, we will apply it to each of the valuations v of U.

For each $v\in U$, (7.1.2) gives a sum

$$\dot{f}'_v(\dot{\psi}'_v) = \sum_{\dot{\sigma}_v} \langle s_{\dot{\psi}}\dot{x}_v, \dot{\sigma}_v\rangle\, \dot{f}_{v,G}(\dot{\sigma}_v), \qquad \dot{x}_v\in\mathcal{S}_{\dot{\psi}_v},\ \dot{f}_v\in\widetilde{\mathcal{H}}(\dot{G}_v),$$

over $\dot{\sigma}_v$ in the finite packet $\widetilde{\Sigma}_{\dot{\psi}_v}$ of linear forms on $\widetilde{\mathcal{H}}(\dot{G}_v)$. We write

$$(\dot{f}'_U)(\dot{\psi}'_U) = \sum_{\dot{\sigma}_U} \langle s_{\dot{\psi}}\dot{x}, \dot{\sigma}_U\rangle\, (\dot{f}_U)_{\dot{G}}(\dot{\sigma}_U), \tag{7.4.9}$$

where $\dot{\sigma}_U$ ranges over the linear forms in the packet

$$\Big\{\dot{\sigma}_U = \bigotimes_{v\in U} \dot{\sigma}_v : \dot{\sigma}_v \in \widetilde{\Sigma}_{\dot{\psi}_v}\Big\}$$

of

$$\dot{\psi}_U = \prod_{v\in U} \dot{\psi}_v$$

and

$$\langle \dot{x}, \dot{\sigma}_U\rangle = \prod_{v\in U} \langle \dot{x}_v, \dot{\sigma}_v\rangle.$$

We temporarily follow the same convention for the supplementary valuations $v \notin U$. We can then write

$$(\dot{f}^{V,U})'((\dot{\psi}^{V,U})') = \sum_{\dot{\sigma}^{U,V}} \langle s_{\dot{\psi}}\dot{x}, \dot{\sigma}^{V,U}\rangle\, (\dot{f}^{V,U})_{\dot{G}}(\dot{\sigma}^{V,U}), \tag{7.4.10}$$

where $\dot{\sigma}^{V,U}$ ranges over elements in the packet

$$\Big\{\dot{\sigma}^{V,U} = \bigotimes_{v\notin V\cup U} \dot{\sigma}_v : \dot{\sigma}_v \in \widetilde{\Sigma}_{\dot{\psi}_v}, \langle\,\cdot\,, \dot{\sigma}_v\rangle = 1 \text{ for almost all } v\Big\}$$

of

$$\dot{\psi}^{V,U} = \prod_{v\notin V\cup U} \dot{\psi}_v,$$

and

$$\langle \dot{x}, \dot{\sigma}^{V,U}\rangle = \prod_{v\notin V\cup U} \langle \dot{x}_v, \dot{\sigma}_v\rangle.$$

We can also write

$$(\dot{f}'_V)(\dot{\phi}'_V) = \sum_{\dot{\sigma}_V} \langle s_{\dot{\psi}}\dot{x}, \dot{\sigma}_V\rangle\, (\dot{f}_V)_{\dot{G}}(\dot{\sigma}_V), \tag{7.4.11}$$

where $\dot{\sigma}_V$ ranges over elements in the packet

$$\Big\{\dot{\sigma}_V = \bigotimes_{v\in V} \dot{\sigma}_v : \dot{\sigma}_v \in \widetilde{\Sigma}_{\dot{\psi}_v}\Big\}$$

of

$$\dot{\psi}_V = \prod_{v\in V} \dot{\psi}_v,$$

and

$$\langle \dot{x}, \dot{\sigma}_V\rangle = \prod_{v\in V} \langle \dot{x}_v, \dot{\sigma}_v\rangle.$$

We will presently specialize the supplementary functions $\dot{f}_V$ and $\dot{f}^{V,U}$, as in the argument of §6.6.

We first substitute the formulas (7.4.9), (7.4.10) and (7.4.11) into the three factors of our linear form

$$\dot{f}'(\dot{\psi}') = \dot{f}'_V(\dot{\psi}'_V)\, \dot{f}'_U(\dot{\psi}'_U)\, (\dot{f}^{V,U})'((\dot{\psi}^{V,U})')$$

in (7.4.8). According to Lemma 4.4.1, the coefficient $\varepsilon'(\dot{\psi}')$ in (7.4.8) equals the value of the character $\varepsilon_{\dot{\psi}} = \varepsilon_{\dot{\psi}}^{\dot{G}}$ on $\mathcal{S}_{\dot{\psi}}$ at $s_{\dot{\psi}}\dot{x}$, with $(\dot{\psi}, \dot{x})$ being the usual image of $(\dot{G}', \dot{\psi}')$. The right hand side of (7.4.8) becomes

$$|\mathcal{S}_{\dot{\psi}}|^{-1} \sum_{x \in \mathcal{S}_{\psi}} \sum_{\dot{\sigma}} \varepsilon_{\dot{\psi}}(s_{\dot{\psi}}\dot{x}) \langle s_{\dot{\psi}}\dot{x}, \dot{\sigma}\rangle \dot{f}_{\dot{G}}(\dot{\sigma}),$$

an expression we write as

$$|\mathcal{S}_{\dot{\psi}}|^{-1} \sum_{x \in \mathcal{S}_{\psi}} \sum_{\dot{\sigma}} \varepsilon_{\dot{\psi}}(\dot{x}) \langle \dot{x}, \dot{\sigma}\rangle \dot{f}_{\dot{G}}(\dot{\sigma}), \tag{7.4.12}$$

where the inner sums are over products

$$\dot{\sigma} = \dot{\sigma}_V \otimes \dot{\sigma}_U \otimes \dot{\sigma}^{V,U}$$

of elements in the packets of $\dot{\psi}_V$, $\dot{\psi}_U$ and $\dot{\psi}^{V,U}$, and

$$\langle \dot{x}, \dot{\sigma}\rangle \dot{f}_{\dot{G}}(\dot{\sigma})$$

equals

$$\langle \dot{x}, \dot{\sigma}_V\rangle \dot{f}_{V,\dot{G}}(\dot{\sigma}_V) \cdot \langle \dot{x}, \dot{\sigma}_U\rangle \dot{f}_{U,\dot{G}}(\dot{\sigma}_U) \cdot \langle \dot{x}, \dot{\sigma}^{V,U}\rangle \dot{f}_{\dot{G}}^{V,U}(\dot{\sigma}^{V,U}),$$

while $\dot{x}$ is the isomorphic image of x in $\mathcal{S}_{\dot{\psi}}$, as usual. Bear in mind that $\dot{\sigma}_V$, $\dot{\sigma}_U$ and $\dot{\sigma}^{V,U}$ at this point are still just linear forms. However, the two supplementary indices $\dot{\sigma}_V$ and $\dot{\sigma}^{V,U}$ both range over linearly independent sets of irreducible, signed characters. This follows from assertions (ii) and (iii)(a) of Proposition 7.2.1, which tell us that the localization $\dot{\psi}_v$ at any $v \notin U$ is dual to a tamely ramified, quadratic, generic parameter. We will now fix elements $\dot{\sigma}_V$ and $\dot{\sigma}^{V,U}$ for which the sign is 1.

Suppose that $\xi \in \widehat{\mathcal{S}}_{\psi}$ is a fixed character on the 2-group $\mathcal{S}_{\psi} \cong \mathcal{S}_{\dot{\psi}}$. We apply condition (iii)(a) of Proposition 7.2.1 to the localization $\dot{\psi}_V$ of $\dot{\psi} = \dot{\psi}_{\dot{M}}$. (Bear in mind that $M = G$ here, since ψ belongs to $\widetilde{\Psi}_2(G)$.) The condition implies that we can fix an element $\dot{\pi}_{V,\xi}$ in the subset

$$\widetilde{\Pi}_{\dot{\psi}_V}^{\dot{G}} = \prod_{v \in V} \widetilde{\Pi}_{\dot{\psi}_v}^{\dot{G}}$$

of the packet of $\dot{\psi}_V$ such that the character $\langle \cdot, \dot{\pi}_{V,\xi}\rangle$ on $\mathcal{S}_{\dot{\psi}}$ equals the product $\varepsilon_{\dot{\psi}}^{-1}\xi^{-1}$. Away from V and U, we take the element

$$\dot{\pi}^{V,U}(1) = \bigotimes_{v \notin V \cup U} \dot{\pi}_v(1)$$

in the subset

$$\widetilde{\Pi}_{\dot{\psi}^{V,U}}^{\dot{G}} = \prod_{v \notin V \cup U} \widetilde{\Pi}_{\dot{\psi}_v}^{\dot{G}_v}$$

of the packet of $\dot{\psi}^{V,U}$ such that for any v, the character $\langle \cdot, \dot{\pi}_v(1)\rangle$ on $\mathcal{S}_{\dot{\psi}_v}$ is 1. We are relying here on Proposition 7.2.1(ii) and Lemma 7.1.2(a), which together imply that $\dot{\pi}_v(1)$ lies in the subset $\widetilde{\Pi}_{\dot{\psi}_v}^{\dot{G}_v}$ of $\widetilde{\Pi}_{\dot{\psi}_v}$. It follows from the definition (7.1.9) that $\dot{\pi}_{v,\xi}$ and $\dot{\pi}^{V,U}(1)$ are both (orbits of) irreducible representations.

We can now specialize the functions $\dot{f}_V$ and $\dot{f}^{V,U}$. We have already noted that the two supplementary indices $\dot{\sigma}_V$ and $\dot{\sigma}^{V,U}$ in (7.4.12) are over linearly independent sets of irreducible signed characters. We choose the corresponding functions $\dot{f}_{V,\dot{G}}(\dot{\sigma}_V)$ and $\dot{f}^{V,U}_{\dot{G}}(\dot{\sigma}^{V,U})$, which represent factors of the summands on the right hand side of (7.4.12), so that they are supported at $\dot{\sigma}_V = \dot{\pi}_{V,\xi}$ and $\dot{\sigma}^{V,U} = \dot{\pi}^{V,U}(1)$. The original formula (7.4.8) then reduces to an identity

$$\sum_{\dot{\pi}_U} n_\psi(\xi, \dot{\pi}_U)\, \dot{f}_{U,\dot{G}}(\dot{\pi}_U) = \sum_{\dot{\sigma}_U} |\mathcal{S}_{\dot{\psi}}|^{-1} \sum_{x \in \mathcal{S}_\psi} \langle \dot{x}, \dot{\sigma}_U \rangle \xi(x)^{-1} \dot{f}_{U,\dot{G}}(\dot{\sigma}_U),$$

where $\dot{\pi}_U$ is summed over $\widetilde{\Pi}(\dot{G}_U)$, $\dot{\sigma}_U$ is summed over the packet of $\dot{\psi}_U$, and

$$n_\psi(\xi, \dot{\pi}_U) = n_{\dot{\psi}}\big(\dot{\pi}_{V,\xi} \otimes \dot{\pi}_U \otimes \dot{\pi}^{V,U}(1)\big). \tag{7.4.13}$$

This is the analogue of the formula (6.6.10) from the generic case in §6.6. Since the groups $\mathcal{S}_{\dot{\psi}_v}$ are all abelian, the function $\langle \dot{x}, \dot{\sigma}_U \rangle$ of $x \in \mathcal{S}_\psi$ attached here to any $\dot{\sigma}_U$ is a linear character. If $\widetilde{\Sigma}_{\dot{\psi}_U}(\xi)$ is the set of $\dot{\sigma}_U$ in the packet $\widetilde{\Sigma}_{\dot{\psi}_U}$ of $\dot{\psi}_U$ such that this character equals ξ, we conclude that

$$\sum_{\dot{\pi}_U \in \widetilde{\Pi}(\dot{G}_U)} n_\psi(\xi, \dot{\pi}_U)\, \dot{f}_{U,\dot{G}}(\dot{\pi}_U) = \sum_{\dot{\sigma}_U \in \widetilde{\Sigma}_{\dot{\psi}_U}(\xi)} \dot{f}_{U,\dot{G}}(\dot{\sigma}_U), \tag{7.4.14}$$

for any $\xi \in \widehat{\mathcal{S}}_\psi$ and $\dot{f}_U \in \widetilde{\mathcal{H}}(\dot{G}_U)$. This is the analogue of (6.6.12)

LEMMA 7.4.2. *For any $\dot{\pi}_U \in \widetilde{\Pi}(\dot{G}_U)$, $n_\psi(\xi, \dot{\pi}_U)$ is a nonnegative integer.*

PROOF. For this lemma, we need a slight embellishment of the proof of its generic counterpart Lemma 6.6.4(a). We recall from (7.4.13) that

$$n_\psi(\xi, \dot{\pi}_U) = |\Psi(\dot{G}, \dot{\psi})|^{-1} \sum_{\dot{\pi}_G \in \Pi(\dot{G}, \dot{\pi})} n_{\dot{\psi}}(\dot{\pi}_G),$$

where

$$\dot{\pi} = \dot{\pi}_{V,\xi} \otimes \dot{\pi}_U \otimes \dot{\pi}^{U,V}(1).$$

Since the summand $n_{\dot{\psi}}(\dot{\pi}_G)$ is a nonnegative integer, the only possible source of difficulty is the inverse of the order

$$|\Psi(G, \psi)| = |\Psi(\dot{G}, \dot{\psi})|$$

of the preimage of ψ in $\Psi(G)$. We can assume that $\widehat{G}$ is orthogonal and that the degrees $N_i = m_i n_i$ of the components $\psi_i = \mu_i \otimes \nu_i$ of ψ are all even, since the order $|\Psi(G, \psi)|$ would otherwise be 1, leaving us with nothing to prove. Suppose that one of the numbers m_i is odd. Then n_i is even, and the corresponding representation ν_i of $SU(2)$ is symplectic. Since the representation μ_i of L_F is orthogonal in this case, ψ_i would then be symplectic. This contradicts the fact that $\psi \in \widetilde{\Psi}_2(G)$ is orthogonal.

We have shown that the degrees m_i of the generic constituents μ_i of ψ_i are all even. Consider the construction of the corresponding global generic constituents $\dot{\mu}_i$ from Proposition 7.2.1. It has the property that for any $v \in V$, the completion $\dot{\mu}_{i,v}$ is a direct sum of 2-dimensional representations of $W_{\dot{F}_v}$. Moreover, for any n, the representation

$$\dot{\mu}_v^n = \bigoplus_{\{i:\, n_i = n\}} \dot{\mu}_{i,v}$$

of $L_{\dot{F}_v}$ is multiplicity free. The completion

$$\dot{\psi}_v = \bigoplus_i (\dot{\mu}_{i,v} \otimes \nu^i) = \bigoplus_n (\dot{\mu}_v^n \otimes \nu^n)$$

of $\dot{\psi}$ therefore belongs to $\widetilde{\Psi}_2(\dot{G}_v)$, and is of the form $\widehat{\phi}_v$, for some $\phi_v \in \widetilde{\Phi}_2(\dot{G}_v)$. Since the dual group of $\dot{G}_v$ is even orthogonal, there are two elements in each of the sets $\Psi(\dot{G}_v, \dot{\psi}_v)$ and $\Phi(\dot{G}_v, \phi_v)$. It then follows from Corollaries 6.6.6 and 6.7.3 that any element $\pi_v \in \widetilde{\Pi}_{\phi_v}$, regarded as an $\widetilde{\mathrm{O}}\mathrm{ut}_N(\dot{G}_v)$-orbit of irreducible representations, contains two representations. The same is therefore true for any element $\dot{\sigma}_v \in \widetilde{\Sigma}_{\dot{\psi}_v}$ by duality, and hence for any component $\dot{\pi}_v$ of the factor $\dot{\pi}_{V,\xi}$ of $\dot{\pi}$ at V. This implies that the group $\widetilde{\mathrm{O}}\mathrm{ut}_N(\dot{G})$ acts freely on the indices of summation $\dot{\pi}_G$ above. Since the original multiplicity $n(\dot{\pi}_G)$ is invariant under the action of any $\dot{F}$-automorphism of $\dot{G}$, the sum itself is an even integer, and is therefore divisible by the order 2 of $\Psi(\dot{G}, \dot{\psi})$. This completes the proof of the lemma. □

It remains to separate the contribution of u to each side of (7.4.14) from that of its complement S_∞^u in U. As in our treatment of (7.3.15) in the last section, we need to recall how we chose the global field $\dot{F}$. If $F = \mathbb{C}$, or $F = \mathbb{R}$ and the simple generic constituents

$$\{\phi_i :\ m_i = 2,\ 1 \leqslant i \leqslant r\} \tag{7.4.15}$$

of ψ are in relative general position, U consists of the one valuation u. In these cases, there is nothing to do. In fact, in dealing with the remaining cases, we can assume inductively that the theorem has been established for (G, ψ) in case $U = \{u\}$. This is where we will use Proposition 7.4.1.

The remaining cases are when F is real but the constituents (7.4.15) of ψ are not in relative general position, or F is p-adic. Then S_∞^u is not empty. It consists of real places v such that the analogues for $\dot{\psi}_v$ of the constituents (7.4.15) *are* in general position. By our induction assumption, we can apply Proposition 7.4.1 to the associated pairs $(\dot{G}_v, \dot{\psi}_v)$. We set

$$\dot{\pi}_\infty^u = \bigotimes_{v \in S_\infty^u} \dot{\pi}_v,$$

where for each v, $\dot{\pi}_v$ is the element in the packet $\widetilde{\Pi}_{\phi_{\dot{\psi}_v}}$ such that the character $\langle \cdot, \dot{\pi}_v \rangle$ on $\mathcal{S}_{\dot{\phi}_v}$ is trivial. The proposition asserts that $\dot{\pi}_v$ belongs to the packet

$\widetilde{\Pi}_{\dot{\psi}_v}$, and has multiplicity 1. We set

$$\dot{f}_\infty^u = \prod_{v \in S_\infty^u} \dot{f}_v,$$

where for each v, $\dot{f}_v$ is a function in $\widetilde{\mathcal{H}}(\dot{G}_v)$ such that

$$\dot{f}_{v,\dot{G}}(\dot{\pi}'_v) = \begin{cases} 1, & \text{if } \dot{\pi}'_v = \dot{\pi}_v, \\ 0, & \text{otherwise,} \end{cases}$$

for any element $\dot{\pi}'_v$ in $\widetilde{\Pi}_{\dot{\psi}_v}$. We shall apply the identity (7.4.14) to the product

$$\dot{f}_U = f \cdot \dot{f}_\infty^u,$$

in which f is a variable function in the space $\widetilde{\mathcal{H}}(G) = \widetilde{\mathcal{H}}(\dot{G}_u)$.

There will be only one term on the right hand side of (7.4.14) at the chosen function $\dot{f}_U$. To describe it, let $\sigma(\xi)$ be the element in the packet $\widetilde{\Sigma}_\psi = \widetilde{\Sigma}_{\dot{\psi}_u}$ with

$$\langle \cdot, \sigma(\xi) \rangle = \xi.$$

It then follows from the definitions, and the fact that each $\dot{\pi}_v$ above has multiplicity 1 in $\widetilde{\Pi}_{\dot{\psi}_v}$, that the right hand side of (7.4.14) reduces simply to $f_G(\sigma(\xi))$. The left hand side of (7.4.14) reduces to a finite sum of irreducible unitary characters in f, with nonnegative integral multiplicities

$$n_\psi(\xi, \pi) = n_\psi(\xi, \dot{\pi}_\infty^u \otimes \pi), \qquad \pi \in \widetilde{\Pi}_{\text{unit}}(G).$$

The identity therefore takes the form

$$\sum_{\pi \in \widetilde{\Pi}_{\text{unit}}(G)} n_\psi(\xi, \pi)\, f_G(\pi) = f_G(\sigma(\xi)), \qquad f \in \widetilde{\mathcal{H}}(G). \tag{7.4.16}$$

Observe that in the earlier cases in which U contained the one element u, (7.4.16) is actually identical to (7.4.14). Our global argument has thus led to a local identity for (G, ψ) that holds in all cases.

Recall that $\sigma(\xi)$ is the element in the provisional packet $\widetilde{\Sigma}_\psi$ attached to the character ξ on $\mathcal{S}_\psi$. It was defined (7.1.2) only as a linear form on $\widetilde{\mathcal{H}}(G)$. However, we see from (7.4.16) that it is actually a finite sum of ($\widetilde{\text{Out}}_N(G)$-orbits of) irreducible characters. We have therefore obtained the following proposition.

PROPOSITION 7.4.3. *For any $\psi \in \widetilde{\Psi}_2(G)$, define*

$$\widetilde{\Pi}_\psi = \coprod_{\xi \in \widehat{\mathcal{S}}_\psi} \widetilde{\Pi}_\psi(\xi), \qquad \xi \in \widehat{\mathcal{S}}_\psi,$$

where $\widetilde{\Pi}_\psi(\xi)$ is the disjoint union over $\pi \in \widetilde{\Pi}_{\text{unit}}(G)$ of multisets consisting of $n(\xi, \pi)$-copies of π. Then

$$f'(\psi') = \sum_{\xi \in \widehat{\mathcal{S}}_\psi} \sum_{\pi \in \widetilde{\Pi}_\psi(\xi)} \xi(s_\psi x)\, f_G(\pi) = \sum_{\pi \in \widetilde{\Pi}_\psi} \langle s_\psi x, \pi \rangle\, f_G(\pi),$$

for any $f \in \widetilde{\mathcal{H}}(G)$ *and* $x \in \mathcal{S}_\psi$, *where*

$$\langle s_\psi x, \pi\rangle = \xi(s_\psi x), \qquad \pi \in \widetilde{\Pi}_\psi(\xi),$$

and as usual, (G', ψ') *maps to the pair* (ψ, x). □

REMARK. Notice that we have not established a ψ-analogue of Corollaries 6.6.6 and 6.7.3, the assertions that the multiplicities $m(\phi)$ and $m(\pi)$ are equal for elements $\phi \in \widetilde{\Phi}_2(G)$ and $\pi \in \widetilde{\Pi}_\psi$. Something that does carry over is the weaker assertion for the elements $\psi \in \widetilde{\Psi}_2(G)$ and $\pi \in \widetilde{\Pi}_\psi$ of Remark 4, which follows the statement of Theorem 2.2.4. In this case, $m(\psi) = 1$ and $m(\pi) = 2$. The assertion is that $\widetilde{m}_\psi(\pi, \widehat{\mathcal{S}}_\psi)$, the multiplicity of the element $\pi \in \widetilde{\Pi}_{\mathrm{unit}}(G)$ over $\widehat{\mathcal{S}}_\psi$, is even. It follows from the proofs of Lemma 7.4.2 and Proposition 7.4.3, and the action of the F-outer automorphism of order 2 on the ψ-component of the automorphic discrete spectrum of G. We do not expect any such elements $\pi \in \widetilde{\Pi}_\psi$ to exist, which is to say that $\widetilde{m}_\psi(\pi, \widehat{\mathcal{S}}_\psi)$ should be zero.

We have now established the endoscopic identity (2.2.6). More precisely, we have shown that the provisional packet $\widetilde{\Sigma}_\psi$ over $\widehat{\mathcal{S}}_\psi$ determines the $\widetilde{\Pi}_{\mathrm{unit}}(G)$-packet $\widetilde{\Pi}_\psi$ that satisfies the conditions of Theorem 2.2.1. This completes the proof of Theorem 2.2.1 and Theorem 1.5.1, for the pair (G, ψ) over F.

In the last two sections, we have established Theorems 2.2.1 and 2.4.1 in general. These are the essential local theorems. We also know that Theorem 1.5.1 is valid. Any of its assertions that are not subsumed in Theorem 2.2.1 follow either from Lemma 7.3.4, or Corollary 6.7.5 and the general classification of tempered reprsentations established in §6.7. This leaves only the twisted local supplements, Theorems 2.2.4 and 2.4.4. They apply to parameters ψ in the subset $\Psi(\widetilde{G})$ of $\widetilde{\Psi}(G)$, where $\widetilde{G}$ is the bitorsor attached to an orthogonal group $G \in \widetilde{\mathcal{E}}_{\mathrm{sim}}(N)$ with N even.

In §6.8, we sketched a proof of the two supplementary theorems for parameters ϕ in the subset $\Phi_{\mathrm{bdd}}(\widetilde{G})$ of $\Psi(\widetilde{G})$. For any such ϕ, the dual parameter $\psi = \widehat{\phi}$ lies in $\Psi(\widetilde{G})$. We shall first describe the analogues for $\widetilde{G}$ of the objects in Lemma 7.1.2.

For any $\phi \in \Phi(\widetilde{G})$, there is a canonical mapping

$$\widetilde{\pi} \longrightarrow \sigma_{\widetilde{\pi}}, \qquad \pi \in \Pi_\phi,$$

where $\widetilde{\pi}$ is an extension of π to $\widetilde{G}(F)$ and $\sigma_{\widetilde{\pi}}$ is a $G(F)$-invariant linear form on $\mathcal{H}(\widetilde{G})$, such that

$$\widetilde{f}'(\psi') = \sum_{\pi \in \Pi_\phi} \langle s_\psi \widetilde{x}, \widetilde{\pi}\rangle \widetilde{f}_{\widetilde{G}}(\sigma_{\widetilde{\pi}}), \qquad \widetilde{f} \in \mathcal{H}(\widetilde{G}).$$

This is the analogue for $\widetilde{G}$ of the case $\psi = \widehat{\phi}$ of (7.1.2) treated prior to the statement of Lemma 7.1.1. In particular, $(\widetilde{G}', \psi')$ is the preimage of a pair

$(\psi, \widetilde{x})$ attached to an element $\widetilde{x} \in \widetilde{\mathcal{S}}_\psi$. We define

$$\Pi_\psi^{\widetilde{G}} = \{\widehat{\pi} \in \Pi_\psi : \sigma_{\widetilde{\pi}} = \widehat{\widehat{\pi}}\}, \tag{7.4.17}$$

where π is the preimage of $\widehat{\pi}$ in Π_ϕ, and $\widehat{\widehat{\pi}}$ is an extension of the representation $\widehat{\pi}$ to $\widetilde{G}(F)$. It is clear that $\widehat{\widehat{\pi}}$ is uniquely determined by $\sigma_{\widetilde{\pi}}$, and hence by the extension $\widetilde{\pi}$ of π. This is the analogue for $\widetilde{G}$ of the set $\widetilde{\Pi}_\psi^G$ in Lemma 7.1.2(a). If ϕ lies in the complement of $\widetilde{\Phi}_2(G)$ in $\widetilde{\Phi}(G)$, we can choose a proper Levi subgroup M of G, with parameters ϕ_M in $\widetilde{\Phi}_2(M, \phi)$ and $\psi_M \in \widetilde{\Psi}_2(M, \psi)$ as in Lemma 7.1.2(b). The analogue for $\widetilde{G}$ of the associated set $\widetilde{\Pi}_{\psi_M}^G$ is the set $\Pi_{\psi_M}^{\widetilde{G}}$ of representations

$$\widehat{\pi}_M, \qquad \pi_M \in \widetilde{\Pi}_{\phi_M},$$

in the packet $\widetilde{\Pi}_{\psi_M}$ such that the analogue

$$\widetilde{\beta}(\phi)(\widetilde{D}\widetilde{f}_G)(\phi, s_\psi \widetilde{u}, \pi_M) = \widetilde{f}_G(\psi, \widetilde{u}, \widehat{\pi}_M), \qquad \widetilde{u} \in \widetilde{\mathfrak{N}}_\phi,\ \widetilde{f} \in \mathcal{H}(\widetilde{G}), \tag{7.4.18}$$

of the identity (7.1.14) holds. The objects $\widetilde{\beta}(\phi)$ and $\widetilde{D}$ are taken with respect to $\widetilde{G}$, which is to say that they are defined in terms of Levi subsets of $\widetilde{G}$ rather than Levi subgroups of G, according to the remarks prior to Lemma 7.1.1.

Suppose that in addition to being of the form $\widehat{\phi}$, the parameter $\psi \in \Psi(\widetilde{G})$ is tamely ramified and quadratic. Then the image of $\Pi_\psi^{\widetilde{G}}$ in $\widehat{\mathcal{S}}_\psi$ generates $\widehat{\mathcal{S}}_\psi$ and contains the trivial character 1, while if M is proper in G, the image of $\Pi_{\psi_M}^{\widetilde{G}}$ in $\widehat{\mathcal{S}}_{\psi_M}$ generates $\widehat{\mathcal{S}}_{\psi_M}$ and contains the trivial character 1. This assertion is the variant of Lemma 7.1.2 for $\widetilde{G}$ mentioned at the end of §7.1. We shall prove it in [**A25**], along with the lemma itself.

Given the analogue for $\widetilde{G}$ of Lemma 7.1.2, we can establish the supplementary Theorems 2.2.4 and 2.4.4, for local parameters $\psi \in \Psi(\widetilde{G})$. In each case, we use Proposition 7.2.1 to construct a family (7.2.7) over the global field $\dot{F}$. If $(\dot{G}, \dot{\psi})$ is the global pair attached to (G, ψ), $\dot{\psi}$ lies in the subset $\Psi(\dot{\widetilde{G}})$ of $\widetilde{\Psi}(\dot{G})$. We have then simply to follow the steps of the proofs of the corresponding Theorems 2.2.1 and 2.4.1. We shall spare the details, since they contain no new arguments. If ψ belongs to the subset $\Psi_2(\widetilde{G})$ of $\Psi(\widetilde{G})$, for example, the reduction of Theorem 2.2.4 to the proof of Theorem 2.2.1 in this section is essentially the same as the generic case $\psi = \phi$ treated at the beginning of §6.8. If ψ belongs to the complement of $\Psi_2(\widetilde{G})$ in $\Psi_{\mathrm{ell}}(\widetilde{G})$, we first observe that the analogue for $\dot{\widetilde{G}}$ of the expression (5.2.8) of Lemma 5.2.1 vanishes. As in the generic case (6.8.6), this follows from the fact that the analogue for $\dot{\widetilde{G}}$ of the other expression (5.2.7) in Lemma 5.2.1 vanishes. The proof of Theorem 2.4.4 for ψ can then be transcribed from the proof of Theorem 2.4.1 in §7.3.

With these brief remarks, we have finished our discussion of Theorems 2.2.4 and 2.4.4. We still have the global induction hypothesis for $\dot{\tilde{\mathcal{F}}}$, the nongeneric family (7.2.7). To resolve it, we must persuade ourselves that the relevant global theorems are valid for parameters $\dot{\psi}$ in the subset $\dot{\tilde{\mathcal{F}}}(N)$ of $\dot{\tilde{\mathcal{F}}}$.

We shall again be brief. The stable multiplicity formula of Theorem 4.1.2 follows from the general reductions of Proposition 4.5.1 and the proofs of Lemmas 7.3.1 and 7.3.2. The actual multiplicity formula is the assertion of Theorem 1.5.2. If $\dot{\psi}$ lies in $\widetilde{\mathcal{F}}_2(\dot{G})$, it follows from the multiplicity formula (7.4.8) (and the local results we have now proved), or alternatively, a direct appeal to the relevant variant of Lemma 4.7.1. If $\dot{\psi}$ lies in the complement of $\widetilde{\mathcal{F}}_2(\dot{G})$ in $\dot{\tilde{\mathcal{F}}}(N)$, it follows as in generic case from (4.4.12), (4.5.5), and the stable multiplicity formula. Theorem 1.5.3(a) is relevant only to generic parameters. Theorem 1.5.3(b) is also an assertion for generic parameters. However, its role in the induction argument is really for nongeneric parameters, insofar as they occur in proof of Lemma 4.3.1 from §4.6. The assertion follows from an application of Lemma 5.1.6 to the nongeneric family (7.2.7). Finally, we have the supplementary global Theorem 4.2.2, for the relevant parameters in $\dot{\tilde{\mathcal{F}}}(N)$. Its main assertion (b) is implicit in the proof of the supplementary local Theorem 2.2.4. We will skip this argument, as we did also in the generic case treated in §6.8. However, we will establish both assertions of Theorem 2.2.4, for general global parameters, in the first section of the next chapter.

We have now resolved the various induction hypotheses for general local parameters ψ, apart from a couple of supplementary arguments we will treat in the next section. With the end of the induction argument, we have established all of the local theorems.

7.5. Some remarks on characters

We conclude this chapter by drawing attention to two character formulas for $GL(N)$. The first applies to the case $F = \mathbb{R}$. It is the formula of Adams and Huang for a Speh character on $GL(N)$. The second is a refinement by Tadic of results of Zelevinsky that are implicit in the discussion from §2.2 of p-adic infinitesimal characters on $GL(N)$. In each case, the formula is an explicit expansion of an irreducible character π_ψ on $GL(N, F)$ in terms of standard characters.

We will need the formulas in the next chapter. They will be used in §8.2 in the critical final stages of the global classification. The formulas might also be of interest in their own right. They apply to untwisted analogues of the integers $\widetilde{n}(\psi', \phi')$ in the double sum (7.4.3), which we recall gives the definition of the coefficients $\lambda(\psi', \pi)$ that determine the packets

$$\widetilde{\Pi}_\psi, \qquad \psi \in \widetilde{\Psi}(G),\ G \in \widetilde{\mathcal{E}}_{\mathrm{sim}}(N).$$

We shall see that the two formulas are parallel. They seem to hint at properties of the packets $\widetilde{\Pi}_\psi$ that are common to the real and p-adic cases.

Suppose that $N = mn$, and that θ is an irreducible, tempered character on $GL(m,F)$. Then θ corresponds to an irreducible, unitary, m-dimensional representation μ of L_F under the Langlands correspondence for $GL(m)$. Let us write

$$\theta^n(f) = \operatorname{tr}\big(\pi_\psi(f)\big) = \operatorname{tr}\big(\pi_{\phi_\psi}(f)\big), \qquad f \in \mathcal{H}\big(GL(N)\big),\ \psi = \mu \otimes \nu^n, \tag{7.5.1}$$

for the irreducible character on $GL(N,F)$ corresponding to μ and the irreducible, n-dimensional representation ν^n of $SU(2)$. We will also consider collections of pairs

$$(m_i, \theta_i), \qquad 1 \leqslant i \leqslant r,$$

where

$$N = m_1 + \cdots + m_r,$$

and θ_i is a *virtual* character on $GL(m_i,F)$. In other words, θ_i is a finite, integral, linear combination of irreducible characters on $GL(m_i,F)$. Let us write

$$\theta_1 \boxplus \cdots \boxplus \theta_r \tag{7.5.2}$$

for the virtual character on $GL(N,F)$ induced from the virtual character $\theta_1 \times \cdots \times \theta_r$ on the Levi subgroup

$$GL(m_1,F) \times \cdots \times GL(m_r,F).$$

If the constituents θ_i are irreducible, (7.5.2) is of course a standard character. The formulas in [**AH**] and [**Tad2**] express a character (7.5.1) explicitly in terms of virtual characters (7.5.2).

Suppose first that $F = \mathbb{R}$. Consider an irreducible representation of W_F of dimension 2. It is then induced from a character

$$z \longrightarrow (z/|z|)^k |z|^\lambda = z^\mu \bar{z}^\nu, \qquad z \in \mathbb{C}^*,\ k \in \mathbb{N},\ \lambda \in \mathbb{C},$$

of the subgroup $\mathbb{C}^*$ of W_F, where $\mu = \frac{1}{2}(k+\lambda)$ and $\nu = \frac{1}{2}(-k+\lambda)$, and $|z| = (z\bar{z})^{\frac{1}{2}}$ is the analytic absolute value. This representation of W_F, which we denote by $\mu(k,\lambda)$, parametrizes a representation in the relative discrete series of $GL(2,F)$, whose character we shall denote by $\theta(k,\lambda)$. Following [**AH**, §5.10], we define $\theta(k,\lambda)$ by coherent continuation for all $k \in \mathbb{Z}$. For fixed λ,

$$\{\theta(k,\lambda) :\ k \in \mathbb{Z}\}$$

is then a coherent family of virtual characters based at λ [**AH**, §4]. In concrete terms, $\theta(k,\lambda)$ can be defined by the natural "continuation" in k from $\mathbb{N}$ to $\mathbb{Z}$ of Harish-Chandra's explicit formula for $\theta(k,\lambda)$. To describe it, let $\eta(\ell,\lambda)$ be the character

$$x \longrightarrow \big(x/|x|\big)^\ell |x|^\lambda, \qquad x \in \mathbb{R}^*,\ \ell = 0,1,\ \lambda \in \mathbb{C},$$

of $\mathbb{R}^*$. Then

$$\theta(0,\lambda) = \eta(0,\lambda) \boxplus \eta(1,\lambda),$$

while if $k \geqslant 1$, the sum

$$\theta(-k, \lambda) + \theta(k, \lambda)$$

equals

$$\big(\eta(0, k + \lambda) \boxplus \eta(1, -k + \lambda)\big) + \big(\eta(1, k + \lambda) \boxplus \eta(0, -k + \lambda)\big),$$

if k is even, and

$$\big(\eta(0, k + \lambda) \boxplus \eta(0, -k + \lambda)\big) + \big(\eta(1, k + \lambda) \boxplus \eta(1, -k + \lambda)\big),$$

if k is odd. These identities are easy consequences of Harish-Chandra's formulas for the characters of representations in the relative discrete series of $GL(2, \mathbb{R})$.

Suppose that $N = 2n$. For any $k \in \mathbb{N}$, we have the unitary, irreducible character

$$\theta(k) = \theta(k, 0)$$

in the relative discrete series for $GL(2, \mathbb{R})$. The corresponding character $\theta^n(k)$ on $GL(N, F)$ defined by (7.5.1) is known as the *Speh character* attached to k and n. According to [**AH**, (5.11)(c)], it has an expansion

$$\theta^n(k) = \sum_{w \in S_n} \operatorname{sgn}(w) \Big(\boxplus_{i=1}^{n} \theta\big(k - (i - wi), (n + 1) - (i + wi)\big) \Big)$$

in terms of virtual characters (7.5.2). Here, wi stands obviously for the image of i under the permutation w in the symmetric group S_n.

Suppose next that F is p-adic. Let r be an irreducible unitary representation of W_F of degree m_r. If

$$r_\lambda(w) = r(w)\, |w|^\lambda, \qquad w \in W_F,$$

is the twist of r by a real number λ, the tensor product

$$\mu_r(k, \lambda) = r_\lambda \otimes \nu^{k+1}, \qquad k \geqslant 0,$$

is an irreducible representation of the Langlands group $L_F = W_F \times SU(2)$ of degree $m = m_r(k + 1)$. By the Langlands correspondence, it indexes a representation in the relative discrete series of $GL(m, F)$, whose character we denote by $\theta_r(k, \lambda)$. As in the case of $\mathbb{R}$ above, we need to define $\theta_r(k, \lambda)$ for all integers k. The value $k = 0$ is already part of the definition here. It corresponds to the case that the character $\theta_r(k, \lambda)$ is supercuspidal. If $k = -1$, we follow the convention of [**Tad2**] of setting $\theta_r(k, \lambda)$ equal to 1, the trivial character on the trivial group $\{1\}$. If $k < -1$, we simply set $\theta_r(k, \lambda) = 0$.

Suppose that

$$N = mn, \qquad m = m_r(k + 1).$$

The character $\theta_r^n(k)$ on $GL(N, F)$ associated to the unitary character

$$\theta_r(k) = \theta_r(k, 0)$$

by (7.5.1) is itself unitary. According to [**Tad2**, Theorem 5.4], it can be written explicitly as a signed sum

$$\theta_r^n(k) = \sum_{w\in S_n} \operatorname{sgn}(w)\Big(\boxplus_{i=1}^{n} \theta_r\big(k-(i-wi),\,(n+1)-(i+wi)\big)\Big)$$

of virtual characters (7.5.2). We have changed the notation of [**Tad2**] slightly in order to make the p-adic formula parallel to the formula for $F=\mathbb{R}$. In [**Tad2**], d equals $(k+1)$, W_n equals S_n, and the sum is taken over the subset

$$\begin{aligned} W_n^{(d)} &= \{w\in W_n:\ w(i)+d\geqslant i,\ 1\leqslant i\leqslant n\}\\ &= \{w\in S_n:\ k-(i-wi)\geqslant -1,\ 1\leqslant i\leqslant n\}\end{aligned}$$

of elements in W_n whose summand in the p-adic formula is nonzero. Moreover, if ρ is the unitary supercuspidal representation corresponding to r, the character of the representation denoted by $\delta([\rho,\nu^k\rho])$ on p. 342 of [**Tad2**] equals $\theta_r\big(k,\frac{k}{2}\big)$ in our notation.

To write the two formulas together, we can always take r to be a unitary 1-dimensional representation of W_F in case $F=\mathbb{R}$. The resulting character of $GL(1,F)$ is of course the only irreducible unitary representation of a general linear group over $F=\mathbb{R}$ that meets the formal definition of *supercuspidal*. The virtual characters $\theta_r(k,\lambda)$ and $\theta_r^n(k)$ then make sense for both real and p-adic F, as do the representations $\mu_r(k,\lambda)$ and

$$\psi_r^n(k) = \mu_r(k)\otimes\nu^n$$

of L_F and $L_F\times SU(2)$ respectively. By including r in the notation, we can then restrict $\lambda\in\mathbb{C}$ to be real, as we did for p-adic F. The case $F=\mathbb{R}$ will still differ slightly from that of p-adic F, in that $\theta_r(k,\lambda)$ will remain unchanged if r is multiplied by the sign character, but this is not serious. Our interest here will generally be confined to the original case that r is trivial. At any rate, the formulas for both real and p-adic F can now be written together as

$$\theta_r^n(k) = \sum_{w\in S_n}\operatorname{sgn}(w)\,\theta_r^w(k), \tag{7.5.3}$$

where

$$\theta_r^w(k) = \boxplus_{i=1}^{n}\theta_r\big(k-(i-wi),\,(n+1)-(i+wi)\big). \tag{7.5.4}$$

The outer automorphism $x\to x^\vee$ of $GL(N)$ gives an involution $\theta\to\theta^\vee$ on the space of virtual characters on $GL(N,F)$. The virtual character in (7.5.3) satisfies

$$\theta_r^n(k)^\vee = \theta_{r^\vee}^n(k),$$

since the associated unitary representation $\psi_r^n(k)$ of $L_F\times SU(2)$ has the same property. To describe the resulting mapping on the summands of (7.5.3), we write

$$w^\vee = w_\ell w^{-1}w_\ell = w_\ell^{-1}w^{-1}w_\ell,\qquad w\in S_n,$$

where

$$w_\ell : i \longrightarrow (n+1) - i, \qquad 1 \leqslant i \leqslant n,$$

is the element of S_n of greatest length.

LEMMA 7.5.1. (i) *The summand in (7.5.3) satisfies*

$$\theta_r^w(k)^\vee = \theta_{r^\vee}^{w^\vee}(k), \qquad w \in S_n.$$

(ii) *Suppose that $k \geqslant n$, and that*

$$\theta_r^w(k) = \theta_r^{w'}(k),$$

for elements $w, w' \in S_n$. Then $w = w'$.

PROOF. (i) It is clear that

$$\theta_r(\ell, \lambda)^\vee = \theta_{r^\vee}(\ell, -\lambda),$$

for any $\ell \in \mathbb{Z}$ and $\lambda \in \mathbb{R}$. We shall check that this gives a bijection between the direct summands in the expansions (7.5.4) of each side of the proposed identity.

Given w, we define a permutation

$$i \longrightarrow i^\vee = w_\ell w i, \qquad 1 \leqslant i \leqslant n,$$

of the indices in the formal direct sum (7.5.4). Then

$$\begin{aligned} k - (i^\vee - w^\vee i^\vee) &= k - (w_\ell w i - w_\ell w^{-1} w_\ell w_\ell w i) \\ &= k - (w_\ell w i - w_\ell i) \\ &= k - \big(((n+1) - wi) - ((n+1) - i)\big) = k - (i - wi). \end{aligned}$$

The permutation therefore stabilizes the first argument

$$\ell = k - (i - wi)$$

of the corresponding two direct summands. Moreover,

$$\begin{aligned} (n+1) - (i^\vee + w^\vee i^\vee) &= (n+1) - (w_\ell w i + w_\ell i) \\ &= (n+1) - \big((n+1) - wi + (n+1) - i\big) \\ &= -\big((n+1) - (i + wi)\big). \end{aligned}$$

The permutation therefore acts as (-1) on the second argument

$$\lambda = (n+1) - (i + wi)$$

of the direct summands. The assertion (i) follows.

(ii) The first condition $k \geqslant n$ implies that

$$k - (i - wi) \geqslant 1, \qquad 1 \leqslant i \leqslant n.$$

In particular, the linear forms $\theta_r^w(k)$ in (7.5.3) are standard characters, with direct summands that lie in the mutually inequivalent family

$$\theta_r(\ell, \lambda), \qquad \ell \in \mathbb{N},\ \lambda \in \mathbb{R},$$

of essentially square integrable characters. The second condition implies that there is a bijection $i \to i'$ of the indices such that

$$k - (i - wi) = k - (i' - w'i')$$

and

$$(n + 1) - (i + wi) = (n + 1) - (i' + w'i').$$

These two equations imply that $i' = i$ and $w'i' = wi$, so that $w' = w$, as claimed. □

This lemma is intended only for perspective. If $k \geqslant n$, it tells us that $\theta_r^w(k)$ is self-dual if and only if $w = w^\vee$. If $k < n$, $\theta_r^w(k)$ is still self-dual if $w = w^\vee$. However, the converse question seems to be more complicated, especially in case $F = \mathbb{R}$. The anti-involution $w \to w^\vee$ of S_n is rather curious here. For it is the involution

$$w \longrightarrow w_\ell w\, w_\ell, \qquad w \in S_n,$$

whose kernel is the Weyl group of the relevant classical group. The proper roles of these two operations are unclear to me.

There are two elements $w \in S_n$ whose summands in (7.5.3) are of special interest. The first is the identity permutation $w = 1$. Then $\theta_r^w(k)$ equals the standard character

$$\phi_r^n(k) = \theta_r^{w_1}(k) = \boxplus_{i=1}^{n} \theta_r\big(k, (n+1) - 2i\big), \qquad w_1 = 1. \tag{7.5.5}$$

This case is characterized by the condition that the direct summands indexed by i are equal, up to twists by unramified quasicharacters. The other element is the longest permutation $w = w_\ell$. Then $\theta_r^w(k)$ equals

$$\theta_r^{w_\ell}(k) = \boxplus_{i=1}^{n} \theta_r\big(k + (n+1) - 2i, 0\big).$$

This case has the property that the factors indexed by i are often tempered. If $k \geqslant n$ as in Lemma 7.5.1(ii), they are all tempered, and $\theta_r^{w_\ell}(k)$ is again a standard character. With this restriction on k, we set $w^* = w_\ell$, and write

$$\phi_r^{n,*}(k) = \theta_r^{w^*}(k) = \theta_r^{w_\ell}(k).$$

The summand of $w = w_\ell$ in (7.5.3) is more complicated if $k < n$. For $\theta_r^{w_\ell}(k)$ is then only a virtual character, which could be 0 if F is p-adic. If F is real, and $k, n \geqslant 1$ are arbitrary, we can set $w^* = w_\ell$ as above, but we define

$$\phi_r^{n,*}(k) = \boxplus_{i=1}^{n} \theta_r^*\big(k + (n+1) - 2i, 0\big), \tag{7.5.6}$$

where

$$\theta_r^*(\ell, \lambda) = \theta_r(|\ell|, \lambda).$$

If F is p-adic, we have to define w^* differently. For any $k, n \geqslant 1$, we set

$$w^* i = \begin{cases} (n+1) - i, & \text{if } i \leqslant k+1, \\ i - (1+k), & \text{if } i \geqslant k+2, \end{cases}$$

for any i in the interval of integers

$$\mathbb{N}[1, n] = \{i \in \mathbb{N} : \ 1 \leqslant i \leqslant n\}.$$

Since w^* restricts to an order reversing bijection from $\mathbb{N}[1, k+1]$ onto $\mathbb{N}[n-k, n]$, and an order preserving bijection from $\mathbb{N}[k+2, n]$ onto $\mathbb{N}[1, n-k-1]$, it is indeed a permutation in S_n. We then set

$$\phi_r^{n,*}(k) = \theta_r^{w^*}(k) = \boxplus_{i=1}^{n^*} \theta_r\big(k + (n+1) - 2i, 0\big), \tag{7.5.7}$$

where $n^* = \min\{n, k+1\}$. The factors of $\phi_r^{n,*}(k)$ are thus (irreducible) tempered characters in all cases. In the p-adic case, we note that the condition $k \geqslant n$, under which we gave a simpler definition for $\phi_r^{n,*}(k)$ above, could be relaxed to $(k+1) \geqslant n$.

According to the definitions (7.5.5) and (7.5.6), $\phi_r^{n,*}(k)$ is induced from an irreducible, tempered, essentially square integrable character of a Levi subgroup of $GL(N, F)$. It is therefore an irreducible tempered character on $GL(N, F)$. The following lemma summarizes the properties that will be needed in §8.2.

LEMMA 7.5.2. *Suppose that $k \geqslant 0$, $n \geqslant 1$, r and N are as above, and that $k \geqslant 1$ in case F is archimedean.*

(i) *The irreducible tempered character $\phi_r^{n,*}(k)$ occurs with multiplicity (± 1) in the decomposition of the irreducible character $\theta_r^n(k)$ into standard characters.*

(ii) *The factor $\theta_r(k+n-1, 0)$ occurs with multiplicity 1 in the general decomposition (7.5.2) of $\phi_r^{n,*}(k)$ into essentially square integrable characters.*

PROOF. (i) Suppose that $F = \mathbb{R}$. The character $\phi_r^{n,*}(k)$ is part of the contribution to $\theta_r^n(k)$ of the summand of $w = w^* = w_\ell$ in (7.5.3). It represents the tempered part of the decomposition of $\theta_r^{w^*}(k)$ into standard characters. This follows from (7.5.6), the definition (7.5.4) in case $w = w^*$, and the decomposition of any direct factor $\theta_r(\ell, \lambda)$ into standard characters of $GL(2, F)$ in case $\ell \leqslant 0$. Similarly, if $w \neq w^* = w_\ell$, the standard characters in the decomposition of $\theta_r^w(k)$ are all nontempered, since

$$(n+1) - (i + wi) \neq 0,$$

for some i, $1 \leqslant i \leqslant n$. Therefore $\phi_r^{n,*}(k)$ is the only tempered character in the decomposition of $\theta_r^n(k)$ into standard characters. It actually occurs with multiplicity 1.

Suppose that F is p-adic. The product

$$\operatorname{sgn}(w^*)\, \phi_r^{n,*}(k) = \operatorname{sgn}(w^*)\, \theta_r^{w^*}(k)$$

is then equal to the summand of $w = w^*$ in (7.5.3). Since the other summands in (7.5.3) are all standard characters (up to a sign), it suffices to show that they are all nontempered. Assume that $w \in S_n$ indexes a summand in (7.5.3) that is nonzero and tempered. We consider the decomposition (7.5.4) of $\theta_r^w(k)$. The index i of any direct summand in (7.5.4) satisfies either

$$wi - i + k = -1, \tag{7.5.8}$$

which means that the factor equals 1, or

$$(n+1) - (i + wi) = 0, \tag{7.5.9}$$

which in the absence of (7.5.8) is necessary for the factor to be tempered. The condition (7.5.8) implies that $i \geqslant k + 2$, since $wi \geqslant 1$. It then follows from (7.5.9) that the restriction of w to $\mathbb{N}[1, k+1]$ equals that of $w^* = w_\ell$, and therefore maps $\mathbb{N}[1, k+1]$ bijectively onto $\mathbb{N}[n-k, n]$. This implies that w^* restricts to a bijection from $\mathbb{N}[k+2, n]$ onto $\mathbb{N}[1, n-k-1]$. But we must have

$$wi \geqslant i - (k+1), \qquad i \in \mathbb{N}[k+2, n],$$

since the factor of i in (7.5.4) would otherwise vanish, contradicting the condition that $\theta_r^w(k)$ is nonzero. This in turn can hold for every $i \in \mathbb{N}[k+2, n]$ only if w maps $\mathbb{N}[k+2, n]$ monotonically onto $\mathbb{N}[1, n-k-1]$. In other words, the restriction of w to the remaining integer interval $\mathbb{N}[k+2, n]$ also equals that of w^*. Therefore w equals w^*, as required.

(ii) The general decomposition (7.5.2) for $\phi_r^{n,*}(k)$ is given explicitly by (7.5.6) or (7.5.7), according to whether F is real or p-adic. In either case, $\theta_r(k+n-1, 0)$ is the factor indexed by $i = 1$. In the case (7.5.6) of $F = \mathbb{R}$, it is clear from the definitions (and the fact that $k \geqslant 1$) that the factor does occur with multiplicity 1. In the case (7.5.7) of p-adic F, the factors are all distinct, and therefore all have multiplicities 1. The lemma follows. □

In the course of proving (i), we established the following additional property.

COROLLARY 7.5.3. *The character $\phi_r^{n,*}(k)$ is the only tempered constituent in the decomposition of $\theta_r^n(k)$ into standard characters.* □

The formula (7.5.3) is an explicit expansion of the irreducible character $\theta_r^n(k)$ of $GL(N, F)$ in terms of standard characters. Our real interest, however, is in the twisted characters for $GL(N, F)$. One would ultimately like a twisted analogue of (7.5.3) for any F, which is to say, an explicit description of the terms on the right hand side of the general formula (2.2.9). For p-adic F, Moeglin and Waldspurger [**MW4**] have established an inductive process, which is both subtle and complex, that leads to such a formula. The fact that (7.5.3) holds uniformly for any F suggests perhaps that a similar process might hold for $F = \mathbb{R}$. However, the twisted nontempered characters of $GL(N, \mathbb{R})$ seem not to have been studied.

In any case, it would be hard to establish an explicit twisted expansion from (7.5.3). The problem, pointed out to me by the referee, is that serious difficulties accumulate from the ambiguity of extensions of self-dual representations. Fortunately, we will need only partial information, which we will in fact be able to extract from (7.5.3). Roughly speaking, we shall show that the coefficients $\widetilde{n}(\psi, \phi)$ in (2.2.9) are congruent modulo 2 to their untwisted analogues.

We fix a parameter $\psi \in \widetilde{\Psi}(N)$ over the local field F. This was the setting of the preamble to Lemma 2.2.2, from which we take our notation. We will also use similar notation for the more elementary untwisted analogues of objects introduced in §2.2. In particular, we have the ψ-subsets

$$\Pi(N, \psi) = \{\pi \in \Pi(N) : \ (\mu_\pi, \eta_\pi) = (\mu_\psi, \eta_\psi), \ \Lambda_\pi \leqslant \Lambda_\psi\}$$

and

$$P(N, \psi) = \{\rho \in P(N) : \ (\mu_\rho, \eta_\rho) = (\mu_\psi, \eta_\psi), \ \Lambda_\rho \leqslant \Lambda_\psi\}$$

of the respective sets $\Pi(N)$ and $P(N)$ of irreducible and standard representations of $GL(N, F)$. We also have untwisted analogues

$$f_N(\rho) = \sum_{\pi \in \Pi(N,\psi)} m(\rho, \pi)\, f_N(\pi), \qquad f \in \mathcal{H}(N), \ \rho \in P(N, \psi) \tag{7.5.10}$$

and

$$f_N(\pi) = \sum_{\rho \in P(N,\psi)} n(\pi, \rho)\, f_N(\rho), \qquad f \in \mathcal{H}(N), \ \pi \in \Pi(N, \psi), \tag{7.5.11}$$

of the twisted expansions (2.2.7) and (2.2.8), or rather their restrictions to the subsets $\widetilde{\Pi}(N, \psi)$ and $\widetilde{P}(N, \psi)$ of $\widetilde{\Pi}(N)$ and $\widetilde{P}(N)$. The expansion (7.5.3) we have just described is an explicit form of (7.5.11), in the special case that ψ is simple and $\pi = \pi_\psi$. To exploit it, we need to find some relation between the general expansions we have just quoted with the twisted expansions of §2.2.

LEMMA 7.5.4. *Given $\psi \in \widetilde{\Psi}(N)$, assume that $\rho = (\widetilde{\rho})^0$ and $\pi = (\widetilde{\pi})^0$ are restrictions to $GL(N, F)$ of representations $\widetilde{\rho} \in \widetilde{P}(N, \psi)$ and $\widetilde{\pi} \in \widetilde{\Pi}(N, \psi)$. Then the coefficients in (2.2.7) and (7.5.10) and in (2.2.8) and (7.5.11) satisfy*

$$m(\widetilde{\rho}, \widetilde{\pi}) \equiv m(\rho, \pi) \quad \text{(modulo 2)}$$

and

$$n(\widetilde{\pi}, \widetilde{\rho}) \equiv n(\pi, \rho) \quad \text{(modulo 2)}.$$

PROOF. According to the notation of §2.2, $\widetilde{P}(N, \psi)$ and $\widetilde{\Pi}(N, \psi)$ are sets of representatives of orbits in the families $\{\widetilde{P}(N)\}$ and $\{\widetilde{\Pi}(N)\}$ from (2.2.7) and (2.2.8). They consist of representations of $\widetilde{G}^+(N, F)$ that are defined by the Whittaker extensions from the beginning of §2.2. In particular, the corresponding coefficients $m(\widetilde{\rho}, \widetilde{\pi})$ and $n(\widetilde{\pi}, \widetilde{\rho})$ are indeed integers. The same is of course true of the general coefficients $m(\rho, \pi)$ and $n(\pi, \rho)$ in (7.5.10) and (7.5.11), whether ρ and π are self-dual (which is to say, of the given form $(\widetilde{\rho})^0$ and $(\widetilde{\pi})^0$) or not. The particular extensions $\widetilde{\rho}$ and $\widetilde{\pi}$ are in fact irrelevant to the lemma. So long as they are actually representations of $\widetilde{G}^+(N,F)$, they are determined up to a sign, which does not affect the assertion of the lemma.

The expansion (7.5.10) is just the decomposition of the standard character ρ of $GL(N,F)$ into irreducible characters π. In particular, $m(\rho,\pi)$ is the multiplicity of π in the π-isotypical subspace $V(\rho,\pi)$ of the space on which ρ acts (or rather the image of this space in the Grothendieck group $\mathcal{K}(G)_{\mathbb{C}}$). The twisted expansion (2.2.7) is the decomposition of the standard character $\widetilde{\rho}$ of $\widetilde{G}^+(N,F)$ into irreducible characters $\widetilde{\pi}$, or rather the restriction of this decomposition to the subset $\widetilde{G}(N,F)$ of $\widetilde{G}^+(N,F)$. It is obtained from (7.5.10) by first removing the summands in (7.5.10) of representations π with $\pi \neq \pi^\vee$, and then for the remaining π, keeping track of the restrictions of $\widetilde{\rho}$ to the subspaces $V(\rho,\pi)$. For any $\pi \in \Pi(N,\psi)$ with $\pi = \pi^\vee$, we thus have

$$m(\widetilde{\rho},\widetilde{\pi}) = m^+(\rho,\pi) - m^-(\rho,\pi),$$

where $m^+(\rho,\pi)$ is the multiplicity of π in the $\widetilde{\rho}$-isotypical subspace of $V(\rho,\pi)$, and $m^-(\rho,\pi)$ is the multiplicity of π in the complementary subspace. Since

$$m(\rho,\pi) = m^+(\rho,\pi) + m^-(\rho,\pi),$$

we see that $m(\widetilde{\rho},\widetilde{\pi})$ is congruent to $m(\rho,\pi)$ modulo 2. This is the first of the required congruence relations.

The second congruence relation will be a little more complicated. We can regard $P(N,\psi)$ and $\Pi(N,\psi)$ as partially ordered sets, defined by the partial order on the linear forms Λ_ρ and Λ_π in $\overline{(\mathfrak{a}_B^*)^+}$. The mapping $\rho \to \pi_\rho$ from $P(N,\psi)$ to $\Pi(N,\psi)$ is then a canonical isomorphism of partially ordered sets. To simplify the notation, we introduce an abstract partially ordered set $K(\psi)$, with corresponding isomorphisms $k \to \rho_k$, and $k \to \pi_k$ onto $P(N,\psi)$ and $\Pi(N,\psi)$. We then write

$$m_{kk'} = m(\rho_k,\pi_{k'}), \qquad k,k' \in K(\psi)$$

and

$$n_{kk'} = n(\pi_k,\rho_{k'}), \qquad k,k' \in K(\psi),$$

for the integral coefficients in the expansions (7.5.10) and (7.5.11). The set $K(\psi)$ inherits an involution $k \to k^\vee$ from the outer automorphism of $GL(N)$, and it follows from the expansions that

$$m_{k^\vee(k')^\vee} = m(\rho_k^\vee,\pi_{k'}^\vee) = m(\rho_k,\pi_{k'}) = m_{kk'}$$

and

$$n_{k^\vee(k')^\vee} = n(\pi_k^\vee,\rho_{k'}^\vee) = n(\pi_k,\rho_{k'}) = n_{kk'},$$

for any k and k' in $K(\psi)$.

Let $\mathcal{U}(\psi)$ be the set of unipotent, integral $\big(K(\psi)\times K(\psi)\big)$-matrices

$$a = (a_{kk'}), \qquad k,k' \in K(\psi),$$

such that

(i) $a_{kk} = 1$,

(ii) $a_{kk'} = 0$, unless $k' \leqslant k$,

and

(iii) $a_{k^\vee (k')^\vee} = a_{kk'}$.

It follows immediately that $\mathcal{U}(\psi)$ is a group under matrix multiplication, in which our two coefficient matrices $(m_{kk'})$ and $(n_{kk'})$ represent inverse elements. To exploit the third condition, we borrow our earlier notation from the slightly different context of §1.4. We write

$$K(\psi) = I(\psi) \amalg J(\psi) \amalg J^\vee(\psi),$$

where

$$I(\psi) = \{i \in K(\psi) : \; i^\vee = i\},$$

while $J(\psi)$ is a fixed set of representatives in $K(\psi)$ of orbits of order two under the involution and $J^\vee(\psi)$ is the complementary set of representatives. We can then introduce a second group. Let $\mathcal{U}_I(\psi)$ be the group of unipotent, integral, $\bigl(I(\psi) \times I(\psi)\bigr)$-matrices

$$a_I = (a_{ii'}), \qquad i, i' \in I(\psi),$$

that satisfy the three conditions above (but with ii' in place of kk'). In this case, there are really only two conditions, since the adjoint condition (iii) is trivial.

There is a natural restriction mapping $a \to a_I$ from $\mathcal{U}(\psi)$ onto $\mathcal{U}_I(\psi)$. However, this projection is not a group homomorphism. To describe the obstruction, consider the image in $\mathcal{U}_I(\psi)$ of a product of matrices a and b in $\mathcal{U}(\psi)$. For any indices i and i' in $I(\psi)$, we can write

$$\begin{aligned}(ab)_{ii'} &= \sum_{k \in K(\psi)} a_{ik}\, b_{ki'} \\ &= \sum_{i_* \in I(\psi)} a_{ii_*}\, b_{i_* i'} + \sum_{j \in J} a_{ij}\, b_{ji'} + \sum_{j^\vee \in J^\vee} a_{ij^\vee}\, b_{j^\vee i'} \\ &= \sum_{i_* \in I(\psi)} a_{ii_*}\, b_{i_* i'} + 2\Bigl(\sum_{j \in J} a_{ij}\, b_{ji'}\Bigr),\end{aligned}$$

since $a_{ij^\vee} = a_{ij}$ and $b_{j^\vee i'} = b_{ji'}$. In particular, the obstruction vanishes if the matrix coefficients are taken modulo 2. Let us therefore write $\overline{\mathcal{U}}(\psi)$ and $\overline{\mathcal{U}}_I(\psi)$ for the two matrix groups defined in the same way as $\mathcal{U}(\psi)$ and $\mathcal{U}_I(\psi)$, but the coefficients in the field $\mathbb{Z}/2\mathbb{Z}$. The restriction mapping $\overline{u} \to \overline{u}_I$ is then a surjective homomorphism from $\overline{\mathcal{U}}(\psi)$ onto $\overline{\mathcal{U}}_I(\psi)$. To put the matter a little more strongly, we write $\overline{x}$ for the reduction modulo 2 of any integral matrix x. It then follows that the mapping

$$\alpha : \; a \longrightarrow \overline{a}_I = \overline{(a_I)}, \qquad a \in \mathcal{U}(\psi),$$

is a surjective homomorphism from $\mathcal{U}(\psi)$ onto $\overline{\mathcal{U}}_I(\psi)$.

We now complete the proof of the lemma. We can of course identify the matrices m and n in $\mathcal{U}(\psi)$ with the coefficient matrices of (7.5.10) and (7.5.11). In particular, n equals the inverse of m. It follows that

$$\overline{n}_I = \alpha(n) = \alpha(m^{-1}) = \alpha(m)^{-1} = (\overline{m}_I)^{-1},$$

since as a homomorphism, α commutes with inverses. Having already established the first congruence relation, we can identify $\overline{m}_I$ with the reduction modulo 2 of the coefficient matrix in (2.2.7). Its inverse $\overline{m}_I^{-1}$ is therefore the reduction modulo 2 of the coefficient matrix in (2.2.8). Since this is in turn equal to $\overline{n}_I$, we have also established the second congruence relation. □

Our interest will be in the special case of (7.5.11) that π equals the Langlands quotient $\pi_\psi = \pi_{\phi_\psi}$ attached to ψ. In this case, (7.5.11) can be written in the form

$$f_N(\psi) = \sum_{\phi\in\Phi(N,\psi)} n(\psi,\phi)\, f_N(\phi), \qquad f \in \mathcal{H}(N), \tag{7.5.12}$$

where

$$\Phi(N,\psi) = \{\phi \in \Phi(N) : \ (\mu_\phi,\eta_\phi) = (\mu_\psi,\eta_\psi), \ \Lambda_\phi \leqslant \Lambda_\psi\}$$

and

$$n(\psi,\phi) = n(\pi_\psi,\pi_\phi).$$

This is the untwisted analogue of the twisted expansion (2.2.9). It is the abstract setting for the explicit character identity (7.5.3).

The formula (7.5.3) actually applies only to case that ψ is simple. However, it is easily extended to any parameter

$$\psi = \ell_1\psi_1 \oplus \cdots \oplus \ell_r\psi_r.$$

When ψ is simple, and corresponds to a pair (r,k) as in (7.5.3), we write

$$\theta(\psi) = \theta_r^n(k).$$

for the right hand side of (7.5.3). For general ψ, we can then set

$$\theta(\psi) = \underbrace{\theta(\psi_1)\boxplus\cdots\boxplus\theta(\psi_1)}_{\ell_1} \boxplus \cdots \boxplus \underbrace{\theta(\psi_r)\boxplus\cdots\boxplus\theta(\psi_r)}_{\ell_r}. \tag{7.5.13}$$

As a character on $GL(N,F)$ induced from an irreducible unitary character, $\theta(\psi)$ is itself irreducible. It is the character of the Langlands quotient π_ψ, whose value at a test function $f \in \mathcal{H}(N)$ equals the left hand side of (7.5.12). The corresponding direct sum of the virtual characters on the right hand side of (7.5.3) is an explicit (though by now somewhat complicated) integral linear combination of standard characters on $GL(N,F)$. Its value at f equals the right hand side of (7.5.12).

The two extremal standard characters in this explicit form of (7.5.12) are

$$\theta(\phi_\psi) = \underbrace{\theta(\phi_{\psi_1})\boxplus\cdots\boxplus\theta(\phi_{\psi_1})}_{\ell_1} \boxplus \cdots \boxplus \underbrace{\theta(\phi_{\psi_r})\boxplus\cdots\boxplus\theta(\phi_{\psi_r})}_{\ell_r}$$

and

$$\theta(\phi_\psi^*) = \underbrace{\theta(\phi_{\psi_1}^*) \boxplus \cdots \boxplus \theta(\phi_{\psi_1}^*)}_{\ell_1} \boxplus \cdots \boxplus \underbrace{\theta(\phi_{\psi_r}^*) \boxplus \cdots \boxplus \theta(\phi_{\psi_r}^*)}_{\ell_r},$$

where if ψ is simple, $\theta(\phi_\psi) = \phi_r^n(k)$ and $\theta(\phi_\psi^*) = \phi_r^{n,*}(k)$, in the earlier notation. In the notation here, it is understood ϕ_ψ and ϕ_ψ^* stand for the Langlands parameters of the standard characters $\theta(\phi_\psi)$ and $\theta(\phi_\psi^*)$. We know very well that

$$n(\psi, \phi_\psi) = 1.$$

But from Lemma 7.5.2(i) we observe also that

$$n(\psi, \phi_\psi^*) = \pm 1.$$

We recall that ϕ_ψ indexes the constituent of (7.5.12) that is most highly non-tempered, while it follows from Corollary 7.5.3 that ϕ_ψ^* indexes the unique constituent of (7.5.12) that is tempered.

With these remarks, we obtain the following corollary of Lemma 7.5.4.

COROLLARY 7.5.5. *For any $\phi \in \widetilde{\Phi}(N, \psi)$, the coefficients in (2.2.9) and (7.5.12) satisfy*

$$\widetilde{n}(\psi, \phi) \equiv n(\psi, \psi) \quad (\text{modulo } 2).$$

In particular,

$$\widetilde{n}(\psi, \phi_\psi^*) \equiv 1 \quad (\text{modulo } 2),$$

so the parameter ϕ_ψ^* has nonzero multiplicity in the twisted expansion (2.2.9). It indexes the unique constituent of (2.2.9) that is tempered. □

Suppose that ψ lies in the subset $\widetilde{\Psi}(G)$ of $\widetilde{\Psi}(N)$ attached to a simple datum $G \in \widetilde{\mathcal{E}}_{\text{sim}}(N)$. The Langlands parameter

$$\phi_\psi : L_F \longrightarrow {}^L G$$

of the standard character $\theta(\phi_\psi)$ on $GL(N, F)$ is defined by

$$\phi_\psi(u) = \psi\big(u, p(u)\big), \qquad u \in L_F,$$

where

$$p(u) = \begin{pmatrix} |u|^{\frac{1}{2}} & 0 \\ 0 & |u|^{-\frac{1}{2}} \end{pmatrix}, \qquad u \in L_F.$$

We can interpret the homomorphism

$$p : L_F \longrightarrow SL(2, \mathbb{C})$$

as the Langlands parameter for the trivial one-dimensional representation of $PGL(2, F)$. As such, it has a dual parameter

$$p^* : L_F \longrightarrow SL(2, \mathbb{C}),$$

which corresponds to the associated square integrable representation of $PGL(2, F)$. More precisely,

$$p^*(w, s) = s, \qquad (w, s) \in L_F = W_F \times SU(2),$$

if F is p-adic, while if $F = \mathbb{R}$, we have

$$p^*(z) = \begin{pmatrix} (z/\overline{z})^{\frac{1}{2}} & 0 \\ 0 & (z/\overline{z})^{-\frac{1}{2}} \end{pmatrix}, \qquad z \in \mathbb{C}^*,$$

and

$$p^*(\sigma_F) = \begin{pmatrix} 0 & 1 \\ -1 & 0 \end{pmatrix},$$

in the notation of §6.1. We can therefore associate a second Langlands parameter

$$\phi^*_\psi : L_F \longrightarrow {}^L G$$

in $\widetilde{\Phi}(G)$ to ψ by defining

$$\phi^*_\psi(u) = \psi(u, p^*(u)), \qquad u \in L_F.$$

This is the Langlands parameter of the standard character $\theta(\phi^*_\psi)$ on $GL(N, F)$. For p-adic F, it plays an important role in the work of Moeglin.

CHAPTER 8

The Global Classification

8.1. On the final step

We are at last in a position to prove the global theorems. We have already had to treat some special cases in order to prove the local theorems. It is now time to work in the opposite direction. We shall use the local theorems we have just established to prove the global theorems in general. We will complete this argument in §8.1 and §8.2. In the last three sections of the chapter, we will discuss some ramifications of the results.

Recall that there are four global theorems. They consist of Theorems 1.5.2 and 1.5.3 (which for their statements implicitly include Theorems 1.4.1 and 1.4.2), and the global supplements Theorems 4.1.2 and 4.2.2. The global intertwining relation of Corollary 4.2.1 is actually a corollary of the associated local intertwining relation, which we have established in all cases. On the other hand, the stable multiplicity formula of Theorem 4.1.2 is really the fundamental global assertion. It represents a foundation on which the others all depend.

The underlying field F will be global throughout this chapter, unless specified otherwise. The theorems apply to parameters ψ in the set $\widetilde{\Psi}$ of self-dual representations that occur in the automorphic spectral decompositions of general linear groups over F. We shall combine the local results we have established with the global formulas of §§5.1–5.3. To this end, we will take the global family $\widetilde{\mathcal{F}}$ of Chapter 5 to be the entire family $\widetilde{\Psi}$. To be consistent with the conventions of §5.1, we must consider the case that $\widetilde{\mathcal{F}}$ equals the subset $\widetilde{\Phi}$ of generic parameters in $\widetilde{\Psi}$, as well as that of $\widetilde{\mathcal{F}} = \widetilde{\Psi}$. However, we will generally write $\widetilde{\Psi}$ in place of $\widetilde{\mathcal{F}}$, distinguishing when necessary between the generic and nongeneric cases. As usual, when working with nongeneric parameters in $\widetilde{\Psi}$, we will assume implicitly that we have already established the global theorems for generic parameters.

We fix a positive integer N, and as in §5.1, assume inductively that the global theorems all hold for any $\psi \in \widetilde{\Psi}$ with $\deg(\psi) < N$. We do not need to add a local hypothesis, of course, since we have now established all the local theorems. We do however have to check that the family satisfies the additional hypothesis of §5.1.

LEMMA 8.1.1. *Suppose that $G \in \widetilde{\mathcal{E}}_{\mathrm{ell}}(N)$ and that ψ belongs to $\widetilde{\Psi}_2(G)$. Then the conditions of Assumption* 5.1.1 *hold for the pair (G, ψ).*

PROOF. This is the analogue of Lemma 5.4.2, for the maximal family $\widetilde{\mathcal{F}} = \widetilde{\Psi}$. However, it will be a little easier for us to prove, now that we have the local theorems in hand. Suppose for example that $G = G_S \times G_O$ is composite. From our induction hypothesis, we observe that any localization $\psi_v = \psi_{S,v} \times \psi_{O,v}$ belongs to the local set $\widetilde{\Psi}(G_v)$. The condition of Assumption 5.1.1 in this case then follows from its local version (2.2.4) in Theorem 2.2.1.

We therefore assume that G is simple. If we can show that any localization ψ_v of ψ still belongs to $\widetilde{\Psi}(G_v)$, we can appeal again to the relevant local assertion (2.2.3) of Theorem 2.2.1. Suppose that ψ is composite. An application of our induction hypothesis to the simple constituents ψ_i of ψ, combined with the inductive definition of the sets $\widetilde{\Psi}(G)$ and $\widetilde{\Psi}(G_v)$, then confirms that ψ_v lies in $\widetilde{\Psi}(G_v)$, and hence that ψ satisfies Assumption 5.1.1. If ψ is simple but not generic, we can apply our induction hypothesis to its simple factor μ. We see again that ψ_v belongs to $\widetilde{\Psi}(G_v)$, and that ψ satisfies Assumption 5.1.1. We can therefore assume that $\phi = \psi$ is simple and generic.

We have reduced the proof of the lemma to the case of a simple generic pair

$$(G, \phi), \qquad G \in \widetilde{\mathcal{E}}_{\mathrm{sim}}(N),\ \phi \in \widetilde{\Phi}_{\mathrm{sim}}(G).$$

The problem is to show that any localization ϕ_v of ϕ belongs to $\widetilde{\Phi}(G_v)$. The reader will recognize in this the assertion of Theorem 1.4.2, the second seed theorem from Chapter 1. However, we are now using the temporary definition of $\widetilde{\Phi}_{\mathrm{sim}}(G)$ from §5.1, rather than the original definition from Theorem 1.4.1 in terms of which Theorem 1.4.2 was stated. The logic of the temporary definition is actually somewhat convoluted. For we took $\widetilde{\Phi}_{\mathrm{sim}}(G)$ to be the subset of elements $\phi \in \widetilde{\Phi}_{\mathrm{sim}}(N)$ that satisfied the condition of Assumption 5.1.1, in addition to the natural global condition $S^G_{\mathrm{disc},\phi} \neq 0$. We then had to reinterpret Assumption 5.1.1 for simple generic pairs as the condition (5.1.6). Our problem, thus modified, is to show that for any $\phi \in \widetilde{\Phi}_{\mathrm{sim}}(N)$, we can find a simple datum $G \in \widetilde{\mathcal{E}}_{\mathrm{sim}}(N)$ such that $S^G_{\mathrm{disc},\phi} \neq 0$ and such that the linear form $\widetilde{f}_N(\phi)$ transfers to G.

For the given $\phi \in \widetilde{\Phi}_{\mathrm{sim}}(N)$, we write

$$\widetilde{I}^N_{\mathrm{disc},\phi}(\widetilde{f}) = \mathrm{tr}\big(\widetilde{R}^N_{\mathrm{disc},\phi}(\widetilde{f})\big) = \widetilde{f}_N(\phi), \qquad \widetilde{f} \in \widetilde{\mathcal{H}}(N).$$

This is one of the various reductions of (4.1.1) (with $\widetilde{G}(N)$ in place of G) with which we are now very familiar. It then follows from the $\widetilde{G}(N)$-analogue (3.3.14) of (4.1.2) that

$$\widetilde{f}_N(\phi) = \sum_{G \in \widetilde{\mathcal{E}}_{\mathrm{sim}}(N)} \widetilde{\iota}(N,G)\, \widehat{S}^G_{\mathrm{disc},\phi}(\widetilde{f}^G), \qquad \widetilde{f} \in \widetilde{\mathcal{H}}(N).$$

Here we are using the fact, obtained by the usual induction argument, that $S^G_{\mathrm{disc},\phi}$ vanishes for any G in the complement of $\widetilde{\mathcal{E}}_{\mathrm{sim}}(N)$ in $\widetilde{\mathcal{E}}_{\mathrm{ell}}(N)$. There must be at least one $G \in \widetilde{\mathcal{E}}_{\mathrm{sim}}(N)$ whose summand does not vanish identically in $\widetilde{f}$. If there is exactly one, we have

$$\widetilde{f}_N(\phi) = \widetilde{\iota}(N,G)\,\widehat{S}^G_{\mathrm{disc},\phi}(\widetilde{f}^G), \qquad \widetilde{f} \in \widetilde{\mathcal{H}}(N),$$

and $\widetilde{f}_N(\phi)$ is the pullback of a unique stable linear form on $\widetilde{\mathcal{H}}(G)$, as required. This is the case for example if N is odd or the character η_ψ is nontrivial.

We can therefore assume that N is even and $\eta_\psi = 1$. The summand of G then vanishes unless G is one of the two split groups in $\widetilde{\mathcal{E}}_{\mathrm{sim}}(N)$, which we denote as usual by G and $G^\vee$. The identity becomes

$$\widetilde{f}_N(\phi) = \widetilde{\iota}(N,G)\,\widehat{S}^G_{\mathrm{disc},\phi}(\widetilde{f}^G) + \widetilde{\iota}(N,G^\vee)\,\widehat{S}^{G^\vee}_{\mathrm{disc},\phi}(\widetilde{f}^{G^\vee}).$$

We can assume that *neither* summand vanishes identically in $\widetilde{f}$, since we would otherwise be in the case settled above. We can also assume that the linear form $\widetilde{f}_N(\phi)$ does not transfer to either G or $G^\vee$, since we would otherwise again be done. We must show that these conditions lead to a contradiction of the identity.

The second condition implies that there are valuations v and $v^\vee$ such that ϕ_v and $\phi_{v^\vee}$ do not lie in the respective local sets $\widetilde{\Phi}(G_v)$ and $\widetilde{\Phi}(G^\vee_{v^\vee})$. Suppose first that $v = v^\vee$. One can then choose a function $\widetilde{f}_v \in \widetilde{\mathcal{H}}_v(N)$ such that $\widetilde{f}^G_v = \widetilde{f}^{G^\vee}_v = 0$, but such that $\widetilde{f}_{v,N}(\phi_v) \neq 0$. We leave the reader to check that this is possible, using the characterization of Proposition 2.1.1 and the fact that ϕ_v must still belong to the local set $\widetilde{\Phi}(G^*_v)$ attached to some $G^*_v \in \widetilde{\mathcal{E}}_v(N)$. Once $\widetilde{f}_v$ is chosen, we set $\widetilde{f} = \widetilde{f}_v \cdot \widetilde{f}^v$, for any function $\widetilde{f}^v \in \widetilde{\mathcal{H}}^v(N)$ such that $\widetilde{f}^v_N(\phi^v) \neq 0$. Then $\widetilde{f}^G = \widetilde{f}^{G^\vee} = 0$, while $\widetilde{f}_N(\phi) \neq 0$. In other words, the right hand side of the last identity vanishes, while the left hand side is nonzero. This is a contradiction. Suppose next that $v \neq v^\vee$. We can then choose functions $\widetilde{f}_v \in \widetilde{\mathcal{H}}_v(N)$ and $\widetilde{f}_{v^\vee} \in \widetilde{\mathcal{H}}_{v^\vee}(N)$ such that $\widetilde{f}^G_v = 0$ and $\widetilde{f}^{G^\vee}_{v^\vee} = 0$, but such that $\widetilde{f}_{v,N}(\phi_v)$ and $\widetilde{f}_{v^\vee,N}(\phi_{v^\vee})$ are both nonzero. We then set $\widetilde{f} = \widetilde{f}_v \cdot \widetilde{f}_{v^\vee} \cdot \widetilde{f}^{v,v^\vee}$, for a complementary function $\widetilde{f}^{v,v^\vee}$ with $\widetilde{f}^{v,v^\vee}_N(\phi^{v,v^\vee}) \neq 0$. Once again, we have $\widetilde{f}^G = \widetilde{f}^{G^\vee} = 0$ and $\widetilde{f}_N(\phi) \neq 0$, and therefore a violation of the last identity. We thus have obtained a contradiction in both cases. Our conclusion is that the linear form $\widetilde{f}_N(\phi)$ does indeed transfer to one of the two groups. Consequently, there is a simple datum $G \in \widetilde{\mathcal{E}}_{\mathrm{sim}}(N)$ such that ϕ lies in the subset $\widetilde{\Phi}_{\mathrm{sim}}(G)$ of $\widetilde{\Phi}_{\mathrm{sim}}(N)$. This completes the proof of the lemma, in the last case of a simple generic parameter. □

REMARK. We see that Assumption 5.1.1, when restricted to the case of simple generic parameters $\phi \in \widetilde{\Phi}_{\mathrm{sim}}(N)$, amounts to the combined assertions of Theorem 1.4.1. The only difference is that it is expressed in terms of our alternate definitions of the sets $\widetilde{\Phi}_{\mathrm{sim}}(N)$ and $\widetilde{\Phi}_v(N)$. In other words, Assumption 5.1.1 (for simple generic ϕ) is tied to the conditions (i) in Corollaries 5.4.7 and 6.8.1, while Theorems 1.4.1 and 1.4.2 are tied to the equivalent conditions (ii) in these corollaries.

We have now placed the maximal family $\widetilde{\mathcal{F}} = \widetilde{\Psi}$ (or $\widetilde{\mathcal{F}} = \widetilde{\Phi}$) within the general framework of §5.1. In particular, the family comes with a set of induction hypotheses on the positive integer N. Our task is to complete the inductive proof by establishing the global theorems for parameters ψ in the set $\widetilde{\mathcal{F}}(N) = \widetilde{\Psi}(N)$ (or $\widetilde{\mathcal{F}}(N) = \widetilde{\Phi}(N)$).

Suppose for a moment that ψ belongs to the complement of the set $\widetilde{\Psi}_{\mathrm{ell}}(N)$. Recall that this means that ψ does not belong to $\widetilde{\Psi}_2(G)$, for any $G \in \widetilde{\mathcal{E}}_{\mathrm{ell}}(N)$. The global theorems reduce in this case to the two vanishing assertions of Proposition 4.5.1. However, the proposition actually imposes a further requirement that ψ also not belong to the larger elliptic set $\widetilde{\Psi}_{\mathrm{ell}}(G)$, if $G \in \widetilde{\mathcal{E}}_{\mathrm{sim}}(N)$ is simple. This second condition does not subsume the first, since it permits ψ to lie in $\widetilde{\Psi}_2(G^*)$ for some $G^* \in \widetilde{\mathcal{E}}_{\mathrm{ell}}(N)$ that is not simple; and indeed, there really was something to establish in this case. We did it in Lemma 5.1.6, which was our setting for a general proof of Theorem 1.5.3(b), and where we also completed the proof of the two vanishing assertions for pairs

$$(G, \psi), \qquad G \in \widetilde{\mathcal{E}}_{\mathrm{sim}}(N),\ \psi \in \widetilde{\Psi}_2(G^*).$$

(The corresponding two assertions do not have to be proved for the pair (G^*, ψ). They follow immediately from our induction hypothesis on N, and the fact that the linear forms $\mathrm{tr}(R^{G^*}_{\mathrm{disc},\psi_v})$ and $S^{G^*}_{\mathrm{disc},\psi}$ for composite G^* decompose into the associated products. We note, incidentally, that the roles of G^* and G here have been interchanged from Lemma 5.1.6.) Proposition 4.5.1 and Lemma 5.1.6, taken together, thus yield the global theorems for any ψ that does not belong to the elliptic set attached to any simple G.

We can therefore assume from this point on that ψ lies in $\widetilde{\Psi}_{\mathrm{ell}}(G)$, for some $G \in \widetilde{\mathcal{E}}_{\mathrm{sim}}(N)$. One sees easily from the general form (4.5.9) of S_ψ that G is uniquely determined by ψ.

Suppose first that ψ lies in the complement $\widetilde{\Psi}^2_{\mathrm{ell}}(G)$ of $\widetilde{\Psi}_2(G)$ in $\widetilde{\Psi}_{\mathrm{ell}}(G)$. Then ψ lies in the complement of $\widetilde{\Psi}_{\mathrm{ell}}(N)$. The global theorems therefore reduce to the two vanishing assertions of Proposition 4.5.1, even though the proposition itself does not apply to this case. We shall instead apply Lemma 5.2.1 or Lemma 5.2.2, according to whether the index r in the decomposition (5.2.4) of ψ is greater than 1 or equal to 1. The global intertwining relation (4.2.6) of Corollary 4.2.1, which we now know is valid, tells us that the given expressions (5.2.8), (5.2.12) and (5.2.13) in these lemmas all vanish. It follows that the remaining expressions (5.2.7) and (5.2.11) in the two cases also vanish. In other words, we have

$$\sum_{G^* \in \widetilde{\mathcal{E}}_{\mathrm{sim}}(N)} \widetilde{\imath}(N, G^*)\, \mathrm{tr}\big(R^*_{\mathrm{disc},\psi}(f^*)\big) = 0,$$

for any r, and any compatible family of functions (5.2.6). It then follows from Proposition 3.5.1 that

$$\operatorname{tr}\big(R^*_{\mathrm{disc},\psi}(f^*)\big) = 0, \qquad G^* \in \widetilde{\mathcal{E}}_{\mathrm{sim}}(N),\ f^* \in \widetilde{\mathcal{H}}(G^*). \tag{8.1.1}$$

This is one of the two vanishing assertions. The second,

$$^0S^*_{\mathrm{disc},\psi}(f^*) = 0, \qquad G^* \in \widetilde{\mathcal{E}}_{\mathrm{sim}}(N),\ f^* \in \widetilde{\mathcal{H}}(G^*), \tag{8.1.2}$$

follows from the first. Indeed, the difference (5.2.9) of the left hand sides of (8.1.1) and (8.1.2) was seen in the proofs of Lemmas 5.2.1 and 5.2.2 to equal one of the expressions (5.2.8), (5.2.12) and (5.2.13) we now know are all equal to 0.

We have thus established the global theorems when ψ does not lie in $\widetilde{\Psi}_2(G^*)$, for any $G^* \in \widetilde{\mathcal{E}}_{\mathrm{sim}}(N)$. This is a major step, even though it amounts only to a pair of vanishing formulas. Theorem 1.5.2 asserts that the contributions from parameters in $\widetilde{\Psi}_2(G)$ exhaust the automorphic discrete spectrum of G. The first formula (8.1.1), applied to $G^* = G$ (and augmented by Corollary 3.4.3), asserts that this is indeed the case. In particular, G has no embedded eigenvalues, in the sense described at the beginning of §4.3. This result has been hard won. The proof we have just completed calls upon much of what we have done since Chapter 1. At its heart are the comparisons from the standard model in §4.3 and §4.4, in which the two vanishing formulas are treated side by side. The second formula (8.1.2) is equivalent to the stable multiplicity formula (4.1.11) for the pair (G, ψ). We have now established that it holds when the group S_ψ attached to (G, ψ) is infinite. This is the case in which the coefficients $\sigma(\overline{S}^0_\psi)$ in (4.1.11) vanish, according to the property (4.1.9).

We come now to the last case, in which ψ is "square integrable". We assume henceforth that ψ lies in $\widetilde{\Psi}_2(G)$, for some $G \in \widetilde{\mathcal{E}}_{\mathrm{sim}}(N)$. In particular,

$$\psi = \psi_1 \boxplus \cdots \boxplus \psi_r$$

is multiplicity free. For this case, there will be assertions to be established from all of the global theorems. If ψ is compound, in the sense that $r > 1$, the techniques we have been using will be quite sufficient. The case $r = 1$ that ψ is simple is considerably harder. It requires something further, which we will discuss in the next section. In both cases, the stable multiplicity formula remains the key result to be proved.

Lemma 5.1.4 gives a criterion for the stable multiplicity formula to be valid for any pair (G, ψ), with $G \in \widetilde{\mathcal{E}}_{\mathrm{sim}}(N)$. It tells us that we need only show that the linear form Λ on $\widetilde{\mathcal{S}}^0(L)$ vanishes. We recall that L and Λ are defined only if N is even and $\eta_\psi = 1$, or equivalently, if there is a second group $G^\vee \in \widetilde{\mathcal{E}}_{\mathrm{sim}}(N)$ with $\eta_{G^\vee} = 1$. Then $L \cong GL(\frac{1}{2}N)$ represents a maximal Levi subgroup of both G and $G^\vee$, and Λ is the obstruction for the stable linear form $S^\vee_{\mathrm{disc},\psi} = S^{G^\vee}_{\mathrm{disc},\psi}$ on $G^\vee$ to vanish as expected. Recall also that if N is odd or $\eta_\psi \neq 1$, we have already established the stable multiplicity formula for each (G^*, ψ) in Corollary 5.1.3.

Assume now that $r > 1$. We apply Lemma 5.3.1 to the supplementary parameter ψ_+. The global intertwining relation, which we now know holds in general, tells us that the expression (5.3.5) vanishes. It then follows from the lemma that the sum

$$\sum_{G^* \in \widetilde{\mathcal{E}}_{\mathrm{sim}}(N_+)} \widetilde{\iota}(N_+, G^*) \operatorname{tr}\big(R^*_{\mathrm{disc},\psi_+}(f^*)\big) + b_+ f^{L_+}(\psi_1 \times \Lambda)$$

vanishes, for a positive constant b_+, and any compatible family of functions (5.3.3). We can now argue exactly as in the proof of Lemma 5.4.5. That is, we replace the supplemental summand on the right by the linear form

$$b_+ f_1^{L_+}(\psi_1 \times \Lambda), \qquad f_1 \in \widetilde{\mathcal{H}}(G_1^\vee),$$

for the composite endoscopic datum $G_1^\vee = G_1 \times G^\vee$ in $\widetilde{\mathcal{E}}_{\mathrm{ell}}(N_+)$, with corresponding function f_1 in (5.3.3). This is a nonnegative linear combination of irreducible characters in f_1, by Corollary 5.1.5 and the application of our induction hypothesis to ψ_1. We can therefore apply Proposition 3.5.1 to the modified sum. This allows us to say that the linear form $\psi_1 \times \Lambda$ attached to the maximal Levi subgroup L_+ of $G_1^\vee$ vanishes. Since the linear form attached to ψ_1 is nonzero by our induction hypothesis, we conclude that Λ vanishes. The stable multiplicity formula thus holds for any pair (G^*, ψ), and for (G, ψ) in particular, whenever the index r for ψ is greater than 1.

Now that we have the stable multiplicity formula for parameters $\psi \in \widetilde{\Psi}_2(G)$ with $r > 1$, we shall establish Theorem 1.5.2 for any such ψ. More precisely, for any $G^* \in \widetilde{\mathcal{E}}_{\mathrm{sim}}(N)$, we shall show that the contribution of ψ to the discrete spectrum of G^* is the summand of ψ on the right hand side of the analogue for G^* of the decomposition (1.5.5) of Theorem 1.5.2.

If $G^* \in \widetilde{\mathcal{E}}_{\mathrm{sim}}(N)$ is distinct from G, ψ does not belong to $\widetilde{\Psi}(G^*)$, by definition. There is consequently no contribution of ψ to the analogue for G^* of (1.5.5). According to the stable multiplicity formula for (G^*, ψ), the stable linear form $S^*_{\mathrm{disc},\psi}$ attached to (G^*, ψ) vanishes. It follows from (4.4.12) that

$$\operatorname{tr}\big(R^*_{\mathrm{disc},\psi}(f^*)\big) = S^*_{\mathrm{disc},\psi}(f^*) = 0, \qquad f^* \in \widetilde{\mathcal{H}}(G^*),$$

so there is also no contribution of ψ to the discrete spectrum of G^*. It therefore suffices to consider the case that $G^* = G$. Since ψ belongs to $\widetilde{\Psi}_2(G)$, its contribution to (1.5.5) is a nontrivial multiplicity formula, rather than zero. But we have already seen how to prove this formula. The proof is the content of our conditional Lemma 4.7.1, from the end of §4.7. The conditions of this lemma (both explicit and implicit) have now all been met. The conclusion (4.7.10) of the lemma, namely that the contribution of ψ to the discrete spectrum of G is indeed the summand of ψ in (1.5.5), is therefore valid. Theorem 1.5.2 therefore holds for the compound parameter $\psi \in \widetilde{\Psi}_2(G)$.

The only other global theorem that is relevant to ψ is Theorem 4.2.2. It concerns the bitorsor $\widetilde{G}$ attached to a datum $G \in \widetilde{\mathcal{E}}_{\mathrm{sim}}(N)$, in case N is even, $\widehat{G} = SO(N, \mathbb{C})$, and ψ lies in $\widetilde{\Psi}_2(G)$. Recall that (4.2.7) provides a canonical extension

$$\widetilde{R}_{\mathrm{disc},\psi} = R^{\widetilde{G}}_{\mathrm{disc},\psi}$$

of the representation $R^G_{\mathrm{disc},\psi}$ to the $G(\mathbb{A})$-bitorsor $\widetilde{G}(\mathbb{A})$. The theorem has two assertions, according to whether ψ lies in the subset $\Psi_2(\widetilde{G})$ of $\widetilde{\Psi}_2(G)$ or not.

The argument is similar to the special case discussed, somewhat hastily, in §6.8. We form the ψ-component

$$\widetilde{I}_{\mathrm{disc},\psi}(\widetilde{f}) = I^{\widetilde{G}}_{\mathrm{disc},\psi}(\widetilde{f}), \qquad \widetilde{f} \in \mathcal{H}(\widetilde{G}),$$

of the discrete part of the twisted trace formula for $\widetilde{G}$, following (3.3.12). According to Hypothesis 3.2.1, it satisfies the analogue

$$\widetilde{I}_{\mathrm{disc},\psi}(\widetilde{f}) = \sum_{\widetilde{G}' \in \mathcal{E}_{\mathrm{ell}}(\widetilde{G})} \iota(\widetilde{G}, \widetilde{G}')\, \widehat{\widetilde{S}}{}'_{\mathrm{disc},\psi}(\widetilde{f}')$$

for $\widetilde{G}$ of (4.1.2). It also satisfies the analogue for $\widetilde{G}$ of (4.1.1). As we have observed many times in the case of a "square integrable" parameter ψ, the terms with $M \neq G$ in (4.1.1) all vanish. The remaining term with $M = G$ is the trace of the operator $\widetilde{R}_{\mathrm{disc},\psi}(\widetilde{f})$, by the general definitions at the end of §3.1. It follows that

$$\mathrm{tr}\big(\widetilde{R}_{\mathrm{disc},\psi}(\widetilde{f})\big) = \sum_{\widetilde{G}' \in \mathcal{E}_{\mathrm{ell}}(\widetilde{G})} \iota(\widetilde{G}, \widetilde{G}')\, \widehat{\widetilde{S}}{}'_{\mathrm{disc},\psi}(\widetilde{f}'). \tag{8.1.3}$$

Suppose that ψ lies in the complement of $\Psi_2(\widetilde{G})$ in $\widetilde{\Psi}_2(G)$. Then ψ is not the imge of any parameter in $\widetilde{\Psi}(\widetilde{G}')$, for any twisted endoscopic datum $\widetilde{G}' \in \mathcal{E}_{\mathrm{ell}}(\widetilde{G})$. It follows from the stable multiplicity formula for $\widetilde{G}'$, or in fact the application to $\widetilde{G}'$ of the more elementary assertion of Proposition 3.4.1, that $\widetilde{S}'_{\mathrm{disc},\psi}$ vanishes. The left hand side of (8.1.3) therefore also vanishes. This is the assertion (a) of Theorem 4.2.2.

Now suppose that ψ belongs to the subset $\Psi_2(\widetilde{G})$ of $\widetilde{\Psi}_2(G)$. The set $\widetilde{\mathcal{S}}_\psi$ is then nonempty, and is a bitorsor over $\mathcal{S}_\psi$. To establish the corresponding assertion (b) of Theorem 4.2.2, we can follow the argument from §6.8 for a generic global parameter with local constraints. Starting with (8.1.3), we obtain a generalization

$$\sum_{\pi \in \Pi(\widetilde{G})} n_\psi(\pi)\, \widetilde{f}_{\widetilde{G}}(\widetilde{\pi}) = |\mathcal{S}_\psi|^{-1} \sum_{\widetilde{x} \in \widetilde{\mathcal{S}}_\psi} \varepsilon'(\psi')\, \widetilde{f}'(\psi'), \tag{8.1.4}$$

of (6.8.4), in which $(\widetilde{G}', \psi')$ maps to $(\psi, \widetilde{x})$, $\Pi(\widetilde{G})$ is the subset of representations in $\widetilde{\Pi}(G)$ that extend to $\widetilde{G}(\mathbb{A})$, and $n_\psi(\pi)$ is the multiplicity of π in $R^G_{\mathrm{disc},\psi}$. On the left hand side, $\widetilde{\pi}$ is the canonical extension of π defined whenever $n_\psi(\pi) \neq 0$.

We noted in §4.8 that the proof of Lemma 4.4.1 in §6.6 is quite general. In particular, it carries over without change to a twisted group $\widetilde{G}$. We can therefore write

$$\sum_{\widetilde{x}\in\widetilde{\mathcal{S}}_\psi} \varepsilon'(\psi')\, \widetilde{f}'(\psi') = \sum_{\widetilde{x}\in\widetilde{\mathcal{S}}_\psi} \varepsilon_\psi(s_\psi\widetilde{x}) \sum_{\pi\in\Pi_\psi} \langle s_\psi\widetilde{x}, \widetilde{\pi}\rangle\, \widetilde{f}_{\widetilde{G}}(\widetilde{\pi}),$$

according to the general argument of §4.7. With a change of variables in the sum over $\widetilde{x}$, we see that the right hand of (8.1.4) equals

$$|\mathcal{S}_\psi|^{-1} \sum_{\widetilde{x}\in\widetilde{\mathcal{S}}_\psi} \varepsilon_\psi^G(\widetilde{x}) \sum_{\pi\in\Pi_\psi} \langle \widetilde{x}, \widetilde{\pi}\rangle\, \widetilde{f}_{\widetilde{G}}(\widetilde{\pi}).$$

This in turn equals

$$\sum_{\pi\in\Pi_\psi(\varepsilon_\psi)} \widetilde{f}_{\widetilde{G}}(\widetilde{\pi}),$$

where $\Pi_\psi(\varepsilon_\psi)$ is the set of representations π in the global packet Π_ψ such that the linear character $\langle\,\cdot\,,\pi\rangle$ on $\mathcal{S}_\psi$ equals $\varepsilon_\psi^{-1} = \varepsilon_\psi = \varepsilon_\psi^G$, while $\widetilde{\pi}$ is the canonical extension of π determined by ε_ψ. To be more precise, we recall that Theorem 2.2.4 assigns an extension $\widetilde{\pi}$ of π from $G(\mathbb{A})$ to $\widetilde{G}(\mathbb{A})$ to any extension $\langle\,\cdot\,,\widetilde{\pi}\rangle$ of the linear form $\langle\,\cdot\,,\pi\rangle$ from $\mathcal{S}_\psi$ to $\widetilde{\mathcal{S}}_\psi$. In the last sum, $\widetilde{\pi}$ corresponds to the canonical extension of $\varepsilon_\psi = \langle\,\cdot\,,\pi\rangle$ to $\widetilde{\mathcal{S}}_\psi$.

The formula (8.1.4) thus takes the form

$$\sum_{\pi\in\Pi(\widetilde{G})} n_\psi(\pi)\, \widetilde{f}_{\widetilde{G}}(\widetilde{\pi}) = \sum_{\pi\in\Pi_\psi(\varepsilon_\psi)} \widetilde{f}_{\widetilde{G}}(\widetilde{\pi}), \qquad \widetilde{f}\in\mathcal{H}(\widetilde{G}).$$

On the left hand side, $\widetilde{\pi}$ represents the extension of π determined by $\widetilde{R}^G_{\mathrm{disc},\psi}$, while on the right hand side, it stands for the extension determined by ε_ψ. The formula tells us that the two extensions match, as asserted in Theorem 4.2.2(b). This completes the proof of Theorem 4.2.2, and hence our study of global parameters $\psi\in\widetilde{\Psi}_2(G)$ with $r>1$.

Suppose now that $r=1$. In other words, $\psi\in\widetilde{\Psi}_{\mathrm{sim}}(G)$ is a simple parameter for the group $G\in\widetilde{\mathcal{E}}_{\mathrm{sim}}(N)$. As we have noted, this case will be more of a challenge. We shall begin the discussion here, leaving the main burden of proof for the next section.

There are two general problems in this case. One remains the stable multiplicity formula, which entails proving that the linear form Λ vanishes. We shall soon see how to reduce this to the case that ψ is not generic. The other is the assertion of Theorem 1.5.3(a) for generic parameters $\psi\in\widetilde{\Psi}_{\mathrm{sim}}(G)$. In §5.3, we formulated it in terms of a sign δ_ψ, which equals 1 or -1 according to whether the assertion is valid for ψ or not. This of course applies only to the case that ψ is generic. There will thus be one problem to solve if ψ is not generic, and another if it is.

We remind ourselves that we are working in the general framework of Chapter 5. In particular, we are following the convention at the beginning of Chapter 5 for the definition of the subset $\widetilde{\Phi}_{\mathrm{sim}}(G)$ of $\widetilde{\Psi}_{\mathrm{sim}}(G)$. Once we have established the stable multiplicity formula in case ψ is generic, we will be able to say that G is uniquely determined by ϕ, and hence that $\widetilde{\Phi}_{\mathrm{sim}}(N)$ is a disjoint union over $G \in \widetilde{\mathcal{E}}_{\mathrm{sim}}(N)$ of the sets $\widetilde{\Phi}_{\mathrm{sim}}(G)$. This leads directly to a proof of Theorem 1.4.1. It will therefore tell us that the temporary definition of $\widetilde{\Phi}_{\mathrm{sim}}(G)$ is equivalent to the original definition in terms of this theorem. We recall that these matters were treated in §5.4, when $\widetilde{\mathcal{F}}$ was the family with local constraints introduced there. We are of course now taking $\widetilde{\mathcal{F}}$ to be either the full set $\widetilde{\Psi}$ or its subset $\widetilde{\Phi}$ of generic parameters.

We begin our investigation of the simple parameter $\psi \in \widetilde{\Psi}_{\mathrm{sim}}(G)$ as we did in the compound case above. That is, we turn to the only result available to us, the relevant lemma from §5.3. This is Lemma 5.3.2, which we apply to the supplementary parameter

$$\psi_+ = \psi \boxplus \psi,$$

and a compatible family of functions (5.3.3).

Consider the terms in Lemma 5.3.2. We first note that the global intertwining relation holds for the pairs (G_+, ψ_+) and $(G_+^\vee, \psi_+)$, since it follows from the local intertwining relation. It gives identities

$$f'_{G_+}(\psi_+, s_{\psi_+} x_1) = f_{G_+}(\psi_+, x_1)$$

and

$$(f^\vee)^{M_+}(\psi) = (f^\vee)'_{G_+^\vee}(\psi_+, x_1) = f^\vee_{G_+^\vee}(\psi_+, x_1)$$

between the terms in the two expressions (5.3.22) and (5.3.23) of the lemma. The difference between (5.3.22) and (5.3.23) therefore equals

$$\tfrac{1}{8}(1-\delta_\psi)(f^\vee)^{M_+}(\psi) - \tfrac{1}{8}(1-\delta_\psi) f_{G_+}(\psi_+, x_1),$$

an expression we can write as

$$\tfrac{1}{8}(1-\delta_\psi)\big(f^{M_+}(\psi) - f_{G_+}(\psi_+, x_1)\big),$$

since

$$(f^\vee)^{M_+}(\psi) = f^{M_+}(\psi).$$

Now $f^{M_+}(\psi)$ represents a unitary character on the group $G_+(\mathbb{A})$. It is induced from the Langlands quotient representation π_ψ of the maximal Levi subgroup $M_+(\mathbb{A}) \cong GL(N, \mathbb{A})$ of $G_+(\mathbb{A})$. By definition, $f_{G_+}(\psi_+, x_1)$ is a difference of two unitary characters on $G_+(\mathbb{A})$ whose sum equals $f^{M_+}(\psi)$. Since δ_ψ equals 1 or -1, the last expression is therefore a nonnegative linear combination of irreducible characters. Lemma 5.3.2 tells us that its sum with

$$\sum_{G^* \in \widetilde{\mathcal{E}}_{\mathrm{sim}}(N_+)} \tilde{\imath}(N_+, G^*)\, \mathrm{tr}\big(R^*_{\mathrm{disc}, \psi_+}(f^*)\big) + \Big(\tfrac{1}{2} f^{L_+}(\Gamma \times \Lambda)\Big),$$

the expression we labelled (5.3.21) in the statement of the lemma, vanishes. Arguing as in the proof of Lemma 5.4.6, we replace the supplemental summand in (5.3.21) by the linear form

$$\tfrac{1}{2} f_1^{L_+}(\Gamma \times \Lambda), \qquad f_1 \in \widetilde{\mathcal{H}}(G_1^\vee),$$

for the composite endoscopic datum $G_1^\vee = G \times G^\vee$ in $\widetilde{\mathcal{E}}_{\mathrm{sim}}(N_+)$, with corresponding function f_1 from the family (5.3.3). By Corollary 5.1.5 and the definition of Γ, this is a nonnegative linear combination of irreducible characters in f_1. The same is of course true of the other summands in (5.3.21).

The terms in Lemma 5.3.2 thus combine to give an expression of the general form (3.5.1). The lemma itself asserts that the expression so obtained vanishes. It then follows from Proposition 3.5.1 that the corresponding coefficients vanish. In particular, we see that

$$R^*_{\mathrm{disc},\psi_+}(f^*) = 0, \qquad G^* \in \widetilde{\mathcal{E}}_{\mathrm{sim}}(N_+),\ f^* \in \widetilde{\mathcal{H}}(G^*), \tag{8.1.5}$$

that

$$\Gamma \times \Lambda = 0,$$

and that

$$(1 - \delta_\psi)\big(f^{M_+}(\psi) - f_{G_+}(\psi_+, x_1)\big) = 0.$$

The second of these equations tells us that either $\Lambda = 0$, which is what we want, or $\Gamma = 0$. In other words, we have

$$\Gamma = 0, \qquad \text{if } \Lambda \neq 0. \tag{8.1.6}$$

The third equation tells us that either $\delta_\psi = 1$, which is what we want, or the complementary difference of linear forms on the left hand side vanishes. In other words, we have

$$f^{M_+}(\psi) = f_{G_+}(\psi_+, x_1), \qquad \text{if } \delta_\psi \neq 1, \tag{8.1.7}$$

for any function $f \in \widetilde{\mathcal{H}}(G_+)$. What is one to make of these conditions?

Recall that $S^G_{\mathrm{disc},\psi}(f)$ equals a scalar multiple of $f^G(\Gamma)$, according to the definition prior to Lemma 5.3.2. If ψ is generic, $S^G_{\mathrm{disc},\psi}$ is nonzero, by our definition of $\widetilde{\Phi}_{\mathrm{sim}}(G)$. It follows that $\Gamma \neq 0$, and therefore that $\Lambda = 0$. We note in passing that this argument could also have been used in the proof of Lemma 5.4.6. The local condition (5.4.1)(a) we used instead is parallel to its nontempered analogue needed for the proof of Lemma 7.3.2. In any case, we have now established the stable multiplicity formula for generic ψ. This is the result we anticipated above. It resolves the temporary definition of $\widetilde{\Phi}_{\mathrm{sim}}(G)$ from §5.1 in terms of the original definition by Theorem 1.4.1, and thus completes our inductive definition of the full set $\widetilde{\Psi}(G)$.

This leaves only the two problems mentioned above. We must show that $\delta_\psi = 1$ if ψ is generic, and that $\Lambda = 0$ if ψ is not generic. At the beginning of the next section, we shall see whether we can solve the two problems by applying the conditions (8.1.7) and (8.1.6).

8.2. Proof by contradiction

The task now is to finish proving the most intractable case of the global theorems. In the last section, we fixed a positive integer N, and imposed the usual induction hypothesis that the theorems all hold for global parameters $\psi \in \widetilde{\Psi}$ with $\deg(\psi) < N$. We then described how our earlier results yield the global theorems for parameters of degree N in the complement of $\widetilde{\Psi}_{\mathrm{sim}}(N)$. It remains to treat the simple global parameters $\psi \in \widetilde{\Psi}_{\mathrm{sim}}(N)$.

In the last section we also obtained information from the application of Lemma 5.3.2. In particular, we proved the stable multiplicity formula in the case that ψ lies in the subset $\widetilde{\Phi}_{\mathrm{sim}}(G)$ of generic parameters in $\widetilde{\Psi}_{\mathrm{sim}}(G)$. It remains only to establish the condition $\delta_\psi = 1$ of Theorem 1.5.3(a) in this case. In the complementary case that $\psi \in \widetilde{\Psi}_{\mathrm{sim}}(G)$ is not generic, $\delta_\psi = 1$ by definition. In this case, it is only the stable multiplicity formula for ψ that needs to be established. We thus have the two separate properties to establish, one when ψ is generic, and the other when it is not. In each case, there is a corresponding condition (8.1.7) or (8.1.6) we can bring to bear on the proof.

Are the conditions (8.1.7) and (8.1.6) we have obtained from the application of Lemma 5.3.2 sufficient to establish the final two properties? Our initial answer to this question is not likely to be encouraging. However, with patience, we will see that it eventually leads to a happy conclusion.

We shall study the question for a general pair

$$(G, \psi), \qquad G \in \widetilde{\mathcal{E}}_{\mathrm{sim}}(N), \ \psi \in \widetilde{\Psi}_{\mathrm{sim}}(G).$$

The most critical case is that of N even and $n_\psi = 1$. This is the case in which G is split, and where we have the auxiliary objects $G^\vee$, L and Λ. We shall discuss these objects in treating the general case, with the implicit understanding that any terms in which they occur vanish if N is odd or $n_\psi \neq 1$.

Consider the condition (8.1.6) that either Λ or Γ must vanish. If it is Λ that vanishes, as required, we see from Lemma 5.1.4 that

$$\begin{aligned} S^G_{\mathrm{disc},\psi}(f) &= |\Psi(G,\psi)|\,|\mathcal{S}_\psi|^{-1}\,\varepsilon^G(\psi)\,f^G(\psi) - \widetilde{o}(G)\,f^L(\Lambda) \\ &= \widetilde{o}(G)\cdot 1 \cdot 1 \cdot f^G(\psi) = \widetilde{o}(G)\,f^G(\psi), \end{aligned}$$

and that

$$S^{G^\vee}_{\mathrm{disc},\psi}(f^\vee) = \widetilde{o}(G^\vee)\,f^{\vee,L}(\Lambda) = 0,$$

for functions $f \in \widetilde{\mathcal{H}}(G)$ and $f^\vee \in \widetilde{\mathcal{H}}(G^\vee)$ that we may as well assume come from a compatible family. These are the stable multiplicity formulas for (G,ψ) and $(G^\vee,\psi)$. In this case, we also obtain the local identity

$$f^G(\psi) = f^G(\Gamma) + f^L(\Lambda) = f^G(\Gamma),$$

from the formula obtained in §5.3 prior to Lemma 5.3.2. On the other hand, if it is Γ that vanishes, we have the local identity

$$f^G(\psi) = f^G(\Gamma) + f^L(\Lambda) = f^L(\Lambda).$$

In particular, the stable distribution on $G(\mathbb{A})$ attached to the original parameter ψ transfers locally from the Levi subgroup $L(\mathbb{A})$. In this case, it follows from Lemma 5.1.4 that

$$\begin{aligned} S^G_{\mathrm{disc},\psi}(f) &= |\Psi(G,\psi)|\, |\mathcal{S}_\psi|^{-1}\, \varepsilon^G(\psi)\, f^G(\psi) - \widetilde{o}(G)\, f^L(\Lambda) \\ &= \widetilde{o}(G)\cdot 1\cdot 1\cdot f^L(\Lambda) - \widetilde{o}(G)\, f^L(\Lambda) = 0, \end{aligned}$$

and that

$$S^{G^\vee}_{\mathrm{disc},\psi}(f^\vee) = \widetilde{o}(G^\vee)\, f^{\vee,L}(\Lambda) = \widetilde{o}(G^\vee)\, f^G(\psi).$$

Consider the case that ψ is generic. Then $\Lambda = 0$, as we have seen. The problem here is to show that $\delta_\psi = 1$. Assume the contrary, namely that $\delta_\psi = -1$. For any function $f \in \widetilde{\mathcal{H}}(G_+)$, we obtain identities

$$\begin{aligned} f^{M_+}(\psi) = f_{G_+}(\psi_+, x_1) &= f'_{G_+}(\psi_+, x_1) \\ &= f^{G\times G}(\psi\times\psi) = f^{G\times G}(\Gamma\times\Gamma), \end{aligned}$$

from (8.1.7), the global intertwining relation, and the definition of f'_{G_+} in terms of the endoscopic transfer $f^{G\times G}$ of f to the endoscopic datum $G'_+ = G\times G$ for G_+. It then follows from Corollary 5.3.3 and (8.1.5) that

$$S^{G_+}_{\mathrm{disc},\psi_+}(f) = -\tfrac{1}{2}\,\widetilde{o}(G_+)\, f^{M_+}(\psi).$$

In addition, if $f^\vee \in \widetilde{\mathcal{H}}(G^\vee_+)$ lies in a compatible family that contains f, we obtain an identity

$$f^{M_+}(\psi) = f^\vee_{G^\vee_+}(\psi_+, x_1),$$

as above. It then follows from Corollary 5.3.3, (8.1.5), and the vanishing of Λ that

$$S^{G^\vee_+}_{\mathrm{disc},\psi_+}(f^\vee) = \tfrac{1}{4}\,\widetilde{o}(G^\vee_+)\, f^{M_+}(\psi).$$

To summarize what we have obtained so far, we take two pairs of functions

$$\{f_1 \in \widetilde{\mathcal{H}}(G),\ f_1^\vee \in \widetilde{\mathcal{H}}(G^\vee)\}$$

and

$$\{f_2 \in \widetilde{\mathcal{H}}(G_+),\ f_2^\vee \in \widetilde{\mathcal{H}}_+(G^\vee_+)\}$$

that are compatible, in the sense that they belong to compatible families for $\widetilde{\mathcal{E}}_{\mathrm{ell}}(N)$ and $\widetilde{\mathcal{E}}_{\mathrm{ell}}(N_+)$ respectively. We then have

$$
\begin{cases}
S^{G}_{\mathrm{disc},\psi}(f_1) = \widetilde{\delta}(G)\, f_1^G(\psi), \\
S^{G^\vee}_{\mathrm{disc},\psi}(f_1^\vee) = 0, \\
S^{G_+}_{\mathrm{disc},\psi_+}(f_2) = -\tfrac{1}{2}\,\widetilde{\delta}(G_+)\, f_2^{M_+}(\psi), \\
S^{G_+^\vee}_{\mathrm{disc},\psi_+}(f_2^\vee) = \tfrac{1}{4}\,\widetilde{\delta}(G_+^\vee)\, f_2^{M_+}(\psi),
\end{cases}
\tag{8.2.1}
$$

under the assumption that $\delta_\psi = -1$ (so that ψ is generic and $\Lambda = 0$).

Suppose next that ψ is not generic. Then $\delta_\psi = 1$ by definition. However, in this case we will still have to show that $\Lambda = 0$. Assume the contrary, namely that $\Lambda \neq 0$, and consequently that $\Gamma = 0$. The first condition implies that N is even and G is split. In particular, we can work with the dual group $\widehat{G}$ in place of ${}^LG$, if we choose. The second condition implies that the invariant linear form on $G(\mathbb{A})$ attached to ψ is the pullback of the invariant linear form Λ on the Levi subgroup $L(\mathbb{A})$. From this property, we shall establish the following consequence of the results of §7.5.

LEMMA 8.2.1. *With our assumption that $\Gamma = 0$, the localization ψ_v of the parameter $\psi \in \widetilde{\Psi}_{\mathrm{sim}}(N)$ at any valuation v factors through the Levi subgroup $\widehat{L}$ of $\widehat{G}$.*

PROOF. We are working with a simple, self-dual parameter ψ in the subset $\widetilde{\Psi}_{\mathrm{sim}}(G)$ of $\widetilde{\Psi}_{\mathrm{sim}}(N)$. It has the usual decomposition

$$\psi = \mu \boxtimes \nu, \qquad \mu \in \widetilde{\Phi}_{\mathrm{sim}}(m),\ \nu = \nu^n,$$

for positive integers m and n with $N = mn$. The unipotent component ν has degree $n > 1$, by our assumption that ψ is not generic. The generic component μ has a determinant η_μ, which is an idèle class character with $\eta_\mu^2 = 1$, since μ is self-dual. It will not be hard to show that the generic degree m is even, and that the character η_μ equals 1. These properties follow from a global argument we shall give at the end of the proof. In the meantime, we shall establish the lemma under the assumption that they hold. This part of the argument is purely local. We shall describe it in local notation that is compatible with the discussion of §7.5.

We therefore take F to be local until near the end of the proof, at which point we will revert to the global notation. We take $G \in \widetilde{\mathcal{E}}_{\mathrm{sim}}(N)$ to be a split endoscopic datum over F, and $\psi \in \widetilde{\Psi}^+(G)$ to be a local parameter. We are assuming that $N = mn$, where $n > 1$ and m is even, and that ψ is of the form

$$\psi = \mu \otimes \nu^n, \qquad \mu \in \widehat{\Psi}^+(m),\ \nu = \nu^n,$$

where ν^n is the irreducible representation of $SU(2)$ of degree n and μ has determinant η_μ equal to 1. The generic factor μ is a self-dual representation

$$\mu = \bigoplus_{i=1}^{r} \ell_i\, \mu_i$$

of L_F, for positive integers ℓ_i and distinct simple elements $\mu_i \in \Phi_{\mathrm{sim}}(m_i)$ such that

$$m = \ell_i\, m_i + \cdots + \ell_r\, m_r.$$

The split group G has a Siegel maximal Levi subgroup $L \cong GL(N/2)$. Our primary assumption on ψ is that the stable character

$$f^G(\psi), \qquad f \in \widetilde{\mathcal{H}}(G),$$

is induced from a linear form on the Levi subgroup $L(F)$ of $G(F)$. We must use this to deduce that the homomorphism

$$\psi : \; L_F \times SU(2) \longrightarrow \widehat{G}$$

factors through the subgroup $\widehat{L}$ of $\widehat{G}$.

The embedding of $\widehat{L}$ into $\widehat{G}$ can of course be taken to be

$$x \longrightarrow (x, x^\vee), \qquad x \in \widehat{L} = GL(N/2, \mathbb{C}).$$

Since μ is self-dual, there is an involution $i \to i^\vee$ on the set of indices i such that $\mu_{i^\vee} = \mu_i^\vee$ and $\ell_{i^\vee} = \ell_i$. An orbit of order 2 contributes a subrepresentation

$$\ell_i(\mu_i \otimes \nu^n) \oplus \ell_i(\mu_i^\vee \otimes \nu^n)$$

of ψ that factors through $\widehat{L}$. Similarly, for any i with $\mu_i^\vee = \mu_i$, the even part of ℓ_i contributes a subrepresentation

$$2\ell_i'(\mu_i \otimes \nu^n), \qquad \ell_i' = [\ell_i/2],$$

that also factors through $\widehat{L}$. If ℓ_i is odd, however, there will be a supplementary subrepresentation $(\mu_i \otimes \nu^n)$ of ψ whose image cannot also be embedded in $\widehat{L}$. We have therefore to show that the set of indices

$$I_- = \{i : \; \mu_i^\vee = \mu_i, \; \ell_i \text{ odd}\}$$

is empty. (This set was discussed under different circumstances in §2.4, where it was denoted $I^{+,o}$.)

Suppose first that $F = \mathbb{C}$. A component μ_i is then a quasicharacter on $\mathbb{C}^*$, which is self-dual if and only if it equals 1. If there is an i with $\mu_i = 1$, the corresponding multiplicity ℓ_i must be even, since m is even, and the other components occur in pairs. Therefore I_- is empty and ψ factors through $\widehat{L}$, as required.

Suppose then that F is real or p-adic. We assume that the subset of indices I_- is not empty. Our task is to obtain a contradiction of the condition on the stable character $f^G(\psi)$. By the remarks above, it suffices to show that if

$$\psi_- = \bigoplus_{i \in I_-} (\mu_i \otimes \nu^n)$$

and

$$N_- = \sum_{i \in I_-} m_i n,$$

and if G_- is the datum in $\widetilde{\mathcal{E}}_{\text{sim}}(N_-)$ such that ψ_- belongs to $\widetilde{\Psi}(G_-)$, then the stable character

$$f_-^{G_-}(\psi_-), \qquad f_- \in \widetilde{\mathcal{H}}(G_-),$$

is not induced from the maximal Levi subgroup L_- of G_-. The answer lies in the decomposition (2.2.12) (with (G_-, ψ_-) in place of (G, ψ)) of this stable character in terms of standard stable characters. Since standard stable characters on G_- are defined directly by the twisted transfer of standard characters from $GL(N_-)$, it suffices to consider the summands in the decomposition (2.2.9) (with (N_-, ψ_-)) in place of (N, ψ) of the twisted irreducible character

$$\widetilde{f}_- \longrightarrow \widetilde{f}_{-,N_-}(\psi_-) = \operatorname{tr}\big(\pi_{\psi_-}(\widetilde{f}_-)\big), \qquad \widetilde{f}_- \in \widetilde{\mathcal{H}}(N_-),$$

into twisted standard characters.

We shall apply the character formulas of §7.5 to the direct summands

$$\psi_{-,i} = \mu_i \otimes \nu^n, \qquad i \in I_-,$$

of ψ_-. As an irreducible representation of L_F of degree m_i, μ_i is self-dual and therefore unitary. If F is p-adic, L_F is of course a product $W_F \times SU(2)$. We write

$$\mu_i = r \otimes \nu^{k+1}, \qquad m_i = m_r(k+1),$$

where r is an irreducible unitary representation of L_F of degree m_r, and k is a nonnegative integer. The irreducible character on $GL(m_i n, F)$ attached to $(\mu_i \otimes \nu^n)$ then equals $\theta_r^n(k)$, in the notation of §7.5. If $F = \mathbb{R}$, m_i equals 1 or 2. In case $m_i = 2$, μ_i parametrizes a self-dual representation of $GL(2, \mathbb{R})$ in the relative discrete series. This in turn is parametrized by a positive integer $k \in \mathbb{N}$. The irreducible character on $GL(2n, F)$ attached to $(\mu_i \otimes \nu^n)$ equals $\theta^n(k)$, in the notation of §7.5. In case $m_i = 1$, there are two possibilities for the self-dual linear character μ_i of $\mathbb{R}^*$, either the trivial character 1 or the sign character $\varepsilon_{\mathbb{R}}$. But it follows from the fact that m is even that there is a second index $i' \in I_-$ such that $\mu_{i'} = \varepsilon_{\mathbb{R}}\mu_i$. The irreducible character on $GL(2n, F)$ attached to $(\mu_i \oplus \mu_{i'}) \otimes \nu^n$ then equals $\theta^n(0)$, again in the notation of §7.5.

We thus obtain a decomposition

$$\theta(\psi_-) = \boxplus_{i \in I_-} \theta(\psi_{-,i}) = \boxplus_{(r,k)} \theta_r^n(k)$$

from (7.5.13) and the surjective correspondence of indices

$$i \longrightarrow (r, k), \qquad i \in I_-, \ k \geqslant 0,$$

where r is understood to be trivial if $F = \mathbb{R}$. The correspondence is almost a bijection, in that the fibre of (r, k) in I_- consists of one element unless $F = \mathbb{R}$ and $k = 0$, in which case it is a pair $\{i, i'\}$. We fix an index (r_1, k_1) with k_1 maximal. If $F = \mathbb{R}$, $(r_1, k_1) = k_1$ is uniquely determined. In this case, we have $k_1 \geqslant 1$. For if $k_1 = 0$, we obtain an identity

$$\eta_\mu = \prod_{i=1}^{r} (\eta_{\mu_i})^{\ell_i} = \prod_{i \in I_-} \eta_{\mu_i} = 1 \cdot \varepsilon_{\mathbb{R}} = \varepsilon_{\mathbb{R}},$$

which contradicts the condition $\eta_\mu = 1$. If F is p-adic, there could be several r paired with k_1. We may as well choose r_1 from among these so that m_{r_1} is maximal.

We shall apply Lemma 7.5.2 and Corollary 7.5.5 to the direct summand $\theta^n_{r_1}(k_1)$ of $\theta(\psi_-)$, or rather the tempered constituent $\phi^{n,*}_{r_1}(k_1)$ of this summand. This gives an extremal factor $\theta_{r_1}(k_1 + n - 1, 0)$ of $\phi_{r_1}(k_1 + n - 1, 0)$, which can be compared with the other factors $\theta_r(\ell, \lambda)$ from the decompositions (7.5.3) and (7.5.4) of any direct summand $\theta^n_r(k)$ of $\theta(\psi_-)$. It would not be hard to see directly that $\theta_n(k_1 + n - 1, 0)$ is distinct from all of these other factors. This would lead us to the required contradiction.

However, it is perhaps more natural (if a little less direct) simply to restrict our attention to the tempered constituent

$$\theta(\phi^*_{\psi_-}) = \boxplus_{i \in I_-} \theta(\phi^*_{\psi_{-,i}}) = \boxplus_{(r,k)} \phi^{n,*}_r(k)$$

of $\theta(\psi_-)$. That this suffices is a consequence of the parametrization of standard characters (ordinary or twisted) for $GL(N_-)$ by the associated set of Langlands parameters (either $\Phi(N_-)$ or $\widetilde{\Phi}(N_-)$), and the fact that these objects form a basis of the space of all virtual characters (ordinary or twisted) for $GL(N_-)$. For it can then be seen, using the germ expansions of twisted characters around singular points for example, that if a given finite linear combination of twisted standard characters $\{\widetilde{\theta}\}$ is induced from a distribution $\widetilde{\Lambda}_-$ on the maximal Levi subset

$$\widetilde{L}_-(F) = \big(GL(N_-/2, F) \times GL(N_-/2, F)\big) \rtimes \widetilde{\theta}(N_-)$$

of $\widetilde{G}(N_-, F)$, the same is true of each $\widetilde{\theta}$. (The symbol $\widetilde{\theta}(N_-)$ here denotes the standard outer automorphism of $GL(N_-)$, or rather, its restriction to the Levi subgroup $L_- = \widetilde{L}^0_-$. It is not to be confused with any of the given twisted characters $\widetilde{\theta}$.) In other words, the Langlands parameter of each $\widetilde{\theta}$ factors through the L-group of L_-.

We are assuming that the twisted character

$$\widetilde{\theta}(\psi_-) : \widetilde{f}_{-,N_-} \longrightarrow \widetilde{f}_{-,N_-}(\psi_-)$$

is induced from such a $\widetilde{\Lambda}_-$. We know from Lemma 7.5.2(i) and Corollary 7.5.5 that the twisted tempered character $\widetilde{\theta}(\phi^*_{\psi_-})$ attached to $\theta(\phi^*_{\psi_-})$ occurs in the decomposition (2.2.9) of $\widetilde{\theta}(\psi_-)$ into twisted standard characters. To obtain a contradiction, we need only show that the self-dual Langlands parameter $\phi^*_{\psi_-}$ of $\theta(\phi^*_{\psi_-})$ does not factor through the L-group of $\widetilde{L}^0_-$.

Lemma 7.5.2(ii) tells us that the factor $\theta_{r_1}(k_1 + n - 1, 0)$ occurs with multiplicity 1 in the direct summand $\phi^{n,*}_{r_1}(k_1)$ of $\theta(\phi^*_{\psi_-})$. Here we are relying on the fact that $k_1 \geqslant 1$ if $F = \mathbb{R}$ (a condition that need not hold if F is p-adic, but which is then superfluous to Lemma 7.5.2(ii)). The integral argument $(n + k_1 - 1)$ in this factor is maximal. It is in fact strictly larger that the arguments in any of the other factors of direct summands $\phi^{n,*}_r(k)$ of $\theta(\phi^*_{\psi_-})$, except for the maximal arguments from pairs

$$(r, k_1), \qquad r \neq r_1,$$

in the p-adic case. But the standard characters $\theta_r(n + k_1 - 1, 0)$ and $\theta_{r_1}(n+k_1-1, 0)$ for $GL\big(m_r(k_1+1)\big)$ are distinct, since the same is true of the corresponding Langlands parameters. It follows that there is a self-dual factor in the decomposition of $\theta(\phi^*_{\psi_-})$ that occurs with multiplicity 1. In other words, there is an irreducible self-dual subrepresentation of the Langlands parameter $\phi^*_{\psi_-} \in \widetilde{\Phi}(N_-)$ that occurs with multiplicity 1. This precludes the possibility that $\phi^*_{\psi_-}$ factor through the L-group of $\widetilde{L}^0_-$, and gives us the required contradiction.

We have shown that the existence of indices in I_- leads to a contradiction. The set I_- is therefore empty. This implies that the parameter $\psi \in \widetilde{\Psi}(G)$ factors through $\widehat{L}$, as we have seen. We have thus established the assertion of the lemma, in local notation, under the assumption that m is even and $\eta_\mu = 1$. This assumption was actually used only in the case that F is archimedean, but of course, we still have to justify it.

We return to the global notation we have been using throughout Chapter 8. Then F, G and $\psi = \mu \boxtimes \nu$ are the global objects from the beginning of the proof. It remains to show that $m = \deg(\mu)$ is even, and that the global idèle class character η_μ equals 1.

Let v be any valuation of F at which G and ψ are unramified. We are free to apply the local argument above to the completion

$$\psi_v = \mu_v \otimes \nu^n$$

of ψ. Since the assumption on m and η_{μ_v} was not needed in this case, the argument tells us that the given condition on the stable character

$$f^G_v(\psi_v), \qquad f_v \in \widetilde{\mathcal{H}}(G_v),$$

namely that it is induced from $L(F_v)$, implies that the local parameter ψ_v factors through the image of $\widehat{L}$ in $\widehat{G}$. This in turn implies that m is even, as required, and that the local determinant η_{μ_v} equals 1. It then follows from Tchebotarev density theorem that the global determinant η_μ equals 1, as required. In particular, $\eta_{\mu_v} = 1$ for any valuation v of F, the condition we used in the archimedean case of the local argument above. This completes the global argument, and hence the proof of the lemma. □

The lemma implies that any localization $\psi_{+,v}$ of ψ_+ factors through the diagonal image of $\widehat{L}$ in the Levi subgroup

$$\widehat{L} \times \widehat{L} \cong GL(N/2,\mathbb{C}) \times GL(N/2,\mathbb{C})$$

of $\widehat{G}_+$. The centralizer of the image of $\widehat{L}$ in $\widehat{G}_+$ can be identified with the connected group

$$GL(2,\mathbb{C}) \times GL(2,\mathbb{C}),$$

which contains the global centralizer

$$O(2,\mathbb{C}) \cong S_{\psi_+}.$$

It follows that S_{ψ_+} maps to the identity component of $\mathcal{S}_{\psi_{+,v}}$. Applying this to the nontrivial point x_1 in S_{ψ_+}, we see from the global intertwining relation that $f_{G_+}(\psi_+, x_1)$ equals $f^{M_+}(\psi)$, for any $f \in \widetilde{\mathcal{H}}(G_+)$. In other words, the condition (8.1.7) holds in the present case, even though $\delta_\psi = 1$. It then follows from Corollary 5.3.3 and (8.1.5) that

$$S^{G_+}_{\mathrm{disc},\psi_+}(f) = \tfrac{1}{4}\,\widetilde{o}(G_+)\,f^{M_+}(\psi).$$

If $f^\vee \in \widetilde{\mathcal{H}}(G^\vee_+)$ belongs to some compatible family that contains f, it again follows from Corollary 5.3.3 and (8.1.5), in combination with the global intertwining relation and the fact that the group $S^\vee_{\psi_+} \cong Sp(2,\mathbb{C})$ is connected, that

$$\begin{aligned} S^{G^\vee_+}_{\mathrm{disc},\psi_+}(f^\vee) &= -\tfrac{1}{4}\,\widetilde{o}(G^\vee_+)\,\big(f^\vee_{G^\vee_+}(\psi_+, x_1) + f^{\vee,L\times L}(\Lambda\times\Lambda)\big) \\ &= -\tfrac{1}{4}\,\widetilde{o}(G^\vee_+)\big(f^{M_+}(\psi) + f^{M_+}(\psi)\big) \\ &= -\tfrac{1}{2}\,\widetilde{o}(G^\vee_+)\,f^{M_+}(\psi). \end{aligned}$$

We thus have a second family of identities

$$\text{(8.2.2)}\qquad \begin{cases} S^G_{\mathrm{disc},\psi}(f_1) = 0, \\ S^{G^\vee}_{\mathrm{disc},\psi}(f_1^\vee) = \widetilde{o}(G^\vee)\,f_1^G(\psi), \\ S^{G_+}_{\mathrm{disc},\psi_+}(f_2) = \tfrac{1}{4}\,\widetilde{o}(G_+)\,f_2^{M_+}(\psi), \\ S^{G^\vee_+}_{\mathrm{disc},\psi_+}(f_2^\vee) = -\tfrac{1}{2}\,\widetilde{o}(G^\vee_+)\,f_2^{M_+}(\psi), \end{cases}$$

under the assumption that $\Lambda \neq 0$ (so that ψ is not generic and $\delta_\psi = 1$), and for functions f_1, $f_1^\vee$, f_2 and $f_2^\vee$ as in (8.2.1). Notice the similarity with (8.2.1). The two families (8.2.1) and (8.2.2) are completely parallel, but with G and G_+ interchanged with their counterparts $G^\vee$ and $G^\vee_+$.

For comparison, we can also write down what the equations in (8.2.1) and (8.2.2) would be if $\delta_\psi = 1$ and $\Lambda = 0$, as expected. We obtain

$$(8.2.3)\qquad \begin{cases} S^G_{\mathrm{disc},\psi}(f_1) = \widetilde{o}(G)\, f^G_1(\psi), \\ S^{G^\vee}_{\mathrm{disc},\psi}(f_1^\vee) = 0, \\ S^{G_+}_{\mathrm{disc},\psi_+}(f_2) = 0, \\ S^{G^\vee_+}_{\mathrm{disc},\psi_+}(f_2^\vee) = -\frac{1}{4}\,\widetilde{o}(G^\vee_+)\, f_2^{M_+}(\psi), \end{cases}$$

under the assumption $\delta_\psi = 1$ and $\Lambda = 0$ (which imposes no further conditions on ψ), and for functions f_1, $f_1^\vee$, f_2 and $f_2^\vee$ as in (8.2.1) and (8.2.2). These follow from Corollary 5.3.3 and discussion above, particularly (8.1.5) and the preceding applications of the global intertwining relation. They would also follow easily from the (as yet unproven) stable multiplicity formula. We are of course trying to establish the conditions $\delta_\psi = 1$ and $\Lambda = 0$. Our task is therefore to focus on the equations (8.2.1) and (8.2.2) that would follow if the conditions are not met.

However, the equations in (8.2.1) and (8.2.2) are vaguely disturbing. They contain nothing that is at odds with anything we have established so far. As a matter of fact, their symmetry seems to give them an air of authority, which is reinforced by their relative simplicity. But the equations are based on premises that we must unequivocally refute! We are left with the uncomfortable feeling that our methods may have run their natural course.

Given the symmetry between (8.2.1) and (8.2.2), we appear to have no choice but to look for a uniform argument that applies to both problems. We shall introduce a second supplementary parameter, which at first sight seems to be unnatural and without promise. The surprise is that it actually works! The new parameter will be seen in fact to be quite natural, and to be exactly what is required to contradict the assumptions $\delta_\psi = -1$ of (8.2.1) and $\Lambda \neq 0$ of (8.2.2).

For the given pair (G, ψ), we define our second supplementary parameter by

$$\psi_{++} = 3\psi = \psi \boxplus \psi \boxplus \psi.$$

Then ψ_{++} belongs to $\widetilde{\Psi}(N_{++})$, for $N_{++} = 3N$. Remember that we are assuming that N is even and $\eta_\psi = 1$, so that G and $G^\vee$ are distinct split groups in $\widetilde{\mathcal{E}}_{\mathrm{sim}}(N)$. Let G_{++} and $G^\vee_{++}$ be the split groups in $\widetilde{\mathcal{E}}_{\mathrm{sim}}(N_{++})$ whose dual groups $\widehat{G}_{++}$ and $(G^\vee_{++})^\wedge$ contain the respective products $\widehat{G} \times \widehat{G} \times \widehat{G}$ and $(G^\vee)^\wedge \times (G^\vee)^\wedge \times (G^\vee)^\wedge$. We then have the maximal Levi subgroups

$$M_{++} = M_+ \times G \cong GL(N) \times G$$

and

$$M^\vee_{++} = M_+ \times G^\vee \cong GL(N) \times G^\vee$$

of G_{++} and $G^{\vee}_{++}$ respectively. The product $\psi \times \psi$ can then be treated as a parameter for M_{++} whose image in $\widetilde{\Psi}(N_{++})$ equals ψ_{++}. Given an associated compatible family of functions

$$\{f^* \in \widetilde{\mathcal{H}}(G^*) : G^* \in \widetilde{\mathcal{E}}_{\text{ell}}(N_{++})\}, \tag{8.2.4}$$

we shall write $f = f^*$ and $f^{\vee} = f^*$ as usual in the cases $G^* = G_{++}$ and $G^* = G^{\vee}_{++}$.

The stage is now set for the lemma that is the key to the final classification. This lemma is founded on the techniques of §5.3, but relies also on the character formulas of §7.5 embodied in Lemma 8.2.1.

LEMMA 8.2.2. *For the given pair*

$$(G, \psi), \qquad G \in \widetilde{\mathcal{E}}_{\text{sim}}(N),\ \psi \in \widetilde{\Psi}_{\text{sim}}(N),$$

assume that N is even and $\eta_\psi = 1$. In addition, assume that either $\delta_\psi = -1$ as in (8.2.1), or $\Lambda \neq 0$ as in (8.2.2). Then the sum

$$\sum_{G^* \in \widetilde{\mathcal{E}}_{\text{sim}}(N_{++})} \operatorname{tr}\big(R^*_{\text{disc},\psi_{++}}(f^*)\big) + \tfrac{1}{2}\, f^{M_{++}}(\psi \times \psi) \tag{8.2.5}$$

vanishes for any compatible family of functions (8.2.4).

PROOF. Once again, we have to modify the arguments from Chapter 4. The proof will be loosely modeled on that of Lemmas 5.3.1 and 5.3.2 for ψ_+. However, our exposition here will be slightly different, since we now have the local theorems at our disposal. In particular, analogues of the terms (5.3.22) and (5.3.23) need not be an explicit part of the discussion here, thanks to the local and global intertwining relations.

We fix a compatible family of functions (8.2.4). The arguments of Chapter 4 would lead ultimately to the vanishing of the sum

$$\sum_{G^* \in \widetilde{\mathcal{E}}_{\text{sim}}(N_{++})} \widetilde{\iota}(N_{++}, G^*) \operatorname{tr}\big(R^*_{\text{disc},\psi_{++}}(f^*)\big), \tag{8.2.6}$$

if we could assume inductively that the global theorems were valid for proper subparameters of ψ_{++}. We cannot do this, of course, because we have assumed that one of the contrary conditions $\delta_\psi = -1$ or $\Lambda \neq 0$ holds. However, we can use this condition, as it is reflected in either (8.2.1) or (8.2.2), to modify the expected formula for (8.2.6).

The modifications of (8.2.6) are of three sorts. They arise from the endoscopic contributions of proper products $G^* = G_1 \times G_2$ in the set $\widetilde{\mathcal{E}}_{\text{ell}}(N_{++})$, from endoscopic contributions of proper products $G' = G_1 \times G_2$ in the sets $\mathcal{E}_{\text{ell}}(G^*)$ attached to simple data $G^* \in \widetilde{\mathcal{E}}_{\text{sim}}(N_{++})$, and from spectral contributions from proper Levi subgroups M^* of the groups $G^* \in \widetilde{\mathcal{E}}_{\text{sim}}(N_{++})$. As we have seen, explicitly in the proof of Lemma 5.3.1 and implicitly in that of Lemma 5.3.2, most such contributions reduce immediately to zero. In the case here they are still direct consequences of Theorem 1.3.2, which will be left to the reader. We will consequently need to consider only one pair of contributions of each of the three kinds. In each case, we will have to subtract the expected value of the contribution from the actual value. The difference will then have to be added or subtracted, as a correction term for (8.2.6), in order that the resulting expression vanish. As in the earlier proofs, the sign of this last operation (addition or subtraction) is dictated by the role of the given contribution in the standard model. Since these signs are critical, it would be a good idea to review again the process by which they are determined.

We know that we can write (8.2.6) as the sum of

$$\sum_{G^* \in \widetilde{\mathcal{E}}_{\mathrm{sim}}(N_{++})} \widetilde{\iota}(N_{++}, G^*) \big(\mathrm{tr}(R_{\mathrm{disc},\psi_{++}}(f^*)) - {}^0S^*_{\mathrm{disc},\psi_{++}}(f^*)\big) \tag{8.2.7}$$

and

$$-\sum_{G^*} \widetilde{\iota}(N_{++}, G^*)\, {}^0S^*_{\mathrm{disc},\psi_{++}}(f^*), \qquad G^* \in \widetilde{\mathcal{E}}^0_{\mathrm{ell}}(N_{++}), \tag{8.2.8}$$

where $\widetilde{\mathcal{E}}^0_{\mathrm{ell}}(N_{++})$ is the complement of $\widetilde{\mathcal{E}}_{\mathrm{sim}}(N_{++})$ in $\widetilde{\mathcal{E}}_{\mathrm{ell}}(N_{++})$, since the condition (5.2.15) of Lemma 5.2.3 holds for ψ_{++}. This is the identity, with (8.2.6) on the left hand side and the sum of (8.2.7) and (8.2.8) on the right, that is to be our starting point.

The first pair of modifications describes the obstruction to the expected vanishing of the summands in (8.2.8). To simplify the right hand side of the identity, we will want to move the obstructions to the left hand side. In other words, we *add* them to (8.2.6), since the summands come with minus signs. For the remaining terms on the right hand side, we will take what we can from Corollaries 4.3.3 and 4.4.3. We can certainly write

$$\mathrm{tr}\big(R_{\mathrm{disc},\psi_{++}}(f^*)\big) - S^*_{\mathrm{disc},\psi_{++}}(f^*) = I'_{\mathrm{end}}(f^*) - I'_{\mathrm{spec}}(f^*), \qquad G^* \in \widetilde{\mathcal{E}}_{\mathrm{sim}}(N_{++}),$$

where I'_{spec} and I'_{end} stand for the proper parts of the respective expansions (4.1.1) and (4.1.2) (namely, the sums taken only over $M \neq G$ and $G' \neq G$ respectively, but of course with (G^*, ψ_{++}) in place of (G, ψ)). Corollaries 4.3.3 and 4.4.3 (and their comparison by Proposition 4.1.1 that in the case of (G, ψ) culminated in (4.5.4)) pertain to the expected values of the terms on the two sides of this last formula. Since the expected value of the trace

$$r^*_{\mathrm{disc},\psi_{++}}(f^*) = \mathrm{tr}\big(R^*_{\mathrm{disc},\psi_{++}}(f^*)\big)$$

is 0, they tell us that

$$\begin{aligned} \mathrm{tr}\big(R^*_{\mathrm{disc},\psi_{++}}(f^*)\big) - {}^0S^*_{\mathrm{disc},\psi_{++}}(f^*) &= {}^0r^*_{\mathrm{disc},\psi_{++}}(f^*) - {}^0s^*_{\mathrm{disc},\psi_{++}}(f^*) \\ &= {}^0I'_{\mathrm{end}}(f^*) - {}^0I'_{\mathrm{spec}}(f^*), \qquad G^* \in \widetilde{\mathcal{E}}_{\mathrm{sim}}(N_{++}), \end{aligned}$$

where ${}^0I'_{\mathrm{end}}(f^*)$ and ${}^0I'_{\mathrm{spec}}(f^*)$ stand for the variance of $I'_{\mathrm{end}}(f^*)$ and $I'_{\mathrm{spec}}(f^*)$ respectively from their expected values. The remaining modifications describe the obstruction to the expected vanishing of these last two terms. The second of the three pairs applies to ${}^0I'_{\mathrm{end}}(f^*)$. To move the resulting obstruction to the left hand side, we must *subtract* it from (8.2.6).

The third pair applies to ${}^0I'_{\mathrm{spec}}(f^*)$. To transfer the resulting obstruction in this last case, we must *add* it to (8.2.6). The expression (8.2.6), thus modified by the three kinds of correction, will then vanish.

The arguments for the two possible conditions $\delta_\psi = -1$ or $\Lambda \neq 0$ are essentially parallel, so we shall only treat one of them in detail. We assume until near the end of the proof that the condition $\Lambda \neq 0$ is in force. Then ψ is not generic, δ_ψ equals 1, and the equations (8.2.2) hold.

The first pair of corrections applies to the linear forms

$$S^*_{\mathrm{disc},\psi_{++}}(f^*) = S^1_{\mathrm{disc},\psi}(f_1)\, S^2_{\mathrm{disc},\psi_+}(f_2), \qquad G^* \in \widetilde{\mathcal{E}}^0_{\mathrm{ell}}(N_{++}), \tag{8.2.9}$$

where $G^* = G_1 \times G_2$ is either of the groups $G \times G_+^\vee$ or $G^\vee \times G_+$ in $\widetilde{\mathcal{E}}^0_{\mathrm{ell}}(N_{++})$, and $f^* = f_1 \times f_2$ is the corresponding function in (8.2.4). If G^* equals $G \times G_+^\vee$, the actual value of (8.2.9) vanishes, according to the first equation in (8.2.2), while the expected value equals

$$-\tfrac{1}{4}\, \widetilde{o}(G)\, \widetilde{o}(G_+^\vee)\, f_1^G(\psi)\, f_2^{M_+}(\psi),$$

by the first and fourth equations in (8.2.3). If G^* equals $G^\vee \times G_+$, the actual value of (8.2.9) is

$$\tfrac{1}{4}\, \widetilde{o}(G^\vee)\, \widetilde{o}(G_+)\, f_1^G(\psi)\, f_2^{M_+}(\psi),$$

by the second and third equations in (8.2.2), while the expected value vanishes by the second (or third) equation in (8.2.3). In each case, we subtract the expected value from the actual value, and then multiply the difference by the coefficient

$$\widetilde{\iota}(N_{++}, G^*) = \tfrac{1}{4}\, \widetilde{o}(G^*)^{-1} = \tfrac{1}{4}\, \widetilde{o}(G_1)^{-1}\, \widetilde{o}(G_2)^{-1}. \tag{8.2.10}$$

We obtain two equal contributions, whose sum

$$\tfrac{1}{8}\, f^{M_{++}}(\psi \times \psi), \tag{8.2.11}$$

must be added as a correction term to (8.2.6).

The second pair of corrections applies to two terms in the analogue for ψ_{++} of (4.1.2). These are the linear forms

$$\begin{gathered} S'_{\mathrm{disc},\psi_{++}}(f') = S^1_{\mathrm{disc},\psi}(f_1)\, S^2_{\mathrm{disc},\psi_+}(f_2), \\ G' \in \mathcal{E}_{\mathrm{sim}}(G^*),\ G^* \in \widetilde{\mathcal{E}}_{\mathrm{sim}}(N_{++}), \end{gathered} \tag{8.2.12}$$

where $G' = G_1 \times G_2$ is either of the groups $G \times G_+ \in \widetilde{\mathcal{E}}_{\mathrm{ell}}(G_{++})$ or $G^\vee \times G_+^\vee \in \mathcal{E}_{\mathrm{ell}}(G_{++}^\vee)$, and $f' = f_1 \times f_2$ is a function in $\widetilde{\mathcal{H}}(G')$ with the same image in $\widetilde{\mathcal{S}}(G')$ as the chosen compatible family. If G' equals $G \times G_+$, the actual value and the expected value of (8.2.12) both vanish, by the first equation in (8.2.2) and the third equation in (8.2.3). If G' equals $G^\vee \times G_+^\vee$, the actual value of (8.2.12) is

$$-\tfrac{1}{2}\, \widetilde{o}(G^\vee)\, \widetilde{o}(G_+^\vee)\, f_1^G(\psi)\, f_2^{M_+}(\psi_+),$$

by the second and fourth equations in (8.2.2), while the expected value vanishes, by the second equation in (8.2.3). In each case, we subtract the expected value from the actual value, and then multiply the difference by the coefficient

(8.2.13)
$$\tilde{\iota}(N_{++}, G^*)\,\iota(G^*, G') = \tfrac{1}{4}\,\tilde{\delta}(G^*)^{-1}\,\mathrm{Out}(G^*, G') = \tfrac{1}{4}\,\tilde{\delta}(G_1)^{-1}\,\tilde{\delta}(G_2)^{-1}.$$

This time we must subtract the result from (8.2.6). The two minus signs cancel, and we obtain another copy of (8.2.11) to be added as a correction term to (8.2.6).

The third pair of corrections applies to two terms in the analogues for ψ_{++} of (4.1.1). These are the linear forms

$$\mathrm{tr}\big(M_{P^*,\psi_{++}}(w^*)\,\mathcal{I}_{P^*,\psi_{++}}(f^*)\big), \qquad G^* \in \widetilde{\mathcal{E}}_{\mathrm{sim}}(N_{++}), \tag{8.2.14}$$

where G^* is either G_{++} or $G_{++}^{\vee}$, M^* is the corresponding Levi subgroup M_{++} or $M_{++}^{\vee}$, and w^* is the associated element in $W_{\psi_{++},\mathrm{reg}}(M^*)$. This requires a little more discussion.

Suppose first that G^* equals G_{++}. Recall that M_{++} is isomorphic to $GL(N) \times G$. The actual discrete spectrum for G and ψ vanishes under our assumption $\Lambda \neq 0$, since the first equation in (8.2.2) tells us that the corresponding stable linear form vanishes. The actual value of (8.2.14) therefore vanishes. To describe the expected value, we note that the centralizer group $\overline{S}_{\psi_{++}}$ is isomorphic to $SO(3, \mathbb{C})$, and that $w^* = w_{++}$ represents the nontrivial element in the Weyl group of $SO(3, \mathbb{C})$. Because ψ is not generic, the induction hypothesis underlying the proof of Lemma 4.3.1 in §4.6 is valid. By a very elementary case of the discussion in §4.6, the global normalizing factor implicit in (8.2.14) therefore equals (-1). If we then combine the first equation in (8.2.3) with the global intertwining relation, we see that the expected value of (8.2.14) is

$$-\tilde{\delta}(G)\,f_{G_{++}}(\psi_{++}, x_1) = -\tilde{\delta}(G)\,f^{M_{++}}(\psi \times \psi),$$

where x_1 stands for the only element in the trivial group $\mathcal{S}_{\psi_{++}}$. We must subtract the expected value from the actual value, multiply the difference by the coefficient

(8.2.15)
$$\tilde{\iota}(N_{++}, G_{++})\,|W(M_{++})|^{-1}\,|\det(w_{++} - 1)|^{-1} = \tfrac{1}{2}\,\tilde{\delta}(G_{++})\,\tfrac{1}{4} = \tfrac{1}{8}\,\tilde{\delta}(G)^{-1},$$

and add the result to (8.2.6). This gives us a third copy of (8.2.11) to be added to (8.2.6).

Suppose finally that G^* equals $G_{++}^{\vee}$. In this case, we cannot quite apply the discussion of §4.6 to the global normalizing factor in (8.2.14), because our assumption $\Lambda \neq 0$ violates the earlier premises. However, we can still represent the normalizing factor as a product of quotients of two automorphic L-functions. One is the Rankin-Selberg L-function $L(s, \psi \times \psi)$. Indeed, it is built out of local Rankin-Selberg factors $L(s, \psi_v \times \psi_v)$ attached to localizations

$$\begin{aligned}\psi_v \times \psi_v : \ L_{F_v} \times SU(2) \longrightarrow (M_{++}^\vee)^\wedge &= GL(N,\mathbb{C}) \times (G^\vee)^\wedge \\ &\subset GL(N,\mathbb{C}) \times GL(N,\mathbb{C})\end{aligned}$$

of $\psi \times \psi$. (We shall recall below how ψ_v can be treated as a local parameter for the group $G^\vee$.) Since ψ is self-dual, it has a pole at $s = 1$. The other is the Langlands-Shahidi L-function attached to the parameter ψ_+ for $G_+^\vee$. Since $S_{\psi_+}^\vee$ is isomorphic to $SL(2,\mathbb{C})$, this L-function also has a pole at $s = 1$. We are of course using our induction hypothesis here that Theorem 1.5.3(a) is valid for the generic component of ψ, as we did in discussing the case $G^* = G_{++}$ above. The two L-functions each contribute a factor (-1) to the product, so the global normalizing factor equals 1 in this case. (By contrast, in the case $G^* = G_{++}$ above, S_{ψ_+} is isomorphic to $O(2,\mathbb{C})$, and the corresponding L-function does not have a pole, and therefore does not contribute a second factor (-1).)

Continuing our analysis of the case $G^* = G_{++}^\vee$, we consider the local normalized intertwining operators in (8.2.14). According to Lemma 8.2.1 and our assumption $\Lambda \neq 0$, any localization ψ_v of ψ factors through the Levi subgroup $\widehat{L}$ of $\widehat{G}$. Since $\widehat{L}$ also represents a Levi subgroup of $(G^\vee)^\wedge$, we can treat ψ_v as a local parameter for $G^\vee$ as well as for G. The image in $(G_{++}^\vee)^\wedge$ of the corresponding localization

$$\psi_{++,v} = \psi_v \boxplus \psi_v \boxplus \psi_v$$

of ψ_{++} is then contained in the threefold diagonal image of $(G^\vee)^\wedge$. The centralizer of this diagonal image, taken modulo the center of $(G_{++}^\vee)^\wedge$, is isomorphic to the connected group $SO(3,\mathbb{C})$. Since $w^* = w^\vee$ represents the nontrivial element in the Weyl group of $SO(3,\mathbb{C})$, we can choose a local representative of $w_{++}^\vee$ in the connected centralizer $(S_{\psi_{++,v}}^\vee)^0$. The local normalized intertwining operator therefore equals the identity.

We have shown that if $G^* = G_{++}^\vee$, the local and global factors of the operator $M_{P^*,\psi_{++}}(w^*)$ in (8.2.14) are all equal to 1. The operator is therefore trivial. In fact, from the global intertwining relation and the second equation in (8.2.2), we see that (8.2.14) itself equals the linear form

$$\widetilde{\sigma}(G^\vee)\, f_{G_{++}^\vee}^\vee(\psi_{++}, x_1) = \widetilde{\sigma}(G^\vee)\, (f^\vee)^{M_{++}^\vee}(\psi \times \psi) = \widetilde{\sigma}(G^\vee)\, f^{M_{++}}(\psi \times \psi).$$

On the other hand, the expected value of (8.2.14) is equal to 0, by the second equation in (8.2.3). Subtracting the expected value from the actual value, and multiplying by the coefficient

(8.2.16)

$$\widetilde{\iota}(N_{++}, G_{++}^\vee)\, |W(M_{++}^\vee)|^{-1}\, |\det(w_{++}^\vee - 1)|^{-1} = \tfrac{1}{2}\, \widetilde{\sigma}(G_{++}^\vee)\, \tfrac{1}{4} = \tfrac{1}{8}\, \widetilde{\sigma}(G^\vee)^{-1},$$

we obtain a fourth copy of (8.2.11) to be added to (8.2.6).

We have obtained four correction terms in all, each of which is equal to (8.2.6). Having left the reader to check that there are no other corrections, we conclude that the sum of (8.2.6) with the total correction term

$$4\big(\tfrac{1}{8}\, f^{M_{++}}(\psi \times \psi)\big) = \tfrac{1}{2}\, f^{M_{++}}(\psi \times \psi)$$

vanishes. Since the sum is equal to the given expression (8.2.2), we have obtained a proof of the lemma in case $\Lambda \neq 0$.

Assume now that $\delta_\psi = -1$. Then ψ is generic, $\Lambda = 0$, and the equations (8.2.1) hold. The structure of the proof is the same in this case, with the only minor differences in detail resulting from our use of the equations (8.2.1) in place of (8.2.2). We can be brief.

Suppose that the group G^* in (8.2.9) equals $G \times G_+^\vee$. Then the actual value of (8.2.9) equals

$$\tfrac{1}{4}\, \widetilde{o}(G)\, \widetilde{o}(G_+^\vee)\, f^{M_{++}}(\psi \times \psi),$$

while the expected value is

$$-\tfrac{1}{4}\, \widetilde{o}(G)\, \widetilde{o}(G_+^\vee)\, f^{M_{++}}(\psi \times \psi),$$

by the first and fourth equations in (8.2.1) and (8.2.3). If G^* equals $G^\vee \times G_+$, the actual and expected values of (8.2.9) both vanish, by the second equations in (8.2.1) and (8.2.3). The product of the total difference with the coefficients (8.2.10) is equal to the linear form (8.2.11) we obtained before.

Suppose that the group G' in (8.2.12) equals $G \times G_+$. Then the actual value of (8.2.12) equals

$$-\tfrac{1}{2}\, \widetilde{o}(G)\, \widetilde{o}(G_+)\, f^{M_{++}}(\psi \times \psi),$$

while the expected value vanishes, by the first and third equations in (8.2.1) and the third equation in (8.2.3). If G' equals $G^\vee \times G_+^\vee$, the actual and expected values of (8.2.12) both vanish, by the second equations in (8.2.1) and (8.2.3). The product of the total difference with the coefficient (8.2.13), when subtracted from (8.2.6), again gives a second copy of the correction term (8.2.9).

Suppose that the groups G^* and M^* in (8.2.14) equal G_{++} and M_{++}. The actual discrete spectrum for M_{++} and $\psi \times \psi$ is equal to what is expected, by the first equations in (8.2.1) and (8.2.3). Since the group $S_{\psi_{++}}$ equals $O(3, \mathbb{C})$, the normalized intertwining operator in (8.2.14) equals 1, as expected. The global normalizing factor implicit in (8.2.14) is therefore the only quantity that differs from its expected value. The expected value equals the sign of $w^* = w_{++}$ in $SO(3, \mathbb{C})$, which is (-1). The actual value equals 1, as in our discussion of $G^* = G_{++}^\vee$ for the case $\Lambda \neq 0$ above, this time using the definition of $\delta_\psi = -1$. Appealing again to the first equations of (8.2.1) and (8.2.3), we see that the actual value of (8.2.12) is

$$\widetilde{o}(G)\, f_{G_{++}}(\psi_{++}, x_1) = \widetilde{o}(G)\, f^{M_{++}}(\psi \times \psi),$$

while the expected value equals (-1) times this quantity. The product of the difference with the coefficient (8.2.15) gives two more copies to (8.2.11) to be added as correction terms to (8.2.6).

Finally, suppose that the groups G^* and M^* in (8.2.14) equal $G_{++}^\vee$ and $M_{++}^\vee$. The discrete spectrum for $M_{++}^\vee$ and $\psi \times \psi$ vanishes, as expected, by the second equations in (8.2.1) and (8.2.3). The actual and expected values of (8.2.14) therefore both vanish in this case, and give no further correction terms for (8.2.6).

The total correction term in this case thus equals

$$2\big(\tfrac{1}{8}\, f^{M_{++}}(\psi \times \psi)\big) + \tfrac{1}{4}\, f^{M_{++}}(\psi \times \psi) = \tfrac{1}{2}\, f^{M_{++}}(\psi \times \psi),$$

as it did in the earlier case. The given expression (8.2.2) therefore again vanishes. We have completed the proof of the lemma in the remaining case $\delta_\psi = -1$. □

With the proof of Lemma 8.2.2, we have finally obtained the result we want. It remains only to appeal one last time to Proposition 3.5.1.

The linear form

$$f^{M_{++}}(\psi \times \psi), \qquad f \in \widetilde{\mathcal{H}}(G_{++}), \tag{8.2.17}$$

in (8.2.5) is an induced character. In fact, it is induced from the nonzero character on the Levi subgroup $M_{++}(\mathbb{A})$ obtained from the local constituents $\psi_v \times \psi_v$ of $\psi \times \psi$. In particular, it is a nonnegative integral combination of irreducible characters. Since the same is true of the linear forms in f^* in (8.2.5), the entire expression (8.2.5) is a nonnegative linear combination of irreducible characters of the general form (3.5.1). If the expression vanishes, we can indeed apply Proposition 3.5.1. It will tell us that the corresponding coefficients all vanish. Since there are nonzero coefficients in (8.2.17), this cannot happen. The expression (8.2.5) therefore does not vanish.

We have arrived at a contradiction to the premise of Lemma 8.2.2. In other words, we have established that $\delta_\psi = 1$ and $\Lambda = 0$. It follows that Theorem 1.5.3(a) holds if $\phi = \psi$ is generic, and that the stable multiplicity formula of Theorem 4.1.2 is valid in the remaining case that ψ is not generic. These are the two assertions left over from the last section.

The remaining global properties for simple parameters $\psi \in \widetilde{\Psi}_{\mathrm{sim}}(N)$ are in the assertions of Theorems 1.5.2 and 4.2.2. They follow immediately. Indeed, given the stable multiplicity formula for ψ, we appeal to (4.4.12) and Lemma 4.7.1 as above to establish the required contribution (4.7.10) to the spectral multiplicity of Theorem 1.5.2. (In case ψ is generic, we are also implicitly relying on the fact that Lemma 8.1.1 includes a proof of Theorem 1.4.2.) If (G, ψ) is as in Theorem 4.2.2, the simple parameter ψ lies in the complement of $\Psi_2(\widetilde{G})$ of $\widetilde{\Psi}_2(G)$. The assertion (b) of this theorem is therefore irrelevant, while assertion (a) again follows from the fact that the right hand side of (8.1.3) vanishes. These are the last assertions to be proved for the simple parameter $\psi \in \widetilde{\Psi}_{\mathrm{sim}}(N)$. Since the simple parameters were all that remained after the last section, we have now established the global theorems for any parameter $\psi \in \widetilde{\Psi}(N)$. This completes the induction argument on N begun in the previous section, and therefore completes our proof of the global theorems.

In summary, we have now resolved all of the induction hypotheses. We have consequently proved all of the theorems. These are the generic local theorems established in Chapter 6, the nongeneric local theorems treated in Chapter 7, and the global theorems proved in this chapter. In particular, the analogue of Corollary 5.4.7 for the full set $\widetilde{\mathcal{F}}_{\mathrm{sim}}(N) = \widetilde{\Phi}_{\mathrm{sim}}(N)$ of simple global generic parameters is valid. We shall restate it as a final corollary of this discussion, for easy reference.

COROLLARY 8.2.3. *Suppose that $\phi \in \widetilde{\Phi}_{\mathrm{sim}}(N)$ is any simple, self-dual, generic parameter of rank N, and that G belongs to $\widetilde{\mathcal{E}}_{\mathrm{sim}}(N)$. Then the following three conditions on the pair (G, ϕ) over our global field F are equivalent.*

(i) *The linear form $S^G_{\mathrm{disc},\phi}$ on $\widetilde{\mathcal{H}}(G)$ does not vanish.*

(ii) *Theorem* 1.4.1 *holds for ϕ, with $G_\phi = G$.*

(iii) *The global quadratic characters η_G and η_ϕ are equal, and the global L-function condition $\delta_\phi = 1$ of Theorem* 1.5.3(a) *holds.*

PROOF. The proof follows from the global theorems we have now established. In particular, it is identical to that of Corollary 5.4.7, the special case for the global family $\widetilde{\mathcal{F}}$ of §5.4. □

Corollary 8.2.3 tells us that any of the three equivalent conditions (i)–(iii) can serve as a definition for the set $\widetilde{\Phi}_{\mathrm{sim}}(G)$ of simple, generic global parameters for G. Thus ends the last ambiguity from the construction of §1.4, which led to the complete set of global parameters $\widetilde{\Psi}(G)$ and the theorems they satisfy. Once again we observe the local-global symmetry, here between the global corollary we have just proved, and its local counterpart Corollary 6.8.1.

We can now go on to other things. In the rest of this chapter, we shall discuss a few ramifications of what we have done. We will then study the representations of inner twists of groups $G \in \widetilde{\mathcal{E}}_{\mathrm{sim}}(N)$ in the final Chapter 9.

8.3. Reflections on the results

Having finally proved the theorems, we shall pause for a few moments of reflection. We will begin the section with a brief overview of the results, taken from the general perspective of characters and multiplicities. We will then pose some questions that are particular to the even orthogonal groups $SO(2n)$. We will complete the section with a proof that local and global L-packets contain generic representations.

We fix one of the quasisplit groups $G \in \widetilde{\mathcal{E}}_{\mathrm{sim}}(N)$ over our field F. We recall that the integer N equals $2n$, $2n+1$ or $2n$, according to whether G equals a simple group $SO(2n+1)$, $Sp(2n)$ or $SO(2n)$ from the infinite family B_n, C_n or D_n, with corresponding dual group $\widehat{G}$ equal to $Sp(2n,\mathbb{C})$, $SO(2n+1,\mathbb{C})$ or $SO(2n,\mathbb{C})$. In the second and third cases, G comes with an arithmetic character η (a character on F^* if F is local or $\mathbb{A}^*/F^*$ if F is global) of order 1 or 2. In the third case, η determines G as a quasisplit inner twist of the corresponding split group. The group G is of course split in the other two cases. The role of η in the second case $G = Sp(2n)$ is to specify the L-embedding of ${}^LG$ into the L-group of $GL(N)$ that determines G as twisted endoscopic datum for $GL(N)$.

Suppose for the moment that F is local. The local Langlands group L_F is then defined as the split extension (1.1.1) of W_F. We have classified the irreducible representations of $G(F)$ in terms of packets $\widetilde{\Pi}_\phi$ indexed by the set $\widetilde{\Phi}(G)$ of equivalence classes of Langlands parameters

$$\phi:\ L_F \longrightarrow {}^LG.$$

Equivalence is defined by $\widehat{G}$-conjugacy as usual when $\widehat{G}$ equals $Sp(2n,\mathbb{C})$ or $SO(2n+1,\mathbb{C})$. In these cases, $\widetilde{\Pi}_\phi$ is the L-packet conjectured by Langlands for any connected reductive group over F. When G equals $SO(2n,\mathbb{C})$, equivalence is defined by conjugacy under the exension $O(2n,\mathbb{C})$ of $\mathbb{Z}/2\mathbb{Z}$ by $SO(2n,\mathbb{C})$. In this case, the packet of ϕ is still an L-packet if the $O(2n,\mathbb{C})$-orbit of ϕ equals its $SO(2n,\mathbb{C})$-orbit . If the $O(2n,\mathbb{C})$-orbit contains two $SO(2n,\mathbb{C})$ orbits, however, $\widetilde{\Pi}_\phi$ is a set of pairs of irreducible representations. We have thus established the full local Langlands correspondence for the split groups $SO(2n+1)$ and $Sp(2n)$, and a slightly weaker form of the correspondence for the quasisplit groups $SO(2n)$.

It is worth emphasizing that the endoscopic classification of representations is by characters. That is, the elements in a packet $\widetilde{\Pi}_\phi$ are represented by their characters. These are defined in terms of the finite 2-group

$$\mathcal{S}_\phi = S_\phi/S_\phi^0\, Z(\widehat{G})^\Gamma, \qquad \Gamma = \Gamma_F = \mathrm{Gal}(\overline{F}/F),$$

and twisted characters for general linear groups. Let us be more explicit.

The distributional character

$$f_G(\pi) = \mathrm{tr}\big(\pi(f)\big), \qquad f \in \mathcal{H}(G),$$

of any irreducible representation $\pi \in \Pi(G)$ of $G(F)$ is a locally integrable function. In other words,

$$f_G(\pi) = \int_{G_{\mathrm{reg}}(F)} \Theta_G(\pi,x)\, f(x)dx = \int_{\Gamma_{\mathrm{reg}}(G)} I_G(\pi,\gamma)\, f_G(\gamma)\, d\gamma,$$

where $\Gamma_{\mathrm{reg}}(G) = \Gamma_{G\text{-reg}}(G)$ is the space of strongly G-regular conjugacy classes in $G(F)$, equipped with its natural measure, and

$$I_G(\pi,\gamma) = |D(\gamma)|^{\frac{1}{2}}\, \Theta_G(\pi,\gamma), \qquad \gamma \in \Gamma_{\mathrm{reg}}(G),$$

is the *normalized character*. This assertion is a fundamental theorem of Harish-Chandra, which is at the heart of harmonic analysis on real and p-adic groups. The endoscopic classification is in terms of tempered representations, or equivalently, their (reducible) analytic continuation to standard representations. For real groups, tempered (and standard) characters have natural formulas, which are based on Harish-Chandra's explicit formulas **[Ha1]** **[Ha2]** for the characters of discrete series. Characters for p-adic groups are more complex, but they have many interesting and well understood qualitative properties.

It is not hard to describe endoscopic transfer in terms of normalized characters. Recall that the tempered representations of $G(F)$ correspond to the subset $\widetilde{\Phi}_{\mathrm{bdd}}(G)$ of parameters $\phi \in \widetilde{\Phi}(G)$ of bounded image. We will need a slightly different convention for treating the constituents of the associated packets $\widetilde{\Pi}_\phi$, since they are elements in the set $\widetilde{\Pi}(G)$ of orbits in $\Pi(G)$ under the group

$$\widetilde{O}(G) = \widetilde{\mathrm{O}}\mathrm{ut}_N(G) = \widetilde{\mathrm{A}}\mathrm{ut}_N(G)/\widetilde{\mathrm{I}}\mathrm{nt}_N(G).$$

For any π in this set, the sum

$$\widetilde{I}_G(\pi,\gamma) = \sum_{\pi_*} I_G(\pi_*,\gamma), \qquad \pi_* \in \Pi(G,\pi),$$

over the $\widetilde{O}(G)$-orbit $\Pi(G,\pi)$ of π in $\Pi(G)$ (of order 1 or 2) depends only on the image of γ in the quotient

$$\widetilde{\Gamma}_{\mathrm{reg}}(G) = \Gamma_{\mathrm{reg}}(G)\,/\,\widetilde{O}(G).$$

We take the quotient measure on $\widetilde{\Gamma}_{\mathrm{reg}}(G)$ inherited from $\Gamma_{\mathrm{reg}}(G)$. We then observe that

$$f_G(\pi) = \int_{\widetilde{\Gamma}_{\mathrm{reg}}(G)} \widetilde{I}_G(\pi,\gamma)\, f_G(\gamma)\, d\gamma, \qquad f \in \widetilde{\mathcal{H}}(G), \tag{8.3.1}$$

for any $\pi \in \widetilde{\Pi}(G)$, and in particular, for π in any packet $\widetilde{\Pi}_\phi$. The stable analogue of this distribution is the linear form $f^G(\phi)$ attached to $\phi \in \widetilde{\Phi}_{\mathrm{bdd}}(G)$. It can be expanded as an integral

$$f^G(\phi) = \int_{\widetilde{\Delta}_{\mathrm{reg}}(G)} \widetilde{S}^G(\phi,\delta)\, f^G(\delta)\, d\delta, \qquad f \in \widetilde{\mathcal{H}}(G),$$

over the space $\widetilde{\Delta}_{\mathrm{reg}}(G) = \widetilde{\Delta}_{G\text{-reg}}(G)$ of $\widetilde{O}(G)$-orbits of strongly G-regular stable conjugacy classes in $G(F)$, equipped with the measure inherited from $\Gamma_{\mathrm{reg}}(G)$. The *normalized stable character* $\widetilde{S}^G(\phi,\delta)$ in the integral is obtained by transfer from $GL(N)$, as we shall see explicitly in a moment.

The function $\widetilde{S}^G(\phi,\delta)$ of course has an analogue

$$\widetilde{S}'(\phi',\delta') = \widetilde{S}^{G'}(\phi',\delta'), \qquad \delta' \in \widetilde{\Delta}_{G\text{-reg}}(G'),$$

for the preimage (G',ϕ') in (1.4.11) of any given pair (ϕ,s). For any $f\in\widetilde{\mathcal{H}}(G)$, we can then write

$$\begin{aligned} f'(\phi') &= \int_{\widetilde{\Delta}_{G\text{-reg}}(G')} \widetilde{S}'(\phi',\delta')\, f'(\delta')\, d\delta' \\ &= \int_{\widetilde{\Delta}_{G\text{-reg}}(G')} \sum_{\gamma\in\widetilde{\Gamma}_{\mathrm{reg}}(G)} \widetilde{S}'(\phi',\delta')\,\Delta(\delta',\gamma)\, f_G(\gamma)\, d\delta' \\ &= \int_{\widetilde{\Gamma}_{\mathrm{reg}}(G)} \sum_{\delta'\in\widetilde{\Delta}_{G\text{-reg}}(G')} \widetilde{S}'(\phi',\delta')\,\Delta(\delta',\gamma)\, f_G(\gamma)\, d\gamma, \end{aligned}$$

by the definition of $f' = f^{G'}$, and a change of variables as, for example, in [**A11**, Lemma 2.3]. On the other hand, if x is the image in $\mathcal{S}_\phi$ of the given (semisimple) point s in the quotient $\overline{S}_{\phi,\mathrm{ss}} = S_{\phi,\mathrm{ss}}/Z(\widehat{G})^\Gamma$, we can also write

$$\begin{aligned} f'(\phi') &= \sum_{\pi\in\widetilde{\Pi}_\phi} \langle x,\pi\rangle\, f_G(\pi) \\ &= \int_{\widetilde{\Gamma}_{\mathrm{reg}}(G)} \sum_{\pi\in\widetilde{\Pi}_\phi} \langle x,\pi\rangle\, \widetilde{I}_G(\pi,\gamma)\, f_G(\gamma)\, d\gamma, \end{aligned}$$

by (2.2.6) and (8.3.1). We thus obtain an identity

$$\sum_{\delta'} \widetilde{S}'(\phi',\delta')\,\Delta(\delta',\gamma) = \sum_{\pi}\langle x,\pi\rangle\, \widetilde{I}_G(\pi,\gamma)$$

of the two integrands. An inversion on the group $\mathcal{S}_\phi$ then yields the general formula

$$\widetilde{I}_G(\pi,\gamma) = |\mathcal{S}_\phi|^{-1} \sum_{x\in\mathcal{S}_\phi} \sum_{\delta'\in\widetilde{\Delta}_{G\text{-reg}}(G')} \langle x,\pi\rangle^{-1}\widetilde{S}'(\phi',\delta')\,\Delta(\delta',\gamma), \tag{8.3.2}$$

in which π and γ lie in $\widetilde{\Pi}_\phi$ and $\widetilde{\Gamma}_{\mathrm{reg}}(G)$ respectively, and (G',δ') corresponds to x according to the usual mappings

$$(G',\phi') \longrightarrow (\phi,s) \longrightarrow (\phi,x), \qquad s\in\overline{S}_{\phi,\mathrm{ss}}.$$

If (G',ϕ') equals (G,ϕ) and δ is the stable conjugacy class of γ, the second last formula reduces to

$$\widetilde{S}^G(\phi,\delta) = \sum_{\pi\in\widetilde{\Pi}_\phi} \widetilde{I}_G(\pi,\gamma), \qquad \gamma\in\widetilde{\Gamma}_{\mathrm{reg}}(G),$$

as might be expected. This is not to be regarded as a means to compute stable characters, however, since stable characters are really the primary objects. We instead use the twisted character

$$\widetilde{f}_N(\phi) = \int_{\widetilde{\Gamma}_{\mathrm{reg}}(N)} \widetilde{I}_N(\widetilde{\pi}_\phi,\widetilde{\gamma})\, \widetilde{f}_N(\widetilde{\gamma})\, d\widetilde{\gamma}, \qquad \widetilde{f}\in\widetilde{\mathcal{H}}(N), \tag{8.3.3}$$

attached to any self-dual parameter $\phi \in \widetilde{\Phi}_{\mathrm{bdd}}(N)$ for $GL(N)$, and in particular, to any $\phi \in \widetilde{\Phi}_{\mathrm{bdd}}(G)$. The left hand side here equals the trace of $\widetilde{\pi}_\phi(\widetilde{f})$, by definition (2.2.1). The domain of integration $\widetilde{\Gamma}_{\mathrm{reg}}(N)$ on the right hand side is the set of strongly regular, twisted conjugacy classes in $GL(N,F)$. The kernel function in the integrand is the normalized twisted character

$$\widetilde{I}_N(\widetilde{\pi}_\phi, \widetilde{\gamma}) = |D(\widetilde{\gamma})|^{\frac{1}{2}} \widetilde{\Theta}_N(\widetilde{\pi}_\phi, \widetilde{\gamma}), \qquad \widetilde{\gamma} \in \widetilde{\Gamma}_{\mathrm{reg}}(N),$$

whose existence has been established in [**Clo1**]. An analysis similar to that of (8.3.2), using (2.2.3) and (8.3.3) in place of (2.2.6) and (8.3.1), yields the general formula

$$\widetilde{S}^G(\phi, \delta) = \sum_{\widetilde{\gamma} \in \widetilde{\Gamma}_{\mathrm{reg}}(N)} \widetilde{I}_N(\widetilde{\pi}_\phi, \widetilde{\gamma}) \, \overline{\Delta(\delta, \widetilde{\gamma})}\,, \tag{8.3.4}$$

for any $\phi \in \widetilde{\Phi}_{\mathrm{bdd}}(G)$ and $\delta \in \widetilde{\Delta}_{N\text{-reg}}(G)$.

The formulas (8.3.2) and (8.3.4) classify the ($\widetilde{O}(G)$-orbits of) irreducible tempered representations of $G(F)$ in terms of twisted, tempered characters for general linear groups. Suppose for example that π lies in the subset $\widetilde{\Pi}_2(G)$ of square integrable (orbits of) representations. Then the parameter ϕ for the packet $\widetilde{\Pi}_\phi$ of π lies in the subset $\widetilde{\Phi}_2(G)$ of $\widetilde{\Phi}_{\mathrm{bdd}}(G)$. The endoscopic data G' that index the terms in the formula (8.3.2) for the character of π consequently lie in the subset $\widetilde{\mathcal{E}}_{\mathrm{ell}}(G)$ of $\mathcal{E}(G)$, and are products

$$G' = G'_1 \times G'_2, \qquad G'_i \in \widetilde{\mathcal{E}}_{\mathrm{sim}}(N'_i),$$

for a partition of N into two integers N'_i. (The notation is as in §1.2, except when G is of type C_n, in which case N'_2 here represents the odd integer (N'_2+1) of §1.2.) The stable characters $\widetilde{S}'(\phi', \delta')$ in the corresponding summands decompose accordingly into products

$$\widetilde{S}'(\phi', \delta') = \widetilde{S}'_1(\phi'_1, \delta'_1)\, \widetilde{S}'_2(\phi'_2, \delta'_2),$$

whose factors can be written in terms of twisted characters on the groups $GL(N'_i, F)$ by the formula (8.3.4). The supplementary coefficients on the right hand sides of (8.3.2) and (8.3.4) are concrete functions. The factor $\langle x, \pi\rangle^{-1}$ in (8.3.2) is just the value at x^{-1} of the character on the 2-group $\mathcal{S}_\phi$ attached to π. The other coefficients $\Delta(\delta', \gamma)$ and $\Delta(\delta, \widetilde{\gamma})$ are transfer factors, which we know are subtle objects, but which are nonetheless defined ([**LS1**], [**KS**]) by explicit formulas.

The local classification embodied in the character formulas (8.3.1)–(8.3.4) was proved by global means. We recall that the proof also had a purely local component. It was the collection of orthogonality relations from §6.5, for elliptic tempered characters on $G(F)$, and twisted, elliptic tempered characters on $GL(N,F)$.

The local theorems apply also to the more general set $\widetilde{\Psi}(G)$ of equivalence classes of nongeneric parameters

$$\psi\colon\ L_F \times SU(2) \longrightarrow {}^L G.$$

The character formulas hold as well, so long as we treat the elements of the packet of ψ as *reducible* representations. In other words, we write $\widetilde{\Sigma}_\psi$ and σ in place of $\widetilde{\Pi}_\psi$ and π, following the notation of §7.1. Then σ represents a sum of elements π of a fibre in $\widetilde{\Pi}_\psi$ over $\widehat{\mathcal{S}}_\psi$. The character formulas (8.3.1)–(8.3.4) remain valid for ψ, provided that the representation $\pi \in \widetilde{\Pi}_\phi$ in (8.3.1) and (8.3.2) is replaced by $\sigma \in \widetilde{\Sigma}_\psi$, and the point x in (8.3.2) is replaced by its translate $s_\psi x$. These lead to explicit character formulas for the elements $\sigma \in \widetilde{\Sigma}_\psi$ in terms of twisted, nontempered characters for general linear groups. The nongeneric local parameters $\psi \in \widetilde{\Psi}(G)$ are not part of the local classification of admissible representations. They serve rather in support of a global goal, the description of nontempered automorphic representations. But since the constituents of their packets are all unitary, the parameters ψ might still have a role in the local classification of unitary representations.

We do have to account for the possible failure of the analogue of Ramanujan's conjecture for $GL(N)$. In other words, we have to work with the larger set

$$\widetilde{\Psi}^+_{\mathrm{unit}}(G) = \widetilde{\Psi}^+(G) \cap \widetilde{\Psi}^+_{\mathrm{unit}}(N)$$

of local parameters in place of $\widetilde{\Psi}(G)$. This is the intermediate set

$$\widetilde{\Psi}(G) \subset \widetilde{\Psi}^+_{\mathrm{unit}}(G) \subset \widetilde{\Psi}^+(G),$$

defined in §1.5 in terms of the irreducible unitary representations of $GL(N,F)$. For any $\psi \in \widetilde{\Psi}^+_{\mathrm{unit}}(G)$, the elements in the packet $\widetilde{\Pi}_\psi$ are the induced representations (1.5.1). Their characters are obtained by analytic continuation of characters (2.2.6) that are provided by our local Theorem 2.2.1. In particular, the discussion above applies to the more general local parameters $\psi \in \widetilde{\Psi}_{\mathrm{unit}}(G)$. We would expect the induced representations in the packets $\widetilde{\Pi}_\psi$ to be irreducible. We shall state this as a conjecture, even though its proof might be quite straightforward.

CONJECTURE 8.3.1. *Assume that F is local, that $G \in \widetilde{\mathcal{E}}_{\mathrm{sim}}(N)$, and that ψ is a parameter in the set $\widetilde{\Psi}^+_{\mathrm{unit}}(G)$. Then the induced representations*

$$\mathcal{I}_P(\pi_{M,\lambda}), \qquad \pi_M \in \widetilde{\Pi}_{\psi_M},$$

in the packet $\widetilde{\Pi}_\psi$ are irreducible and unitary.

Rather than attempt a proof of the assertion, let me just append a couple of brief remarks. There is a bijection $\psi \to \pi_\psi$ from the set $\widetilde{\Psi}^+_{\mathrm{unit}}(N)$ attached to $GL(N)$ onto $\widetilde{\Pi}_{\mathrm{unit}}(N)$. The irreducible representation $\pi_\psi \in \widetilde{\Pi}_{\mathrm{unit}}(N)$ of $GL(N,F)$ can be regarded either as the Langlands quotient attached to the Langlands parameter ϕ_ψ, or an irreducible induced representation $\mathcal{I}_{P(N)}(\pi_{M(N),\lambda})$, for a Levi subgroup $M(N)$ of $GL(N)$ and the Langlands quotient $\pi_{M(N)}$ of a parameter $\psi_{M(N)} \in \Psi\big(M(N)\big)$. The classification of the unitary dual $\Pi_{\mathrm{unit}}(N)$ in [**Tad1**] and [**V2**] is in terms of such induced representations. In particular, the irreducibility is built into the construction. The problem is to show that the irreducibility property is preserved under the transfer from $GL(N)$ to G. One would have to show that the singularities of the relevant intertwining operators for $G(F)$ are dominated by those for $GL(N,F)$. Such singularities are governed by exponents of representations [**BoW**, p. 113] for p-adic F, and by some variant of the conditions of [**SV**], when F is archimedean. In the case that ψ lies in the subset

$$\widetilde{\Phi}_{\mathrm{unit}}(G) = \widetilde{\Psi}^{+}_{\mathrm{unit}}(G) \cap \widetilde{\Phi}(G)$$

of generic parameters in $\widetilde{\Psi}^{+}_{\mathrm{unit}}(G)$, the two assertions of the conjecture are not hard to check. □

Suppose now that F is global. If the fundamental local objects are irreducible characters, the fundamental global objects are Hecke eigenfamilies $c = \{c_v\}$. The global theorems classify the Hecke eigenfamilies attached to automorphic representations of G in terms of those for general linear groups. But they do more. Just as the local theorems sort irreducible characters into local packets for G in relating them to $GL(N)$, so the global theorems characterize the automorphic representations for G attached to eigenfamilies in terms of a global packet, and hence in terms of tensor products of local characters, even as they relate the eigenfamilies to those of $GL(N)$. The global process could be regarded as a description of the ramification properties of automorphic representations that are hidden in the unramified data of Hecke eigenfamilies. This is all summarized compactly in the multiplicity formula of Theorem 1.5.2.

Recall that $\mathcal{A}_{\mathrm{cusp}}(N)$ denotes the set of equivalence classes of unitary, cuspidal automorphic representations of $GL(N)$. When we began in §1.3, we agreed that this was the basic global object. It contains the family $\widetilde{\mathcal{A}}_{\mathrm{cusp}}(N)$ of self-dual such representations, which we have also denoted by $\widetilde{\Phi}_{\mathrm{sim}}(N)$ when thinking in terms of global parameters. The disjoint unions

$$\widetilde{\mathcal{A}}_{\mathrm{cusp}} = \coprod_{N} \widetilde{\mathcal{A}}_{\mathrm{cusp}}(N) \ \subset \ \mathcal{A}_{\mathrm{cusp}} = \coprod_{N} \mathcal{A}_{\mathrm{cusp}}(N)$$

are then the foundation for all of the global results. Theorem 1.3.2 implies that these sets map bijectively onto the corresponding sets of Hecke eigenfamilies in the disjoint unions

$$\widetilde{\mathcal{C}}_{\mathrm{sim}} = \coprod_{N} \widetilde{\mathcal{C}}_{\mathrm{sim}}(N) \ \subset \ \mathcal{C}_{\mathrm{sim}} = \coprod_{N} \mathcal{C}_{\mathrm{sim}}(N), \tag{8.3.5}$$

where

$$\widetilde{\mathcal{C}}_{\mathrm{sim}}(N) = \{c \in \mathcal{C}_{\mathrm{sim}}(N) : \ c^{\vee} = c\},$$

and

$$\mathcal{C}_{\mathrm{sim}}(N) = \mathcal{C}_{\mathrm{cusp}}(N)$$

is our fundamental set of simple Hecke eigenfamilies from (1.3.11). The more concrete data in (8.3.5) can thus also serve as a foundation for the global results.

Ideally, one would like to be able to formulate general global properties explicitly in terms of data from (8.3.5). Sometimes one can do so in an elementary way. For example, $\widetilde{\mathcal{C}}_{\mathrm{sim}}(N)$ is the subset of families $c = \{c_v\}$ in $\mathcal{C}_{\mathrm{sim}}(N)$ that are equal to their corresponding dual families $c^\vee = \{c_v^\vee\}$. The same goes for the description in §1.3 of the larger sets of Hecke eigenfamilies

$$c(\psi) = c(\mu) \otimes c(\nu) = \{c_v(\psi) = c_v(\mu)\, q_v^{\frac{n-1}{2}} \oplus \cdots \oplus c_v(\mu)\, q_v^{-\frac{n-1}{2}}\},$$

which are attached to elements in the set $\Psi_{\mathrm{sim}}(N) \cong \mathcal{A}_2(N)$ that parametrizes the automorphic discrete spectrum of $GL(N)$. These are obtained in an elementary way from families $c(\mu)$ in the basic sets $\mathcal{C}_{\mathrm{sim}}(m)$. Sometimes one must use transcendental means. This is the case for the description of the subset $\mathcal{C}_{\mathrm{sim}}(G)$ of families $c \in \widetilde{\mathcal{C}}_{\mathrm{sim}}(N)$ attached to the given G, according to which of the partial L-functions

$$L^S(s, c, S^2) = \prod_{v \notin S} \det\left(I - S^2(c_v)\, q_v^{-s}\right)$$

or

$$L^S(s, c, \Lambda^2) = \prod_{v \notin S} \det\left(I - \Lambda^2(c_v)\, q_v^{-s}\right)$$

has a pole at $s = 1$ (since the partial L-functions have the same behaviour at $s = 1$ as the completed L-functions of Theorem 1.5.3). Sometimes, however, the ties are less direct. For example, it would be hard to claim that the local data at ramified places v can be obtained from the unramified data in this explicit way. Theorems 1.5.1 and 1.5.2 do reduce the question to the case of $GL(N)$. One might then argue that the answer is in the theory of Rankin-Selberg L-functions, specifically the completion of a partial L-function $L^S(s, c_1 \times c_2)$ to a full L-function $L(s, \pi_1 \times \pi_2)$ with the appropriate functional equation.

At any rate, the global theorems are ultimately statements about data from (8.3.5). In particular, Theorem 1.5.2 describes the automorphic discrete spectrum of G in terms of parameters $\psi \in \widetilde{\Psi}_2(G)$ attached to compound Hecke eigenfamilies $c(\psi)$. These in turn are derived from simple families $\mu_i \in \widetilde{\mathcal{C}}_{\mathrm{sim}}(m_i)$, as in the special case of the subset $\widetilde{\Psi}_{\mathrm{sim}}(G)$ above. It is the multiplicity formula (1.5.3) that provides the quantitative link between global data and local properties. Through the parameters $\psi \in \widetilde{\Psi}_2(G)$, their completions $\psi_v \in \widetilde{\Psi}(G_v)$, and the corresponding mappings

$$\mathcal{S}_\psi \longrightarrow \prod_v \mathcal{S}_{\psi_v}$$

of centralizers, it characterizes the automorphic discrete spectrum explicitly in terms of local characters for G. These are then classified in terms of twisted characters for general linear groups by the formulas (8.3.1)–(8.3.4), and their generalizations for nongeneric parameters ψ_v.

The global theorems of course have other consequences. Since some of these were discussed as they came up in the text, we will not spend more time on them here. We recall only that they include a general theory of Rankin-Selberg L-functions and ε-factors for pairs of representations of G that is inherited from $GL(N)$, the special properties of L-functions and ε-factors given by Theorem 1.5.3, the relation between multiplicities and symplectic ε-factors implicit in the sign character ε_ψ of Theorem 1.5.2, and the absence of embedded eigenvalues for G established in Chapter 4.

Our proof of the local and global theorems is reminiscent of a fundamental theorem from the very beginnings of the subject. This is the analytic classification by Hermann Weyl of the representations of compact, simply connected Lie groups [**We1**]. Weyl's proof, which includes the Weyl character formula, is by invariant harmonic analysis. It entails a felicitous interplay of characters and multiplicities, linked by the orthogonality relations satisfied by irreducible characters. For a proof in the elementary case of $U(N)$ (which we could consider the analogue of our reductive group $GL(N)$), see [**We2**, 377–385].

These general observations on characters and multiplicities are all implicit in the statements of the various theorems. They have served us here as a pretext to collect our thoughts. They also lead us naturally to the next topic of discussion, that of even orthogonal groups. This will be divided into two parts, according to whether the integer $m(\psi)$ attached to a parameter ψ equals 1 or 2. We shall see that each part comes also with a broader philosophical question.

We fix an even orthogonal group

$$G \in \widetilde{\mathcal{E}}_{\mathrm{sim}}(N), \qquad \widehat{G} = SO(N,\mathbb{C}),\ N = 2n,$$

over the local or global field F. The group $\widetilde{O}(G)$ is of order 2 in this case, and G is either split or a quasisplit outer twist by the nontrivial element $\widetilde{\theta}$ in $\widetilde{O}(G)$. For any $\psi \in \widetilde{\Psi}(G)$, we write $m(\psi)$ as in Theorem 1.5.2 for the number of $\widehat{G}$-orbits in the associated $\widetilde{\mathrm{Aut}}_N(G)$-orbit of L-homomorphisms attached to ψ. Then $\widetilde{\Psi}(G)$ is the disjoint union of the subset

$$\Psi(\widetilde{G}) = \{\psi \in \widetilde{\Psi}(G) :\ m(\psi) = 1\}, \qquad \widetilde{G} = G \rtimes \widetilde{\theta},$$

and its complement

$$\widetilde{\Psi}'(G) = \{\psi \in \widetilde{\Psi}(G) :\ m(\psi) = 2\}.$$

We write $\Psi_2(\widetilde{G})$, $\Phi_{\mathrm{bdd}}(\widetilde{G})$, $\Phi_2(\widetilde{G})$ and $\widetilde{\Psi}'_2(G)$, $\widetilde{\Phi}'_{\mathrm{bdd}}(G)$, $\widetilde{\Phi}'_2(G)$, etc., for the obvious subsets of $\Psi(\widetilde{G})$ and $\widetilde{\Psi}'(G)$.

Consider, for example, a parameter

$$\psi = \psi_1 \boxplus \cdots \boxplus \psi_r, \qquad \psi_i \in \widetilde{\Psi}_{\mathrm{sim}}(G_i),\ G_i \in \widetilde{\mathcal{E}}_{\mathrm{sim}}(N_i),$$

in the subset $\widetilde{\Psi}_2(G)$ of $\widetilde{\Psi}(G)$. Then ψ lies in the subset $\widetilde{\Psi}'_2(G)$ of $\widetilde{\Psi}_2(G)$ if and only if each of the integers N_i is even. This is the most important case. A general parameter $\psi \in \widetilde{\Psi}(G)$ is a self-dual sum (1.4.1) of simple parameters with higher multiplicities ℓ_i. It is then easy to see that ψ belongs to the subset $\widetilde{\Psi}'(G)$ if and only if N_i is even for each i in the subset I_ψ of indices that correspond to self dual simple parameters. The condition can also be expressed in terms of the general subparameter $\psi_- \in \widetilde{\Psi}_2(G_-)$ we have attached to ψ, in which the associated endoscopic datum $G_- \in \widetilde{\mathcal{E}}_{\mathrm{sim}}(N_-)$ is now an even orthogonal group. We write

$$\psi = \psi^* \boxplus \psi_-, \qquad \psi^* \in \widetilde{\Psi}(G^*),\ G^* \in \widetilde{\mathcal{E}}_{\mathrm{sim}}(N^*),$$

where G^* is also an even orthogonal group, and ψ^* factors through a Levi subgroup of G^* that is a product of general linear groups. Then ψ lies in $\widetilde{\Psi}'(G)$ if and only if ψ^* and ψ_- lie in the corresponding subsets $\widetilde{\Psi}'(G^*)$ and $\widetilde{\Psi}'_2(G_-)$ of $\widetilde{\Psi}(G^*)$ and $\widetilde{\Psi}_2(G_-)$ (with the convention that $\widetilde{\Psi}'(G) = \widetilde{\Psi}(G)$ in case $N = 0$). The remarks here, with the accompanying notation, obviously extend to the larger family of parameters $\widetilde{\Psi}^+(G)$.

We shall limit our discussion here to the subset $\widetilde{\Phi}_{\mathrm{bdd}}(G)$ of generic parameters in $\widetilde{\Psi}(G)$. (If F is global, the subscript *bdd* here is a little misleading. It anticipates our being able at some point to replace the complex algebraic group $\mathcal{L}_\psi$ by a locally compact group.) The first part of the discussion concerns parameters ϕ in the subset $\Phi_{\mathrm{bdd}}(\widetilde{G})$ of $\widetilde{\Phi}_{\mathrm{bdd}}(G)$. It is an extension of our remarks on characters and multiplicities to twisted endoscopy for G.

Suppose that F is local. For a local parameter $\phi \in \Phi_{\mathrm{bdd}}(\widetilde{G})$, Theorem 2.2.4(b) tells us that the elements in $\widetilde{\Pi}_\phi$ are actually irreducible representations, rather than nontrivial $\widetilde{O}(G)$-orbits. In other words, the packet $\Pi_\phi = \widetilde{\Pi}_\phi$ is an actual L-packet, as we noted at the beginning of the section. The character formulas (8.3.1)–(8.3.4) are of course still in force. However, there are supplementary character formulas that come from Theorem 2.2.4(a). They apply to the $\widetilde{G}$-twisted characters

$$\tilde{f}_{\widetilde{G}}(\tilde{\pi}) = \int_{\Gamma_{G\text{-reg}}(\widetilde{G})} I_{\widetilde{G}}(\tilde{\pi}, \tilde{\gamma})\, \tilde{f}_{\widetilde{G}}(\tilde{\gamma})\, d\tilde{\gamma}, \qquad \tilde{f} \in \mathcal{H}(\widetilde{G}), \tag{8.3.6}$$

of extensions $\tilde{\pi}$ to $\widetilde{G}(F)$ of representations $\pi \in \Pi_\phi$. An analysis similar to that of (8.3.2), with (2.2.17) and (8.3.6) in place of (2.2.6) and (8.3.1), yields the general formula

$$I_{\widetilde{G}}(\tilde{\pi}, \tilde{\gamma}) = |\mathcal{S}_\phi|^{-1} \sum_{\tilde{x} \in \widetilde{\mathcal{S}}_\phi} \ \sum_{\tilde{\delta}' \in \Delta_{G\text{-reg}}(\widetilde{G}')} \langle \tilde{x}, \tilde{\pi} \rangle^{-1}\, \widetilde{S}'(\tilde{\phi}', \tilde{\delta}')\, \Delta(\tilde{\delta}', \tilde{\gamma}), \tag{8.3.7}$$

for any $\phi \in \Phi_{\mathrm{bdd}}(\widetilde{G})$, $\pi \in \Pi_\phi$ and $\tilde{\gamma} \in \Gamma_{G\text{-reg}}(\widetilde{G})$ and for $\tilde{\pi}$, $\widetilde{\mathcal{S}}_\phi$ and $(\widetilde{G}', \tilde{\phi}')$ as in Theorem 2.2.4(a).

Given that the L-packet Π_ϕ has already been defined by the original formula, one might ask whether (8.3.7) is superfluous. It is not. Suppose for example that ϕ lies in the subset $\Phi_2(\widetilde{G})$ of $\Phi_{\mathrm{bdd}}(\widetilde{G})$. The endoscopic data $\widetilde{G}'$ that index the terms in (8.3.7) then lie in the subset $\mathcal{E}_{\mathrm{ell}}(\widetilde{G})$ of $\mathcal{E}(G)$, and are products

$$\widetilde{G}' = \widetilde{G}'_1 \times \widetilde{G}'_2, \qquad \widetilde{G}'_i \in \widetilde{\mathcal{E}}_{\mathrm{sim}}(\widetilde{N}'_i),$$

for a partition of N into two odd integers $\widetilde{N}'_i$, as in §1.2. The stable characters in the corresponding summands decompose accordingly into products

$$\widetilde{S}'(\widetilde{\phi}', \widetilde{\delta}') = \widetilde{S}'(\widetilde{\phi}'_1, \widetilde{\delta}'_1)\, \widetilde{S}'(\widetilde{\phi}'_2, \widetilde{\delta}'_2)$$

to which we can apply (8.3.4). The formula (8.3.6) thus expresses twisted characters on $\widetilde{G}(F)$ in terms of twisted characters of general linear groups of *odd* rank $\widetilde{N}'_i$. The original formula (8.3.2) expresses ordinary characters on $G(F)$ in terms of twisted characters of general linear groups of *even* rank N'_i (except when G is of type C_n, in which case the second integer N'_2 is odd). Therefore, (8.3.7) represents a separate reciprocity law, which gives new local information.

Suppose now that F is global. A global parameter $\phi \in \Phi_2(\widetilde{G})$ has a global L-packet Π_ϕ, whose contribution to the discrete spectrum is given explicitly by Theorem 1.5.2. Since its completions ϕ_v lie in the corresponding local subsets $\Phi(\widetilde{G}_v)$, the representations in the local packets Π_{ϕ_v} satisfy the supplementary twisted character formulas (8.3.7). What more is there to say? The extra piece of global information is given by Theorem 4.2.2(b). It asserts that two canonical extensions to $\widetilde{G}(\mathbb{A})$ of any $\pi \in \Pi_\phi$ in the discrete spectrum, one obtained from (4.2.7), the other from Theorem 1.5.2 and the extensions $\widetilde{\pi}_v$ in (8.3.7), are the same. Viewed as a signed multiplicity formula for the extension (4.2.7), it tells us that the sign in question equals 1.

Suppose again that each completion ϕ_v of ϕ lies in $\Phi(\widetilde{G}_v)$, but that ϕ itself belongs to the complementary global set $\widetilde{\Phi}'_2(G)$. This is the case in which any representation $\pi \in \widetilde{\Pi}_\phi$ that contributes to the discrete spectrum does so with multiplicity 2. Theorem 4.2.2(a) implies that for any $\widetilde{f} \in \mathcal{H}(\widetilde{G})$, the restriction of the operator $R^{\widetilde{G}}_{\mathrm{disc},\phi}(\widetilde{f})$ to the π-isotypical subspace L^2_π of the discrete spectrum has vanishing trace. The eigenspaces of the operator $R^{\widetilde{G}}_{\mathrm{disc},\phi}(\widetilde{\theta})$ then give a canonical decomposition

$$L^2_\pi = L^2_{\pi,+} \oplus L^2_{\pi,-}, \qquad \pi \in \Pi_\phi, \tag{8.3.8}$$

of L^2_π into invariant subspaces on which the restriction $R^G_{\mathrm{disc},\phi}$ is equivalent to π. In particular, π does occur with multiplicity 2.

This phenomenon is related to a general philosophical question posed by A. Beilinson. Guided by his work on the geometric Langlands program, Beilinson has asked whether spaces of automorphic forms of higher multiplicity might have natural bases. Examples from [**LL**] lend support to his suggestion. The decomposition (8.3.8), which we have obtained from twisted endoscopy for G, can be regarded as further evidence. It would obviously be interesting to investigate this question heuristically in terms of parameters for general groups. (For some partial results, including the case here of even orthogonal groups, see [**Lar**].)

The other half of our discussion of the even orthogonal group G is for the complementary subset $\widetilde{\Phi}'_{\mathrm{bdd}}(G)$ of $\Phi_{\mathrm{bdd}}(\widetilde{G})$. This is not implicit in the earlier theorems. It concerns the question of constructing genuine L-packets Π_{ϕ_*} from the coarser packets $\widetilde{\Pi}_\phi$, and will be the setting for two new theorems in the next section. Our purpose here is to give some motivation and lay the groundwork for these theorems.

Suppose that F is local. Regarding $\phi \in \widetilde{\Phi}'_{\mathrm{bdd}}(G)$ as an $\widetilde{\mathrm{Aut}}_N(G)$-orbit of L-homomorphisms, we write $\Phi(\phi)$ for the corresponding pair of $\widehat{G}$-orbits of L-homomorphisms ϕ_*. We also write Π_ϕ for the preimage of $\widetilde{\Pi}_\phi$ in $\Pi(G)$ under the projection of $\Pi(G)$ onto $\widetilde{\Pi}(G)$. Then Π_ϕ is a disjoint union of $\widetilde{O}(G)$-torsors in $\Pi(G)$, which is to say, transitive $\widetilde{O}(G)$-orbits (of order 2) of irreducible representations of $G(F)$. We would like to separate it into two disjoint subsets $\Pi_{\phi,\star}$ of representatives of the $\widetilde{O}(G)$-orbits, which are compatible with endoscopic transfer, and are canonically parametrized by the two elements $\phi_\star$ in the $\widetilde{O}(G)$-torsor $\Phi(\phi)$. We will eventually realize this goal, apart from the canonical parametrization by $\Phi(\phi)$. The idea is to replace $\Phi(\phi)$ by a second $\widetilde{O}(G)$-torsor $T(\phi)$ constructed directly in terms of representations.

For the formal definition of $T(\phi)$, consider first a simple parameter $\phi \in \widetilde{\Phi}_{\mathrm{sim}}(G)$. Notice that this automatically implies that ϕ lies in $\widetilde{\Phi}'_{\mathrm{bdd}}(G)$, since N is even. According to our weaker version of the local Langlands correspondence for G, ϕ corresponds to an $\widetilde{O}(G)$-orbit $\pi_\phi \in \widetilde{\Pi}(G)$ of irreducible representations in $\Pi_{\mathrm{temp}}(G)$. By Corollary 6.7.3, the order $m(\pi_\phi)$ of the orbit equals 2. We define $T(\phi)$ in this case to be the $\widetilde{O}(G)$-torsor π_ϕ. Consider next a general element

$$\phi = \phi_1 \oplus \cdots \oplus \phi_r, \qquad \phi_i \in \widetilde{\Phi}_{\mathrm{sim}}(G_i),\ G_i \in \mathcal{E}_{\mathrm{sim}}(N_i),\ N_i \text{ even},$$

in $\widetilde{\Phi}'_2(G)$. In this case, we take

$$T(\phi) = \{t = t_1 \times \cdots \times t_r :\ t_i \in T(\phi_i)\} / \sim \tag{8.3.9}$$

to be a set of equivalence classes in the product over i of the sets $T(\phi_i)$. The equivalence relation is defined simply by writing $t' \sim t$ if the set

$$\{i :\ t'_i \neq t_i\}$$

is even. Finally, suppose that ϕ lies in the complement of $\widetilde{\Phi}'_2(G)$ in $\widetilde{\Phi}'_{\mathrm{bdd}}(G)$. If ϕ equals its subparameter ϕ^*, in the notation above, ϕ is the image of a product of local Langlands parameters for general linear groups, and there is no ambiguity in identifying the two Langlands parameters in $\Phi(\phi)$ with two L-packets. In this case, we simply set $T(\phi) = \Phi(\phi)$. We can therefore assume that ϕ is a proper direct sum $\phi^* \oplus \phi_-$. Having already defined $T(\phi_-)$

for the parameter $\phi_- \in \widetilde{\Phi}_2(G_-)$, we set

$$T(\phi) = \{t = t^* \times t_- \in T(\phi^*) \times T(\phi_-)\} / \sim, \tag{8.3.10}$$

where $t' \sim t$ if the components of t' and t are either the same or both distinct from each other.

With our reference to Corollary 6.7.3 in the definition of $T(\phi)$, we have implicitly taken F to be p-adic. The same definition of course holds if F is archimedean. In fact, by the general form of the Langlands correspondence for real groups, there is a canonical $\widetilde{O}(G)$-isomorphism between the torsors $\Phi(\phi)$ and $T(\phi)$. For p-adic F, we do not have a canonical bijection. Our goal, which we will pursue in the next section, will be to construct the two L-packets for ϕ using $T(\phi)$ in place of $\Phi(\phi)$.

Suppose that F is global, and that ϕ is a global parameter in $\widetilde{\Phi}'_{\text{bdd}}(G)$. We can then define an $\widetilde{O}(G)$-torsor in a rather similar fashion. Suppose first that ϕ lies in the subset $\widetilde{\Phi}_{\text{sim}}(G)$ of $\widetilde{\Phi}'_2(G)$. For any valuation v on F, we take π_v to be the representation in the packet $\widetilde{\Pi}_{\phi_v}$ corresponding to the trivial character on $\mathcal{S}_{\phi_v}$. In the case of archimedean v, where the mapping from $\widetilde{\Pi}_{\phi_v}$ to $\widehat{\mathcal{S}}_{\phi_v}$ is not surjective, we appeal to Shelstad's characterization [**S6**, Theorem 11.5] of the elements in $\widetilde{\Pi}_{\phi_v}$ with Whittaker model for the existence of π_v. The global product $\pi = \bigotimes \pi_v$, which is to be regarded as an orbit under the group

$$\widetilde{\mathrm{O}}\mathrm{ut}_N(G_{\mathbb{A}}) = \widetilde{\bigotimes_v} \big(\widetilde{\mathrm{O}}\mathrm{ut}_N(G_v)\big),$$

then occurs in the global discrete spectrum. More precisely, there are exactly two representations $\pi_\star$ in the orbit of π that occur in the discrete spectrum of G. They each occur with multiplicity 1, and together, form a transitive orbit under the diagonal subgroup $\widetilde{O}(G)$ of $\widetilde{\mathrm{O}}\mathrm{ut}_N(G_{\mathbb{A}})$. This follows directly from the proof of Proposition 6.7.2, or alternatively, the general assertion of Theorem 1.5.2. We can therefore define $T(\phi)$ to be the $\widetilde{O}(G)$-torsor $\{\pi_\star\}$, in case ϕ is simple. Suppose next that ϕ is a general parameter in the subset $\widetilde{\Phi}'_2(G)$, with simple components $\{\phi_i\}$. We then define the global $\widetilde{O}(G)$-torsor $T(\phi)$ by the local construction (8.3.9). Finally, if ϕ lies in the complement of $\widetilde{\Phi}'_2(G)$ in $\widetilde{\Phi}'_{\text{bdd}}(G)$, we follow the local prescription (8.3.10) to define the global torsor $T(\phi)$.

We note that if $\phi \in \widetilde{\Phi}_{\text{sim}}(G)$ is simple, and ϕ_v lies in the subset $\Phi(\widetilde{G}_v)$ of $\widetilde{\Phi}(G_v)$ for every v, the two representations $\pi_\star$ above are actually equivalent. But as we saw in the decomposition (8.3.8) obtained from the action of $\widetilde{\theta}$, we can indeed treat them as two separate representations in the definition. At any rate, our interest in global torsor $T(\phi)$ attached to $\phi \in \Phi_{\text{bdd}}(G)$ will be confined to the case that some ϕ_v lies in the corresponding local set $\widetilde{\Phi}'(G_v)$, since we have treated the other case above.

In the next section, we will use the local torsor $T(\phi)$ to construct the two L-packets attached to a general local parameter $\phi \in \widetilde{\Phi}'_{\mathrm{bdd}}(G)$. We shall use the global torsor $T(\phi)$ to establish corresponding multiplicity formulas for global parameters $\phi \in \widetilde{\Phi}'_2(G)$. This will give the refined endoscopic classification for generic parameters for our even orthogonal group G. These results represent a different perspective, in at least one sense. The characters of individual representations $\pi_\star$ in a local packet $\Pi_{\phi,\star}$ cannot be defined in terms of twisted characters for general linear groups. They do satisfy internal endoscopic relations analogous to (8.3.2), but nothing akin to (8.3.4). From this point of view, the representations $\pi_\star$ are less accessible than their $\widetilde{O}(G)$-orbits π.

Of interest also is what we do not obtain. We will not be able to define a canonical parametrization of the two L-packets $\Pi_{\phi,\star}$ by the two local Langlands parameters $\phi_\star \in \Phi(\phi)$. The difficulty seems to be deeply entrenched, at least insofar as it pertains to the methods of this volume. The obstruction can be likened to what in physics might be called a $(\mathbb{Z}/2\mathbb{Z})$-symmetry. The philosophical question is whether there might be any endoscopic way to resolve it. Is there any "experiment" within the conjectural theory of endoscopy that would determine the predicted bijection between the two $\widetilde{O}(G)$-torsors?

The third topic for this section is that of Whittaker models. We shall prove a strong form of the generic packet conjecture, for any group $G \in \widetilde{\mathcal{E}}_{\mathrm{sim}}(N)$. To do so, we have only to combine our classification of local and global L-packets for G with the results of Cogdell, Kim, Piatetskii-Shapiro and Shahidi, and Ginzburg, Rallis and Soudry on the transfer of automorphic representations with global Whittaker models. The basic idea of applying the work that culminated in [**CKPS2**] and [**GRS2**] to the general existence of Whittaker models is not new. I learned of it from Rallis, but the idea was also known to Shahidi, and no doubt others as well.

We should recall the underlying global definition. Assume for a moment that F is global, and that G is a general quasisplit, connected group over F. Suppose also that $(B, T, \{X_k\})$ is a Γ_F-stable splitting of G over F, and that ψ_F is a nontrivial additive character on $\mathbb{A}/F$. The function

$$\chi(u) = \psi_F(u_1 + \cdots + u_n), \qquad u \in N_B(\mathbb{A}),$$

where $\{u_k\}$ are the coordinates of u with respect to the simple root vectors $\{X_k\}$, is a nondegenerate, left $N_B(F)$-invariant character on $N_B(\mathbb{A})$. It represents a global Whittaker datum (B, χ) for G. Given χ, one defines the Whittaker functional

$$\omega(h) = \int_{N_B(F)\backslash N_B(\mathbb{A})} h(n)\, \chi(n)^{-1}\, dn$$

on the space of smooth functions h on $G(F)\backslash G(\mathbb{A})$. Suppose that π is an irreducible cuspidal automorphic representation of $G(\mathbb{A})$, acting on a closed invariant subspace V of $L^2_{\mathrm{cusp}}\big(G(F)\backslash G(\mathbb{A}), \chi_G\big)$ (where χ_G denotes a central character datum). We say that (π, V) is (globally) (B, χ)-generic if the restriction of ω to the space of smooth functions in V is nonzero. If this is so, the restriction ω_v of ω to any local constituent π_v of π is a local (B_v, χ_v)-Whittaker functional.

We assume that F is either local or global, and that G is a fixed group in $\widetilde{\mathcal{E}}_{\text{sim}}(N)$ over F. We can then fix (B, χ) by fixing ψ_F, and taking $(B, T, \{X_k\})$ to be the standard splitting. If F is global, and (G_v, F_v) is a localization of (G, F), we can take (B_v, χ_v) to be the corresponding localization of (B, χ).

PROPOSITION 8.3.2. (a) *Suppose that F is local, that $\phi \in \widetilde{\Phi}_{\text{bdd}}(G)$ is a bounded generic parmeter, and that π represents the $\widetilde{\text{Out}}_N(G)$-orbit in the local packet $\widetilde{\Pi}_\phi$ such that the linear character $\langle \cdot, \pi \rangle$ on $\mathcal{S}_\phi$ is trivial. Then π is locally (B, χ)-generic.*

(b) *Suppose that F is global, and that ϕ is a generic parameter in the larger global set $\widetilde{\Psi}_2(G)$. Then there is a globally (B, χ)-generic element π in the global packet $\widetilde{\Pi}_\phi$ such that the linear character $\langle \cdot, \pi \rangle$ on $\mathcal{S}_\phi$ is trivial.*

PROOF. (a) If F is archimedean, the result is known for any quasisplit G, and for the general L-packet attached to a Langlands parameter $\phi \in \Phi_{\text{bdd}}(G)$. For it follows from [**V2**, Theorem 6.2] and [**Kos**, p. 105] that there is exactly one (B, χ)-generic representation π in the packet Π_ϕ. It then follows from [**S6**, Theorem 11.5] that with the (B, χ)-Whittaker normalization of transfer factors we are using, π is indeed the representation in $\widetilde{\Pi}_\phi$ such that $\langle \cdot, \pi \rangle$ equals 1.

We can therefore assume that F is p-adic. Suppose first that ϕ belongs to the complement of $\widetilde{\Phi}_2(G)$. The argument in this case is purely local. We choose a local parameter $\phi_M \in \Phi_2(M, \phi)$, for a proper standard Levi subgroup M of G. Let π_M be the representation in $\widetilde{\Pi}_{\phi_M}$ such that the linear form $\langle \cdot, \pi_M \rangle$ on $\mathcal{S}_{\phi_M}$ is trivial. We can assume inductively that π_M is (B_M, χ_M)-generic, since M is proper in G. Let π be the unique (B, χ)-generic component of $\mathcal{I}_P(\pi_M)$. It then follows from (2.5.3), together with the definition (2.4.17) of the character $\langle \cdot, \pi \rangle$ in terms of its twisted analogue $\langle \cdot, \widetilde{\pi}_M \rangle$ for π_M and the self-intertwining operators $R_P(w, \widetilde{\pi}_M)$, that the character is trivial. This gives the proposition for ϕ.

Suppose next that ϕ belongs to $\widetilde{\Phi}_2(G)$. In this case we use a global argument, specifically the global construction of §6.3. From Proposition 6.3.1, together with the discussion at the beginning of this section in case ϕ is simple, we obtain global objects $(\dot{F}, \dot{G}, \dot{\phi})$ from the given local objects (F, G, ϕ). Then $\dot{\phi}$ is a generic parameter in the subset $\widetilde{\Phi}_2(\dot{G})$ of $\widetilde{\Phi}_{\text{ell}}(N)$. It corresponds to an automorphic representation $\pi_{\dot{\phi}}$ of $GL(N)$, which is induced from a self-dual, cuspidal automorphic representation of a Levi subgroup. It follows by induction, or by Lemma 5.4.6 in case $\dot{\phi}$ is simple, that Theorem 1.5.3 is valid for the simple components $\dot{\phi}_i \in \widetilde{\Phi}_{\text{sim}}(\dot{G}_i)$ of $\dot{\phi}$. This is the required condition for the global descent theorem of [**GRS1**], [**So**] and [**GRS2**] (which depends for its proof on [**CKPS2**]). The theorem states that there is a globally $(\dot{B}, \dot{\chi})$-generic cuspidal automorphic representation $\dot{\pi}$ of $\dot{G}(\dot{\mathbb{A}})$ whose near equivalence class (in the language of [**So**]) maps to that of $\pi_{\dot{\phi}}$. In other words, $\dot{\pi}$ satisfies the condition

$$\xi_{\dot{\phi}}\big(c(\dot{\pi})\big) = c(\pi_{\dot{\phi}}) = c(\dot{\phi}),$$

and thus contributes to the subspace of the automorphic discrete spectrum of $\dot{G}$ determined by $\dot{\phi}$. As we noted following the proof of Corollary 6.7.4, $\dot{\pi}$ therefore represents an element in the global packet $\widetilde{\Pi}_{\dot{\phi}}$ we now have at our disposal, for which associated linear character

$$\langle \dot{x}, \dot{\pi} \rangle = \prod_v \langle \dot{x}_v, \dot{\pi}_v \rangle, \qquad \dot{x} \in \mathcal{S}_{\dot{\phi}}, \tag{8.3.11}$$

equals 1.

Since $\dot{\pi}$ is globally $(\dot{B}, \dot{\chi})$-generic, each of its components $\dot{\pi}_v$ is locally $(\dot{B}_v, \dot{\chi}_v)$-generic. If $v \notin S_\infty(u)$, we know that $\dot{\pi}_v$ is characterized uniquely as the element in the local packet $\widetilde{\Pi}_{\dot{\phi}_v}$ such that the local linear character $\langle\,\cdot, \dot{\pi}_v\rangle$ on $\mathcal{S}_{\dot{\phi}_v}$ equals 1. This follows from the discussion based on the group $Sp(2)$ at the end of §2.5, which led to the proof of Lemmas 6.4.1 and 6.6.2. We have just seen that the same condition holds for the archimedean places $v \in S_\infty$. This leaves only the p-adic place u of $\dot{F}$, which we recall has the property that $(\dot{F}_u, \dot{G}_u, \dot{\phi}_u)$ equals (F, G, ϕ). The component $\dot{\pi}_u$ of $\dot{\pi}$ at u lies in the local packet $\widetilde{\Pi}_\phi = \widetilde{\Pi}_{\dot{\phi}_u}$. It follows from the product formula (8.3.11) that the corresponding linear character $\langle\,\cdot, \dot{\pi}_u\rangle$ on the group $\mathcal{S}_{\dot{\phi}_u} = \mathcal{S}_\phi$ equals 1, and consequently that $\dot{\pi}_u$ equals the given representation π of $G(F)$. But we also know that $\dot{\pi}_u$ is $(\dot{B}_u, \dot{\chi}_u)$-generic. The given representation is therefore (B, χ)-generic, as required.

(b) The existence of π is proved in the same way as that of the automorphic representation $\dot{\pi}$ in (a). That is to say, it follows directly from the global descent theorem of [**GRS1**], [**So**] and [**GRS2**]. The assertion on the character $\langle \cdot, \pi\rangle$ is just the condition that π occurs in the discrete spectrum. □

REMARKS. 1. We have identified the irreducible representation π with the corresponding $\widetilde{\mathrm{O}}\mathrm{ut}_N(G)$-orbit of representations in the packet $\widetilde{\Pi}_\phi$. This is just our usual convention in the local case (a), and applies also to the global case (b) if π is taken to be the $\widetilde{\mathrm{O}}\mathrm{ut}_N(G)$-orbit of representations in the associated $\widetilde{\mathrm{O}}\mathrm{ut}_N(G_{\mathbb{A}})$-orbit that actually occur in the discrete spectrum. The group of F-automorphisms of G with which we identified $\widetilde{\mathrm{O}}\mathrm{ut}_N(G)$ in Chapter 1 stabilizes (B, χ). We can therefore regard a (B, χ)-Whittaker functional for the representation π as an $\widetilde{\mathrm{O}}\mathrm{ut}_N(G)$-orbit of functionals for the corresponding $\widetilde{\mathrm{O}}\mathrm{ut}_N(G)$-orbit of representations.

2. In the local case (a), it is expected that π is the *unique* (B, χ)-generic element in the packet $\widetilde{\Pi}_\phi$. In fact, we had to use the known analogue of this property for valuations $v \neq u$ in the proof. The property for G follows from

[**JiS1**] (and the existence of the packet $\widetilde{\Pi}_\phi$), in the case $G = SO(2n+1)$. (See also [**JiS2**], for a proof of the local Langlands correspondence for generic representations of the p-adic group $SO(2n+1)$.) Extensions of [**JiS1**] to the other groups G were sketched in [**JiS3**], while the group $G = Sp(2n)$ was recently treated in detail in [**Liu**]. Taking these for granted, we see that the automorphic representation π in (b) is the (unique) element in the global packet $\widetilde{\Pi}_\phi$ whose local components are the elements $\pi_v \in \widetilde{\Pi}_{\phi_v}$ whose linear characters $\langle \cdot, \pi_v \rangle$ on the corresponding groups $\mathcal{S}_{\phi_v}$ are all trivial.

3. Consider again the local case (a). As we noted in §2.5, Konno [**Kon**] proved that the packet $\widetilde{\Pi}_\phi$, once constructed, would contain a (B,χ)-generic element. His methods are purely local. Their refinement, based on a study of the Whittaker normalization of transfer factors in [**KS**, (5.3)], would presumably lead to a local proof of the proposition in case char$(F) \neq 2$, as well as a proof of the uniqueness of π.

8.4. Refinements of generic parameters for $SO(2n)$

We shall now continue the discussion of even orthogonal groups begun in the last section. Our goal is to construct genuine L-packets from the cruder packets $\widetilde{\Pi}_\phi$ obtained in the main theorems. To see that this might be possible, we need only observe that the full force of the stabilized trace formula has yet to be exploited. We have used it so far just for test functions in the symmetric global Hecke algebra. We will now apply it as it was stated in the general formula (3.3.6) of Lemma 3.3.1, for arbitrary functions in the Hecke algebra. The discussion will be rather abstract, making this section one of the more difficult parts of the volume.

As in the last section, we fix an even orthogonal group

$$G \in \widetilde{\mathcal{E}}_{\text{sim}}(N), \qquad \widehat{G} = SO(N,\mathbb{C}),\ N = 2n,$$

over the field F. We are interested in parameters ϕ that lie in the subset

$$\widetilde{\Phi}'(G) = \{\phi \in \widetilde{\Phi}(G):\ m(\phi) = 2\}$$

of $\widetilde{\Phi}(G)$. For any such ϕ, the associated pair

$$\Phi(\phi) = \Phi(G,\phi)$$

of parameters is a torsor under the group

$$\widetilde{O}(G) = \widetilde{\text{O}}\text{ut}_N(G) = \widetilde{\text{A}}\text{ut}_N(G)/\widetilde{\text{I}}\text{nt}_N(G)$$

of order 2. In the last section, we defined another $\widetilde{O}(G)$-torsor $T(\phi)$ in terms of irreducible representations. The L-packets we construct from $\widetilde{\Pi}_\phi$ will be parametrized by $T(\phi)$ instead of $\Phi(\phi)$.

If ϕ lies in the complement $\Phi(\widetilde{G})$ of $\widetilde{\Phi}'(G)$ in $\widetilde{\Phi}(G)$, the set $\Phi(\phi)$ consists of ϕ alone. A formal application of the earlier definition of $T(\phi)$ to this case yields a set that also contains only one element. With this understanding, we note that for any $\phi \in \widetilde{\Phi}(G)$, the set $T(\phi)$ behaves well under endoscopic transfer. Suppose for example that s belongs to the subset $\overline{S}_{\phi,\text{ell}}$ of elliptic elements in $\overline{S}_\phi$. The endoscopic preimage of the pair (ϕ, s) is then a pair (G', ϕ'), where

$$G' = G'_1 \times G'_2, \qquad G'_i \in \widetilde{\mathcal{E}}_{\text{sim}}(N'_i),$$

belongs to $\mathcal{E}_{\text{ell}}(G)$, and

$$\phi' = \phi'_1 \times \phi'_2, \qquad \phi'_i \in \widetilde{\Phi}(G'_i),$$

belongs to $\widetilde{\Phi}(G')$. The finite group

$$\widetilde{O}(G') = \widetilde{O}(G'_1) \times \widetilde{O}(G'_2)$$

is an extension

$$1 \longrightarrow \text{Out}_G(G') \longrightarrow \widetilde{O}(G') \longrightarrow \widetilde{O}(G) \longrightarrow 1,$$

which acts transitively on the product

$$T(\phi') = T(\phi'_1) \times T(\phi'_2).$$

It is clear that there is a surjective mapping $T(\phi') \to T(\phi)$, which is compatible with the actions of $\widetilde{O}(G')$ and $\widetilde{O}(G)$. In particular, the mapping $(G', \phi') \to (\phi, s)$ extends to a surjective mapping

$$(G', \phi', t') \longrightarrow (\phi, s, t), \qquad t \in T(\phi),\ t' \in T(\phi').$$

We are really only interested in the case that ϕ lies in $\widetilde{\Phi}'(G)$ (or even the smaller set $\widetilde{\Phi}'_{\text{bdd}}(G)$), so that $T(\phi)$ has order 2 and is indeed an $\widetilde{O}(G)$-torsor. The set $T(\phi')$ then typically has order 4, and is an $\widetilde{O}(G')$-torsor. In this case, the fibres of the mapping $T(\phi') \to T(\phi)$ are $\text{Out}_G(G')$-torsors.

Suppose that F is local. If $\widetilde{\Pi}$ represents a subset of $\widetilde{\Pi}_{\text{unit}}(G)$, we shall write Π for its preimage in $\Pi_{\text{unit}}(G)$. In particular, Π_ϕ is the preimage in $\Pi_{\text{unit}}(G)$ of the packet $\widetilde{\Pi}_\phi$ of $\widetilde{O}(G)$-orbits in $\widetilde{\Pi}_{\text{unit}}(G)$ we have attached to any ϕ. Assume that ϕ lies in the subset $\widetilde{\Phi}'_{\text{bdd}}(G)$ of $\widetilde{\Phi}(G)$. It follows from the results of Chapter 6 that the $\widetilde{O}(G)$-orbit of any $\pi \in \widetilde{\Pi}_\phi$ has order 2. The fibres of the mapping

$$\Pi_\phi \longrightarrow \widetilde{\Pi}_\phi$$

are thus $\widetilde{O}(G)$-torsors. The following theorem posits among other things a canonical $\widetilde{O}(G)$-orbit of sections

$$\widetilde{\Pi}_\phi \longrightarrow \Pi_\phi$$

parametrized by $T(\phi)$. The images in $\Pi_{\text{unit}}(G)$ of the two sections in the orbit represent the true L-packets attached to the two Langlands parameters in the set $\Phi(\phi)$.

THEOREM 8.4.1. *Suppose that F is local, and that ϕ lies in $\widetilde{\Phi}'_{\text{bdd}}(G)$. Then there is an $\widetilde{O}(G)$-equivariant bijection*

$$t \longrightarrow \phi_t, \qquad t \in T(\phi),$$

from $T(\phi)$ onto a pair of stable linear forms

$$f \longrightarrow f^G(\phi_t), \qquad f \in \mathcal{H}(G),\ t \in T(\phi),$$

on $\mathcal{H}(G)$, and an $\widetilde{O}(G)$-equivariant bijection

$$(\pi, t) \longrightarrow \pi_t, \qquad \pi \in \widetilde{\Pi}_\phi,\ t \in T(\phi),$$

from $\widetilde{\Pi}_\phi \times T(\phi)$ onto Π_ϕ, such that

$$f^G(\phi) = f^G(\phi_t), \qquad f \in \widetilde{\mathcal{H}}(G),\ t \in T(\phi), \tag{8.4.1}$$

and

$$f_G(\pi) = f_G(\pi_t), \qquad f \in \widetilde{\mathcal{H}}(G),\ t \in T(\phi), \tag{8.4.2}$$

and such that for any $s \in S_{\phi,ss}$ with image x in $\mathcal{S}_\phi$, the identity

$$f'(\phi'_{t'}) = \sum_{\pi \in \widetilde{\Pi}_\phi} \langle x, \pi\rangle f_G(\pi_t), \qquad f \in \mathcal{H}(G),\ t \in T(\phi), \tag{8.4.3}$$

is valid for any endoscopic preimage (G', ϕ', t') of (ϕ, s, t).

REMARKS. 1. The formula (8.4.1) tells us that the action ϕ_t of $T(\phi)$ on the stable distributions in question is compatible with the action of $\widetilde{O}(G)$ on corresponding parameters. The formula (8.4.2) asserts that the action π_t of $T(\phi)$ on the relevant characters is also compatible with that of $\widetilde{O}(G)$. The formula (8.4.3) is a refinement (for generic parameters) of the main assertion (2.2.6) of the local Theorem 2.2.1.

2. Given (8.4.1) and (8.4.2), one sees without difficulty that the proposed mappings $t \to \phi_t$ and $t \to \pi_t$ are uniquely determined by (8.4.3).

3. Suppose that the theorem holds for F. It then extends formally to standard representations attached to parameters in the larger set $\widetilde{\Phi}'(G)$. In particular, the theorem is valid for the packet $\widetilde{\Pi}_\phi$ attached to any parameter in the subset

$$\widetilde{\Phi}'_{\mathrm{unit}}(G) = \widetilde{\Phi}'(G) \cap \widetilde{\Psi}^+_{\mathrm{unit}}(G)$$

of $\widetilde{\Phi}'(G)$.

We shall use global methods to establish this local theorem. Before doing so, we may as well first state the associated global theorem. The burden of proof for both theorems will then fall on the stabilized trace formula for G.

Suppose that F is global, and that $\phi \in \widetilde{\Phi}_{\mathrm{bdd}}(G)$ is a generic global parameter. We can then write

$$I_{\mathrm{disc},\phi}(f) = \sum_{G' \in \mathcal{E}_{\mathrm{ell}}(G)} \iota(G, G')\, \widehat{S}'_{\mathrm{disc},\phi}(f'), \qquad f \in \mathcal{H}(G), \tag{8.4.4}$$

in the notation (3.3.15). The essential point here is that the formula holds for general functions in $\mathcal{H}(G)$, rather than just the functions in the symmetric subalgebra $\widetilde{\mathcal{H}}(G)$. To make the argument work, we will need a nonsymmetric version of the stable multiplicity formula, at least for some ϕ, to express the terms on the right.

Any localization ϕ_v of ϕ lies in the generic subset $\widetilde{\Phi}_{\text{unit}}(G_v)$ of $\widetilde{\Psi}^{+}_{\text{unit}}(G_v)$. Assume that Theorem 8.4.1 is valid for F_v, and that ϕ_v lies in the subset $\widetilde{\Phi}'_{\text{unit}}(G_v)$ of $\widetilde{\Phi}_{\text{unit}}(G_v)$. The assertions of Theorem 8.4.1 are then valid for ϕ_v, as we noted above in Remark 3. This allows us to define an isomorphism

$$t \longrightarrow t_v$$

between the torsors $T(\phi)$ and $T(\phi_v)$. To describe it, suppose first that $\phi \in \widetilde{\Phi}_{\text{sim}}(G)$ is simple. According to the definition in §8.3, an element $t \in T(\phi)$ is represented by an automorphic representation $\pi_\star = \pi_t$ attached to the unique $\widetilde{O}(G)$-orbit $\pi = \pi_\phi$ in $\widetilde{\Pi}_\phi$. We use the main assertion (8.4.3) of the theorem to define t_v as the element in $T(\phi_v)$ such that

$$\pi_{v,t_v} = \pi_{t,v}.$$

We extend the construction from this basic case to more general ϕ, first in $\widetilde{\Phi}'_2(G)$ and then in the complement of $\widetilde{\Phi}'_2(G)$ in $\widetilde{\Phi}'_{\text{bdd}}(G)$, directly from the definitions.

Suppose now that Theorem 8.4.1 holds for *every* completion F_v of F. The mappings $t \to t_v$ then allow us to globalize the two constructions of the local theorem. We define a global $\widetilde{O}(G)$-equivariant mapping

$$t \longrightarrow \phi_t, \qquad t \in T(\phi),$$

from $T(\phi)$ to the space of linear forms on the global space $\mathcal{S}(G)$ by setting

$$\phi_t = \bigotimes_v (\phi_{v,t_v}).$$

If ϕ_v lies in the complement of $\widetilde{\Phi}'_{\text{unit}}(G_v)$, t_v of course represents the lone element in the set $T(\phi_v)$. In this case, ϕ_{v,t_v} is understood to be the $\widetilde{O}(G_v)$-invariant linear form ϕ_v we identified in (2.2.2) with the corresponding parameter. To describe the second global construction, we extend its accompanying local notation in the natural way. Namely, we write $\Pi_\star$ for the preimage in $\Pi_{\text{unit}}(G)$ of any subset $\widetilde{\Pi}_\star$ of the restricted tensor product

$$\widetilde{\Pi}_{\text{unit}}(G) = \widetilde{\bigotimes_v} \big(\widetilde{\Pi}_{\text{unit}}(G_v)\big).$$

We then define an $\widetilde{O}(G)$-equivariant mapping

$$(\pi, t) \longrightarrow \pi_t, \qquad \pi \in \widetilde{\Pi}_\phi,\ t \in T(\phi),$$

from $\widetilde{\Pi}_\phi \times T(\phi)$ to Π_ϕ by setting

$$\pi_t = \bigotimes_v \pi_{v,t_v}, \qquad \pi \in \widetilde{\Pi}_\phi,\ t \in T(\phi).$$

This is compatible with our definition of t_v in the special case that ϕ lies in $\widetilde{\Phi}_{\text{sim}}(G)$.

THEOREM 8.4.2. *Suppose that F is global, and that ϕ is a parameter in $\widetilde{\Phi}'_2(G)$ such that the canonical mapping*

$$\mathcal{S}_\phi \longrightarrow \mathcal{S}_{\phi_{\mathbb{A}}} = \prod_v \mathcal{S}_{\phi_v}$$

is injective. Then

$$S_{\mathrm{disc},\phi}(f) = |\mathcal{S}_\phi|^{-1} \sum_{t \in T(\phi)} f^G(\phi_t), \qquad f \in \mathcal{H}(G), \tag{8.4.5}$$

and

$$\mathrm{tr}\big(R_{\mathrm{disc},\phi}(f)\big) = \sum_{\pi \in \widetilde{\Pi}_\phi(1)} \sum_{t \in T(\phi)} f_G(\pi_t), \qquad f \in \mathcal{H}(G). \tag{8.4.6}$$

REMARKS. 1. The assertions are refinements (for certain generic, square-integrable parameters) of the global Theorems 4.1.2 and 1.5.2. By not requiring f to lie in the symmetric Hecke algebra $\widetilde{\mathcal{H}}(G)$, they place the group G of type D_n on an equal footing with our groups of type B_n and C_n for these parameters.

2. We recall that the left hand side of (8.4.6) was defined in (3.4.5). The indexing set $\widetilde{\Pi}_\phi(1)$ on the right hand side is the set of representations π in the global packet $\widetilde{\Pi}_\phi$ such that the character $\langle \cdot, \pi \rangle$ on $\mathcal{S}_\phi$ is trivial.

3. The supplementary condition on ϕ amounts to an imposition of local constraints. It is inductive, in the sense that if the condition holds for (G, ϕ), it is valid for any of the pairs (G', ϕ') that index the summands in (8.4.4). It can obviously be formulated equivalently with $\phi_{\mathbb{A}}$ replaced by ϕ_V, for some finite set of valuations V. We would of course be happier not to impose local restrictions on a global theorem. The theorem should certainly be valid without the condition, but its proof in this generality seems to call for new methods.

The proofs of Theorems 8.4.1 and 8.4.2 have a common core. It begins with the preliminary interpretation of the stabilized trace formula (8.4.4). To describe this, we assume inductively that both theorems hold if N is replaced by any positive integer $N' < N$.

Suppose that F and ϕ are as in the global Theorem 8.4.2. The analogue of (8.4.5) is then valid for the summand of any $G' \neq G$ on the right hand side of (8.4.4). Making the appropriate substitution, and recalling how we treated these terms in (6.6.4) for example, we find that the right hand side of (8.4.4) equals

$$S^G_{\mathrm{disc},\phi}(f) + |\mathcal{S}_\phi|^{-1} \sum_{\substack{x \in \mathcal{S}_\phi \\ x \neq 1}} \sum_{t \in T(\phi)} f'(\phi'_{t'}), \tag{8.4.7}$$

where (G', ϕ', t') is a preimage of (ϕ, x, t). The left hand side of (8.4.4) reduces to the left hand side of (8.4.6). For the present, we write it simply as

$$\sum_{\pi \in \widetilde{\Pi}_{\mathrm{unit}}(G)} \; \sum_{\pi_\star \in \Pi(\pi)} n_\phi(\pi_\star)\, f_G(\pi_\star), \tag{8.4.8}$$

where

$$\Pi(\pi) = \Pi(G, \pi)$$

is the preimage of π in $\Pi_{\mathrm{unit}}(G)$, and $n_\phi(\pi_\star)$ is the multiplicity of π_* in the discrete spectrum.

The identity of (8.4.7) and (8.4.8) will be a starting point for the proofs of both theorems. We shall establish the two theorems in succession.

PROOF OF THEOREM 8.4.1. We are assuming that F is local, and that ϕ lies in $\widetilde{\Phi}'_{\mathrm{bdd}}(G)$. If F is archimedean, the theorem follows from the endoscopic Langlands classification of Shelstad. In fact, the Langlands classification in this case provides a canonical $\widetilde{O}(G)$-bijection between the torsors $\Phi(\phi)$ and $T(\phi)$, which is compatible with the assertions of the theorem. It is the lack of this bijection in the p-adic case that forces us to introduce $T(\phi)$ as a substitute for $\Phi(\phi)$.

We therefore assume that the local field F is nonarchimedean. To deal with this case, we shall use the global methods of Chapter 6. In particular, we shall apply Proposition 6.3.1 to the local parameter ϕ. Having established the local theorems of Chapter 6, we do not need the induction hypothesis $N_i < N$ from Proposition 6.3.1 that allowed us to apply Corollary 6.2.4. In other words, we do not have to rule out the case that ϕ is simple.

The essential case is that of a parameter (6.6.1) in the subset $\widetilde{\Phi}'_2(G)$ of $\widetilde{\Phi}'_{\mathrm{bdd}}(G)$. With this assumption on ϕ, we obtain global objects $\dot{F}$, $\dot{G} \in \widetilde{\mathcal{E}}_{\mathrm{sim}}(N)$ and $\dot{\phi} \in \widetilde{\Phi}'_2(\dot{G})$ from the local objects F, G and ϕ, which satisfy the conditions of Proposition 6.3.1. In particular, we have

$$(\dot{F}_u, \dot{G}_u, \dot{\phi}_u) = (F, G, \phi),$$

for a p-adic valuation u on the global field $\dot{F}$. We identify the local group $\widetilde{O}(G) = \widetilde{O}(\dot{G}_u)$ (of order 2) with its global counterpart $\widetilde{O}(\dot{G})$. From the construction of Proposition 6.3.1, together with the fact that the simple degrees $N_i = \deg(\dot{\phi}_i) = \deg(\phi_i)$ are even, we see that $\dot{\phi}_v$ belongs to $\widetilde{\Phi}'_2(\dot{G}_v)$ for each archimedean valuation $v \in S_\infty$. In particular, the set $\Phi(\dot{\phi}_v)$ is an $\widetilde{O}(\dot{G}_v)$-torsor. This allows us to identify the group $\widetilde{O}(G) = \widetilde{O}(\dot{G})$ also with its analogue $\widetilde{O}(\dot{G}_v)$ for any $v \in S_\infty$, and thereby treat $\Phi(\dot{\phi}_v)$ as an $\widetilde{O}(G)$-torsor.

There is a canonical $O(G)$-bijection $t \to \dot{t}$ from $T(\phi)$ to $T(\dot{\phi})$. In the case that ϕ is simple, this follows from the fact that a global representation $\dot{\pi}$ in $T(\dot{\phi})$ is uniquely determined by its component $\pi_\star = \dot{\pi}_{\star,u}$ at u. For arbitrary ϕ, the bijection is then a consequence of the definition of the two parallel equivalence relations for $T(\phi)$ and $T(\dot{\phi})$. Its construction thus does not require the validity of Theorem 8.4.1 for the completion $\phi = \dot{\phi}_u$.

We do know that the theorem holds for the completion $\dot{\phi}_v$ of $\dot{\phi}$ at any archimedean valuation $v \in S_\infty$. The earlier construction then gives us an $O(G)$-bijection $t \to \dot{t}_v$ from $T(\phi)$ to $T(\dot{\phi}_v)$. This in turn can be composed with the isomorphism between $T(\dot{\phi}_v)$ and any of the other $\widetilde{O}(\dot{G}_v)$-torsors attached to $\dot{\phi}_v$. We thus obtain an $\widetilde{O}(G)$-isomorphism $t \to \dot{\phi}_{v,t}$ from $T(\phi)$ to $\Phi(\dot{\phi}_v)$, and an $\widetilde{O}(G)$-isomorphism $t \to \dot{\pi}_{v,t}$ from $T(\phi)$ to $T(\dot{\pi}_v)$, for any $\dot{\pi}_v \in \widetilde{\Pi}_{\dot{\phi}_v}$.

We are carrying the earlier induction hypothesis that both theorems hold if N is replaced by a positive integer $N' < N$. The global parameter $\dot{\phi}$ for $\dot{G}$ satisfies the supplementary condition of Theorem 8.4.2 (where we can take the set V mentioned in Remark 3 above equal to the set $S_\infty \cup \{u\}$, or for that matter, just S_∞). The expressions (8.4.7) and (8.4.8), obtained from the stabilized trace formula (8.4.4) but with $(\dot{F}, \dot{G}, \dot{\phi})$ in place of the earlier triplet (F, G, ϕ), are therefore equal. To be consistent with the notation here, we need also to write $\dot{f}$ for the global test function in $\mathcal{H}(\dot{G})$ that was earlier denoted by f. We fix its component

$$\dot{f}^{\infty,u} = \prod_{v \notin S_\infty(u)} \dot{f}_v$$

away from $S_\infty(u)$ by choosing each $\dot{f}_v \in \widetilde{\mathcal{H}}(\dot{G}_v)$ to be a symmetric, spherical function so that $\dot{f}'_v(\dot{\phi}'_v) = 1$, where $(\dot{G}'_v, \dot{\phi}'_v)$ is the v-component of the preimage $(\dot{G}', \dot{\phi}')$ of the pair attached to any $x \in \mathcal{S}_\phi$. We can then regard the equality of (8.4.7) and (8.4.8) as an identity of linear forms in the complementary component

$$\dot{f}_{\infty,u} = \dot{f}_\infty \cdot \dot{f}_u = \dot{f}_\infty \cdot f, \qquad \dot{f}_\infty \in \mathcal{H}(\dot{G}_\infty),\ f \in \mathcal{H}(G),$$

of $\dot{f}$.

The summand of x and t in (8.4.7) equals

$$\dot{f}'(\dot{\phi}'_{t'}) = \dot{f}'_\infty(\dot{\phi}'_{\infty,t'}) \cdot f'(\phi'_{t'}),$$

since

$$(\dot{f}^{\infty,u})'(\dot{\phi}^{\infty,u}) = \prod_{v \notin S_\infty(u)} \dot{f}'_v(\dot{\phi}'_v) = 1.$$

The expression (8.4.7) therefore equals

$$S^{\dot{G}}_{\mathrm{disc},\dot{\phi}}(\dot{f}) + |\mathcal{S}_\phi|^{-1} \sum_{\substack{x \in \mathcal{S}_\phi \\ x \neq 1}} \sum_{t \in T(\phi)} \dot{f}'_\infty(\dot{\phi}'_{\infty,t'}) f'(\phi'_{t'}). \tag{8.4.9}$$

The description of (8.4.8) requires a little more discussion.

For any character $\xi \in \widehat{\mathcal{S}}_\phi$, we have the $\widetilde{O}(G)$-orbit of representations $\pi(\xi) \in \widetilde{\Pi}_\phi$ such that $\langle \cdot, \pi(\xi) \rangle = \xi$. We also have the set $\widetilde{\Pi}_{\dot{\phi}_\infty,\xi}$ of representations

$$\dot{\pi}_\infty = \bigotimes_{v \in S_\infty} \dot{\pi}_v, \qquad \dot{\pi}_v \in \widetilde{\Pi}_{\dot{\phi}_v},$$

in the packet $\widetilde{\Pi}_{\dot{\phi}_\infty}$ such that the character

$$\langle \dot{x}, \dot{\pi}_\infty \rangle = \prod_{v \in S_\infty} \langle \dot{x}_v, \dot{\pi}_v \rangle, \qquad x \in \mathcal{S}_\phi,$$

on $\mathcal{S}_\phi$ equals $\xi(x)^{-1}$. It follows from Proposition 6.3.1(iii)(a) that this set is not empty. In general, the elements $\dot{\pi}_\infty$ in $\widetilde{\Pi}_{\dot{\phi}_\infty}$ represent simply transitive orbits in $\Pi_{\text{unit}}(\dot{G}_\infty)$ under the product

$$\widetilde{O}(\dot{G}_\infty) = \prod_{v \in S_\infty} \widetilde{O}(\dot{G}_v).$$

For any $\dot{\pi}_\infty \in \widetilde{\Pi}_{\dot{\phi}_\infty,\xi}$ and any local representation $\pi_\star(\xi) \in \Pi\big(\pi(\xi)\big)$, there is a unique global representation

$$\dot{\pi}_\star = \dot{\pi}_{\infty,\star} \otimes \pi_\star(\xi) \otimes \dot{\pi}_\star^{\infty,u},$$

where $\dot{\pi}_{\infty,\star}$ belongs to the set

$$\Pi(\dot{G}_\infty, \dot{\pi}_\infty) = \bigotimes_{v \in S_\infty} \big(\Pi(\dot{G}_v, \dot{\pi}_v)\big)$$

and $\dot{\pi}_\star^{\infty,u} \in \Pi(\dot{G}^{\infty,u})$ is spherical, such that the coefficient $n_{\dot{\phi}}(\dot{\pi}_\star)$ on (8.4.8) is nonzero. This follows from Proposition 6.3.1(ii) and either the proof of Propositions 6.6.5 and 6.7.2 or simply the general assertion of Theorem 1.5.2. In particular, the representation $\dot{\pi}_{\infty,\star}$ is uniquely determined by $\dot{\pi}_\infty$ and $\pi_\star(\xi)$. We can then write

$$\dot{\pi}_{\infty,\star} = \dot{\pi}_{\infty,\dot{t}_\infty(\star)} = \bigotimes_{v \in S_\infty} \dot{\pi}_{v,\dot{t}_v(\star)},$$

for a direct product

$$\dot{t}_\infty(\star) = t_\infty\big(\dot{\pi}_\infty, \pi_\star(\xi)\big) = \prod_{v \in S_\infty} t_v\big(\dot{\pi}_\infty, \pi_\star(\xi)\big)$$

of uniquely determined elements $\dot{t}_v(\star) = t_v\big(\dot{\pi}_\infty, \pi_\star(\xi)\big)$ in $T(\phi)$. Since the nonzero coefficient $n_{\dot{\phi}}(\dot{\pi}_\star)$ actually equals 1, the expression (8.4.8) becomes

$$\sum_{\xi \in \widehat{\mathcal{S}}_\phi} \sum_{\dot{\pi}_\infty \in \widetilde{\Pi}_{\dot{\phi}_\infty,\xi}} \sum_{\pi_\star(\xi) \in \Pi(\pi(\xi))} \dot{f}_{\infty,\dot{G}}(\dot{\pi}_{\infty,\star})\, f_G\big((\pi_\star(\xi)\big). \tag{8.4.10}$$

LEMMA 8.4.3. *For any $\pi \in \widetilde{\Pi}_\phi$, there is an $\widetilde{O}(G)$-isomorphism $t \to \pi_t$ from $T(\phi)$ onto $\Pi(\pi)$ such that for any $\xi \in \widehat{\mathcal{S}}_\phi$ and $\dot{\pi}_\infty \in \widetilde{\Pi}_{\dot{\phi}_\infty,\xi}$, we have*

$$t_v\big(\dot{\pi}_\infty, \pi_t(\xi)\big) = t, \qquad t \in T(\phi), \;\; v \in S_\infty.$$

PROOF. This lemma is at the heart of the argument. We have to show that for any ξ and $\pi_\star(\xi)$, the $T(\phi)$-components

$$t_v\big(\dot{\pi}_\infty, \pi_\star(\xi)\big), \qquad v \in S_\infty,$$

of the representations $\dot{\pi}_{\infty,\star} = \dot{\pi}_{\infty,\dot{t}_\infty(\star)}$ in (8.4.10) are independent of $\dot{\pi}_\infty$ and v. The main point is to show that $t_v\big(\dot{\pi}_\infty, \pi_\star(\xi)\big)$ is independent of the first argument $\dot{\pi}_\infty$. The proof of this will be based on the identity of (8.4.9) and (8.4.10) that we have obtained from the stabilized trace formula.

We first impose a constraint on the function $\dot{f}_\infty \in \mathcal{H}(\dot{G}_\infty)$ that forces the summands on the right hand side of (8.4.9) to vanish. We require that for any $x \neq 1$ in $\mathcal{S}_\phi$ and any $t \in T(\phi)$, the linear form

$$\dot{f}'_\infty(\dot{\phi}'_{\infty,t'}) = \sum_{\dot{\pi}_\infty \in \widetilde{\Pi}_{\dot{\phi}_\infty}} \langle \dot{x}, \dot{\pi}_\infty \rangle \dot{f}_{\infty,\dot{G}}(\dot{\pi}_{\infty,t})$$

vanishes. In other words, for any $t \in T(\phi)$, the sum

$$\dot{f}_{\infty,\dot{G}}(\xi, t) = \sum_{\dot{\pi}_\infty \in \widetilde{\Pi}_{\dot{\phi}_\infty,\xi}} \dot{f}_{\infty,\dot{G}}(\dot{\pi}_{\infty,t}), \qquad \xi \in \widehat{\mathcal{S}}_\phi,$$

is independent of ξ. This forces the sum over x in (8.4.9) to vanish. We then fix the indices ξ and $\pi_\star(\xi)$ in (8.4.10), and take $f \in \mathcal{H}(G)$ to be a pseudocoefficient $f_\star(\xi)$ of the representation $\pi_\star(\xi) \in \Pi_2(G)$. This simplifies the triple sum in (8.4.10) to a simple sum over $\widetilde{\Pi}_{\dot{\phi}_\infty,\xi}$. The equality of (8.4.9) and (8.4.10) becomes

$$S^{\dot{G}}_{\mathrm{disc},\dot{\phi}}(\dot{f}_\infty \cdot f_\star(\xi) \cdot \dot{f}^{\infty,u}) = \sum_{\dot{\pi}_\infty \in \widetilde{\Pi}_{\dot{\phi}_\infty,\xi}} \dot{f}_{\infty,\dot{G}}(\dot{\pi}_{\infty,*}). \tag{8.4.11}$$

Consider the abstract spectral expansion of the left hand side of (8.4.11), as a stable linear form in $\dot{f}_\infty$. Using the right hand to bound its spectral support, and keeping in mind the spectral criterion for stability in terms of archimedean Langlands parameters, we obtain a linear combination

$$S^{\dot{G}}_{\mathrm{disc},\dot{\phi}}(\dot{f}_\infty \cdot f_\star(\xi) \cdot \dot{f}^{\infty,u}) = \sum_{\dot{t}_\infty} c(\dot{t}_\infty)\, \dot{f}^{\dot{G}}_\infty(\dot{\phi}_{\infty,\dot{t}_\infty}) \tag{8.4.12}$$

over indices

$$\dot{t}_\infty = \prod_{v \in S_\infty} t_v, \qquad t_v \in T(\phi),$$

with complex coefficients $c(\dot{t}_\infty) = c(\dot{t}_\infty, \pi_\star(\xi))$.

The formulas (8.4.11) and (8.4.12) were derived with the condition that $\dot{f}_{\infty,\dot{G}}(\xi, t)$ be independent of ξ. We can actually treat the two right hand sides as linear forms on the space of functions on the preimage $\Pi_{\dot{\phi}_\infty,\xi}$ of $\widetilde{\Pi}_{\dot{\phi}_\infty,\xi}$ in $\Pi(\dot{G}_\infty)$. To be precise, we claim that if the coefficients $c(\dot{t}_\infty)$ are rescaled by the factor

$$|\Pi_{\dot{\phi}_\infty}|\,|\Pi_{\dot{\phi}_\infty,\xi}|^{-1} = |\widetilde{\Pi}_{\dot{\phi}_\infty}|\,|\widetilde{\Pi}_{\dot{\phi}_\infty,\xi}|^{-1},$$

the two formulas remain valid if the condition in $\dot{f}_\infty$ is replaced by the requirement that the function $\dot{f}_{\infty,\dot{G}}$ on $\Pi(\dot{G}_\infty)$ be supported on $\Pi_{\dot{\phi}_\infty,\xi}$. This is because the restriction mapping, from the subspace of functions $\dot{f}_\infty \in \mathcal{H}(\dot{G}_\infty)$ that satisfy the first condition onto the space of functions $\dot{f}_{\infty,\dot{G}}$ on $\Pi_{\dot{\phi}_\infty,\xi}$, is surjective. The claim follows easily from this fact, together with the form of the terms on the two right hand sides. The two formulas can then be combined as an identity

$$(8.4.13) \qquad \sum_{\dot{\pi}_\infty \in \widetilde{\Pi}_{\dot{\phi}_\infty,\xi}} \dot{f}_{\infty,\dot{G}}(\dot{\pi}_{\infty,\star}) = \sum_{\dot{t}_\infty} c(\dot{t}_\infty)\, \dot{f}_\infty^{\dot{G}}(\dot{\phi}_{\infty,\dot{t}_\infty}),$$

for complex coefficients $c(\dot{t}_\infty) = c\big(\dot{t}_\infty, \pi_\star(\xi)\big)$, and any $\dot{f}_\infty \in \mathcal{H}(\dot{G}_\infty)$ such that the function $\dot{f}_{\infty,\dot{G}}$ on $\Pi(\dot{G}_\infty)$ is supported on $\Pi_{\dot{\phi}_\infty,\xi}$. The representation $\dot{\pi}_{\infty,\star}$ on the left hand side is the image of $\dot{\pi}_\infty$ under the section from $\widetilde{\Pi}_{\dot{\phi}_\infty,\xi}$ to $\Pi_{\dot{\phi}_\infty,\xi}$ determined by $\pi_\star(\xi)$, while the index of summation $\dot{t}_\infty$ on the right hand is over elements in the direct product of $|S_\infty|$-copies of $T(\phi)$.

The left hand side of (8.4.13) does not vanish, since we can always take $\dot{f}_\infty$ to be a pseudocoefficient of some $\dot{\pi}_{\infty,\star}$. We can therefore choose an index $\dot{t}_\infty$ so that the corresponding coefficient $c(\dot{t}_\infty)$ on the right is nonzero. Suppose that $\dot{\pi}_\infty \in \widetilde{\Pi}_{\dot{\phi}_\infty,\xi}$ is arbitrary, and that $\dot{f}_\infty$ is a pseudocoefficient of the representation $\dot{\pi}_{\infty,\dot{t}_\infty}$. The left hand side then equals 1 if $t_\infty\big(\dot{\pi}_\infty, \pi_\star(\xi)\big)$ equals $\dot{t}_\infty$, and vanishes otherwise. But the right hand side reduces in this case to the nonzero coefficient $c(\dot{t}_\infty) = c\big(\dot{t}_\infty, \pi_\star(\xi)\big)$. It follows that the product

$$t_\infty\big(\dot{\pi}_\infty, \pi_\star(\xi)\big) = \prod_{v \in S_\infty} t_v\big(\dot{\pi}_\infty, \pi_\star(\xi)\big)$$

equals the index

$$\dot{t}_\infty = \prod_{v \in S_\infty} \dot{t}_v.$$

But $\dot{\pi}_\infty$ was taken to be any element in $\widetilde{\Pi}_{\dot{\phi}_\infty,\xi}$. It follows that for any $v \in S_\infty$, $t_v\big(\dot{\pi}_\infty, \pi_\star(\xi)\big)$ equals $\dot{t}_v$, and is therefore independent of $\dot{\pi}_\infty$.

We now have to check that $\dot{t}_v = t_v\big(\dot{\pi}_\infty, \pi_\star(\xi)\big)$ is independent of v. If ϕ is simple, ξ equals 1, and $\widetilde{\Pi}_\phi$ consists of one element π. In this case, the definitions give isomorphisms

$$t = \pi_t \longrightarrow \dot{\pi}_t \longrightarrow \dot{\pi}_{t,v} = \dot{\pi}_{v,t} \longrightarrow \dot{\phi}_{v,t}$$

of $\widetilde{O}(G)$-torsors. (We recall that for simple ϕ, $T(\phi)$ is just defined to be the $\widetilde{O}(G)$-torsor Π_ϕ.) The global representation $\dot{\pi}_t$ is determined by π_t and a particularly chosen factor $\dot{\pi}_\infty$ in the set $\widetilde{\Pi}_{\dot{\phi}_\infty} = \widetilde{\Pi}_{\dot{\phi}_\infty,1}$. It follows that

$$t_v(\dot{\pi}_\infty, \pi_t) = t,$$

by the definition of the left hand side. Having already shown that $t_v(\dot{\pi}_\infty, \pi_t)$ is independent of $\dot{\pi}_\infty$, we conclude that it is also independent of v. For general ϕ, we have an isomorphism $t \to \dot{\phi}_{v,t}$ of $\widetilde{O}(G)$-torsors $T(\phi)$ and $\Phi(\dot{\phi}_v)$. The independence of $t_v\big(\dot{\pi}_\infty, \pi_\star(\xi)\big)$ on v then follows from the definitions, and the case we have just established for the simple constituents of ϕ.

We can now construct the required isomorphism. Suppose that $\pi \in \widetilde{\Pi}_\phi$. Then $\pi = \pi(\xi)$, for a unique character $\xi \in \widehat{\mathcal{S}}_\phi$. For any representation $\pi_\star(\xi) \in \Pi(\pi(\xi))$, we now know that the point

$$t = t_v\big(\dot{\pi}_\infty, \pi_\star(\xi)\big), \qquad \dot{\pi}_\infty \in \widetilde{\Pi}_{\dot{\phi}_\infty,\xi},\ v \in S_\infty,$$

in $T(\phi)$ is independent of $\dot{\pi}_\infty$ and v, and therefore depends only on $\pi_\star(\xi)$. We thus obtain an isomorphism $\pi_\star(\xi) \to t$ between the $\widetilde{O}(G)$-torsors $\Pi(\pi(\xi))$ and $T(\phi)$. Its inverse $t \to \pi_t$ then satisfies the conditions of the lemma. □

We return to the proof of Theorem 8.4.1. There are two mappings to construct. The required $\widetilde{O}(G)$-equivariant isomorphism $(\pi, t) \to \pi_t$ is provided by the last lemma. We define the other proposed mapping by setting

$$f^G(\phi_t) = \sum_{\pi \in \widetilde{\Pi}_\phi} f_G(\pi_t), \qquad t \in T(\phi),\ f \in \mathcal{H}(G). \tag{8.4.14}$$

The required identities (8.4.1) and (8.4.2) then follow from the definitions. We must still show that the right hand side of (8.4.14) is stable in f in order to justify the notation on the left hand side, and of course also because it is one of the requirements of the theorem. We have also to establish (8.4.3).

The mapping $(\pi, t) \to \pi_t$ is defined by the formula

$$t_\infty\big(\dot{\pi}_\infty, \pi_t(\xi)\big) = t, \qquad \dot{\pi}_\infty \in \widetilde{\Pi}_{\dot{\phi}_\infty,\xi},\ t \in T(\phi),$$

according to the discussion at the end of the proof of Lemma 8.4.3. The expression (8.4.10) can then be written as

$$\sum_{\xi \in \widehat{\mathcal{S}}_\phi} \sum_{t \in T(\phi)} \Big(\sum_{\dot{\pi}_\infty \in \widetilde{\Pi}_{\dot{\phi}_\infty,\xi}} \dot{f}_{\infty,\dot{G}}(\dot{\pi}_{\infty,t}) \Big) f_G\big(\pi_t(\xi)\big). \tag{8.4.15}$$

The inner sum in brackets is what we denoted by $\dot{f}_{\infty,\dot{G}}(\xi, t)$ near the beginning of the proof of the lemma. As before, let us assume that $\dot{f}_\infty$ is such that this is independent of ξ. The inner sum can then be written as the product of $|\mathcal{S}_\phi|^{-1}$ with

$$\sum_{\xi \in \widehat{\mathcal{S}}_\phi} \dot{f}_{\infty,\dot{G}}(\xi, t) = \sum_{\dot{\pi}_\infty \in \widetilde{\Pi}_{\dot{\phi}_\infty}} \dot{f}_{\infty,\dot{G}}(\dot{\pi}_{\infty,t}) = \dot{f}_\infty^{\dot{G}}(\dot{\phi}_{\infty,t}).$$

Our expression (8.4.15) for (8.4.10) reduces to the simpler expression

$$|\mathcal{S}_\phi|^{-1} \sum_{t \in T(\phi)} \dot{f}_\infty^{\dot{G}}(\dot{\phi}_{\infty,t}) \Big(\sum_{\xi \in \widehat{\mathcal{S}}_\phi} f_G\big(\pi_t(\xi)\big) \Big),$$

which in turn equals

$$|\mathcal{S}_\phi|^{-1} \sum_{t \in T(\phi)} \dot{f}_\infty^{\dot{G}}(\dot{\phi}_{\infty,t})\, f^G(\phi_t),$$

according to the definition (8.4.14). On the other hand, our condition on $\dot{f}_\infty$ implies that the summands in (8.4.9) with $x \neq 1$ all vanish. The equality of (8.4.9) and (8.4.10) becomes

$$S^{\dot{G}}_{\mathrm{disc},\dot{\phi}}(\dot{f}) = |\mathcal{S}_{\dot{\phi}}|^{-1} \sum_{t \in T(\phi)} \dot{f}^{\dot{G}}_\infty(\dot{\phi}_{\infty,t})\, f^G(\phi_t). \tag{8.4.16}$$

The left hand side of this formula is stable in the variable function $f \in \mathcal{H}(G)$. The coefficients $\dot{f}^{\dot{G}}_\infty(\dot{\phi}_{\infty,t})$ may be chosen arbitrarily by varying $\dot{f}_\infty$. It follows that the linear form $f^G(\phi_t)$ defined by (8.4.14) is stable, as required.

It remains to prove (8.4.3). The formula (8.4.16) is still valid without the condition on $\dot{f}_\infty$ under which it was derived. This follows from the stability of each side and the fact that the subspace of functions $\dot{f}_\infty \in \mathcal{H}(\dot{G}_\infty)$ that satisfy the condition maps *onto* $\mathcal{S}(\dot{G}_\infty)$. We substitute it for the leading term on the right hand side (8.4.9) of the stabilized trace formula (8.4.4). The right hand side of (8.4.4) becomes

$$|\mathcal{S}_\phi|^{-1} \sum_{x \in \mathcal{S}_\phi} \sum_{t \in T(\phi)} \dot{f}'_\infty(\dot{\phi}'_{\infty,t'})\, f'(\phi'_{t'}),$$

for any $\dot{f}_\infty \in \mathcal{H}(\dot{G}_\infty)$ and $f \in \mathcal{H}(G)$, an expression we write as

$$|\mathcal{S}_\phi|^{-1} \sum_x \sum_t \Big(\sum_{\dot{\pi}_\infty \in \widetilde{\Pi}_{\dot{\phi}_\infty}} \langle \dot{x}, \dot{\pi}_\infty \rangle\, \dot{f}_{\infty,\dot{G}}(\dot{\pi}_{\infty,t}) \Big)\, f'(\phi'_{t'}). \tag{8.4.17}$$

This consequently equals the expression (8.4.15) we have obtained for the left hand side (8.4.10) of (8.4.4). We choose the function $\dot{f}_\infty$ to be a pseudo-coefficient of a representation $\dot{\pi}_{\infty,t} \in \Pi_{\dot{\phi}_\infty}(\xi)$, for fixed ξ and t. The expression (8.4.15) reduces to $f_G\big(\pi_t(\xi)\big)$. The other expression (8.4.17) becomes

$$|\mathcal{S}_\phi|^{-1} \sum_{x \in \mathcal{S}_\phi} \xi(x)^{-1}\, f'(\phi'_{t'}),$$

since $\langle \dot{x}, \dot{\pi}_\infty \rangle = \xi(x)^{-1}$. Applying Fourier inversion on $\mathcal{S}_\phi$ to each side of the resulting identity, we obtain

$$f'(\phi'_{t'}) = \sum_{\xi \in \widehat{\mathcal{S}}_\phi} \xi(x)\, f_G\big(\pi_t(\xi)\big) = \sum_{\pi \in \widetilde{\Pi}_\phi} \langle x, \pi \rangle\, f_G(\pi_t).$$

This is the required formula (8.4.3).

We have completed the proof of Theorem 8.4.1 in case ϕ is a parameter (6.6.1) in $\widetilde{\Phi}'_2(G)$. It remains to consider a parameter ϕ in the complement of $\widetilde{\Phi}'_2(G)$ in $\widetilde{\Phi}'_{\mathrm{bdd}}(G)$. Then ϕ is of the general form (6.3.1), and is the image of a parameter $\phi_M \in \widetilde{\Phi}_2(M)$, for a proper Levi subgroup M of G. We shall say only a few words about this case since it is much simpler.

We define the mappings $t \to \phi_t$ and $(\pi, t) \to \pi_t$ directly from their analogues for M, guided by the requirement that they be compatible with induction. In this way, we also obtain general analogues $f'_G(\phi_t, x)$ and $f_G(\phi_t, x)$

of the two sides of the local intertwining relation (2.4.7). The required identities (8.4.1) and (8.4.2) follow directly from the definitions. It remains to establish the transfer identity (8.4.3). As we saw in the proof of Proposition 2.4.3, this in turn will follow from the local intertwining relation. In other words, it suffices to establish the following analogue of Theorem 2.4.1.

PROPOSITION 8.4.4. *Suppose that F is local, and that ϕ lies in the complement of $\widetilde{\Phi}'_2(G)$ in $\widetilde{\Phi}'_{\mathrm{bdd}}(G)$. Then*

$$f'_G(\phi_t, s) = f_G(\phi_t, u), \qquad f \in \mathcal{H}(G),\ t \in T(\phi), \tag{8.4.18}$$

for u and s as in Theorem 2.4.1.

PROOF. Recalling the discussion at the beginning of §6.4, and noting that there is nothing new to be established in the two exceptional cases (4.5.11) and (4.5.12), we see that it is enough to prove (8.4.18) for a parameter ϕ of the special form (6.4.1), with $r > q$, and N_i even whenever $q < i \leqslant r$. We can then take $u = s$ to be an element x in the set we denoted by $\mathcal{S}_{\phi,\mathrm{ell}}$. We shall be brief.

Suppose first that $F = \mathbb{R}$. By the results of Shelstad and the multiplicativity of intertwining operators, the generalization

$$f_G(\phi_t, x) = \sum_{\pi \in \widetilde{\Pi}_\phi} \varepsilon_{\pi_{M,t}}(x) \langle x, \pi \rangle f_G(\pi_t), \qquad t \in T(\phi),\ f \in \mathcal{H}(G),$$

of (6.1.5) remains valid. If $f \in \widetilde{\mathcal{H}}(G)$, this equals

$$f_G(\phi, x) = f'_G(\phi, x) = \sum_{\pi \in \widetilde{\Pi}_\phi} \langle x, \pi \rangle f_G(\pi),$$

by the special case of (8.4.18) given by Theorem 2.4.1. We can choose $f \in \widetilde{\mathcal{H}}(G)$ so that

$$f_G(\pi_t^*) = f_G(\pi^*) = \begin{cases} 1, & \text{if } \pi^* = \pi, \\ 0, & \text{otherwise,} \end{cases}$$

for any $\pi^* \in \widetilde{\Pi}_\phi$. It then follows that

$$\varepsilon_{\pi_{M,t}}(x) = 1, \qquad x \in \mathcal{S}_{\phi,\mathrm{ell}},\ t \in T(\phi).$$

Therefore

$$f_G(\phi_t, x) = \sum_{\pi \in \widetilde{\Pi}_\phi} \langle x, \pi \rangle f_G(\pi_t) = f'_G(\phi_t, x),$$

for any t, x and $f \in \mathcal{H}(G)$, as required.

Suppose now that F is p-adic. In this case, we use Proposition 6.3.1 to attach global objects $\dot{F}$, $\dot{G}$, $\dot{\phi}$, $\dot{M}$ and $\dot{\phi}_M$ to F, G, ϕ, M and ϕ_M, as at the beginning of §6.4. We then have the analogue

$$\sum_{t \in T(\phi)} \sum_{x \in \mathcal{S}_{\phi,\mathrm{ell}}} \big(\dot{f}'_{\dot{G}}(\dot{\phi}_t, \dot{x}) - \dot{f}_{\dot{G}}(\dot{\phi}_t, \dot{x})\big) = 0, \qquad \dot{f} \in \mathcal{H}(\dot{G}), \tag{8.4.19}$$

of (6.4.2). This follows from an application of the standard model to $(\dot{G}, \dot{\phi})$, as in the relevant part of the proof of Lemma 5.2.1, and the fact that $\dot{R}^{\dot{G}}_{\mathrm{disc},\dot{\phi}}(\dot{f})$ vanishes, as we have shown for symmetric functions $\dot{f} \in \widetilde{\mathcal{H}}(\dot{G})$ (and consequently any $\dot{f}$). We can fix a symmetric function $\dot{f}^{\infty,u} \in \widetilde{\mathcal{H}}(\dot{G}^{\infty,u})$ away from $S_\infty(u)$ so that if

$$\dot{f} = \dot{f}_\infty \cdot f \cdot \dot{f}^{\infty,u}, \qquad \dot{f}_\infty \in \widetilde{\mathcal{H}}(\dot{G}_\infty),\ f \in \mathcal{H}(G),$$

the left hand side of (8.4.19) decomposes

$$\sum_{t\in T(\phi)} \sum_{x\in\mathcal{S}_{\phi,\mathrm{ell}}} \big(\dot{f}'_{\infty,\dot{G}}(\dot{\phi}_{\infty,t}, \dot{x})\, f'_G(\phi_t, x) - \dot{f}_{\infty,\dot{G}}(\dot{\phi}_{\infty,t}, \dot{x})\, f_G(\phi_t, x)\big).$$

It then follows from the archimedean case we have proved that

$$\sum_t \sum_x \dot{f}_{\infty,\dot{G}}(\dot{\phi}_{\infty,t}, \dot{x})\big(f'_G(\phi_t, x) - f_G(\phi_t, x)\big) = 0.$$

Since the linear forms

$$\dot{f}_{\infty,\dot{G}}(\dot{\phi}_{\infty,t}, \dot{x}), \qquad \dot{f}_\infty \in \mathcal{H}(\dot{G}_\infty),\ t \in T(\phi),\ x \in \mathcal{S}_{\phi,\mathrm{ell}},$$

are linearly independent, we conclude that

$$f'_G(\phi_t, x) = f_G(\phi_t, x), \qquad f \in \mathcal{H}(G),\ t \in T(\phi),\ x \in \mathcal{S}_{\phi,\mathrm{ell}},$$

as required. This completes our sketch of the proof of Proposition 8.4.4, and hence also of the remaining part of the proof of Theorem 8.4.1. □

We write

$$\Pi_{\phi,t} = \{\pi_t : \pi \in \widetilde{\Pi}_\phi\}, \qquad t \in T(\phi), \tag{8.4.20}$$

for any $\phi \in \widetilde{\Phi}'_{\mathrm{bdd}}(G)$, and π_t as in Theorem 8.4.1. The sets $\Pi_{\phi,t}$ parametrized by the two elements $t \in T(\phi)$ are then the two L-packets attached to ϕ. If ϕ belongs to the complementary set $\Phi_{\mathrm{bdd}}(\widetilde{G})$, $T(\phi)$ consists of one element, and the assertions of Theorem 8.4.1 hold trivially with $\phi_t = \phi$ and $\pi_t = \pi$. If we extend the definition (8.4.20) to this case, $\Pi_{\phi,t}$ is simply the L-packet $\Pi_\phi = \widetilde{\Pi}_\phi$ we have already attached to ϕ.

COROLLARY 8.4.5. (a) *The set $\Pi_{\mathrm{temp}}(G)$ of irreducible, tempered representations of $G(F)$ is a disjoint union of the L-packets*

$$\Pi_{\phi,t}, \qquad \phi \in \widetilde{\Phi}_{\mathrm{bdd}}(G),\ t \in T(\phi).$$

(b) *The distributions*

$$\phi_t, \qquad \phi \in \widetilde{\Phi}_{\mathrm{bdd}}(G),\ t \in T(\phi),$$

form a basis of the space of stable linear forms on $\mathcal{H}(G)$.

PROOF. (a) For any $\phi \in \widetilde{\Phi}_{\mathrm{bdd}}(G)$, the preimage Π_ϕ of $\widetilde{\Pi}_\phi$ in $\Pi_{\mathrm{temp}}(G)$ is the disjoint union over $t \in T(\phi)$ of the packets $\Pi_{\phi,t}$. The assertion follows from its analogue for $\widetilde{\Pi}_{\mathrm{temp}}(G)$, which was stated in Theorem 1.5.1(a) and proved in Corollary 6.7.5.

(b) We can assume that F is nonarchimedean, as we have done implicitly in (a), since the assertions for archimedean F follow from the general endoscopic classification of Shelstad. In particular, Shelstad's archimedean results imply that the spectral characterization of stability (in terms of Langlands parameters) matches the geometric characterization (in terms of stable conjugacy classes), a property we have already used in the derivation of (8.4.12). This is what must be verified for p-adic F.

Theorem 8.4.1 tells us that the distributions ϕ_t are stable. Part (a) above implies that they are linearly independent. What remains is to show that they span the space of all stable linear forms on $\mathcal{H}(G)$. This is the analogue for G of the assertion of Proposition 2.1.1 for $\widetilde{G}(N)$, alluded to at the beginning of the proof of the proposition. It is implicit in the main results [**A11**] (namely, Theorems 6.1 and 6.2, together with Proposition 3.5), and will be made explicit in [**A24**]. □

PROOF OF THEOREM 8.4.2. We assume now that F is global. We have established that Theorem 8.4.1 holds for the localizations ϕ_v of the given global parameter $\phi \in \widetilde{\Phi}_2'(G)$. However, we must still retain what is left of the original induction hypothesis, namely that Theorem 8.4.2 holds if N is replaced by any integer $N' < N$. The expressions (8.4.7) and (8.4.8) are therefore equal.

The main point is to prove the stable multiplicity formula (8.4.5). The argument will be similar to a part of the proof of Theorem 8.4.1, specifically Lemma 8.4.3, so we can be brief. We shall first consider the other assertion, the actual multiplicity formula (8.4.6).

One advantage we have here is that the right hand side of (8.4.5) is actually defined. It equals the contribution of $x = 1$ that is missing from (8.4.7). Following a convention from §4.4, we write ${}^0S^G_{\mathrm{disc},\phi}(f)$ for the difference between the two sides of (8.4.5). Since (8.4.7) equals the left hand side of (8.4.6), namely the trace of $R^G_{\mathrm{disc},\phi}(f)$, we see that the difference

$$\mathrm{tr}\big(R^G_{\mathrm{disc},\phi}(f)\big) - {}^0S^G_{\mathrm{disc},\phi}(f), \qquad f \in \mathcal{H}(G),$$

equals

$$|\mathcal{S}_\phi|^{-1} \sum_{x \in \mathcal{S}_\phi} \sum_{t \in T(\phi)} f'(\phi'_{t'}).$$

Another advantage is that we can now apply the formula (8.4.3) to the local components of the global summands $f'(\phi'_{t'})$. From a simplified variant of the remarks at the end of §4.7, we deduce that the last expression equals

$$\sum_{\pi \in \widetilde{\Pi}_\phi(1)} \sum_{t \in T(\phi)} f_G(\pi_t).$$

This is just the right hand side of (8.4.6). Writing ${}^0 r^G_{\mathrm{disc},\phi}(f)$ for the difference of the two sides of (8.4.6), we conclude that

$$(8.4.21) \qquad {}^0 r^G_{\mathrm{disc},\phi}(f) = {}^0 S^G_{\mathrm{disc},\phi}(f), \qquad f \in \mathcal{H}(G).$$

In particular, the stable multiplicity formula (8.4.5) implies the actual multiplicity formula (8.4.6).

Suppose that the group $\mathcal{S}_\phi$ is trivial. The formula (8.4.6) then reduces to our definition in §8.3 of the global torsor $T(\phi)$. Since the other formula (8.4.5) follows from (8.4.21), the theorem holds in this case. We can therefore assume that the group $\mathcal{S}_\phi$ is nontrivial.

Let V be a large finite set of valuations of F such that the mapping

$$\mathcal{S}_\phi \to \mathcal{S}_{\phi_V} = \prod_{v \in V} \mathcal{S}_{\phi_v}$$

in injective. Having fixed V, we choose a function

$$f = f_V f^V,$$

where $f_V \in \mathcal{H}(G_V)$, and $f^V \in \widetilde{\mathcal{H}}(G^V)$ is the analogue of the function $\dot{f}^{\infty,u}$ from the proof of Theorem 8.4.1. For example, we can take f^V to be the characteristic function of the underlying hyperspecial maximal compact subgroup K^V of $G(\mathbb{A}^V)$. We will apply the identity (8.4.21) to f.

The left hand side ${}^0 r^G_{\mathrm{disc},\phi}(f)$ of (8.4.21) is defined as the difference of the two sides of the putative identity (8.4.6). The left hand side of (8.4.6) in turn equals the double sum (8.4.8). Arguing as in the derivation of (8.4.10) from (8.4.8) in the proof of Theorem 8.4.1, we see that this equals

$$\sum_{\pi_V \in \widetilde{\Pi}_{\phi_V}(1)} \sum_{\pi_{V,\star}} f_{V,G}(\pi_{V,\star}),$$

where $\widetilde{\Pi}_{\phi_V}(1)$ is the subset of elements $\pi_V \in \widetilde{\Pi}_{\phi_V}$ such that the character $\langle \cdot, \pi_V \rangle$ on $\mathcal{S}_\phi$ is trivial, and $\pi_{V,\star}$ ranges over some $\widetilde{O}(G)$-orbit of order 2 in the set $\Pi(\pi_V)$. In particular, we have

$$\pi_{V,\star} = \pi_{V,t_V(\star)} = \bigotimes_{v \in V} \pi_{v,t_V(\star)},$$

for a direct product

$$t_V(\star) = t_V(\pi_V) = \prod_{v \in V} t_v(\pi_V)$$

of points in $T(\phi)$ that is determined up to the diagonal action of $\widetilde{O}(G)$. The right hand side of (8.4.6) is just equal to

$$\sum_{\pi_V \in \widetilde{\Pi}_{\phi_V,\xi}} \sum_{t \in T(\phi)} f_{V,G}(\pi_{V,t}),$$

where

$$\pi_{V,t} = \bigotimes_{v \in V} \pi_{v,t}, \qquad t \in T(\phi).$$

The difference of these last two expressions therefore equals the right hand side ${}^0S^G_{\mathrm{disc},\psi}(f_V f^V)$ of (8.4.21). Supplementing the argument used to derive (8.4.12) in the proof of Lemma 8.4.3 with the criterion for stability at places $v \in V$ in Corollary 8.4.5(b), we write the stable linear form ${}^0S^G_{\mathrm{disc},\psi}(f_V f^V)$ as a linear combination

$$\sum_{t_V} c(t_V)\, f^G_V(\phi_{V,t_V})$$

over indices

$$t_V = \prod_{v \in S_V} t_v, \qquad t_v \in T(\phi_v),$$

with complex coefficients $c(t_V)$. The identity (8.4.21) becomes

$$\sum_{\pi_V \in \widetilde{\Pi}_{\phi_V}(1)} \big(\overline{f}_{V,G}(\pi_{V,t_V(\star)}) - \overline{f}_{V,G}(\pi_{V,t})\big) = \sum_{t_V} c(t_V)\, f^G_V(\phi_{V,t_V}), \tag{8.4.22}$$

where

$$\overline{f}_V(x) = \sum_{\alpha \in \widetilde{O}(G)} f_V(\alpha(x)).$$

The right hand side of (8.4.22) is a linear combination of linear forms, indexed by products t_V. The problem is to show that the coefficients $c(t_V)$ all vanish.

It follows from Corollary 8.4.5 that the linear forms on the right hand side of (8.4.22) are linearly independent. In fact, we can do better. Writing

$$f_{V,G}(\phi_{V,t_V}) = \sum_{\pi_V \in \widetilde{\Pi}_{\phi_V}} f_{V,G}(\pi_{V,t_V}),$$

we decompose the right hand side of (8.4.22) into a finer linear combination indexed by pairs (t_V, π_V), with coefficients $c(t_V)$ that depend only on t_V. These are still linearly independent. We claim that we can find a representation

$$\pi^*_V = \bigotimes_{v \in V} \pi^*_v, \qquad \pi^*_v \in \widetilde{\Pi}_{\phi_v},$$

in $\widetilde{\Pi}_{\phi_V}$ such that the character

$$\langle \cdot, \pi^*_V \rangle = \prod_{v \in V} \langle \cdot, \pi^*_v \rangle$$

on $\mathcal{S}_\psi$ is nontrivial. Indeed, the given condition on ϕ and V implies that the projection

$$\widehat{\mathcal{S}}_{\phi_V} = \prod_{v \in V} \widehat{\mathcal{S}}_{\phi_v} \longrightarrow \widehat{\mathcal{S}}_\phi$$

is surjective. At an archimedean valuation v, we recall that the injection $\Pi_{\phi_v} \longrightarrow \widehat{\mathcal{S}}_{\phi_v}$ need not be bijective. However, its image generates $\widehat{\mathcal{S}}_{\phi_v}$, according to Lemma 6.1.2, and it certainly contains the trivial character.

The claim follows. For any t_V, $f_{V,G}(\pi^*_{V,t_V})$ lies outside the span of the complementary linear forms in our expansion of the right hand side of (8.4.22). It also lies outside the span of the linear forms

$$f_V \longrightarrow \overline{f}_{V,G}(\pi_{V,t_V(*)}) - \overline{f}_{V,G}(\pi_{V,t_V}), \qquad \pi_V \in \widetilde{\Pi}_{\phi_V}(1),$$

on the left hand side of (8.4.22). This is because π^*_V is distinct from any representation $\pi_V \in \widetilde{\Pi}_\phi(1)$, since it maps to a nontrivial character on $\mathcal{S}_\phi$. The coefficient $c(t_V)$ of π^*_{V,t_V} therefore vanishes. Since t_V represents an arbitrary index, all of the coefficients on the right hand side of (8.4.22) vanish, as required.

We have established that the right hand side of (8.4.22) equals 0. In other words, the linear form ${}^0S^G_{\mathrm{disc},\phi}(f)$ on the right hand side of (8.4.21) vanishes. This gives the first formula (8.4.5) of the theorem. The second formula (8.4.6) then follows from the vanishing of the linear form ${}^0r^G_{\mathrm{disc},\phi}(f)$ on the left hand side of (8.4.21).

Notice that in proving Theorem 8.4.2, we appear to have used the condition of local constraints on ϕ in only a minor way. Is the condition too strong? I have not checked it carefully, but we should also remember that we needed the inductive property of the condition in our earlier derivation of (8.4.7) □

We have completed our study of local and global L-packets for the even orthogonal group G. In particular, we have obtained a refined local endoscopic classification of admissible representations, and a refined global endoscopic classification of the representations in the discrete spectrum of Ramanujan type. We have not tried to treat the packets attached to nongeneric parameters ψ. The global methods of this section undoubtedly apply to such ψ. However, they would have to be combined with a finer understanding of the constituents of their packets. It would be interesting to apply the p-adic results of Moeglin to this question.

As we noted in §8.3, the refined endoscopic classification we have established for p-adic F does not quite imply a refined Langlands correspondence. The impediment is the internal $(\mathbb{Z}/2\mathbb{Z})$-symmetry implicit in the lack of a canonical bijection between two sets of order 2. In principle, there will be one copy of the group $(\mathbb{Z}/2\mathbb{Z})$ for each local parameter $\phi \in \widetilde{\Phi}'_{\mathrm{bdd}}(G)$. The obstruction is actually a little more modest. Suppose that for each even N, and each even orthogonal group $G \in \widetilde{\mathcal{E}}_{\mathrm{sim}}(N)$ over the p-adic field F, the $(\mathbb{Z}/2\mathbb{Z})$-symmetry for any simple parameter $\phi \in \widetilde{\Phi}_{\mathrm{sim}}(G)$ has been "broken". In other words, the bijection between the $\widetilde{O}(G)$-torsors $\Phi(\phi)$ and $T(\phi)$ predicted by the local Langlands conjecture has been defined. The same then holds for any of the more general parameters $\phi \in \widetilde{\Phi}'_{\mathrm{bdd}}(G)$, which are attached to any N and G. This is a direct consequence of the definitions. Notice that the question here is purely local. Our makeshift global group

$\mathcal{L}_\psi$ is too coarse at this point even to consider a global analogue of the problem.

8.5. An approximation of the "regular part" of L_F

We would like to be able to streamline the definitions of §1.4. The goal would be to replace the complex groups $\mathcal{L}_\psi$ in terms of which we formulated the global theorems by a locally compact group that is independent of ψ. We shall conclude Chapter 8 with a step in this direction. We shall introduce a group that characterizes many of the global parameters we have studied.

The new group will be closer in spirit to the hypothetical Langlands group L_F. As an approximation it is still pretty crude. It serves as a substitute for L_F only so far as to describe the global endoscopic problems we have now solved. In fact, the group treats just a part of the problem, in that it pertains only to global parameters we shall call regular. Its missing complement depends on the solution of a couple of supplementary problems related to base change. There is no point in attempting to formulate it here, even though these problems might well be within reach. Our aim for now is merely to give a partial description, which illustrates what we can hope is a general construction in the simplest of situations.

There are two steps. The first will be to introduce a group that characterizes automorphic representations of general linear groups. We will then describe a subgroup that accounts also for automorphic representations of orthogonal and symplectic groups.

For any $N \geqslant 1$, we have the basic set $\mathcal{C}_{\mathrm{sim}}(N)$, defined in §1.3 and discussed further in §8.3. We recall that it consists of simple families

$$c = \{c_v : \ v \notin S\}$$

of semisimple conjugacy classes in $GL(N, \mathbb{C})$, taken up to the equivalence relation $c' \sim c$ if $c'_v = c_v$ for almost all v. As usual, $S \subset \mathrm{val}(F)$ represents a finite set of valuations of F that contains the set S_∞ of archimedean places. The special case

$$\mathcal{D} = \mathcal{C}_{\mathrm{sim}}(1) = \mathcal{C}(1)$$

is an abelian group under pointwise multiplication. It has an action

$$d: \ c \longrightarrow dc, \qquad d \in \mathcal{D}, \ c \in \mathcal{C}_{\mathrm{sim}}(N),$$

on $\mathcal{C}_{\mathrm{sim}}(N)$, given by pointwise scalar multiplication on families of $(N \times N)$-matrices. The complex determinant

$$GL(N, \mathbb{C}) \longrightarrow GL(1, \mathbb{C}),$$

which is of course dual to the central embedding

$$GL(1) \longrightarrow GL(N),$$

determines a second operation. Its pointwise application gives a mapping

$$\widehat{\varepsilon}: \ \mathcal{C}_{\mathrm{sim}}(N) \longrightarrow \mathcal{D}$$

such that

$$\widehat{\varepsilon}(dc) = d^N\,\widehat{\varepsilon}(c).$$

We shall use these two operations to define an endoscopic Langlands group that treats "most" automorphic representations of general linear groups.

We first recall that the two operations have natural analogues for automorphic representations. We know that there is a canonical bijection

$$\pi \longrightarrow c(\pi), \qquad \pi \in \mathcal{A}_{\mathrm{cusp}}(N),$$

from the set of unitary cuspidal automorphic representations of $GL(N)$ onto $\mathcal{C}_{\mathrm{sim}}(N)$. This is the theorem of strong multiplicity one, which can be regarded as the specialization of Theorem 1.3.2 to $\mathcal{A}_{\mathrm{cusp}}(N)$, and is the basis for our definition of the set $\mathcal{C}_{\mathrm{sim}}(N)$. The abelian group

$$\mathfrak{X} = \mathcal{A}_{\mathrm{cusp}}(1) = \mathcal{A}(1)$$

of idèle class characters acts in the usual way

$$\chi: \pi \longrightarrow \chi\pi = \pi \otimes (\chi \circ \det), \qquad \chi \in \mathfrak{X},\ \pi \in \mathcal{A}_{\mathrm{cusp}}(N),$$

on $\mathcal{A}_{\mathrm{cusp}}(N)$. The central character then gives a second operation, a mapping

$$\varepsilon: \mathcal{A}_{\mathrm{cusp}}(N) \longrightarrow \mathfrak{X}$$

such that

$$\varepsilon(\chi\pi) = \chi^N\,\varepsilon(\pi).$$

These are the automorphic analogues of the two operations for $\mathcal{C}_{\mathrm{sim}}(N)$. We shall write

$$c \longrightarrow \pi_c, \qquad c \in \mathcal{C}_{\mathrm{sim}}(N),$$

for the inverse of the bijection $\pi \to c(\pi)$, and also

$$d \longrightarrow \chi_d, \qquad d \in \mathcal{D},$$

for its specialization to the case $N = 1$. We then have relations

$$\pi_{dc} = \chi_d \pi_c, \qquad d \in \mathcal{D},\ c \in \mathcal{C}_{\mathrm{sim}}(N),$$

and

$$\varepsilon(\pi_c) = \chi_c, \qquad c \in \mathcal{C}_{\mathrm{sim}}(N),$$

for the idèle class character

$$\chi_c = \chi_{\widehat{\varepsilon}(c)}.$$

For any $c \in \mathcal{C}_{\mathrm{sim}}(N)$, we have isomorphic subgroups

$$\mathcal{D}_c = \{d \in \mathcal{D}:\ dc = c\}$$

and

$$\mathfrak{X}_c = \{\chi \in \mathfrak{X}:\ \chi\pi_c \cong \pi_c\} = \{\chi_d:\ d \in \mathcal{D}_c\}$$

of $\mathcal{D}$ and $\mathfrak{X}$, which consist of elements of order dividing N. We shall say that c and π_c are *regular* if these two subgroups are trivial, which is to say that the associated representations

$$\pi_c \otimes (\chi \circ \det), \qquad \chi \in \mathfrak{X},$$

are mutually inequivalent. We then write $\mathcal{C}_{\text{sim,reg}}(N)$ for the set of regular elements in $\mathcal{C}_{\text{sim}}(N)$. The complement of $\mathcal{C}_{\text{sim,reg}}(N)$ in $\mathcal{C}_{\text{sim}}(N)$ should then be sparse in $\mathcal{C}_{\text{sim}}(N)$. We shall introduce a locally compact variant L_c of the L-group of $GL(N)$ for every $c \in \mathcal{C}_{\text{sim,reg}}(N)$.

The unitary group $U(N)$ is a compact real form of the complex dual group $GL(N,\mathbb{C})$. For any $c \in \mathcal{C}_{\text{sim,reg}}(N)$, we define an extension

$$1 \longrightarrow K_c \longrightarrow L_c \longrightarrow W_F \longrightarrow 1 \tag{8.5.1}$$

of W_F by the special unitary group $K_c = SU(N)$ by setting

$$L_c = \{g \times w \in U(N) \times W_F : \ \det(g) = \chi_c(w)\}.$$

We have identified the central character χ_c of π_c here with a character on the Weil group. The group L_c is a locally compact extension of W_F by the compact simply connected group $SU(N)$, which need not be split. By itself, L_c depends only on χ_c. It has the property that for any unitary, N-dimensional representation of W_F with determinant equal to χ_c, the corresponding global Langlands parameter factors through the image of the L-embedding

$$L_c \longrightarrow {}^L(GL(N)) = GL(N,\mathbb{C}) \times W_F.$$

We shall make L_c more rigid by attaching the local Langlands parameters associated to c.

The group $SU(N)$ is of course a compact real form of the complex special linear group $SL(N,\mathbb{C})$. To account for the possible failure of the generalized Ramanujan conjecture for $GL(N)$, we will need to work with the complexification $L_{c,\mathbb{C}}$ of L_c. This is the extension

$$1 \longrightarrow K_{c,\mathbb{C}} \longrightarrow L_{c,\mathbb{C}} \longrightarrow W_F \longrightarrow 1 \tag{8.5.2}$$

of W_F by $K_{c,\mathbb{C}} = SL(N,\mathbb{C})$ defined by

$$L_{c,\mathbb{C}} = \{g \times w \in GL(N,\mathbb{C}) \times W_F : \ \det(g) = \chi_c(w)\}.$$

It has the property that any L-homomorphism

$$\phi : \ L_c \longrightarrow {}^L G,$$

for a connected reductive group G over F, extends analytically to $L_{c,\mathbb{C}}$. Our interest in $L_{c,\mathbb{C}}$ is in the local Langlands parameters

$$L_{F_v} \longrightarrow {}^L(GL(N)_v) = GL(N,\mathbb{C}) \times W_{F_v}$$

attached to the local components $\pi_{c,v}$ of the cuspidal automorphic representation π_c of c. They provide the left hand vertical arrows in the diagrams

$$\begin{array}{ccc} L_{F_v} & \longrightarrow & W_{F_v} \\ \downarrow & & \downarrow \\ L_{c,\mathbb{C}} & \longrightarrow & W_F \end{array}, \qquad v \in \text{val}(F), \tag{8.5.3}$$

and are defined up to conjugacy by the subgroup $K_{c,\mathbb{C}}$ of $L_{c,\mathbb{C}}$.

We would like to treat L_c as an object that depends only on the $\mathcal{D}$-orbit of c. Since $\mathcal{D}$ acts freely on $\mathcal{C}_{\mathrm{sim,reg}}(N)$, any point c' in the orbit of c equals dc, for a unique element $d \in \mathcal{D}$. The mapping

$$g \times w \longrightarrow g\chi_d(w) \times w, \qquad g \times w \in L_c,$$

is then a canonical L-isomorphism from L_c to $L_{c'}$, which is compatible with the associated diagrams (8.5.3).

Before we formalize the last property in our definitions, we should build in the group $\widetilde{O}(N)$ of outer automorphisms of $GL(N)$. This is the group of order 2 generated by our standard automorphism $\widetilde{\theta}(N)$. Its canonical actions on $\mathcal{D}$ and $\mathfrak{X}$ give isomorphic semidirect products $\mathcal{D} \rtimes \widetilde{O}(N)$ and $\mathfrak{X} \rtimes \widetilde{O}(N)$. The first group $\mathcal{D} \rtimes \widetilde{O}(N)$ acts in the obvious way on $\mathcal{C}_{\mathrm{sim}}(N)$. If c' is any point in the corresponding orbit of c, we can still choose a canonical element α in $\mathcal{D} \rtimes \widetilde{O}(N)$ such that $c' = \alpha c$. We do need to be a little careful if the orbit of c is self-dual, in the sense that the family $c^\vee = \widetilde{\theta}(N)(c)$ equals dc, for some $d \in \mathcal{D}$, since the stability group of c in $\mathcal{D} \rtimes \widetilde{O}(N)$ then has order 2. In this case, we simply take α to be the unique element in the normal subgroup $\mathcal{D}$ of $\mathcal{D} \rtimes \widetilde{O}(N)$. The second group $\mathfrak{X} \rtimes \widetilde{O}(N)$ acts on the L-group of $GL(N)$ according to the automorphisms

$$^L_\chi\theta = {}^L\theta_{\chi^{-1}}, \qquad \chi \in \mathfrak{X},\ \theta \in \widetilde{O}(N),$$

defined at the beginning of §3.2. In other words,

$$(\chi \rtimes \theta) :\ g \times w \longrightarrow \theta(g)\,\chi(w) \times w, \qquad g \in GL(N, \mathbb{C}),\ w \in W_F.$$

In particular, if c' equals $(d \rtimes \theta)c$, the restriction of the automorphism defined by $(\chi_d \rtimes \theta)$ maps L_c to $L_{c'}$. It follows that for any point c' in the $\big(\mathcal{D} \rtimes \widetilde{O}(N)\big)$-orbit of c, we have a *canonical* L-isomorphism from L_c to $L_{c'}$, which is compatible with the associated diagrams (8.5.3).

We can therefore treat c as an element in the set $\mathcal{C}^*_{\mathrm{sim,reg}}(N)$ of $\big(\mathcal{D} \rtimes \widetilde{O}(N)\big)$-orbits in $\mathcal{C}_{\mathrm{sim,reg}}(N)$. With this interpretation of c, we can still form a canonical extension (8.5.1), with complexification (8.5.2) and local diagrams (8.5.3). It comes with a canonical isomorphism onto the associated concrete object we have attached to any c' in the class c, and is defined as a formal inverse limit over the points c', as for example in Kottwitz's definitions in [**K3**, (1.1)].

We are now working with the family

$$\mathcal{C}^*_{\mathrm{sim,reg}} = \coprod_{N \geqslant 1} \mathcal{C}^*_{\mathrm{sim,reg}}(N)$$

of orbits of the group

$$\bigoplus_{N \geqslant 1} \big(\mathcal{D} \rtimes \widetilde{O}(N)\big)$$

in the subset

$$\mathcal{C}_{\mathrm{sim,reg}} = \coprod_{N \geqslant 1} \big(\mathcal{C}_{\mathrm{sim,reg}}(N)\big)$$

of

$$\mathcal{C}_{\mathrm{sim}} = \coprod_{N \geqslant 1} \big(\mathcal{C}_{\mathrm{sim}}(N)\big).$$

We amalgamate the groups we have attached to elements of the family into a fibre product

$$L^*_{F,\mathrm{reg}} = \prod_{c \in \mathcal{C}^*_{\mathrm{sim,reg}}} (L_c \longrightarrow W_F) \tag{8.5.4}$$

over W_F. We thus obtain a locally compact extension

$$1 \longrightarrow K^*_{F,\mathrm{reg}} \longrightarrow L^*_{F,\mathrm{reg}} \longrightarrow W_F \longrightarrow 1 \tag{8.5.5}$$

of W_F by a direct product $K^*_{F,\mathrm{reg}}$ of compact, simply connected groups. The extension comes with a "complexification"

$$1 \longrightarrow K^*_{F,\mathrm{reg},\mathbb{C}} \longrightarrow L^*_{F,\mathrm{reg},\mathbb{C}} \longrightarrow W_F \longrightarrow 1,$$

where $K^*_{F,\mathrm{reg},\mathbb{C}}$ is a restricted direct topological product of complex special linear groups, and an associated family of conjugacy classes of local embeddings

$$\begin{array}{ccc} L_{F_v} & \longrightarrow & W_{F_v} \\ \downarrow & & \downarrow \\ L^*_{F,\mathrm{reg},\mathbb{C}} & \longrightarrow & W_F \end{array}, \qquad v \in \mathrm{val}(F). \tag{8.5.6}$$

Consider a (continuous, semisimple) unitary representation

$$\phi:\ L^*_{F,\mathrm{reg}} \longrightarrow GL(N,\mathbb{C})$$

of $L^*_{F,\mathrm{reg}}$. In case ϕ is irreducible, we shall call it *standard* if it factors to an irreducible, unitary representation of a quotient L_c of $L^*_{F,\mathrm{reg}}$, whose restriction to the subgroup $K_c = SU(N)$ is standard, in the extended sense that it is equivalent to either the standard embedding of $SU(N)$ into $GL(N,\mathbb{C})$ *or* its dual. An irreducible standard representation thus amounts to a tensor product of a standard representation of $SU(N)$ with a unitary character of W_F. In particular, it is trivial on the commutator subgroup W_F^{c} of W_F. In the general case, we shall say that ϕ is *standard* if it is a (finite) direct sum of irreducible standard representations. We then write $\Phi^*_{\mathrm{reg,bdd}}(N)$ for the set of standard representations of $L^*_{F,\mathrm{reg}}$, and $\Psi^*_{\mathrm{reg}}(N)$ for the set of equivalence classes of unitary representations

$$\psi:\ L^*_{F,\mathrm{reg}} \times SU(2) \longrightarrow GL(N,\mathbb{C})$$

whose restriction to $L^*_{F,\mathrm{reg}}$ belongs to $\Phi^*_{\mathrm{reg,bdd}}(N)$.

We have constructed a formal analogue $L^*_{F,\mathrm{reg}}$ of the global Langlands group L_F. It governs the basic representation theory of $GL(N)$, which is to say, the theory encompassed by the three theorems stated in §1.3. Of course, we have really to preface this statement with the adjective *regular*. Let $\mathcal{A}_{\mathrm{reg}}(N)$ be the subset of representations in $\mathcal{A}(N)$, the set of automorphic representations of $GL(N)$ introduced in §1.3 that occur in the spectral decomposition, whose cuspidal components are regular. It then follows from the definitions, together with the three theorems, that there is a canonical bijection

$$\psi \longrightarrow \pi_\psi, \qquad \psi \in \Psi^*_{\mathrm{reg}}(N),$$

from $\Psi^*_{\mathrm{reg}}(N)$ onto $\mathcal{A}_{\mathrm{reg}}(N)$ such that

$$\pi_{\psi,v} = \pi_{\psi_v}, \qquad v \in \mathrm{val}(F).$$

Here, $\psi_v \in \Psi_v(N)$ is the local parameter attached to the complexification of ψ by (8.5.6), while $\pi_{\psi,v}$ is obviously the local component of the automorphic representation π_ψ.

The only role of the "complexification" $L^*_{F,\mathrm{reg},\mathbb{C}}$ of $L^*_{F,\mathrm{reg}}$ is to extend the domain of parameters $\phi \in \Phi^*_{\mathrm{reg,bdd}}(N)$. Any such ϕ factors to a representation of an extension of W_F by a finite-dimensional quotient of $K^*_{F,\mathrm{reg}}$, which in turn extends analytically to the extension of W_F by the corresponding quotient of $K^*_{F,\mathrm{reg},\mathbb{C}}$. We recall again that the complexification $L^*_{F,\mathrm{reg},\mathbb{C}}$ is needed only because we do not know the generalized Ramanujan conjecture for cuspidal automorphic representations of $GL(N)$.

The group $L^*_{F,\mathrm{reg}}$ is adapted to the theory of (standard) endoscopy for general linear groups, which amounts to the three theorems stated in §1.3, together with the theory of Eisenstein series for $GL(N)$. We shall now introduce a subgroup of $L^*_{F,\mathrm{reg}}$ that is adapted to the more sophisticated theory of endoscopy we have established for orthogonal and symplectic groups. Consider the subset

$$\widetilde{\mathcal{C}}_{\mathrm{sim}} = \coprod_{N\geqslant 1} \widetilde{\mathcal{C}}_{\mathrm{sim}}(N) = \coprod_{N\geqslant 1} \{c \in \mathcal{C}_{\mathrm{sim}}(N):\ c^\vee = c\}$$

of self-dual families in $\mathcal{C}_{\mathrm{sim}}$. For any $c \in \widetilde{\mathcal{C}}_{\mathrm{sim}}$, we write

$$\widetilde{\mathcal{D}}_c = \{d \in \mathcal{D}:\ dc \in \widetilde{\mathcal{C}}_{\mathrm{sim}}\}.$$

Since

$$(dc)^\vee = d^\vee c^\vee = d^{-1}c,$$

we see that

$$\widetilde{\mathcal{D}}_c = \{d \in \mathcal{D}:\ d^2 \in \mathcal{D}_c\}.$$

We shall again restrict our attention to the corresponding subset

$$\widetilde{\mathcal{C}}_{\mathrm{sim,reg}} = \coprod_{N\geqslant 1} \widetilde{\mathcal{C}}_{\mathrm{sim,reg}}(N) = \coprod_{N\geqslant 1} \big(\widetilde{\mathcal{C}}_{\mathrm{sim}}(N) \cap \mathcal{C}_{\mathrm{sim,reg}}(N)\big)$$

of regular elements. For any c in this subset, the associated subset

$$\widetilde{\mathcal{D}} = \widetilde{\mathcal{D}}_c = \{d \in \mathcal{D}:\ d^2 = 1\}$$

of $\mathcal{D}$ is the subgroup of elements of order dividing 2, which is of course independent of c. We form the corresponding family

$$\widetilde{\mathcal{C}}^*_{\mathrm{sim,reg}} = \coprod_{N\geqslant 1} \big(\widetilde{\mathcal{C}}^*_{\mathrm{sim,reg}}(N)\big)$$

of $\widetilde{\mathcal{D}}$-orbits in $\widetilde{\mathcal{C}}_{\mathrm{sim,reg}}$. There is then an injection

$$\widetilde{\mathcal{C}}^*_{\mathrm{sim,reg}} \longrightarrow \mathcal{C}^*_{\mathrm{sim,reg}},$$

whose image is the set of $\big(\mathcal{D} \rtimes \widetilde{O}(N)\big)$-orbits in $\mathcal{C}^*_{\mathrm{sim,reg}}$ that have representatives in $\widetilde{\mathcal{C}}_{\mathrm{sim,reg}}$, or equivalently, just the set of self-dual orbits in $\mathcal{C}^*_{\mathrm{sim,reg}}$.

It follows from Theorem 1.4.1 that

$$\widetilde{\mathcal{C}}_{\mathrm{sim,reg}}(N) = \coprod_{G \in \widetilde{\mathcal{E}}_{\mathrm{sim}}(N)} \big(\widetilde{\mathcal{C}}_{\mathrm{sim,reg}}(G)\big),$$

where $\widetilde{\mathcal{C}}_{\mathrm{sim,reg}}(G)$ is the subset of families $c \in \widetilde{\mathcal{C}}_{\mathrm{sim,reg}}(N)$ such that $c = c(\pi)$, for some π in the set $\mathcal{A}_2(G)$. If N is even, the free action of $\widetilde{\mathcal{D}}$ on $\widetilde{\mathcal{C}}_{\mathrm{sim,reg}}(N)$ stabilizes the associated subsets $\widetilde{\mathcal{C}}_{\mathrm{sim,reg}}(G)$, as one sees from the criterion of Theorem 1.5.3. If N is odd, it is the opposite that holds. In this case $\widetilde{\mathcal{D}}$ acts simply transitively on the data $G \in \widetilde{\mathcal{E}}_{\mathrm{sim}}(N)$. Recall however that these different endoscopic data all have the same endoscopic group $G = Sp(N-1)$, with dual group $\widehat{G} = SO(N,\mathbb{C})$.

Suppose that c belongs to the subset $\widetilde{\mathcal{C}}_{\mathrm{sim,reg}}(G)$ of $\widetilde{\mathcal{C}}_{\mathrm{sim,reg}}(N)$ attached to a given $G \in \widetilde{\mathcal{E}}_{\mathrm{sim}}(N)$. We write $\widetilde{K}_c$ for the standard maximal compact subgroup of $\widehat{G}$. Thus, $\widetilde{K}_c$ equals the special unitary quaternion subgroup $SU(n,\mathbb{H})$ of $\widehat{G} = Sp(2n,\mathbb{C})$ if $G = SO(2n+1)$, the compact orthogonal subgroup $SO(2n+1,\mathbb{R})$ of $\widehat{G} = SO(2n+1,\mathbb{C})$ if $G = Sp(2n)$, and the compact orthogonal subgroup $SO(2n,\mathbb{R})$ of $\widehat{G} = SO(2n,\mathbb{C})$ if $G = SO(2n)$. In the first two cases, W_F acts trivially on $\widetilde{K}_c$, while in the third case, it acts according to the mapping of W_F to the group of outer automorphisms defined by χ_c. We define an extension

$$1 \longrightarrow \widetilde{K}_c \longrightarrow \widetilde{L}_c \longrightarrow W_F \longrightarrow 1 \tag{8.5.7}$$

of W_F by $\widetilde{K}_c$ simply by setting

$$\widetilde{L}_c = \widetilde{K}_c \rtimes W_F.$$

We can think of $\widetilde{L}_c$ as a compact real form of the group ${}^LG$. In other words, its complexification is the extension

$$1 \longrightarrow \widetilde{K}_{c,\mathbb{C}} \longrightarrow \widetilde{L}_{c,\mathbb{C}} \longrightarrow W_F \longrightarrow 1 \tag{8.5.8}$$

of W_F by the complex group $\widehat{K}_{c,\mathbb{C}} = \widehat{G}$ defined by the L-group $\widetilde{L}_{c,\mathbb{C}} = {}^LG$. Our notation here is obviously designed to be compatible with that of the groups L_c and $L_{c,\mathbb{C}}$ above. In particular, we have local diagrams

$$\begin{array}{ccc} \widetilde{L}_{F_v} & \longrightarrow & W_{F_v} \\ \big\downarrow & & \big\downarrow \\ \widetilde{L}_{c,\mathbb{C}} & \longrightarrow & W_F \end{array}, \qquad v \in \mathrm{val}(F), \tag{8.5.9}$$

in which the left hand vertical arrows are defined up to conjugacy by the subgroup $\widetilde{K}_{c,\mathbb{C}}$ of $\widetilde{L}_{c,\mathbb{C}}$. This follows from Theorem 1.4.2, but with one caveat. In the case that $\widetilde{K}_{c,\mathbb{C}} = \widehat{G} = SO(2n+1,\mathbb{C})$, we must take c to be the element in its $\widetilde{\mathcal{D}}$-orbit that $\chi_c = 1$ in order that the images of the mappings of the groups L_{F_v} actually be contained in $\widetilde{L}_{c,\mathbb{C}}$.

Following our convention for the group L_c above, we can treat $\widetilde{L}_c$ as an object that depends only on the image of c in the set $\widetilde{\mathcal{C}}^*_{\mathrm{sim,reg}}(G)$ of $\widetilde{\mathcal{D}}$-orbits in $\widetilde{\mathcal{C}}_{\mathrm{sim,reg}}(G)$. In the case $\widehat{G} = SO(2n+1,\mathbb{C})$ we have just flagged, this is part of the definition. In either of the other two cases, suppose that $c' = dc$ is the element in the orbit of c attached to $d \in \widetilde{\mathcal{D}}$. The mapping

$$g \times w \longrightarrow g\chi_d(w) \times w, \qquad g \times w \in \widetilde{L}_c,$$

is then a canonical L-ismorphism from $\widetilde{L}_c$ to $\widetilde{L}_{c'}$, which is compatible with the associated diagrams (8.5.9). We can therefore regard c as a $\widetilde{\mathcal{D}}$-orbit in all three cases. With this interpretation, we still have a canonical extension (8.5.7), with complexification (8.5.8) and local diagrams (8.5.9). It again comes with a canonical isomorphism onto the concrete object we have attached to any c' in the class c.

We can now define our subgroup of $L^*_{F,\mathrm{reg}}$. We have assigned a group $\widetilde{L}_c$ with complexification $\widetilde{L}_{c,\mathbb{C}}$ to each $c \in \widetilde{\mathcal{C}}^*_{\mathrm{sim,reg}}$. Let us identify $\widetilde{\mathcal{C}}^*_{\mathrm{sim,reg}}$ with its injective image in $\mathcal{C}^*_{\mathrm{sim,reg}}$. For any c in its complement, we just set $\widetilde{L}_c = L_c$ and $\widetilde{L}_{c,\mathbb{C}} = L_{c,\mathbb{C}}$. With these conventions, we amalgamate the groups as a fibre product

$$\widetilde{L}^*_{F,\mathrm{reg}} = \prod_{c \in \mathcal{C}^*_{\mathrm{sim,reg}}} (\widetilde{L}_c \longrightarrow W_F) \tag{8.5.10}$$

over F. We thus obtain a locally compact extension

$$1 \longrightarrow \widetilde{K}^*_{F,\mathrm{reg}} \longrightarrow \widetilde{L}^*_{F,\mathrm{reg}} \longrightarrow W_F \longrightarrow 1 \tag{8.5.11}$$

of W_F by a direct product $\widetilde{K}^*_{F,\mathrm{reg}}$ of compact, connected groups. It comes with a "complexification"

$$1 \longrightarrow \widetilde{K}^*_{F,\mathrm{reg},\mathbb{C}} \longrightarrow \widetilde{L}^*_{F,\mathrm{reg},\mathbb{C}} \longrightarrow W_F \longrightarrow 1,$$

where $\widetilde{K}^*_{F,\mathrm{reg},\mathbb{C}}$ is a restricted direct product of complex dual groups, and an associated family of conjugacy classes of embeddings

$$\begin{array}{ccc} L_{F_v} & \longrightarrow & W_{F_v} \\ \downarrow & & \downarrow \\ \widetilde{L}^*_{F,\mathrm{reg},\mathbb{C}} & \longrightarrow & W_F \end{array}, \qquad v \in \mathrm{val}(F). \tag{8.5.12}$$

To embed $\widetilde{L}^*_{F,\mathrm{reg}}$ as a subgroup of $L^*_{F,\mathrm{reg}}$, consider a class $c \in \widetilde{\mathcal{C}}^*_{\mathrm{sim,reg}}$. By choosing suitable representatives of c in $\widetilde{\mathcal{C}}_{\mathrm{sim,reg}}$ and $\mathcal{C}_{\mathrm{sim,reg}}$, we obtain a canonical L-embedding of $\widetilde{L}_c$ into L_c from the standard embedding of $\widetilde{K}_c$

into K_c. If c belongs to the complement of $\widetilde{\mathcal{C}}^*_{\text{sim,reg}}$ in $\mathcal{C}^*_{\text{sim,reg}}$, the groups $\widetilde{L}_{c,\text{reg}}$ and $L_{c,\text{reg}}$ are equal by definition. It follows from the definitions (8.5.4) and (8.5.10) that there is a canonical L-embedding

$$\widetilde{L}^*_{F,\text{reg}} \hookrightarrow L^*_{F,\text{reg}}, \tag{8.5.13}$$

which is compatible with the local diagrams (8.5.6) and (8.5.12).

Consider an L-homomorphism

$$\widetilde{\phi}:\ \widetilde{L}^*_{F,\text{reg}} \longrightarrow {}^LG, \qquad G \in \widetilde{\mathcal{E}}_{\text{sim}}(N).$$

We shall say that $\widetilde{\phi}$ is *standard* if it can be embedded into a commutative diagram

$$\begin{array}{ccc} \widetilde{L}^*_{F,\text{reg}} & \xrightarrow{\ \widetilde{\phi}\ } & {}^LG \\ \downarrow & & \downarrow \\ L^*_{F,\text{reg}} & \xrightarrow{\ \phi\ } & GL(N,\mathbb{C}) \end{array},$$

where the left hand vertical arrow is the L-embedding we have just constructed, the right hand vertical arrow is the representation determined by G as a twisted endoscopic datum, and ϕ is a standard representation of $L^*_{F,\text{reg}}$ defined as above. We write $\widetilde{\Phi}^*_{\text{reg,bdd}}(G)$ for the set of standard L-homomorphisms $\widetilde{\phi}$, taken up to the equivalence relation defined by conjugation of ${}^LG$ by the group $\widetilde{\text{Aut}}_N(G)$. We define $\widetilde{\Psi}^*_{\text{reg}}(G)$ to be the set of L-homomorphisms

$$\widetilde{\psi}:\ \widetilde{L}^*_{F,\text{reg}} \times SU(2) \longrightarrow {}^LG$$

whose restriction to $\widetilde{L}^*_{F,\text{reg}}$ belongs to $\widetilde{\Phi}^*_{\text{reg,bdd}}(N)$, taken up to the same equivalence relation.

The point of course is that one could use the single group $\widetilde{L}^*_{F,\text{reg}}$ in place of the family of ad hoc groups $\mathcal{L}_\psi$ attached to the subset $\widetilde{\Psi}_{\text{reg}}(G)$ of "regular" parameters in our earlier set $\widetilde{\Psi}(G)$. A parameter (1.4.1) in $\widetilde{\Psi}(G)$ would again be called regular if the cuspidal automorphic representations $\mu_i \in \widetilde{\mathcal{A}}_{\text{cusp}}(m_i)$ in its simple constituents $\psi_i = \mu_i \otimes \nu_i$ are regular. One sees from the definitions that there is a bijection from $\widetilde{\Psi}^*_{\text{reg}}(G)$ onto $\widetilde{\Psi}_{\text{reg}}(G)$, which is compatible with the local diagrams (8.5.12) and (1.4.14), and such that

$$\psi(L^*_{F,\text{reg},\mathbb{C}}) \subset \mathcal{L}_\psi, \qquad \psi \in \widetilde{\Psi}^*_{\text{reg}}(G),$$

for suitable representatives of the corresponding parameters within their $\widetilde{\text{Out}}_N(G)$-orbits. Moreover, the bijection preserves the global centralizers S_ψ, their localizations S_{ψ_v}, and the associated homomorphisms

$$\mathcal{S}_\psi \longrightarrow \prod_v \mathcal{S}_{\psi_v}.$$

Therefore our theorems, insofar as they apply to regular parameters, can indeed be formulated in terms of the group $\widetilde{L}^*_{F,\mathrm{reg}}$. In particular, we can simply identify the two parameter sets, writing $\widetilde{\Psi}_{\mathrm{reg}}(G)$ in place of $\widetilde{\Psi}^*_{\mathrm{reg}}(G)$. This is in keeping with the convention in §4.8.

We have reserved the symbols L^*_F and $\widetilde{L}^*_F$ for the larger locally compact groups that could be expected if one included the complementary, "non-regular" representations in the construction. If c lies in the complement of $\mathcal{C}_{\mathrm{sim,reg}}(N)$ in $\mathcal{C}_{\mathrm{sim}}(N)$, $\mathcal{D}_c$ should be a cyclic group, whose order ℓ_K then divides N. In this case, c would be obtained by automorphic induction from the subgroup W_K of W_F attached to $\mathfrak{X}_c$ by class field theory. One would define L_c as an extension (8.5.1) of W_F by the smaller simply connected group

$$K_c = \underbrace{SU(N_K) \times \cdots \times SU(N_K)}_{\ell_K}, \qquad \ell_K N_K = N,$$

on whose factors W_F acts by permutation through its cyclic quotient $W_K \backslash W_F$. For any element $c' = dc$ in the $\mathcal{D}$-orbit of c, the L-isomorphism from L_c to $L_{c'}$ would then depend only on the image of d in $\mathcal{D}/\mathcal{D}_c$, and would therefore be canonical. This would lead to a canonical group L_c that depends only on the image of c in the set $\mathcal{C}^*_{\mathrm{sim}}(N)$ of $\big(\mathcal{D} \rtimes O(N)\big)$-orbits in $\mathcal{C}_{\mathrm{sim}}(N)$. Similar properties could be expected of any element c in the complement of $\widetilde{\mathcal{C}}_{\mathrm{sim,reg}}(N)$ in $\widetilde{\mathcal{C}}_{\mathrm{sim}}(N)$. The would lead to a canonical group $\widetilde{L}_c$ that depends only on the image of c in the set $\widetilde{\mathcal{C}}^*_{\mathrm{sim}}(N)$ of $\widetilde{\mathcal{D}}$-orbits in $\widetilde{\mathcal{C}}_{\mathrm{sim}}(N)$. We would then be able to define the larger groups L^*_F and $\widetilde{L}^*_F$ as fibre products (8.5.4) and (8.5.10), but taken over the larger set

$$\mathcal{C}^*_{\mathrm{sim}} = \coprod_N \mathcal{C}^*_{\mathrm{sim}}(N).$$

The detailed construction of the larger groups L^*_F and $\widetilde{L}^*_F$ should be possible, given the global results of §8.1 and §8.2 and perhaps an assumption that Statement B of [**LR**] for $GL(N)$ is valid. (See also [**Ra**].) However, it would take us beyond the methods of this volume. We add here only the remark that the two groups will come with an L-embedding

$$\widetilde{L}^*_F \hookrightarrow L^*_F,$$

and two larger families of parameters $\Psi^*(N)$ and $\widetilde{\Psi}^*(G)$. These would allow us to formulate our global results entirely in terms of the natural approximations L^*_F and $\widetilde{L}^*_F$ of the Langlands group L_F. We hope to return to these matters elsewhere.

CHAPTER 9

Inner Forms

9.1. Inner twists and automorphisms

This last chapter might be more elementary than some of the others. My original intention was to extend the classification to general orthogonal and symplectic groups. However, it became clear that the complete proofs would take considerably more than one chapter. For there are interesting new phenomena that arise from the inner forms of quasisplit orthogonal and symplectic groups. These all require further discussion, and sometimes more complex proofs. Our goal for the last chapter has therefore to be more modest. We will be content to look at nonquasisplit inner forms from the perspective of earlier chapters, and then simply to state analogues of the results for we have obtained quasisplit groups. In this sense, the last chapter is parallel to the first. The difference is that the assertions we make here will have to be proved elsewhere [**A28**].

The first three sections of the chapter will each be built around a distinctive property of inner forms that bears on their representation theory. The first section is devoted to a review of the classification of inner forms. The new phenomenon here is the possible failure of an outer automorphism to have a representative (in its outer class) that is defined over the ground field. Among other things, this is responsible for the failure of the Hasse principle for groups of type D_n. In the second section, we will discuss the nonabelian groups that must be used in place of the abelian 2-groups $\mathcal{S}_\psi$ attached to local parameters ψ for quasisplit groups. They lead ultimately to higher global multiplicities for automorphic representations. The third section concerns the noncanonical nature of the local transfer factors in [**LS1**]. We shall describe how to rectify the problem, according to ideas of Kaletha and Kottwitz. We will then be able to state the local results in §9.4, and the global results in §9.5.

A high point in the theory of algebraic groups is the classification of reductive groups over F (our local or global field of characteristic 0). We shall review this part of the theory, as it applies to orthogonal and symplectic groups. We described the quasisplit orthogonal and symplectic groups G^* in §1.2. (From this point on, we will be writing G^* for the groups that were earlier denoted by G, since they now assume only a secondary role.) What remains to discuss are the inner forms of these groups. This part of the classification is best formulated in terms of the finer classification of inner twists of G^*.

An *inner twist* of the quasisplit group $G^* \in \widetilde{\mathcal{E}}_{\mathrm{sim}}(N)$ over our field F is a pair (G, ψ), where G is a connected reductive group over F, and ψ is an isomorphism from G to G^* such that for every element σ in the Galois group Γ_F, the automorphism

$$\alpha(\sigma) = \psi\sigma(\psi)^{-1}$$

of G^* is inner. An *isomorphism* from (G, ψ) to a second inner twist (G_1, ψ_1) is an isomorphism $\theta_1\colon G \to G_1$ over F such that the automorphism $\psi_1\theta_1\psi^{-1}$ of G^* is inner. (See [**KS**, p. 141].) The correspondence that sends (G, ψ) to the 1-cocycle $\alpha(\sigma) \in Z^1(F, G^*_{\mathrm{ad}})$ then descends to a bijection between the set of isomorphism classes of inner twists of G^* and the Galois cohomology set $H^1(F, G^*_{\mathrm{ad}})$.

An *inner form* of G^* can be defined simply as a connected reductive group over F that is the first component of some inner twist of G^*. There is consequently a surjective mapping from the set of isomorphism classes of inner twists onto the set of isomorphism classes of inner forms. The fibres of the mapping correspond to the orbits in $H^1(F, G^*_{\mathrm{ad}})$ of the group $\mathrm{Out}(G^*)$ of outer automorphisms of G^*, under an action we can denote by

$$\theta^* : \alpha(\sigma) \longrightarrow \theta^*\alpha(\sigma)\sigma(\theta^*)^{-1}, \qquad \theta^* \in \mathrm{Out}(G^*),\ \alpha(\sigma) \in Z^1(F, G^*_{\mathrm{ad}}),$$

or equivalently, by

$$\tilde{\theta}^* : \alpha(\sigma) \longrightarrow \tilde{\theta}^*\big(\alpha(\sigma)\big) = \tilde{\theta}^* \cdot \alpha(\sigma) \cdot (\tilde{\theta}^*)^{-1},$$

where $\tilde{\theta}^*$ is the F-automorphism of G^* in the F-outer class θ^* that preserves a given F-splitting. We will return to these matters in the discussion of Lemma 9.1.1 below, once we have described the classification. We note in passing that our remarks here remain valid if G^* is replaced by any quasisplit group over F. In general, the extra structure of a quasisplit inner twist ψ for a group G is somewhat like a "co-ordinate system" for G.

Thus, although the classification of inner twists of G^* is finer than that of inner forms, its fibres are quite transparent. Accordingly, we will often let G stand for a given inner twist (G, ψ) of G^* taken up to isomorphism, in the spirit of our earlier convention for endoscopy, in addition to the orthogonal or symplectic group that is its first component. We will consider only the subgroup $\widetilde{\mathrm{Out}}_N(G^*)$ of $\widetilde{\mathrm{Out}}(G^*)$ in keeping with our original agreement to avoid triality automorphisms. This just means that if G^* equals the split group $SO(8)$, the only case in which $\widetilde{\mathrm{Out}}_N(G^*)$ is proper in $\mathrm{Out}(G^*)$, there will be a nonstandard isomorphism between inner forms in two different $\widetilde{\mathrm{Out}}_N(G^*)$-orbits of inner twists. We recall that the group $\widetilde{\mathrm{Out}}_N(G^*)$ is trivial unless G^* is of the form $SO(2n)$, in which case it equals $\mathbb{Z}/2\mathbb{Z}$.

There are really three classifications of inner twists. One is by Galois cohomology, which can be based on the elegant formulation of Kottwitz [**K5**, §1–2], with the results in [**Se**, III.35–III.37] taken as an archimedean supplement. This is the best suited to our purposes. A second classification is by the Tits indices in [**Ti**]. It is actually a classification of inner forms G (rather than inner twists), which comes with a transparent description of the Levi subgroups M of G. The third is the explicit classification in terms of bilinear forms, about which we will say little.

Given a group $G^* \in \widetilde{\mathcal{E}}_{\text{sim}}(N)$ over F, we may as well write $\widehat{Z}_{\text{sc}} = Z(\widehat{G}^*_{\text{sc}})$ (rather than $\widehat{Z}^*_{\text{sc}}$) for the centre of the simply connected dual group $\widehat{G}^*_{\text{sc}}$. Then we have

$$(9.1.1)\qquad \widehat{Z}_{\text{sc}} \cong \begin{cases} \mathbb{Z}/2\mathbb{Z}, & \text{if } G^* = SO(2n+1),\\ \mathbb{Z}/2\mathbb{Z}, & \text{if } G^* = Sp(2n),\\ (\mathbb{Z}/2\mathbb{Z}) \times (\mathbb{Z}/2\mathbb{Z}), & \text{if } G^* = SO(2n), \text{ with } n > 1 \text{ even},\\ \mathbb{Z}/4\mathbb{Z}, & \text{if } G^* = SO(2n), \text{ with } n > 1 \text{ odd}. \end{cases}$$

We also have the subgroup $\widehat{Z}^{\Gamma}_{\text{sc}}$ of elements in $\widehat{Z}_{\text{sc}}$ fixed by the Galois group $\Gamma = \Gamma_F$. If G^* is split, Γ of course acts trivially on $\widehat{Z}_{\text{sc}}$, and $\widehat{Z}^{\Gamma}_{\text{sc}} = \widehat{Z}_{\text{sc}}$. If G^* is not split, we recall that G is of the form $SO(2n)$, and Γ acts on $\widehat{Z}_{\text{sc}}$ through the quadratic extension E/F. The nontrivial element in the corresponding quotient of Γ acts by permutation of the two factors of $\widehat{Z}_{\text{sc}}$ if n is even, and by the automorphism $z \to z^{-1}$ if n is odd. It follows that $\widehat{Z}^{\Gamma}_{\text{sc}}$ equals $\mathbb{Z}/2\mathbb{Z}$ if G^* is not split.

Suppose first that F is local. Then there is a canonical morphism of pointed sets

$$(9.1.2)\qquad K^*_{\text{ad}} = K_{G^*_{\text{ad}}} : H^1(F, G^*_{\text{ad}}) \longrightarrow \Pi(\widehat{Z}^{\Gamma}_{\text{sc}}),$$

where $\Pi(\widehat{Z}^{\Gamma}_{\text{sc}})$ is the group of (linear) characters on $\widehat{Z}^{\Gamma}_{\text{sc}}$. If F is p-adic, the map is bijective. If F is archimedean, we have

$$\ker(K^*_{\text{ad}}) = \text{im}\big[H^1(F, G^*_{\text{sc}}) \longrightarrow H^1(F, G^*_{\text{ad}})\big]$$

and

$$\text{im}(K^*_{\text{ad}}) = \ker\big[\Pi(\widehat{Z}^{\Gamma}_{\text{sc}}) \longrightarrow \Pi(\widehat{Z}_{\text{sc}})\big],$$

where the kernel on the right is induced from the norm homomorphism from $\widehat{Z}_{\text{sc}}$ to $\widehat{Z}^{\Gamma}_{\text{sc}}$ given by the action of Γ_F in $\widehat{Z}_{\text{sc}}$. In particular, K^*_{ad} is neither injective nor surjective if $F = \mathbb{R}$. The set $H^1(F, G^*_{\text{ad}})$ is of course trivial if $F = \mathbb{C}$. (See the general local result [**K5**, Theorem 1.2], in which G^*_{ad} is replaced by an arbitrary connected reductive group G over F, and $\Pi(\widehat{Z}^{\Gamma}_{\text{sc}})$ is replaced by the finite character group

$$A(G) = \Pi\big(Z(\widehat{G})^{\Gamma}/(Z(\widehat{G})^{\Gamma})^0\big).$$

Suppose next that F is global. Then the canonical mapping

$$H^1(F, G^*_{\text{ad}}) \longrightarrow \bigoplus_v H^1(F_v, G^*_{\text{ad}})$$

is injective. Its image is the kernel of the composition

$$\bigoplus_v H^1(F_v, G^*_{\mathrm{ad}}) \longrightarrow \bigoplus_v \Pi(\widehat{Z}^{\Gamma_v}_{\mathrm{sc}}) \longrightarrow \Pi(\widehat{Z}^{\Gamma}_{\mathrm{sc}}).$$

(See the general global results [**K5**, Theorem 2.2 and Proposition 2.6]. If G^*_{ad} is replaced by a general group G, the mapping need not be injective, though it does have an easily described kernel [**K3**, §4]. In addition, the groups $\Pi(\widehat{Z}^{\Gamma_v}_{\mathrm{sc}})$ and $\Pi(\widehat{Z}^{\Gamma}_{\mathrm{sc}})$ in the composition above have to be replaced by $A(G_v)$ and $A(G)$ respectively.)

We are generally letting G stand for an isomorphism class of inner twists (G, ψ) over F (as well as the associated connected reductive group). If F is local, we can write

$$\widehat{\zeta}_G(z) = \langle z, G\rangle, \qquad z \in \widehat{Z}^{\Gamma}_{\mathrm{sc}}, \tag{9.1.3}$$

for the character on $\widehat{Z}^{\Gamma}_{\mathrm{sc}}$ obtained from K^*_{ad} and the image of G in $H^1(F, G^*_{\mathrm{ad}})$. If F is global, we have a mapping $z \to z_v$ from $\widehat{Z}^{\Gamma}_{\mathrm{sc}}$ to the group $\widehat{Z}^{\Gamma_v}_{\mathrm{sc}}$ attached to any completion F_v of F. The isomorphism classes of inner twists over F are then bijective with the set of families

$$G = \{G_v : \ v \in \mathrm{val}(F)\}$$

of (isomorphism classes of) local inner twists such that $G_v = G^*_v$ for almost all v, and such that the character

$$\langle z, G\rangle = \prod_v \langle z_v, G_v\rangle, \qquad z \in \widehat{Z}^{\Gamma}_{\mathrm{sc}}, \tag{9.1.4}$$

on $\widehat{Z}^{\Gamma}_{\mathrm{sc}}$ is trivial. The analogy with the deeper pairings of Theorems 1.5.1 and 1.5.2 is clear.

This classification of inner twists by Galois cohomology is easy to work with. However, the description we have just given is incomplete. It does not explicitly characterize the inner twists over the local field $F = \mathbb{R}$. A related gap is the lack of an immediate description of the Levi subgroups of G over F, especially in case F equals $\mathbb{R}$ or is global.

It is easy to describe the image of K^*_{ad} when $F = \mathbb{R}$. If G equals $SO(2n+1)$ or $Sp(2n)$, G^* is split. The norm mapping from the group $\widehat{Z}^{\Gamma}_{\mathrm{sc}} = \widehat{Z}_{\mathrm{sc}} = \mathbb{Z}/2\mathbb{Z}$ to itself is trivial, and K^*_{ad} is surjective. The third case that $G = SO(2n)$ depends on whether n is even or odd. If n is even, one checks that K^*_{ad} is surjective if G^* is split, and trivial otherwise. If n is odd, K^*_{ad} is surjective if G^* is not split, and maps onto the subgroup (of index 2) in $\Pi(\widehat{Z}^{\Gamma}_{\mathrm{sc}})$ of elements of order 1 or 2 if G^* is split. In particular, we observe that in all cases K^*_{ad} is surjective if and only if $G(\mathbb{R})$ has square integrable representations. The problem still is that our description does not account for the fibres of the mapping. There are two ways we could refine the discussion so as to fill the gap.

One modification would be to let G be what we called a K-group in [**A12**, §2] (for F local) and [**A14**, §4] (for F global), following ideas of

Kottwitz and Vogan. This would have the effect of compressing the fibres of $K^*_{\rm ad}$ into single objects. It would also account for the Levi subgroups of G [**A12**, Lemma 2.1], [**A14**, Lemma 4.1] (by redefining the problem as much as actually solving it). However, there is no point letting G stand here for anything other than the connected reductive group it has been up until now. The other refinement would be to include Serre's formulation of the archimedean cohomology sets $H^1(\mathbb{R}, G^*_{\rm ad})$ [**Se**, III.35–III.37]. This would also lead to a description of the Levi subgroups of the connected group G. However, we will not give the details, since they would take us too far afield.

We shall instead simply list the relevant Tits indices from [**Ti**]. More precisely, we list the Tits indices for our groups of type B_n, C_n and D_n, specialized to the local or global field F. They are taken directly from Table II of [**Ti**], but adjusted slightly so that they represent inner twists G of the groups $G^* \in \widetilde{\mathcal{E}}_{\rm sim}(N)$ rather than inner forms. (The distinction is only relevant to indices of type ${}^1D_n$ below, specifically the subdiagrams (9.1.5) and (9.1.6).) Together with a few supplementary details we include without proof, the indices allow one to interpolate the missing information.

We refer the reader to the explanations in [**Ti**, p. 39,54]. Each index is built on the Coxeter-Dynkin diagram of the group G^*. It comes with two supplementary integers, in addition to the number n of vertices in the diagram: the F-rank $r \in \{0, 1, \dots, n\}$ of G and a degree $d \in \{1, 2\}$. Vertices that lie in the same Galois orbit (which occur only for the quasisplit, nonsplit groups G^* of type ${}^2D_n$ below) are written side by side. Vertex orbits that correspond to isotropic simple roots of G are then circled. At the suggestion of Bill Casselman, we have also included *broken* loops to designate vertex orbits of order two that are *anisotropic* (and that are hence not enclosed in a solid loop).* The minimal Levi subgroup M_0 of G (or rather its derived group, which is called the anisotropic kernel in [**Ti**]) is represented by the diagram obtained by removing the isotropic vertex orbits. In particular, the split rank r of G over F equals the number of circled vertex orbits. The standard Levi subgroups M of G are in bijection with subsets of isotropic vertex orbits. They are represented by the diagrams obtained by deleting these subsets.

If F is local, the inner twist G is determined by its index (with the interpretations below for the subdiagrams (9.1.5) and (9.1.6)). In this case, we state without proof a description of the character $\widehat{\zeta}_G$ on $\widehat{Z}^{\Gamma}_{\rm sc}$ in terms of its index. If F is global, the inner twist G is not determined by the index. It is characterized instead by its completions (G_v), which is to say their local indices, subject to the condition (9.1.4). In this case, the set of isotropic vertex orbits in the global index can be seen to be the intersection over v of the sets of local isotropic vertex orbits (suitably interpreted in case G is of the type ${}^2D_n$ below, and some G_v of type ${}^1D_n$ contains a diagram of

*I thank Casselman for his composition of all of these diagrams.

the form (9.1.5)). We can thus construct a global index explicitly from the associated family of local indices.

Type $\mathbf{B_n}$

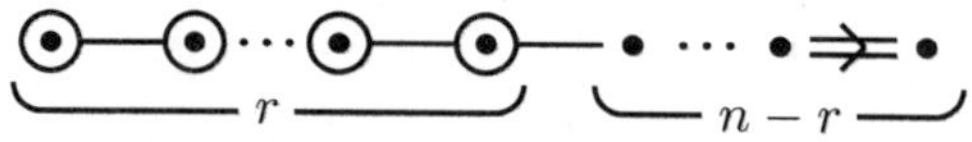

, $d = 1$.

(The degree $d = 2$ does not occur in this case.)

Further conditions on r and d

$F = \mathbb{R}$: none.

F p-adic: $n - r \in \{0, 1\}$.

F global: none.

The group G

$G^* = SO(2n + 1)$ is split.

$G = SO(2n + 1, q_r)$, where q_r is a (nondegenerate) quadratic form of index r over F.

$\widehat{G} = Sp(2n, \mathbb{C}) = \widehat{G}^*$.

The character $\widehat{\zeta}_G$ (F local)

$$\widehat{\zeta}_G = (\widehat{\varepsilon}_{\text{sc}})^{n-r},$$

where $\widehat{\varepsilon}_{\text{sc}}$ is the nontrivial character on the group $\widehat{Z}_{\text{sc}}^{\Gamma} = \mathbb{Z}/2\mathbb{Z} = \widehat{Z}_{\text{sc}}$.

Type $\mathbf{C_n}$

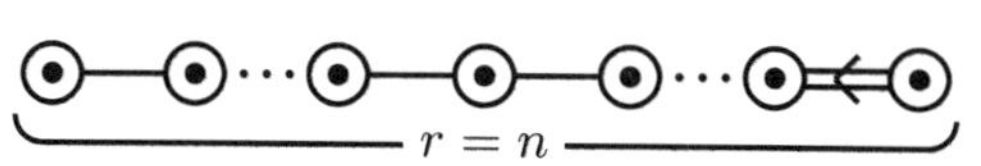

, $d = 1$.

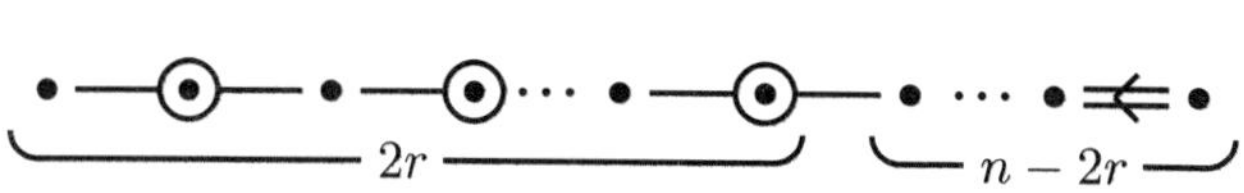

, $d = 2$.

Further conditions on r and d

$F = \mathbb{R}$: $n - r = 0$ if $d = 1$ (as in the diagram).
No further conditions if $d = 2$.

F p-adic: $n - r = 0$ if $d = 1$ (as in the diagram).
$n - 2r \in \{0, 1\}$, if $d = 2$.

F global: $n - r = 0$ if $d = 1$ (as in the diagram).
No further conditions if $d = 2$.

The group G

$G^* \cong Sp(2n)$ is split.

$G = G^* = Sp(2n)$, if $d = 1$; $G = SU(n, h_r)$, where h_r is a (nondegenerate) Hermitian form of index r over a quaternion algebra Q over F, if $d = 2$.

$\widehat{G} = \widehat{G}^* = SO(2n+1, \mathbb{C})$.

The character $\widehat{\zeta}_G$ (F local)

$$\widehat{\zeta}_G = (\widehat{\varepsilon}_{\mathrm{sc}})^{d-1},$$

where $\widehat{\varepsilon}_{\mathrm{sc}}$ is again the nontrivial character on the group $\widehat{Z}_{\mathrm{sc}}^{\Gamma} = \mathbb{Z}/2\mathbb{Z} = \widehat{Z}_{\mathrm{sc}}$.

Type $^1\mathrm{D_n}$

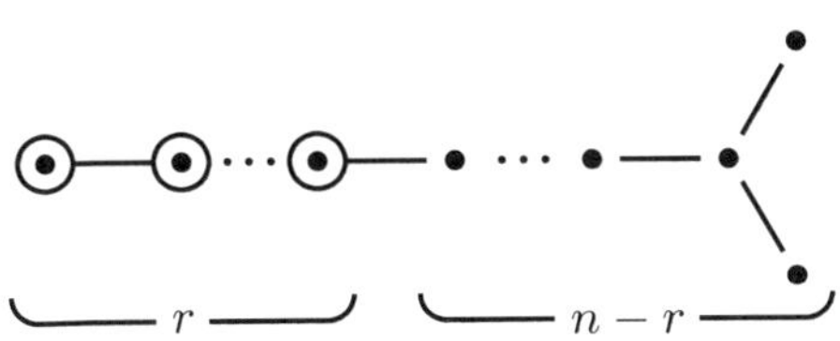

, $d = 1$.

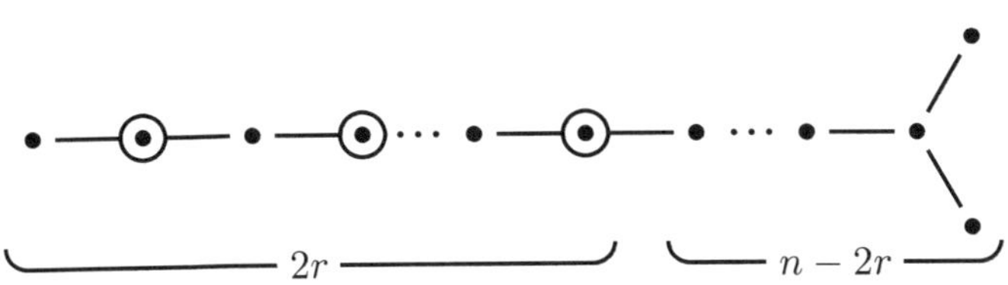

, $d = 2$.

If $d = 2$ and $n - 2r = 0$, the right hand end is one of the two diagrams

(9.1.5(i))

or

,

when $n > 2$, and one of the two diagrams

(9.1.5(ii))

or

,

when $n = 2$ (in which case these are of course the full diagrams). If $d = 2$ and $n-2r$ is odd, $n-2r$ equals 3 and F is either global or p-adic, according to the further conditions below. In this case, we take the right hand end somewhat artificially to be one of two formal copies of the diagram

(9.1.6)  ,

which we identify with one of the two anisotropic, central simple algebras of degree 4 that are implicit in G as a group over F. In each of these cases, the associated two diagrams represent the two inner twists G in the fibre of an inner form. (Compare with the diagrams at the bottom of p. 56 in [**Ti**].)

Further conditions on r and d

$F = \mathbb{R}$: $n - r$ is even if $d = 1$.
$n - 2r = 0$ if $d = 2$ (so in particular, n is even).

F p-adic: $n - r \in \{0, 2\}$ if $d = 1$.
$n - 2r \in \{0, 3\}$ if $d = 2$ (so n can be even or odd).

F global: $n - r$ is even if $d = 1$.
$n - 2r$ is even or equal to 3 if $d = 2$.

The group G

$G^* \cong SO(2n)$ is split.

$G \cong SO(2n, q_r)$, where q_r is a (nondegenerate) quadratic form of discriminant 1 and index r over F, if $d = 1$; $G \cong SU(n, h_r)$, where h_r is a nondegenerate anti-Hermitian form of discriminant 1 and index r over a quaternion algebra Q over F, if $d = 2$.

$\widehat{G} = SO(2n, \mathbb{C}) = \widehat{G}^*$.

The character $\widehat{\zeta}_G$ (F local)

$$\widehat{\zeta}_G = \begin{cases} (\widehat{\varepsilon}_{\mathrm{sc}})^{\frac{1}{2}(n-r)}, & \text{if } d = 1, \\ \widehat{\chi}_{\mathrm{sc}}, & \text{if } d = 2, \end{cases}$$

where $\widehat{\varepsilon}_{\mathrm{sc}}$ is the nontrivial character on the group

$$\widehat{Z}_{\mathrm{sc}}^{\Gamma} = \widehat{Z}_{\mathrm{sc}} = \begin{cases} (\mathbb{Z}/2\mathbb{Z}) \times (\mathbb{Z}/2\mathbb{Z}), & n \text{ even}, \\ (\mathbb{Z}/4\mathbb{Z}), & n \text{ odd}, \end{cases}$$

that is fixed by the standard automorphism, and $\widehat{\chi}_{\mathrm{sc}}$ is one of the remaining two nontrivial characters. The latter is determined by the diagram (9.1.5) attached to G if n is even, and the anisotropic, central simple algebra of degree 6 attached to G and the diagram (9.1.6) if n is odd (and F is p-adic).

Type $^2\mathrm{D_n}$

, $d = 1.$

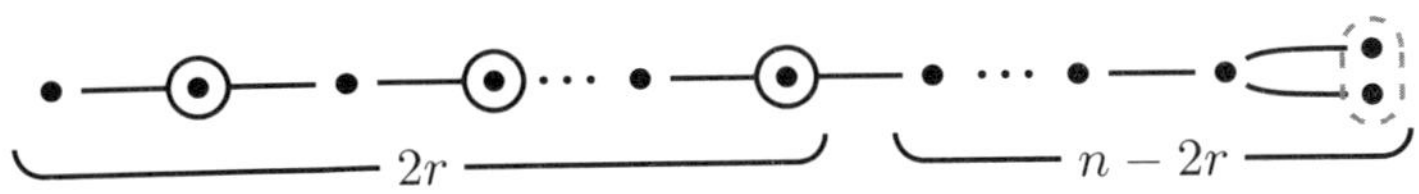

, $d = 2.$

If $d = 1$ and $n - r = 1$, the right hand end is the diagram

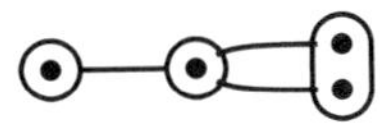

,

and G equals G^*. If $d = 2$ and $n-2r = 1$, the right hand end is the diagram

.

Further conditions on r and d

$F = \mathbb{R}$: $n - r$ is odd if $d = 1$.
$n - 2r = 1$ if $d = 2$ (so in particular, n is odd).

F p-adic: $n - r = 1$ if $d = 1$.
$n - 2r \in \{1, 2\}$ if $d = 2$ (so n can be even or odd).

F global: none.

The group G

$G^* \cong SO(2n, \eta_E)$ is the nonsplit, quasisplit group attached to a quadratic extension E/F.

G is as in ^1D_n, except that the forms q_r and h_r in question now have discrimimant $\alpha \in F^*/(F^*)^2$ such that $E = F(\sqrt{\alpha})$.

$\widehat{G} = SO(2n, \mathbb{C}) = \widehat{G}^*$, with the nontrivial L-action of the Galois group $\Gamma_{E/F}$.

The character $\widehat{\zeta}_G$ (F local)

$$\widehat{\zeta}_G = (\widehat{\varepsilon}_{\mathrm{sc}})^{d-1},$$

where $\widehat{\varepsilon}_{\mathrm{sc}}$ is the nontrivial character on $\widehat{Z}_{\mathrm{sc}}^{\Gamma} = \mathbb{Z}/2\mathbb{Z} \subset \widehat{Z}_{\mathrm{sc}}$.

To focus the discussion, we have agreed to consider one property in each of the first three sections that makes inner twists qualitatively different from quasisplit groups, and that complicates the classification of their representations. The question for this section is the possible nonexistence of rational automorphisms. Suppose that G^* is isomorphic to one of the quasisplit groups $SO(2n)$ over F. We shall now write $\widetilde{\theta}^*$ for the F-automorphism of G^* that was denoted by $\widetilde{\theta} = \mathrm{Int}\big(\widetilde{w}(N)\big)$ in (1.2.5). This of course represents the nontrivial element in $\mathrm{Out}(G^*)$. Given an inner twist (G, ψ) with associated 1-cocycle $\alpha(\sigma)$, we can ask whether there is a corresponding F-automorphism of G. In other words, is there an F-automorphism of G, which we now denote by $\widetilde{\theta}$, that transforms under ψ to an automorphism in the inner class of $\widetilde{\theta}^*$?

LEMMA 9.1.1. *The following conditions on the objects* (G, ψ) *and* $\widetilde{\theta}^*$ *attached to* $G^* \in \widetilde{\mathcal{E}}_{\mathrm{sim}}(N)$, *the given quasisplit group of type* $\mathbf{D}_n$ *over our local or global field* F, *are equivalent.*

(i) *There is an* F-*automorphism* $\widetilde{\theta}$ *of the group* G *in the inner class of the automorphism* $\psi^{-1}\widetilde{\theta}^*\psi$.

(ii) *The inner twist* $(G, \widetilde{\theta}^*\psi)$ *is isomorphic to* (G, ψ).

(iii) *The cocycles* $\alpha(\sigma)$ *and*

$$\widetilde{\theta}^*\big(\alpha(\sigma)\big) = \widetilde{\theta}^* \cdot \alpha(\sigma) \cdot (\widetilde{\theta}^*)^{-1}, \qquad \sigma \in \Gamma_F,$$

have the same image in $H^1(F, G^*_{\mathrm{ad}})$.

(iv) *There is an element* $\gamma^* \in \mathrm{Int}(G^*)$ *such that*

$$\widetilde{\theta}^*\big(\alpha(\sigma)\big) = (\gamma^*)^{-1}\alpha(\sigma)\sigma(\gamma^*), \qquad \sigma \in \Gamma_F.$$

PROOF. The lemma is a direct consequence of the definitions. For example, we can regard ψ as an F-isomorphism from G to G^* if we equip G^* with the twisted Galois action

$$\sigma_G = \alpha(\sigma)\sigma = \psi\sigma(\psi)^{-1}\sigma, \qquad \sigma \in \Gamma_F.$$

The first condition (i) becomes the existence of an element $\gamma^* \in \mathrm{Int}(G^*)$ such that

$$\sigma_G(\gamma^*\widetilde{\theta}^*) = \gamma^*\widetilde{\theta}^*.$$

This amounts in turn to the existence of an F -automorphism

$$\widetilde{\theta} = \psi^{-1}\gamma^*\widetilde{\theta}^*\psi \tag{9.1.7}$$

of G such that the automorphism

$$\gamma^* = (\psi\widetilde{\theta})(\widetilde{\theta}^*\psi)^{-1}$$

of G^* is inner. In other words, $\widetilde{\theta}$ is an isomorphism from $(G, \widetilde{\theta}^*\psi)$ to (G, ψ), the existence of which is the second condition (ii). Conditions (i) and (ii) are therefore equivalent. The conditions (iii) and (iv) are just alternate ways of stating (ii), so they are also equivalent to the first two conditions. □

REMARKS. 1. What happens when the conditions of the lemma are not met? Given (G, ψ) and $\widetilde{\theta}^*$ as in the lemma, let $(G^\vee, \psi^\vee)$ be any inner twist of G^* such that

$$\alpha^\vee(\sigma) \stackrel{\text{def}}{=\!=} \psi^\vee\sigma(\psi^\vee)^{-1} = \widetilde{\theta}^*\big(\alpha(\sigma)\big), \qquad \sigma \in \Gamma_F.$$

The isomorphism

$$\widetilde{\theta}^\vee = (\psi^\vee)^{-1}\widetilde{\theta}^*\psi : G \longrightarrow G^\vee \tag{9.1.8}$$

is then defined over F, since we have

$$\begin{aligned}\sigma(\widetilde{\theta}^\vee) &= \sigma(\psi^\vee)^{-1}\sigma(\widetilde{\theta}^*)\sigma(\psi) = (\psi^\vee)^{-1}\alpha^\vee(\sigma)\widetilde{\theta}^*\alpha(\sigma)^{-1}\psi \\ &= (\psi^\vee)^{-1}\widetilde{\theta}^*\alpha(\sigma)\alpha(\sigma)^{-1}\psi = (\psi^\vee)^{-1}\widetilde{\theta}^*\psi = \widetilde{\theta}^\vee,\end{aligned}$$

for any $\sigma \in \Gamma_F$. For example, we could take

$$(G^\vee, \psi^\vee) = (G, \widetilde{\theta}^*\psi) \text{ and } \widetilde{\theta}^\vee = I.$$

Alternatively, we could define

$$(G^\vee, \psi^\vee) = (\widetilde{\theta}^*G, \psi) \text{ and } \widetilde{\theta}^\vee = \psi^{-1}\widetilde{\theta}^*\psi,$$

where $\widetilde{\theta}^*G$ is the group G with the twisted Galois action for which the isomorphism $\psi^{-1}\widetilde{\theta}^*\psi$ from G to $\widetilde{\theta}^*G$ is defined over F.

Suppose that the 1-cocycles $\alpha(\sigma)$ and $\alpha^\vee(\sigma)$ have distinct images in $H^1(F, G^*_{\mathrm{ad}})$. Then $\widetilde{\theta}^\vee$ is not an isomorphism of the inner twists (G, ψ) and $(G^\vee, \psi^\vee)$, as we see directly from the fact that the automorphism

$$\widetilde{\theta}^* = \psi^\vee\widetilde{\theta}^\vee\psi^{-1}$$

of G^* is not inner. Thus, $\widetilde{\theta}^\vee$ is an isomorphism between G and $G^\vee$ as groups over F, but *not as inner twists*. A second observation is that $\widetilde{\theta}$ does not exist as an *F-automorphism* of G. For there is no canonical way to identify G and $G^\vee$ over F other than by $\widetilde{\theta}^\vee$. These two related points are both important, obvious as they may be.

2. The lemma of course deals with the case that $\alpha(\sigma)$ has the same image in $H^1(F, G^*_{\mathrm{ad}})$ as the 1-cocycle $\alpha^\vee(\sigma) = \widetilde{\theta}^*\big(\alpha(\sigma)\big)$ (condition (iii)). We can then choose a corresponding "intertwining operator" $\gamma^* \in \mathrm{Int}(G^*)$ (condition (iv)), from which we define $\widetilde{\theta}$ as a product (9.1.7). Consequently $\widetilde{\theta}$ exists, as an isomorphism from the inner twist $(G^\vee, \psi^\vee) = (G, \widetilde{\theta}^*\psi)$ to (G, ψ) (condition (iii)), and as a nontrivial F-automorphism of G (condition (i)).

We return to the classification, as illustrated in the table of indices above. We are assuming for the rest of this section that G^* is of type D_n, so that $\widetilde{\theta}^*$ exists as an automorphism of G^* of order 2. It acts by permutation on the indices of G^*, which are by definition the family of markings on its Coxeter-Dynkin diagram in the list above, by interchanging the two right hand vertices in the diagram. We see by inspection that a $\widetilde{\theta}^*$-orbit of indices for G^* has order 1 unless G is of type ${}^1D_n$ and the integer d of its index equals 2, in which case the orbit is of order 2.

Suppose that F is local. The inner twist G is then completely determined by its index. Therefore $\widetilde{\theta}^*$ transfers to an F-automorphism $\widetilde{\theta}$ of G in this case unless G is of type ${}^1D_n$ and $d = 2$, or equivalently, unless the index of G contains a subdiagram (9.1.5) or (9.1.6). This is to be regarded as the exceptional case. It occurs only when G^* is split (of type D_n), and then only for one $\widetilde{\theta}^*$-orbit of indices if F is p-adic. If $F = \mathbb{R}$, it occurs for one orbit of indices if n is even, and none if n is odd. We see from these remarks that $\widetilde{\theta}^*$ transfers to an F-automorphism of G (as either a group or an inner twist) if and only if

$$\widetilde{\theta}^*\widehat{\zeta}_G = \widehat{\zeta}_G, \tag{9.1.9}$$

for the character $\widehat{\zeta}_G$ on the group $\widehat{Z}_{\mathrm{sc}}^{\Gamma} = \widehat{Z}_{\mathrm{sc}}$ attached to G (as an inner twist). The exceptional case thus occurs when $\widetilde{\theta}^*\widehat{\zeta}_G$ is distinct from $\widehat{\zeta}_G$. In this case, we can attach the complementary inner twist $(G^\vee, \psi^\vee)$ to the given inner twist (G, ψ), as in Remark 1 above. Then $\widetilde{\theta}^*\widehat{\zeta}_G = \widehat{\zeta}_{G^\vee}$, and we have the F isomorphism $\widetilde{\theta}^\vee$ from G to $G^\vee$ as groups, but not as inner twists.

Suppose that F is global. Then the inner twist G is determined by the indices of its localizations (G_v), subject to the condition (9.1.4) on the characters $\widehat{\zeta}_{G_v}$. In this case, $\widetilde{\theta}^*$ transfers to an F-automorphism of G (as a group as an inner twist) if and only if

$$\widetilde{\theta}_v^*\widehat{\zeta}_{G_v} = \widehat{\zeta}_{G_v}, \tag{9.1.10}$$

for all v. This follows, for example, from the application of any of the conditions of Lemma 9.1.1 to G and to each G_v. The exceptional case occurs when $\widetilde{\theta}_v^*\widehat{\zeta}_{G_v}$ is distinct from $\widehat{\zeta}_{G_v}$, for some v. The complementary inner twist $(G^\vee, \psi^\vee)$ over F is determined by the property that $\widetilde{\theta}_v^*\widehat{\zeta}_{G_v} = \widehat{\zeta}_{G_v^\vee}$ for all v. We again have an F-isomorphism $\widetilde{\theta}^\vee$ from G to $G^\vee$ as groups, but not as inner twists. But in the global case, there are other inner twists $(G^{\cdot}, \psi^{\cdot})$ over F that are related to (G, ψ). Suppose that $(G^{\cdot}, \psi^{\cdot})$ is such that its localizations $(G_v^{\cdot})$ satisfy (9.1.4), and such that one of the identities $\widehat{\zeta}_{G_v^{\cdot}} = \widehat{\zeta}_{G_v}$ or $\widehat{\zeta}_{G_v^{\cdot}} = \widetilde{\theta}^*\widehat{\zeta}_{G_v}$ holds for any v, but such that neither of the identities holds for all v. Then there is an F_v-isomorphism between the groups G_v and $G_v^{\cdot}$, but no F-isomorphism between the groups G and $G^{\cdot}$. In other words, the groups G and $G^{\cdot}$ over F are locally isomorphic, but not globally isomorphic. In this sense, they violate the Hasse principle. However, the only inner twist $(G^{\cdot}, \psi^{\cdot})$ over F that is locally isomorphic to (G, ψ) (at all places v) is (G, ψ) itself. In other words, the Hasse principle, in its usual interpretation for the Galois cohomology set $H^1(F, G^*)$, remains valid.

Although it is perhaps redundant, let us illustrate these ideas with a simple example of type ${}^1\mathbf{D}_2$. The underlying quasisplit group G^* is split, and equals

$$G^* = Sp(2) \times Sp(2)/\{\pm 1\},$$

where the group $\{\pm 1\}$ is embedded diagonally.

Suppose first that F is local, and that the integer d equals 2. This is the case of the general classification that corresponds to the $\widetilde{\theta}^*$-orbit of indices (9.1.5(ii)). The two indices parametrize the two groups

$$G_1 = Sp(2) \times Q^1/\{\pm 1\}$$

and

$$G_2 = Q^1 \times Sp(2)/\{\pm 1\},$$

where $Q^1(F)$ is the group of elements of norm 1 in a quaternion algebra over F. We assume implicitly that we have identified $Q^1(\overline{F})$ with $Sp(2, \overline{F})$. Then G_1 and G_2 are defined as inner twists by the isomorphisms

$$\psi_1 = \psi_2 = 1 \times 1.$$

It follows that

$$(G_2, \psi_2) = (G_1^\vee, \psi_1^\vee) = (\widetilde{\theta}^* G, \psi),$$

in the notation of Remark 1 above, while the F-isomorphism $\widetilde{\theta}^\vee$ in (9.1.8) is equal to the permutation $\widetilde{\theta}$ of the two factors. This is obviously the only way to identify G_1 and G_2, since the identity isomorphism is not defined over F.

Suppose next that F is global. If S is a finite, nonempty, even set of valuations of F, we write $Q^1(S)$ for the group represented by the elements of norm 1 in a quaternion algebra over F that is ramified at S and unramified outside of S. We fix S, and take

$$G_1 = Q^1(S_1) \times Q^1(S_1')$$

and

$$G_2 = Q^1(S_2) \times Q^1(S_2'),$$

for two partitions

$$S = S_1 \coprod S_1' = S_2 \coprod S_2'$$

of S into nonempty, even subsets. For any $v \in S$, the mapping

$$\widetilde{\theta}_v^\vee = \begin{cases} I, & \text{if } v \in (S_1 \cap S_2) \cup (S_1' \cap S_2'), \\ \widetilde{\theta}, & \text{if } v \in (S_1 \cap S_2') \cup (S_1' \cap S_2), \end{cases}$$

is then an F_v-isomorphism from G_1 to G_2. If $v \notin S$, we obviously have two F_v-isomorphisms from G_1 to G_2, either the identity I or the permutation mapping $\widetilde{\theta}$. Therefore G_1 and G_2 are locally isomorphic as groups over F. On the other hand they are not globally isomorphic unless $S_1 = S_2$ or $S_1 = S_2'$. They are of course neither locally or globally isomorphic as inner twists, except if $S_1 = S_2$. We thus have a clear illustration of the failure of the Hasse principle for inner forms of type ${}^1\mathbf{D}_n$, even though it remains valid for inner twists.

This completes our discussion of inner twists. For the transfer of automorphisms, our special topic for the section, we will keep in mind the summary from Remark 1 following Lemma 9.1.1, especially the two points at the end. We have concentrated on the first of these, the difference between isomorphisms of groups and isomorphisms of inner twists, and its relation to the Hasse principle. The other is the difference between F-isomorphisms and F-automorphisms. Its main implication is that there are inner forms G for which we cannot define the symmetric Hecke algebra $\widetilde{\mathcal{H}}(G)$. This means that the formulations of the theorems for quasisplit groups in §1.5 do not carry over as stated to the inner twist G.

9.2. Parameters and centralizers

In this section we consider a second point of departure for general orthogonal and symplectic groups. If $G = G^*$ is quasisplit, our description of the associated representations has been in terms of parameters $\psi \in \widetilde{\Psi}(G)$ and centralizers S_ψ. If G is not quasisplit, we need to modify these objects. In particular, we need to consider a finite extension of the abelian quotient $\mathcal{S}_\psi$ of S_ψ that is often nonabelian. We shall look at the structure of these groups, and describe their irreducible characters.

We fix G, which is to say the inner twist (G, ψ) of $G^* \in \widetilde{\mathcal{E}}_{\text{sim}}(N)$ over F that G represents. Following a convention from [**LS1**, (1.2)], we will generally use ψ to identify the L-groups of G and G^*. In case G is itself quasisplit, we might as well simply take $G^* = G$ and $\psi = 1$. In general, it makes sense to suppress ψ from the notation whenever we can, since we will again be using this symbol to represent a local or global parameter.

Having identified the L-groups of G and G^*, we write $\widetilde{\Psi}(G)$ for the subset of parameters $\psi \in \widetilde{\Psi}(G^*)$ that are locally G-relevant. For F local, this means that ψ has a representative with the usual property that if its image is contained in a parabolic subgroup ${}^LP$ of ${}^LG$, then ${}^LP$ corresponds to a parabolic subgroup P of G that is defined over F. For F global, it means that any localization ψ_v of ψ is G_v relevant. In other words, ψ belongs to the global subset $\widetilde{\Psi}(G)$ of $\widetilde{\Psi}(G^*)$ if and only if each localization ψ_v belongs to the local subset $\widetilde{\Psi}(G_v)$ of $\widetilde{\Psi}(G_v^*)$. We remark in passing that this definition is not entirely standard. The usual definition applies (at least for local F) to the finer set $\Psi(G)$, in which $\widetilde{\Psi}(G)$ represents the set of $\widetilde{\text{Out}}_N(G^*)$-orbits. Our definition of $\widetilde{\Psi}(G)$ here would be more satisfactory if we treated G itself as an $\widetilde{\text{Out}}_N(G^*)$-orbit of inner twists. We shall touch on this point again in §9.4, in discussing difficulties raised by the questions of §9.1.

We are interested in the irreducible representations of $G(F)$ if F is local, and the automorphic representations of G if F is global. Our first eight chapters were devoted to quasisplit G. We saw that in this case, the contribution of a parameter ψ to the representation theory of G is closely related to the (nonconnected) complex reductive group

$$\overline{S}_\psi = S_\psi / Z(\widehat{G})^\Gamma.$$

For general G, we shall attach a slightly larger group to any (locally G-relevant) parameter.

For any $\psi \in \widetilde{\Psi}(G)$, $\overline{S}_\psi$ is contained in the adjoint group $\widehat{G}_{\mathrm{ad}}$ of $\widehat{G}$. We shall write $S_{\psi,\mathrm{sc}}$ for the preimage of $\overline{S}_\psi$ in the simply connected cover $\widehat{G}_{\mathrm{sc}}$ of $\widehat{G}_{\mathrm{ad}}$. Then $S_{\psi,\mathrm{sc}}$ contains the group $\widehat{Z}_{\mathrm{sc}}$, and is a central extension

$$1 \longrightarrow \widehat{Z}_{\mathrm{sc}} \longrightarrow S_{\psi,\mathrm{sc}} \longrightarrow \overline{S}_\psi \longrightarrow 1 \tag{9.2.1}$$

of $\overline{S}_\psi$ by $\widehat{Z}_{\mathrm{sc}}$. For quasisplit G, the component group $\mathcal{S}_\psi = \overline{S}_\psi / \overline{S}_\psi^0$ governs the endoscopic packet of representations attached to ψ. For general G, the larger component group

$$\mathcal{S}_{\psi,\mathrm{sc}} = S_{\psi,\mathrm{sc}} / S_{\psi,\mathrm{sc}}^0$$

is what is needed for the associated endoscopic packet. It is a central extension

$$1 \longrightarrow \widehat{Z}_{\psi,\mathrm{sc}} \longrightarrow \mathcal{S}_{\psi,\mathrm{sc}} \longrightarrow \mathcal{S}_\psi \longrightarrow 1 \tag{9.2.2}$$

of $\mathcal{S}_\psi$ by the group

$$\widehat{Z}_{\psi,\mathrm{sc}} = \widehat{Z}_{\mathrm{sc}} / \widehat{Z}_{\mathrm{sc}} \cap \widetilde{S}_{\psi,\mathrm{sc}}^0.$$

In general, $\mathcal{S}_{\psi,\mathrm{sc}}$ is a nonabelian finite group. We shall write

$$\mathcal{Z}_{\psi,\mathrm{sc}} = Z(\mathcal{S}_{\psi,\mathrm{sc}})$$

for its centre, and

$$\mathcal{Z}_\psi = \mathcal{Z}_{\psi,\mathrm{sc}} / \widehat{Z}_{\psi,\mathrm{sc}}$$

for the image of $\mathcal{Z}_{\psi,\mathrm{sc}}$ in the abelian quotient $\mathcal{S}_\psi$ of $\mathcal{S}_{\psi,\mathrm{sc}}$. The fact that the irreducible characters of $\mathcal{S}_{\psi,\mathrm{sc}}$ need not be one-dimensional obviously adds interest to the representation theory of G. The group $\mathcal{S}_{\psi,\mathrm{sc}}$ is of Heisenberg type, so its representation theory is still quite simple. In this section, we shall describe the group explicitly in terms of ψ, and review how to classify its irreducible characters. The subgroup $\mathcal{Z}_\psi$ plays an essential role in this classification.

It is really the local case that is of interest here, since we will see that the global theory descends to the original group $\mathcal{S}_\psi$. We may as well therefore assume that F is local. The groups $S_{\psi,\mathrm{sc}}$ and $\mathcal{S}_{\psi,\mathrm{sc}}$ were introduced for p-adic F and parameters ϕ in the subset

$$\Phi(G) = \Psi(G) \cap \Phi(G^*)$$

in [**A19**] (where $\mathcal{S}_{\phi,\mathrm{sc}}$ was denoted by $\widetilde{\mathcal{S}}_\psi$). The purpose of [**A19**] was to state a conjectural parametrization of the local packet attached to ϕ in terms of representations of the associated nonabelian group $\mathcal{S}_{\psi,\mathrm{sc}}$. The stated parametrization [**A19**, §3] depends on a noncanonical choice of an extension of the character $\widehat{\zeta}_G$ on $\widehat{Z}_{\rm sc}^\Gamma$ to the larger group $\widehat{Z}_{\rm sc}$. With this in mind, we fix a section

$$\Pi(\widehat{Z}_{\rm sc}^\Gamma) \hookrightarrow \Pi(\widehat{Z}_{\rm sc}) \tag{9.2.3}$$

for the restriction mapping from $\Pi(\widehat{Z}_{\rm sc})$ to $\Pi(\widehat{Z}_{\rm sc}^\Gamma)$ that depends only on the quasisplit group $G^* \in \widetilde{\mathcal{E}}_{\rm sim}(N)$. (We are now writing $\Pi(\widehat{Z}_{\rm sc}^\Gamma)$ and $\Pi(\widehat{Z}_{\rm sc})$ for the groups of characters on the abelian groups $\widehat{Z}_{\rm sc}^\Gamma$ and $\widehat{Z}_{\rm sc}$.) For our given inner twist G, the section then gives an extension to $\widehat{Z}_{\rm sc}$ of the associated character $\widehat{\zeta}_G \in \Pi(\widehat{Z}_{\rm sc}^\Gamma)$, which we continue to denote by $\widehat{\zeta}_G$.

The group G^* belongs to one of three general families. If it is of type $\mathbf{B}_n$, G^* is isomorphic to the split group $SO(2n+1)$, and its dual group is isomorphic to $Sp(2n,\mathbb{C})$. In this case, $\widehat{G}^* = \widehat{G}^*_{\rm sc}$ and $S_{\psi,{\rm sc}} = S_\psi$, and we have nothing to do. We suppose therefore that G^* is of type $\mathbf{C}_n$ or $\mathbf{D}_n$. Then G^* is isomorphic to either a split group $Sp(2n)$ or a quasisplit group $SO(2n)$, and its dual group is isomorphic to either $SO(2n+1,\mathbb{C})$ or $SO(2n,\mathbb{C})$. We have therefore to deal with parameters $\psi \in \widetilde{\Psi}(G)$, for an inner twist G of a group $G^* \in \widetilde{\mathcal{E}}_{\rm sim}(N)$ with $\widehat{G} = \widehat{G}^* = SO(N,\mathbb{C})$. The essential case is that of a parameter ψ in the subset

$$\widetilde{\Psi}_2(G) = \widetilde{\Psi}(G) \cap \widetilde{\Psi}_2(G^*)$$

of square integrable parameters. The group $S_{\psi,{\rm sc}}$ is then finite, and the short exact sequence (9.2.2) reduces to (9.2.1). For any such ψ, we would like to describe the structure of the finite group $S_{\psi,{\rm sc}} = \mathcal{S}_{\psi,{\rm sc}}$. This reduces to a lemma on complex spin groups, which was suggested to me by Steve Kudla.

It will be convenient to modify the earlier convention from §1.2 slightly by taking the traditional embedding of $O(N,\mathbb{C})$ into $GL(N,\mathbb{C})$. In other words, we treat $O(N,\mathbb{C})$ as the group of complex $(N\times N)$-matrices that preserve the dot product on $\mathbb{C}^n$. For any partition

$$\Pi = (N_1,\dots,N_r), \qquad N = N_1 + \cdots + N_r,$$

of N, the product

$$O(\Pi) = O(N_1,\mathbb{C}) \times \cdots \times O(N_r,\mathbb{C})$$

then represents the corresponding group of block diagonal matrices in $O(N,\mathbb{C})$.

Set

$$A(N) = \underbrace{\big(O(1,\mathbb{C}) \times \cdots \times O(1,\mathbb{C})\big)}_{N} \cap SO(N,\mathbb{C}),$$

and let $B(N)$ be the preimage of $A(N)$ in the extension $\mathrm{Spin}(N,\mathbb{C})$ of $SO(N,\mathbb{C})$. Then $B(N)$ is an extension

$$1 \longrightarrow \{\pm1\} \longrightarrow B(N) \longrightarrow A(N) \longrightarrow 1 \tag{9.2.4}$$

of $A(N)$ by the multiplicative group $\{\pm1\}$. We can treat $A(N)$ as the group of elements $a(S)$, parametrized by *even* subsets S of $\{1,\dots,N\}$, with multiplication

$$a(S)a(T) = a(S\triangle T)$$

defined by the divided difference

$$S\triangle T = (S \cup T) - (S \cap T).$$

Indeed, we identify $a(S)$ with an element in $SO(N, \mathbb{C})$ by setting

$$a(S)e_j = \begin{cases} -e_j, & \text{if } j \in S, \\ e_j, & \text{if } j \notin S, \end{cases}$$

for the standard orthonormal basis $\{e_1, \ldots, e_N\}$ of $\mathbb{C}^n$. Given $a(S)$, we introduce a formal product

$$b(S) = e_{i_1} \cdots e_{i_s}, \qquad S = \{i_1 < i_2 < \cdots < i_s\}.$$

We then define a group law on the set

$$\{b(S), -b(S) : \; S \subset \{1, \ldots, N\} \text{ even}\} \tag{9.2.5}$$

by writing

$$b(S)b(T) = \varepsilon(S,T)b(S\triangle T), \tag{9.2.6}$$

for the 2-cocycle

$$\varepsilon(S,T) = \prod_{i \in S} (-1)^{|T_i|}, \qquad T_i = \{j \in T : \; j < i\}.$$

It is of course understood that the element $(-1) = -b_{\varnothing}$ lies in the centre of the group (9.2.5). It is then clear that (9.2.5) represents a central extension of the group $A(N)$ by $\{\pm 1\}$.

LEMMA 9.2.1. *There is a canonical isomorphism from the group* (9.2.5), *as an extension of* $A(N)$ *by* $\{\pm 1\}$, *onto the group* $B(N)$ *in* (9.2.4). *Moreover, the elements in* (9.2.5) *satisfy the supplementary relations*

$$\begin{cases} b(S)^2 = (-1)^{\frac{|S|(|S|-1)}{2}}, \\ b(S)b(T) = (-1)^{|S \cap T|} b(T)b(S), \end{cases} \tag{9.2.7}$$

for even subsets S *and* T *of* $\{1, \ldots, N\}$.

PROOF. The basis vectors e_i embed into the Clifford algebra $C(Q)$ attached to the standard quadratic form $Q(v) = v \cdot v$ on the complex vector space $V = \mathbb{C}^N$. We can then regard the products $b(S)$ as elements in the even part $C^0(Q)$ of $C(Q)$. The lemma is a consequence of the construction of $C(Q)$, and its relation to the complex spin group $\mathrm{Spin}(N, \mathbb{C})$. Let us review these notions, for the sake of the author* if not the reader.

The Clifford algebra $C(Q)$ is the associative $\mathbb{C}$-algebra generated by the orthonormal basis vectors $\{e_i\}$, with relations

$$\begin{cases} e_i^2 = (e_i \cdot e_i) = 1, \\ e_i e_j + e_j e_i = 2(e_i \cdot e_j) = 0, \; \text{if } i \neq j. \end{cases} \tag{9.2.8}$$

---

*I will follow the elementary discussion of an article *Clifford algebra* in Wikipedia, as it appeared on June 10, 2011.

It has a $(\mathbb{Z}/2\mathbb{Z})$-grading, defined by the algebra involution generated by the isomorphism $v \to (-v)$ of V, and hence a decomposition

$$C(Q) = C^0(Q) \oplus C^1(Q)$$

into even and odd parts. This is a $\mathbb{C}$-algebra grading, which is not to be confused with the natural $\mathbb{Z}$-grading of $C(Q)$ as a vector space. The Clifford algebra also has a transpose anti-involution generated by the mapping

$$(v_1 \cdots v_k) \longrightarrow (v_1 \cdots v_k)^t = v_k \cdots v_1, \qquad v_i \in V.$$

This is used to define the Clifford scalar product

$$Q(x) = (x^t x)_0, \qquad x \in C(Q),$$

where $(\cdot)_0$ denotes the projection of $C(Q)$ onto $\mathbb{C}$ given by 0-components of its $\mathbb{Z}$-grading. The Clifford scalar product reduces to the original quadratic form on the subspace V of $C(Q)$.

The spin group $\mathrm{Spin}(N, \mathbb{C})$ lies in a chain of embedded subgroups

$$\mathrm{Spin}(N, \mathbb{C}) \subset G\,\mathrm{Spin}(N, \mathbb{C}) \subset C^0(Q)^* \subset C(Q)^*,$$

where as usual, $C^0(Q)^*$ and $C(Q)^*$ denote the groups of units in the underlying algebras. The *general spin group* $G\,\mathrm{Spin}(N, \mathbb{C})$ is identified with the subgroup of elements $x \in C^0(Q)^*$ such that the conjugation mapping

$$y \longrightarrow xyx^{-1}, \qquad y \in C(Q),$$

stabilizes the subspace V of $C(Q)$. The resulting representation of $G\,\mathrm{Spin}(N, \mathbb{C})$ on V maps this group onto the subgroup $SO(N, \mathbb{C})$ of $GL(N, \mathbb{C})$, and defines a short exact sequence

$$1 \longrightarrow \mathbb{C}^* \longrightarrow G\,\mathrm{Spin}(N, \mathbb{C}) \longrightarrow SO(N, \mathbb{C}) \longrightarrow 1.$$

The spinor norm is the restriction

$$x \longrightarrow x^t x = (x^t x)_0, \qquad x \in G\,\mathrm{Spin}(N, \mathbb{C}),$$

of the Clifford scalar product. It is a homomorphism from $G\,\mathrm{Spin}(N, \mathbb{C})$ onto $\mathbb{C}^*$. Its kernel is the subgroup $\mathrm{Spin}(N, \mathbb{C})$, while the kernel of its restriction to the central subgroup $\mathbb{C}^*$ of $G\,\mathrm{Spin}(N, \mathbb{C})$ is the subgroup $\{\pm 1\}$ of $\mathbb{C}^*$. We thus have a canonical embedding of the extension

$$1 \longrightarrow \{\pm 1\} \longrightarrow \mathrm{Spin}(N, \mathbb{C}) \longrightarrow SO(N, \mathbb{C}) \longrightarrow 1$$

into the group of units $C^0(Q)^*$ in the even Clifford algebra.

The lemma is now a direct consequence of the embedding of $\mathrm{Spin}(N, \mathbb{C})$ into $C^0(Q)^*$. The multiplication law (9.2.6) for elements $b(S)$ in (9.2.5) reduces to its analogue (9.2.8) for general elements in $C(Q)$. The supplementary relations (9.2.7) follow in turn from (9.2.6), as the reader can easily check. □

We return to our inner twist G of the quasisplit group $G^* \in \widetilde{\mathcal{E}}_{\text{sim}}(N)$, with

$$\widehat{G} = \widehat{G}^* = SO(N, \mathbb{C}).$$

Suppose that

$$\psi = \psi_1 \oplus \cdots \oplus \psi_r, \qquad \psi_i \in \widetilde{\Psi}_{\text{sim}}(N_i),$$

is a parameter in $\widetilde{\Psi}_2(G)$. Given ψ, we have a decomposition

$$I = I_e \coprod I_o = I_{\psi,e} \coprod I_{\psi,o}$$

of its set of indices

$$I = I_\psi = \{1, \dots, r\}$$

into two disjoint subsets, consisting of those indices k whose associated degrees N_k are even or odd. Let A_ψ be the group of elements a_σ, parametrized by subsets $\sigma \subset I$ of indices such that the intersection

$$\sigma_o = \sigma \cap I_o$$

is *even*, with multiplication

$$a_\sigma a_\tau = a_{\sigma \triangle \tau}.$$

We identify A_ψ with a subgroup of $A(N)$ under the injective homomorphism $a_\sigma \to a(S_\sigma)$, where S_σ is the (disjoint) union over k in σ of the subsets

$$\{N_1 + \cdots + N_{k-1} + 1, \dots, N_1 + \cdots + N_k\} \subset \{1, \dots, N\}$$

of the original indexing set. In particular, A_ψ is a subgroup of $SO(N, \mathbb{C})$. Its preimage in $\text{Spin}(N, \mathbb{C})$ is the subgroup

$$B_\psi = \{\pm b_\sigma : \ \sigma \subset I, \ \sigma_o \text{ even}\}$$

of $B(N)$, where $b_\sigma = b(S_\sigma)$. We thus obtain a subextension

$$1 \longrightarrow \{\pm 1\} \longrightarrow B_\psi \longrightarrow A_\psi \longrightarrow 1$$

of (9.2.4).

We will need to know when B_ψ is abelian. More generally, we consider the quotient

$$B_\psi / Z_\psi = \overline{B}_\psi / \overline{Z}_\psi,$$

where $Z_\psi = Z(B_\psi)$ is the centre of B_ψ, and $\overline{Z}_\psi = Z_\psi / \widehat{Z}_{\text{sc}}$ is its image in $\overline{B}_\psi = B_\psi / \widehat{Z}_{\text{sc}}$. Observe that

$$|B_\psi| = 2|A_\psi| = 2^{r+\varepsilon_\psi},$$

where

$$\varepsilon_\psi = \begin{cases} 1, & \text{if } |I_o| = 0, \\ 0, & \text{if } |I_o| \neq 0. \end{cases}$$

If N is odd, the set I_o has an odd number of elements. In particular, it is nonempty, and $|B_\psi| = 2^r$. In this case, the centre $\widehat{Z}_{\text{sc}}$ of $\text{Spin}(N, \mathbb{C})$ equals $\{\pm 1\}$. If N is even, the set I_o has an even number of elements. It can be empty, so that $|B_\psi|$ can equal either 2^{r+1} or 2^r. In this case, we have elements $a_I \in A_\psi$ and $b_I \in B_\psi$ attached to the maximal set $\sigma = I$, and the center of $\mathrm{Spin}(N, \mathbb{C})$ equals

$$\widehat{Z}_{\mathrm{sc}} = \{\pm 1, \pm b_I\}.$$

It follows in general that

$$|\overline{B}_\psi| = |B_\psi|/|\widehat{Z}_{\mathrm{sc}}| = 2^{r+\varepsilon_\psi-\delta_\psi}, \tag{9.2.9}$$

where

$$\delta_\psi = \delta_N = \begin{cases} 1, & \text{if } N \text{ is odd,} \\ 2, & \text{if } N \text{ is even.} \end{cases}$$

As for the center of B_ψ, we claim that

$$Z_\psi = \{b_\sigma : \sigma \subset I_e\} \cdot \widehat{Z}_{\mathrm{sc}}. \tag{9.2.10}$$

To check this, we appeal to the specialization

$$b_\tau^{-1} b_\sigma b_\tau = (-1)^{|\sigma\cap\tau|} b_\sigma = (-1)^{|\sigma_o\cap\tau_o|} b_\sigma \tag{9.2.11}$$

of the second relation in (9.2.7), which holds for any subsets σ and τ of I with σ_o and τ_o even. It follows that Z_ψ contains the right hand side of (9.2.10). Conversely, suppose that b_σ lies in Z_ψ, but that σ is not contained in I_e. It is then easy to verify from (9.2.11) and the various definitions that $\sigma_o = I_o$, and that $|I_o| = 2$. This implies that N is even, and that b_σ is the product of an element $(\pm b_I) \in \widehat{Z}_{\mathrm{sc}}$ with an element $b_\tau \in B_\psi$ such that τ is contained in I_e. The claim follows.

It follows from (9.2.10) that

$$|Z_\psi| = 2^{|I_e|+\delta_\psi-\varepsilon_\psi},$$

and that

$$|\overline{Z}_\psi| = |Z_\psi|/|\widehat{Z}_{\mathrm{sc}}| = 2^{|I_e|-\varepsilon_\psi} \tag{9.2.12}$$

We have now only to combine (9.2.12) with (9.2.9). This gives a formula

$$|B_\psi/Z_\psi| = 2^{|I_o|+2\varepsilon_\psi-\delta_\psi} \tag{9.2.13}$$

for the order of the quotient in question. In particular, we see from the definitions of ε_ψ and δ_ψ that B_ψ is abelian if and only if $|I_o| \leqslant 2$. In the case that B_ψ is not abelian, (9.2.11) implies that its derived group satisfies

$$B_{\psi,\mathrm{der}} = \{\pm 1\} \tag{9.2.14}$$

Suppose for example that $F = \mathbb{R}$, and that $\phi = \psi$ lies in the subset

$$\widetilde{\Phi}_2(G) = \widetilde{\Psi}_2(G) \cap \widetilde{\Phi}(G^*)$$

of $\widetilde{\Psi}_2(G)$. The simple components ϕ_k of ϕ must be distinct and self-dual, and their degrees N_k are either 2 or 1. There can be at most two ϕ_k with $N_k = 1$, the trivial character $1_{\mathbb{R}}$ on $\mathbb{R}^*$ and the sign character $\varepsilon_{\mathbb{R}}$. It follows that $|I_o| \leqslant 2$ for any such ϕ, and hence that the group B_ϕ is abelian. This property, which is valid for any G, is a basic tenet of the work of Shelstad

[**S6**]. We note that if $F = \mathbb{C}$, the only simple, self-dual, generic parameter ϕ_k is the trivial character $1_{\mathbb{C}}$. In this case, we have

$$|I_o| = |I| = 1 = N,$$

and the group $B_\phi = \{1\}$ is trivial.

We shall use the relations we have established among the various groups

$$\{\pm 1\} \subset \widehat{Z}_{\mathrm{sc}} \subset Z_\psi \subset B_\psi$$

to characterize the set $\Pi(B_\psi)$ of equivalence classes of irreducible characters of B_ψ. It suffices to treat the subset $\Pi(B_\psi, \widehat{\zeta}_\psi)$ of characters in $\Pi(B_\psi)$ whose central character on $\widehat{Z}_{\mathrm{sc}}$ equals a fixed character $\widehat{\zeta}_\psi$. We can assume that B_ψ is nonabelian, since we would otherwise be dealing with a set of one-dimensional characters. We can also assume that the value of $\widehat{\zeta}_\psi$ on the element (-1) in $\widehat{Z}_{\mathrm{sc}}$ is (-1), since the quotient of B_ψ by $\{\pm 1\}$ is the abelian group A_ψ.

LEMMA 9.2.2. *Assume that the group B_ψ attached to $\psi \in \widetilde{\Psi}_2(G)$ is nonabelian, and that the given character $\widehat{\zeta}_\psi$ on $\widehat{Z}_{\mathrm{sc}}$ satisfies $\widehat{\zeta}_G(-1) = -1$. Then there is a bijective correspondence*

$$\xi \longrightarrow \langle b, \xi \rangle, \qquad b \in B_\psi,\ \xi \in \Pi(Z_\psi, \widehat{\zeta}_\psi),$$

from the set of (abelian) characters $\Pi(Z_\psi, \widehat{\zeta}_\psi)$ on Z_ψ whose restriction to $\widehat{Z}_{\mathrm{sc}}$ equals $\widehat{\zeta}_G$, onto the corresponding nonabelian character set $\Pi(B_\psi, \widehat{\zeta}_\psi)$ on B_ψ, such that

$$\langle b, \xi \rangle = \begin{cases} d_\psi\, \xi(b), & \text{if } b \in Z_\psi, \\ 0, & \text{if } b \notin Z_\psi, \end{cases} \tag{9.2.15}$$

where

$$d_\psi = \begin{cases} 2^{(|I_o|-1)/2}, & \text{if } N \text{ is odd}, \\ 2^{(|I_o|-2)/2}, & \text{if } N \text{ is even}, \end{cases}$$

is the degree of any character in $\Pi(B_\psi, \widehat{\zeta}_\psi)$.

PROOF. As a finite nonabelian group, B_ψ is particularly simple. It is a 2-stage nilpotent group of Heisenberg type, whose representation theory is elementary and well known. We shall just quote what we need.

The quotient B_ψ / Z_ψ is a finite abelian 2-group. It has a well defined pairing

$$[x, y] = \widetilde{x}\widetilde{y}\widetilde{x}^{-1}\widetilde{y}^{-1}, \qquad x, y \in B_\psi / Z_\psi,$$

for preimages $\widetilde{x}$ and $\widetilde{y}$ in B_ψ of x and y, which takes values in $\{\pm 1\}$ by (9.2.14), and is nondegenerate by (9.2.11). With this structure, we can regard B_ψ / Z_ψ as a 2-dimensional symplectic vector space over the field of

2-elements, in which addition (in both the vector space and the field) is written multiplicatively. As a vector space, B_ψ/Z_ψ has dimension equal to

$$|I_o| + 2\varepsilon_\psi - \delta_\psi = |I_o| - \delta_\psi = \begin{cases} |I_o| - 1, & \text{if } N \text{ is odd,} \\ |I_o| - 2, & \text{if } N \text{ is even,} \end{cases}$$

by (9.2.13), and our condition that B_ψ is nonabelian. Since $|I_o|$ has the same parity as N, this integer is even, which of course is also a consequence of the existence of the symplectic form $[\cdot,\cdot]$. Let U/Z_ψ be a maximal isotropic subspace of B_ψ/Z_ψ. This is a subspace of dimension one-half that of B_ψ/Z_ψ. Its preimage in B_ψ is a maximal abelian subgroup U, which embeds into the chain

$$\{\pm 1\} \subset Z_\psi \subset U \subset B_\psi.$$

The putative degree, defined in the statement of the theorem as a power of 2, then satisfies

$$d_\psi = 2^{\frac{1}{2}\dim(B_\psi/Z_\psi)} = 2^{\dim(U/Z_\psi)} = |B_\psi/U|.$$

Suppose that ξ belongs to $\Pi(Z_\psi, \widehat{\zeta}_\psi)$. We first extend ξ as an abelian character to the subgroup U, and then form the induced representation

$$\rho_\xi = \mathrm{Ind}_U^{B_\psi}(\xi)$$

of B_ψ. It is an elementary consequence of the theory of induced representations that the representation ρ_ξ is irreducible, that its character vanishes on the complement of Z_ψ, and that

$$\mathrm{tr}\big(\rho_\xi(z)\big) = \langle z, \xi\rangle = |B_\psi/U|\,\xi(z) = d_\psi\,\xi(z),$$

for any $z \in Z_\psi$. In particular, the equivalence class of ρ_ξ is independent of the choice of U and the extension of ξ. The lemma follows. □

We have been working with a square integrable parameter $\psi \in \widetilde{\Psi}_2(G)$. It is clear that the definitions of the groups in the lemma match those of the centralizer groups defined at the beginning of the section. We have $S_\psi = A_\psi$, $S_{\psi,\mathrm{sc}} = B_\psi$, $\overline{S}_\psi = \mathcal{S}_\psi = B_\psi/\widehat{Z}_{\mathrm{sc}}$, $\mathcal{Z}_{\psi,\mathrm{sc}} = Z_\psi$ and $\mathcal{Z}_\psi = \overline{Z}_\psi$. Lemma 9.2.2 therefore gives us a description of the set $\Pi(S_{\psi,\mathrm{sc}}, \widehat{\zeta}_G)$ of irreducible characters on $S_{\psi,\mathrm{sc}}$ with central character $\widehat{\zeta}_G = \widehat{\zeta}_\psi$ on $\widehat{Z}_{\mathrm{sc}}$.

More generally, suppose that ψ is an arbitrary parameter in $\widetilde{\Psi}(G)$. We recall that the original centralizer group $\mathcal{S}_\psi$ comes with a short exact sequence

$$1 \longrightarrow \mathcal{S}_\psi^1 \longrightarrow \mathcal{S}_\psi \longrightarrow R_\psi \longrightarrow 1.$$

What is the analogue of this for the extension $\mathcal{S}_{\psi,\mathrm{sc}}$ of $\mathcal{S}_\psi$ in (9.2.2)?

Since ψ is relevant to G, we can choose a Levi subgroup M of G with respect to which ψ is square integrable. In other words, we can choose a pair (M, ψ_M) such that ψ_M lies in the set $\widetilde{\Psi}_2(M,\psi)$. The group $\mathcal{S}_{\psi_M}$ is then canonically isomorphic to the subgroup $\mathcal{S}_\psi^1$ of $\mathcal{S}_\psi$, a property that we have used repeatedly since it first arose in §2.4 (and in the global case of §4.2), and which follows from the well known fact that

$$Z(\widehat{M})^{\Gamma} = Z(\widehat{G})^{\Gamma}\big(Z(\widehat{M})^{\Gamma}\big)^{0}.$$

(See [**A12**, Lemma 1.1], for example.) The last formula holds also for the preimage $\widehat{M}_{\mathrm{sc}}$ of $\widehat{M}$ in $\widehat{G}_{\mathrm{sc}}$, where we interpret it as an identity

$$\pi_0\big(Z(\widehat{M}_{\mathrm{sc}})^{\Gamma}\big) = \widehat{Z}_{\mathrm{sc}}^{\Gamma}/\widehat{Z}_{\mathrm{sc}}^{\Gamma} \cap \big(Z(\widehat{M}_{\mathrm{sc}})^{\Gamma}\big)^{0}$$

between two finite groups. On the other hand, the character $\widehat{\zeta}_G$ on $\widehat{Z}_{\mathrm{sc}}^{\Gamma}$ attached to G is trivial on the intersection of $\widehat{Z}_{\mathrm{sc}}^{\Gamma}$ with the identity component of $Z(\widehat{M}_{\mathrm{sc}})^{\Gamma}$. (See for example the proof of [**A12**, Lemma 2.1], specifically the remarks at the end of the second paragraph on p. 219 of [**A12**].) We have agreed to write $\widehat{\zeta}_G$ also for its extension (9.2.3) to a character on $\widehat{Z}_{\mathrm{sc}}$, a larger group which is still contained in $Z(\widehat{M}_{\mathrm{sc}})$. Since

$$\widehat{Z}_{\mathrm{sc}} \cap \big(Z(\widehat{M}_{\mathrm{sc}})^{\Gamma}\big)^{0} = \widehat{Z}_{\mathrm{sc}}^{\Gamma} \cap \big(Z(\widehat{M}_{\mathrm{sc}})^{\Gamma}\big)^{0},$$

we can identify $\widehat{\zeta}_G$ with a character on the quotient $\widehat{Z}_{\psi,\mathrm{sc}}$ of $\widehat{Z}_{\mathrm{sc}}$ by this intersection.

According to (9.2.2), $\widehat{Z}_{\psi,\mathrm{sc}}$ is the kernel of the projection of $\mathcal{S}_{\psi,\mathrm{sc}}$ onto $\mathcal{S}_{\psi}$. Since elements in $\widehat{Z}_{\mathrm{sc}}$ commute with the connected group

$$\big(Z(\widehat{M}_{\mathrm{sc}})^{\Gamma}\big)^{0} = S_{\psi,\mathrm{sc}}^{0},$$

the quotient $\widehat{Z}_{\psi,\mathrm{sc}}$ embeds into $\mathcal{S}_{\psi,\mathrm{sc}}^{1}$. It is in fact the kernel of the projection of $\mathcal{S}_{\psi,\mathrm{sc}}^{1}$ onto $\mathcal{S}_{\psi}^{1}$. As a consequence of these various remarks, we obtain a commutative diagram

(9.2.16)

$$\begin{array}{ccccccccc}
 & & 1 & & 1 & & & & \\
 & & \downarrow & & \downarrow & & & & \\
 & & \widehat{Z}_{\psi,\mathrm{sc}} & = & \widehat{Z}_{\psi,\mathrm{sc}} & & & & \\
 & & \downarrow & & \downarrow & & & & \\
1 & \longrightarrow & \mathcal{S}_{\psi,\mathrm{sc}}^{1} & \longrightarrow & \mathcal{S}_{\psi,\mathrm{sc}} & \longrightarrow & R_{\psi} & \longrightarrow & 1 \\
 & & \downarrow & & \downarrow & & \| & & \\
1 & \longrightarrow & \mathcal{S}_{\psi}^{1} & \longrightarrow & \mathcal{S}_{\psi} & \longrightarrow & R_{\psi} & \longrightarrow & 1 \\
 & & \downarrow & & \downarrow & & & & \\
 & & 1 & & 1 & & & &
\end{array}$$

of short exact sequences. The upper horizontal sequence is the analogue for $\mathcal{S}_{\psi,\mathrm{sc}}$ of the lower horizontal sequence for $\mathcal{S}_{\psi}$.

We can regard M as an inner twist of a Levi subgroup M^* of G^* that is a product (2.3.4) of general linear groups and a group G^*_- of the same type as G^*. From the list of indices in the last section, we observe that M is a product of general linear groups, all over either F or a fixed F-quaternion algebra, with an inner form G_- of the group G^*_-. The group $\mathcal{S}^1_{\psi,\mathrm{sc}}$ in (9.2.16) satisfies

$$\mathcal{S}^1_{\psi,\mathrm{sc}} = \mathcal{S}_{\psi_M,\mathrm{sc}} = \mathcal{S}_{\psi_-},$$

where $\psi_- \in \widetilde{\Psi}_2(G_-)$ is the subparameter of ψ attached to G_-. Its irreducible characters are given by Lemma 9.2.2. The irreducible characters of the larger group $\mathcal{S}_{\psi,\mathrm{sc}}$ are then defined in terms of those of $\mathcal{S}^1_{\psi,\mathrm{sc}}$, and the characters on the abelian R-group R_ψ, by a simple process analogous to that in the proof of Proposition 2.4.3. We will establish the analogue for G of Proposition 2.4.3 in [**A28**], as part of the local intertwining relation for G.

9.3. On the normalization of transfer factors

In the earlier chapters, we classified the representations of quasisplit orthogonal or symplectic groups G^* in terms of those of general linear groups. The purpose of this chapter is to describe the representations of arbitrary orthogonal or symplectic groups G in terms of those of groups G^*. In the first section, we distinguished between inner forms and inner twists of a given $G^* \in \widetilde{\mathcal{E}}_{\mathrm{sim}}(N)$. The reason for working with the latter is that the parametrization of representations of an inner form G really depends on the extra structure of a quasisplit inner twist ψ.

The representation theory of a given group depends on something else as well. Its quasisplit inner twist ψ must be supplemented by a system of transfer factors for its endoscopic groups G' [**LS1**, (3.7)]. (We are assuming at this point, and until further notice, that the underlying field F is local.) If G is not quasisplit, its transfer factors cannot be normalized by Whittaker data as in [**KS**, §5.4] (or by F-splittings as in [**LS1**, (3.7)]). According to the definitions in [**LS1**, (3.7)], a transfer factor Δ for (G, G') is determined only up to a multiplicative constant $u \in U(1)$ of absolute value 1.

This is our topic for the third section. We shall first discuss the abstract problem of normalizing general transfer factors. We will then describe the natural normalizations established in recent work of Kaletha [**Kal1**], [**Kal2**], [**Kal4**].

We fix G, which according to our convention represents an inner twist (G, ψ) of a (connected) quasisplit orthogonal or symplectic group $G^* \in \widetilde{\mathcal{E}}_{\mathrm{sim}}(N)$ over F. We still have the notion of an *isomorphism* α' of endoscopic data G'_1 and G'_2 for G (which in this case represent triplets (G'_1, s'_1, ξ'_1) and (G'_2, s'_2, ξ'_2)). As earlier, we follow the definition of [**KS**]. Then α' is identified with an element $g' = g(\alpha')$ in $\widehat{G}$ such that $\mathrm{Int}(g')^{-1}$ restricts to an L-isomorphism between the L-subgroups $\xi'_2({}^L G'_2)$ and $\xi'_1({}^L G'_1)$ of ${}^L G$, and such that $g'(s'_1)(g')^{-1}$ belongs to the set

$$Z(\widehat{G})^{\Gamma} \cdot \mathrm{Cent}\big(\xi'_2({}^L G'_2), \widehat{G}\big)^0 \cdot s'_2.$$

The symbol α' serves also to represent a dual F-isomorphism between the quasisplit groups G'_1 and G'_2. The logic might be a little clearer here if we define G'_1 and G'_2 to be *strongly isomorphic* if there is an isomorphism (of endoscopic data) G'_1 and G'_2 with $g(\alpha') = 1$. A general α' can be treated as an isomorphism of strong isomorphism classes of endoscopic data. It is then uniquely determined by the element $g' \in \widehat{G}$, and in fact by the image $\mathrm{Int}(g')$ of g' in $\widehat{G}_{\mathrm{ad}}$.

We are in the practice of writing $\mathcal{E}(G)$ for the set of isomorphism classes of endoscopic data for G. For any $G' \in \mathcal{E}(G)$, let us also write $E(G, G')$ for the set of strong isomorphism classes in the general isomorphism class G'. With this understanding, $E(G, G')$ is a complex algebraic variety on which the group $\widehat{G}$ acts transitively. The stabilizer in $\widehat{G}$ of any $G'_1 \in E(G, G')$ is a reductive subgroup of $\widehat{G}$, whose group of connected components is the finite group we have denoted by $\mathrm{Out}_G(G'_1)$. We can write

$$E(G) = \coprod_{G' \in \mathcal{E}(G)} E(G, G')$$

for the variety of all such classes of data for G. When there is no danger of confusion, we will continue to let G' stand interchangeably for an endoscopic datum (which we will now usually identify with its strong isomorphism class) or a corresponding isomorphism class. For example, the finite group $\mathrm{Out}_G(G')$ makes sense whether we treat G' as an element in $E(G)$ or a representative in $E(G)$ of a class in $\mathcal{E}(G)$.

We do, however, need to exercise some care. As we have noted, the transfer factors defined in [**LS1**] and [**KS**] do not come with a natural normalization if G is not quasisplit. Let us write

$$T(G) = \coprod_{G' \in \mathcal{E}(G)} T(G, G')$$

for the set of transfer factors for G. (Again, we apologize for an overlap with earlier notation, namely the set $T(G)$ introduced in §3.5 in which T was meant to serve as an upper case τ.) An element in $T(G)$ is then a pair (G', Δ), where G' belongs to $E(G)$, and Δ is a transfer factor for (G, G'). For a given G', Δ is determined only up to a scalar multiple in $U(1)$. In other words, the natural projection

$$T(G) \longrightarrow E(G)$$

is a principal $U(1)$-bundle. The point is that this bundle does not generally have a continuous section, let alone a canonical section. To describe the obstruction, we recall some remarks from [**A19**, §3].

If α' is an isomorphism between two data G_1' and G_2' in $E(G)$, and Δ is a transfer factor for (G, G_1'), the function

$$(\alpha'\Delta)(\delta_2', \gamma) = \Delta\big((\alpha')^{-1}\delta_2', \gamma\big), \qquad \delta_2' \in \Delta_{G\text{-reg}}(G_2'),\ \gamma \in \Gamma_{\text{reg}}(G),$$

is a transfer factor for (G, G_2'). In particular, we obtain an action of the finite group $\mathrm{Out}_G(G')$ on the fibre in $T(G)$ of a point $G' \in E(G)$. Given $G' \in E(G)$ and $\alpha' \in \mathrm{Out}_G(G')$, let s_{sc} and g_{sc} be preimages in $\widehat{G}_{\text{sc}}$ of the projections onto $\widehat{G}_{\text{ad}}$ of the points s' and g' in $\widehat{G}$ attached to G' and α'. We can then write

$$g_{\text{sc}} s_{\text{sc}} g_{\text{sc}}^{-1} = s_{\text{sc}} z(\alpha'),$$

where

$$\alpha' \longrightarrow z(\alpha')$$

is a homomorphism from $\mathrm{Out}_G(G')$ into $\widehat{Z}_{\text{sc}}^{\Gamma}$. It is then not hard to show that

$$(\alpha'\Delta) = \widehat{\zeta}_G\big(z(\alpha')\big)\Delta, \qquad \alpha' \in \mathrm{Out}_G(G'),\ G' \in E(G), \tag{9.3.1}$$

where Δ is any element in the fibre of G' in $T(G)$, and $\widehat{\zeta}_G$ is the character on $\widehat{Z}_{\text{sc}}^{\Gamma}$ attached in §9.1 to the inner twist G of G^*. (See [**A24**].)

The action of $\mathrm{Out}_G(G')$ on the fibre of G' in $T(G)$ is therefore not trivial. (This point was overlooked in several of the papers leading up to the stable trace formula [**A16**], as we noted in [**A19**, §3]. The necessary modifications are minor, and will appear elsewhere.) It follows from (9.3.1) that we cannot assign transfer factors to pairs (G, G') in a consistent way as G' ranges over elements in $E(G)$ in a given isomorphism class. To describe a formal resolution, it is perhaps suggestive to write

$$S' = S_{G'} = \mathrm{Cent}({}^L G', \widehat{G}) = Z(\widehat{G}')^{\Gamma}$$

for the centralizer in $\widehat{G}$ of the L-subgroup ${}^L G'$ of ${}^L G$, in analogy with the group S_ψ attached to a parameter ψ. We then write S_{sc}' for the preimage in $\widehat{G}_{\text{sc}}'$ of the quotient

$$\overline{S}' = S'/Z(\widehat{G})^{\Gamma}.$$

This gives us an exact sequence

$$1 \longrightarrow \widehat{Z}_{\text{sc}} \longrightarrow S_{\text{sc}}' \longrightarrow \overline{S}' \longrightarrow 1$$

analogous to (9.2.1). It follows from the definitions that the point $s' \in \widehat{G}$ attached to G' lies in S', and therefore projects to a point $\overline{s}'$ in $\overline{S}'$.

With this notation, we define

$$E_{\text{sc}}(G) = \{(G', s_{\text{sc}}) :\ G' \in E(G)\},$$

where s_{sc} ranges over the preimages in S_{sc}' of the point $\overline{s}' \in \overline{S}'$. Then $E_{\text{sc}}(G)$ is a covering space over $E(G)$, or more precisely, a principal $\widehat{Z}_{\text{sc}}$-bundle over $E(G)$. The fibre product $T_{\text{sc}}(G)$ of $T(G)$ and $E_{\text{sc}}(G)$ over $E(G)$ is the set of triplets

$$\{(G', \Delta, s_{\text{sc}}) :\ (G', \Delta) \in T(G),\ (G', s_{\text{sc}}) \in E_{\text{sc}}(G)\}.$$

It is clear that $T_{\mathrm{sc}}(G)$ is a principal $U(1)$-bundle over $E_{\mathrm{sc}}(G)$, which we can display over the original $U(1)$-bundle in a commutative diagram

$$\begin{array}{ccccccccc} 1 & \longrightarrow & U(1) & \longrightarrow & T_{\mathrm{sc}}(G) & \longrightarrow & E_{\mathrm{sc}}(G) & \longrightarrow & 1 \\ & & \| & & \downarrow & & \downarrow & & \\ 1 & \longrightarrow & U(1) & \longrightarrow & T(G) & \longrightarrow & E(G) & \longrightarrow & 1. \end{array} \tag{9.3.2}$$

This second principal bundle does have a section. Its existence amounts to the following elementary lemma.

LEMMA 9.3.1. *There is a continuous mapping*

$$\rho:\ T_{\mathrm{sc}}(G) \longrightarrow U(1)$$

such that

$$\rho(u\Delta, s_{\mathrm{sc}}z) = u\rho(\Delta, s_{\mathrm{sc}})\widehat{\zeta}_G(z)^{-1}, \qquad u \in U(1),\ z \in \widehat{Z}_{\mathrm{sc}}, \tag{9.3.3}$$

and

$$\rho(\alpha'\Delta, \mathrm{Int}(g')s_{\mathrm{sc}}) = \rho(\Delta, s_{\mathrm{sc}}), \qquad g' = g(\alpha'),\ \alpha' \in \mathrm{Out}_G(G'), \tag{9.3.4}$$

for any point

$$(\Delta, s_{\mathrm{sc}}) = (G', \Delta, s_{\mathrm{sc}})$$

in $T_{\mathrm{sc}}(G)$.

PROOF. Suppose that (G', s_{sc}) is the image in $E_{\mathrm{sc}}(G)$ of a point $(\Delta, s_{\mathrm{sc}})$ in $T_{\mathrm{sc}}(G)$, and that g' equals the point $g(\alpha')$ attached to an outer automorphism $\alpha' \in \mathrm{Out}_G(G')$ of G'. It follows from (9.3.1) that

$$\rho(\alpha'\Delta, \mathrm{Int}(g')s_{\mathrm{sc}}) = \rho(\widehat{\zeta}_G(z(\alpha'))\Delta, s_{\mathrm{sc}}z(\alpha')),$$

for any $U(1)$-valued function ρ on $T_{\mathrm{sc}}(G)$. The condition (9.3.4) is then a consequence of (9.3.3) in this case. Since the two conditions are otherwise independent, they are compatible with each other. We can therefore define the required section $\rho(\Delta, s_{\mathrm{sc}})$ by (9.3.3). Indeed, we have only to specify its values at any set of points $\{(\Delta, s_{\mathrm{sc}})\}$ that projects bijectively onto the set of isomorphism classes $G' \in \mathcal{E}(G)$. □

REMARK. The isomorphism classes of endoscopic data $G' \in \mathcal{E}(G)$ parametrize the connected components $T_{\mathrm{sc}}(G, G')$ of $T_{\mathrm{sc}}(G)$. It is clear that the section $\rho(\Delta, s_{\mathrm{sc}})$ of the lemma is unique up to a $U(1)$-multiple on each component $T_{\mathrm{sc}}(G, G')$.

Consider the Langlands-Shelstad transfer correspondence

$$f \longrightarrow f' = f'_\Delta, \qquad f \in \mathcal{H}(G),\ (G', \Delta) \in T(G).$$

We recall that for fixed (G', Δ), it is the mapping from $\mathcal{H}(G)$ to $\mathcal{S}(G')$ defined by the geometric transform

$$f'_\Delta(\delta') = \sum_\gamma \Delta(\delta', \gamma) f_G(\gamma), \qquad \delta' \in \Delta_{G\text{-reg}}(G'),\ \gamma \in \Gamma_{\mathrm{reg}}(G),$$

introduced in §2.1. For fixed f and δ', $f'_\Delta(\delta')$ can be regarded as a section of a bundle, specifically the line bundle attached to $T(G, G')$ by the action of its structure group $U(1)$ on $\mathbb{C}$. Rather than regard endoscopic transfer as a mapping that varies with a choice of Δ, we could normalize it with a section $\rho(\Delta, s_{\text{sc}})$ from Lemma 9.3.1. Given ρ and f, we define f'_ρ to be the function in $\mathcal{S}(G')$ whose value at any δ' equals

$$f'_\rho(\delta') = \rho(\Delta, s_{\text{sc}})^{-1} f'_\Delta(\delta'). \tag{9.3.5}$$

Observe that the function f'_ρ remains unchanged if Δ is replaced either by a $U(1)$-multiple $u\Delta$ or an $\text{Out}_G(G')$-image $\alpha'\Delta$, by the properties of $\rho(\Delta, s_{\text{sc}})$ in Lemma 9.3.1. It is therefore independent of Δ. It does, however, depend on the implicit choice of a preimage s_{sc} of $\bar{s}'$ in S'_{sc}.

The transfer f'_ρ of course also depends on the normalizing section ρ. However, Kaletha has introduced a finer normalization, following ideas of Kottwitz. Specifically, he has shown how to normalize the transfer factor Δ for (G, G') (or equivalently, how to choose a section ρ) explicitly in terms of a transfer factor Δ^* for the quasisplit pair (G^*, G') [**Kal1**, §2.2], [**Kal2**, §2.2]. I shall describe how Kaletha's general constructions apply to the case at hand. Since some of his work is still in progress (and my own understanding of it leaves something to be desired), parts of the following discussion might for the moment be best taken conditionally. In any case, I am indebted to Kottwitz for describing his original unpublished ideas to me.

The field F remains local. Kaletha's constructions depend on an inner twist with some extra structure. A *pure inner twist* over F is a triplet (G, ψ, z), where (G, ψ) is an inner twist of a connected, quasisplit group G^* over F, which we temporarily take to be arbitrary, and $z \in Z^1(F, G^*)$ is a Galois 1-cocycle in G^* such that

$$\psi\sigma(\psi)^{-1} = \text{Int}\big(z(\sigma)\big), \qquad \sigma \in \Gamma_F.$$

In contrast to ordinary inner twists, pure inner twists do not necessarily exist. That is, an inner form G of G^* need not be the first component of a pure inner twist. Suppose, however, that G satisfies the condition that its center $Z(G)$ is connected (so that $\widehat{G}_{\text{der}} = \widehat{G}_{\text{sc}}$), and the further condition that the Γ-invariant part $Z(\widehat{G})^\Gamma$ of the center of $\widehat{G}$ is discrete. The inner form G of G^* can then be inflated to a pure inner twist. With this structure, we of course have $Z(G) = Z(G^*)$ and $Z(\widehat{G})^\Gamma = Z(\widehat{G}^*)^\Gamma$. The two conditions can thus be stated in terms of G^*, and therefore apply uniformly to all G.

Assume that G represents a pure inner twist (G, ψ, z) of G^*. Suppose also that $\gamma \in G(F)$ and $\gamma^* \in G^*(F)$ are strongly G-regular elements that are stably conjugate, in the sense that

$$\psi(\gamma) = g^*\gamma^*(g^*)^{-1},$$

for some $g^* \in G^*(\overline{F})$. Then the 1-cocycle

$$\sigma \longrightarrow g^*z(\sigma)\sigma(g^*)^{-1}, \qquad \sigma \in \Gamma_F,$$

takes values in the maximal F-torus $T^* = G^*_{\gamma^*}$ of G^*. Its image $\mathrm{inv}(\gamma, \gamma^*)$ in $H^1(F, T^*)$ is the *invariant* of γ and γ^* (with respect to (ψ, z)), defined by Kaletha in [**Kal1**, §2.1]. Suppose also that G' represents an endoscopic datum $(G', \mathcal{G}', s', \xi')$, and that $\delta' \in G'(F)$ is an image of γ^*, in the sense of [**LS1**, p. 236]. There is then an admissible embedding [**LS1**, p. 236] from the centralizer $T' = G'_{\delta'}$ into G^* that maps δ' to γ^* (and that therefore maps T' isomorphically onto T^*). Let $s^* = s_{T^*}$ be the preimage in $(\widehat{T}^*)^\Gamma$ of the element s' in the subgroup $Z(\widehat{G}')^\Gamma$ of $(\widehat{T}')^\Gamma$, under the dual isomorphism from $\widehat{T}^*$ to $\widehat{T}'$. Then s^* maps to a component in the group $\pi_0\big((\widehat{T}^*)^\Gamma\big)$, which we denote also by s^*. The Tate-Nakayama pairing between $H^1(F, T^*)$ and $\pi_0\big((\widehat{T}^*)^\Gamma\big)$ [**K2**, §1.1] then provides a complex number

$$\langle \mathrm{inv}(\gamma, \gamma^*), s^* \rangle, \tag{9.3.6}$$

which depends on the point δ' (as well as the cocycle z).

This pairing is one ingredient in Kaletha's normalization of the transfer factors for G. The other is the choice of a family of transfer factors Δ^* for the quasisplit group G^*, which we can assume are normalized by a fixed Whittaker datum for G^*. Given Δ^*, we write

$$\Delta_K(\delta', \gamma) = \Delta^*(\delta', \gamma^*)\langle \mathrm{inv}(\gamma, \gamma^*), s^* \rangle^{-1}, \tag{9.3.7}$$

for γ, g^* and δ' as above. This is Kaletha's normalization. He shows that it is independent of γ^* in [**Kal1**, Lemma 2.2.1], and proves that it is actually a transfer factor for (G, G^*) in [**Kal1**, §2.3–§2.5].

The construction (9.3.7) imposes the two restrictions above on G (or G^*). For p-adic F, Kaletha removes the condition that $Z(\widehat{G})^\Gamma$ be discrete in [**Kal2**] by using Kottwitz's theory of isocrystals with additional structure [**K4**], [**K7**]. Kottwitz has pointed out to me that his results can be stated equivalently in terms of algebraic cohomology of Weil groups. This alternate formulation has the advantage that it applies also to the case that F is archimedean. I thank Kottwitz for the following description, which I hope I have interpreted correctly.

Suppose that T is a torus over the local field F. The Tate-Nakayama pairing that provides the definition of (9.3.6) is a canonical isomorphism

$$H^1(F, T) \stackrel{\mathrm{def}}{=\!=} H^1(\Gamma_{K/F}, T) \xrightarrow{\sim} \pi_0(\widehat{T}^\Gamma)^* \stackrel{\mathrm{def}}{=\!=} X^*\big(\pi_0(\widehat{T}^{\Gamma_{K/F}})\big),$$

where K/F is a large finite Galois extension and $X^*(\cdot)$ denotes the finite abelian group of (algebraic) characters on the finite abelian (diagonalizable) group $\pi_0(\widehat{T}^\Gamma)$. Its analogue for the Weil group is a canonical isomorphism

$$H^1(W_F, T) \stackrel{\mathrm{def}}{=\!=} H^1_{\mathrm{alg}}(W_{K/F}, T) \xrightarrow{\sim} (\widehat{T}^\Gamma)^* \stackrel{\mathrm{def}}{=\!=} X^*(\widehat{T}^{\Gamma_{K/F}}),$$

where $H^1_{\mathrm{alg}}(\cdot)$ denotes the cohomology group of classes represented by 1-cocycles that are algebraic as morphisms of varieties over F, and $X^*(\cdot)$ now represents the finitely generated abelian group of (algebraic) characters on the complex diagonalizable group $\widehat{T}^\Gamma$. More generally, suppose that G is a connected reductive group over F. Then the Tate-Nakayama pairing is a canonical mapping

$$H^1(F,G) \stackrel{\text{def}}{=\!=} H^1(\Gamma_{K/F},G) \longrightarrow \pi_0\big(Z(\widehat{G})^\Gamma\big)^* \stackrel{\text{def}}{=\!=} X^*\big(\pi_0\big(Z(\widehat{G})^{\Gamma_{K/F}}\big)\big),$$

which is a bijection when F is p-adic. (See [**K5**, §1.2].) Its Weil group analogue is a canonical mapping

$$H^1(W_F,G) \stackrel{\text{def}}{=\!=} H^1_{\text{alg}}(W_{K/F},G) \longrightarrow \big(Z(\widehat{G})^\Gamma\big)^* \stackrel{\text{def}}{=\!=} X^*\big(Z(\widehat{G})^{\Gamma_{K/F}}\big),$$

which is again a bijection if F is p-adic. In this case, the definition of $H^1_{\text{alg}}(W_{E/F},G)$ includes the supplementary condition that the underlying 1-cocycles map the subgroup E^* of $W_{E/F}$ to the subgroup $Z(G)$ of G.

Kaletha defines an *extended pure inner twist* to be a triplet (G,ψ,z) as above, but with the 1-cocycle appropriately weakened. In our terms, z is an element in the set of 1-cocycles

$$Z^1(W_F,G^*) \stackrel{\text{def}}{=\!=} Z^1_{\text{alg}}(W_{K/F},G^*)$$

whose quotient is the Weil group cohomology set $H^1(W_F,G^*)$ we have just described. If G represents an extended pure inner twist (G,ψ,z), the pairing (9.3.6) makes sense. In [**Kal2**, Propositon 2.1], Kaletha shows that the corresponding product (9.3.7) is a transfer factor for (G,G'). This removes the condition that $Z(\widehat{G})^\Gamma$ be discrete, at least if F is p-adic. We take for granted the analogous result for archimedean F.

The other restriction is that $Z(G)$ be connected. To remove it, Kaletha embeds a general group G into a group G_1 with connected center whose representation theory is essentially the same as that of G. Following an idea of Kottwitz, he defines a *z-embedding* of G to be an F-extension

$$1 \longrightarrow G \longrightarrow G_1 \longrightarrow C_1 \longrightarrow 1 \tag{9.3.8}$$

of an induced torus C_1 over F by G such that $Z(G_1)$ is connected, and such that the mapping

$$H^1\big(F,Z(G)\big) \longrightarrow H^1\big(F,Z(G_1)\big) \tag{9.3.9}$$

is an isomorphism. The reader will note that this is reminiscent of the well known definition [**K1**, §1] of a *z-extension*, which would in fact be dual to that of a z-embedding were it not for the extra condition on (9.3.9). The properties of z-embeddings are part of the work in progress [**Kal4**] of Kaletha.* We shall briefly describe some of them.

One of the simplest properties is that the restriction of representations from $G_1(F)$ to $G(F)$ gives a *surjective* mapping $\pi_1 \to \pi$ from $\Pi(G_1)$ to $\Pi(G)$. This follows from the fact that

$$G_1(F) = G(F)Z(G_1,F),$$

---

*I thank Kaletha for making his notes available to me.

where $Z(G_1, F) = \big(Z(G_1)\big)(F)$ is the centre of $G(F)$, a consequence in turn of the injectivity of (9.3.9) and the long exact cohomology sequence attached to the short exact sequence

$$1 \longrightarrow Z(G) \longrightarrow Z(G_1) \longrightarrow C_1 \longrightarrow 1, \tag{9.3.10}$$

which together imply that the mapping from $Z(G_1, F)$ to $C_1(F)$ is surjective. Since the group $\Pi(C_1)$ of characters on $C_1(F)$ acts transitively on the fibres of the mapping $\pi_1 \to \pi$, the representation theory of $G_1(F)$ is indeed essentially the same as that of $G(F)$. Kaletha also establishes finer endoscopic relations among the representations of the two groups. They arise naturally from the short exact sequence

$$1 \longrightarrow \widehat{C}_1 \longrightarrow \widehat{G}_1 \longrightarrow \widehat{G} \longrightarrow 1, \tag{9.3.11}$$

of dual groups, equipped with their compatible L-actions of W_F. In particular, the projection of $\widehat{G}_1$ onto $\widehat{G}$ gives a mapping of Langlands parameters $\phi_1 \to \phi$ that takes $\Phi(G_1)$ to $\Phi(G)$. Kaletha observes that this mapping is surjective, that its fibres are torsors for the action of $H^1(W, \widehat{C}_1)$ by multiplication by cocycles, and that the centralizer groups of corresponding parameters are related by a short exact sequence

$$1 \longrightarrow \widehat{C}_1^{\Gamma} \longrightarrow S_{\phi_1} \longrightarrow S_{\phi} \longrightarrow 1. \tag{9.3.12}$$

Similar relations apply to the parameter sets $\Psi(G_1)$ and $\Psi(G)$.

There are parallel relations among endoscopic data. Given an endoscopic datum $(G', s', \mathcal{G}', \xi')$ for G, Kaletha constructs an endoscopic datum $(G'_1, s'_1, \mathcal{G}_1, \xi'_1)$ for G_1 by defining $s'_1 \in \widehat{G}_1$ to be any preimage of s' in $\widehat{G}_1$, and $(\mathcal{G}'_1, \xi'_1)$ to be the pullback (fibre product) in the larger diagram

$$\begin{array}{ccccccccc} 1 & \longrightarrow & \widehat{C}_1 & \longrightarrow & {}^L G_1 & \longrightarrow & {}^L G & \longrightarrow & 1 \\ & & \Vert & & \uparrow{\scriptstyle \xi'_1} & & \uparrow{\scriptstyle \xi'} & & \\ 1 & \longrightarrow & \widehat{C}_1 & \longrightarrow & \mathcal{G}'_1 & \longrightarrow & \mathcal{G}' & \longrightarrow & 1. \end{array}$$

The quasisplit group G'_1 is then the pushout (fibre sum) in the diagram

$$\begin{array}{ccc} Z(G) & \longrightarrow & Z(G_1) \\ \downarrow & & \downarrow \\ G' & \longrightarrow & G'_1 \end{array}$$

Conversely, given an endoscopic datum $(G'_1, s'_1, \mathcal{G}'_1, \xi'_1)$ for G_1, Kaletha takes the F-surjection $G'_1 \to C_1$ whose dual injection $\widehat{C}_1 \to \widehat{G}'_1$ is the restriction of ξ'_1 to the subgroup $\widehat{C}_1$ of $\xi'_1(\mathcal{G}'_1)$. This gives the quasisplit group

$$G' = \ker(G'_1 \to C_1)$$

over F. He then inflates G' to an endoscopic datum $(G', s', \mathcal{G}', \xi')$ for G by setting s' equal to the image of s'_1 under the mapping $Z(\widehat{G}'_1) \to Z(\widehat{G}')$

attached to

$$\widehat{G}' = \mathrm{im}(\widehat{G}'_1 \to \widehat{G}),$$

and $(\mathcal{G}', \xi')$ equal to the composition of three maps

$$\mathcal{G}' \hookrightarrow \mathcal{G}'_1 \xrightarrow{\xi'_1} \widehat{G}_1 \to \widehat{G},$$

with $\mathcal{G}'$ defined in the natural way. Finally, he shows that these two correspondences give mutually inverse bijections

$$(G', s', \mathcal{G}', \xi') \leftrightarrow (G'_1, s'_1, \mathcal{G}'_1, \xi'_1)$$

between the strong isomorphism classes (as defined above) of endoscopic data for G and G_1.

In the course of comparing endoscopic data, Kaletha observes that the mapping $G' \to G'_1$ satisfies all the conditions of a z-embedding *except* for the connectedness of $Z(G'_1)$. He calls such a mapping a *pseudo-z-embedding*. The properties of G_1 we have just described all remain valid if the original z-embedding is weakened to a pseudo-z-embedding. In particular, a pseudo-z-embedding of G into G_1 (with quotient C_1) still gives a bijective correspondence $G' \leftrightarrow G'_1$ between the associated endoscopic data, taken up to strong isomorphism, and equipped with pseudo-z-embeddings $G' \to G'_1$ (with the same quotient C_1). The supplementary condition on (9.3.9) required for a z-embedding is used only for the normalization (9.3.7) of transfer factors.

Kaletha notes that z-embeddings (or pseudo-z-embeddings) can be chosen uniformly for inner twists. Suppose for example that G^* is quasisplit over F, and that $G^* \to G^*_1$ is a z-embedding. Suppose also that $\psi\colon G \to G^*$ is an inner twist, with

$$\psi\sigma(\psi)^{-1} = \mathrm{Int}\big(u(\sigma)\big), \qquad \sigma \in \Gamma_F,$$

for a 1-cocycle $u \in Z^1(F, G^*_{\mathrm{ad}})$. Then there is a z-embedding $G \to G_1$, and an inner twist $\psi_1\colon G_1 \to G^*_1$, with

$$\psi_1\sigma(\psi_1)^{-1} = \mathrm{Int}\big(u(\sigma)\big), \qquad \sigma \in \Gamma_F,$$

that fits into the diagram

$$\begin{array}{ccccccccc} 1 & \longrightarrow & G & \longrightarrow & G_1 & \longrightarrow & C_1 & \longrightarrow & 1 \\ & & {\scriptstyle\psi}\downarrow & & {\scriptstyle\psi_1}\downarrow & & \| & & \\ 1 & \longrightarrow & G^* & \longrightarrow & G^*_1 & \longrightarrow & C_1 & \longrightarrow & 1. \end{array}$$

The goal is to normalize the transfer factors for (G, ψ) in terms of the normalized transfer factors for an extended inner twist (G_1, ψ_1, z_1).

Let us first describe how Kaletha constructs z-embeddings. To be concrete, we shall treat only orthogonal and symplectic groups. We return therefore to our original setting, in which G represents an inner twist of $G^* \in \widetilde{\mathcal{E}}_{\mathrm{sim}}(N)$. It suffices to construct a z-embedding of the quasisplit group G^*. If G^* is of type $\mathbf{B}_n$, it is isomorphic to a split group $SO(2n+1)$. In this case the center $Z(G^*)$ is trivial, and we can take $G_1^* = G^*$. We suppose therefore that G^* is of type $\mathbf{C}_n$ or $\mathbf{D}_n$, and is consequently isomorphic to a split group $Sp(2n)$ or a quasisplit group $SO(2n)$. In these cases, the center $Z(G^*)$ has order 2.

The main ingredient of Kaletha's construction, as it applies to our group G^* of type $\mathbf{C}_n$ or $\mathbf{D}_n$, is a short exact sequence

$$1 \longrightarrow Z(G^*) \longrightarrow Z_1 \longrightarrow C_1 \longrightarrow 1,$$

where C_1 is the induced torus of the putative z-embedding, and Z_1 is an F-torus that will become the centre $Z(G_1^*)$ of G_1^*. Let K/F be the compositum of the finite set of quadratic extensions of F, an abelian extension of F with Galois group

$$\Gamma_{K/F} \cong (F^*)^2 \backslash F^*.$$

We take the induced torus to be the restriction of scalars

$$C_1 = \mathrm{Res}_{K/F}(\mathbb{G}_{m,K}).$$

The norm $N_{K/F}$ then represents an F-homomorphism from C_1 to $\mathbb{G}_m = \mathbb{G}_{m,F}$. We take Z_1 to be the fibre product

$$\begin{array}{ccc} Z_1 & \longrightarrow & C_1 \\ \big\downarrow & & \big\downarrow{\scriptstyle N_{K/F}} \\ \mathbb{G}_m & \xrightarrow{\ p\ } & \mathbb{G}_m, \end{array}$$

where $p = p_2$ is the homomorphism $x \rightarrow x^2$. It fits into a commutative diagram of short exact sequences

$$\begin{array}{ccccccccc} 1 & \longrightarrow & Z(G^*) & \longrightarrow & Z_1 & \longrightarrow & C_1 & \longrightarrow & 1 \\ & & \big\| & & \big\downarrow & & \big\downarrow{\scriptstyle N_{K/F}} & & \\ 1 & \longrightarrow & Z(G^*) & \longrightarrow & \mathbb{G}_m & \xrightarrow{\ p\ } & \mathbb{G}_m & \longrightarrow & 1. \end{array}$$

This in turn gives rise to the commutative diagram

$$\begin{array}{ccccccccc} 1 & \longrightarrow & Z(G^*) & \longrightarrow & Z_1(F) & \longrightarrow & C_1(F) & \longrightarrow & H^1\big(F, Z(G^*)\big) & \longrightarrow & \cdots \\ & & \big\| & & \big\downarrow & & \big\downarrow{\scriptstyle N_{K/F}} & & \big\| & & \\ 1 & \longrightarrow & Z(G^*) & \longrightarrow & F^* & \xrightarrow{\ p\ } & F^* & \longrightarrow & H^1\big(F, Z(G^*)\big) & \longrightarrow & \cdots \end{array}$$

of long exact sequences of cohomology.

The group Z_1 is connected, since it is a torus over F. We have already noted that C_1 is an induced torus, so in particular, $H^1(F, C_1) = \{1\}$. It follows from the continuation of the upper long exact sequence that the mapping

$$H^1\big(F, Z(G^*)\big) \longrightarrow H^1(F, Z_1)$$

is surjective. To prove that the mapping is injective, it suffices to show that the image of $C_1(F)$ in $H^1\big(F, Z(G^*)\big)$ in the upper sequence is trivial. But the image of $C_1(F)$ in upper sequence equals the image of $N_{K/F}\big(C_1(F)\big)$

in the lower sequence. By local class field theory, $N_{K/F}(C_1(F))$ equals $(F^*)^2$, which is just the image of p in F^*. The image of $N_{K/F}(C_1(F))$ in $H^1(F, Z(G^*))$ is therefore trivial. It follows that the mapping above is an injection and hence an isomorphism, as will be required of (9.3.9).

We have been following the argument of Kaletha, which is valid for any group over F. In our case that G^* is of type $\mathbf{C}_n$ or $\mathbf{D}_n$, one sees explicitly that

$$Z_1 = \{(z, c_1, \dots, c_k) \in \mathbb{G}_m^{k+1} : \; z^2 t_1 \cdots t_k = 1\}, \tag{9.3.13}$$

and that

$$\begin{aligned}\widehat{Z}_1 &= \{(\widehat{z}, \widehat{c}_1, \dots, \widehat{c}_k) \in (\mathbb{C}^*)^{k+1}\}/\{(z^2, z, \dots, z) : \; z \in \mathbb{C}^*\} \\ &\cong (\mathbb{C}^*/\{\pm 1\}) \times ((\mathbb{C}^*)^k/\mathbb{C}^*),\end{aligned}$$

where the subscripts index the Galois group

$$\Gamma_{K/F} = \{\sigma_1 = 1, \; \sigma_2, \dots, \sigma_k\},$$

and the twisted Galois action on Z_1 and $\widehat{Z}_1$ is provided by right translation of $\Gamma_{K/F}$ on these indices. Using the relations

$$H^1(F, Z_1) = \pi_0(\widehat{Z}_1^{\Gamma})^* \cong \pi_0\big(((\mathbb{C}^*)^k/\mathbb{C}^*)^{\Gamma_{K/F}}\big)^* \cong \Pi(\Gamma_{K/F}),$$

one can then see directly that the mapping above is an isomorphism.

Once we have the torus Z_1, we can define G_1^* as the fibre sum of Z_1 and G^* over $Z(G^*)$. This is easily seen to be an F-extension of C_1 by G^*, which fits into the commutative diagram

$$\begin{array}{ccccccccc} 1 & \longrightarrow & Z(G^*) & \longrightarrow & Z_1 & \longrightarrow & C_1 & \longrightarrow & 1 \\ & & \downarrow & & \downarrow & & \| & & \\ 1 & \longrightarrow & G^* & \longrightarrow & G_1^* & \longrightarrow & C_1 & \longrightarrow & 1. \end{array}$$

The mapping from Z_1 to G_1^* is an injection that identifies Z_1 with the centre $Z(G_1^*)$ of G_1^*. The mapping of G^* into G_1^* then satisfies the required conditions of a z-embedding. This is what we set out to describe. As we have seen, it in turn gives a canonical z-embedding of our inner twist G of G^*, as well as a canonical pseudo-z-embedding of any endoscopic group G' of G, both as above.

The z-embedding for G we have just described gives us a group G_1 with connected center. This was the condition for Kaletha's construction (9.3.7) of normalized transfer factors. Suppose that (G_1, ψ_1, z_1) is an inflation of the inner twist (G_1, ψ_1) to an extended pure inner twist, and that $\{\Delta_1^*\}$ is the family of transfer factors for the quasisplit group G_1^* normalized by a fixed Whittaker datum. The construction (9.3.7) then gives a family of normalized transfer factors

$$\Delta_{1,K}(\delta_1', \gamma_1)$$

for G_1. These functions are of course defined on subsets of products $G_1'(F) \times G_1(F)$, parametrized by endoscopic data G_1' for G_1. Suppose that G' is an endoscopic datum for G that corresponds to G'_1 in the manner described above. We can then consider the restriction

$$\Delta_{1,K}(\delta', \gamma)$$

of $\Delta_{1,K}$ to the intersection of its domain with the subset $G'(F) \times G(F)$ of $G'_1(F) \times G_1(F)$. Is it a transfer factor for (G, G')?

Kaletha answers the question affirmatively in [**Kal4**] with a comparison of the *relative* transfer factors for (G_1, G'_1) and (G, G'). He proves that these objects satisfy the identity

$$\Delta_1(\delta', \gamma; \overline{\delta}', \overline{\gamma}) = \Delta(\delta', \gamma; \overline{\delta}', \overline{\gamma}),$$

where $(\overline{\delta}', \overline{\gamma})$ is any second pair in the intersection of the domain of $\Delta_{1,K}$ with $G'(F) \times G(F)$. Since $\Delta_{1,K}(\delta', \gamma)$ is a transfer factor for (G_1, G'_1), it equals a product of the left hand side of the identity with a factor in $U(1)$ that is independent of (δ', γ). It therefore equals the product of the right hand side by the same factor, and is consequently a transfer factor for (G, G').

Our description of the normalized transfer factor $\Delta_{1,K}$, which Kaletha constructs for any connected reductive group, is the culmination of this section. We can now go back to the discussion leading up to Lemma 9.3.1. The z-embedding for G is given by the short exact sequence (9.3.8). The center Z_1 of G_1 defines a second short exact sequence

$$1 \longrightarrow Z_1 \longrightarrow G_1 \longrightarrow G_{\mathrm{ad}} \longrightarrow 1. \tag{9.3.14}$$

Since Z_1 is connected, the derived group of $\widehat{G}_1$ is simply connected. It equals the simply connected cover $\widehat{G}_{\mathrm{sc}}$ of $\widehat{G}$. This becomes the first term in the dual short exact sequence

$$1 \longrightarrow \widehat{G}_{\mathrm{sc}} \longrightarrow \widehat{G}_1 \longrightarrow \widehat{Z}_1 \longrightarrow 1, \tag{9.3.15}$$

which obviously bears the same relation to (9.3.11) as (9.3.14) bears to (9.3.8).

Suppose that we start with G', which represents an endoscopic datum $(G', s', \mathcal{G}', \xi')$ for G'. The corresponding endoscopic datum $(G'_1, s'_1, \mathcal{G}'_1, \xi'_1)$ for G_1 was defined by taking s'_1 to be any preimage of s' in $\widehat{G}_1$. In particular, we can take $s'_1 = s_{\mathrm{sc}}$ to lie in the subgroup $\widehat{G}_{\mathrm{sc}}$ of $\widehat{G}_1$. We then define

$$\rho_K(\Delta, s_{\mathrm{sc}}) = \Delta(\delta', \gamma)\Delta_{1,K}(\delta', \gamma)^{-1}, \tag{9.3.16}$$

for *any* transfer factor Δ for (G, G'). Since $\Delta_{1,K}(\delta, \gamma)$ is a transfer factor for (G, G'), $\rho_K(\Delta, s_{\mathrm{sc}})$ is independent of (δ', γ).

LEMMA 9.3.2. *The function*

$$\rho(\Delta, s_{\mathrm{sc}}) = \rho_K(\Delta, s_{\mathrm{sc}})$$

satisfies the two conditions of Lemma 9.3.1.

We shall leave the proof for [**A28**], even though it is not difficult. The main point is the $\widehat{Z}_{\mathrm{sc}}$-equivariance condition in (9.3.3). This depends on our choice of the extension (9.2.3) of the character $\widehat{\zeta}_G$ from $\widehat{Z}_{\mathrm{sc}}^{\Gamma}$ to $\widehat{Z}_{\mathrm{sc}}$, in the case that G^* is not split. We note here only that in verifying (9.3.3) for $\rho_K(\Delta, s_{\text{sc}})$, one sees that the extension (9.2.3) is actually imposed on us by the chosen 1-cocycle $z_1 \in Z^1(W_{K/F}, G_1)$ in (G_1, ψ_1, z_1).

The normalizing section $\rho_K(\Delta, s_{\text{sc}})$ allows us to formulate theorems directly for G. It would be interesting to see how unique it is, beyond say, its dependence on the extension (9.2.3) and the transfer factors $\{\Delta^*\}$ for G^*. Questions of this nature have been examined by Kaletha, and will perhaps be discussed in [**Kal4**].

9.4. Statement of the local classification

We shall now formulate our local assertions for inner twists. Taken together, they represent a collective extension of Theorems 1.5.1 and 2.2.1 that encompasses the local refinements from Section 8.4. The first applies to generic parameters ϕ. It amounts to the local classification of representations conjectured by Langlands. But as we know, the local classification must be supplemented by further assertions for the general parameters ψ in order to establish the local framework for the global classification.

The field F will be local throughout this section. As before, G stands for a fixed inner twist (G, ψ) of a quasisplit group $G^* \in \widetilde{\mathcal{E}}_{\text{sim}}(N)$ over F. We can think of the local classification for G as the specialization of general local theorems to the subset $\widetilde{\Phi}_{\text{bdd}}(G)$ of Langlands parameters in $\widetilde{\Psi}(G)$. We will need to formulate it from the perspective reached in Theorem 8.4.1.

Suppose for the moment that the group $\widehat{G}^*$ is of the form $SO(2n, \mathbb{C})$. As we explained in §8.3, there remains a "$\mathbb{Z}/2\mathbb{Z}$ symmetry" between the Langlands parameters $\phi^* \in \Phi_{\text{bdd}}(G^*)$ and the associated stable characters on $G^*(F)$. In order to sidestep this ambiguity, we shall state the classification for G directly in terms of stable distributions. The set

$$\Phi'_{\text{dis}}(G^*) = \{\phi^*_t : \ \phi^* \in \widetilde{\Phi}'_{\text{bdd}}(G^*), \ t \in T(\phi^*)\},$$

defined in the notation of Theorem 8.4.1, indexes certain stable tempered distributions on $G^*(F)$. We shall still denote these objects by ϕ^*, since they are bijective with the subset $\Phi'_{\text{bdd}}(G^*)$ of Langlands parameters in $\Phi_{\text{bdd}}(G^*)$ whose stabilizers in the group $\widetilde{O}(G^*) = \widetilde{\text{O}}\text{ut}_N(G^*) = \mathbb{Z}/2\mathbb{Z}$ are trivial. It is this bijection that is determined only up to the free action of $\widetilde{O}(G^*)$ on $\Phi'_{\text{bdd}}(G^*)$. The complement $\Phi_{\text{bdd}}(\widetilde{G}^*)$ of $\Phi'_{\text{bdd}}(G^*)$ in $\Phi_{\text{bdd}}(G^*)$ is *canonically* bijective with the associated family of stable distributions, a set we will denote by $\Phi_{\text{dis}}(\widetilde{G}^*)$. We then write

$$\Phi_{\text{dis}}(G^*) = \Phi_{\text{dis}}(\widetilde{G}^*) \coprod \Phi'_{\text{dis}}(G^*).$$

If $\widehat{G}^*$ is not of the form $SO(2n, \mathbb{C})$, $\Phi_{\text{bdd}}(G^*)$ equals the set $\widetilde{\Phi}_{\text{bdd}}(G^*)$. By Theorem 2.2.1(a), it is again canonically bijective with a family $\Phi_{\text{dis}}(G^*)$ of stable distributions on $G^*(F)$. In all cases then, $\Phi_{\text{dis}}(G^*)$ represents a set $\{\phi^*\}$ of stable tempered distributions on $G^*(F)$. We denote them in the usual way by

$$f^* \longrightarrow f^*(\phi^*), \qquad f^* \in \mathcal{S}(G^*),\ \phi^* \in \Phi_{\mathrm{dis}}(G^*).$$

This set is in fact a basis of the subspace of stable, tempered distributions on $G^*(F)$ that are *admissible*, in the natural sense inherited from representation theory.

Our interest is in the group G. We write $\Phi_{\mathrm{dis}}(G)$ for the subset of distributions in $\Phi_{\mathrm{dis}}(G^*)$ that are G-relevant, in the sense that they do not vanish on the open subset of strongly regular, stable conjugacy classes in $G^*(F)$ that come from $G(F)$. They are bijective with the subset $\Phi_{\mathrm{bdd}}(G)$ of G-relevant parameters in $\Phi_{\mathrm{bdd}}(G^*)$. We shall sometimes regard $\Phi_{\mathrm{dis}}(G)$ as a separate set, equipped with an injection $\phi \to \phi^*$ into $\Phi_{\mathrm{dis}}(G^*)$. This leaves us free to identify any $\phi \in \Phi_{\mathrm{dis}}(G)$ with the stable distribution

$$f^G(\phi) = e(G) f^*(\phi^*), \qquad f \in \mathcal{H}(G), \tag{9.4.1}$$

on $G(F)$, where $f \to f^*$ is the Langlands-Shelstad transfer mapping from G to G^*, and

$$e(G) = \begin{cases} (-1)^{q(G^*)-q(G)}, & \text{if } F \text{ is archimedean,} \\ (-1)^{r(G^*)-r(G)}, & \text{if } F \text{ is } p\text{-adic,} \end{cases}$$

is the Kottwitz sign [**K2**]. We recall that $q(G)$ equals one-half the dimension of the symmetric space attached to G over F, and $r(G)$ equals the F-rank of the derived group of G (which is G itself in the case at hand).

We have used the correspondence (1.4.11) (in both its local and global forms) repeatedly throughout the earlier chapters. It carries over to the present setting. If ϕ^* belongs to $\Phi_{\mathrm{dis}}(G^*)$, we write S_{ϕ^*} for the usual centralizer in $\widehat{G}^*$ of the image of a corresponding parameter. Then (1.4.11) can be interpreted as a correspondence

$$(G', \phi') \longrightarrow (\phi^*, s^*), \qquad G' \in \mathcal{E}(G),\ s^* \in S_{\phi^*,\mathrm{ss}},$$

in which ϕ' and ϕ^* are now distributions in $\Phi_{\mathrm{dis}}(G')$ and $\Phi_{\mathrm{dis}}(G^*)$ respectively. This version follows from (1.4.11) itself, its variant at the beginning of §8.4, and the definitions above. Similarly, we have a correspondence

$$(G', \phi') \longrightarrow (\phi, s), \qquad G' \in \mathcal{E}(G),\ s \in S_{\phi,\mathrm{ss}}, \tag{9.4.2}$$

for any $\phi \in \Phi_{\mathrm{dis}}(G)$, and its centralizer $S_\phi = S_{\phi^*}$.

Suppose that ϕ belongs to $\Phi_{\mathrm{dis}}(G)$. Its centralizer group S_ϕ has a quotient $\overline{S}_\phi$, which comes with the extension $S_{\phi,\mathrm{sc}}$ by $\widehat{Z}_{\mathrm{sc}}$ defined as at the end of §9.2. The component group $\mathcal{S}_{\phi,\mathrm{sc}} = \pi_0(S_{\phi,\mathrm{sc}})$ of $S_{\phi,\mathrm{sc}}$ is in turn an extension of $\mathcal{S}_\phi = \pi_0(\overline{S}_\phi)$ by the quotient $\widehat{Z}_{\phi,\mathrm{sc}}$ of $\widehat{Z}_{\mathrm{sc}}$, again as in §9.2. The local classification for G requires a packet $\Pi_\phi \subset \Pi_{\mathrm{temp}}(G)$ of tempered representations for each $\phi \in \Phi_{\mathrm{dis}}(G)$, and an irreducible character $\langle \cdot, \pi \rangle$ for each $\pi \in \Pi_\phi$. When $G = G^*$ is quasisplit, $\langle \cdot, \pi \rangle$ is a linear character on the abelian group $\mathcal{S}_\phi$, or equivalently, a $\widehat{Z}_{\phi,\mathrm{sc}}$-invariant linear character on $\mathcal{S}_{\phi,\mathrm{sc}}$. For general G, however, $\langle \cdot, \pi \rangle$ will have to be an irreducible character on

$\mathcal{S}_{\phi,\text{sc}}$ that is equivariant under the pullback of $\widehat{\zeta}_G$ to $\widehat{Z}_{\phi,\text{sc}}$. For example, if ϕ belongs to the subset

$$\{\Phi_{\text{dis},2}(G) = \{\phi \in \Phi_{\text{dis}}(G) : \ |S_\phi| < \infty\}$$

of "square integrable" elements, $S_{\phi,\text{sc}}$ equals $\mathcal{S}_{\phi,\text{sc}}$, and $\widehat{Z}_{\phi,\text{sc}}$ equals $\widehat{Z}_{\text{sc}}$. Since the finite group $S_{\phi,\text{sc}}$ is often nonabelian, $\langle \cdot, \pi \rangle$ will generally be an irreducible character of higher degree, of the kind we described in §9.2.

We discussed the Langlands-Shelstad transfer mapping for G in the last section. For a given transfer factor $\Delta \in T(G, G')$, f'_Δ is a function in $\mathcal{S}(G')$ that can be paired with any of the stable distributions $\phi' \in \Phi_{\text{dis}}(G')$. A given ϕ' thus provides a linear form $f'_\Delta(\phi')$ in $f \in \mathcal{H}(G)$. We have implicitly agreed to normalize the transfer factors according to the ideas of Kaletha and Kottwitz. We therefore fix the section $\rho = \rho_K$ of Lemma 9.3.1 by the formula (9.3.16) that is based on the constructions in §9.3. This gives us the normalization

$$f' = f'_\rho = \rho(\Delta, s_{\text{sc}})^{-1} f'_\Delta \tag{9.4.3}$$

of the Langlands-Shelstad transfer mapping. We are interested in the corresponding normalization

$$f \longrightarrow f'(\phi') = \rho(\Delta, s_{\text{sc}})^{-1} f'_\Delta(\phi'), \qquad f \in \mathcal{H}(G), \tag{9.4.4}$$

of the linear form attached to any $\phi' \in \Phi_{\text{dis}}(G')$.

Recall that the pair (Δ, s_{sc}) in (9.4.4) belongs to the covering space $T_{\text{sc}}(G)$ in (9.3.2), or more precisely, to the fibre $T_{\text{sc}}(G, G')$ of $G' \in E(G)$ in $T_{\text{sc}}(G)$. The right hand side of (9.4.4) remains unchanged if Δ is replaced either by a $U(1)$-multiple of $t\Delta$ or an $\text{Out}_G(G')$-image of $\alpha'\Delta$, by the properties of $\rho(\Delta, s_{\text{sc}})$ in Lemma 9.3.1. It is therefore independent of Δ. It does depend on the preimage s_{sc} of the point $\bar{s}'$ attached to G', as does its geometric counterpart (9.3.5). For if s_{sc} is replaced by the translate zs_{sc} by an element $z \in \widehat{Z}_{\psi,\text{sc}}$, the right hand side of (9.4.4) is replaced by its product with $\widehat{\zeta}_G(z)$, again by Lemma 9.3.1. We would like to describe the decomposition of the $\widehat{\zeta}_G$-equivariant linear form (9.4.4) into irreducible characters.

The local classification for G is formulated as the following theorem. It represents an analogue for G of the assertions of Theorems 1.5.1 and 2.2.1(b), specialized to generic parameters ϕ but with the local refinements of §8.4.

THEOREM 9.4.1. *Assume that F is local, and that G is an inner twist of the quasisplit group $G^* \in \widetilde{\mathcal{E}}_{\text{sim}}(N)$ over F.*

(a) *For each $\phi \in \Phi_{\text{dis}}(G)$, there is a finite packet $\Pi_\phi \subset \Pi_{\text{temp}}(G)$ of irreducible tempered representations of $G(F)$, together with a mapping*

$$\pi \longrightarrow \langle \cdot, \pi \rangle, \qquad \pi \in \Pi_\phi,$$

from Π_ϕ to the set $\Pi(\mathcal{S}_{\phi,\text{sc}}, \widehat{\zeta}_G)$ of irreducible, $\widehat{\zeta}_G$-equivariant characters on $\mathcal{S}_{\phi,\text{sc}}$, with the following property. If s_{sc} is a semisimple element in the group

$S_{\phi,\mathrm{sc}}$ with images s in S_ϕ and x_{sc} in $\mathcal{S}_{\phi,\mathrm{sc}}$, and (G',ϕ') is the preimage of the pair (ϕ, s) under the correspondence (9.4.2), *then*

$$f'(\phi') = \sum_{\pi\in\Pi_\phi} \langle x_{\mathrm{sc}}, \pi\rangle f_G(\pi), \qquad f \in \mathcal{H}(G). \tag{9.4.5}$$

(b) *For any $\phi \in \Phi_{\mathrm{dis}}(G)$, the mapping from Π_ϕ to $\Pi(\widehat{\mathcal{S}}_{\phi,\mathrm{sc}}, \widehat{\zeta}_G)$ is injective, and bijective if F is nonarchimedean. Moreover, any representation in $\Pi_{\mathrm{temp}}(G)$ occurs in exactly one packet Π_ϕ.*

REMARKS. 1. As we have explained, the left hand side $f'(\phi')$ of the spectral identity (9.4.5) depends on the underlying quasisplit inner twist of the group G, and the normalizing section $\rho_K(\Delta, s_{\mathrm{sc}})$ for the corresponding transfer factors. Once these data have been fixed, it is clear that (9.4.5) characterizes the packet Π_ϕ and the pairing $\langle x_{\mathrm{sc}}, \pi\rangle$. We can think of the coefficients $\langle x_{\mathrm{sc}}, \pi\rangle$ in this formula as "spectral transfer factors", normalized by the functions $\rho_K(\Delta, s_{\mathrm{sc}})$.

2. The theorem asserts that $\Pi_{\mathrm{temp}}(G)$ is a disjoint union over $\phi \in \Phi_{\mathrm{dis}}(G)$ of finite subsets Π_ϕ, which are in turn parametrized by irreducible characters in the sets $\Pi(\mathcal{S}_{\phi,\mathrm{sc}}, \widehat{\zeta}_G)$. It does therefore represent a classification of the set $\Pi_{\mathrm{temp}}(G)$ of irreducible tempered representations of $G(F)$.

3. If F is archimedean, the assertions of the theorem are no doubt implicit in the results **[S4]**–**[S6]** of Shelstad. She actually establishes the $\mathbb{R}$-analogue of Theorem 9.4.1 (for general groups) with an abstract normalizing section $\rho(\Delta, s_{\mathrm{sc}})$ as in **[A19]** (and Lemma 9.3.1), rather than the explicit factor $\rho_K(\Delta, s_{\mathrm{sc}})$ of Kaletha. We will need to assume the more precise version of her results for our group G in the global proof **[A28]** of the theorem for p-adic F.

4. As is well known, the endoscopic classification of tempered representations leads directly to the classification of the set $\Pi(G)$ of all irreducible representations. (See for example the remarks for quasisplit groups $G^* \in \widetilde{\mathcal{E}}_{\mathrm{sim}}(N)$ at the beginning of §7.4 or in the discussion following (1.5.1).) For the group G here, we would extend the definition of the set $\Phi_{\mathrm{dis}}(G)$ to a larger set $\Phi^+_{\mathrm{dis}}(G)$ that is (noncanonically) bijective with the set $\Phi(G)$ all Langlands parameters. Theorem 9.4.1 implies that $\Pi(G)$ is a disjoint union over $\phi^+ \in \Phi^+_{\mathrm{dis}}(G)$ of finite packets of Langlands quotients Π_{ϕ^+} that are in canonical bijection with the associated packets Π_ϕ in $\Pi_{\mathrm{temp}}(G)$ given by the theorem, and a suitable correspondence $\phi^+ \to \phi$ from $\Phi^+_{\mathrm{dis}}(G)$ to $\Phi_{\mathrm{dis}}(G)$. However, the identity (9.4.5) is no longer valid in this case. (We recall that, alternatively, the identity (9.4.5) would remain intact if Π_{ϕ^+} were replaced by the corresponding packet R_{ϕ^+} of reducible standard representations, rather than irreducible Langlands quotients.) This circumstance can be seen as a local reflection of the need for a set $\Psi_{\mathrm{dis}}(G)$ that lies between $\Phi_{\mathrm{dis}}(G)$ and $\Phi^+_{\mathrm{dis}}(G)$.

The results of §8.4 apply only to generic parameters ϕ. We have therefore to be content to work here with set $\widetilde{\Psi}(G)$ of ($\widetilde{O}(G)$-orbits of) general parameters for G, which we introduced in §9.2. However, there is a problem that arises with this set. It concerns the transfer of automorphisms discussed in §9.1, and occurs specifically for indices in the local classification that contain subdiagrams (9.1.5) and (9.1.6). For in those cases, we cannot define a G-analogue of the set $\widetilde{\Pi}_{\text{unit}}(G^*)$ that we would use to formulate the general local theorem.

Accordingly, we shall say that the inner twist G is *symmetric* if any outer automorphism θ^* of G^* (other than a triality automorphism for the split group $G = SO(8)$) transfers to an F-automorphism θ of G. This is automatic if G is of type $\mathbf{B}_n$ or $\mathbf{C}_n$. For the case of type $\mathbf{D}_n$, it means that G and $\theta^* = \widetilde{\theta}^*$ satisfy the equivalent conditions of Lemma 9.1.1. In general, then, G is symmetric if and only if its index from the tables of §9.1 does not contain a subdiagram (9.1.5) or (9.1.6).

Assume that G is symmetric. If it is of type $\mathbf{D}_n$, we write $\widetilde{\theta}$ for a fixed F-automorphism of G of order 2 obtained by transfer of the automorphism $\widetilde{\theta}^*$ of G^*. If G is of type $\mathbf{B}_n$ or $\mathbf{C}_n$, we simply set $\widetilde{\theta} = 1$. In all cases, we then write $\widetilde{\mathcal{H}}(G)$ for the subalgebra of $\widetilde{\theta}$-invariant functions in the Hecke algebra $\mathcal{H}(G)$. We can also write $\widetilde{\Pi}(G)$, $\widetilde{\Pi}_{\text{unit}}(G)$ and $\widetilde{\Pi}_{\text{temp}}(G)$ for the families of $\widetilde{\theta}$-orbits in the sets $\Pi(G)$, $\Pi_{\text{unit}}(G)$ and $\Pi_{\text{temp}}(G)$ of irreducible representations of $G(F)$. We are of course already familiar with the quotients $\widetilde{\Phi}(G)$, $\widetilde{\Psi}(G)$ and $\widetilde{\Phi}_{\text{bdd}}(G)$ of the parameter sets $\Phi(G)$, $\Psi(G)$ and $\Phi_{\text{bdd}}(G)$. They are the orbits for the dual automorphism $\widehat{\widetilde{\theta}} = \widehat{\widetilde{\theta}}^*$ of $\widehat{G} = \widehat{G}^*$, which are defined for all of our G. We note that the symmetric, stable Hecke algebra $\widetilde{\mathcal{S}}(G^*)$ for the quasisplit group G^* can be defined in terms of either $\widehat{\widetilde{\theta}}^*$ (and Langlands parameters ϕ^*) or the original automorphism $\widetilde{\theta}^*$ of G^* (and stable conjugacy classes δ^*). Similarly, the symmetric algebras $\widetilde{\mathcal{S}}(G')$ for endoscopic groups G', which played such a key role in the first eight chapters, are defined in terms of either the automorphism $\widetilde{\theta}' = \widetilde{\theta}^{G'}$ of G' or its dual.

The F-automorphism $\widetilde{\theta}$ of G is not uniquely determined by the original outer automorphism $\widetilde{\theta}^*$ of G^*. Any choice of $\widetilde{\theta}$ can always be modified by composition with an inner automorphism in $\mathrm{Int}\big(G_{\text{ad}}(F)\big)$. The normalizing section ρ_K is also not uniquely determined, and can be modified in a similar fashion. We will assume that they are both chosen so that for any $G' \in \mathcal{E}(G)$, the transfer mapping $f \to f'$ satisfies the identity

$$(f \circ \widetilde{\theta}) = f' \circ \widetilde{\theta}', \qquad f \in \mathcal{H}(G), \tag{9.4.6}$$

and therefore takes the symmetric subalgebra $\widetilde{\mathcal{H}}(G)$ into $\widetilde{\mathcal{S}}(G')$. The linear form

$$f'(\psi'), \qquad f \in \widetilde{\mathcal{H}}(G),$$

on $\widetilde{\mathcal{H}}(G)$ is then defined for any $\widetilde{\theta}'$-orbit $\psi' \in \widetilde{\Psi}(G')$.

I am labelling the following assertion as a conjecture, since I have not written down the details of its proof. However, the methods of proof are parallel to those of the quasisplit case treated in Chapter 7, and should pose no new problems.

CONJECTURE 9.4.2. *Assume that F is local, and that G is a symmetric inner twist of the quasisplit group $G^* \in \widetilde{\mathcal{E}}_{\mathrm{sim}}(N)$ over F. Assume also that $\widetilde{\theta}$ and ρ_K are fixed so that for any $G' \in \mathcal{E}(G)$, the transfer mapping $f \to f'$ satisfies (9.4.6). Then for each $\psi \in \widetilde{\Psi}(G)$, there is a finite set $\widetilde{\Pi}_\psi$ over $\widetilde{\Pi}_{\mathrm{unit}}(G)$, together with a mapping*

$$\pi \longrightarrow \langle \cdot, \pi \rangle, \qquad \pi \in \widetilde{\Pi}_\psi,$$

from $\widetilde{\Pi}_\psi$ to the set $\Pi(\mathcal{S}_{\psi,\mathrm{sc}}, \widehat{\zeta}_G)$, such that

$$f'(\psi') = \sum_{\pi \in \widetilde{\Pi}_\psi} \langle s_\psi x_{\mathrm{sc}}, \pi \rangle f_G(\pi), \qquad f \in \widetilde{\mathcal{H}}(G), \tag{9.4.7}$$

for $s_{\mathrm{sc}} \in S_{\psi,\mathrm{sc}}$, $s \in S_\psi$, and $x_{\mathrm{sc}} \in \mathcal{S}_{\psi,\mathrm{sc}}$ as in the statement of Theorem 9.4.1, and (G', ψ') the preimage of (ψ, s).

REMARKS. 1. If F is archimedean, we again expect the packets $\widetilde{\Pi}_\psi$ to be special cases for G of the general packets constructed in [**ABV**]. However, this is not presently known.

2. If F is p-adic, at least some of the packets $\widetilde{\Pi}_\psi$ are among those constructed by Moeglin, by quite different methods. As in the the quasisplit case, she proves the fundamental theorem that these packets are multiplicity free, and are therefore simply subsets of $\widetilde{\Pi}(G)$ [**M4**].

3. Observe that, as in the earlier quasisplit case, the theorem includes the assertion that the irreducible representations that comprise the $\widetilde{\theta}$-orbits in the packets $\widetilde{\Pi}_\psi$ are unitary. This is one point of difference between the assertion here and the sharper results described in Remarks 1 and 2. Another is the fact that the proof of Conjecture 9.4.2 includes an extension to G of the local intertwining relation, which will in turn be required for the interpretation of the global trace formula.

4. The method for proving Conjecture 9.4.2 is global (as is that of Theorem 9.4.1). For this reason, both proofs are conditional on a certain global hypothesis that has not yet been established for the local normalizations of Kaletha. We shall state the hypothesis formally in the next section.

5. The point s_ψ in (9.4.5) represents the canonical preimage in $\mathcal{S}_{\psi,\mathrm{sc}}$ of the original point $s_\psi \in \mathcal{S}_\psi$ in (2.2.6). It is defined in terms of the lifting of a homomorphism whose domain is the simply connected group $SL(2, \mathbb{C})$. If $\phi = \psi$ lies in the subset $\widetilde{\Phi}_{\mathrm{bdd}}(G)$ of $\widetilde{\Psi}(G)$, we remind ourselves that the point s_ϕ (regarded as an element in either $\mathcal{S}_\phi$ or $\mathcal{S}_{\phi,\mathrm{sc}}$) is trivial. The spectral identity (9.4.7) then reduces formally to its analogue (9.4.5) from the previous theorem, although the terms in the earlier identity of course have slightly different meanings.

The conjecture does not account for the groups G that are not symmetric. These are the inner forms of a split even orthogonal group $G^* = SO(N)$

that do not extend to inner forms of the full orthogonal group $O(N)$. The case is of considerable interest, for the theory of Shimura varieties among other things. Theorem 9.4.1 does apply to these groups, and gives a construction of their tempered L-packets Π_ϕ. It would obviously be desirable to modify Conjecture 9.4.2 so that it applies as well. We shall describe a strategy for doing so, with the hope of carrying it out in [**A28**].

The problem is that we cannot define the symmetric Hecke subalgebra $\widetilde{\mathcal{H}}(G)$ if G itself is not symmetric, since $\widetilde{\theta}$ is then not an automorphism of G. If f is a general function in $\mathcal{H}(G)$, its endoscopic transfer f' does not belong to the stable symmetric image $\widetilde{\mathcal{S}}(G')$ of $\widetilde{\mathcal{H}}(G')$, and cannot be paired with the stable linear form ψ' given by Theorem 2.2.1(a). In other words, the left hand side $f'(\psi')$ of (9.4.7) is undefined. It ϕ belongs to the subset $\widetilde{\Phi}_{\mathrm{bdd}}(G)$ of $\widetilde{\Psi}(G)$, we do have associated linear forms $\phi' \in \Phi_{\mathrm{dis}}(G')$ on the full space $\mathcal{S}(G')$. This was one of the main points of §8.4, and is the reason that the left hand side $f'(\phi')$ of (9.4.5) is defined. But we have not extended the methods of §8.4 to general parameters $\psi \in \widetilde{\Psi}(G)$. Let us consider instead how we might extend Conjecture 9.4.2, as an assertion for symmetric objects. We can start by trying to reconcile the assertion of Conjecture 9.4.2 with that of Theorem 9.4.1(a), in the case of a parameter $\psi = \phi$ that is generic.

We should first say a word about the transform represented by the formula (9.4.5) of Theorem 9.4.1. It maps a given function $f \in \mathcal{H}(G)$ to the function

$$f_G^{\mathcal{E}}(G', \phi') = f_G^{\mathcal{E}}(\phi, s_{\mathrm{sc}}),$$

defined by either of the two sides of the identity

$$f'(\phi') = \sum_{\pi \in \Pi_\phi} \langle x_{\mathrm{sc}}, x \rangle f_G(\pi).$$

Recall that ϕ belongs to $\Phi_{\mathrm{dis}}(G)$, that s_{sc} is a semisimple element in $S_{\phi,\mathrm{sc}}$ with images $s \in S_\phi$ and x_{sc} in $\mathcal{S}_{\phi,\mathrm{sc}}$, and that (G', ϕ') is the preimage of (ϕ, s). As the notation suggests, the image $f_G^{\mathcal{E}}$ of f depends only on the projection f_G of f onto the invariant Hecke algebra $\mathcal{I}(G)$. Let us write $\mathcal{I}^{\mathcal{E}}(G)$ for the image of $\mathcal{I}(G)$ under this mapping, regarded as a space of functions $f_G^{\mathcal{E}}$ of either the variable (G', ϕ') or the variable (ϕ, x_{sc}). Using the trace Paley-Wiener theorem for G, one can describe $\mathcal{I}^{\mathcal{E}}(G)$ explicitly in terms of the appropriate Paley-Wiener spaces. The mapping

$$T_G^{\mathcal{E}} : f_G \longrightarrow f_G^{\mathcal{E}}, \qquad f_G \in \mathcal{I}(G), \tag{9.4.8}$$

then becomes a topological isomorphism from $\mathcal{I}(G)$ onto $\mathcal{I}^{\mathcal{E}}(G)$.

How does the mapping (9.4.8) relate to automorphisms? If G is symmetric, we have the F-automorphism $\widetilde{\theta}$. It acts on $G(F)$, and allows us to define the symmetric Hecke $\widetilde{\mathcal{H}}(G)$ above. It also descends to a linear (topological) automorphism $\widetilde{\theta}_G$ of $\mathcal{I}(G)$ such that

$$\widetilde{\theta}_G f_G = (f \circ \widetilde{\theta})_G, \qquad f \in \mathcal{H}(G),$$

where each side can be treated as a function of either $\gamma \in \Gamma_{\mathrm{reg}}(G)$ or $\pi \in \Pi_{\mathrm{temp}}(G)$. This allows us to define the symmetric invariant Hecke algebra $\widetilde{\mathcal{I}}(G)$, as for example the image of $\widetilde{\mathcal{H}}(G)$ in $\mathcal{I}(G)$. But what is pertinent to the discussion here is that we also have the dual automorphism $\widehat{\widetilde{\theta}} = \widehat{\widetilde{\theta}}^*$ of $\widehat{G} = \widehat{G}^*$, and this exists in all cases. It acts on the set $\Phi_{\mathrm{bdd}}(G^*)$ of bounded Langlands parameters for G^*, and therefore on the set of stable distributions $\Phi_{\mathrm{dis}}(G^*)$.

Suppose that G is symmetric. Then $\widehat{\widetilde{\theta}}$ stabilizes the image of $\Phi_{\mathrm{dis}}(G)$ in $\Phi_{\mathrm{dis}}(G^*)$. It therefore acts both on the set of pairs (G', ϕ') and the set of pairs (ϕ, s_{sc}), either of which serves as the domain for the space of functions $\mathcal{I}^{\mathcal{E}}(G)$. This gives a linear automorphism $\widetilde{\theta}_G^{\mathcal{E}}$ of $\mathcal{I}^{\mathcal{E}}(G)$. One sees from (9.4.6) that $\widetilde{\theta}_G^{\mathcal{E}}$ is the transfer to $\mathcal{I}^{\mathcal{E}}(G)$ of the automorphism $\widetilde{\theta}_G^{-1}$ (which equals $\widetilde{\theta}_G$ in this case). In other words,

$$\widetilde{\theta}_G^{\mathcal{E}} = (\mathcal{T}_G^{\mathcal{E}})\widetilde{\theta}_G^{-1}(\mathcal{T}_G^{\mathcal{E}})^{-1}.$$

We emphasize that on the right hand side of this relation, $\widetilde{\theta}_G^{-1}$ is defined in terms of the automorphism $\widetilde{\theta}^{-1}$ of G, while its counterpart $\widetilde{\theta}_G^{\mathcal{E}}$ on the left hand side is obtained from the dual automorphism $\widehat{\widetilde{\theta}}$ of $\widehat{G}$. The relation, which is only for motivation, allows us to compare Conjecture 9.4.2 and Theorem 9.4.1 in the special cases that $\psi = \phi$ is generic and G is symmetric.

Suppose now that G is not symmetric, the case at which the discussion is aimed. We would still like to define a symmetric subspace $\widetilde{\mathcal{I}}(G)$ of $\mathcal{I}(G)$.

According to Remark 1 following Lemma 9.1.1, we have an F-isomorphism $\widetilde{\theta}^\vee\colon\ G \to G^\vee$ of groups, in which $G^\vee$ represents a second inner twist $(G^\vee, \psi^\vee)$ of G^* that is not isomorphic (as an inner twist) to (G, ψ). The remarks above can be extended to the isomorphism $\widetilde{\theta}^\vee$. They lead to topological isomorphisms

$$\widetilde{\theta}_G^\vee : \mathcal{I}(G^\vee) \longrightarrow \mathcal{I}(G)$$

and

$$\widetilde{\theta}_G^{\vee,\mathcal{E}} : \mathcal{I}^{\mathcal{E}}(G) \longrightarrow \mathcal{I}^{\mathcal{E}}(G^\vee),$$

such that

$$\widetilde{\theta}_G^{\vee,\mathcal{E}} = (\mathcal{T}_{G^\vee}^{\mathcal{E}})(\widetilde{\theta}_G^\vee)^{-1}(\mathcal{T}_G^{\mathcal{E}})^{-1}. \tag{9.4.9}$$

It has been our convention to identify $\widehat{G}^*$ with $\widehat{G}$ by means of the dual isomorphism $\widehat{\psi}$ of the original inner twist ψ. Let us agree also to identify $\widehat{G}^\vee$ with $\widehat{G}$ by using the dual $\widehat{\widetilde{\theta}}^\vee$ of $\widetilde{\theta}^\vee$. What then remains is the dual isomorphism $\widehat{\psi}^\vee$ from $\widehat{G}^*$ to $\widehat{G}^\vee$ of the second inner twist $\psi^\vee$. As an automorphism of $\widehat{G} = \widehat{G}^* = \widehat{G}^\vee$, it equals the dual $\widehat{\widetilde{\theta}}^*$ of $\widetilde{\theta}^*$. It gives rise to a topological isomorphism

$$\widetilde{\theta}_G^{*,\mathcal{E}} : \mathcal{I}^{\mathcal{E}}(G^\vee) \longrightarrow \mathcal{I}^{\mathcal{E}}(G),$$

since the dual automorphism $\widehat{\widetilde{\theta}}{}^*$ is all that is needed to apply the original construction above. The composition

$$\widetilde{\theta}_G^{\mathcal{E}} = \widetilde{\theta}_G^{*,\mathcal{E}} \cdot \widetilde{\theta}_G^{\vee,\mathcal{E}} \tag{9.4.10}$$

is then a topological linear automorphism of $\mathcal{I}^{\mathcal{E}}(G)$. It in turn gives a topological linear automorphism

$$\widetilde{\theta}_G = (T_G^{\mathcal{E}})^{-1}(\widetilde{\theta}_G^{\mathcal{E}})^{-1}(T_G^{\mathcal{E}}) :\ \mathcal{I}(G) \longrightarrow \mathcal{I}(G) \tag{9.4.11}$$

of the original invariant Hecke algebra $\mathcal{I}(G)$.

It is not hard to see explicitly how the two isomorphisms $\widetilde{\theta}_G^{\vee,\mathcal{E}}$ and $\widetilde{\theta}_G^{*,\mathcal{E}}$ transform functions in $\mathcal{I}^{\mathcal{E}}(G^\vee)$ and $\mathcal{I}^{\mathcal{E}}(G)$ on their respective domains $\{(\phi^\vee, x_{\mathrm{sc}}^\vee)\}$ and $\{(\phi, x_{\mathrm{sc}})\}$. This exercise, which we leave to the reader, leads to concrete descriptions of the automorphisms $\widetilde{\theta}_G^{\mathcal{E}}$ and $\widetilde{\theta}_G$ of (9.4.10) and (9.4.11). In particular, one sees that $\widetilde{\theta}_G$ acts through an involution

$$\pi \longrightarrow \widetilde{\theta}\pi, \qquad \pi \in \Pi_{\mathrm{temp}}(G)$$

on $\Pi_{\mathrm{temp}}(G)$ that is given explicitly in terms of the endoscopic classification of Theorem 9.4.1. We can therefore define the symmetric subalgebra $\widetilde{\mathcal{I}}(G)$ of $\mathcal{I}(G)$, and the corresponding quotient $\widetilde{\Pi}_{\mathrm{temp}}(G)$ of orbits in $\Pi_{\mathrm{temp}}(G)$, even though we do not have an underlying F-automorphism $\widetilde{\theta}$ of G. This is what we were looking for. It allows us to extend the assertion of Conjecture 9.4.2 to the nonsymmetric group G simply by using an invariant function $f_G \in \widetilde{\mathcal{I}}(G)$ in place of $f \in \widetilde{\mathcal{H}}(G)$.

The extension of Conjecture 9.4.2 will also require a modification for the set $\widetilde{\Pi}_{\mathrm{unit}}(G)$, since the involution $\widetilde{\theta}$ described above is defined only for tempered representations. The simplest solution would be to postulate a natural extension of $\widetilde{\theta}$ to $\Pi_{\mathrm{unit}}(G)$ as part of the conjecture. However, this could be quite hard to prove without further hypotheses. The methods for proving Conjecture 9.4.2, as it is stated for symmetric groups, will no doubt lead to some result for the nonsymmetric group G, once we have the symmetric invariant Hecke algebra $\widetilde{\mathcal{I}}(G)$. But the packet $\widetilde{\Pi}_\psi$ of orbits of irreducible unitary representations might then have to be replaced by something weaker.

Could we expect to be able to define an involution on $\Pi_{\mathrm{unit}}(G)$ (or $\Pi_{\mathrm{temp}}(G)$) directly? There is a natural involution on the full set $\Pi(G)$, the contragredient

$$\pi \longrightarrow \pi^\vee, \qquad \pi \in \Pi(G).$$

Is it possible to relate it to the involution on $\Pi_{\mathrm{unit}}(G)$ we seek, and in particular, the involution $\widetilde{\theta}$ on $\Pi_{\mathrm{temp}}(G)$ we have just described? D. Prasad [**Pr**] has given a conjectural description of the contragredient involution for any group in terms of the Langlands parametrization, which of course is also still conjectural in general. Kaletha [**Kal3**] has recently proved Prasad's conjecture for general real groups (following earlier work of Adams and Vogan [**AV**] and Shelstad's endoscopic classification [**S6**]), and for quasisplit orthogonal and symplectic groups (using the version of the Langlands parametrization established in §8.4). We are still thinking of a nonsymmetric inner twist G, necessarily of type $\mathbf{D}_n$, for which there is no F-automorphism $\widetilde{\theta}$ we can use to define our involution. The answer to the questions then depends on the parity of n.

Suppose that n is odd. The index of G then has a subdiagram (9.1.6). In this case, $\widehat{\widetilde{\theta}}^*$ equals the opposition involution [**Ti**, 1.5.1] (also called the Chevalley or duality involution), as an outer automorphism of $\widehat{G} = \widehat{G}^*$. Prasad's conjecture describes the contragredient involution on $\Pi(G)$ in terms of the action of the opposition involution on $\mathcal{I}^{\mathcal{E}}(G)$. In our notation, it amounts to an assertion

$$\widetilde{\theta}\pi = \eta\pi^\vee, \qquad \pi \in \Pi_{\text{temp}}(G),$$

where η is an explicit element in the finite group

$$\text{Int}_F(G)/\text{Int}\big(G(F)\big) = \text{Int}\big(G_{\text{ad}}(F)\big)/\text{Int}\big(G(F)\big)$$

that is independent of π, and

$$(\eta\pi^\vee)(x) = \pi^\vee\big(\eta^{-1}(x)\big), \qquad x \in G(\mathbb{A}).$$

Kaletha's proof of Prasad's conjecture in the p-adic case applies only to the quasisplit group G^*. His methods could presumably be combined with Theorem 9.4.1 to establish the conjecture for G. This would give a direct formula for our involution in terms of the contragredient of π. Its extension from $\Pi_{\text{temp}}(G)$ to $\Pi(G)$, a set that of course contains $\Pi_{\text{unit}}(G)$, ought then to follow from the familiar properties of Langlands quotients. It thus seems likely that Conjecture 9.4.2 and its proof will extend directly to the nonsymmetric groups with n odd.

Suppose now that n is even. The index of G then has a subdiagam of type (9.1.5). This is the unique case among all inner twists of G^* in which the subset $\Phi(G)$ of Langlands parameters in $\Phi(G^*)$ is *not* $\widehat{\widetilde{\theta}}^*$-stable. In this case, $\widehat{\widetilde{\theta}}^*$ is not equal to the opposition involution on $\widehat{G} = \widehat{G}^*$, since the latter is trivial (as an outer automorphism). It follows from Prasad's conjecture that our involution $\widetilde{\theta}$ on $\Pi_{\text{temp}}(G)$ has nothing to do with the contragredient. Is there anything more that one can say? I have no answer, other than to recall the example with $n = 2$ from the end of §9.1. Set

$$G = G_1 = Sp(2) \times Q^1/\{\pm 1\}$$

in the earlier notation, and suppose that

$$\pi = \sigma \otimes \tau$$

is a representation in $\Pi_2(G)$, with factors $\sigma \in \Pi_2\big(Sp(2)\big)$ and $\tau \in \Pi_2(Q^1)$ that both have trivial central character. The automorphism $\widehat{\widetilde{\theta}}^*$ acts on

$$\widehat{G} = \widehat{G}^* = Sp(2,\mathbb{C}) \times Sp(2,\mathbb{C})/\{\pm 1\}$$

by interchanging the two $Sp(2,\mathbb{C})$ factors. It is then not hard to see that

$$\widetilde{\theta}\pi = \tau^* \otimes \sigma_*,$$

where τ^* is the Jacquet-Langlands "lift" of τ from Q^1 to $Sp(2)$, and σ_* is the Jacquet-Langlands "descent" of σ from $Sp(2)$ to Q^1. This is a rather subtle operation. It is hard to see how its generalization could lead to a direct way of characterizing the involution $\widetilde{\theta}$ on $\Pi_{\mathrm{temp}}(G)$.

We have discussed at some length how we might extend Conjecture 9.4.2 to all inner twists G of G^*. We now complete the section with a few words on the possibility of generalizing Theorem 9.4.1(a) to all parameters for G. For this, we may as well just assume that G is of type $\mathbf{D}_n$, since the answer is otherwise just the assertion of Conjecture 9.4.2.

Theorem 9.4.1 is based on the results of §8.4 for generic parameters $\phi^* \in \widetilde{\Phi}_{\mathrm{bdd}}(G^*)$. To generalize the theorem, one would need to establish similar results for nongeneric parameters $\psi^* \in \widetilde{\Psi}(G^*)$. For the moment, we assume that $G = G^*$ is quasisplit. Then $G \in \widetilde{\mathcal{E}}_{\mathrm{sim}}(2n)$ is a quasisplit, even orthogonal group over the general local field F.

Recall that $\widetilde{\Pi}_{\mathrm{unit}}(G)$ is the set of orbits in $\Pi_{\mathrm{unit}}(G)$ of the group

$$\widetilde{O}(G) = \widetilde{\mathrm{O}}\mathrm{ut}_N(G) = \mathbb{Z}/2\mathbb{Z}.$$

Any subset $\widetilde{\Pi}$ of $\widetilde{\Pi}_{\mathrm{unit}}(G)$ has a preimage Π in $\Pi_{\mathrm{unit}}(G)$. More generally, if $\widetilde{\Pi}$ is a set *over* $\widetilde{\Pi}_{\mathrm{unit}}(G)$, we can write Π for the corresponding set over $\Pi_{\mathrm{unit}}(G)$. For any $\psi \in \widetilde{\Psi}(G)$, the packet $\widetilde{\Pi}_\psi$ of Theorem 1.5.1(a) (which was constructed in §7.4) is a finite set over $\widetilde{\Pi}_{\mathrm{unit}}(G)$. Its preimage Π_ψ over $\Pi_{\mathrm{unit}}(G)$ is then the fibre product of $\widetilde{\Pi}_\psi$ and $\Pi_{\mathrm{unit}}(G)$ over $\widetilde{\Pi}_{\mathrm{unit}}(G)$. The problem is to construct a section

$$\widetilde{\Pi}_\psi \longrightarrow \Pi_\psi \tag{9.4.12}$$

that is compatible with endoscopic transfer for G. This is essentially the content of the assertions of Theorem 8.4.1 of §8.4, for the case that $\psi = \phi$ lies in the subset $\widetilde{\Phi}_{\mathrm{bdd}}(G)$ of generic local parameters. In the generic case, it is an easy consequence of the structure of the original packet $\widetilde{\Pi}_\phi$ that the section is uniquely determined up to the action of the group $\widetilde{O}(G)$. More precisely, the set of such sections is a single orbit under the group $\widetilde{O}(G)$. For a general parameter $\psi \in \widetilde{\Psi}(G)$, one can construct a section (9.4.12) by following the global proof of Theorem 8.4.1. However, its uniqueness requires more information about the packet $\widetilde{\Pi}_\psi$. This is a serious matter, since without the uniqueness, the global applications of the refinements (9.4.12) will fail.

It is not hard to impose conditions on ψ that would imply the uniqueness of (9.4.12). Chief among these is the requirement that elements in the packet $\widetilde{\Pi}_\psi$ occur with multiplicity 1, or in other words, that $\widetilde{\Pi}_\psi$ be a subset of $\widetilde{\Pi}_{\mathrm{unit}}(G)$. For p-adic F, Moeglin [**M4**] has established that any packet $\widetilde{\Pi}_\psi$

has multiplicity 1. This is a deep theorem, which describes a fundamental structural property of the packets. But if $F = \mathbb{R}$, the property has so far been elusive, even though one might expect it to be true. In any case, we can formally introduce a canonical set

$$\Psi_{\mathrm{dis}}(G) = \coprod_{\psi} \Psi_{\mathrm{dis}}(\psi), \tag{9.4.13}$$

where $\psi \in \widetilde{\Psi}(G)$ ranges over parameters that satisfy the conditions of uniqueness, and $\Psi_{\mathrm{dis}}(\psi)$ is the set (of order 1 or 2) of $O(G)$-orbits of sections (9.4.12). For any ψ_* in $\Psi_{\mathrm{dis}}(\psi)$, we define Π_{ψ_*} to be the image of ψ_* in Π_ψ, regarded as a packet of invariant linear forms on $\mathcal{H}(G)$. The main point is that Π_{ψ_*} satisfies the appropriate refinements of the endoscopic character relations for $\widetilde{\Pi}_\psi$. In particular, the packet comes with a canonical stable distribution

$$f \longrightarrow f^G(\psi_*) = \sum_{\pi \in \Pi_{\psi_*}} \langle s_{\psi_*}, \pi \rangle f_G(\pi), \qquad f \in \mathcal{H}(G),$$

which we identify with ψ_* itself.

We can now go back to the setting at the beginning of the section. We let G revert to a general inner twist of the quasisplit group $G^* \in \widetilde{\mathcal{E}}_{\mathrm{sim}}(N)$ of type $\mathbf{D}_n$. We can then write $\Psi_{\mathrm{dis}}(G)$ for the subset of distributions in $\Psi_{\mathrm{dis}}(G^*)$, which we now denote by ψ, that correspond to G-relevant parameters for G^*. The various definitions that precede the statement of Theorem 9.4.1 extend to elements $\psi \in \Psi_{\mathrm{dis}}(G)$. So does the assertion of Theorem 9.4.1(a), but with two obvious changes. The packet $\Pi_\phi \subset \Pi_{\mathrm{temp}}(G)$ is replaced by a packet $\Pi_\psi \subset \Pi_{\mathrm{unit}}(G)$, and the identity (9.4.5) becomes

$$f'(\psi') = \sum_{\pi \in \Pi_\psi} \langle s_\psi x_{\mathrm{sc}}, \pi \rangle f_G(\pi), \qquad f \in \mathcal{H}(G).$$

Let us call this new assertion a conjecture, though the techniques for proving it should again be available. Keep in mind, however, that the set $\Psi_{\mathrm{dis}}(G)$ is defined by conditions that have not been proved for all parameters in $\widetilde{\Psi}(G)$, so the result would still have significant limitations.

This completes the discussion. We have described possible extensions of Conjecture 9.4.2 and Theorem 9.4.1(a). In principle, a generalization of the theorem would be stronger than the extension of the conjecture. However, there are reasons to consider both. Some extension of Conjecture 9.4.2 ought to be within reach, as I have suggested above, while a complete generalization of Theorem 9.4.1(a) will require information that is not presently accessible. Moreover, the characters treated in Conjecture 9.4.2 satisfy local reciprocity laws that are more elementary, to the extent that they are explicitly related to twisted characters on $GL(N, F)$. The characters of Theorem 9.4.1 are defined in terms of stable characters on $G^*(F)$, which one can see from §8.4 are in general only indirectly related to twisted characters on $GL(N, F)$. We reiterate that the discussion has all been directed at groups of type $\mathbf{D}_n$. If G is of type $\mathbf{B}_n$ or $\mathbf{C}_n$, there is nothing further to say. In these cases, Conjecture 9.4.2 already represents the generalization of Theorem 9.4.1 from the set $\Phi_{\mathrm{bdd}}(G) = \widetilde{\Phi}_{\mathrm{bdd}}(G)$ to the larger set $\Psi(G) = \widetilde{\Psi}(G)$.

9.5. Statement of a global classification

We come finally to the global setting. We shall state global analogues of Theorem 9.4.1 and Conjecture 9.4.2. The first is an extension of Theorem 8.4.2. It amounts to a refined classification for the subspace of the discrete spectrum attached to certain global Langlands parameters ϕ. The full classification should be governed by the general parameters ψ, objects whose origins are unambiguously global. We shall formulate it as a conjecture that, as in the local case, applies to most but not all groups.

The field F will be global throughout this final section. We fix a quasisplit group $G^* \in \widetilde{\mathcal{E}}_{\mathrm{sim}}(N)$. Our general interest is then in global inner twists (G, ψ) of G^*. We will have to formulate the full classification of the global conjecture in terms of the locally symmetric subalgebra of the global Hecke algebra. This of course is analogous to Theorem 1.5.2. But for the inner forms G here, we will consequently be forced to restrict ourselves to groups that are locally symmetric, in the sense suggested by the last section.

The global assertions will be formulated under the assumption that the global hypothesis of [**L10**, p. 149] (which was established in [**LS1**, §6.4]) extends appropriately to the local normalized transfer factors of Kaletha. We have of course used the global results of [**LS1**] in our study of quasisplit groups. The reader will recall that in §3.2, we normalized the relevant local transfer factors by Whittaker data obtained by localization of a fixed global Whittaker datum. What we need here is a product formula for the normalizing sections

$$\rho_v(\Delta_v, s_{\mathrm{sc},v}) = \rho_{v,K}(\Delta_v, s_{\mathrm{sc},v}), \qquad (\Delta_v, s_{\mathrm{sc},v}) \in T_{\mathrm{sc}}(G_v), \tag{9.5.1}$$

attached to the localizations of a global inner twist (G, ψ). The points

$$(\Delta_v, s_{\mathrm{sc},v}) = (G'_v, \Delta_v, s_{\mathrm{sc},v})$$

depend implicitly on local endoscopic data $G'_v \in \mathcal{E}(G_v)$. They could, for example, be the localizations of a fixed global point

$$(\Delta, s_{\mathrm{sc}}) = (G', \Delta, s_{\mathrm{sc}}), \qquad G' \in \mathcal{E}(G),\ s_{\mathrm{sc}} \in S'_{\mathrm{sc}}, \tag{9.5.2}$$

in which Δ is the canonical adelic transfer factor for (G, G') [**LS1**, (6.3)], [**KS**, (6.3)].

HYPOTHESIS 9.5.1. *Suppose that (G, ψ) is a global inner twist of G^* over F. Then one can choose the normalizing sections* (9.5.1) *for the completions (G_v, ψ_v) of (G, ψ) so that*

$$\prod_v \rho_v(\Delta_v, s_{\mathrm{sc},v}) = 1, \tag{9.5.3}$$

if $(\Delta_v, s_{\mathrm{sc},v})$ *is the localization of a global point* (9.5.2).

One would expect the sections to equal 1 at places v with G_v quasisplit, so in particular, the product (9.5.3) would be over a finite set. In general, there are four sets of canonical data that ought to be part of the solution: the global transfer factor Δ for (G, G'), its localizations Δ_v at the quasisplit places v, the global transfer factor Δ^* for (G^*, G'), and its localizations Δ_v^* at all places. I have not tried to analyze these objects with the methods of [**LS1**, (4.2), (6.3)–(6.4)], partly for a lack of understanding of how to use the results of Kaletha at the remaining places v at which G is not quasisplit. I do not know whether Kaletha or Kottwitz are considering the global situation, but it seems reasonable to suppose that the questions will be resolved by the time the article [**A28**] has been written. In any case, we assume from now on that Hypothesis 9.5.1 holds. More precisely, we assume implicitly that the normalizing sections (9.5.1) for the localizations G_v of G have been chosen to satisfy (9.5.3) and of course also to be given (9.3.16) by Kaletha's local constructions. The local assertions of Theorem 9.4.1 and Conjecture 9.4.2 are also to be regarded as conditional on the hypothesis, since as we noted at the time, their proofs will be by global means.

Consider a global parameter $\psi \in \widetilde{\Psi}(G)$ and a point $s_{\mathrm{sc}} \in S_{\psi,\mathrm{sc}}$ with a semisimple image s in S_ψ. Let (G', ϕ') be the preimage of the pair (ψ, s). On one hand, the global transfer mapping

$$f'(\psi') = f'_\Delta(\psi'), \qquad f \in \mathcal{H}(G),$$

from G to G' is defined by the canonical adelic transfer factor Δ. On the other, we have agreed to define the local transfer mapping

$$f'_v(\psi'_v) = f'_{v,\rho_v}(\psi'_v), \qquad f_v \in \mathcal{H}(G_v),$$

in terms of a fixed transfer section $\rho_v = \rho_{v,K}$. We are assuming that as v varies, the local transfer sections $\{\rho_v\}$ satisfy the condition (9.5.3) of Hypothesis 9.5.1. It then follows from (9.3.3) that

$$f' = \prod_v f'_{v,\rho_v}, \tag{9.5.4}$$

for a product $f = \prod f_v$. In particular, although the factors on the right depend on the point s_{sc}, the product does not.

Before describing the finer structure of the discrete spectrum for G, we need a coarse decomposition analogous to that of the quasisplit case in Chapter 3. Extending the definition prior to Corollary 3.4.3 in §3.4, we set

$$L^2_{\mathrm{disc},\psi}\big(G(F)\backslash G(\mathbb{A})\big) = L^2_{\mathrm{disc},t(\psi),c(\psi)}\big(G(F)\backslash G(\mathbb{A})\big),$$

for our inner twist G and a general global parameter $\psi \in \widetilde{\Psi}(N)$. The next proposition is the generalization to G of the assertion (3.4.4) of Corollary 3.4.3. It can be proved with modest changes in the methods of §3.4–§3.5.

PROPOSITION 9.5.2. *The discrete spectrum for our inner twist (G, ψ) of $G^* \in \widetilde{\mathcal{E}}_{\mathrm{sim}}(N)$ over F has the following decomposition*

$$L^2_{\mathrm{disc}}\big(G(F)\backslash G(\mathbb{A})\big) = \bigoplus_{\psi \in \widetilde{\Psi}(N)} L^2_{\mathrm{disc},\psi}\big(G(F)\backslash G(\mathbb{A})\big).$$

What remains to describe is an explicit decomposition of any of the invariant subspaces $L^2_{\mathrm{disc},\psi}\big(G(F)\backslash G(\mathbb{A})\big)$ into irreducible representations. As in the last section, our first main assertion applies to the subset $\widetilde{\Phi}(G)$ of generic parameters. We need to formulate it from the perspective of the global Theorem 8.4.2.

For any global parameter $\phi^* \in \widetilde{\Phi}(G^*)$ for the quasisplit group G^*, we have the set $T(\phi^*)$ defined in §8.3. It is the domain of the mapping

$$t \longrightarrow \phi^*_t = \bigotimes_v \phi^*_{v,t_v}, \qquad t \in T(\phi^*),$$

from $T(\phi^*)$ to the space of linear forms on the global Hecke algebra $\mathcal{S}(G^*)$, defined prior to the statement of Theorem 8.4.2. Following our local convention from the last section, we now write

$$\Phi(G^*) = \big\{\phi^*_t : \phi^* \in \widetilde{\Phi}(G^*),\ t \in T(\phi^*)\big\}.$$

Then $\Phi(G^*)$ represents a family of stable linear forms on the global Hecke algebra $\mathcal{H}(G^*)$, equipped with mappings $\phi^*_t \to \phi^*_{v,t_v}$ to families $\Phi^+_{\mathrm{dis}}(G^*_v)$ of stable linear forms on the local Hecke algebras $\mathcal{H}(G^*_v)$. It is to be regarded as a refined substitute for our set of global Langlands parameters for G^*, since elements in the set $\widetilde{\Phi}(G^*)$ that have served us until now represent only $\widetilde{O}(G)$-orbits of parameters. There is no need to append the subscript "dis" to the new set, as there is no set of true Langlands parameters to distinguish it from, the global Langlands group L_F not having been defined. These remarks are of course only of interest in the case that G^* is of type $\mathbf{D}_n$, since the global sets $\Phi(G^*)$ and $\widetilde{\Phi}(G^*)$ are otherwise the same.

For our global inner twist G, we define $\Phi(G)$ to be the subset of stable linear forms in $\Phi(G^*)$ that do not vanish on the Hecke algebra $\mathcal{H}(G)$. We then have localization mappings

$$\phi \longrightarrow \phi_v, \qquad \phi \in \Phi(G),$$

from $\Phi(G)$ to the local sets $\Phi^+_{\mathrm{dis}}(G_v)$. Recall from Remark 4 following the statement of Theorem 9.4.1 that $\Phi^+_{\mathrm{dis}}(G_v)$ is a set of stable distributions on $G(F_v)$ that is in (noncanonical) bijection with the set $\Phi(G_v)$ of all local Langlands parameters. Since ϕ_v comes from a parameter ϕ in the global subset $\Phi(G)$ of $\Phi(G^*)$, it actually lies in the local subset

$$\Phi^+_{\mathrm{dis,unit}}(G_v) = \Phi^+_{\mathrm{dis}}(G_v) \cap \Phi^+_{\mathrm{dis,unit}}(G^*_v)$$

of $\Phi^+_{\mathrm{dis}}(G_v)$ attached to representations of $GL(N, F_v)$ that are unitary. (See the definitions following the statement of Theorem 1.5.1.) The point of this distinction is that elements in $\Phi^+_{\mathrm{dis,unit}}(G_v)$ should satisfy a version of

Conjecture 8.3.1, which would imply that representations in the local packet Π_{ϕ_v} are irreducible (and unitary). A parameter $\phi \in \Phi(G)$ has its global centralizer quotient $\mathcal{S}_\phi$, with its extension $\mathcal{S}_{\phi,\mathrm{sc}}$ by the quotient $\widehat{Z}_{\phi,\mathrm{sc}}$ of $\widehat{Z}_{\mathrm{sc}}$. It too has localization mappings

$$x_{\mathrm{sc}} \longrightarrow x_{\mathrm{sc},v}, \qquad x_{\mathrm{sc}} \in \mathcal{S}_{\phi,\mathrm{sc}},$$

from $\mathcal{S}_{\phi,\mathrm{sc}}$ to the local centralizer extensions $\mathcal{S}_{\phi_v,\mathrm{sc}}$. Given Theorem 9.4.1, we can define the global packet

$$\Pi_\phi = \Big\{\pi = \bigotimes_v \pi_v : \ \pi_v \in \Pi_{\phi_v}, \ \langle \cdot, \pi_v\rangle = 1 \text{ for almost all } v\Big\}$$

of representations of $G(\mathbb{A})$ attached to ϕ. Any representation $\pi = \bigotimes \pi_v$ in this packet then determines a character

$$\langle x, \pi\rangle = \prod_v \langle x_{\mathrm{sc},v}, \pi_v\rangle, \qquad x_{\mathrm{sc}} \in \mathcal{S}_{\phi,\mathrm{sc}}, \tag{9.5.5}$$

on $\mathcal{S}_{\phi,\mathrm{sc}}$. It follows easily from Hypothesis 9.5.1 that, as the notation suggests, $\langle x, \pi\rangle$ depends only on the image x of x_{sc} in $\mathcal{S}_\phi$.

THEOREM 9.5.3. *Assume that F is global, and that G is an inner twist of the quasisplit group $G^* \in \widetilde{\mathcal{E}}_{\mathrm{sim}}(N)$ over F. Suppose that $\psi = \widetilde{\phi}$ lies in the subset*

$$\widetilde{\Phi}_2(G) = \widetilde{\Phi}(G) \cap \widetilde{\Phi}_{\mathrm{ell}}(N)$$

of generic, square-integrable parameters for G in $\widetilde{\Psi}(N)$, and that $\psi = \widetilde{\phi}$ satisfies the condition of local constraints from Theorem 8.4.2, in case G is of type $\mathbf{D}_n$. Then there is an $\mathcal{H}(G)$-module isomorphism

$$L^2_{\mathrm{disc},\psi}(G(F)\backslash G(\mathbb{A})) \cong \bigoplus_{\phi \in \Phi(\psi)} \bigoplus_{\pi \in \Pi_\phi} m_\phi(1,\pi)\pi, \tag{9.5.6}$$

where $\Phi(\psi)$ is the preimage of ψ in $\Phi(G)$ (of order 1 or 2), and

$$m_\phi(1,\pi) = |\mathcal{S}_\phi|^{-1} \sum_{x \in \mathcal{S}_\phi} \langle x, \pi\rangle$$

is the multiplicity of the trivial representation of $\mathcal{S}_\phi$ in the irreducible decomposition of the character $\langle \cdot, \pi\rangle$.

The theorem applies to the generic part of the discrete spectrum. This is the subspace of the automorphic discrete spectrum of G whose irreducible constituents are believed to satisfy the analogue of Ramanujan's conjecture, in the sense that they are locally tempered. The theorem extends the generic case of Theorem 1.5.2 in two ways. It generalizes the earlier assertion to inner twists G of the quasisplit group G^*, and it refines it for the parameters in question to the full global Hecke algebra $\mathcal{H}(G)$. This is of course parallel to the relation that Theorem 9.4.1 bears to the generic part of the assertion of Theorem 2.2.1(b).

For general parameters ψ, we need a global version of Conjecture 9.4.2. It applies to the case of a global inner twist G that is *locally symmetric*.

In other words, each completion G_v of G is symmetric, in the sense of the last section. As in the local case, this is automatic if G is type $\mathbf{B}_n$ or $\mathbf{C}_n$. If G is of type $\mathbf{D}_n$, it means that each completion G_v of G satisfies the equivalent conditions (i)–(iv) of Lemma 9.1.1. From the condition (iii) and the injectivity of the mapping

$$H^1(F, G^*_{\mathrm{ad}}) \longrightarrow \bigoplus_v H^1(F_v, G^*_{\mathrm{ad}}),$$

we see that G itself satisfies the four equivalent conditions of the lemma. In particular, G is locally symmetric if and only if the automorphism $\widetilde{\theta}^*$ of G^* transfers to an F-automorphism $\widetilde{\theta}$ of G, in the sense of the condition (i) of Lemma 9.1.1.

Assume that G is locally symmetric. For a given choice of F-automorphism $\widetilde{\theta}$ of G (in which we take $\widetilde{\theta} = 1$ if G is of type $\mathbf{B}_n$ or $\mathbf{C}_n$), we have the usual subspace

$$\widetilde{\mathcal{H}}(G) = \widetilde{\bigotimes_v} \big(\widetilde{\mathcal{H}}(G_v)\big)$$

of locally $\widetilde{\theta}_v$-invariant functions in the global Hecke algebra $\mathcal{H}(G)$. We also have the the quotient

$$\widetilde{\Pi}(G) = \widetilde{\prod_v} \big(\widetilde{\Pi}(G_v)\big)$$

of (restricted tensor products of) local $\widetilde{\theta}_v$-orbits of representations in $\Pi(G)$, as well as corresponding quotients $\widetilde{\Pi}_{\mathrm{unit}}(G)$ and $\widetilde{\Pi}_{\mathrm{temp}}(G)$ of the subsets $\Pi_{\mathrm{unit}}(G)$ and $\Pi_{\mathrm{temp}}(G)$. As always, we obtain a well defined pairing

$$(f, \pi) \longrightarrow f_G(\pi) = \mathrm{tr}\big(\pi(f)\big), \qquad f \in \widetilde{\mathcal{H}}(G),\ \pi \in \widetilde{\Pi}(G),$$

in which the image is independent of the representative in $\Pi(G)$ of the orbit π.

The next assertion will again be stated as a conjecture, even though I do not expect any serious difficulties with the proof. However, there is one further point to mention. We are assuming that the local normalizing sections (9.5.1) are chosen to satisfy Hypothesis 9.5.1. We want to assume in addition that they are compatible with the localizations $\widetilde{\theta}_v$ of the chosen global F-automorphism $\widetilde{\theta}_v$, in the sense that the local transfer condition (9.4.6) holds for each v. I have not verified that this is possible, so it will have to be regarded as part of the conjecture.

The assertion is formulated in terms of the objects that we now know well. For our locally symmetric inner twist G, we have the localization mappings

$$\psi \longrightarrow \psi_v, \qquad \psi \in \widetilde{\Psi}(G),$$

from the set $\widetilde{\Psi}(G)$ of general global parameters to the local sets $\widetilde{\Psi}^+_{\mathrm{unit}}(G_v)$, defined as in §1.5. A parameter $\psi \in \widetilde{\Psi}(G)$ has its centralizer quotient $\mathcal{S}_\psi$, with localization mappings

$$x_{\mathrm{sc}} \longrightarrow x_{\mathrm{sc},v}, \qquad x_{\mathrm{sc}} \in \mathcal{S}_{\psi,\mathrm{sc}},$$

from the extension $\mathcal{S}_{\psi,\mathrm{sc}}$ of $\mathcal{S}_\psi$ to the corresponding local extensions $\mathcal{S}_{\psi_v,\mathrm{sc}}$. Assuming Conjecture 9.4.2, we form the global packet

$$\widetilde{\Pi}_\psi = \Big\{\pi = \bigotimes_v \pi_v : \ \pi_v \in \widetilde{\Pi}_{\psi_v}, \ \langle \cdot, \pi_v\rangle = 1 \text{ for almost all } v\Big\}$$

of (adelic orbits of) representations in $\widetilde{\Pi}_{\mathrm{unit}}(G)$ attached to ψ. Any element $\pi = \bigotimes_v \pi_v$ in this packet then determines a character

$$\langle x, \pi\rangle = \prod_v \langle x_{\mathrm{sc},v}, \pi_v\rangle, \qquad x_{\mathrm{sc}} \in \mathcal{S}_{\psi,\mathrm{sc}}, \tag{9.5.7}$$

on $\mathcal{S}_{\psi,\mathrm{sc}}$, which depends only on the image x of x_{sc} in $\mathcal{S}_\psi$.

CONJECTURE 9.5.4. *Assume that F is global, and that G is a locally symmetric inner twist of the quasisplit group $G^* \in \widetilde{\mathcal{E}}_{\mathrm{sim}}(N)$ over F. Assume also that the local transfer sections of Hypothesis* 9.5.1 *are chosen so that they are compatible with the localizations $\widetilde{\theta}_v$ of the fixed F-automorphism $\widetilde{\theta}$ of G. Then there is an $\widetilde{\mathcal{H}}(G)$-module isomorphism*

$$L^2_{\mathrm{disc}}\big(G(F)\backslash G(\mathbb{A})\big) \cong \bigoplus_{\psi \in \widetilde{\Psi}_2(G)} \bigoplus_{\pi \in \widetilde{\Pi}_\psi} \big(m_\psi \cdot m_\psi(\varepsilon_\psi, \pi)\big)\pi, \tag{9.5.8}$$

where $m_\psi = \{1,2\}$ and $\varepsilon_\psi \colon \mathcal{S}_\psi \to \{\pm 1\}$ are as defined after the statement of Theorem 1.5.2, *and*

$$m_\psi(\varepsilon_\psi, \pi) = |\mathcal{S}_\psi|^{-1} \sum_{x \in \mathcal{S}_\psi} \varepsilon_\psi(x)\langle x, \pi\rangle \tag{9.5.9}$$

is the multiplicity of the one-dimensional representation ε_ψ of $\mathcal{S}_\psi$ in the irreducible decomposition of the character $\langle \cdot, \pi\rangle$.

The proof of the conjecture will be considerably simpler than that of Theorem 1.5.2. This is because we already have the stable multiplicity formula of Theorem 4.1.2 for G^*. In particular, we can apply Corollary 4.1.3 to the quasisplit endoscopic groups G' for G^*. We are also free to apply Theorem 1.5.3, as well as the sign Lemmas 4.3.1 and 4.4.1, since they are the same for G and G^*. The proof should consequently amount to an adaptation of the earlier parts of the discussion of §4.3 and §4.4, or if one prefers, the specialization to G of the general arguments of §4.8.

We remind ourselves that the assertion of Conjecture 9.5.4 applies to all global inner twists G of type $\mathbf{B}_n$ or $\mathbf{C}_n$. It also applies to any G of type $\mathbf{D}_n$ whose local indices at places v (which we can of course restrict to the finite set of places at which G is not quasisplit) do not contain subdiagrams (9.1.5) or (9.1.6). The tables in §9.1, together with the global reciprocity law (9.1.4), allow us to quantify the remaining exceptional set of global inner twists. Conjecture 9.5.4 places any G in its complement on the same footing as that attained by the quasisplit group G^* with the proof of Theorem 1.5.2.

Theorem 9.5.3 and Conjecture 9.5.4 of course require their local analogues, Theorem 9.4.1 and Conjecture 9.4.2, even to state. At the end of the last section, we discussed how these local assertions might possibly be strengthened to include all parameters for all inner twists of type $\mathbf{D}_n$. This would govern how the global assertions could also be strengthened. The supplementary discussion for Conjecture 9.4.2, for example, concerned how to extend the local assertion (and its expected proof) to inner twists G_v that are not symmetric. The corresponding extension of Conjecture 9.5.4 would then apply to all global inner twists G whose localizations G_v either are locally symmetric, or satisfy such an extended local assertion. The closing remarks on Theorem 9.4.1 concerned the possibility of generalizing part (a) of the theorem (and its proof) to local parameters ψ_v in the set $\widetilde{\Psi}(G_v)$. The corresponding generalization of Theorem 9.5.3 would be more complicated. In principle, it ought to apply to global parameters $\psi \in \widetilde{\Psi}_2(G)$ whose localizations ψ_v each satisfy the generalized local assertion. But we would also be restricted to global parameters ψ that satisfy the injectivity condition of Theorem 8.4.2. With Corollary 8.4.5, this becomes a condition of local constraints, in the generic case of §8.4. Some analogue of the corollary would seem to be needed for the nongeneric case here. This is an interesting question to be sure, but obviously not for now. We remind ourselves in any case that these last remarks apply only to groups of type $\mathbf{D}_n$.

An interesting aspect of global inner twists G is the presence of higher multiplicities in the discrete spectrum. Consider the generic case represented by Theorem 9.5.3. It consists of an explicit multiplicity formula (9.5.6) for the contribution to the automorphic discrete spectrum of a generic global parameter

$$\psi = \psi_1 \boxplus \cdots \boxplus \psi_r, \qquad \psi_i \in \widetilde{\Phi}_{\mathrm{sim}}(G_i), \ G_i \in \mathcal{E}_{\mathrm{sim}}(N_i),$$

in $\widetilde{\Phi}_2(G)$, or if one prefers, of the automorphic family $c = c(\psi)$ of semisimple conjugacy classes in ${}^LG$ attached to any such ψ. The right hand side of the formula is a double (direct) sum over $\phi \in \Phi(\psi)$ and $\pi \in \Pi_\phi$. Let us review constituents of these sums in explicit terms.

The sum over ϕ in (9.5.6) is quite similar to its quasisplit analogue discussed in §8.3. The indexing set $\Phi(\psi)$ is of order 1 unless G is of type $\mathbf{D}_n$ and the degrees N_i of the irreducible constituents ψ_i of ψ are all even, in which case $\Phi(\psi)$ has order 2. In this latter case, the sum over ϕ contributes a factor of 2 to the multiplicities of the interior terms, unless one of the associated local sets $\Phi(\psi_v)$ has order 2. If some $\Phi(\psi_v)$ does have order 2, the set $\Phi(\psi)$ indexes two distinct global L-packets Π_ϕ, and therefore does not increase the inner multiplicities. The conditions for a local set $\Phi(\psi_v)$ to have order 2 are that the simple factors $\psi_{v,i}$ of ψ_v all have even degree, *and* the two parameters ϕ_v^* for G_v^* in $\Phi(\psi_v^*)$ both be relevant to G_v. The second of these conditions here is automatic unless the local index of G_v contains a subdiagram of the form (9.1.5).

The inner sum over $\pi \in \Pi_\phi$ in (9.5.6) is more interesting here than in the quasisplit case. This is because the function $\langle \cdot, \pi \rangle$ in the summand is the restriction to the abelian 2-group $\mathcal{S}_\phi$ of an irreducible character $\prod_v \langle \cdot, \pi_v \rangle$ on a group that could be nonabelian, namely the product of the local extensions $\mathcal{S}_{\phi_v,\mathrm{sc}}$. The multiplicity $m_\phi(1,\pi)$ of the trivial representation in this restricted character could then be greater than 1.

We discussed the groups $\mathcal{S}_{\phi_v,\mathrm{sc}}$ in §9.2. If $G^* = SO(2n+1)$ is of type $\mathbf{B}_n$, the dual group $\widehat{G} = Sp(2n,\mathbb{C}) = Sp(N,\mathbb{C})$ is simply connected. Each group $\mathcal{S}_{\phi_v} = \mathcal{S}_{\phi_v,\mathrm{sc}}$ is then abelian, and there are no higher multiplicities. If $G^* = Sp(2n)$ is of type $\mathbf{C}_n$, or $G^* = SO(2n)$ is of type $\mathbf{D}_n$, $\widehat{G}$ equals an orthogonal group $SO(N,\mathbb{C})$, where N is respectively equal to $(2n+1)$ or $2n$. In these cases, $\mathcal{S}_{\phi_v,\mathrm{sc}}$ is a nontrivial extension of $\mathcal{S}_{\phi_v}$ that is often nonabelian. Suppose for simplicity that

$$\phi_v = \phi_{v,1} \oplus \cdots \oplus \phi_{v,r_v}, \qquad \phi_{v,i} \in \Phi_{\mathrm{dis,sim}}(G_{v,i}),\ G_{v,i} \in \widetilde{\mathcal{E}}_{\mathrm{sim}}(N_{v,i}),$$

belongs to the subset $\Phi_{\mathrm{dis},2}(G_v)$ of $\Phi_{\mathrm{dis}}(G_v)$. As we then recall from §9.2, the group $\mathcal{S}_{\phi_v,\mathrm{sc}} = S_{\phi_v,\mathrm{sc}}$ is nonabelian if and only if the set

$$I_{v,o} = \{i : \ N_{v,i} \text{ is odd}\}$$

has order greater than 2, a condition that precludes the case that F_v is archimedean. Assume that $\mathcal{S}_{\phi_v,\mathrm{sc}}$ *is* nonabelian. Assume also that the character $\widehat{\zeta}_{G_v}$ on $\widehat{Z}_{\mathrm{sc}}$, which then determines G_v as a p-adic inner twist, is nontrivial on the derived group

$$(S_{\phi_v,\mathrm{sc}})_{\mathrm{der}} = \{\pm 1\}.$$

It then follows from Lemma 9.2.2 that the degree of any irreducible representation in the set $\Pi(S_{\phi_v,\mathrm{sc}}, \widehat{\zeta}_{G_v})$ equals the positive integer

$$d_{\phi_v} = \begin{cases} 2^{(|I_{v,o}|-1)/2}, & \text{if } N \text{ is odd}, \\ 2^{(|I_{v,o}|-2)/2)}, & \text{if } N \text{ is even}. \end{cases}$$

This is the contribution of v to the degree of the character $\langle \cdot, \pi \rangle$. A modest extension of the formula, which we did not include in §9.2, applies to any localization ϕ_v of ϕ. Our conclusion is that the multiplicity $m_\phi(1,\pi)$ on the right hand side of (9.5.6) is equal to a product

$$m_\phi(1,\pi) = \Big(\prod_v d_{\phi_v}\Big) e(\pi), \tag{9.5.10}$$

where v is taken over the finite set of places v such that $|I_{o_v}| \geqslant 3$ and $\widehat{\zeta}_{G_v}(-1) = -1$, d_{ϕ_v} is given by an explicit formula like that above, and $e(\pi) \in \{0,1\}$ is defined by a global reciprocity law that is entirely parallel to that of the quasisplit case (from the right hand sum in (1.5.5), with $\varepsilon_\psi = 1$).

The analysis of multiplicities for arbitrary parameters $\psi \in \widetilde{\Psi}_2(G)$ will be similar, if slightly more complicated. We do not presently have general results that are as sharp as Theorem 9.5.3, but suppose that the multiplicity formula (9.5.8) of Conjecture 9.5.4 has been established. In principle, any element $\pi_v \in \widetilde{\Pi}_{\text{unit}}(G_v)$ in the local packet $\widetilde{\Pi}_{\psi_v}$ of a completion of ψ could occur in the packet with higher multiplicity $\mu_{\psi_v}(\pi_v)$. This would break into a sum

$$\mu_{\psi_v}(\pi_v) = \sum_{\xi} \mu_{\psi_v}(\xi, \pi_v)$$

of multiplicities in the fibres of $\widetilde{\Pi}_{\psi_v}$ over the set $\Pi(\mathcal{S}_{\psi_v}, \widehat{\zeta}_G)$ of irreducible characters, parametrized as in §9.2. These summands would in turn contribute appropriately to the sum over π in the global multiplicity formula (9.5.8). Suppose, however, that none of the local multiplicities $\mu_{\psi_v}(\pi_v)$ is greater than one, as seems likely. Then (9.5.8) will simplify to something close to the formula we have described for the right hand side of (9.5.6). In particular, the coefficient m_ψ in (9.5.8) could no doubt be replaced by a sum over the set $\Psi(\psi)$, like the outer sum in (9.5.6). The only remaining difference would then be the sign character ε_ψ in (9.5.8), which we recall was defined in §1.5 in terms of symplectic ε-factors, in place of the trivial character from (9.5.6). The analysis of its multiplicity $m_\psi(\varepsilon_\psi, \pi)$ would be identical to the discussion of $m_\phi(1, \pi)$ above that led to (9.5.10).

Higher multiplicities for global inner forms of the group $Sp(2)$ appeared in the seminal paper [**LL**] by Labesse and Langlands on the stabilization of the trace formula for this and related groups. (See also [**S1**, §12, §15].) Higher multiplicities for more general orthogonal and symplectic groups were later discovered by Jian-Shu Li [**Li1**], [**Li2**].* He found many such examples by using the theta correspondence. They arise in Li's constructions from a phenomenon we described in §9.1, the existence of nonisomorphic groups over F that are locally isomorphic.

Suppose that G and G' are inner forms of the split groups $Sp(2n)$ and $SO(2n)$. We then have the usual embeddings

$$\widehat{G}' = SO(2n, \mathbb{C}) \subset SO(2n+1, \mathbb{C}) = \widehat{G}$$

of their dual groups. If G' were itself split, it would of course be an endoscopic group for G. In general, (G', G) is a dual reductive pair, which comes with a theta correspondence of automorphic representations. As we saw in §9.1, we can sometimes choose G' so that it belongs to a larger family $\{G'_*\}$ of nonisomorphic groups over F that are all locally isomorphic. Li chooses G, and certain families $\pi'_* \in \Pi_2(G'_*)$ of locally isomorphic, automorphic representations of the groups G'_*, such that the theta correspondence for the various pairs (G'_*, G) takes the representations π'_* to distinct, locally isomorphic, automorphic representations $\pi_* \in \Pi_2(G)$ of the group G. This, if I have understood it, is the source of Li's constructions of automorphic representations $\pi \in \Pi_2(G)$ that occur with higher multiplicities in the discrete spectrum of G.

---

*I thank Steve Kudla for these references, and for conversations on their contents.

For us, the inner forms G'_* simply represent different groups over F. For any G'_*, an adelic isomorphism from G'_* to G' gives a bijection between the associated sets $\Pi\big(G(\mathbb{A})\big)$ and $\Pi\big(G'_*(\mathbb{A})\big)$ of irreducible adelic representations. This does not a priori have anything to say about the associated automorphic representations. The implicit question here amounts to a comparison of automorphic spectra for the locally compact group $G'(\mathbb{A})$ (identified with $G'_*(\mathbb{A})$ by the adelic isomorphism) relative to a pair $G'(F)$ and $G'_*(F)$ of different discrete subgroups. The group G' is not locally symmetric, according to the remarks at the end of §9.1, so it is not included in the classification of automorphic representations of Conjecture 9.5.4. However, one can hope that the extensions of Conjecture 9.5.4 proposed above and in the last section would lead to an isomorphism between the automorphic discrete spectra of G' and G'_*.

In any event, it seems quite surprising that the two methods for establishing higher multiplicities are so different. In Theorem 9.5.3 and Conjecture 9.5.4, they arise from a local phenomenon, the fact that the local centralizer groups $\mathcal{S}_{\psi_v,\mathrm{sc}}$ can be nonabelian. In [**Li1**] and [**Li2**], they are a consequence of the existence of locally isomorphic groups that are not globally isomorphic, a global property. The two general methods of course have quite different sources, the trace formula on one hand, and the theta correspondence on the other. It would be very interesting to relate the classification of local and global representations of this volume with the explicit constructions provided by the theta correspondence and its generalizations. A full answer is probably still far away. For some preliminary results, a reader can consult the paper [**Ji**] of Jiang.

Bibliography

[ABV] J. Adams, D. Barbasch, and D. Vogan, *The Langlands Classification and Irreducible Characters for Real Reductive Groups*, Progr. Math. **104**, Birkhauser, Boston, 1992.

[AH] J. Adams and J.-S. Huang, *Kazhdan-Patterson lifting for $GL(n, \mathbb{R})$*, Duke Math. J. **89** (1997), 423–443.

[AJ] J. Adams and J. Johnson, *Endoscopic groups and packets for non-tempered representations*, Compositio Math. **64** (1987), 271–309.

[AV] J. Adams and D. Vogan, *The contragredient*, arXiv: 1201.0496, 2012.

[AS] J. Adler and L. Spice, *Supercuspidal characters of reductive p-adic groups*, Amer. J. Math. **131** (2009), 1137–1210.

[A1] J. Arthur, *Eisenstein series and the trace formula*, in *Automorphic Forms, Representations and L-functions*, Proc. Sympos. Pure Math. **33** (1979), Part 1, Amer. Math. Soc., 253–274.

[A2] ______, *On a family of distributions obtained from Eisenstein series I: Application of the Paley-Wiener theorem*, Amer. J. Math. **104** (1982), 1243–1288.

[A3] ______, *On a family of distributions obtained from Eisenstein series II: Explicit formulas*, Amer. J. Math. **104** (1982), 1289–1336.

[A4] ______, *The local behaviour of weighted orbital integrals*, Duke Math. J. **56** (1988), 223–293.

[A5] ______, *The invariant trace formula II. Global theory*, J. Amer. Math. Soc. **1** (1988), 501–554.

[A6] ______, *The L^2-Lefschetz numbers of Hecke operators*, Invent. Math. **97** (1989), 257–290.

[A7] ______, *Intertwining operators and residues I. Weighted characters*, J. Funct. Anal. **84** (1989), 19–84.

[A8] ______, *Unipotent automorphic representations: conjectures*, Astérisque **171–172** (1989), 13–71.

[A9] ______, *Unipotent Automorphic Representations: Global Motivation*, in *Automorphic Forms, Shimura Varieties and L-functions*, vol. I, Academic Press, 1990, 1–75.

[A10] ______, *On elliptic tempered characters*, Acta. Math. **171** (1993), 73–138.

[A11] ______, *On local character relations*, Selecta Math. **2**, No. 4 (1996), 501–579.

[A12] ______, *On the transfer of distributions: weighted orbital integrals*, Duke Math. J. **99** (1999), 209–283.

[A13] ______, *Stabilization of a family of differential equations*, Proc. Sympos. Pure Math. **158** (2000), 77–98.

[A14] ______, *A stable trace formula I. General expansions*, Journal of the Inst. of Math. Jussieu **1** (2002), 175–277.

[A15] ______, *A stable trace formula II. Global descent*, Invent. Math. **143** (2001), 157–220.

[A16] ______, *A stable trace formula III. Proof of the main theorems*, Annals of Math. **158** (2003), 769–873.

[A17] ———, *A note on the automorphic Langlands group*, Canad. Math. Bull. **45** (2002), 466–482.

[A18] ———, *An introduction to the trace formula*, in *Harmonic Analysis, the Trace Formula, and Shimura Varieties*, Clay Mathematics Proceedings, vol. 4, 2005, 1–263.

[A19] ———, *A note on L-packets*, Pure and Applied Math Quarterly **2** (2006), 199–217.

[A20] ———, *Induced representations, intertwining operators and transfer*, Contemp. Math. **449** (2008), 51–67.

[A21] ———, *Problems for real groups*, Contemp. Math. **472** (2008), 39–62.

[A22] ———, *Report on the trace formula*, in *Automorphic forms and L-functions* I. *Global aspects*, Contemp. Math. **488** (2009), 1–12.

[A23] ———, *The embedded eigenvalue problem for classical groups*, in *On Certain L-functions*, Clay Mathematics Proceedings, vol. 13, 2011, 19–31.

[A24] ———, *Endoscopy and singular invariant distributions*, in preparation.

[A25] ———, *Duality, Endoscopy and Hecke operators*, in preparation.

[A26] ———, *A nontempered intertwining relation for $GL(N)$*,

[A27] ———, *Transfer factors and Whittaker models*, in preparation.

[A28] ———, *Automorphic representations of inner twists*, in preparation.

[AC] J. Arthur and L. Clozel, *Simple Algebras, Base Change, and the Advanced Theory of the Trace Formula*, Ann. of Math. Studies **120**, Princeton Univ. Press, Princeton, N.J., 1989.

[Au] A.-M. Aubert, *Dualité dans le groupe de Grothendieck de la catégorie des représentations lisses de longueur finie d'un groupe réductif p-adique*, Trans. Amer. Math. Soc. **347** (1995), 2179–2189, and Erratum, Ibid. **348** (1996), 4687–4690.

[Ba] A. Badalescu, *Global Jacquet-Langlands correspondence, multiplicity one and classification of automorphic representations*, Invent. Math. **172** (2008), 383–438.

[Ban] D. Ban, *The Aubert involution and R-groups*, Ann. Sci. École Norm. Sup. **35** (2002), 673–693.

[BV] D. Barbasch and D. Vogan, *Unipotent representations of complex semisimple Lie groups*, Ann. of Math. **121** (1985), 41–110.

[Be] J. Bernstein, *P-invariant distributions on $GL(n)$*, in *Lie Group Representations* II, Lecture Notes in Math., vol. 1041, Springer, New York, 1984, 50–102.

[BDK] J. Bernstein, P. Deligne and D. Kazhdan, *Trace Paley-Wiener theorem for reductive p-adic groups*, J. d'Analyse Math. **47** (1986), 180–192.

[BZ] J. Bernstein and A. Zelevinsky, *Induced representations of reductive p-adic groups* I, Ann. Sci. École Norm. Sup. **10** (1977), 441–472.

[Bo] A. Borel, *Automorphic L-functions*, in *Automorphic Forms, Representations and L-functions*, Proc. Sympos. Pure Math. vol. 33, Part 2, Amer. Math. Soc., 1979, 27–62.

[BoW] A. Borel and N. Wallach, *Continuous Cohomology, Discrete Subgroups, and Representations of Reductive Groups*, Ann. of Math. Studies **94**, Princeton Univ. Press, Princeton, N.J., 1980.

[C] W. Casselman, *The unramified principal series of p-adic groups* I. *The spherical function*, Compositio Math. **40** (1980), 387–406.

[CS] W. Casselman and J. Shalika, *The unramified principal series of p-adic groups* II. *The Whittaker function*, Compositio Math. **41** (1980), 207–231.

[Cassels] J. Cassels, *Global Fields*, in *Algebraic Number Theory*, Thompson Book Company, Washington, 1967, 42–84.

[CL1] P.-H. Chaudouard and G. Laumon, *Le lemme fondamental pondéré* I: *constructions géométriques*, Compositio Math. **146** (2010), 1416–1506.

[CL2] ______, *Le lemme fondamental pondéré* II: *Énoncés cohomologiques*, Annals of Math. **176** (2012), 1647–1781.

[CC] G. Chenevier and L. Clozel, *Corps de nombres peu ramifiés et formes automorphes autoduales*, J. Amer. Math. Soc. **22** (2009), 467–519.

[Clo1] L. Clozel, *Characters of nonconnected, reductive p-adic groups*, Canad. J. Math. **39** (1987), 149–167.

[Clo2] ______, *The fundamental lemma for stable base change*, Duke Math. J. **61** (1990), 255–302.

[CD] L. Clozel and P. Delorme, *Le théorème de Paley-Wiener invariant pour les groupes de Lie réductif* II, Ann. Scient. Éc. Norm. Sup., 4e série, 23 (1990), 193–228.

[CLL] L. Clozel, J.-P. Labesse and R.P. Langlands, *Morning Seminar on the Trace Formula*, Lecture Notes, Institute for Advanced Study, Princeton, N.J., 1984.

[CHL] R. Cluckers, T. Hales, and F. Loeser, *Transfer principle for the fundamental lemma. On the stabilization of the trace formula*, 309–347, Stab. Trace Formula Shimura Var. Arith. Appl. 1, Int. Press, Somerville, MA, 2011.

[CKPS1] J. Cogdell, H. Kim, I. Piatetski-Shapiro, and F. Shahidi, *On lifting from classical groups to* GL_N, Publ. Math. Inst. Hautes Études Sci. **93** (2001), 5–30.

[CKPS2] ______, *Functoriality for the classical groups*, Publ. Math. Inst. Hautes Études Sci. **99** (2004), 163–233.

[CST] J. Cogdell, F. Shahidi, and T-L. Tsai, *Local Langlands correspondence for* GL_n *and the exterior and symmetric square* ε*-factors*, with an appendix by H. Jacquet, in preparation.

[CH] C. Cunningham and T. Hales, *Good Orbital Integrals*, Represent. Theory **8** (2004), 414–457.

[DR] S. DeBacker and M. Reeder, *Depth-zero supercuspidal L-packets and their stability*, Ann. of Math. **169** (2009), 795–901.

[D1] P. Deligne, *Les constantes des équations fonctionnelles des fonctions L*, in *Modular Forms of One Variable* II, Lecture Notes in Math. 349, Springer, New York, 1973, 501–597.

[D2] ______, *Les constantes locales de l'équation fonctionnelle de la fonction L d'Artin d'une représentation orthogonale*, Invent. Math. **35** (1976), 299–316.

[DM] P. Delorme and P. Mezo, *A twisted invariant Paley-Wiener theorem for real reductive groups*, Duke Math. J. **144** (2008), 341–380.

[F] D. Flath, *Decomposition of representations into tensor products*, in *Automorphic Forms, Representations and L-functions*, Proc. Sympos. Pure Math. vol. 33, Part 1, Amer. Math. Soc., 1979, 179–184.

[FLN] E. Frenkel, R. Langlands and B.C. Ngo, *Formule des traces et functorialité: le début d'un programme*, Ann. Sci. Math. Québec **34** (2010), 199–243.

[FQ] A. Frohlich and J. Queyrot, *On the functional equations of the Artin L-function for characters of real representations*, Invent. Math. **20** (1973), 125–138.

[G] D. Goldberg, *Reducibility of induced representations for* $Sp(2n)$ *and* $SO(n)$, Amer. J. Math. **116** (1994), 1101–1151.

[GRS1] D. Ginzburg, S. Rallis, and D. Soudry, *Generic automorphic forms on* $SO(2n+1)$*: functorial lift to* $GL(2n)$*, endoscopy and base change*, Internat. Math. Res. Notices **14** (2001), 729–764.

[GRS2] ______, *The descent map from automorphic representations of* $GL(n)$ *to classical groups*, World Scientific Publishing Co. Pte. Ltd., Hackensack, NJ, 2011.

[GKM1] M. Goresky, R. Kottwitz, and R. MacPherson, *Homology of affine Springer fibres in the unramified case*, Duke Math. J. **121** (2004), 509–561.

[GKM2] ______, *Purity of equivalued affine Springer fibres*, Representation Theory **10** (2006), 130–146.

[Hal] T. Hales, *On the fundamental lemma for standard endoscopy: reduction to unit elements*, Canad. J. Math. **47** (1995), 974–994.

[Ha1] Harish-Chandra, *Discrete series for semisimple Lie groups.* II. *Explicit determination of the characters*, Acta Math. **116**, 1–111.

[Ha2] ———, *Harmonic analysis on real reductive groups.* I. *The theory of the constant term*, J. Funct. Anal. **19**, 104–204.

[Ha3] ———, *Harmonic analysis on real reductive groups* II. *Wave packets in the Schwartz space*, Invent. Math. **36** (1976), 1–55.

[Ha4] ———, *Harmonic analysis on real reductive groups* III. *The Moass-Selberg relations and the Plancherel formula*, Ann. of Math. **104** (1976), 117–201.

[Ha5] ———, *The Plancherel formula for reductive p-adic groups*, in *Collected Papers*, vol. IV, Springer-Verlag, 353–367

[HT] M. Harris and R. Taylor, *On the Geometry and Cohomology of Some Simple Shimura Varieties*, Ann. of Math. Studies **151**, Princeton Univ. Press, Princeton, N.J., 2001.

[He1] G. Henniart, *Une preuve simple des conjectures de Langlands de $GL(n)$ sur un corps p-adique*, Invent. Math. **139** (2000), 439–455.

[He2] ———, *Correspondence de Langlands et fonctions L des carrés extérieur et symétrique*, Int. Math. Res. Not. **2010**, no. 4, 633–673.

[Hi] K. Hiraga, *On functoriality of Zelevinski involutions*, Compositio Math. **140** (2004), 1625–1656.

[JL] H. Jacquet and R. Langlands, *Automorphic Forms on $GL(2)$*, Lecture Notes in Math., vol. 114, Springer, New York, 1970.

[JPS] H. Jacquet, I. Piatetski-Shapiro, and J. Shalika, *Rankin-Selberg convolutions*, Amer. J. Math. **105** (1983), 367–464.

[JS] H. Jacquet and J. Shalika, *On Euler products and the classification of automorphic representations* II, Amer. J. Math. **103** (1981), 777–815.

[Ji] D. Jiang, *Automorphic integral transforms for classical groups* I: *endoscopy correspondences*, to appear in *Conference on Automorphic Forms and Related Geometry: Assessing the Legacy of I.I. Piatetskii-Shapiro.*

[JiS1] D. Jiang and D. Soudry, *The local converse theorem for $SO(2n+1)$ and applications*, Annals of Math. **157** (2003), 743–806.

[JiS2] ———, *Generic representations and local Langlands reciprocity law for p-adic $SO(2n+1)$*, in *Contributions to Automorphic Forms, Geometry and Number Theory*, Johns Hopkins Univ. Press, Baltimore, MD, 2004, 457–519.

[JiS3] ———, *On local descent from $GL(n)$ to classical groups*, Amer. J. Math. **134** (2012), 767–772.

[Kal1] T. Kaletha, *Endoscopic character identities for depth-zero supercuspidal L-packets*, Duke Math. J. **158** (2011), 161–224.

[Kal2] ———, *Supercuspidal L-packets via isocrystals*, preprint.

[Kal3] ———, *Genericity and contragredience in the local Langlands correspondence*, preprint.

[Kal4] ———, *Depth-zero supercuspidal L-packets for tamely-ramified groups*, in preparation.

[Ka] D. Kazhdan, *Cuspidal geometry on p-adic groups* (Appendix), J. Analyse Math. **47** (1980), 1–36.

[KL] D. Kazhdan and G. Lusztig, *Proof of the Deligne-Langlands conjecture for Hecke algebras*, Invent. Math. **87** (1987), 153–215.

[KeS] D. Keys and F. Shahidi, *Artin L-functions and normalization of intertwining operators*, Ann. Scient. Ec. Norm. Sup. 4e série, t. **21** (1988), 67–89.

[KM] J.-L. Kim and F. Murnaghan, *Character expansions and unrefined minimal K-types*, Amer. J. Math. **125** (2003), 1199–1234.

[Kn] A. Knapp, *Commutativity of intertwining operators.* II, Bull. Amer. Math. Soc. **82** (1976), 271–273.

[KnS] A. Knapp and E. Stein, *Intertwining operators for semisimple groups* II, Invent. Math. **60** (1980), 9–84.

[KnZ1] A. Knapp and G. Zuckerman, *Normalizing factors and L-groups*, in *Automorphic Forms, Representations and L-functions*, Proc. Sympos. Pure Math. vol. 33, Part 1, Amer. Math. Soc., 1979, 93–106.

[KnZ2] ———, *Classification of irreducible tempered representations of semisimple groups*, Ann. of Math. **116** (1982), 389–455.

[Kon] T. Konno, *Twisted endoscopy and the generic packet conjecture*, Israel J. Math. **129** (2002), 253–289.

[Kos] B. Kostant, *On Whittaker vectors and representation theory*, Invent. Math. **48** (1978), 101–184.

[K1] R. Kottwitz, *Rational conjugacy classes in reductive groups*, Duke Math. J. **49** (1982), 785–806.

[K2] ———, *Sign changes in harmonic analysis on reductive groups*, Trans. Amer. Math. Soc. **278** (1983), 289–297.

[K3] ———, *Stable trace formula: cuspidal tempered terms*, Duke Math. J. **51** (1984), 611–650.

[K4] ———, *Isocrystals with additional structure*, Compositio Math. **56** (1985), 201–220.

[K5] ———, *Stable trace formula: elliptic singular terms*, Math. Ann. **275** (1986), 365–399.

[K6] ———, *Shimura Varieties and λ-adic Representations*, in *Automorphic Forms, Shimura Varieties and L-functions*, vol. 1, Academic Press, 1990, 161–209.

[K7] ———, *Isocrystals with additional structure*, II, Compositio Math. **109** (1997), 255–339.

[KR] R. Kottwitz and J. Rogawski, *The distributions in the invariant trace formula are supported on characters*, Canad. J. Math. **52** (2000), 804–814.

[KS] R. Kottwitz and D. Shelstad, *Foundations of Twisted Endoscopy*, Astérisque, vol. 255.

[LL] J.-P. Labesse and R. Langlands, *L-indistinguishability for $SL(2)$*, Canad. J. Math. **31** (1979), 726–785.

[LW] J.-P. Labesse, and J.-L. Waldspurger, *La formule des traces tordues d'apres le Friday Morning Seminar*, CRM Monograph Series **31**, Amer. Math. Soc., 2013.

[L1] R. Langlands, *Eisenstein series*, in *Algebraic Groups and Discontinuous Subgroups*, Proc. Sympos. Pure Math., vol. 9, Amer. Math. Soc., 1966, 235–252.

[L2] ———, *Problems in the theory of automorphic forms*, in *Lectures in Modern Analysis and Applications,* Lecture Notes in Math. 170, Springer, New York, 1970, 18–61.

[L3] ———, *On Artin's L-function*, Rice University Studies **56** (1970), 23–28.

[L4] ———, *Euler Products*, Yale University Press, 1971.

[L5] ———, *On the Functional Equations Satisfied by Eisenstein Series*, Lecture Notes in Math. 544, Springer, New York, 1976.

[L6] ———, *On the notion of an automorphic representation. A supplement to the preceding paper*, in *Automorphic Forms, Representations and L-functions*, Proc. Sympos. Pure Math. vol. 33, Part 1, Amer. Math. Soc., 1979, 203–208.

[L7] ———, *Automorphic representations, Shimura varieties, and motives. Ein Märchen*, in *Automorphic Forms, Representations and L-functions*, Proc. Sympos. Pure Math. vol. 33, Part 2, Amer. Math. Soc., 1979, 205–246.

[L8] ——— *Stable conjugacy: definitions and lemmas*, Canad. J. Math. **31** (1979), 700–725.

[L9] ———, *Base Change for $GL(2)$*, Ann. of Math. Studies 96, Princeton Univ. Press, Princeton, N.J., 1980.

[L10] ———, *Les débuts d'une formule des traces stables*, Publ. Math. Univ. Paris VII **13**, 1983.

[L11] ———, *On the classification of irreducible representations of real algebraic groups*, in *Representation Theory and Harmonic Analysis on Semisimple Lie Groups*, AMS Mathematical Surveys and Monographs, vol. 31, 1989, 101–170.

[L12] ———, *Representations of abelian algebraic groups*, Pacific J. Math. (1997), 231–250.

[L13] ———, *Beyond endoscopy*, in *Contributions to Automorphic Forms, Geometry, and Number Theory*, Johns Hopkins University Press, 2004, 611–698.

[L14] ———, *Un nouveau point de repère des formes automorphes*, Canad. Math. Bull. **50** (2007), 243–267.

[LS1] R. Langlands and D. Shelstad, *On the definition of transfer factors*, Math. Ann. **278** (1987), 219–271.

[LS2] ———, *Descent for transfer factors, The Grothendieck Festschift*, Vol. II, Birkhäuser, Boston, 1990, 485–563.

[Lap] E. Lapid, *On the root number of representations of orthogonal type*, Compositio Math. **140** (2004), 274–286.

[LR] E. Lapid and J. Rogawski, *On twists of cuspidal representations of $GL(2)$*, Forum Math. **10** (1998), 175–197.

[Lar] M. Larsen, *On the conjugacy of element-conjugate homomorphisms*, Israel J. Math. **88** (1994), 253–277.

[LN] G. Laumon and B.C. Ngo, *Le lemme fondamental pour les groupes unitaires*, Annals of Math. **168** (2008), 477–573.

[Li1] J-S. Li, *Singular Automorphic Forms*, Proceedings of the International Congress of Mathematicians, Vol. 1, Birkhäuser, Basel, 1995, 790–799.

[Li2] ———, *Automorphic forms with degenerate Fourier coefficients*, Amer. J. Math. **119** (1997), 523–578.

[Liu] B. Liu, *Genericity of Representations of p-Adic Sp_{2n} and Local Langlands Parameters*, Canad. J. Math. **63** (2011), 1107–1136.

[Lu1] G. Lusztig, *Some examples of square integrable representations of semisimple p-adic groups*, Trans. Amer. Math. Soc. **277** (1983), 623–652.

[Lu2] ———, *Classification of unipotent representations of simple p-adic groups*, Int. Math. Res. Not. **1995**, no. 11, 517–589.

[Lu3] ———, *Classification of unipotent representations of simple p-adic groups* II, Represent. Theory **6** (2002), 243–289.

[Ma] I.G. Macdonald, *Spherical Functions on a Group of p-Adic Type*, Publications of the Ramanujan Institute, Madras, 1971.

[Me] P. Mezo, *Character identities in the twisted endoscopy of real reductive groups*, to appear in Memoires of the AMS.

[M1] C. Moeglin, *Sur certaines paquets d'Arthur et involution d'Aubert-Schneider-Stuhler généralisée*, Represent. Theory **10** (2006), 89–129.

[M2] ———, *Paquets d'Arthur discrets pour un groupe classique p-adique*, Contemp. Math. **489** (2009), 179–257.

[M3] ———, *Paquets d'Arthur pour les groupes classiques p-adiques; point de vue combinatoire*, in *Automorphic forms and L-functions* II. *Local aspects*, Contemp. Math. **489** (2009), 179–257.

[M4] ———, *Multiplicité* 1 *dans les paquets d'Arthur aux places p-adiques*, in *On Certain L-functions*, Clay Mathematics Proceedings, vol. 13, 2011, 233–274.

[MW1] C. Moeglin and J.-L. Waldspurger, *Modèles de Whittaker dégénérés pour des groupes p-adiques*, Math. Z. **1996** (1987), 427–452.

[MW2] ______, *Le spectre résiduel de $GL(n)$*, Ann. Scient. Éc. Norm. Sup. 4e série **22** (1989), 605–674.

[MW3] ______, *Pacquets stables de représentations tempérés et réductions unipotent pour $SO(2n+1)$*, Invent. Math. **152** (2003), 461–623.

[MW4] ______, *Sur le transfert des traces d'un groupe classique p-adique à un groupe linéaire tordu*, Select Math. **12** (2006), 433–515.

[Mu] W. Müller, *The trace class conjecture in the theory of automorphic forms*, Ann. of Math. **130** (1989), 473–529.

[Mur] F. Murnaghan, *Characters of supercuspidal representations of classical groups*, Ann. Sci. École Norm. Sup. (4) **29** (1996), 49–105.

[N] B.C. Ngô, *Le lemme fondamental pour les algèbres de Lie*, Publ. Math. Inst. Hautes Études Sci. **111** (2010), 1–269.

[Pr] D. Prasad, *A "relative" local Langlands conjecture*, partial preprint.

[Ra] C.S. Rajan, *On the image and fibres of solvable base change*, Math. Res. Lett. **9** (2002), 499–508.

[Re] D. Renard, *Endoscopy for real reductive groups. On the stabilization of the trace formula*, 95–141, Stab. Trace Formula Shimura Var. Arith. Appl., 1, Int. Press, Somerville, MA, 2011.

[Rod1] F. Rodier, *Whittaker models for admissible representations of reductive p-adic split groups*, Proc. Sympos. Pure Math. **26** (1974), Amer. Math. Soc., 425–430.

[Rod2] ______, *Modèle de Whittaker et caractères de représentations*, in *Non-Commutative Harmonic Analysis*, Lecture Notes in Math. **466**, Springer, New York, 1975, 151–171.

[Ro1] J. Rogawski, *The trace Paley-Wiener theorem in the twisted case*, Trans. Amer. Math. Soc. **309** (1988), 215–229.

[Ro2] ______, *Automorphic Representations of Unitary Groups in Three Variables*, Ann. of Math. Studies **123**, Princeton Univ. Press, Princeton, N.J., 1990.

[Ro3] ______, *The multiplicity formula for A-packets*, in *The Zeta Functions of Picard Modular Surfaces*, Les Publications CRM, Montreal, 1992, 395–420.

[Sa] I. Satake, *Classification Theory of Semisimple Algebraic Groups*, notes prepared by D. Schattschneider, M. Dekker, New York, 1971.

[ScS] P. Schneider and U. Stuhler, *Representation theory and sheaves on the Bruhat-Tits building*, Pub. Math. IHES **85** (1997), 97–191.

[Sch] P. Scholze, *The local Langlands correspondence for GL_n over p-adic fields*, to appear in Invent. Math.

[Se] J.-P. Serre, *Cohomologie Galoisienne*, Lecture Notes in Math., vol. 4, Springer, New York, 1970.

[Sha1] F. Shahidi, *On certain L-functions*, Amer. J. Math. **103** (1981), 297–355.

[Sha2] ______, *Local coefficients as Artin factors for real groups*, Duke Math. J. **52** (1985), 973–1007.

[Sha3] ______, *On the Ramanujan conjecture and finiteness of poles for certain L-functions*, Annals of Math. **127** (1988), 547–584.

[Sha4] ______, *A proof of Langlands' conjecture on Plancherel measures; Complementary series for p-adic groups*, Annals of Math. **132** (1990), 273–330.

[Sha5] ______, *Eisenstein series and automorphic L-functions*, American Mathematical Society Colloquium Publications, **58**, 2010.

[Shal] J. Shalika, *The multiplicity one theorem for GL_n*, Annals of Math. **100** (1974), 171–193.

[S1] D. Shelstad, *Notes on L-indistinguishability (based on a lecture by R.P. Langlands)*, in *Automorphic Forms, Representations and L-functions*, Proc. Sympos. Pure Math., vol. 33, Part 2, Amer. Math. Soc., 1979, 193–204.

[S2] ———, *Orbital integrals and a family of groups attached to a real reductive group*, Ann. Sci. École Norm. Sup. **12** (1979), 1–31.

[S3] ———, *L-indistinguishability for real groups*, Math. Ann. **259** (1982), 385–430.

[S4] ———, *Tempered endoscopy for real groups* I: *geometric transfer with canonical factors*, Contemp. Math., **472** (2008), 215–246.

[S5] ———, *Tempered endoscopy for real groups* II: *spectral transfer factors*, in *Automorphic Forms and the Langlands Program*, 236–276, Adv. Lect. Math. (ALM), 9, Int. Press, Somerville, MA, 2010.

[S6] ———, *Tempered endoscopy for real groups* III: *inversion of transfer and L-packet structure*, Representation Theory **12** (2008), 369–402.

[S7] ———, *On geometric transfer in real twisted endoscopy*, Ann. of Math. **176** (2012), 1919–1985.

[S8] ———, *On spectral factors in real twisted endoscopy*, preprint.

[Si1] A. Silberger, *Introduction to Harmonic Analysis on Reductive p-adic Groups*, Mathematical Notes, Princeton University Press, 1979.

[Si2] ———, *The Knapp-Stein dimension theorem for p-adic groups*, Proc. Amer. Math. Soc. **76** (1978), 243–246; *Correction*, Proc. Amer. Math. Soc. **76** (1979), 169–170.

[So] D. Soudry, *On Langlands functoriality from classical groups to* GL_n, Astérisque **298** (2005), 335–390.

[SV] B. Speh and D. Vogan, *Reducibility of generalized principal series representations*, Acta. Math. **145** (1980), 227–299.

[Sp] T. Springer, *Linear Algebraic Groups*, Prog. Math. **9**, 1981, Birkhäuser, Boston, Basel, Stuttgart.

[SpS] T. Springer and R. Steinberg, *Conjugacy classes*, in *Seminar on Algebraic Groups and Related Finite Groups*, Lecture Notes in Math., vol. 131, 1970, 167–266.

[Tad1] M. Tadic, *Topology of unitary dual of non-archimedean* $GL(n)$, Duke Math. J. **55** (1987), 385–422.

[Tad2] ———, *On characters of irreducible unitary representations of general linear groups*, Abh. Math. Sem. Univ. Hamburg **65** (1995), 341–363.

[T1] J. Tate, *Fourier Analysis in Number Fields and Hecke's Zeta Functions*, in *Algebraic Number Theory*, Thompson, Washington, D.C., 1967, 305–347.

[T2] ———, *Number theoretic background*, in *Automorphic Forms, Representations and L-functions*, Proc. Sympos. Pure Math. vol. 33, Part 2, Amer. Math. Soc., 1979, 3–26.

[Ti] J. Tits, *Classification of Algebraic Semisimple Groups*, in *Algebraic Groups and Discontinuous Subgroups*, Proc. Sympos. Pure Math., vol. 9, Amer. Math. Soc., 1966, 33–62.

[V1] D. Vogan, *Gelfand-Kirillov dimension for Harish-Chandra modules*, Invent. Math. **48** (1978), 75–98.

[V2] ———, *The unitary dual of* $GL(n)$ *over an archimedean field*, Invent. Math. **83** (1986), 449–505.

[V3] ———, *The local Langlands conjecture*, Contemp. Math. **145** (1993), 305–379.

[W1] J.-L. Waldspurger, *Le lemme fondamental implique le transfer*, Compositio Math. **105** (1997), 153–236.

[W2] ———, *Représentations de réduction unipotente pour* $SO(2n+1)$*: quelques conséquences d'une article de Lusztig*, Contributions to automorphic forms, geometry, and number theory, 803–910, Johns Hopkins University Press, Baltimore, MD, 2004.

[W3] ———, *Endoscopie et changement de caractéristique*, J. Inst. Math. Jussieu **5** (2006), 423–525.

[W4] ———, *Le groupe* GL_N *tordu, sur un corps p-adique.* I, Duke Math J. **137** (2007), 185–234.

[W5] ———, *Le groupe GL_N tordu, sur un corps p-adique.* II, Duke Math J. **137** (2007), 235–336.

[W6] ———, *L'endoscopie tordue n'est pas si tordue*, Memoirs of AMS **908** (2008).

[W7] ———, *A propos du lemme fondamental pondéré tordu*, Math. Ann. **343** (2009), 103–174.

[W8] ———, *Endoscopie et changement de charactéristique: intégrales orbitales pondérées*, Ann. Inst. Fourier **59**(5) (2009), 1753–1818.

[W9] ———, *Préparation à la stabilisation de la formule des traces tordue* I: *endoscopie tordue sur un corps local*, preprint.

[W10] ———, *Préparation à la stabilisation de la formule des traces tordue* II: *integrales orbitales et endoscopie sur un corps non-archimédien*, preprint.

[We1] H. Weyl, *Theorie der Darstellung kontinuierlicher halb-einfacher Gruppen durch lineare Transformationen*, Gesammelte Abhandlungen, Bd. II, **68**, 543–647.

[We2] ———, *The Theory of Groups and Quantum Mechanics*, 1931, rept. Dover Publications, 1950.

[Wh] D. Whitehouse, *The twisted weighted fundamental lemma for the transfer of automorphic forms from $GSp(4)$ to $GL(4)$*, Astérisque **302** (2005), 291–436.

[Z] A. Zelevinsky, *Induced representations of reductive p-adic groups* II. *On irreducible representations of $GL(n)$*, Ann. Sci. École Norm. Sup. **13** (1980), 165–210.

Notational Index

This index is considerably more detailed than necessary. It is meant to account for some of the unnamed objects not in the Subject Index. It might even convey some feeling for the internal relationships among different parts of the volume. The index can of course also be used strictly for reference.

Subject Index